Corn growing in the field.

A flower bed brightens the sidewalk.

Ornamental grass in the landscape.

Tulip exhibit at the Holland Tulip Festival in Michigan.

Plant parts encased in ice can be protected against injury.

HORTICULTURE
PRINCIPLES AND PRACTICES

HORTICULTURE
PRINCIPLES AND PRACTICES

George Acquaah
LANGSTON UNIVERSITY

PRENTICE HALL, Upper Saddle River, New Jersey 07458

Library of Congress Cataloging-in-Publication Data

Acquaah, George.
   Horticulture: principles and practices   /   George Acquaah.
      p.   cm.
   Includes bibliographical references and index.
   ISBN 0–13–518275–1
   1. Horticulture.   I. Title.
  SB318.A3   1999
  635—dc21

99–18886
CIP

Acquisitions Editor: **Charles Stewart**
Assistant Editor: **Kate Linsner**
Production Editor: **Lori Harvey/Carlisle Publishers Services**
Production Liaison: **Eileen M. O'Sullivan**
Director of Manufacturing & Production: **Bruce Johnson**
Managing Editor: **Mary Carnis**
Production Manager: **Marc Bove**
Marketing Manager: **Melissa Bruner**
Cover Designer: **Bruce Kenselaar**
Formatting/page make-up: **Carlisle Communications, Ltd.**

 © 1999 by Prentice-Hall, Inc.
Simon & Schuster/A Viacom Company
Upper Saddle River, New Jersey 07458

Printed in the United States of America

10 9 8 7 6 5 4 3 2 1

ISBN 0-13-518275-1

Prentice-Hall International (UK) Limited, *London*
Prentice-Hall of Australia Pty. Limited, *Sydney*
Prentice-Hall Canada Inc., *Toronto*
Prentice-Hall Hispanoamericana, S.A., *Mexico*
Prentice-Hall of India Private Limited, *New Delhi*
Prentice-Hall of Japan, Inc., *Tokyo*
Simon &Schuster Asia Ptc. Ltd., *Singapore*
Editora Prentice-Hall do Brasil, Ltda., *Rio de Janeiro*

With love to Theresa, quarterback; Parry, wide receiver; Kwasi, running back; Bozuma, Homecoming Queen; and Tina, cheerleader extraordinaire. In my book you will always be winners!

George Acquaah

Brief Contents

Detailed Contents

Part 2   Protecting Horticultural Plants   207

Part 3   Propagating Horticultural Plants   281

Part 4 Growing Plants Indoors   339

Part 5　Growing Plants Outdoors　439

Part 7 Miscellaneous Topics 627

Preface

Horticulture is the area of plant science that caters to the needs of a broad range of people, from the small backyard farmer in the urban area to the large-scale producer. Horticulture's adaptability to the home situation makes it attractive to people from all walks of life, including those who may not wish to study agriculture or be identified as farmers in the way society defines them but are willing to grow and care for plants. Horticulture is thus a popular instructional program and part-time activity indulged in by many people. The horticulture industry, as explained in the text, is making it increasingly more attractive for nonprofessional plant growers to participate in plant culture at various levels and for various needs.

The purpose of this text is to provide a resource for use in instruction in the fundamentals of horticulture and as a reference for hobbyists and professionals. As an instructional text, *Horticulture: Principles and Practices* is designed for use at the undergraduate level. Emphasis is placed on instruction in the basic principles and practices of horticulture, thereby minimizing regional and national biases.

Horticulture is presented as a science, an art, and a business. However, large-scale production is not described in detail but is adequately discussed as appropriate for this level of presentation. A format with unique features is adopted throughout the text. First, the text is divided into parts, within which related topics are treated as chapters. Chapters on broad topics are subdivided into appropriate modules. Each chapter opens with a stated purpose or objective, followed by a list of expected outcomes upon completion of the chapter. The key words and terminologies encountered in the text are listed next, providing an opportunity for the reader to evaluate his or her understanding of the material in the text. An overview is designed to introduce the subjects to be discussed and to define the scope of presentation. Each subject is discussed under clearly defined headings and subheadings. Key words and terminologies in the text are highlighted in italics and defined or explained. The reader is also frequently referred to other places in the text where certain key terminologies or concepts are presented in detail. The reader is thus able to refresh his or her memory, if need be, to facilitate the learning of the current material. A brief summary at the end of each chapter reviews the main message for emphasis. A list of literature is presented at the end of each chapter to acknowledge the sources consulted by the author in preparing the text and to suggest sources for further information on the topics discussed. If one has difficulty in defining or explaining any terminology or key word, the glossary at the end of the text may be used as a quick reference. Practical activities to enhance the understanding of the material discussed are suggested at the end of the chapters. Finally, the student is provided an opportunity to assess whether the material in the chapter was really understood. The outcomes assessment is conducted at three levels. The first part requires the reader to simply agree or disagree with a statement. The second part is designed to test the understanding of terminologies and concepts and requires the student to provide specific information. Lastly, part three of the outcomes assessment requires the student to think a little harder and to discuss, explain, or describe events, concepts, principles, or methodologies and to communicate effectively in writing.

The presentation includes many photographs, line drawings, and tables to facilitate the comprehension of the material and for a quick reference. The materials included in this textbook were chosen to provide the student and the user a complete introduction to the four general areas of horticulture: ornamental horticulture, fruit culture, vegetable culture, and landscape architecture. Part 1 is devoted to describing the underlying science. The amount of time spent on these chapters depends on the background of the student. The presentation is such that the reader clearly sees the relevance of the science in horticulture. Topics are presented from the point of view of the horticulturist. The chapters take the reader through a review of pertinent topics in plant taxonomy, plant anatomy, plant growth environment, plant physiology, and plant genetics and improvement. The role of these disciplines of science in the horticultural industry and how they are applied or manipulated to increase the performance of plants are also discussed.

Part 2 is devoted to a discussion of how horticultural plants are protected. In this section, the student learns about biological enemies of horticultural plants and the principles and methods of disease and pest control. Part 3 presents a discussion on plant propagation, discussing the characteristics of sexual and asexual methods of propagating plants.

Growing plants indoors forms the theme for part 4. Horticulture can be conducted in an open area or under a controlled environment where growth factors are manipulated for the optimal performance of plants. The student learns how the greenhouse is designed and used in the production of plants. Hydroponics is discussed in some detail. In this section, the student is instructed in the science and art of growing plants in containers in the home and office.

Part 5 discusses growing plants outdoors including detailed coverage of the installation, use, and maintenance of plants in the landscape. A discussion on the establishment and maintenance of a lawn, as well as pruning and landscape maintenance tools, are also presented.

Part 6 focuses on the culture of plants for food, especially in the home garden. Selected garden crops and herbs are discussed. Methods of growing fruit trees and small fruits are presented, and a selected number of small fruits are discussed in detail. Part 7 includes discussions of a variety of specific and very important topics that have a bearing on the other subjects treated in preceding parts. The subjects include postharvest handling of horticultural products, organic farming, cut flowers and flower arranging, computers in horticulture, growing succulents, terrarium culture, and the art of bonsai.

ACKNOWLEDGMENTS

I am very grateful to Jeanne Bronston, whose persistence inspired me to undertake this project. I extend sincere thanks to Mr. Larry Acker of Langston University for providing all the black and white photos; also the management of the TLC Greenhouse of Edmond, Oklahoma; the Homeland Store of Guthrie, Oklahoma; and the Oklahoma State University horticulture greenhouse, for graciously providing the subjects for those photos. Similarly, I thank Mr. Isaac Sithole for providing a dozen sketches that were later refined for line drawings. Deep appreciation is extended to Ms. Gail Latimer and Mr. Brent Pannell of the Department of Agriculture and Natural Resources at Langston University for their clerical assistance. Finally, I thank Nana Nyame for her guidance, support, and help throughout the entire project.

George Acquaah

HORTICULTURE
PRINCIPLES AND PRACTICES

INTRODUCTION

What Is Horticulture?

PURPOSE

The introduction to this book is devoted to discussing the operational and scientific boundaries and the importance of horticulture to society. Horticulture is presented as a science, art, and business. This discussion is preceded by a brief history of horticulture.

EXPECTED OUTCOMES

After studying this chapter, the student should be able to

1. Define the term *horticulture*.
2. Briefly discuss the history of horticulture.
3. Describe the boundaries of horticulture in relation to other applied sciences.
4. Discuss the importance of horticulture in society.
5. List 10 jobs that require training in horticulture.
6. List and describe four horticulture-related industries (service industries).

KEY TERMS

Arboriculture
Floriculture
Florist

Landscape architect
Olericulture

Ornamental horticulture
Pomology

OVERVIEW

Horticulture is a very important branch of plant science. It accounts for food from three major sources: vegetables, fruits, and nuts. Apart from food, it plays a significant role in other aspects of society. It provides employment and also beautifies the environment. In this introduction, the divisions of professional horticulture are discussed, along with the field's importance to society. Horticulture is presented as an art, a science, and a business. Scientists use knowledge from genetics, physiology, botany, chemistry, and other disciplines to produce elite cultivars of plants and prescribe the best cultural practices to use for success in their production. Horticulture is supported by a variety of service industries that develop and provide chemicals, machinery, and implements for its numerous activities. Many people who do not care to grow food crops often enjoy growing flowers outdoors or indoors for aesthetic purposes.

1.1: WHAT IS HORTICULTURE?

Horticulture
Science and art of cultivating, processing, and marketing of fruits, vegetables, nuts, and ornamental plants.

The term *horticulture* is derived from the Latin *hortus* (garden) and *cultura* (cultivation), which means garden cultivation. Modern horticulture is the science and art of cultivating fruits, vegetables, and ornamental plants (figure 1). Certain institutions of higher learning such as colleges and universities have educational programs in horticulture for training as well as conducting research to advance the area of study. Some of these academic programs are general in their scope of coverage, whereas others are devoted to in-depth research and training in a specific aspect of horticulture. Modern horticulture is also big business. It provides employment for people with a wide variety of skills and is supported by an equally large number of service industries. When considered from both the science and business perspectives, horticulture can be more broadly defined as the "science and art of cultivating, processing, and marketing of fruits, vegetables, nuts, and ornamental plants."

Horticulture is related to other plant sciences (figure 2). From the four divisions of the horticulture industry shown in figure 3, it is clear that horticulture has two main goals—to provide food and to impact the environment. The relationship between horticulture and other plant sciences is evidenced by the fact that plants cannot be confined strictly to one category distinguished from others by features such as use and cultural practices. For example, an oak tree has great ornamental value in the landscape, but as a forest tree, oaks are excellent sources of lumber for high-quality furniture. Similarly, bermudagrass may be cultivated as an agronomic crop for feeding livestock and also makes an excellent turfgrass in the landscape.

Generally, growing horticultural plants is more production intensive than growing agronomic and forest plants. The returns on investment per unit area of production are also generally higher for horticultural plants. Further, horticultural plants are largely utilized fresh or as living materials (as ornamentals in the landscape), whereas agronomic and forestry products are generally utilized in the nonliving state (e.g., as grain, fiber, and timber).

1.2: A BRIEF HISTORY

The deliberate use of plants by humans for aesthetic and functional purposes has its origin in antiquity. The hanging gardens of Babylon were hailed as one of the seven wonders of the ancient world. As society evolved, deliberate cultivation and domestication of edible plants replaced the less-efficient food-gathering habits of primitive societies. Agriculture, and for that matter horticulture, is therefore not a modern-day invention but one that continues to be transformed as society advances technologically.

In terms of food production, ancient civilizations, notably that of Egypt, pioneered the basic crop production methods still in use today with modification and modernization. Land was set aside and prepared by plowing; crops were provided with supplemental irrigation for increased productivity in cultivation; crops received appropriate plant husbandry for the best results. Postharvest storage and processing (e.g., drying, fermenting, and milling) were employed to increase the shelf life of the otherwise highly perishable horticultural products. Most of the valued ancient crops are still of interest today. They include fruits (e.g., date, fig, grape, pomegranate, and olive), vegetables (e.g., garlic, melon, radish, lentil, artichoke, and chicory), oil and fiber crops, and medicinal herbs. For aesthetic uses, gardeners were employed to manicure the formal gardens of ancient royalty. As already mentioned, the gardens and landscape designs of the Babylonians were proverbial.

The foundation of agriculture was built upon by civilizations that followed the Egyptians. The Greeks and Romans were next to impact practical agriculture, with the Greeks

(a)

CACTUS
& SUCCULENTS

(b)

(c)

(d)

(e)

(f)

(g)

(h)

FIGURE 1 The many faces of horticulture. Horticulture's role in society is diverse: (a) horticultural produce is found in the grocery store; (b) greenhouses provide employment and plants for various uses; (c) landscaping enhances the premises of commercial buildings; (d) potted plants are used to enhance the interior and exterior decor of homes; (e) researchers and teachers of horticulture train students and develop improved cultivars of plants; (f) florists cater to a variety of needs in the community where flowers play a role and provide jobs; and (g) tropical and (h) temperate fruits are nutritious.

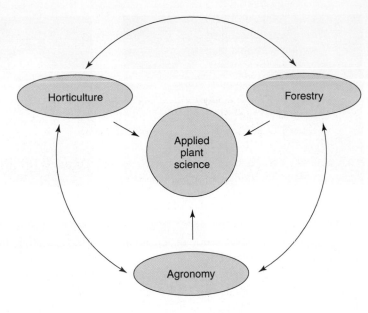

FIGURE 2 Horticulture is interrelated with other disciplines of applied sciences.

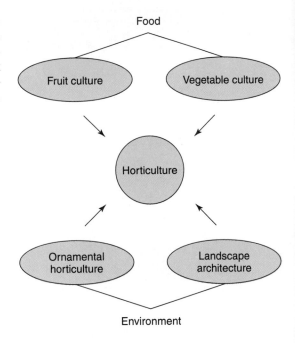

FIGURE 3 The four divisions of professional horticulture. The role of fruit and vegetable culture is to provide food; ornamental horticulture and landscape architecture impact the environment aesthetically or functionally, depending on the goal of the designer.

being noted for their contributions to early botany, as chronicled in some ancient writings—notably those of Theophrastus of Eresos. However, the Romans were more agriculturally oriented than the Greeks. They refined some of the technologies of their Egyptian predecessors. Records show that the Romans used horticultural practices such as grafting, budding, fertilization, and crop rotation, which are still in use today. In fact, their use of structures that functioned like greenhouses for forcing vegetables to grow indicates that they understood the principles of controlled-environment agriculture. With the advent of the slave trade, commercial production on plantations or large tracts of land was introduced. It was during this era that ornamental horticulture received great attention; the wealthy and nobility maintained elaborate gardens on their lavish estates. Growing plants in containers (flower boxes) and topiary (plant sculpting) became part of the landscape of these estates, some of which included swimming pools.

During the Dark Ages, monasteries became a significant factor in the preservation and advancement of horticulture. Fruit and vegetable gardens became an integral part of monastic life. One of the most significant contributions to science was made by Gregor Mendel, an

Augustinian monk, while working with plants in a garden monastery. Mendel's laws provide us with an understanding of how traits are transmitted from one generation to another. With the Renaissance came a resurgence in the interest in horticulture. Gardening not only became popular but also formalized. Landscaping was pursued with great creativity and diligence, resulting in some of the most magnificent designs ever produced, as exemplified by the gardens of the Palace of Versailles in France, designed by Andre LeNoire.

Horticulture was further advanced with the discovery of the New World. This advancement came not only in the areas of new and improved technology and an increase in knowledge but also in the introduction of new crops and improvement in trade. New World crops included fruits and nuts (e.g., cashew, avocado, pecan, pineapple, cranberry, and black walnut) and vegetables (e.g., kidney bean, lima bean, tomato, maize, potato, and sweet potato). With the expansion in trade routes and increased profitability of trading in horticultural products, coupled with greater diversity in plant types, various centers of production of specific horticultural crops were established around the globe. For example, the bulb industry flourished in Holland, while the cacao industry blossomed from the introduction into West Africa of this New World crop.

On the American horticultural scene, a number of pioneer practitioners played significant roles in the establishment of the horticulture industry. Robert Prince is credited with establishing the first nursery in the United States in the early 1730s in New York. This nursery introduced the Lombardy poplar plant in 1784, a plant that later became the most common tree in America during the post-Revolutionary era. Andrew J. Downing revolutionized the art of landscaping in the early eighteenth century by emphasizing simplicity, nature, and permanency of exhibits in the landscape. This concept of landscape design was studied by his students, the most famous being Frederick Law Olmstead, whom many acknowledge as the father of landscape architecture. The work of Olmstead is exemplified by the still-popular Central Park in New York City. In the twentieth century, the name that stands out among American horticulturalists is Liberty H. Bailey. He contributed significantly to horticulture in the areas of nomenclature and taxonomy, among others. His outstanding publications include *The Manual of Cultivated Plants* and *How Plants Get Their Names*.

Up until this period in history, horticultural production benefited primarily from improvements and changes in the production environment. Evidence suggests that genetic improvement (breeding) was conducted in the Dark Ages. The simplest method of breeding entails visual selection and saving seeds from a plant with desirable characteristics for planting in the next planting cycle. Knowledge of categorization or classification of plants according to use and other characteristics existed. With time, additional discoveries were made about the nature of things. Curiosity about the nature and response of plants to their environment led to an interest in the practical application of the existing knowledge through formal experimentation such as hybridization. Plant classification was improved with the systematic method developed by Carolus Linnaeus. Advances, including modern machinery and equipment for planting and harvesting crops and chemicals to protect crops from harmful pests and to provide supplemental nutrition, have been made through the accumulation of knowledge in a diversity of disciplines.

Modern horticulture continues to see advances in the way crops and other plants are produced. Productivity per unit area has increased, and mechanization makes it possible to grow large acreages of plants. Plant diversity has increased through advanced breeding practices. Modern horticulture also enjoys tremendous support from academic programs in institutions of higher learning, research in public and private sectors, and industry. Advanced processing and storage techniques have extended the shelf lives of products.

However, with advances in development and application of technology have come a variety of issues of great social concern. For example, increased use of agricultural chemicals has produced serious environmental consequences such as groundwater pollution. Production of crops under controlled environments (greenhouses) has expanded production of plants, making it possible to produce plants in and out of season. New production techniques such as hydroponics, tissue culture, and other biotechnological advances promise to take horticulture to a new height. However, biotechnology is embroiled in a variety of safety and ethical debates.

Pomology
The science and practice of fruit culture.

Olericulture
The science and practice of growing vegetables.

Ornamental Horticulture
The branch of horticulture that deals with the cultivation of plants for their aesthetic value.

Floriculture
The science and practice of cultivating and arranging ornamental flowering plants.

Arboriculture
The science of growing and caring for ornamental trees.

Landscape Architect
A professional who designs landscape plans.

Horticultural activities may be divided into several broad categories based mainly on the kinds of plants involved (figure 3). These divisions form the basis of certain academic programs in horticulture:

1. *Fruit culture.* Fruits vary in numerous ways (1.3.4). Some are borne on trees, others on bushes. Some fruits are succulent and juicy, whereas others are dry. Growing fruits is a long-term operation. *Fruit trees* take a long time (several years) to come into bearing. They also require more growth space per plant than vegetables. An area of land on which fruit trees are grown in a significant concentration is called an *orchard.* The branch of horticulture involved with the production (including growing, harvesting, processing, and marketing) of fruit trees (including nuts) is called *pomology.* Fruit trees such as apple, orange, and pear are operationally distinguished from *small fruits* such as grape, blueberry, and strawberry.

2. *Vegetable culture.* Vegetable production is one of the most popular horticultural activities indulged in by homeowners, often in the backyard or private section of the property. The branch of horticulture involved with the production of vegetables is called *olericulture.* Some vegetable plants are grown for their fruits (e.g., tomato), leaves (e.g., spinach), roots (e.g., carrot), or pods (e.g., bean). Unlike fruits, vegetables are generally short-duration plants that need to be restarted each growing season. Vegetables may be harvested and used fresh. However, they are also processed in a variety of ways.

3. *Ornamental horticulture.* The production and use of ornamentals is the branch of horticulture generally called *ornamental horticulture.* The term *use* is included in the definition because it is an integral part of this branch of horticulture. Ornamentals may be cultivated in open space (or landscape) or in indoor containers. They may also be grown, arranged, and displayed in a variety of ways. Subdivisions of this branch of horticulture involve distinct activities. *Floriculture* is the production and use of flowering plants and one of the areas most readily identified with when horticulture is mentioned. An important aspect of the landscape is the ground covering, which is usually grass. *Turfgrass science* has developed into a full-fledged program at many colleges. A lawn is the basic landscape element in most cases. Other plants are then added to this ground cover. Turfgrasses can be found on football fields, golf courses, playgrounds, and home grounds. Flowers in the landscape may be herbaceous or woody. The branch of horticulture involved with the production of trees is called *arboriculture.* Trees are perennial elements in a landscape design. They usually are large in size and hence require more space than annual plants.

4. *Landscape architecture. Landscaping* is the use of ornamental plants in conjunction with other elements to beautify a given area. The professionals who design such plans are called *landscape architects.* Since landscaping can enhance a property, it has become an integral part of home construction. Commercial facilities and other public areas are also appropriately landscaped. Malls, playgrounds, boulevards, and parks are examples of public places where ornamental plants are used to enhance the environment aesthetically and make it more functional. The use of plants indoors is called *interioscaping* (as opposed to landscaping).

1.4: Allied Horticultural Industries

As previously stated, horticulture as an industry is supported by other allied industries. The major ones are the *nursery industry* and *seed industry.*

I.4.1 THE NURSERY INDUSTRY

The growth in the horticultural industry today is attributable in part to the growth in the nursery industry (chapter 13). Nurseries provide seedlings for growers who do not want to raise plants from scratch and prefer to take advantage of their convenience. In fact, certain plants are difficult to propagate without special conditions that the homeowner ordinarily cannot provide. Nurseries also grow and sell mature plants in containers for use indoors and outdoors. Nurseries facilitate the work of landscape architects and contractors by providing materials that are ready to be installed on-site (13.6), enabling a bare ground to be instantly transformed into a lawn with trees and other ornamental plants. Commercial nurseries are equipped to provide ideal conditions for plant growth. By growing plants under a controlled environment, nurseries provide growers a head start on plant production for the season. They start the plants in the greenhouse (11.1) in winter when growing them outside is impossible. These plants are timed to be ready for transplanting into the field when spring conditions arrive. Nurseries produce a variety of plants—bushes, trees, tubers, roots, and other succulent and woody plants. They can handle tropical and temperate (warm- and cool-season) plants because they are equipped to control the plant growth environment. The small-scale home grower can purchase portable plant growth chambers for use at home.

I.4.2 THE SEED INDUSTRY

Researchers (geneticists and breeders) are continually developing new plant cultivars (5.9). These new types may be higher yielding, more resistant to environmental stresses (such as moisture, temperature, and light) and diseases, higher in nutritional value, or aesthetically more pleasing, among other qualities. Seeds from the research domain reach the consumer after going through several steps in the seed release process (8.2). Once certified and released as a cultivar, seed growers in the seed industry become responsible for multiplying the seed of the new cultivar, processing it, and packaging it for sale. The role of the seed industry is crucial to the success that the horticultural industry currently enjoys. Seed packets come with instructions about how the plant should be raised to maturity. These instructions are of tremendous help, especially to novice growers. The seed industry has eliminated the need for growers to produce their own seed for planting, unless they so desire. The price of commercial seed is reasonable, and mail-order purchases are possible in many cases. Seed production is usually concentrated in areas where the growing season is most favorable for cropping.

I.5: HORTICULTURE AND SOCIETY

Horticulture is important to society in a variety of ways, including as a source of food, ornamentals, and jobs.

I.5.1 SOURCE OF FOOD

Society depends on horticulture for a substantial portion of its food needs in the form of vegetables, fruits, and nuts (figure 4). These types of food sources are high in complex carbohydrates and rich in vitamins and minerals (figure 5). Leguminous vegetable plants are high in both carbohydrates and protein. Horticultural products are hence part of a balanced diet for humans.

Commercial producers account for most of the horticultural products in the nation. However, numerous homeowners are engaged in gardening on their property as a source of fresh produce for the table and as a hobby for recreation and exercise. Horticultural products may be purchased fresh or processed. For example, certain vegetables (such as carrot, lettuce, tomato, and pepper) may be eaten fresh and raw in salads, and fruits (such as apple, orange, and grape) may be eaten fresh or processed into beverages.

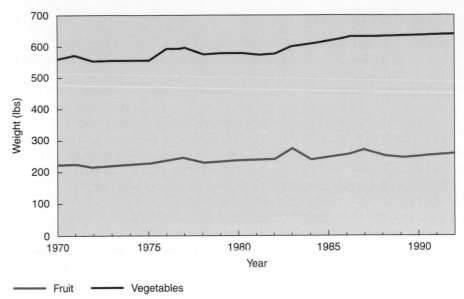

FIGURE 4 Per capita consumption of fruits and vegetables in the United States. (*Source:* Figure drawn from USDA data)

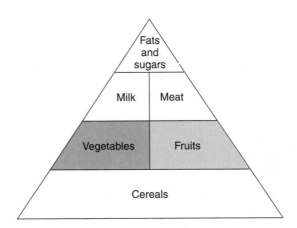

FIGURE 5 The food pyramid. The importance of horticultural produce in healthful nutrition is evidenced by the role of fruits and vegetables in the food pyramid. (*Source:* Figure modified from the original USDA diagram)

Horticultural crops are an important source of revenue through exports for a country (table 1). The United States exports both fresh and processed horticultural crops worldwide (table 2).

I.5.2 ORNAMENTALS

As previously indicated, landscaping has become an integral part of home construction. Ornamentals are found both indoors and outdoors (figure 6). Plants in the landscape include trees, shrubs, bedding plants, and grasses. Ornamentals are formally displayed for public enjoyment in places such as arboretums, parks, and botanical gardens. Botanical gardens are designed to exhibit a large variety of plant types for the pleasure and education of visitors. Public areas such as malls, playgrounds, and cemeteries are places where plants are displayed for specific purposes. In many societies, certain flowers are associated with specific social events. For example, roses are associated with Valentine's Day, carnations with graduation and Mother's Day, poinsettias with Christmas, and lilies with Easter, which is not to say that these flowers cannot be used for other purposes. Flowers feature very prominently

Table 1 Top-10 U.S. Export Markets in 1993 for Agricultural Products

Rank	Foreign Market	Export Value ($ Billion)
1.	Japan	8.4
2.	European Union	7.0
3.	Canada	5.2
4.	Mexico	3.6
5.	South Korea	2.0
6.	Taiwan	2.0
7.	Former Soviet Union	1.6
8.	Hong Kong	0.9
9.	Egypt	0.8
10.	Philippines	0.5
Subtotal		32.0
Total U.S. agricultural exports		42.5

Source: USDA data (1994 agricultural fact book)

Table 2 Top-15 U.S. Agricultural Exports for 1993

Rank	Agricultural Product	Export Value ($ Billion)
1.	Wheat	4.7
2.	Soybean	4.6
3.	Corn	4.3
4.	Red meats	3.3
5.	**Processed fruits and vegetables, juices**	**2.1**
6.	Feeds and fodders	1.7
7.	**Fresh fruits**	**1.7**
8.	Cotton	1.5
9.	Soybean meal and oil	1.5
10.	Tobacco	1.4
11.	Hides and skins	1.3
12.	Poultry meat	1.0
13.	**Fresh vegetables**	**1.0**
14.	Snack foods (excluding nuts)	1.0
15.	**Tree nuts**	**0.9**
Subtotal		32.0
Total U.S. agricultural exports		42.5

Horticultural products are in bold.
Source: USDA data (1994 agricultural fact book)

(a)

(b)

FIGURE 6 Potted plants (a) are used indoors while trees (b) are planted outdoors for aesthetic and functional purposes.

at funerals and weddings. The golf course industry is a major horticultural enterprise that involves not only turf grasses but also a wide variety of trees, shrubs, bedding plants, and other ornamentals.

I.5.3 JOBS

The ornamental industry provides a wide variety of jobs for many categories of people, directly or indirectly. Conducting a computer search via the Internet (chapter 23) reveals a wide variety of advertised jobs.

Direct Jobs

Florist
One who sells or grows for sale flowers and ornamental plants.

A large number of jobs require knowledge and training in horticulture. The level of training could be vocational or at the college level. The work may be indoors or outdoors (figure 7). Intense manual labor or paperwork in the office may be involved. Many jobs in horticulture require a high school diploma and a short course in horticulture or agriculture. A college education provides more in-depth knowledge of the field and offers job opportunities at supervisory or managerial levels and to conduct research. The following are selected categories of jobs that require varying degrees of familiarity with horticulture:

1. Greenhouse manager or worker
2. Nursery manager or worker
3. Florist
4. Golf course manager or worker
5. Landscape designer or architect
6. Tree surgeon
7. Groundskeeper
8. Garden center manager or technician
9. Vegetable grower

(a)   (b)

FIGURE 7 Jobs in the field of horticulture may be outdoors (a) as for example in landscape maintenance or indoors (b) as for example a greenhouse worker.

10. Fruit grower
11. Flower grower
12. Researcher
13. Extension officer
14. Sales or marketing officer
15. Teacher
16. Farm manager

As mentioned earlier, certain jobs do not require any familiarity with horticulture by way of formal training. For example, one can find numerous jobs in the greenhouse that require only an ability to follow directions and instructions and a sense of responsibility. Many workers in the greenhouse perform jobs such as watering, transplanting, filling pots with media, harvesting produce, and so on. Job prospects for those who pursue formal training in agriculture or horticulture are very bright.

Indirect Jobs

The ornamental industry has spawned a number of supporting or service industries, including the following:

1. *Research.* Many scientists are engaged in developing new and improved types of vegetables, fruits, and ornamentals. These new cultivars may have wider and better adaptation, be higher yielding and of higher nutritional quality, and have other qualities depending on breeding objectives. Research is conducted in both private and public sectors (at universities, research institutes, and research companies) to find solutions to problems in the horticultural industry. College-level training (often graduate level) is required to adequately prepare for a career in research. Research institutes invest a great amount of human and financial resources in developing new cultivars, which is why commercial seed companies sell their improved seed (such as *hybrid* seed) at premium prices. Apart from improving the agronomic and nutritional qualities of plants, horticultural scientists also devote considerable time to improving the aesthetics of ornamentals and the quality of products.

2. *Chemical industry.* The horticulture industry depends on large amounts and varieties of chemicals, including fertilizers, pesticides, and growth hormones. Many companies are involved in producing chemicals that are used to enhance plant production and the quality of produce. Chemicals are an integral part of modern high-input production practices. The increasing trend toward ensuring a safer environment has been the impetus for the creation and enforcement of laws and guidelines for the judicious and safe use of chemicals. Crop production using little or no chemicals, called *organic farming,* is gradually gaining popularity.

3. *Machinery.* Engineers design and produce tools and machinery for use in the production of horticultural plants. Machinery and implements are available for preparing land, planting, cultivating, spraying, harvesting, storing, and packaging. These aids enable large-scale production of horticultural plants to be undertaken. Home garden versions of some of this machinery and equipment are available.

4. *Distribution.* Horticultural products are transported from the areas of production to marketing outlets. Because of their largely perishable nature, horticultural produce and products require special handling in transportation to retain their quality for a long time. Certain items require refrigeration during storage. Horticulture has spawned an elaborate transportation and distribution network. Because most horticultural products are harvested and used fresh, the ability to preserve quality in transit is critical to the industry. In certain cases, the produce is harvested before it ripens in order to increase its shelf life. Home gardeners have the advantage of ready access to vine-ripened and fresh produce.

Numerous jobs are available in these four general areas at various levels. These jobs can be obtained by persons trained in fields other than horticulture, such as basic science, engineering, economics, marketing, agribusiness, genetics, and postharvest physiology.

1.6: HORTICULTURE THERAPY

Horticulture is known to have therapeutic value that can be derived by participating in it or simply enjoying what has been created by others. Walking through a botanical garden can be very relaxing and healthy.

People with emotional and mental problems have been helped when they were deliberately exposed to ornamental plants. For the visually impaired, horticulture can be enjoyed by touching the plant parts and enjoying the sweet scents.

Gardening can be undertaken by people to keep fit or to relieve boredom and other negative emotional feelings. People who are incarcerated or severely limited in their movement are prone to frustration. For such individuals, horticultural activities can be helpful in better managing their emotions.

SUMMARY

Horticulture (garden cultivation) is the branch of agricultural plant sciences that deals with the production of fruits, vegetables, nuts, and ornamentals. It is a major source of food and employment in society. Operationally, there are several divisions of horticulture: olericulture (vegetable production), floriculture (flower production), turf culture (turf production), pomology (tree fruit production), arboriculture (tree production), and landscape architecture (design and use of plants in the landscape). Horticultural foods are rich in minerals and vitamins. Horticulture can be undertaken on a small scale by homeowners on their property, producing flowers and food plants. As an industry, horticulture is supported by a large number of service providers that supply equipment, chemicals, and implements. Nurseries provide plant materials for growers. Numerous jobs are available to persons with formal training in horticulture, but an equally large number of jobs in this branch of agriculture require little or no formal training.

REFERENCES AND SUGGESTED READING

Janick, J. 1986. Horticultural science, 4th ed. San Francisco: W. H. Freeman.

USDA. 1977. Yearbook of agriculture: Gardening for food and fun. Washington, D.C.: U.S. Government Printing Office.

PRACTICAL EXPERIENCE

1. Consult the commercial edition of your local telephone book (the yellow pages) and list up to 10 businesses each in the following categories:
   a. Florists
   b. Nurseries or greenhouses
2. Visit a local business in each of the following categories. During the visit, obtain information about the following: plants cultivated (or plants and flowers used), educational level or training of the owner or manager, size of the operation, and profitability.
   a. Florist shop
   b. Greenhouse
   c. Vegetable farm
   d. Fruit farm

3. Visit the library for the following exercise.
   a. Select a local periodic publication. For a period (e.g., three months), count the number of advertised horticultural and allied jobs.
   b. Find the job description of each of the following jobs from, for example, the *Occupational Outlook* publication:
      1. Golf course manager
      2. Greenhouse manager
      3. Three other horticultural jobs
4. Internet search. Search the Internet to find out the number of jobs advertised in the area of horticulture. You may want to search according to categories (e.g., teaching jobs, greenhouse jobs, floriculture, turfgrass, and so on).

OUTCOMES ASSESSMENT

PART A

Please answer true (T) or false (F) for the following statements.

1. T F Horticulture is a branch of agriculture.
2. T F Olericulture is the production of vegetables.
3. T F A tree surgeon is a person who treats sick trees.
4. T F A college degree is required for all jobs in the horticultural enterprise.
5. T F Turf culture is concerned with lawns.
6. T F Arboriculture is concerned with the production of trees.

PART B

Please answer the following questions.

1. The production of fruit trees is called _____.
2. List three specific ways in which the nursery industry supports the general horticultural industry. _____
   _____
   _____
3. If there is a horticultural program on a university campus for training in the design and use of plants in beautifying the environment, it is probably called _____.
4. Horticultural products are rich in _____ and _____.
5. List two different industrial enterprises that provide services to the horticultural industry. _____
   _____

PART C

1. Describe how horticulture can be used in therapy.
2. In what ways is horticulture important to society?
3. Briefly trace the history of horticulture.

PART 1

THE UNDERLYING SCIENCE

CHAPTER 1

Classifying and Naming Horticultural Plants

PURPOSE

This chapter is designed to show the need and importance of a universal system for classifying and naming plants and to describe the various methods currently in use for accomplishing this goal.

EXPECTED OUTCOMES

After studying this chapter, the student should be able to

1. Define the term *taxonomy.*
2. Explain the need for a universal nomenclature.
3. Describe the binomial nomenclature.
4. Describe various operational systems of classification of plants based on growth form, fruits, life cycle, use, stem type, leaf characteristics, adaptation, and flower type.

KEY TERMS

Annual	Genus	Perennial
Biennial	Family	Species
Binomial nomenclature	Kingdom	Taxonomy
Broadleaf	Narrowleaf	Variety
Cultivar	Order	

OVERVIEW

Nature is characterized by diversity. No two individuals are exactly alike. Some individuals have identical *genomes* (arrays of genes). However, in appearance, even *clonal populations* (chapter 9) or identical twins exhibit subtle differences. Every culture has a system for grouping individuals for a variety of practical purposes; names are attached to the groups and the component types of which they are comprised. As long as a culture remained closed to the outside world, there was no problem with the culture-based nomenclature. However, as cultures merged with each other and plant materials were moved across cultural and geographic lines, it became necessary, for effective cross-cultural communication, to have a universal system of naming plants. This system ensured that corn, even though called maize in another culture, would have a neutral name and mean the same crop to all people.

Some superficial differences automatically place organisms into distinct classes. For example, there are plants and there are animals. In plants, some bear flowers, others do not; some have broad leaves, others have narrow leaves; some bear fleshy fruits, others bear grains; and so on. These *natural systems* of classification are arbitrary and reflect the uses human cultures have for plants. Other forms of nomenclature are based on scientific principles that have universal application. This chapter explores the origin and nature of these different systems for grouping and naming plants.

1.1: SCIENTIFIC AND BOTANICAL SYSTEMS OF CLASSIFICATION

Taxonomy
The science of identifying, naming, and classifying plants.

Scientific systems of classification go beyond the superficial or natural system by employing a number of criteria that include morphological, anatomical, ultrastructural, physiological, phytochemical, cytological, and evolutionary (phylogenetical) criteria. Pyrame de Candole is credited with the introduction of the term *taxonomy* as the science of classifying and naming plants. Taxonomy is sometimes used synonymously with *systematics*. The latter, however, is a field of biology involved with the study of diversity among organisms to establish their natural (evolutionary) relationships, making taxonomy a discipline of systematics.

In plant identification, individuals are assigned to a descending series of related plants, based on their known common characteristics. For example, a marigold plant is first placed in a more distant group with plants that have seed, then among seed plants with flowers, and eventually in the most closely associated groups of varieties of marigold. In terms of *botanical nomenclature* (naming plants), Carolus Linnaeus is credited with developing the current Latin-based system called the *binomial nomenclature* (because an individual is given two names, as opposed to the polynomial system, which was more descriptive). The international body that sets the rules for naming plants by this system publishes the *International Code of Botanical Nomenclature (ICBN)* to provide guidelines for standardizing the naming of plants. These rules are revised as new scientific evidence becomes available.

Binomial Nomenclature
A system of naming plants whereby a plant is given a two-part name representing the genus and species.

1.2: TAXONOMIC GROUPS

Kingdom
The highest taxonomic category.

Seven general classification categories have been defined in plants. These classifications can be arranged in order from the most inclusive group (*kingdom*) to the least inclusive group (*species*) (figure 1-1). Each of these groups constitutes a *taxon* (plural: *taxa*). In addition to these basic groups, subcategories are used in certain cases. These include levels such as *subdivision, subclass, suborder, subspecies,* and *variety* (or *cultivar*). An example of plant classification is presented in table 1-1.

According to the binomial nomenclature, each individual has a two-part name; the first part is called the *genus* (plural: *genera*) and the second part is called a *specific epithet* or *species*. This system is equivalent to surnames and first names in the naming of people.

Species
A category of biological classification ranking immediately below the genus or subgenus.

1.2.1 KINGDOMS

Placing organisms into groups is a work in progress. Traditionally, living organisms are recognized as belonging to one of two categories or kingdoms: the *plant kingdom* or the *animal kingdom*. However, as science advances and knowledge increases, this scheme periodically comes under review. From the two-kingdom scheme evolved the three-, four-, and five-kingdom classifications of organisms. Even the five-kingdom scheme is deemed inadequate by

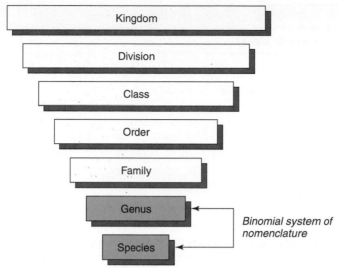

FIGURE 1-1   Major classification categories of living organisms.

Table 1-1   An Example of Scientific Classification of Plants

Taxon	Example	Common Name
Kingdom	Plantae	Plant
Division	Magnoliophyta	Flowering plant
Class	Liliopsida	Monocot
Order	Liliales	Lily order
Family	Liliaceae	Lily family
Genus	*Allium*	
Species	*Allium cepa*	Onion

certain scientists, who propose a further breakdown into more kingdoms. The five-kingdom classification proposed by R. H. Whittaker in 1969 is presented in table 1-2. The criteria for the classification are cellular structure (complexity) and forms of nutrition (photosynthesis, ingestion, or absorption of food in solution).

Even though horticulture focuses on organisms in Kingdom Plantae, other kingdoms are directly or indirectly important to the field. Horticulture exists because of humans (Kingdom Animalia). The field was developed by humans to serve humans. Organisms in the Kingdoms Monera, Protoctista, and Fungi include those that are pests of horticultural plants, namely, bacteria, fungi, and viruses.

1. *Kingdom Monera.* Monera is a kingdom of unicellular (one-celled) organisms (2.1.1). They are called prokaryotes and have no nuclear membrane or compartmentalization into distinct organelles. They reproduce primarily by cell division and are mostly heterotrophic (cannot make organic compounds and thus feed on materials made by others). Bacteria are classified under this kingdom.

2. *Kingdom Protoctista.* The Kingdom Protoctista includes algae (green, brown, and red), slime molds, and eukaryotes (cells with a nuclear membrane and compartmentalization) (2.1.2).

3. *Kingdom Fungi.* Fungi are filamentous eukaryotes that lack plastids and the photosynthetic pigment (chlorophyll) (4.3.1). Thus, they feed on dead or living organisms. Most plant diseases are caused by fungi.

4. *Kingdom Animalia.* Kingdom Animalia consists of multicellular organisms that are eukaryotes but without cell walls, plastids, and capacity for photosynthesis (processing

Table 1-2 The Five Kingdoms of Organisms as Described by Whittaker

Monera (Have Prokaryotic Cells)
Bacteria
Protoctista (Have Eukaryotic Cells)
Algae Slime molds Flagellate fungi Protozoa Sponges
Fungi (Absorb Food in Solution)
True fungi
Plantae (Produce Own Food by the Process of Photosynthesis)
Bryophytes Vascular plants
Animalia (Ingest Their Food)
Multicellular animals

of food production from the sun by plants) (4.3.1). Animals generally ingest their food and reproduce primarily by sexual means. Animals have the highest level of organization and tissue differentiation of any organism in any kingdom. They have complex sensory and neuromotor systems.

5. *Kingdom Plantae.* Organisms in the Kingdom Plantae are photosynthetic (make food from inorganic materials; few plants are heterotrophic, that is, feeding on organic material from other sources). They are multicellular, have cell walls (2.1.2), and live on land.

1.2.2 DIVISIONS OF KINGDOM PLANTAE

Several divisions are recognized in the Kingdom Plantae (table 1-3). These divisions can be divided into two major categories: *bryophytes* (nonvascular plants—the mosses, hornworts, and liverworts) and *vascular plants.* Vascular plants are large bodied and have three primary vegetative organs, *stem, leaves,* and *roots,* and also *conducting tissues (vascular tissues)* (2.1.2). Vascular plants may produce seeds or be seedless. Most plants of horticultural interest are vascular plants.

In terms of relative abundance, more than 80 percent of all species in the plant kingdom are flowering plants. Even though *gymnosperms* (seed plants whose seeds are not enclosed within an ovary during development) make up only 0.2 percent of species in the plant kingdom, conifers (e.g., pines) occur on about one-third of forested lands of the world.

1.2.3 VARIETY VERSUS CULTIVAR

> **Variety**
> Any of various groups of plants or animals ranking below a species.

> **Cultivar**
> Derived from the words cultivated and variety, often designating a product of plant breeding.

The lowest and least-inclusive taxon is the species, as already indicated. Species may be subdivided into specific categories. A botanical variety is a naturally occurring variant of the species that is significantly different from the general species originally described. Botanical varieties may differ in subtle or more visible ways, such as in color, shape, size, chemical quality, or some other traits. Instead of two names, as expected in the binomial nomenclature, a variety requires the use of a third name after the introduction of the abbreviation *var.* (for *variety*). For example, broccoli is called *Brassica oleraceae* var. *botrytis.*

Through plant breeding, humans sometimes create new variants that are maintained under human supervision (as opposed to being naturally maintained, as is the case in varieties). The product of plant breeding is called a *cultivar,* a contraction of two terms—*culti*vated and *var*iety.

Table 1-3 The Divisions of the Kingdom Plantae

	Divisions	Common Name
Bryophytes	Hepaticophyta	Liverworts
	Anthocerotophyta	Hornworts
	Bryophyta	Mosses
Vascular plants		
Seedless	Psilotophyta	Whisk ferns
	Lycophyta	Club mosses
	Sphenophyta	Horsetails
	Pterophyta	Ferns
Seeded	Pinophyta	Gynosperms
	Subdivision: Cycadicae	Cycads
	Subdivision: Pinicae	
	Class: Ginkgoatae	*Ginkgo*
	Class: Pinatae	Conifers
	Subdivision: Gneticae	*Gnetum*
	Magnoliophyta	Flowering plants
	Class: Liliopsida	Monocots
	Class: Magnoliopsida	Dicots

Cultivars are maintained as clones in vegetatively propagated (increasing the number of plants by using plant parts other than seed) (chapter 9) species and as lines in species propagated by seed under specific conditions. Many flowers and vegetables have cultivars that are propagated by seed, whereas others are *hybrids* (F_1 seed from a cross of two different parents) (5.7.5).

1.2.4 RULES IN CLASSIFICATION

In plant taxonomy, the ending of a name is often characteristic of the taxon. Classes often end in *-opsida* (e.g., Magnoliopsida). Names ending in *-ae* are subclasses of class names (depending on the classification system). Exceptions include several families such as Compositae (now called Asteraceae). Plant orders end in *-ales* (e.g., Rosales [roses]), while family names end in *-aceae* (e.g., Rosaceae).

These higher-order taxa are not routinely encountered, unless one is conducting taxonomic studies. The binomial names (genus and species) are the most frequently encountered. When you walk through a botanical garden, or even a college campus where there is a good horticulture program or a good grounds and gardens department, you may find that some plants in the landscape are labeled with the correct binomial name or scientific name, as well as the common name (figure 1-2). The family name is quite frequently indicated.

The rules for writing names become more stringent at the binomial level. The main rules are as follows:

1. The binary name must be underlined or written in italics (to indicate that such names are non-English names).

2. The genus name starts with an uppercase letter, and the species name is written in lowercase throughout. The term *species* is both singular and plural. It may be shortened to "spp," for the plural "species."

3. In technical writing, an initial *L.* may follow the species, indicating that Linnaeus first named the plant. Other abbreviations may be encountered in the literature. An example of a full binary name for corn, for example, is <u>Zea mays</u> L., or *Zea mays* L. The genus may be abbreviated (e.g., *Z. mays* L.). Some plants may have a subspecies and hence have a third name added to the binary name. In such a case, the third name is also underlined or italicized.

4. Whereas the generic name can be written alone to refer to individuals in the group, the specific epithet cannot be used by itself (i.e., *Zea* but not *mays*).

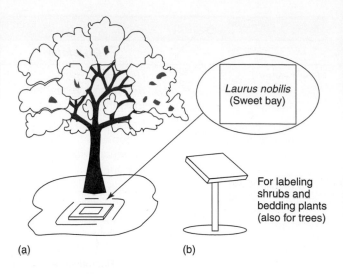

FIGURE 1-2 Labeling plants in the landscape. (a) Trees may be identified by using a metallic plate label set in concrete, usually showing both the scientific and common names. (b) A stake-mounted label may also be used to label plants.

Laurus nobilis (Sweet bay)

For labeling shrubs and bedding plants (also for trees)

(a)   (b)

5. At the bottom of the taxa hierarchy is variety, which is the naturally occurring and very closely related variant. As previously indicated, the binomial name is followed by the abbreviation *var.* and then the variety name (1.2.3). Cultivar names are not underlined or italicized (e.g., *Lycopersicon esculentum* Mill. cultivar 'Big Red' or L. esculentum cv. 'Big Red,' or L. esculentum 'Big Red').

Specific epithets are adjectival in nature. Many genera can have the same specific epithet. Some of them indicate color, such as *alba* (white), *variegata* (variegated), *rubrum* (red), and *aureum* (golden). Examples of frequently encountered epithets are *vulgaris* (common), *esculentus* (edible), *sativus* (cultivated), *tuberosum* (bearing tubers), and *officinalis* (medicinal).

In developing new horticultural cultivars, plant breeders employ a variety of techniques (5.7). The conventional techniques involve crossing or hybridizing plants that differ in desirable characteristics. In terms of taxonomic hierarchy, hybridization can be routinely performed at the base of the hierarchy (i.e., among varieties or cultivars of the same species). Crossing at other levels such as among species (*interspecies hybridization* [5.12]) leads to genetic complications. Such a cross is problematic and has limited success, requiring the use of additional techniques, such as *embryo rescue* (9.13), in some cases.

1.3: OTHER CLASSIFICATION SYSTEMS (OPERATIONAL)

A number of operational classification systems are employed simultaneously in the field of horticulture. The following are some of the major systems.

1.3.1 SEASONAL GROWTH CYCLE

Plants may be classified into three general groups based on growth cycle (figure 1-3). Growth cycle refers to the period from first establishment (e.g., by seed) to when the plant dies. The three categories are as follows:

> **Annual**
> *A plant that completes its life cycle in one growing season or one year.*

1. *Annuals.* An annual plant lives through only one growing season, completing its life cycle (seed, seedling, flowering, fruiting, and death) in that period. This group includes many weeds, garden flowers, vegetables, and wild flowers. The duration of a life cycle is variable and may be a few weeks to several months, depending on the species. Annuals are the basis of a major horticultural production group called annual bedding plants (13.7). These plants are produced largely for use in the landscape and

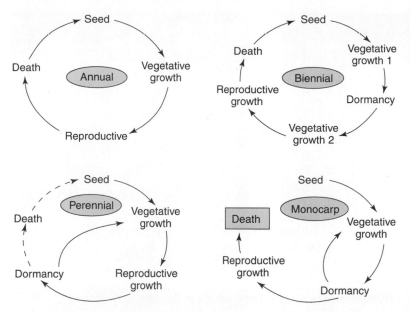

FIGURE 1-3 Classification of flowering plants according to the duration of their growth cycle from seed to seed. Variations occur within each category, even for the same species, due in part to the activities of plant breeders.

also the vegetable garden. Popular annual flowers are geranium (*Geranium* spp.), zinnia (*Zinnia elegans*), marigold (*Tagetes* spp.), and pansy (*Viola tricolor*). In cultivation, certain vegetables such as tomato (*Lycopersicon esculentum*) are produced on an annual cycle.

2. *Biennials.* A biennial is a plant that completes its life cycle in two growing seasons. In the first season, it produces only basal leaves; it grows a stem, produces flowers and fruits, and dies in the second season. The plant usually requires some special environmental condition or treatment such as exposure to a cold temperature (*vernalization* [4.5.2]) to be induced to enter the reproductive phase. Examples of biennials are sugar beet (*Beta vulgaris*) and onion (*Allium cepa*). Even though annuals and biennials rarely become woody in temperate regions, these plants may sometimes produce secondary growth in their stems and roots.

3. *Perennials.* Perennials may be herbaceous or woody. They persist year-round through the adverse weather of the nongrowing seasons (winter or drought) and then flower and fruit after a variable number of years of vegetative growth beyond the second year. Herbaceous perennials survive the unfavorable season as dormant underground structures (e.g., roots, rhizomes, bulbs, and tubers [Chapter 9]) that are modified primary vegetative parts of the plant. Examples of herbaceous perennials are turfgrasses such as bermudagrass (*Cynodon dactylon*) and flowers such as daylilies (*Lilium* spp.) and irises (*Iris* spp.).

   Woody perennials may be vines, shrubs, or trees. These plants do not die back in adverse seasons but usually suspend active growth. Although some perennials may flower in the first year of planting, woody perennials flower only when they become adult plants. This stage may be attained within a few years or even after 100 years. Woody perennials may be categorized into two types:
   a. *Evergreen.* Evergreen perennials maintain green leaves year-round. Some leaves may be lost, but not all at one time. Examples of evergreen perennials are citrus (*Citrus* spp.) and pine (*Pinus* spp.).
   b. *Deciduous.* Deciduous plants shed their leaves at the same time during one of the seasons of the year (dry, cold). New leaves are developed from dormant buds upon the return of favorable growing conditions. Examples of deciduous perennials are oak (*Quercus* spp.) and elm (*Ulmus* spp.). It should be mentioned that intermediate conditions occur in which some plants do not lose all of their leaves (*semideciduous*).

> **Biennial**
> A plant that completes its life cycle in two cropping seasons, the first involving vegetative growth and the second flowering and death.

> **Perennial**
> A plant that grows year after year without replanting.

> **Monocarp**
> A plant (e.g., bromeliad, century plant) that lives for many years but flowers only once in a lifetime and then dies, new plants arising on the roots of the old plant.

FIGURE 1-4 An example of a herb or herbaceous plant. Stems can also be herbaceous.

FIGURE 1-5 A shrub showing the typical multiple stems arising from the ground.

1.3.2 KINDS OF STEMS

There are three general classes of horticultural plants based on stem type. However, intermediates do occur between these classes.

1. *Herbs.* Herbs are plants with soft, nonwoody stems (figure 1-4). They have primary vegetative parts. Examples include corn (*Zea mays*), many potted plants, many annual bedding plants, and many vegetables. In another usage, the term *herbs* is associated with spices (plants that are aromatic or fragrant and used to flavor foods or beverages) (18.1).

2. *Shrubs.* A shrub has no main trunk. Branches arise from the ground level on a shrub (figure 1-5). It is woody and has secondary tissue. Shrubs are perennials and usually smaller than trees. Examples of shrubs are dogwood (*Cornus* spp.), kalmia (*Kalmia* spp.), and azalea (*Rhododendron* spp.).

3. *Trees.* Trees are large plants characterized by one main trunk (figure 1-6). They branch on the upper part of the plant, are woody, and have secondary tissue. Examples include pine (*Pinus* spp.), oak (*Quercus* spp.), cedar (*Cedrus* spp.), and orange (*Citrus sinensis*).

1.3.3 COMMON STEM GROWTH FORMS

The criterion for classification is how the stem stands in relation to the ground (figure 1-7). There are several types of stem growth forms, the most common ones including the following.

1. *Erect.* A stem is erect if, without artificial support, it stands upright (stands at a 90-degree angle to the ground level). Because of the effect of strong winds and other environmental factors, an erect plant may incline slightly. Trees have erect stems. To adapt crop plants to mechanized harvesting, plant breeders have developed what are called "bush" cultivars. These plants have strong stems and stiff branches.

FIGURE 1-6 A typical tree showing a well-defined, woody central axis. Certain species produce or can be manipulated to produce several stems.

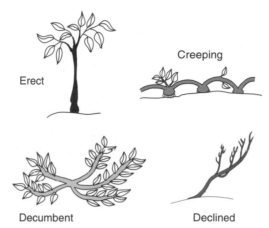

Erect

Creeping

Decumbent

Declined

FIGURE 1-7 Examples of the variations in the direction or method of stem growth.

2. *Decumbent.* The stems of decumbent plants are extremely inclined, with the tips raised. An example is peanut (*Arachis hypogaea*).

3. *Creeping (or repent).* A plant is described as creeping when it crawls on the ground, producing adventitious roots at specific points on the stem. Stems that grow horizontally in this fashion are called *stolons* (2.3.3). The strawberry plant (*Fragaria* spp.) has creeping stems.

4. *Climbing.* Climbers are vines that, without additional support, will creep on the ground. There are three general modes of climbing (figure 1-8). *Twiners* are climbing plants that simply wrap their stringy stems around their support, as occurs in sweet potato (*Ipomea batatas*). Another group of climbers develop cylindrical structures called *tendrils* that are used to coil around the support on physical contact. An example of a plant that climbs by this method is the garden pea (*Pisum sativum*). The third mode of climbing is by *adventitious roots* formed on aerial parts of the plant, as found in the English ivy (*Hedera helix*) and *Philodendron*.

1.3.4 CLASSIFICATION OF FRUITS

Fruits can be classified on a botanical basis and for several operational purposes.

Botanical Classification

Fruits exhibit a variety of apparent differences that may be used for classification. Some fruits are borne on herbaceous plants and others on woody plants. A very common operational way of classifying fruits is according to fruit succulence and texture on maturity and ripening. On this basis there are two basic kinds of fruits—*fleshy fruits* and *dry fruits*. However, anatomically, fruits are distinguished by the arrangement of the carpels from which they developed. A carpel is sometimes called the *pistil* (consisting of a stigma, style, and ovary), the female reproductive structure (2.8).

FIGURE 1-8 A climbing plant. To climb, such plants have various structural adaptations for holding onto nearby physical support.

Fruit
A mature ovary.

A fruit is a mature ovary. The ovary may have one or more carpels. Even though the fruit is a mature ovary, some fruits include other parts of the flower and are called *accessory fruits.* Combining carpel number, succulence characteristics, and anatomical features, fruits may be classified into three kinds, *simple, multiple,* or *aggregate* (figure 1-9).

Simple fruits develop from a single carpel or sometimes from the fusing together of several carpels. This group of fruits is very diverse. When mature and ripe, the fruit may be soft and fleshy, dry and woody, or have a papery texture. There are three types of fleshy fruits.

1. *Fleshy fruits.*
    a. *Drupe.* A drupe may comprise one to several carpels. Usually, each carpel contains one seed. The endocarp (inner layer) of the fruit is hard and stony and is usually highly attached to the seed (figure 1-10). Examples are cherry (*Prunus* spp.), olive, coconut (*Cocos nucifera*), peach (*Prunus persica*), and plum (*Prunus domestica*).
    b. *Berry.* A berry is a fruit characterized by an inner pulp that contains a few to several seeds but not pits. It is formed from one or several carpels. Examples are tomato (*Lycopersicon esculentum*), grape (*Vitis* spp.), and pepper (*Capsicum anuum*) (figure 1-11). If the *exocarp* (skin) is leathery and contains oils, as in the citrus fruits (e.g., orange [*Citrus sinensis*], lemon [*Citrus lemon*], and grapefruit [*Citrus paradisi*]), the berry is called a *hesperidium* (figure 1-12). Some berries have a thick rind, as in watermelon (*Citrullus vulgaris*), cucumber (*Cucumis sativus*), and pumpkin (*Cucurbita pepo*) (figure 1-13). This type of a berry is called a *pepo.*
    c. *Pome.* A pome is a pitted fruit with a stony interior. The pit usually contains one seed chamber and one seed. This very specialized fruit type develops from the ovary, with most of the fleshy part formed from the receptacle tissue (the enlarged base of the perianth) (figure 1-14). Pomes are characteristic of one subfamily of the family Rosaceae (rose family). Examples of pomes are apple (*Pyrus malus*), pear (*Pyrus communis*), and quince (*Cydonia oblonga*).

2. *Dry Fruits.* Dry fruits are not juicy or succulent when mature and ripe. When dry, they may split open and discharge their seeds (called *dehiscent fruits*) or retain their seeds (called *indehiscent fruits*).

FIGURE 1-9 A classification of fruits.

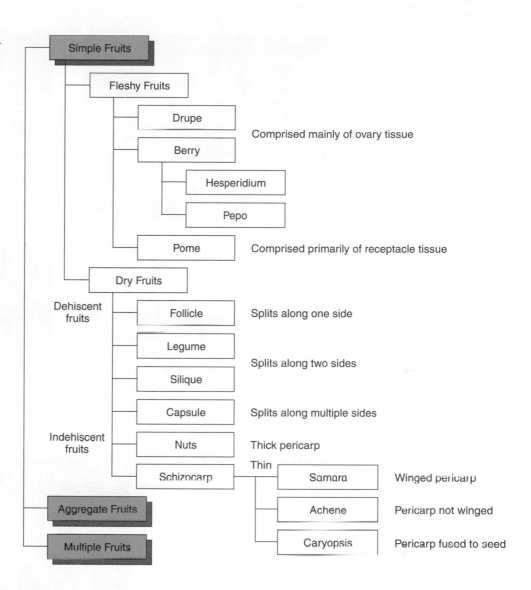

Simple Fruits

Fleshy Fruits

Drupe

Berry — Comprised mainly of ovary tissue

Hesperidium

Pepo

Pome — Comprised primarily of receptacle tissue

Dry Fruits

Dehiscent fruits

Follicle — Splits along one side

Legume — Splits along two sides

Silique

Capsule — Splits along multiple sides

Indehiscent fruits

Nuts — Thick pericarp

Schizocarp — Thin

Samara — Winged pericarp

Achene — Pericarp not winged

Caryopsis — Pericarp fused to seed

Aggregate Fruits

Multiple Fruits

FIGURE 1-10 A drupe, represented by coconut (*Cocos nucifera*).

FIGURE 1-11 A berry, represented by tomato (*Lycopsersicon esculentum*).

FIGURE 1-12 A hesperidium, represented by citrus (*Citrus* spp.).

FIGURE 1-13 A pepo, represented by watermelon (*Citrullus vulgaris*).

FIGURE 1-14 A pome, represented by apple (*Pyrus malus*).

a. *Dehiscent Fruits.* A fruit developed from a single carpel may split from only one side at maturity to discharge its seeds. Such a fruit is called a *follicle.* Examples are columbine (*Aquilegia* spp.), milkweed (*Asclepias* spp.), larkspur (*Delphinium* spp.), and magnolia (*Magnolia* spp.). Sometimes, the splitting of the ovary occurs along two seams, with seeds borne on only one of the halves of the split ovary. Such a fruit is called a legume (figure 1-15), examples being pea (*Pisum sativum*), bean (*Phaseolus vulgaris*), and peanut (*Arachis hypogaea*). In a third type of dehiscent fruit, called *silique* or *silicle,* seeds are attached to a central structure, as occurs in radish (*Raphanus sativus*) and mustard (*Brassica campestris*) (figure 1-16). The most common dehiscent simple fruit is the capsule, which develops from a compound ovary. In some species, seeds are discharged when the capsule splits longitudinally. In others, seeds exit through holes near the top of the capsule, such as in lily (*Lilium* spp.), iris (*Iris* spp.), and poppy (*Papaver* spp.).

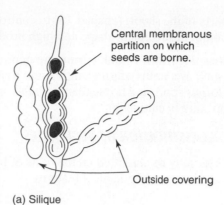

Central membranous partition on which seeds are borne.

Outside covering

(a) Silique

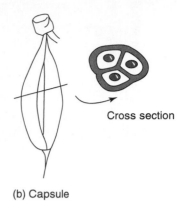

Cross section

(b) Capsule

FIGURE 1-16   A silique and a capsule.

FIGURE 1-15   A legume or pod, represented by garden bean (*Phaseolus vulgaris*).

b.   *Indehiscent Fruits.* Some indehiscent fruits may have a hard *pericarp* (exocarp + mesocarp + endocarp [2.10]). This stony fruit wall is cracked in order to reach the seed. Such fruits are called *nuts,* as found in chestnut (*Castanea* spp.) and hazelnut (*Corylus* spp.). Nuts develop from a compound ovary. Sometimes the pericarp of the fruit is thin and the ovaries occur in pairs, as found in dill (*Anethum graveolens*) and carrot (*Daucus carota*). This fruit type is called a *schizocarp.* In maple (*Acer* spp.), ash (*Fraxinus* spp.), elm (*Ulmus* spp.), and other species, the pericarp has a wing and is called a *samara.* Where the pericarp is not winged but the single seed is attached to the pericarp only at its base, the fruit is called an *achene.* Achenes are the most common indehiscent fruits. Examples are the buttercup family (Ranunculaceae) and sunflower. In cereal grains (Poaceae or grass family), the seed, unlike in an achene, is fully fused to the pericarp. This fruit type is called a *caryopsis* or *grain.*

Other Operational Classifications

Fruits may also be classified according to other operational uses.

1.   *Temperate fruits* or *tropical fruits.* Temperate fruits are fruits from plants adapted to cool climates, and tropical fruits are produced on plants adapted to warm climates. For example, apple (*Pyrus malus*), peach (*Prunus persica*), and plum (*Prunus domestica*) are temperate fruits, whereas mango (*Mangifera indica*) and coconut (*Cocos nucifera*) are tropical fruits.

2.   *Fruit trees.* Tree fruits are fruits borne on trees, such as apple (*Pyrus malus*) and mango (*Mangifera indica*).

3.   *Small fruits.* Small fruits are predominantly woody, perennial, dicot angiosperms. They are usually vegetatively propagated and bear small- to moderate-sized fruits on herbs, vines, or shrubs. Examples are grape (*Vitis* spp.), strawberry (*Fragaria* spp.), and blackberry (*Rubus* spp.). Small fruits require training and pruning (removal of

parts of the shoot) (chapter 15) to control growth and remove old canes (branches) to obtain desired plant shape and high productivity.

4. *Bramble fruits.* Bramble fruits are nontree fruits that usually require physical support (such as a trellis) during cultivation. Examples are raspberry (*Rubus* spp.), blackberry (*Rubus* spp.), and boysenberry (*Rubus* spp.). Bramble fruits also require training and pruning in cultivation.

1.3.5 CLASSIFICATION OF VEGETABLES

Vegetables may be classified on the basis of life cycle, edible or economic parts of the plant (use), adaptation, and botanical features.

1. *Life cycle.* Based on life cycle, vegetables may be classified as annuals, biennials, or perennials.
   a. *Annual.* Most vegetable garden crops are true annuals, such as corn (*Zea mays*), or are cultivated as annuals, such as tomato (*Lycopersicon esculentum*). These plants are selected for either fall or summer gardening (17.1). They require a few weeks to several months to maturity, depending on the cultivar.
   b. *Biennial.* Few popular vegetable garden crops are biennials, and, even then, they are frequently cultivated as annuals and replanted each season. Examples are sugar beet (*Beta vulgaris*) and carrot (*Daucus carota*).
   c. *Perennial.* Whenever perennial vegetable garden crops are cultivated, they must be strategically located so as not to interfere with seasonal land preparation activities needed for planting annual crops. These plants may be pruned to control growth or to remove dead tissue. Examples are asparagus (*Asparagus officinalis*) and horseradish (*Rorippa armoracia*).

2. *Edible or economic parts.* Vegetables may be operationally classified according to the parts of the plant harvested for food or other uses.
   a. *Pods.* Pods are legumes that are harvested prematurely, cooked, and eaten with the seeds inside. When harvesting is delayed, pods develop fiber and become

FIGURE 1-17 A root, represented by sugar beet (*Beta vulgaris*).

FIGURE 1-18 A bulb, represented by onion (*Allium cepa*).

stringy and undesirable for fresh use. Examples are green bean and okra (see figure 1-15).

b. *Roots.* Sometimes primary plant parts (stem, root, and leaf) may become modified as storage organs for food (9.11). Roots may become enlarged as a result of the accumulation of stored food (figure 1-17). The roots are dug and eaten baked, boiled, or fried. An example is the sweet potato (9.11.2).

c. *Bulbs.* Like roots, bulbs are modified stems and leaves, as found in onions (figure 1-18). The stem is highly compressed to form what is called a basal plate, while the leaves are storage organs (9.11).

d. *Tubers.* Tubers look like modified roots. The difference between them is that tubers are swollen stems, whereas roots are swollen roots (figure 1-19) (9.11).

e. *Greens.* Greens are vegetable crops whose leaves are usually picked at tender stages to be used for food. The leaves are generally cooked before being eaten.

3. *Adaptation.* Just like fruits, certain vegetable species prefer cool temperatures during production, and others prefer warm temperatures. Based on seasons in which they grow best, vegetables may be classified into two groupings.

a. *Cool season.* Cool-season crops require monthly temperatures of 15–18°C (60–65°F). Examples are sugar beet (*Beta vulgaris*) and cabbage (*Brassica oleracea*).

b. *Warm season.* Warm-season crops prefer monthly temperatures of 18–27°C. Examples are okra (*Hibiscus esculentus*), eggplant (*Solanum melongena*), corn (*Zea mays*), and shallot (*Allium cepa*).

    It should be mentioned that plant breeders have developed cultivars with wide adaptation for many crop species. For example, popular garden crops including corn, tomato, and pepper are grown over a wide range of climates. Even though cultivars with cold or heat tolerance may have been bred for different crops, commercial large-scale production occurs in regions of best adaptation of these crops, unless production is under a controlled environment (greenhouse) (11.1).

4. *Botanical features.* Vegetables may be classified according to specific botanical characteristics they share in common.

> **Cool-season Plant**
> *A plant that grows best at daytime temperatures of between 15°-18°C.*

> **Warm-season Plant**
> *A plant that grows best at daytime temperatures of between 18°-27°C.*

FIGURE 1-19 A tuber, represented by Irish potato (*Solanum tuberosum*).

(a)

(b)

FIGURE 1-20 Cole crops, represented by (a) broccoli (*Brassica oleracea* var. *botrytis*) and (b) cabbage (*Brassica oleracea*).

a. *Vines.* Vines are plants with stems that need physical support; without it they creep on the ground or climb onto other nearby plants in cultivation. Examples are squash, pumpkin, and cucumber.
b. *Solanaceous plants.* Solanaceous plants belong to the family Solanaceae. Examples are eggplant, tomato, and pepper (see figure 1-11).
c. *Cole crops.* Cole plants belong to the Brassica family. Examples are cabbage, cauliflower, and broccoli (*Brassica oleracea* var *botrytis*) (figure 1-20).

1.4: CLASSIFICATION OF ORNAMENTAL PLANTS

Ornamental plants may also be classified based on stem type, growth cycle, leaf form, use, and other characteristics.

1.4.1 HERBACEOUS ORNAMENTAL PLANTS

Herbaceous plants are nonwoody. Many horticultural plants, especially those grown indoors, are nonwoody. They have a wide variety of uses in landscapes as well. They may be classified in various ways.

1. *Growth cycle*
    a. *Annuals.* Annual ornamentals are planted each season. Flowering annuals are prominent in the landscape in favorable seasons, providing most of the color (figure 1-21). Versatile landscape plants can be used to fill in gaps, provide color in bulb beds after the bulbs have bloomed, and create colorful flower beds, hanging baskets (flowers grown in containers and hung) (10.2.7), and cut flowers (flowers grown and cut for use) (11.7). Examples include petunia (*Petunia* spp.), zinnia (*Zinnia elegans*), and marigold (*Tagetes* spp.).
    b. *Biennials.* Biennial ornamentals are vegetative in their first year of growth and bloom in the next season. Examples are foxglove (*Digitalis* spp.) and hollyhock.
    c. *Perennials.* Since perennials live for a long time in the landscape, locating them requires a great deal of thought and planning. Perennials may be flowering or nonflowering.
       Flowering. Flowering perennials may be planted in flower beds in the fall season to provide early blooms, after which annuals may be planted. Examples are geranium (*Geranium* spp.), lily (*Lilium* spp.), and tulip (*Tulipa* spp.).

FIGURE 1-21 Flats of annual bedding plants. Annual plants provide a tremendous amount of variety in color in the landscape.

FIGURE 1-22 A foliage plant. Foliage plants are usually green, but certain species have streaks of color or variegation.

Foliage. Foliage, or nonflowering, perennials are popular for indoor use in houses and offices as potted plants (figure 1-22). Examples are coleus (*Coleus blumei*), sansevieria (*Sansevieria* spp.), and dumbcane (*Dieffenbachia*).

2. *Other operational classifications.* Herbaceous plants may be used in a variety of other ways, both indoors and outdoors.

    a. *Bedding plants.* Bedding plants (11.6) are annual plants raised for planting outdoors in flower beds. They are usually started from seed indoors in the off-season and transplanted later in the growing season. Examples include petunia (*Petunia* spp.), zinnia (*Zinnia elegans*), pansy (*Viola tricolor*), and marigold (*Tagetes* spp.).

    b. *Hanging plants.* Hanging basket plants (10.2.7) are plants, annual or perennial, flowering or foliage, that are grown in decorative containers and hung by equally elegant ropes from the ceiling in the patio, in the doorway area, or from decorative plant poles. Examples are geranium (*Geranium* spp.) and spider plant (*Chlorophytum comosum*).

    c. *Houseplants.* Indoor plants or houseplants (10.1), are plants adapted to indoor conditions. They are grown in containers, are usually slow growing, and may be flowering or foliage plants. Examples are sansevieria (*Sansevieria* spp.) and Indian rubber plant (*Ficus elastica*).

1.4.2 WOODY ORNAMENTAL PLANTS

Woody ornamentals differ in size and growth pattern. Some shed their leaves and are called *deciduous,* whereas others maintain fresh leaves year-round and are called *evergreen.* Some are shrubs, and others are trees. Woody ornamentals may be grown in a perennial garden along with herbaceous perennials. When choosing these plants, attention should be paid to growth habit, color, texture, shape, and adaptation.

1. *Shrubs.* As previously described, a shrub is a perennial woody plant of relatively low stature and usually produces multiple stems that arise from the ground or very close to it. Shrubs, which may be used as hedge plants or ground cover (13.12), can be classified as deciduous or evergreen.

    a. *Deciduous shrubs.* Deciduous shrubs shed their leaves at some point in the year. Examples are lilac (*Sylinga vulgaris*), honeysuckle (*Lonicera japonica*), and barberry (*Berberis* spp.).

    b. *Evergreen shrubs.* Evergreen shrubs may be further divided into two groups according to leaf size.

FIGURE 1-23 A narrowleaf plant.

FIGURE 1-24 A broadleaf plant.

> **Narrowleaf**
> A group of evergreen plants having needlelike leaves.

> **Broadleaf**
> A group of evergreen plants having a large leaf lamina.

Narrowleaf. Narrowleaf shrubs have needlelike leaves (figure 1-23), as in pine (*Pinus* spp.) and juniper (*Juniperus* spp.).
Broadleaf. Broadleaf shrubs have large leaf lamina (figure 1-24), as in gardenia (*Gardenia* spp.) and rhododendron.

2. *Trees.* Trees are the largest plant materials in the landscape and thus should be located with care. They can overwhelm a house in the adult stage if inappropriate trees species are planted. Trees can also modify the local climate (e.g., as shade trees or wind breaks).

 a. *Deciduous trees.* Examples of deciduous trees are birch (*Betula papyrifera*), elm (*Ulmus* spp.), and willow (*Salix* spp.).

 b. *Evergreen trees.*
  Narrowleaf. Examples of narrowleaf evergreen trees are spruce (*Picea* spp.) and red cedar (*Juniperus virginiana*).
  Broadleaf. Examples of broadleaf evergreen trees are holly (*Ilex opaca*) and citrus (*Citrus* spp.).

3. *Vines.* Vines are climbers (figure 1-25) and can be manipulated to create a variety of structures and for various functions in the landscape. Like trees and shrubs, vines can be deciduous or evergreen.

 a. *Deciduous vines.* An example of a deciduous vine is clematis (*Clematis recta*).

 b. *Evergreen vines.* An example of an evergreen vine is English ivy (*Hedera helix*).

1.4.3 CLASSIFICATION BASED ON HARDINESS (ADAPTATION)

Plants can be classified according to their hardiness or adaptation to local climate. Certain trees are able to thrive under cold temperatures, whereas others prefer warm or tropical conditions. The U.S. Department of Agriculture (USDA) Plant Hardiness Zone map divides the United States into 10 zones (3.2.1). Zone 1 is the coldest, and zone 10 is the warmest. Plants' adaptive ranges may be narrow or broad. For example, a shrub such as cinquefoil (*Potentilla fruticosa*) is adaptable to zones 2 through 7, and lantana (*Lantana camara* 'Nivea' L.) is adapted to zones 9 and 10. All plants may be placed into one or more of these hardiness zones.

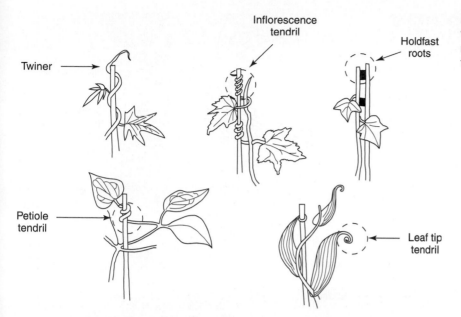

Inflorescence
tendril

Holdfast
roots

Twiner

Petiole
tendril

Leaf tip
tendril

FIGURE 1-25 Twiners. Twining species have a variety of adaptive features for attaching to nearby physical supports. These are modifications of stems, leaves, or roots.

1.5: FLOWERS IN CLASSIFICATION

Flowers are described in detail in chapter 2. They play a major role in the classification of flowering plants (angiosperms) because flowers have very stable plant parts in different environments.

Flower characteristics may vary within certain families. However, specific flowers are characteristic of certain families. For example, the legume family (Fabaceae) is characterized by an irregular flower with a keel petal, two wing petals, and a banner petal (figure 1-26). These flowers develop into a fruit, the legume (2.10). The grass family (Poaceae) is also characterized by a flower with a spike inflorescence (figure 1-27). In the nightshade family (Solanaceae), the petals of the flower are fused into a corolla tube with stamens (male flower parts) fused to the corolla (figure 1-28).

The sunflower family (Asteraceae) has a compact inflorescence in which numerous tiny flowers (florets) are arranged in a manner to resemble a single large flower (figure 1-29). The spurge family (Euphorbiaceae) has a rather unique inflorescence (figure 1-30).

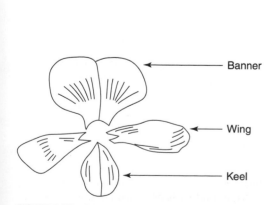

Banner

Wing

Keel

FIGURE 1-26 Petals of a typical legume flower.

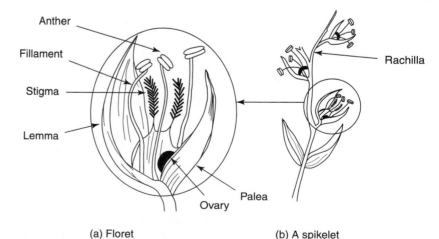

Anther

Filament

Stigma

Lemma

Rachilla

Palea

Ovary

(a) Floret

(b) A spikelet

FIGURE 1-27 Parts of a typical grass flower: (a) a floret and (b) a spikelet.

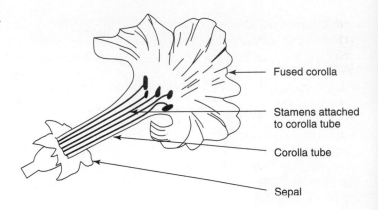

FIGURE 1-28 Fused petals of the family Solanaceae (nightshade family).

Fused corolla

Stamens attached to corolla tube

Corolla tube

Sepal

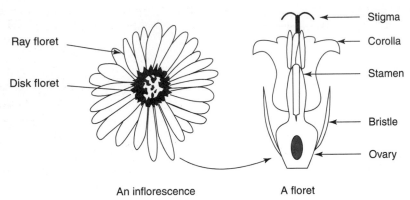

Stigma

Corolla

Stamen

Bristle

Ovary

Ray floret

Disk floret

An inflorescence

A floret

FIGURE 1-29 An inflorescence and single floret of the family Asteraceae (sunflower family).

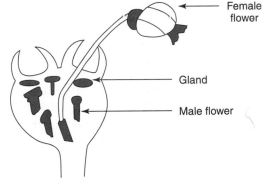

Female flower

Gland

Male flower

FIGURE 1-30 A cyathium of the family Euphorbiaceae (spurge family).

1.6: SOME FAMILIES OF HORTICULTURAL IMPORTANCE

The following is a partial listing and brief discussion of important crop families of horticultural interest.

1.6.1 ANGIOSPERMS (FLOWERING PLANTS)

Monocotyledons (Monocots)

1. *Poaceae (grass family).* In terms of numbers, the grass family is the largest of flowering plants. It is also the most widely distributed. This family includes all of the cereals (e.g., wheat, barley, oats, rice, and corn). In horticulture, most grasses are ornamental and especially noted for their role as lawn material in the landscape. Important grasses include the fescues (*Festuca* spp.) and bluegrass (*Poa* spp.). A popular grass used as a vegetable is sweet corn (*Zea mays*).

2. *Liliaceae (lily family).* Lilies are frequently characterized by large flowers whose parts occur in multiples of three. They are also known by their underground structures (bulbs, rhizomes, and other storage organs). They are mostly ornamental or medicinal in use. Asparagus (*Asparagus officinalis*) is the rare example of a plant in this family used as a food source for humans. Important ornamentals include *Lilium, Aspidistra,* and *Hemerocallis* genera. The *Aloe* genus provides species used as both ornamentals and medicinals.

3. *Orchidaceae (orchid family).* The orchid family is very large, with diverse characteristics. Many of this family's members are epiphytes found on tree bark. Some are terrestrial and

others aquatic. The flowers, which can range in size from barely conspicuous to gigantic, are characterized by three sepals and three petals. The distinguishing feature is a unique formation of one of the petals into a structure described as the lip. The widely used vanilla flavoring is obtained from an orchid, *Vanilla planifolia*. Orchids are also used in the cut flower industry, where important genera include *Dendrobium* and *Cattleya*.

4. *Amaryllidaceae (amaryllis family).* The amaryllis family is characterized by plants with tunicate bulbs (9.11.1). Plants in this family are adapted to temperate and warm regions of South America and South Africa. One of the most important genera is the *Allium*, which includes important species such as onion, garlic, and chives. *Narcissus* and *Amaryllis* are popular ornamental genera.

5. *Aracaceae (palm family).* The palm family is tropical and subtropical in adaptation. Its members make elegant landscape plants; the most popular ones include the royal palm (*Roystonea regia*). Certain palms are shrubs that can be grown indoors. Food-producing palms include the oil palm (*Elaeis guineensis*) and the coconut (*Cocos nucifera*).

Dicotyledons (Dicots)

1. *Brassicaceae (mustard family).* The mustard family is noted for its pungent herbs. This family contains some of the most popular garden crops, such as cabbage, brussels sprout, broccoli, radish, turnip, cauliflower, rutabaga, and horseradish. Apart from pungency, this family is also characterized by fruit types called siliques or silicles. Certain ornamental types of the vegetable species occur.

2. *Fabaceae (legume family).* In terms of number, the legume family ranks third behind the sunflower and orchid families. Fabaceae is characterized by flowers that may be *regular* (radially symmetrical) or *irregular* (bilaterally symmetrical) in shape. Irregular (i.e., capable of being divided into two symmetrical halves only by a single lengthwise plane passing through the axis) flowers have two wing petals, one large banner petal, and a keel petal that is boat shaped. Legumes are important sources of foods that are high in protein. They are used in improving the protein quality and quantity of forage for livestock. Important garden legumes include pea, lima bean, garbanzo bean, and mung bean.

3. *Cactaceae (cactus family).* Cacti are native to North and South America. They vary in size from the pinheadlike forms to the giant saguaro forms, which may be as tall as 15 meters (50 feet) and weigh in excess of 4.5 metric tons (5 tons). They have large and brightly colored flowers. Cacti grow very slowly and require little care when grown as houseplants. Their leaves are usually small, and their fleshy stems may be flattened or cylindrical, with the capacity to photosynthesize.

4. *Lamiaceae (mint family).* Most plants in the mint family produce aromatic oils in their leaves and stems. They are also characterized by their angular stems, opposite leaves, and irregular flowers. The mint family includes popular herbs such as rosemary, sage, thyme, marjoram, basil, catnip, peppermint, lavender, and spearmint. Apart from being useful for medicinal and culinary purposes, mints are used as ornamentals in landscaping.

5. *Solanaceae (nightshade family).* The nightshade family is noted for the poisonous alkaloids many of them produce (e.g., the deadly drug complex called *belladonna* that is extracted from the nightshade plant). Other drugs produced by this family are atropine, scopolamine, nicotine, solanine, and hyoscyamine. Important vegetable plants in this family are tomato (*Lycopersicon esculentum*), eggplant (*Solanum melongena*), pepper (*Capsicum* spp.), and potato (*Solanum tuberosum*). Other important plants are tobacco (*Nicotiana tabacum*) and petunia (*Petunia hybrida*). When potato tubers are exposed to the sun, they produce a green color at the surface. These green areas are known to contain toxins.

6. *Apiaceae (carrot family).* Plants in the carrot family frequently produce numerous tiny flowers that are arranged in umbels (2.20). The plants of horticultural importance include vegetables and herbs such as parsley, carrot, celery, dill, coriander, and parsnip.

7. *Cucurbitaceae (pumpkin family).* The pumpkin or gourd family is characterized by prostrate or climbing herbaceous vines with tendrils and large, fleshy fruits containing numerous seeds. Important plants include pumpkin (*Cucurbita maxima*), melon (*Cucumis melo*), watermelon (*Citrullus lunatus*), and cucumber (*Cucumis sativus*).

8. *Asteraceae (sunflower family).* The sunflower family has the second largest number of flowering plant species. Flowers in this family occur in a compact inflorescence or head (figure 1-29). Some of the members are edible and others ornamental. Important plants include sunflower (*Helianthus annuus*), marigold (*Tagetes* spp.), *Dahlia* spp., *Chrysanthemum* spp., *Aster* spp., and edible plants such as lettuce (*Lactuca sativa*), Jerusalem artichoke (*Helianthus tuberosus*), and endive (*Cichorium intybus*). The common dandelion, (*Taraxacum officinale*), a noxious weed in lawns, belongs to this family.

9. *Euphobiaceae (spurge family).* Most members of the spurge family produce milky latex, and the family includes a number of poisonous species. The largest genus in this family is the *Euphorbia*. Important plants include the Christmas plant, or poinsettia (*Euphorbia pulcherrima*); a root crop, cassava (*Manihot esculenta*); and the castor bean (*Ricinus communis*).

10. *Rutaceae (rue family).* Most of the species in the rue family are aromatic shrubs or trees. An important and popular genus is the *Citrus,* which includes plants such as mandarin, lemon, lime, grapefruit, and sweet orange.

11. *Ericaceae (heath family).* The heath family consists of shrubs that are adapted to acidic soils. The genera of horticultural importance include *Rhododendron* (rhododendron and azalea) and *Vaccinium* (blueberry and cranberry).

1.6.2 GYMNOSPERMS

Gymnosperms
Plants that bear seeds that are not within fruits (naked).

Gymnosperms have naked seed. There are four divisions of gymnosperms with living representatives: Cycadophyta (cycads), Ginkgophyta (ginkgo, maidenhair tree), Coniferophyta (conifers) and Gnetophyta (gnetophytes). The most widespread of these divisions is the Coniferophyta which consists of about 50 genera and 550 species. The most familiar of all conifers are the pines of the family Pinaceae. The important genera of conifers other than the pines are the firs (*Abies*), spruces (*Picea*), hemlocks (*Tsuga*), Douglas firs (*Pseudotsuga*), cypresses (*Cupressus*), and junipers (*Juniperus*). These predominantly evergreen trees and shrubs occur primarily in temperate areas.

SUMMARY

The science of classifying and naming plants is called *plant taxonomy.* Carolus Linnaeus developed the current Latin-based binomial nomenclature in which plants are given two names, the first name called the *genus* and the second name called the *species.* The International Code of Botanical Nomenclature (ICBN) provides guidelines for the naming of plants. The taxonomic groups in order of descending hierarchy are kingdom, division, class, subclass, order, family, genus, species, and variety. Kingdom is the most genetically divergent level, and variety represents the level at which individuals are most similar in genotype and external features.

There are three plant kingdoms—Monera, Protoctista, and Fungi. Most plants of horticultural interest are vascular plants (having conducting tissues—xylem and phloem). Apart from classifying plants on a scientific basis, there are a number of operational classifications based on (1) stem type (herbs, shrubs, and trees), (2) kind of herb (annual, biennial, or perennial), (3) stem growth form (erect, decumbent, creeping, or climbing), (4) fruit type (herbaceous fruiting plants versus woody fruiting plants), (5) adaptation (cool season or warm season), and (6) edible parts (roots, bulbs, pods, greens, and so forth).

REFERENCES AND SUGGESTED READING

Benson, L. 1979. Plant classification, 2d ed. Lexington, Mass.: Heath.

Esau, K. 1977. Anatomy of seed plants, 2d ed. New York: John Wiley & Sons.

Glendfill, D. 1989. The names of plants. 2d ed. New York: Cambridge University Press.

Hartman, H. T., A. M. Kofranek, V. E. Rubatzky, and W. J. Flocker. 1988. Plant science, 2d ed. Englewood Cliffs, N.J.: Prentice-Hall.

Radford, A. E. 1986. Fundamentals of plant systematics. New York: Harper & Row.

Rice, L. W., and P. R. Rice Jr. 1993. Practical horticulture, 2d ed. Englewood Cliffs, N.J.: Prentice-Hall.

Stern, K. R. 1997. Introductory plant biology, 2d ed. Dubuque, Iowa: Wm. C. Brown Publishers.

PRACTICAL EXPERIENCE

1. Visit a grocery store and purchase different horticultural produce in the following classes:
   a. Fruits: drupes
              berries
              nuts
   b. Roots
   c. Leaves
   d. Bulbs
2. Cut across (transverse section) the fruits to expose the inside and compare them.
3. Complete the plant taxonomic classification of sweet corn, following the example given in section 1.2.
4. Field trip
   a. Take a walk on campus, bringing along copies of figures to use in identifying the variety of characteristics that are the basis of plant taxonomy. Record as many distinguishing features as are represented in the plants on campus (e.g., flower arrangement, leaf shape, and margin).
   b. Take a trip to a botanical garden and repeat the exercises in part a.

OUTCOMES ASSESSMENT

PART A

Please answer true (T) or false (F) for the following statements.

1. T F   de Candole is the father of taxonomy.
2. T F   If a taxon has a name ending in *aceae,* it categorizes a plant family.
3. T F   In a descending hierarchy the taxon *genus* is lower than *species*.
4. T F   According to the rules of the binomial nomenclature, the species name always begins with an uppercase letter.
5. T F   A pumpkin is a drupe.
6. T F   Pepos have a leathery rind.

PART B

Please answer the following questions.

1. What does the acronym ICBN stand for? _____
2. Write the binomial name FRAGARIA VIRGINIANA (scarlet strawberry) in lower-case letters according to standard rules. _____
3. The binomial nomenclature was developed by _____.
4. The science of classifying and naming plants is called _____.
5. Give two examples for each of the following:
   a. annual vegetable _____
   b. evergreen ornamental _____
   c. deciduous fruit tree _____
   d. cool-season vegetable _____
   e. bulb _____
   f. cole crop _____
   g. bedding plant _____

PART C

1. Why is the function of the ICBN important to scientists?
2. Certain plants can be classified as both annuals and biennials or (perennials). Explain.
3. Most breeding work occurs at the variety level in the taxonomic hierarchy. Explain.

CHAPTER 2

Plant Anatomy

PURPOSE

This chapter is designed to describe the physical and structural organization of higher plants and to show how plant anatomy is used as a basis of classifying horticultural plants.

EXPECTED OUTCOMES

After studying this chapter, the student should be able to

1. Describe the levels of eukaryotic organization.
2. Describe the cell structure and function of major organelles.
3. List and describe the primary tissues of higher plants and their functions.
4. Describe various plant organs and how they are used as a basis for classifying horticultural plants.

KEY TERMS

Annual rings	Heartwood	Primary vegetative body
Complete flower	Hypogeous germination	Reaction wood
Complex tissue	Imperfect flower	Sapwood
Compound leaves	Incomplete flower	Sclerenchyma cells
Compression wood	Inflorescence	Simple fleshy fruits
Conducting tissue	Meristems	Simple leaves
Dicot	Monocot	Simple tissue
Differentiation	Multiple fruits	Softwood
Epidermis	Organelles	Tension wood
Epigeous germination	Parenchyma cells	Turgor pressure
Foliage leaves	Parthenocarpy	Xerophytes
Hardwood	Plasma membrane	

OVERVIEW

The scientific discipline of *plant anatomy* deals with cataloging, describing, and understanding the function of plant structures. The functional aspect of the study of anatomy overlaps with *plant physiology* (chapter 4). Plant anatomy can be studied at various levels of eukaryotic organization, the most fundamental being the molecular level, which deals with macromolecules (nucleic acids, proteins, and carbohydrates) (figure 2-1). These basic molecules are organized into *organelles* (such as mitochondria, chloroplasts, and nuclei). The next level of complexity of organization is *tissues* (such as xylem and phloem). Cellular substructures are not visible to the naked eye, requiring the aid of magnifying instruments (such as the light microscope) to be seen. Even though organs and whole-plant structures are the most readily visible to the naked eye, it is important to know that what we see are products of subcellular function involving the effects of physiological processes.

A good understanding of plant anatomy helps horticultural scientists in manipulating plants for increased productivity and aesthetic value. Before nudging nature, one needs to know and understand the norm, how it responds to change, and how to effect change.

2.1: CELL

2.1.1 THE UNIT OF ORGANIZATION OF LIVING THINGS

The *cell* is the unit of organization of living things. Some organisms have only one cell *(unicellular),* and others are made up of many cells *(multicellular).* Unicellular organisms are also called *prokaryotes,* or lower organisms, an example being the algae. They lack a distinct nucleus due to the absence of a nuclear membrane. Higher organisms, or *eukaryotes,* have distinct nuclei and cells that are compartmentalized by means of membranes such that each compartment has a different function. They may be unicellular or multicellular. Plants of horticultural interest are eukaryotes, or higher plants.

Under appropriate laboratory conditions, a single cell may be nurtured to grow and develop to produce the entire plant from which it was derived. This capability is called *totipotency* and results because each cell has the complete *genome* (the complete set of genes for the particular organism) to direct the development of the whole plant (5.19). This capacity is exploited in propagating certain horticultural plants (9.12) and manipulating the genetic structure of others to produce new and improved types.

Cells, like all living things, grow and age. There are different sizes and shapes of cells. Through the process of *cell division,* a single cell rapidly divides and multiplies to produce a uniform mass of cells. These cells subsequently undergo changes through the process of *differentiation* to perform specific functions in the plant, as needed. For example, some cells change to produce strengthening tissues, while others produce flower or leaf buds. During

FIGURE 2-1 Levels of organization of eukaryotes. The study of eukaryotes below the level of organs usually requires the aid of special equipment such as microscopes to observe and manipulate.

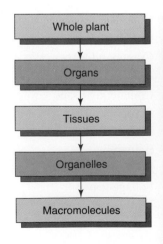

this differentiation process, the shapes and sizes of cells are modified appropriately, as is their structural strength. Horticultural products are harvested in time to obtain products that have the optimal quality desired by consumers. A delay in harvesting a product may reduce its quality and consequently the market value. As cells age, their physical structures change such that products that should be juicy and succulent, for example, become less juicy and more fibrous.

2.1.2 CELL STRUCTURE

The plant cell may be divided into three parts—the *outer membrane* (*plasma membrane* or *plasmalemma*), the *cytoplasm,* and the *nucleus.* The plasma membrane functions as a selective barrier to the transport of substances into and out of the protoplast.

The content of a living plant cell, excluding the wall, is called the protoplasm (or protoplast). Embedded in the protoplasm are discrete bodies called organelles. The most prominent of cellular organelles is the nucleus, the organelle that houses most of the cell's genetic material (deoxyribonucleic acid, or DNA). DNA is responsible for directing cellular functions (chapter 5). Some old cells lack nuclei. The area outside of the nucleus is called the cytoplasm, and it contains the other organelles (figure 2-2). Some of these organelles are described in the following sections.

> **Plasma Membrane**
> The membrane that surrounds the entire protoplast.

> **Organelle**
> A specialized region in a cell that is bound by a membrane.

The Cell Wall

All plant cells except the sperm and some egg cells have walls. When the protoplast dies and degenerates, what is left is the *cell wall,* which constitutes the bulk of woody plants (wood and bark [2.6 and 2.7]). Chemically, the cell wall consists of *cellulose, hemicellulose, protein,* and *pectic substances* (4.2.1). Cellulose, the most abundant cell wall component, is a polysaccharide (a polymer of sucrose molecules) and nutritionally of little value to humans since it is not digestible. Pectins are acidic polysaccharides (4.2.1) (polymers of galacturonic acid). At a particular stage in the life of a cell, deposition of *lignin* (a complex mixture of polymers of phenolic acid) occurs to harden the cell wall, rendering it rigid and inelastic. This process is called *lignification.*

The thickness and other structural features of a cell wall vary according to the age and type of cell. All cells have a standard *primary cell wall,* which is the first to form when a cell is developing. This type of cell wall is found where cells are actively growing and dividing. Chemically, it is composed predominantly of cellulose and pectic substances. When cell growth ceases, a *secondary cell wall* is deposited inside of the primary cell wall. With this new layer of cellulose and lignin, the cell wall becomes rigid. Adjacent cell walls are held together

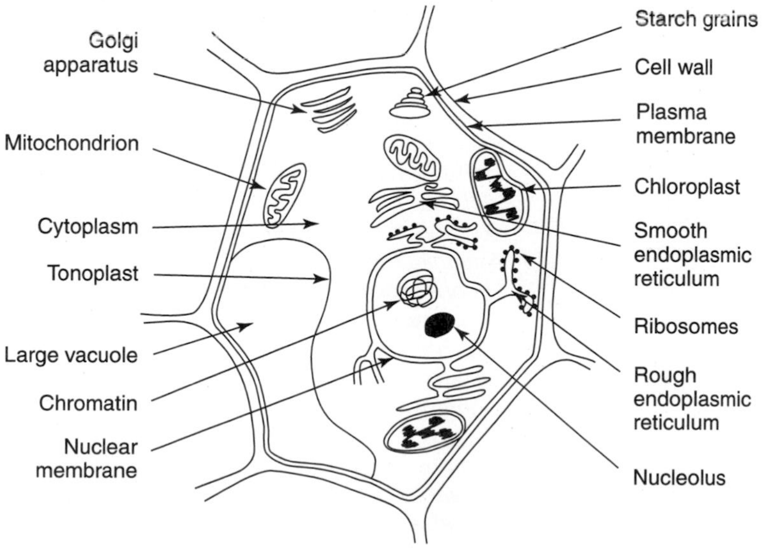

FIGURE 2-2 The parts of a plant cell.

by a pectin-rich material called the *middle lamella.* When fruits rot under fungal attack, the middle lamella breaks down into the characteristic slimy fluid associated with rotting. Cell-to-cell interconnections are produced by structures called *plasmodesmata* (cytoplasmic strands).

Nucleus

Upon staining a cell in the resting stage (interphase), the *nucleus* usually shows up as a spheroidal and densely stained body. This structure is the fundamental organelle of a cell since it is the primary repository of genetic information for the control and maintenance of cellular structure and function. The nucleus is composed of DNA, ribonucleic acid (RNA), proteins, and water. The DNA (5.2.4) occurs in defined structures called *chromosomes* that allow it to be replicated accurately (except in occasional alterations called mutations [5.10]).

Chromosomes are visible as strands when the DNA-histone complex coils and appears to condense. Chromosomes stain differentially to reveal dark and light sections. The dark sections are called *heterochromatin* and represent DNA-containing genes that are not actually directing the synthesis of RNA. The light-staining sections are called *euchromatin* and contain active genes. The chromosome number per cell is characteristic of the species (table 2-1). The number of chromosomes in the *gametic cell* (sex cell, such as pollen) is half that of the *somatic cell* (body cell). When two *homologous pairs* (identical mates) occur in a cell, it is called a *diploid.* Sometimes, in certain plant species, cells may contain multiple copies beyond the diploid number, a condition called *polyploidy* (5.11).

Vacuoles

Vacuoles are cavities in cells that contain a liquid called the *vacuolar sap,* or *cell sap,* within the vascular membrane called *tonoplast.* The sap consists mainly of water, but other substances such as salt, sugars, and dissolved proteins occur, according to the physiological state of the cell. Vacuoles also store water-soluble pigments called *anthocyanins.* These pigments are responsible for the red and blue colors of many flowers (e.g., geranium, rose, and delphinium), fruits (e.g., cherry, apple, and grape), and vegetables (e.g., cabbage, turnip, and onion). Anthocyanins are also involved in the fall colors of some leaves. Vacuoles vary in size: In meristematic cells (young and actively dividing) (2.2.2), the vacuoles in a single cell are small in size but numerous. In mature cells, however, these numerous small vacuoles usually fuse into large cavities that may occupy about 90 percent of the cell, pushing the re-

> **Chromosome**
> A highly organized nuclear body that contains DNA.

Table 2-1 The Number of Chromosomes Possessed by a Variety of Plant Species

Species	Scientific Name	Chromosome Number (n)
Carrot	*Daucus carrota*	18
Garden pea	*Pisum sativum*	7
Evening primrose	*Oenothera biennis*	7
Broad bean	*Vicia faba*	6
Potato	*Solanum tuberosum*	24
Snapdragon	*Antirrhinum majus*	8
Tomato	*Lycopersicon esculentum*	12
Corn	*Zea mays*	10
Lettuce	*Lactuca sativa*	18
Garden onion	*Allium cepa*	8
White oak	*Quercus alba*	12
Yellow pine	*Pinus ponderosa*	12
Cherry	*Prunus cerasus*	16
Bean	*Phaseolus vulgare*	11
Cabbage	*Brassica oleracea*	9
Cucumber	*Cucumis sativus*	7

mainder of the protoplasm against the cell wall. Vacuoles absorb water to create the *turgor pressure* (4.1.1) required for physical support in plants. Plants under moisture stress wilt for lack of turgor pressure. A variety of rapid movements in plants such as flower opening, leaf movement in response to touch (e.g., in *Memosa pudica*), and the opening and closing of guard cells are attributed to functions of the vascular vacuoles. Vacuoles have some digestive functions similar to those of lysosomes in animal cells; macromolecules are broken down in vacuoles and their components recycled within the cell.

Plastids

Plastids are very dynamic plant cell organelles capable of dividing, growing, and differentiating into different forms, each of which has a different structure and function. Plastids contain their own DNA. They are said to be *semiautonomous* because they synthesize some of their own proteins. The genes they contain are not inherited according to Mendelian laws (they have extrachromosomal inheritance, meaning that it does not occur in chromosomes in the nucleus [2.1.2]). *Chloroplasts* are plastids that contain *chlorophyll,* the green pigment that gives plants their characteristic green color and, more importantly, is involved in photosynthesis (4.3.1). A chloroplast contains saclike vesicles called *thylakoids,* which are stacked in units called *grana* (singular: *granum*). The grana are suspended in a fluid called *stroma.* Chloroplasts occur only in plants, not animals. During cell division, no special mechanism ensures equal distribution of plastids. Consequently, certain cells may receive no plastids at all; the parts of the leaf that have cells without chloroplasts develop no green color. Instead, they produce white, pink, or purple coloration. Leaves showing such patches of color are said to be *variegated* (figure 2-3). Colorless plastids are generally called *leucoplasts.*

Whereas nonangiosperms are generally not colorful, being predominantly green, angiosperms (flowering plants whose seeds develop within ovaries that mature into fruits) have certain plastids with the capacity to produce large amounts of *carotenoids* (bright yellow or orange and red pigments). These plastids are called *chromoplasts* (*chroma* means color). More than 30 different types of pigments have been found in the chromoplasts of pepper (*Capsicum* spp.). The pigments are diverse in their composition and the colors they produce. Flowers display a spectacular array of colors. Colors found in petals and fruits are caused by plastids. Although leaves are predominantly green, they may also exhibit other colors. The presence of chlorophyll overwhelms other colors and masks their expression; however, under the right conditions (such as occurs seasonally in fall), the chlorophyll breaks down, allowing the masked colors to be expressed as beautiful colors during the fall season in temperate regions. All types of plastids, especially chloroplasts and chromoplasts, are interconvertible. When plants are grown in darkness, chloroplasts change into *etioplasts,* resulting in a deformation called *etiolation* (spindly growth due to excessive elongation of internodes) (3.2.1). Light is required to reverse this abnormal growth. To prevent etiolation,

> **Turgor Pressure**
> The pressure on a cell wall that is created from within the cell by the movement of water into it.

> **Plastid**
> An organelle that is bound by a double membrane and associated with different pigments and storage products.

FIGURE 2-3 Variegation of the leaf of dumbcane (*Dieffenbachia* spp.).

plants grown under conditions of insufficient daylight (such as houseplants and greenhouse plants) are provided with supplemental light from an artificial source.

Mitochondria

A cell may survive without plastids, but all cells must have *mitochondria,* organelles that provided the energy (adenosine triphosphate, ATP [4.3.2]) required for plant processes. These organelles are bound by a double membrane. The inner membrane is folded into projections called cristae; this extreme folding increases the internal surface area of mitochondria for biochemical reactions. Mitochondria have their own DNA, just like chloroplasts. They are sites of respiration, the cellular process responsible for producing energy for living organisms (4.3.2).

Ribosomes

Ribosomes are sites of protein synthesis (the process of translation of the genetic message) (5.2.4). Tiny structures, they consist of approximately equal amounts of RNA and protein. Ribosomes may occur freely in the cytoplasm or attached to the endoplasmic reticulum (described next). When engaged in protein synthesis, ribosomes tend to form clusters called *polyribosomes* or *polysomes.*

Endoplasmic Reticulum

The *endoplasmic reticulum* is a membranous structure distributed throughout the cytoplasm as a system of interconnected, flattened tubes and sacs called *cisternae.* The extent of folding and the amount of endoplasmic reticulum depend on the cell type and function and the cell's stage of development. A transverse section of this organelle shows two parallel membranes with a space between them. When ribosomes are attached to the surface, it is called *rough endoplasmic reticulum;* otherwise it is called *smooth endoplasmic reticulum.* This organelle also serves as a channel for the transport of substances such as proteins and lipids to different parts of the cell and as the principal site of membrane synthesis.

Golgi Apparatus

The *Golgi apparatus* is a collective term for structures called *Golgi bodies* or *dictyosomes,* which consist of a stack of about four to eight flattened sacs, or cisternae. In higher plants, dictyosomes have secretory functions. They secrete new cell wall precursors and other substances.

Microbodies

Microbodies are single-membraned, spherical bodies that have roles in metabolic processes. A group of microbodies associated with mitochondria and chloroplasts are called *peroxisomes.* These microbodies contain the enzyme glycolate oxidase and function in glycolate oxidation during photorespiration (light-dependent production of glycolic acid in chloroplasts and its subsequent oxidation in peroxisomes (4.3.1). Another group of microbodies called *glyoxysomes* contain enzymes involved in the breakdown of lipids to fatty acids. The fatty acids are converted to carbohydrates that are used for growth and development during the germination of many seeds.

Ergastic Substances

Ergastic substances are miscellaneous substances in the cell that include waste products and storage products such as starches, anthocyanin, resins, tannins, gums, and protein bodies. These substances may be classified as either *primary metabolites* (e.g., starch and sugars that have a basic role in cell metabolism) or *secondary metabolites* (e.g., resins and tannins that have no role in primary metabolism). Secondary metabolites are known to play a role in protecting the plant from herbivores and insect attack. Tannins are toxic to animals. For example, the rhubarb leaf blade contains calcium oxalate crystals, but the petiole does not; thus, while the petioles are edible, the leaf blade is toxic to animals. Similarly, *Philodendron* stems and shoots and *Dieffenbachia* (dumbcane) leaves contain calcium oxalate in the raphides (sharp needles on the plant) that is very irritating to the throat. Noncrystalline ergastic substances such as silica deposits occur in the cell walls of some grasses and sedges. These deposits strengthen the tissue and also protect the plant against insects and other pests.

2.2: TYPES OF PLANT CELLS AND TISSUES

Different types of cells are used to construct the different tissues and organs to meet the variety of functional needs of the plant. These cells can be classified into three basic types: *parenchyma, collenchyma,* and *sclerenchyma.* The three types of cells differ from each other in their cell wall characteristics. Simple cells aggregate in certain characteristic patterns to form *tissues.* When the tissue consists of one type of cell, it is described as a *simple tissue;* when more than one type of cell is present, the tissue is called a *complex tissue.*

> **Tissue**
> *A set of cells that function together.*

2.2.1 SIMPLE TISSUES

1. *Parenchyma.* Parenchyma cells are characterized by their thin wall (figure 2-4). The tissue they form is described as *parenchyma (parenchymatous) tissue.* This cell type occurs extensively in herbaceous plants. Functionally, parenchyma cells are found in actively growing regions of plants called *meristems* (2.2.2). Meristematic cells are undifferentiated. The parenchyma cells in meristematic regions are also called *meristematic parenchyma.* Some parenchyma cells have synthetic functions, such as in the chloroplasts, where they are called *photosynthetic parenchyma,* or *chlorenchyma* cells, and function in photosynthesis. Some parenchyma cells have secretory roles and are called *secretory parenchyma.* The fleshy and succulent parts of fruits and other swollen parts such as roots and tubers consist of large amounts of parenchyma tissue.

> **Meristem**
> *A cell or region of specialized tissue whose principal function is to undergo cell division.*

2. *Collenchyma.* Collenchyma cells have a thick primary wall that plays a mechanical role in the plant support system by strengthening tissues (figure 2-5). This role is confined to regions of the plant where active growth occurs (i.e., where thin-wall meristematic cells occur) so as to provide the plant some protection from damage. Collenchymatous tissue occurs in the stem below the epidermis (outermost layer of the cells in the particular plant part) (2.2.2); in leaves, it occurs in the petiole (or leaf stalk), leaf margins, and main veins of the leaf blade or lamina. Fruit rinds that are soft and edible contain collenchyma tissue. As these cells age, they accumulate hardening substances and become unevenly sclerified, or thickened.

3. *Sclerenchyma.* Sclerenchyma cells have two walls, primary and secondary, the latter being thicker (figure 2-6). They also have a mechanical function in plants, serving as reinforcement for tissues. Sclerenchyma cells have elasticity and resiliency and thus can bend without snapping. Naturally, therefore, they are found in places in the plant,

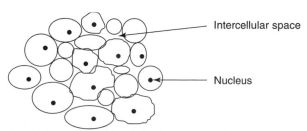

Intercellular space

Nucleus

FIGURE 2-4 Parenchyma cells are thin walled and appear undifferentiated.

FIGURE 2-5
Collenchyma cells have thicker walls than parenchyma cells and may have a thin layer of lignin in certain cases.

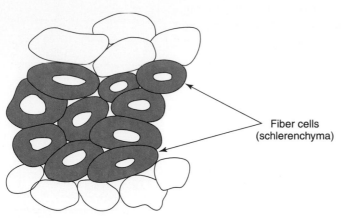

FIGURE 2-6 Schlerenchyma cells have secondary walls that are thickened for added strength.

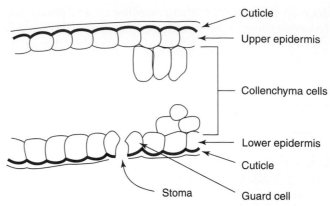

FIGURE 2-7 The leaf epidermis may be homogeneous or have trichomes or glands in addition to stomata. It is overlaid by a layer of cuticle.

such as the leaf petiole, where bending and movement occur. There are two basic types of sclerenchyma cells—short cells, called *sclereids,* and long cells, called *fibers.* The primary cell wall is made up of cellulose, hemicellulose, and other pectic substances. The secondary wall is formed from large deposits of lignin; cells with deposits of lignin are said to be lignified. Sclerenchyma occurs, for example, in the stones of fruits, around the seeds, or in immature fruits. Sclerenchyma cells abound in plants that yield fiber such as kenaf, flax, and hemp.

2.2.2 COMPLEX TISSUES

The three basic cell types may aggregate separately or in combination to form complex tissues that perform a variety of functions in the plant. Some of these tissues are *epidermis, secretory tissue,* and *conducting tissue.*

Epidermis

> **Epidermis**
> *The outermost layer of cells on all parts of the primary body—stems, leaves, roots, flowers, fruits, and seed—but absent from root tips and apical meristems.*

The *epidermis* is the outermost layer of the plant that separates its internal structures from its external environment (figure 2-7). Since a plant's environment changes, its epidermis should possess developmental plasticity or flexibility such that it adapts to a wide range of conditions. Such flexibility may involve physiological, structural, and anatomical variability.

By virtue of its position, the primary function of the epidermis is to regulate water and gas movement into and out of the plant. Some plants have a waterproof epidermis, whereas others are permeable to moisture and gases. Waterproofing is caused by the occurrence of a hydrophobic (water-repelling) substance called *cutin* (polymerized fatty acids) that is deposited on the outside of the epidermal wall. Waterproofing occurs in most rain forest epiphytes, such as cactus, orchid, and philodendron, where protection against leaching of minerals is critical. The resulting layer is called a *cuticle* and is waxy in nature. The epidermis protects the plant against sunlight. Intense sunlight can overheat the protoplasm and bleach the chlorophyll. Sometimes orchardists paint the exposed trunks of fruit trees to reflect sunlight and thus prevent damage from excessive heat. The epidermal layer also has a protective role, resisting the intrusion of biological pests such as bacteria and fungi. Some layers protect the plant against chewing insects, and others have hairlike structures (pubescence) called *trichomes* that interfere with oviposition (deposition of eggs). Some of these epidermal outgrowths secrete a variety of substances for many different purposes.

In the green parts of plants, especially in leaves, the epidermis has pores called *stomatal pores,* or *stomata* (singular: *stoma*) (2.4.2). The pores occur predominantly on the abaxial (lower surface) part of the leaf and are bordered by structures called *guard cells.* The guard cells control the opening of the stoma. These pores function in gaseous (carbon dioxide and oxygen) and moisture exchange. In some horticultural practices, certain chemicals are administered to plants by foliar application. Such chemicals enter the plant through the stomata.

Secretory Tissue

According to the nature of the material secreted, secretory systems may be classified according to where they are found (i.e., whether outside or inside of the plant).

Secretion
The movement, either by diffusion or active transport, of materials out of a plant or into a space where it can accumulate for storage.

Found Outside of the Plant

1. *Nectaries.* Nectaries are found on parts of the plant. When they occur in flowers they are called floral nectaries, and when they occur elsewhere they are called extrafloral nectaries. They secrete a fluid called *nectar* that consists of sugars (especially glucose, sucrose, and fructose) and numerous other organic compounds. Certain insects including butterflies and bees feed on nectar and in the process aid in the pollination of the flowers.

2. *Hydathodes.* Secretory structures called hydathodes secrete almost pure water. They are thus thought to play a role in the transport of minerals to young tissues, in addition to the role played by transpiration (the loss of water vapor by plant parts) (4.3.3). Under conditions of moist soil, high humidity, and cool air, the leaves of many plants, especially grasses, are known to produce droplets of water along their margins that appear similar to dew. However, the moisture is due to a special secretory process called *guttation* (4.3.4).

3. *Salt glands.* Salt glands, which secrete inorganic salts, occur in plants that grow in desert and brackish areas that are high in salts. These salts may accumulate on the leaves of certain plants and thereby make them unattractive to herbivores.

4. *Osmophores.* Osmophores are fragrance-secreting glands found in flowers. They secrete odors and perfumes (predominantly oily compounds belonging to the class of volatile, small terpenes). The repulsive odor of aroids is attributed to the amines and ammonia produced by osmophores.

5. *Digestive glands.* Digestive glands are found in insect-eating (insectivorous) plants. They secrete enzymes used in digesting the animal materials trapped by such plants (e.g., pitcher plant).

6. *Adhesive cells.* Adhesive cells secrete materials that allow for attachment between host and parasite. The strong attachment due to adhesive material helps parasites during penetration of their host.

Found Inside of the Plant

1. *Resin ducts.* Resin ducts are long canals that contain sticky resin. They are most abundant in the wood and leaves of conifers.

2. *Mucilage cells.* Mucilage cells are slimy secretions high in carbohydrate and water content. The mucilage secreted by the growing root tip, called mucigel, is believed to be important in lubricating the passage of the root through the soil.

3. *Oil chambers.* Certain glands secret compounds that are commonly deposited in large cavities in the plant. However, these oils are moved outside of the plant, where they are aromatic.

4. *Gum ducts.* Cell wall modification results in the production of gums in certain tree species.

5. *Laticifers.* Laticifers are latex-secreting glands (turbid and usually milky secretions). They occur in species such as milkweed, poppy, and euphorbia. The latex may contain diverse compounds including carbohydrates, lipids, tannins, rubber, protein, and crystals. Even though the secretion from poinsettia (*Euphorbia pulcherrima*) has been found to be nontoxic to humans, "plant milk" should not be ingested.

6. *Myrosin cells.* Myrosin cells contain a neutral enzyme called myrosinase. However, when this harmless protein is mixed with its substrate (thioglucosides), a toxic mustard oil (isothiocyanate) is produced. This mixing occurs when myrosin cells are ruptured during chewing by insects or other animals.

Since all cells are capable of transporting materials across their boundaries, all cells may be said to have secretory functions. Secretory functions are needed by cells for a variety of reasons. Accumulation of metabolic waste in the cell may be toxic, and so the waste must be excreted. Compounds that are actively transported out of the cell include sugars, inorganic salts, hormones, nitrogenous compounds, and sulfur-containing compounds. However, deposits in the cell wall such as cuticle, suberin, lignin, and waxes are products of secretory activities that are desired for specific roles in the plant, as previously mentioned. In carnivorous plants, secretions are needed to digest trapped animals.

Apart from secretory activities that are related to cellular metabolism, cells secrete other chemicals. Some glands in flowering plants secrete scents to attract pollinators. Some flowers produce very sweet fragrances that people enjoy. Other plant secretions are substances that repel animals that may be pests.

Conducting Tissues

Vascular plants have an elaborate system of vessels used in conducting organic and inorganic solutes from place to place in the plant. These systems are complex tissues consisting of a variety of cell types. There are two conducting tissues in plants: *xylem* and *phloem.*

Xylem Tissue   The xylem tissue conducts water and solutes from the roots up to the leaf, where food is manufactured by the process of photosynthesis (4.3.1). Since it consists of sclerenchyma cells, it also provides structural support to the plant as a whole, and it stores nutrients or new materials for photosynthesis. The conducting cells of the xylem are of two types, *tracheids* and *vessel elements.* These two types of cells are collectively called *tracheary elements,* which vary widely in size, shape, and types of secondary walls (figure 2-8).

Water is moved up the xylem tissue by water potential. This movement is due to passive transport because the xylem cells have no protoplasm, just the cell wall, and hence function essentially as dead cells. Tracheids tend to be long and spindle shaped; vessel elements may be narrow or wide. Both cell types have lateral perforations (or *pits*) to permit flow of cell sap from cell to cell. The elements may have large holes in the primary wall and thus are classified as tracheids, or they may lack such holes and be called vessel elements. Tracheids are the only conducting elements in gymnosperms. In angiosperms, vessel elements are good conductors when water is abundant and flows in high volume. Wood is predominantly xylem tissue (2.6). In nonwoody (herbaceous) plants, xylem occurs in amounts necessary for the conduction of water.

Phloem Tissue   Structurally, phloem tissue may be made up of parenchyma cells or a combination of parenchyma and sclerenchyma cells. The conducting elements are called *sieve elements* and are of two types, *sieve cells* (primarily parenchyma) and *sieve tube membrane* (figure 2-9). In angiosperms, sieve elements are closely associated with spindle cells called *companion cells.* In nonangiosperms, these cells are called *albuminous cells.* They are believed to be involved in *phloem loading* (in which newly synthesized sugars are loaded for export to other plant parts) (4.3.5). This loading occurs in the minor veins of leaves. Phloem cells, of necessity, function as living cells. The general function of phloem tissue is to move food from the leaves, where it is manufactured, to other parts of the plant, where it is used or stored. A significant difference between xylem and phloem tissue function is that the plant actively controls the distribution of photosynthates as conducted by the phloem tissue system but a passive role in the movement of raw materials up to the leaves. The active (under the plant's control) movement of food by plants enables food to be transported to the parts of the plant where it is needed and similarly be extracted in other areas (*source-sink* relationship) (4.3.5). This dynamic nature of the movement of sugars, amines, and other nutrients en-

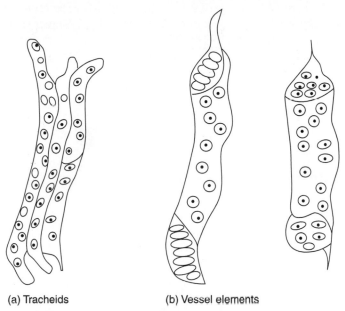

(a) Tracheids   (b) Vessel elements

FIGURE 2-8 Tracheary elements of the xylem tissue. Tracheids (a) are usually long and spindle shaped while vessel elements (b) are connected end to end to form a vessel. The two structures have secondary walls on the inside and are pitted.

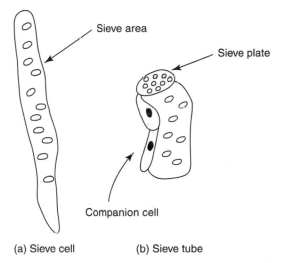

Sieve area

Sieve plate

Companion cell

(a) Sieve cell   (b) Sieve tube

FIGURE 2-9 Sieve elements of the phloem tissue. Sieve cells are usually tapered (a) while sieve tube members are wider (b). Both cell types are largely parenchymatous and are pitted.

ables the plant to reallocate resources as it grows through various phases. Phloem cells are not durable and must be replaced constantly.

Apical Meristems

In animals, growth occurs by means of the *diffuse growth* process. In this process, growth occurs throughout the entire individual, all parts growing simultaneously. In plants, however, the means of growth is by the *localized growth* process, whereby growth is limited to certain regions called meristems.

Meristems are areas of active growth where cells are dividing rapidly. The cells in these regions are undifferentiated (sources of *developmental plasticity*). In organs such as the root and the shoot, the meristems are located at the tip (apex) and hence are called *apical meristems* (figure 2-10). Some meristematic cells occur in the leaf axil and are called *axillary*

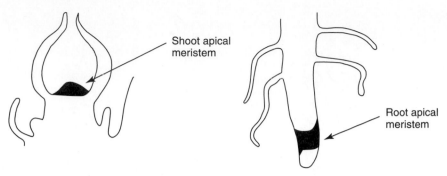

FIGURE 2-10   Sites of the apical meristems in plants.

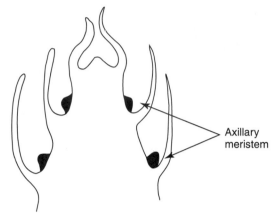

FIGURE 2-11   Sites of axillary meristems.

meristems (figure 2-11). Meristems also occur in other parts of the plant (e.g., *basal, lateral,* and *intercalary*). The localized growth process makes it possible for juvenile and mature adult cells to coexist in a plant provided good environmental conditions exist. The plant can continue to grow while certain organs and tissues are fully mature and functional and can grow indefinitely without any limit on final size. This growth pattern is called *open* or *indeterminate.* In reality, many plants appear to have predictable sizes that are characteristic of the species, but this feature is believed to be largely an environmental and statistical phenomenon. Under controlled environmental conditions in which optimal growth conditions prevail, many annual plants have been known to grow perennially.

In some species, the apical meristem dies after it has produced a certain number of leaves. The plant then ceases to grow and is said to be *determinate* (as opposed to indeterminate). In flowering species, plants go through a vegetative growth phase of variable duration (depending on the species) before flowering.

2.3: STEM

The *stem* is the central axis of the shoot of a plant.

2.3.1 FUNCTIONS

The functions of the stem include the following:

1.   Stems produce and provide mechanical support for holding up the branches, leaves, and reproductive structures. Leaves of plants need to be displayed such that they intercept light for photosynthesis.

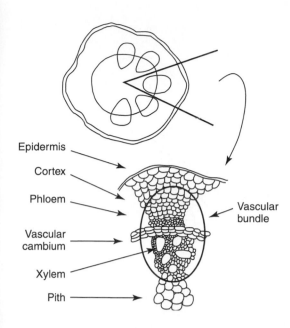

FIGURE 2-12   A cross section of a dicot stem.

Epidermis
Cortex
Phloem
Vascular
cambium
Xylem
Pith

Vascular
bundle

> **Vascular Cambium**
> A sheet-like fundamental type of
> meristem that produces secondary
> xylem and phloem.

2. Stems move water and minerals through their conducting vessels (xylem vessels) up to the leaves for the manufacture of food and then conduct manufactured food down from the leaves (through the phloem vessels) to other parts of the plant.

3. Stems may be modified to serve as storage organs for food, water, and minerals (9.11.1). Succulents such as cacti have stems that are designed for storage. The Irish potato is a swollen stem.

4. Certain plants are propagated through asexual means (chapter 9) by using pieces of the stem as cuttings. These cuttings are rooted and then planted as seedlings to raise new plants.

2.3.2   STEM TYPES

There are two basic types of stems in vascular plants: *dicots* (*dicotyledons,* or two cotyledons) and *monocots* (*monocotyledon,* or one cotyledon). They differ in how the primary tissues (xylem and phloem) are arranged. The outermost layer of the stem is the *epidermis,* which borders the internal part called the *cortex.* The cortex is usually composed predominantly of parenchyma cells. It may contain sclerenchyma cells in some plants. The cortex is usually narrow, except in herbaceous stems of monocots, in which it is often extensive. The cortex surrounds the *vascular tissues,* which form a central cylinder called the *stele.* The stele consists of *vascular bundles* and is made up of xylem and phloem tissues. In dicots and gymnosperms, vascular bundles are arranged in a ring (figure 2-12). In the center of the stem lies a region of purely parenchyma cells called the *pith.* In monocots, the bundles are distributed throughout the cortex, and there is no pith (figure 2-13).

2.3.3   MODIFIED STEMS

Stems do not always grow upright or vertically. Other forms of stems occur either above or below ground.

1. *Crowns.* The crown may be likened to a compressed stem, as found in bulbs. Leaf and flower buds occur on the crown and give rise to leaves and flowers. In plants such as asparagus, the crown may be further modified into a food storage organ.

2. *Stolons.* Stolons are stems that grow horizontally above ground and occur in plants such as strawberry and bermudagrass (figure 2-14). Roots may arise at the nodes of the stem as it creeps on the surface of the soil. When stolons have long internodes that originate at the base of the crown of the plant (as in strawberry), the stolon is called a *runner.*

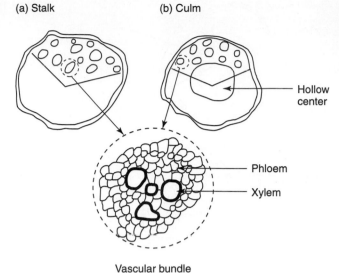

FIGURE 2-13 Cross sections of monocot stems: (a) a stalk and (b) a culm with a hollow center. A corn plant has a stalk and a bamboo has a culm.

(a) Stalk (b) Culm

Hollow center

Phloem

Xylem

Vascular bundle

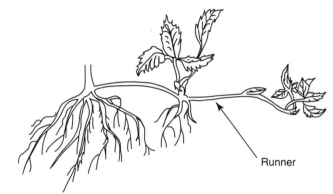

FIGURE 2-14 Runner of strawberry.

Runner

3. *Spurs.* Spurs are found on branches of woody plants such as pear and apple. They are stems whose growth has been severely restricted due to shortened internodes (figure 2-15). Spurs may resume normal growth at a later stage.

4. *Rhizomes.* Plants such as bamboo, banana, and canna produce horizontally growing underground stems called rhizomes (figure 2-16). Rhizomes differ in size, as in the species ginger (*Zingiber officinale*), in which they perform storage functions.

5. *Corms.* Corms are underground structures that are compressed and thickened stems (figure 2-17). They occur in only some monocots. Ornamentals including crocus and gladiolus produce corms.

6. *Bulbs.* Like corms, bulbs have compressed stems. They are so highly compressed that the prominent part of the structure is not the stem but rather modified leaves that are attached to the stem and wrapped up into a round structure called a bulb. These scale leaves store food. Plants including tulip, lily, onion, and hyacinth produce bulbs. There are two basic types of bulbs. A bulb is described as tunicate (as in onion) when the modified leaves completely cover the stem in concentric layers (figure 2-18). In Easter lily, the attachment of the leaves is only partial, not concentric, and irregular. This type of bulb is called a scaly bulb (figure 2-19).

7. *Tubers.* In plants such as yam (*Dioscorea* spp.) and white (or Irish) potato (*Solanum tuberosum*), the underground stems are highly enlarged as storage organs (figure 2-20). Whereas Irish potato tubers are generally small in size, yam tubers vary widely in size and shape, according to the species, some attaining lengths of several feet.

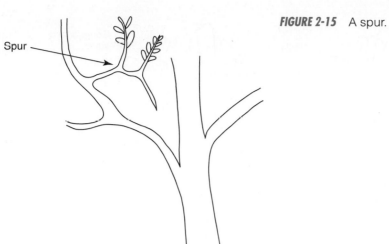

FIGURE 2-15 A spur.

Spur

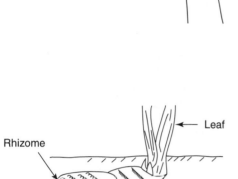

Leaf

Rhizome

(a) Thick rhizome

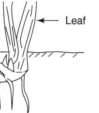

(b) Thin rhizome

Rhizome

FIGURE 2-16 Rhizomes. These underground structures differ in size and may be significant storage organs of the plant where they are thick, as in ginger *(Zingiber officinale)*.

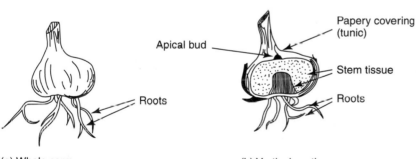

Papery covering (tunic)

Apical bud

Stem tissue

Roots

Roots

(a) Whole corm

(b) Vertical section

FIGURE 2-17 A corm.

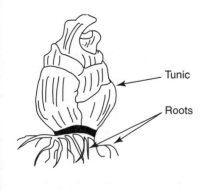

Tunic

Roots

Scale

Flower bud

Tunic

Basal plate

Roots

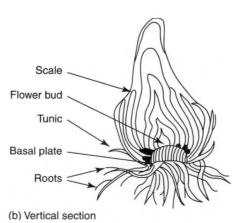

(a) Whole bulb

(b) Vertical section

FIGURE 2-18 A whole tunicate bulb (a) and cross section of a bulb (b).

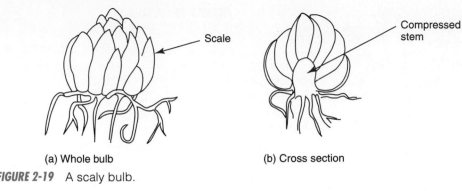

(a) Whole bulb (b) Cross section

FIGURE 2-19 A scaly bulb.

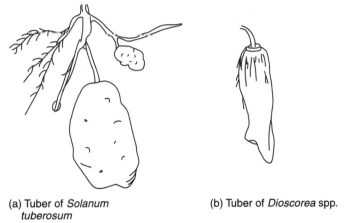

(a) Tuber of *Solanum tuberosum* (b) Tuber of *Dioscorea* spp.

FIGURE 2-20 Tubers. These swollen stems vary in size and shape depending on the species. The flesh color is also variable and may be creamish, whitish, yellowish, or some other color.

2.4: LEAF

There are five types of leaves: *foliage leaves, bud scales, floral bracts, sepals,* and *cotyledons.*

2.4.1 FUNCTIONS

The major functions of leaves are the following:

1. *Food manufacture.* The most widely known function of the leaf is photosynthesis. This function is performed by *foliage leaves,* which are the most readily visible type of leaf.

2. *Protection.* Buds (vegetative or floral) are delicate tissues that need protection while developing. Several nonfoliage leaves assume the important role of protecting developing buds:
   a. *Bud scales.* Also called *cataphylls,* bud scales are modified leaves that protect buds (apical or axillary) while in the resting stage (dormancy) or vegetative buds.
   b. *Floral bracts.* Floral bracts, also called *hysophylls,* protect the inflorescence during development. They perform a role similar to that of cataphylls.
   c. *Sepals.* Collectively called *calyx,* sepals operate like hysophylls. Sepals are not durable, however, and often either senesce or abscise after the flower has fully expanded to its mature size. In certain flowers, sepals may be brightly colored like petals and may exude fragrances and produce nectar.

3. *Storage.* Storage as a function of the leaf is often overlooked. *Cotyledons* (or seed leaves) store food that the embryo depends on early in life while the seed germinates (8.10) until the seedling has developed roots and sometimes leaves to start photosynthesis.

2.4.2 STRUCTURE

Foliage Leaves

Most leaves consist of a flat and thin structure called the *leaf blade,* or *lamina,* that is attached to the stem or branch by a narrow stalk called a *petiole* (figure 2-21). The lamina may be *simple* or *compound* (2.4.5). Leaves lacking this stalk and attached directly to the stem are said to be *sessile.* The leaf is *dorsoventral* (flattened top to bottom), and as such the epidermis on the *adaxial* (upper) and *abaxial* (lower) surfaces experiences different environmental conditions. The upper epidermal layer of a leaf has a thicker cuticle than the lower layer. In some families, such as Portulacaceae and Begoniaceae, certain species have no *stomata* (openings or pores in the epidermis) on the adaxial epidermis. These pores are bordered by specialized cells called *guard cells* (figure 2-22). In floating species of families such as Ranuculaceae, abaxial stomata are lacking; in completely submerged plants, such as certain species of Nymphaeaceae, stomata may be absent altogether.

Most stomata occur on the abaxial surface, as do trichomes. The internal part of the leaf, the *mesophyll,* is equivalent to the cortex in the stem (figure 2-23). Directly below the upper epidermis are columns of cells called *palisade parenchyma.* There may be more than one row of this tissue if the plant is exposed to intense sunlight. Palisade tissue cells contain chloroplasts used in photosynthesis. Next to the lower epidermis is a layer of widely spaced cells called *spongy mesophyll.* This tissue provides flexibility of the lamina as it moves in the wind.

The vascular system of the stem extends to the leaf. Dicot leaves generally have a single, large *central vein,* or *midrib,* from which secondary and tertiary veins branch out into the lamina. This vascular system provides the framework of the leaf. The pattern of veins (called *venation)* is a basis for classifying plants. Dicots have *reticulate venation* (weblike), while monocots have *parallel venation* (figure 2-24).

Leaves drop after a period of being on the stem. In dicots, a zone called the *abscision zone,* located at the base of the petiole, is responsible for the dropping of leaves as they age or as a result of adverse environmental conditions. The base of the petiole is a swollen structure called a *pulvinus,* enlarged as a result of water imbibed by the parenchyma cells in that region. Under moisture stress, the pulvinus cells lose water and collapse, resulting in drooping of the petiole. When the pulvinus becomes turgid again, the leaf is lifted up.

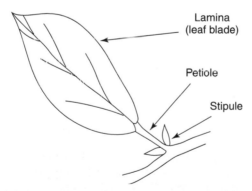

FIGURE 2-21   A typical broad leaf. The size, shape, and other features vary widely among species.

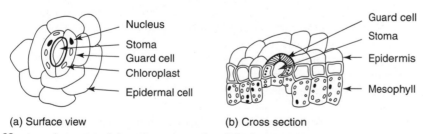

(a) Surface view   (b) Cross section

FIGURE 2-22   A surface view (a) and cross section (b) of a stoma.

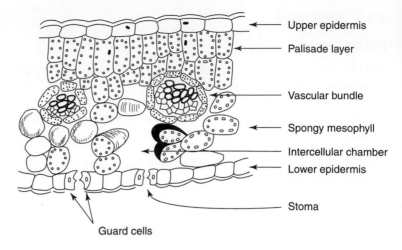

FIGURE 2-23 A transverse section of a typical dicot leaf showing internal structure.

Upper epidermis
Palisade layer
Vascular bundle
Spongy mesophyll
Intercellular chamber
Lower epidermis
Stoma
Guard cells

FIGURE 2-24 Leaf venation in a monocot (a) and dicot (b).

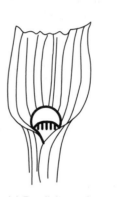 

(a) Parallel venation   (b) Reticulate venation

Monocot leaves lack a petiole (figure 2-25). Instead, they have a base (*sheath*) and a lamina with parallel venation. The junction where the lamina attaches to the sheath may have a structure called a *ligule*. An *auricle* or *stipule* may also be present.

Leaf organization in ferns is similar to that of dicots. However, the lamina is usually compound (figure 2-26). The venation of mature fern leaves is open and dichotomously branched. Few ferns have reticulate venation. The leaves of conifers are always simple and perennial. They are leathery, tough, and sclerified. The epidermis consists of thick-walled cells. In the pine family (Pinaceae), the leaves are long and needlelike (figure 2-27). *Transfusion tissue* occurs around the conducting tissue, and resin canals are also characteristic of conifer leaves.

Modified Foliage Leaves

> **Xerophyte**
> *A plant adapted for growth in arid conditions.*

Xeromorphic Foliage Leaves   Certain plants have leaf anatomy and growth forms that are adapted to desert conditions. These plants are called *xerophytes* and are characterized by a thick-walled epidermis and hypodermis. The epidermis is covered by a dense and waxy cuticle. A large number of trichomes are found, and many have salt glands. To decrease transpirational loss under xeric conditions, the external surface of xeromorphic leaves is small. The leaves may be small in size, cylindrical, and succulent. The internal packaging of cells is also tight so as to reduce the surface area for moisture loss. The spongy mesophyll may be completely absent in some cases. Water storage cells usually occur in xerophytes. In certain species, leaves are shed after only a few weeks. This strategy reduces the danger of excessive transpiration in the event of a severe drought. Such species, however, are capable of producing a fresh flush of leaves when the rains return.

Submerged Foliage Leaves   Aquatic plants have submerged foliage leaves. These plants are called *hydrophytes* (e.g., *Eleocharis*, *Najas*, and *Sagittaria*). Since they do not need to conserve mois-

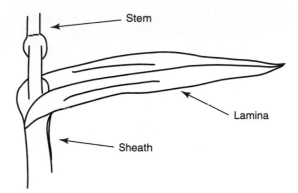

FIGURE 2-25 A typical grass leaf.

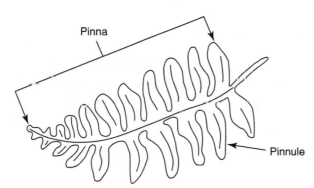

FIGURE 2-26 A fern leaflet (pinna). The fern leaf, called a frond, is typically dissected. The leaf blade is usually segmented into pinnae, which are attached to the main axis (rachis) of the frond. The lower surfaces of the pinnae may be covered with clusters of sporangia enclosed in structures called indusia (singular: *indusium*). These structures give a rust-colored appearance to the pinnae.

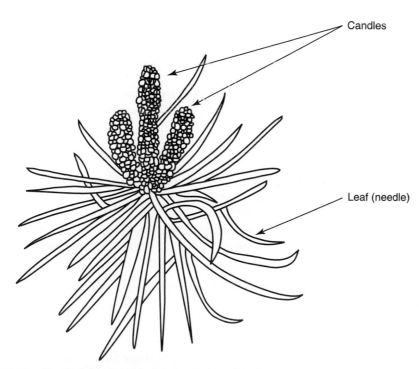

FIGURE 2-27 Pine leaves are also referred to as needles. If a number of leaves are held together, each cluster or fascicle of needles forms a cylindrical rod.

FIGURE 2-28 A leaf of water lily. This circular-shaped leaf may grow to a large enough size to support the weight of a small person.

ture, their leaves have thin cuticles. Similarly, other cell walls tend to be thin. *Gas chambers* in the spongy mesophyll trap internally generated gases, making the leaves buoyant (figure 2-28).

Bud Scales

Bud scales (cataphylls) are leaves designed for protecting buds (apical or axillary). This type of leaf is absent in annual plants in which there are no terminal resting buds. Plants that grow continuously or have only brief resting periods (such as tropical plants) also lack bud scales. However, some types of plants such as mango (*Mangifera indica*) have resting buds protected by bud scales. Bud scales are especially critical in temperate perennial species for protection against desiccation from winter winds and insect damage. The epidermal layer may be composed of cells with thickened walls. More commonly, however, the epidermis forms a protective layer of corky bark. Since there is little need for conduction, vascular tissue may occur in limited amounts. Similarly, stomata are very uncommon and if present eventually are lost when cork forms.

Floral Bracts

Floral bracts (hyposophylls) are leaves designed to protect the inflorescence during development, as previously stated. They are similar to bud scales but are weaker. Usually green in color and thus capable of photosynthesis, they are less resistant than bud scales to environmental factors.

2.4.3 GENERAL LEAF MODIFICATIONS

In some species, modified leaf forms occur along with the normal forms on the same plant. Some of these modifications bear no resemblance to leaves when viewed casually (figure 2-29). Modified leaves serve a variety of functions:

1. Some are glands for secretion of various substances.

2. The spines or thorns found on some plants protect the plant against pests and animals.

3. Some modified stems (such as bulbs) have leaves that store food.

4. Under arid or xeric conditions, xerophytes develop a thick-walled epidermis and hypodermis. These structures are covered with wax that resists the attack of chewing insects and also protects the leaves from excessive light.

5. Plants that grow under submerged conditions (hydrophytes) have very thin cuticles and cell walls.

2.4.4 FOLIAGE LEAF FORMS

The form of a leaf refers to the shape of the lamina. Leaves range from narrow needles (as in pines) to circular shapes such as in water lily (*Victoria amazonia*). The most common leaf forms are shown in figure 2-30. Each shape is representative of only the particular class, since

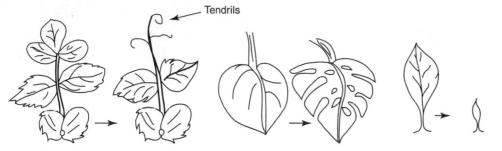

(a) Tendrils of *Pisum*

(b) Fenestrated leaf of *Monstera*

(c) *Berberis* leaf reduced to spine

FIGURE 2-29 Modified leaves. (a) The terminal leaflets of a pea compound leaf may change to become stringy tendrils. (b) When grown under intense light, the solid leaf of *Monstera* develops holes. (c) In certain species, the leaf lamina may be drastically reduced to become a spine.

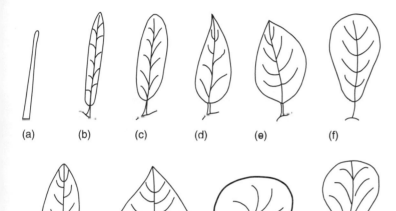

(a)   (b)   (c)   (d)   (e)   (f)

(g)   (h)   (i)   (j)

FIGURE 2-30 Selected common leaf forms: (a) filiform, (b) linear, (c) elliptic, (d) lanceolate, (e) ovate, (f) obvate, (g) hastate, (h) cordate, (i) peltate, (j) spatulate.

there are degrees of expression as well as size in each group. Certain leaf forms involve incomplete or partial separation of the lamina into parts called lobes. Certain species have deeply lobed leaves.

2.4.5 COMPOUND LEAF SHAPES

Simple leaves occur individually with one lamina (single leaf). A *compound leaf* consists of two to many small leaves (leaflets or *pinnae*) arranged on either side of the midrib or rachis in a variety of patterns (figure 2-31). Compound leaves with this arrangement are called *pinnate leaves*. In certain species (e.g., ferns), the pinna is further subdivided into secondary segments, or *pinnules*. Leaves with secondary segments are called *bipinnate leaves*. Further subdivision produces *tri-* and *quadripinnate leaves*.

2.4.6 LEAF MARGINS

A leaf may have an unindented margin or border or an indented one. In the latter case, there are also degrees of expression, some being more deeply incised than others (figure 2-32). Some leaf margins or edges are smooth, whereas others are jagged or serrated.

2.4.7 LEAF ARRANGEMENTS

The three basic leaf arrangements are *alternate, opposite,* and *whorl* (figure 2-33). An alternate arrangement involves leaves set on opposite sides of the branch or stem in a staggered pattern. In opposite arrangement, the placement of leaves is in opposite pairs; in a whorl arrangement, leaves are placed around the stem at each node.

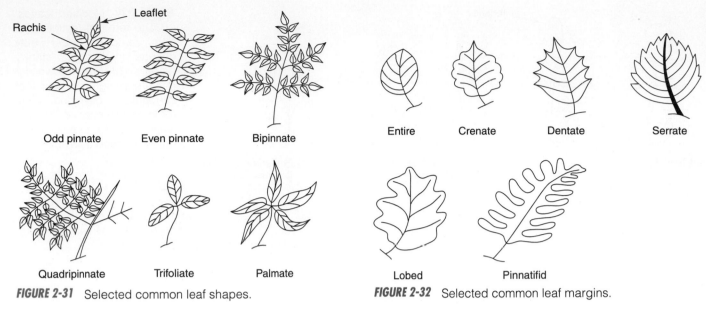

FIGURE 2-31 Selected common leaf shapes.

FIGURE 2-32 Selected common leaf margins.

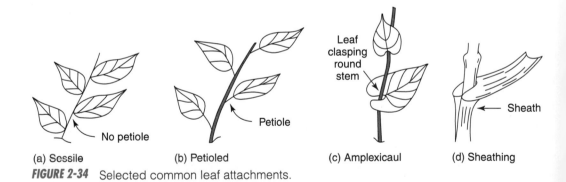

FIGURE 2-33 Selected common leaf arrangements.

FIGURE 2-34 Selected common leaf attachments.

2.4.8 LEAF ATTACHMENT

Figure 2-34 shows a variety of leaf attachments in plants. Grasses have sheathing *(sheath)* attachment to the stem of the plant, as in the case of corn, in which a tubular structure protectively surrounds the stem. Some leaves arise directly from the plant stem and are called *sessile*. Other are attached to the stem by a stalk called a *petiole*.

2.4.9 LEAF TIPS AND BASES

Plants exhibit a wide variety of shapes in the leaf tip and base (figure 2-35). Tips may be pointed or rounded. In the leaf base, certain species have an indented lamina at the petiole-lamina junction (e.g., *cordate*), and others are straight (e.g., *hastate*).

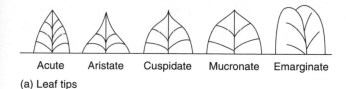

FIGURE 2-35 Selected common leaf tips and bases.

Acute   Aristate   Cuspidate   Mucronate   Emarginate

(a) Leaf tips

Cunneate   Attenuate   Obtuse   Cordate   Haste

(b) Leaf bases

2.5: ROOTS

The root is the underground vegetative organ of a plant.

2.5.1 FUNCTIONS

Roots have several functions, some of which are universal and others that are limited to certain species.

General Functions

1. *Anchorage.* Roots hold up or anchor the stem and other above-ground plant parts to the soil. If not properly anchored in the soil, a plant can be toppled easily by the wind.

2. *Nutrient and water absorption.* Most of the nutrients and water required for plant growth are obtained from the soil. These are absorbed into the plant through its roots.

3. *Hormone synthesis.* Some hormones (cytokinins and gibberelins) (4.6) required for shoot development and growth are synthesized in the roots.

Specialized Functions

1. *Storage of carbohydrates.* Some species (e.g., sweet potato) have swollen roots that store carbohydrates (figure 2-36). The plant falls back on such food reserves during times of limited food.

2. *Aerial support.* Even though roots are mostly underground, some species have *aerial roots*. In some grasses, such as corn, modified roots called *prop roots* provide additional anchorage for the plant to its growing medium (figure 2-37). In climbers, including some ivy, aerial roots enable plants to cling to walls and other structures that they climb (figure 2-38).

> **Prop Roots**
> *Adventitious roots that originate from the shoot and pass through the air before entering the soil.*

2.5.2 TYPES OF ROOTS AND ROOT SYSTEMS

Roots that develop from a seed are called *seminal roots*. Seminal roots are called the true roots of the plant. A germinated seed produces a young root called a *radicle*. The radicle grows to become the *primary root* of the plant from which *lateral roots* (or *secondary roots*) emerge (figure 2-39). Any "root" (other than the true root) that originates from other parts of the plant is said to be *adventitious* (e.g., the prop roots in corn or nodal roots or crown roots in other plants). There are two basic root systems.

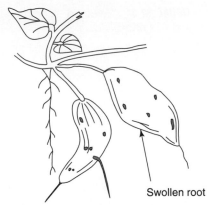

FIGURE 2-36 A swollen root. This root modification may assume a conical and elongated shape, as in carrot *(Daucus carota),* or a round shape, as in radish *(Raphanus sativus).*

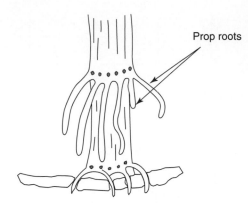

FIGURE 2-37 Prop roots of corn *(Zea mays).*

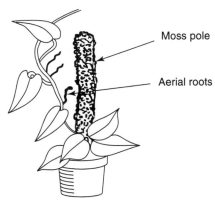

FIGURE 2-38 Aerial roots are used in certain species to aid in the climbing of vines onto nearby physical supports.

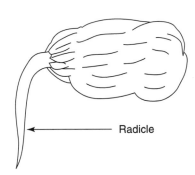

FIGURE 2-39 A radicle of a germinating seed.

Tap Root
The radicle is more prominently enlarged than any of the laterals.

1. *Tap root.* The tap root system, also called the *primary root* system, is characterized by a large central axis that is larger than the lateral roots that develop from it (figure 2-40). A tap root grows deep into the soil as the conditions of the medium permit and provides a strong anchor for the plant. The tap root system is seen in dicots and gymnosperms.

2. *Fibrous root.* The fibrous root system is found in monocots and is characterized by the lack of a single dominant root (figure 2-41). A dominant root is absent because the radicle of the embryo dies immediately after germination. Instead, a number of lateral roots develop, but they do not penetrate the soil as deeply as the tap root.

The transverse sections of a tap root of a dicot root and monocot root are presented in figure 2-42. A root exhibits endodermis in the inner layer, which is rare in stems. The conducting vessels in the roots are connected with the vessels in the stem and other plant parts. The tip of the root is protected by the *root cap* (figure 2-43). As the root pushes its way through the soil, it sheds its peripheral cells. These cells are replaced as they are shed. The root epidermis is usually one cell layer thick. One feature of the root system is the occurrence of *root hairs,* structures often associated with absorption of water and minerals from the soil (figure 2-44). The density of these hairs within the soil depends on the environmental conditions. Roots usually produce more hairs under dry conditions and less under moist conditions.

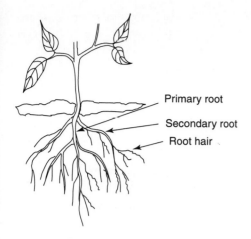

FIGURE 2-40 Tap root system. A tap root may be a swollen root, as found in carrot (*Daucus carota*) and radish (*Raphanus sativus*).

Primary root
Secondary root
Root hair

FIGURE 2-41 Fibrous root system.

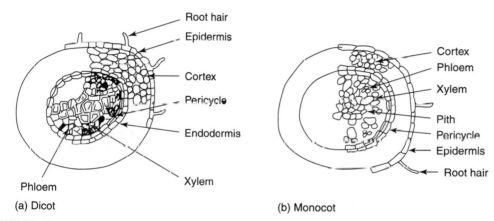

Root hair
Epidermis
Cortex
Pericycle
Endodermis
Xylem
Phloem

(a) Dicot

Cortex
Phloem
Xylem
Pith
Pericycle
Epidermis
Root hair

(b) Monocot

FIGURE 2-42 Transverse sections of dicot and monocot roots.

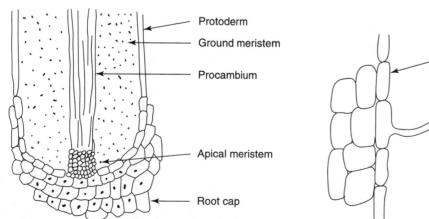

Protoderm
Ground meristem
Procambium
Apical meristem
Root cap

FIGURE 2-43 A longitudinal section of a root cap region.

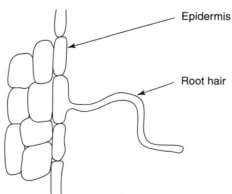

Epidermis
Root hair

FIGURE 2-44 A root hair.

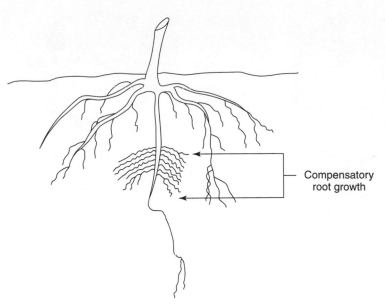

FIGURE 2-45 Compensatory root growth.

Root hairs are relatively short lived. When roots grow through a medium of differential nutritional quality and moisture, the parts in more favorable conditions (high moisture, high fertility) assume an accelerated growth period called *compensatory growth*, while the growth of other parts slows down (figure 2-45).

2.6: STRUCTURE OF WOOD

All plants consist of what is called a *primary vegetative body*. This body is made up of the three basic organs—stem, leaf, and root. These organs, however, occur in their soft form (i.e., they have no wood). In herbaceous dicots, ferns, and most monocots, this primary vegetative phase persists throughout the lifetime of the plant. However, nonherbaceous dicots and gymnosperms are able to initiate a secondary body within the primary one using *secondary tissues*. The result is a woody plant with a large body. For example, instead of being like an herb, a tree is the result of secondary growth. It is important to note that even in woody plants, primary vegetative tissue may occur in certain parts.

The tissue responsible for part of the secondary growth in woody plants is called the *vascular cambium*. This tissue occurs as a continuous ring of several layers of cells located between the xylem and the phloem tissues. The region where it occurs is called the *cambial region*. In plants such as cactus and euphorbia, the cambial layer is less pronounced and confined to the vascular bundles.

In active plants, the cells in the cambial region divide to produce cells that differentiate into conducting tissues. The secondary growth that occurs to the interior of the vascular cambium produces the secondary xylem, or *wood*. The activity of the vascular cambium is influenced by the environment, specifically by moisture and temperature. Under favorable conditions the cambium is active, but it is dormant under adverse weather conditions such as drought and cold temperature. Once formed, the secondary xylem remains in the stem. As the new layer develops, the older layer is pushed outward in a radial manner. This process makes the plant grow larger and stronger. In situations in which cambial activity is influenced by the environment (especially temperature), a cyclical pattern develops such that *annual rings* representing the previous secondary xylem tissues are observable (figure 2-46). In such cases, the age of woody plants can be estimated from the number of such rings found in the wood. However, in species such as ebony (*Diospyros ebenum*) that grow in benign or seasonless tropical regions, annual rings are not formed. On the other hand, plants that grow under con-

Wood
Secondary xylem.

Annual Rings
Cylinders of secondary xylem added to the stem of a woody plant in successive years.

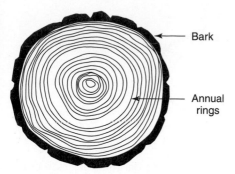

FIGURE 2-46 Annual rings of a woody dicot stem. A ring is equivalent to one year's growth of the xylem tissue.

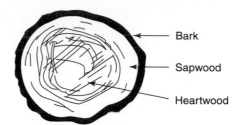

FIGURE 2-47 Heartwood and sapwood. Heartwood consists of nonconducting tissue; sapwood is still-functioning xylem tissue.

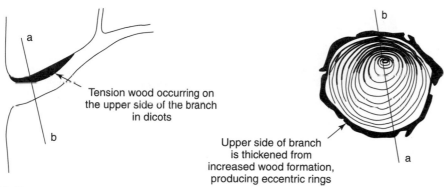

FIGURE 2-48 Tension wood. This reaction wood of dicots is found on the upper side of the branch and consists of large amounts of gelatinous fibers.

ditions of erratic moisture, such as that in arid and semiarid regions, may produce more than one growth ring per year in response to moisture patterns. The science of studying growth rings is called *dendrochronology.*

Wood may be classified according to the kind of plant that produces it. Nonflowering plants found in temperate regions (e.g., spruce, pine, fir, and larch) produce wood that is relatively homogeneous. This wood, called *softwood,* is composed predominantly of tracheids without vessels. Because softwoods are readily penetrable by nails, they are widely used in construction. They contain large amounts of lignin, which makes them desirable for use as lumber, because lignin stabilizes the wood and reduces warping.

The wood of dicots found in temperate and tropical zones is composed predominantly of fibers and vessels. These structures make the wood stronger and denser, the resultant wood being called *hardwood.* Examples of hardwood include walnut, oak, maple, ash, and hickory. These woods are characteristically hard to nail and hence not preferred for construction.

Wood may also be classified on the basis of location and function. Conduction of sap occurs only in outer secondary xylem where the wood is relatively weak. This part of the wood is called *sapwood* and is light and pale colored (figure 2-47). The center of the wood is dry, dark colored, and dense as a result of the deposits of metabolites such as gums, tannins, and resins. This wood is called *heartwood.* Sapwood converts into heartwood as the plant grows older.

Branches of trees are attached to the trunk at a variety of angles. They sway in the wind and are weighed down by gravity. These external factors cause the plant to respond by developing a special kind of wood called *reaction wood.* In dicots, this specialized wood is called *tension wood* and occurs on the upper side of the stem (figure 2-48). This wood contains gelatinous fibers made from cellulose, making the wood brittle and difficult to cut. Reaction wood in conifers is called *compression wood* and occurs on the lower side of the branch (figure 2-49). The amount of wood in the compressed area increases with time and contains a large amount of lignin.

Softwood
Wood produced by a conifer.

Reaction Wood
Wood produced in response to a stem that has lost its vertical position.

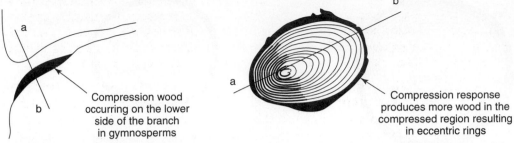

Compression wood
occurring on the lower
side of the branch
in gymnosperms

Compression response
produces more wood in the
compressed region resulting
in eccentric rings

FIGURE 2-49 Compression wood. In conifers the reaction wood forms on the lower side of the branch. This wood is heavier and more brittle than normal wood.

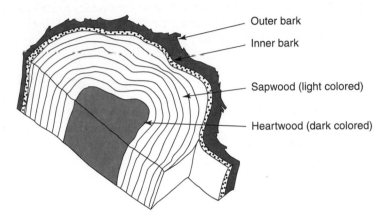

Outer bark

Inner bark

Sapwood (light colored)

Heartwood (dark colored)

FIGURE 2-50 Tree bark, which is comprised of the tissues outside the cambium, including the phloem.

2.7: BARK

> **Bark**
> *The part of the stem or root exterior to the vascular cambium.*

As previously stated, the secondary xylem produces wood. The secondary phloem is a major part of the *bark* of a tree (figure 2-50). The bark develops in perennial plants as the epidermis is replaced by a structure called the *periderm.* The periderm is composed of a mixture of cells, some of which are meristematic and are called the *phellogen,* or *cork cambium.* The bark includes the cork cambium and the secondary phloem tissue.

Bark thickness is variable (figure 2-51). Some plants, such as *Betula platyphylla,* have very thin bark that peels off easily. Others, such as the cork oak (*Quercus suber*), may have several centimeters (3 centimeters, or 1.2 inches, or more thick) of bark. The texture of the bark varies widely. Some barks are made of short fibers that are not elastic, so that as the bark grows and stretches, it cracks deeply, as in the willow tree, and breaks into large chunks. Where the fibers are long, as in the case of junipers, the bark peels off in long pieces. Some barks, such as those found in the madrone (*Arbutus xalapensis*), are flaky. Barks are rare in monocots (e.g., *Aloe dichotoma*). The function of the bark is to protect the plant against hazards in the plant's environment. However, a thick bark that is waterproof prevents gaseous exchange between the plant and environment. Therefore, woody plants have openings in their bark called *lenticels* that allow exchange of gases.

2.8: FLOWERS

The *flower* is the part of the plant most readily associated with the field of horticulture. It contains the reproductive organs of flowering plants (angiosperms). A typical flower has four parts: *sepal, petal, stamen,* and *pistil.* A developing flower bud is protected by leaflike struc-

(a)

(b)

(c)

FIGURE 2-51 Tree bark differs in thickness and roughness. Some are thin and flaky, whereas others are thick and firm.

tures called *sepals*. All of the sepals of a flower are collectively called *calyx*. The most showy parts of the flower are the petals, which collectively are called a *corolla* (figure 2-52). Petals have color and fragrance that attract pollinators to the flower. The texture of petals may be smooth, or the epidermis may have trichomes (hairs). Usually, petals drop soon after the flower has been pollinated.

The petals in some flowers are of about the same size and shape, such as those found in magnolia. These flowers are said to be *actinomorphic* (figure 2-53). In families that have different types of petals on one flower (such as occurs in clover) the plants are said to be *zygomorphic*. Sometimes the corolla is made up of individual petals, and the flower is described as *apopetalous*. Some flowers (e.g., honeysuckle) have a fused corolla (*sympetalous*), forming a *corolla tube*.

The collective term for the male reproductive organ parts is *androecium*. The stamen is comprised of a stalk called the *filament* that is capped by a structure called an *anther,* the sack that contains the pollen grains (figure 2-54). The female reproductive parts are collectively called *gynoecium*. The *carpel* consists of three structures: the *style,* which is similar to the filament, is tubular and capped by the *stigma,* a receptacle for receiving pollen grains; the third structure is the *ovary,* which contains ovules (figure 2-55). Gymnosperms have no flowers. Instead, their reproductive structures are called *strobili* (singular: *strobilus*).

If a flower has both male and female parts it is said to be *perfect*. When all of the four parts of a flower are present, a *complete flower* exists. If one or more parts are missing, the flower is said to be *incomplete*. Certain flowers are either male or female and therefore *imperfect*. In plants described as *monoecious,* both male and female flowers occur in one plant

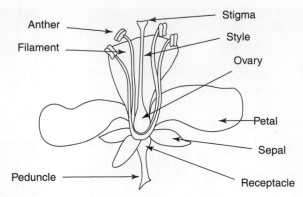

FIGURE 2-52   Parts of a typical flower.

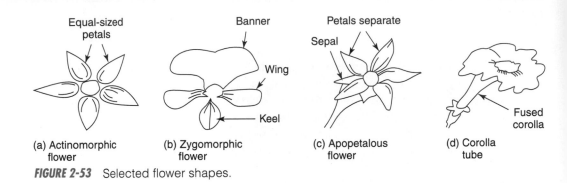

(a) Actinomorphic flower

(b) Zygomorphic flower

(c) Apopetalous flower

(d) Corolla tube

FIGURE 2-53   Selected flower shapes.

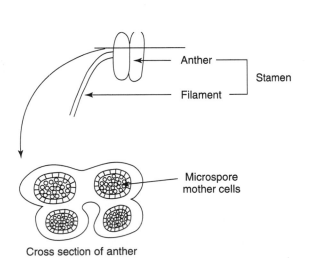

Cross section of anther

FIGURE 2-54   Whole and cross-sectional view of male flower parts.

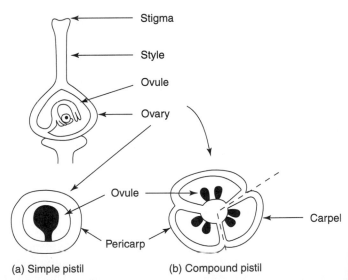

(a) Simple pistil

(b) Compound pistil

FIGURE 2-55   Whole and cross-sectional view of female flower parts.

but are physically located on different parts. In sweet corn, the male flowers (*tassel*) occur at the terminal parts while the female flowers (*silk*) occur on the middle region on the leaf axil. Cucumber, walnut, and pecan are also monoecious, and as such both sexes are required for fruiting to occur in cultivation. In *dioecious* plants such as date palm, holly, and asparagus, however, one plant is exclusively either male or female.

Flowers may occur individually (*solitary*) or in a bunch or cluster (*inflorescence,* as in urn plant, lupine, and snapdragon). The main stalk of an inflorescence is called a *peduncle;* the smaller stalks are called *pedicels.* There are three basic types of inflorescence: *head* (e.g.,

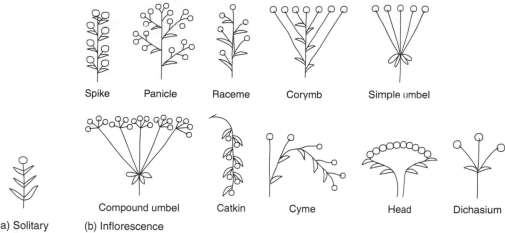

FIGURE 2-56 Inflorescence type.

daisy and sunflower), *spike* (e.g., gladiolus and wheat), and *umbel* (e.g., onion and carrot). These and other types of flower clusters are shown in figure 2-56.

The physical structure and display of certain flowers make them capable of *self-pollination* (pollen grains from the anther are deposited on the stigma of the same flower). Since the mating system excludes foreign pollen, the species tends to be genetically pure, or homozygous. Imperfect flowers have no choice but to engage in *cross-pollination* (pollen transferred from one flower and deposited on a different flower). Insects, birds, mammals, and wind are all agents of this process that promotes heterozygosity by permitting foreign pollen to be deposited.

2.9: SEED

Horticultural plants may be seed bearing (gymnosperms and angiosperms) or nonseed bearing (*cryptogams*). Seeds remain dormant until the proper environmental conditions for germination prevail (8.6). A seed contains an *embryo,* or miniature plant (sporophyte), that is encased in a seed coat, or *testa,* in dicots and a pericarp in monocots. In dicots, the embryo is sandwiched between two structures called cotyledons, which function as storage organs (figure 2-57). Monocots have one cotyledon, which is called a *scutellum* in grasses (figure 2-58). In grasses, the *endosperm* is the storage organ. The cotyledons provide nutrients for the growing embryo until the seedling is able to photosynthesize.

The dicot seed has a tiny opening called a *hilum,* through which water and air enter to initiate germination. In dicots, the cotyledons are pushed above the soil during germination (*epigeous germination*), whereas in monocots, the cotyledon remains underground (*hypogeous germination*) (8.10).

> **Monocot**
> *A type of angiosperm characterized by seeds with only one seed leaf or cotyledon.*

> **Dicot**
> *A type of angiosperm characterized by seeds with two seed leaves or cotyledons.*

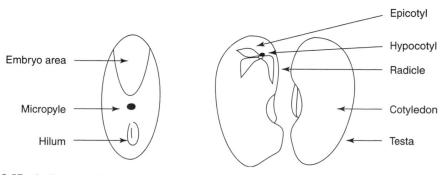

FIGURE 2-57 A dicot seed.

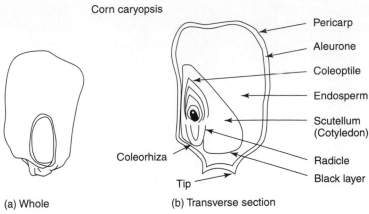

Corn caryopsis

(a) Whole (b) Transverse section

FIGURE 2-58 A monocot caryopsis.

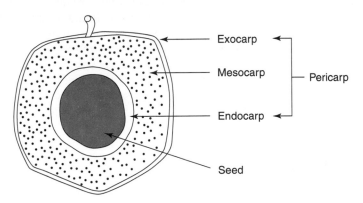

FIGURE 2-59 Parts of a typical fruit.

2.10: FRUITS

The mature ovary together with the associated parts form the *fruit*. In most species, the fruit bears *seeds* (2.9 and 8.1). In others, the fruit may develop without fertilization, a phenomenon called *parthenocarpy* (5.15). Parthenocarpic fruits (e.g., Cavendish banana, Washington navel orange, and many fig cultivars) are seedless. The natural function of the fruit is to protect the seed; however, animals and humans most desire the fruit.

A typical fruit has three regions: the *exocarp* (which is the outer covering or skin), the *endocarp* (which forms a boundary around the seed and may be hard and stony or papery), and the *mesocarp* (the often fleshy tissue that occurs between the exocarp and the endocarp) (figure 2-59).

The anatomy of fruits is discussed in great detail in chapter 1, where such description was necessary to help in the classification of fruits.

> **Parthenocarpy**
> Development of fruit without sexual fertilization.

SUMMARY

Plant anatomy is the science of cataloging, describing, and understanding the functions of plant structures. Eukaryotes have various levels of structural organization. The cell is the fundamental unit of organization of living things. It is totipotent, grows, and ages. Cells consist of a variety of organelles such as mitochondria, chloroplasts, and nuclei, each with specific functions. The mitochondria are involved in energy-related functions, and the chloroplasts in photosynthesis, or food manufacture. The nucleus houses the chromosomes that contain most of the genetic material of the plant.

Plant cells undergo differentiation to produce a variety of types, namely, parenchyma, collenchyma, and sclerenchyma. These types differ in cell wall characteristics and have specific structural roles. Cells aggregate to form tissues with specific functions. Parenchyma cells are thin walled and occur in tissues that have secretory, photosynthetic, and growth functions. Collenchyma and sclerenchyma are thicker walled and have mechanical roles in plant-strengthening tissues. Apart from simple tissues, cells aggregate to form complex tissues such as the epidermis (protective role) and those with secretory functions as found in nectaries. Some complex tissues are involved in the movement of organic and inorganic solutes through the plant. The xylem tissues move water and minerals from the roots to the leaves, where they are used in photosynthesis. The photosynthates are transported from leaves to other parts of the plant through the phloem tissue. Unlike animals, in which all parts of the organism grow simultaneously, growth in plants is limited to regions called meristems that occur in the apex or axils of the plant.

The primary vegetative body of a plant consists of three organs—root, stem, and leaf. The root is the underground organ involved in anchorage of the plant and absorption of nutrients from the soil. In some species, the roots act as storage organs for carbohydrates. The tap root system, with one large central axis with laterals, is found in dicots, and the fibrous root system, characterized by several dominant roots, is seen in monocots. Tap roots penetrate the soil more deeply than fibrous roots. The stem is the central axis of the shoot of a plant. It holds up the foliage; conducts water, minerals, and food; and sometimes acts as a storage organ. It may be modified to be a vine or may creep (as in stolon), grow horizontally (as in rhizome), or store food (as in bulb, tuber, and corm). The leaf is the primary photosynthetic apparatus of the plant and is usually green in color. It may also be modified to be a storage organ. It varies in shape, size, form, margin, and arrangement.

In dicots, secondary growth in nonherbaceous species produces wood, or secondary xylem. The inner layers of the wood that have lost conducting ability constitute the heartwood, and the outer layer forms the sapwood. Gymnosperms produce softwood because they lack certain strengthening fibers. The secondary phloem produces a tough outer layer called bark that replaces the epidermis. The bark in species such as the oak may be several centimeters thick.

The flower, showy and colorful, is the reproductive organ of the flowering plant. The male organ comprises the filament and the anther (contains pollen grains). The female organ, the carpel, consists of a style that is capped by a stigma and an ovary that contains ovules. After fertilization, a seed is produced. It contains an embryo, or a miniature plant. In dicot seeds, the embryo is sandwiched between two storage organs called cotyledons. Only one cotyledon occurs in monocots. When dicot seeds are planted, they germinate by pushing the cotyledons above the soil surface (epigeous germination). Monocots leave one cotyledon below the ground (hypogeous germination).

REFERENCES AND SUGGESTED READING

Esau, K. 1977. Anatomy of seed plants, 2d ed. New York: John Wiley & Sons.

Hartmann, H. T., A. M. Kofrnek, V. E. Rubatzky, and W. J. Flocker. 1988. Plant science: Growth, development, and utilization of cultivated plants, 2d ed. Englewood Cliffs, N.J.: Prentice-Hall.

Hayward, H. E. 1967. The structure of economic plants. New York: Lubrect & Crammer.

Moore, R., and W. D. Clark. 1994. Botany: Form and function. Dubuque, Iowa: Wm. C. Brown Publishers.

Stern, K. R. 1997. Introductory plant biology, 2d ed. Dubuque, Iowa: Wm. C. Brown Publishers.

Wilson, L., and W. E. Loomis. 1967. Botany, 4th ed. New York: Holt, Rinehart & Winston.

LABORATORY

1. Obtain prepared slides of transverse sections of dicot stems and roots. Compare and contrast the two kinds of stems and the two kinds of roots.
2. Obtain samples of modified roots:
    a. rhizome
    b. corm
    c. stolon
    d. bulb
3. Obtain samples of leaves from different species showing a variety of forms, arrangement, and margin types.

GREENHOUSE

Plant a legume seed and a grass seed. Compare and contrast germination types, the root systems, leaf characteristics, and other anatomical differences.

FIELD TRIP

1. Take a walk across your campus and observe the variety of types of tree bark.
2. Visit a botanical garden or a greenhouse to observe the following:
    a. flower types
    b. leaf types—arrangement, form, and margins

OUTCOMES ASSESSMENT

PART A

Please answer true (T) or false (F) for the following statements.

1. T F   The cell is the fundamental unit of organization of living things.
2. T F   Plants of horticultural interest are eukaryotes.
3. T F   Amyloplasts are plastids that store protein.
4. T F   Meristematic regions of plants consist of sclerenchyma cells.
5. T F   Xylem tissue conducts food from the leaves down to other parts of the plant.
6. T F   Tubers are swollen stems.
7. T F   *Entire* is a description of a kind of leaf margin.
8. T F   Sepals are the showy parts of flowers.
9. T F   Monocots have two cotyledons.
10. T F   Tap roots are found in dicots but not in monocots.

Please answer the following questions.

1. The scientific discipline of cataloging, describing, and understanding the function of plant structures is called_____.
2. List three plastids and give their functions in plants. _____
   _____
3. What are the three basic plant cell types?_____
   _____
4. Give two examples of places where collenchyma cells may be found in plants.
   _____
5. Areas of active growth in plants where cells are rapidly dividing are called _____.
6. Give two functions for each of the following:
   a. stem   _____ _____
   b. root   _____ _____
   c. leaf   _____ _____
7. Stems that grow horizontally on the surface of the ground are called_____.
8. Give two examples each of plants with the following characteristics:
   a. bulbs   _____ _____
   b. tubers   _____ _____
   c. corms   _____ _____
   d. rhizomes   _____ _____
   e. stolons   _____ _____
9. Collectively, all petals of a flower are called a _____.
10. What are the components of a carpel? _____
11. Dicots have reticulate leaf venation, whereas monocots have _____.

1. Distinguish between plant and animal cells.
2. Describe and explain how certain plant foliage changes color in the fall season.
3. Describe and explain the phenomenon of variegation as it occurs in leaves.
4. Distinguish between dicot and monocot stems.
5. Describe and distinguish between the process of seed germination in dicots and monocots.
6. Describe how annual rings are formed in woody plants.

CHAPTER 3

Plant Growth Environment

PURPOSE

The purpose of this chapter is to list and discuss the nature and roles of the essential plant growth environmental factors and how they can be manipulated to enhance the performance of plants.

EXPECTED OUTCOMES

After studying this chapter, the student should be able to

1. List the important plant growth factors in both the above-ground and below-ground environments.
2. Discuss the roles of each environmental factor in plant growth and development.
3. Describe how each environmental factor may be managed for better plant performance.

KEY TERMS

Abiotic factors
Biotic factors
Broadcast application
Chlorosis
Climate
Compound fertilizer
Conservation tillage
Conventional tillage
Day-neutral plant
Drip irrigation
Etiolation
Fertilizer analysis
Fertilizer grade
Hopkins bioclimatic law

Listing
Long-day plant
Macroclimate
Microclimate
Minimum tillage
Peat moss
Perlite
Photoperiodism
Primary tillage
Secondary tillage
Sidedressing
Short-day plant
Slow-release fertilizer

Smog
Soil horizons
Soilless mixes
Soil profile
Sphagnum moss
Sprinkler irrigation
Starter fertilizer
Straight fertilizer
Textural triangle
Tilth
Vermiculite
Weather
Zero tillage

OVERVIEW

Plant growth and development do not occur in a vacuum but in an environment. The genetic blueprint (chapter 5) directs the development of the plant within an environment. Two individuals of identical genotype are likely to manifest differences in appearance (phenotype) if they develop in different environments. In other words, one can alter the course of a plant's growth and development by manipulating its environment. Such manipulation of plants can be advantageous for humans. The implication and practical application for horticulture are that the output, in terms of yield and quality of produce of a plant, to some extent is within the control of the grower. To manipulate the plant's environment, it is important first to know what elements of the general environment affect plant growth and how they exert their influence.

Factors that affect plant growth may be divided into two broad categories, namely, those that occur in the *above-ground environment* and those that occur in the *below-ground environment*. Some factors occur in both categories.

3.1: CLIMATE, WEATHER, AND HORTICULTURE

Climate is a combination of above-ground environmental factors—*temperature, moisture, sunlight,* and *air*—and is characteristic of a region. It determines what crops can be cultivated in a given area. Some regions receive 0 to 25 centimeters (0 to 9.9 inches) of precipitation per annum and are said to be *arid,* or dry, whereas other regions may receive 75 to 100 centimeters (29.5 to 39.4 inches) of precipitation and are called *humid* regions. The immediate environment of plants is called the *microclimate.* This environment plays a role in plant processes such as *evaporation* (loss of moisture from any surface) and *transpiration* (loss of moisture by plants) (4.3.3) and also in the incidence of disease.

Weather is the composite effect of the interplay of temperature, precipitation, wind, light, and relative humidity as it pertains to a specific locality. These weather factors are dynamic, having daily, weekly, monthly, and seasonal patterns of variation. These local patterns are repeated year after year, creating the *climate* of the specific area. Regional weather patterns are affected by factors including altitude, latitude, and geographic features such as mountains and large bodies of water. The higher you go, the cooler it becomes. Oceans and large lakes moderate temperature extremes of nearby land masses. Vegetables and fruit crops are delicate and require stable climates for optimal production. They are grown on the leeward side of large lakes; grain crops, which perform well under drier and less-stable climatic conditions, are grown on the windward side. Large bodies of water also store heat in fall and are cold reservoirs in spring. The stored heat has a warming effect on the land around the lake by delaying the onset of frost on the leeward side (figure 3-1). Valley floors are colder than hillsides because cold, dense air sinks to lower levels. A belt of warm air above the cold air in the valley is called a *thermal belt* and is a region where conditions are conducive to fruit production (figure 3-2).

3.2: ABOVE-GROUND ENVIRONMENT

The above-ground environmental factors may be classified into two types—*abiotic* (nonliving) and *biotic* (involving living organisms).

3.2.1 ABIOTIC FACTORS

The abiotic factors in the environment are air, water, temperature, and light.

Air

Air is a gaseous environment that consists of many components—most importantly, nitrogen, carbon dioxide, hydrogen, and oxygen. Oxygen is required for respiration (4.3.2), the process

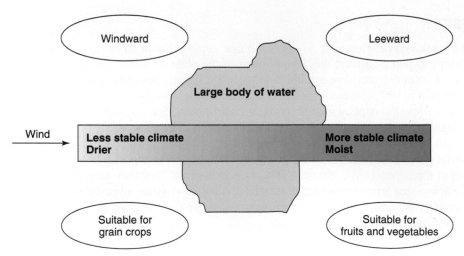

FIGURE 3-1 The role of large bodies of water in altering the climate.

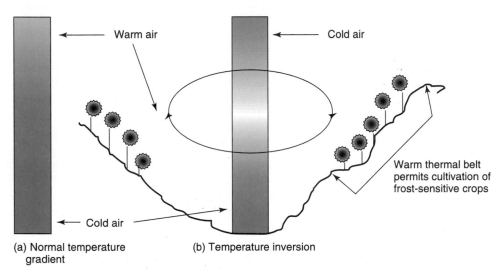

FIGURE 3-2 The formation of thermal belts. Colder and heavier air occurring at higher altitudes moves down and pushes the warmer and lighter air upward. This air convection leaves the higher band of land warmer. This thermal belt is warm and permits the culture of frost-sensitive crops on certain parts of slopes in areas that are normally too cold for growing crops.

Relative Humidity
The ratio of the weight of water vapor in a given quantity of air to the total weight of water vapor that quantity of air can hold at a given temperature, expressed as a percentage.

Evapotranspiration
The total loss of water by evaporation from the soil surface and by transpiration from plants, from a given area, and over a specified period.

by which energy is released from stored plant food for use by plants. The other role of air in plant growth is in the amount of water it holds. The water content of air is called *humidity* and is measured in units of *relative humidity (RH)* by using an instrument called a *psychrometer.* Humidity depends on *vapor pressure* (concentration of water vapor in the air) and temperature. Relative humidity decreases when temperature increases and water vapor remains constant. The amount of water needed by a plant for normal growth is directly related to the humidity or water content of the air. Plants lose moisture through the process of transpiration, and soil loses moisture through evaporation. A combination of these two processes is called *evapotranspiration.* When the air is dry, evaporation occurs at a much more rapid rate than when the air is moist. Although soil-stored moisture is depleted slowly, high humidity encourages the incidence of disease.

Air movement is also important to crop production. Strong winds can damage trees in the landscape as well as food crops. *Wind breaks* (12.7) may be erected in windy regions to reduce wind velocity so that crops can be produced successfully. Windy conditions, combined with high temperatures, increase moisture loss by transpiration. Strong winds adversely affect the success of pollination involving insects and may damage tender leaves and young fruits and consequently reduce crop yield.

Air Pollution   Environmental pollutants are a problem, especially in areas where heavy industrialized activities occur. At significant levels, air pollutants inhibit photosynthetic processes, resulting in reduced crop yield. The major air pollutants are *fluoride, sulfur dioxide, ozone, peroxyacetyl nitrate (PAN),* and *pesticides.*

Fluoride occurs in rocks and is also used in the steel industry. When present in toxic amounts, the photosynthetic surfaces of plants are reduced through *chlorosis* (yellowing) or *necrosis* (tissue death). Sensitive horticultural plants such as pine, gladiolus (*Gladiolus* spp.), grape, and corn experience retarded growth when exposed to fluoride. Sulfur dioxide is a problem where coal is burned, such as in the ore-smelting industry (iron, copper, zinc, and others). The general reaction is

$$2CuS + 3O_2 \rightarrow 2CuO + 2SO_2$$

The effects of sulfur dioxide are similar to those of fluoride.

Smog (*sm*oke and f*og*) is not a problem only in heavily industrialized areas because the exhaust of automobiles is a source of such pollution. When sunlight acts on exhaust fumes, the photochemical action causes the hydrocarbons and nitrogen oxides present to be converted into certain toxic compounds, especially ozone (O_3) and PAN. These products are injurious to horticultural plants. Ozone injury often shows up as bleached patches on leaves, leading to smaller photosynthetic surfaces and eventually reduced crop yield and inferior produce quality. PAN is more toxic than ozone but occurs in much lower concentration in the atmosphere. A class of compounds called *chlorofluorocarbons (CFCs)* release chlorine upon breaking down. This chlorine rises and, upon reaching the upper atmosphere, reacts with ozone, reducing it to oxygen gas. The ozone layer thus becomes depleted, allowing more harmful ultraviolet radiation to reach the earth. An example of CFC is *freon*, a refrigerant and propellant in aerosol cans. Aerosols used in homes (coolants in refrigerators and some cosmetic sprays) also can deplete the ozone layer. Pesticides are sources of environmental pollution. Arsenic-based insecticides, for example, cause chlorosis and necrosis, resulting in poor growth and quality of plant products.

> **Ozone**
> A form of oxygen found in the stratosphere that is more effective than ordinary oxygen in shielding living organisms from the adverse effects of intense ultraviolet radiation.

Acid Rain   Acid rain is a consequence of air pollution, because the pollutants in the atmosphere are brought down in the various kinds of precipitation (rain, ice, and snow). When the pH of rain is below 5.6, it is described as acid rain. Acid rain is produced when sulfur oxide and nitrogen oxide react with water in the atmosphere to form sulfuric acid and nitric acid, respectively. Normal (unpolluted) rain has a pH of about 6.0. Acid rain with a pH of less than 3 has been recorded in heavily industrialized and polluted parts of Scotland, Norway, and Ireland. The eastern United States and southeastern Canada record an average pH of 4 to 4.5 in rain water. Pollutants can remain airborne for long distances. Thus, nonindustrial areas can experience acid rain because of a drift effect from polluting sources in neighboring countries.

> **Acid Precipitation**
> Precipitation (e.g., rain, snow, sleet, and hail) with a pH of less than 5.6.

Mychorrhizal fungi are known to be affected adversely by acid rain. Acidified lakes have shown dramatically reduced fish populations. This decline is attributed in part to aluminum toxicity. Aluminum is a trace element in plant nutrition but comprises about 5 percent of the earth's crust. It is not soluble under alkaline conditions. Acid rain provides the low pH required to dissolve soil aluminum and other heavy metals such as lead, mercury, and cadmium. Acid rain is also suspected in chemical damage to certain forests.

Water

Role   Water is required for germination (8.10), the first step in plant growth. Water from the above-ground environment comes from precipitation (including rain and snow) and evaporation. In terms of plant growth needs, the distribution of precipitation throughout the crop growth period is as important as the total amount. Water is needed for photosynthesis (4.3.1), the process by which plants manufacture food. Water is the medium by which minerals and photosynthates are transported through the plant (4.3.4). Plants are cooled through the process of transpiration (4.3.3).

Moisture Stress The effect of moisture stress on horticultural plants is discussed more fully later in this chapter. Lack of moisture in the above-ground environment makes the air less humid, thereby increasing its drying power. The rates of plant processes such as transpiration, diffusion, and evaporation are affected directly by the vapor pressure of the air (the part of the total air pressure attributable to the water molecules present in the air). As previously indicated, if air temperature is increased but the amount of water vapor in the air stays the same, the relative humidity of the air decreases. Excessive moisture in the microclimate of plants predisposes them to disease. Horticultural plants grown indoors are sometimes given a misty spray of water to increase the humidity of the microclimate, especially in winter when the heaters are turned on to warm the building.

Temperature

Temperature, the intensity factor of heat energy, is important in all plant biological, chemical, and physiological processes. It regulates the rate of chemical reactions and consequently regulates the rate of plant growth. As a contributing factor to climate, it plays a major role in plant adaptation and the length of the growing season. The kind of horticultural plants that can be grown in an area therefore depends on temperature (in conjunction with rainfall, light, and air movements). As previously indicated, the U.S. Department of Agriculture (USDA) has developed a *plant hardiness zone* map that shows crop adaptation in the United States (figure 3-3). Most plants can live and grow within temperature ranges of 0 to 50°C (32 to 122°F). However, sensitivity to cold temperature limits the regions in which crops can be cultivated successfully. High temperatures may kill plants outright or reduce production when they coincide with flowering and fruiting periods. Plants representative of those adapted to the various hardiness zones are presented in table 3-1.

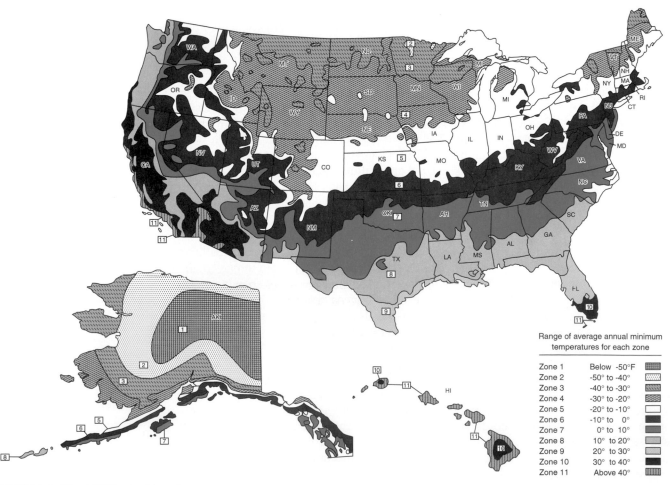

Range of average annual minimum temperatures for each zone

Zone 1	Below -50°F
Zone 2	-50° to -40°
Zone 3	-40° to -30°
Zone 4	-30° to -20°
Zone 5	-20° to -10°
Zone 6	-10° to 0°
Zone 7	0° to 10°
Zone 8	10° to 20°
Zone 9	20° to 30°
Zone 10	30° to 40°
Zone 11	Above 40°

FIGURE 3-3 The USDA plant hardiness zone map. (*Source:* USDA)

Table 3-1 Climatic Adaptation of Selected Plants

Cool-Temperature Plants

Plant	Scientific Name
Aspidistra	*Aspidistra* spp.
Azalea	*Rhododendron* spp.
Cyclamen	*Cyclamen* spp.
English ivy	*Hedera helix*
Fatshedera	*Fatshedera lizei*
Geranium	*Pelargonium* spp.
Hydrangea	*Hydrangea* spp.
Impatiens	*Impatiens* spp.
Primrose	*Primula* spp.
Beet	*Beta vulgaris*
Broccoli	*Brassica oleraceae* var. *botrytis*
Carrot	*Daucus carota*
Cabbage	*Brassica oleraceae*
Lettuce	*Lactuca sativa*
Radish	*Raphanus sativa*

Warm-Temperature Plants

Plant	Scientific Name
African violet	*Saintpaulia* spp.
Agave	*Agave* spp.
Aluminum plant	*Pilea cadierei*
Begonia	*Begonia* spp.
Bromeliads	Many species
Cacti	Many species
Caladium	*Caladium* spp.
Chinese evergreen	*Aglaonema costatum*
Coleus	*Coleus blumei*
Croton	*Codiaeum* spp.
Corn plant	*Dracaena fragraus* 'Massangeana'
Dwarf banana	*Musa cavendishii*
Dumbcane	*Dieffenbachia* spp.
Jade plant	*Crassula argenta*
Palms	Many species
Pineapple	*Ananas comosus*
Petunia	*Petunia* spp.
Corn	*Zea mays*
Cucumber	*Cucurbita sativus*
Eggplant	*Solanum melongena*
Melon	*Cucumis melo*
Tomato	*Lycopersicon esculentum*
Sweet potato	*Ipomea batatas*

The damage caused depends on the physiological state of the plant. If actively growing, succulent tissue in plant parts such as flower buds are more susceptible than dormant tissue to cold damage. *Frost damage* is critical when flower buds start to open. Warm-season crops are more prone to frost damage. Although low temperatures may harm some plants, others flower only after receiving cold treatment to break dormancy (4.4). Chilling is part of the culture of some ornamental bulbs such as daffodil (*Narcissus*) and hyacinth (*Hyacinthus*). Biennial plants frequently require cold treatment or winter chilling to flower.

To improve plants' response to cold temperatures, growers may put them through a process called *hardening* (8.12). This process involves exposing plants gradually to an increasingly harsh environment so that they slowly acquire resistance to the adverse conditions.

> **Hardening Off**
> Adapting plants to outdoor conditions by withholding water, lowering the temperature, or altering the nutrient supply.

Protection against Cold Temperature–Related Losses The best protection against frost damage is to plant *after* the threat of frost has passed. However, this period is not always predictable and it may not be economical to wait that long. Most crop damage from frost occurs when the temperature drops unexpectedly in early spring. Just-opened flower buds and young seedlings are most vulnerable. Growers obtain premium prices if they have produce available for sale early in the season. Rather than waste time, many growers prefer to take the risk to plant early and adopt a variety of strategies to minimize frost damage when it occurs. Such strategies include the following:

<table>
<tr><td></td><td></td></tr>
</table>

> **Hardy Plants**
> Plants adapted to cold temperatures or other adverse climatic conditions of an area.

1. *Use frost forecast.* USDA publications of maps and tables showing predicted patterns of frost occurrence throughout the nation are available. These charts show average dates for the last killing frost (in spring and in fall), after which frost damage is not likely. Sometimes, certain plants in the locality may act as indicator plants to guide growers as to when it is safe to plant crops outside. For example, when the American dogwood blooms, it is safe to plant outside.

2. *Protect plants.*
   a. *Hot caps.* Hot caps are dome-shaped, moisture-resistant paper caps used to cover plants individually in the field. As plants grow taller, the tops can be torn off to give more room for plant growth. When the threat of frost is over, the caps are removed. Because of its high cost, this protective measure is cost-effective only in the production of high-premium crops such as tomato, summer squash, and pepper.

   b. *Sprinkler application.* Growers may provide additional heat for frost protection by sprinkling water on plants. The principle behind this strategy is that water releases *latent heat* for fusion when it changes state from liquid to solid (ice). This heat energy is enough to protect some plants against frost damage. The amount of heat generated is even greater if warmer water (above 0°C or 32°F) is applied. Further protection of the buds is derived from insulation provided by ice as it encases the young bud. Sprinkling is necessary throughout the duration of the frost period. However, too much sprinkling may lead to excessive ice formation on plants that may cause limbs to break under the weight of the ice. The soil may also be in danger of flooding under conditions of continuous sprinkling.

   c. *Plastic mulching.* Spreading polyethylene sheets over the seedbed provides warmth for germination and seedling growth. Opaque sheets minimize weed problems, since weeds receive partial light while germinated seeds receive full light.

   d. *Row covers.* Polyethylene sheets or other fabric may be used to cover rows of crops for protection from frost. These materials may be laid directly on plants in certain cases or they may be supported with wire hoops. Row covers also protect plants against certain insect pests. Plants under the cover are warmed and experience early and increased growth.

   e. *Polyethylene tunnels.* Polyethylene tunnels are a kind of row cover, but the top is designed to be opened (if necessary) during the daytime when temperatures are too high.

   f. *Wind machines.* Wind machines are like giant fans erected in crop fields to be used for mixing up the colder bottom (near the soil) layer of air with the warmer top layer of air. The air is colder near the soil surface because the soil radiates heat into the atmosphere at night. The condition in which colder air underlies warmer air is called *air inversion.* The temperature differential (between the colder and warmer air) may be small such that normal daytime conditions can eliminate the temperature inversion. However, when the temperature near the surface of the soil is very low, it may be necessary to use mechanical processes such as a strong draft from a fan to mix the air.

g. *Heaters.* Orchards may be heated by using portable burners (gas burners). This practice is not common because of the high cost of fuel and the pollution that results from its combustion.

Temperature has a diurnal pattern; that is, it varies between the daytime and nighttime, rising in the day and falling at night. In some horticultural operations, especially under controlled environments in the greenhouse, success depends on maintaining a certain nighttime temperature.

As a general rule, planting dates for horticultural crops are delayed as one moves northward. This generalization is embodied in the *Hopkins bioclimatic law,* which states that crop production activities (such as planting and harvesting) and specific morphological developments are delayed four days for every one degree of latitude, five degrees of longitude, and 122 meters (400 feet) of altitude as one moves northward, eastward, and upward. Another important generalization is that within the normal temperature range for plant growth and development, the growth rate is doubled for every 10°C (50°F) increase in temperature. Plant metabolic rate is slowed as temperatures decrease and accelerated as temperatures rise. Therefore, plant growth is slower in the cool season.

Temperature Stress   Biochemical reactions have an optimal temperature at which they occur. Photosynthesis declines as the temperature rises to excessive levels, with negative consequences (4.3.1). High temperatures cause plants to transpire excessively. The moisture in the soil is lost rapidly through evaporation, causing moisture stress to plants.

Dormancy   Low temperatures are required for purposes other than flower induction in plant growth and development. Horticultural plants with corms or tubers and many flowering shrubs and fruit trees require low temperatures to break dormancy. Dormancy is discussed in detail later in the text (4.4, 8.6).

Other Temperature Effects   Cool temperature is required by bulb plants such as narcissus, tulip, and hyacinth for good flower development. Table 3-2 presents some plants adapted to various temperature conditions.

Heat Units   Temperature may be used to quantify the amount of growth that occurs in a plant because the two factors are correlated. This relationship is used by scientists to predict the harvest dates of crops and also to determine the adaptability of plants to various climatic zones. Plant development can be measured in *heat units.* A heat unit is the number of degrees Fahrenheit by which the mean daily temperature exceeds a base minimum growth temperature. It is calculated using the following formula:

$$\text{heat unit} = [(\text{daily minimum temperature} + \text{daily maximum temperature})/2] - \text{base temperature (°F)}$$

The base temperature used in the calculation of heat units varies among species. For example, a value of 50°F (10°C) is used for corn and many fruit trees. A plant requires a certain number of hours of warmth for a specific growth phase to occur. Dormant buds on temperate fruit trees require winter chilling and a specific heat unit for the buds to break. Species such as high-bush blueberry have high (long) winter chilling and low heat unit requirements, whereas others such as pecan have low chilling and high heat unit requirements.

Light

Light for plant growth comes primarily from the sun. The role of light in the growth and development of horticultural plants depends on its *quality, quantity,* and *daily duration.* When

Table 3-2 Adaptation of Plants to Light

Full Sunlight–Loving Plants

Plant	Scientific Name
African lily	*Agapanthus* spp.
Agave	*Agave* spp.
Aluminum plant	*Pilea cadierei*
Avocado	*Persea* spp.
Cactus	Many species
Cape jasmine	*Gardenia jasminoides*
Calla lily	*Zantedeschia aethiopica*
Coleus	*Coleus blumei*
Dwarf banana	*Musa cavendishii*
Hibiscus	*Hibiscus rosa-sinensis*
Hen and chickens	*Sempervivum tectorum*
Geranium	*Perlargonium* spp.
Corn	*Zea mays*
Melon	*Cucumis melo*
Tomato	*Lycopersicon esculentum*
Squash	*Cucurbita mixta*
Pepper	*Capsicum anuum*
Marigold	*Tagetes* spp.
Rose	*Rosa* spp.

Partial Shade–Loving Plants

Plant	Scientific Name
Azalea	*Rhododendron* spp.
Bird's-nest fern	*Aspleniun nidus*
Boston fern	*Nephrolepis exaltata*
Chrysanthemum	*Chrysanthemum* spp.
Fatsia	*Fatsia japonica*
Fatshedera	*Fatshedera lizei*
Prayer plant	*Maranta leuconeura*
Primrose	*Primula* spp.
Spider plant	*Chlorophytum comosum*
Sansevieria	*Sansevieria* spp.

plants are grown indoors, artificial lighting is required. The most readily recognized role of light is in photosynthesis, but it also has other important functions, such as seed germination in some horticultural species. *Solar radiation* is electromagnetic in nature. Radiant energy is described by its *wavelength* and *frequency*. The shorter the wavelength and higher the frequency, the higher the energy transmitted. Cosmic rays have the most energy, and radio waves have the least energy (figure 3-4).

Because of the curvature of the earth's surface, incoming solar radiation strikes the earth directly at the equator but obliquely toward the pole (figure 3-5). The rays at the poles are spread over a wider surface and pass through more air mass and thus are more filtered than at the equator. Hence, polar radiation has less energy (colder). The duration of the radiation reaching the earth varies with the season, since day length is also seasonally variable. Sunrise and sunset patterns differ from one season to another. The amount and duration of sunlight are affected by the angle of the sun (figure 3-6). The angles are wider in summer than in winter. A knowledge of these seasonal changes in the sun angles is important in the orientation of a greenhouse (11.1) and other solar collectors for maximum exposure to sunlight. Cloud cover can also reduce effective solar radiation. To increase light interception by plants, growers may use closer spacing (increase plant density). In row crops, provided it is convenient and slope is not a factor, rows should be oriented in the east-west direction. Plants in

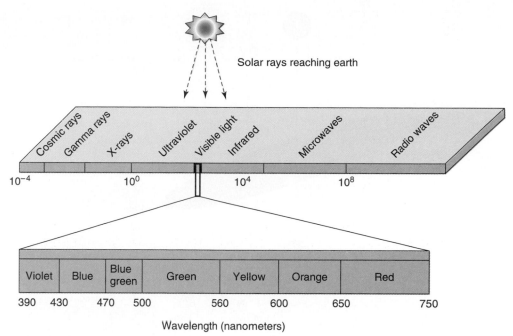

Violet	Blue	Blue green	Green	Yellow	Orange	Red

390   430   470   500   560   600   650   750

Wavelength (nanometers)

FIGURE 3-4 The radiant energy spectrum. Visible radiation or light constitutes a small proportion of the range of wavelengths of electromagnetic radiation, occurring between about 400 and 735 nanometers.

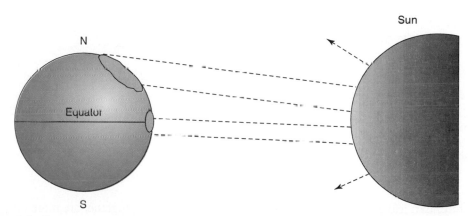

FIGURE 3-5 Sunlight hits the earth directly at the equator and is more intense there than at higher latitudes, where it hits the earth at an angle and after traveling a much longer distance.

north-south–oriented rows are self-shading as the sun moves. Shade cast by plants in east-west rows falls to the ground (figure 3-7).

Light Intensity   Sunlight *intensity* (brightness or quantity) is highest at noon. The sun may shine brightly, but the percentage intercepted by the plant's photosynthetic apparatus (mainly leaves) is what is important. Leaves in the upper parts of plants may be able to fully intercept light while shading those below. Horticultural practices such as *pruning* (chapter 16) open up the plant canopy for penetration of light to lower leaves for increased photosynthetic efficiency. By adopting an appropriate row spacing and plant density and using cultivars with desirable plant architecture (erect plant with leaves displayed at less than a 90-degree angle), more light can be intercepted for increased photosynthesis and subsequent higher yield.

Sunlight intensity at midday is about 10,000 foot-candles. Of this quantity, many plants can effectively utilize only about 50 percent for photosynthesis (4.3.1). A relatively low light intensity is more efficiently utilized by plants for photosynthesis than high-intensity light.

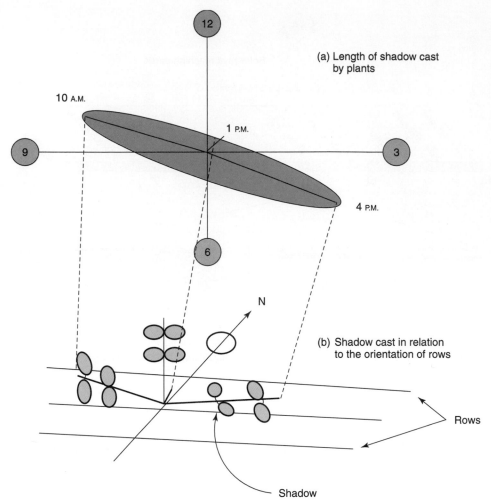

(a) Length of shadow cast by plants

(b) Shadow cast in relation to the orientation of rows

Rows

Shadow

FIGURE 3-6 The length of shadow of objects at various times of day. (a) The longest shadow is cast at around midmorning (10 A.M.) and late afternoon (4 P.M.). (b) By orienting plant rows in the east-west direction, the gardener minimizes the effect of overshadowing from taller plants. The long shadows fall relatively harmlessly between the rows.

Therefore, two leaves, each receiving 50 percent of full sunlight, together are more efficient in photosynthesizing than one leaf receiving full sunlight. Most houseplants and trees such as maple and oak do not increase their photosynthetic rate significantly as light intensity increases, which occurs with grasses such as corn and bermudagrass (figure 3-8). Therefore, many ornamental houseplants need subdued light to survive and develop properly. A selected list of plants for various light intensities is presented in table 3-3.

ETIOLATION Even though a relatively low-intensity light is utilized more efficiently than a high-intensity light, very low light intensity can be detrimental to the growth and development of plants. Plants grown in the dark or dim light (shade) may exhibit difficulties, including growing tall and spindly with yellowing of leaves, a condition called *etiolation* (figure 3-9). Plant *hormones* (4.6) play a role in etiolation. *Auxins* (plant growth hormones, especially indoleacetic acid [IAA]) accumulate under shaded or dark conditions; sunlight destroys IAA. The high concentration of IAA induces accelerated growth, leading to the rapid gain in height. Plants grown in dense populations shade each other, leading to etiolated growth.

COLOR DEVELOPMENT In apples, fruits formed on the inner branches do not develop to their full intense color because the reduced light intensity experienced inside the canopy is not adequate for the anthocyanin pigments to develop. Chlorophyll development is light dependent.

> **Etiolation**
> *Abnormal elongation of stems caused by insufficient light.*

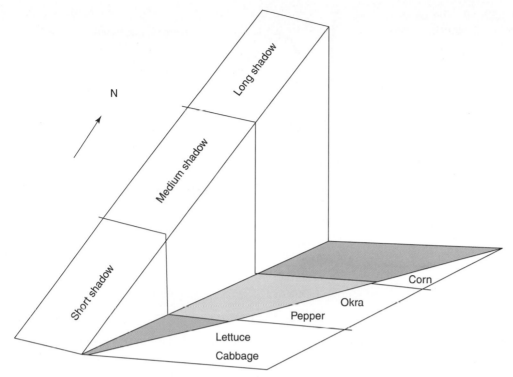

FIGURE 3-7 The variations in the shadow lengths of plants. Plants with long shadows should be located on the north sides of gardens.

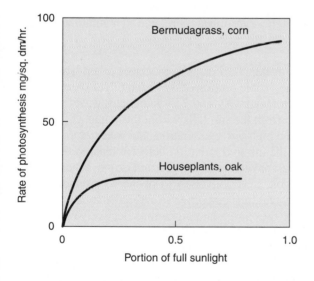

FIGURE 3-8 Light utilization of plants in photosynthesis.

Table 3-3	Selected Short-Day Plants

Plant	Scientific Name
Chrysanthemum	*Chrysanthemum x morifolium*
Gardenia	*Gardenia jasminoides*
Poinsettia	*Euphorbia pulcherrima*
Kalanchoe	*Kalancho blossfeldiana*
Bryophyllum	*Bryophyllum pinnatum*
Orchid	*Cattleya trianae*
Strawberry	*Fragaria x ananasia*
Violet	*Viola papilionaceae*

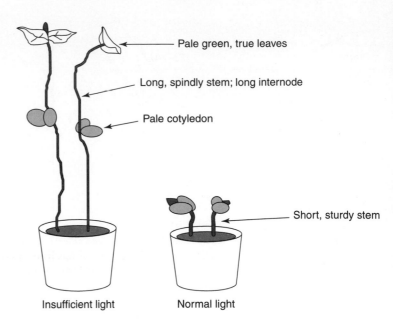

FIGURE 3-9 The effect of low light intensity or darkness on plants.

Pale green, true leaves

Long, spindly stem; long internode

Pale cotyledon

Short, sturdy stem

Insufficient light

Normal light

In the absence of light, asparagus and celery become blanched (white). These white products are preferred in certain cultures and markets; therefore, some growers in such areas deliberately cover the base of the plants by mounding with soil or using paper to block out light.

TILLERING AND BRANCHING Plants such as grasses produce additional shoots or secondary stems from the crown, called *tillers,* when the IAA concentration in the plant falls below the optimum for vegetative elongation. Unless the IAA level is reduced, plants will not tiller, branch, or produce seed. The role of light in these responses is related to regulating the concentration of IAA.

Light Quality   Light quality relates to its wavelength. For photosynthesis, plant leaf chlorophyll is able to absorb light wavelength in the visible portion of the radiant light spectrum (see figure 3-4). The visible light ranges between about 390- and 735-nanometer wavelengths. In fact, most of the radiation reaching the earth from the sun falls within this range. Oxygen and ozone in the atmosphere filter out much of the high-energy radiation, and water vapor and carbon dioxide effectively screen out much of the infrared radiation before it reaches the earth's surface. This radiation is involved in the interconversion of oxygen and ozone. When oxygen absorbs short-wavelength ultraviolet radiation, it is converted to ozone. Conversely, when ozone absorbs long-wavelength ultraviolet waves, it is converted to carbon dioxide.

   Of the minute fraction of light reaching the earth, plant pigments selectively absorb various wavelengths. A green leaf absorbs light through the visible spectrum but most strongly in the blue (approximately 430 nanometers) and red (approximately 660 nanometers) areas. It absorbs most poorly in the green (approximately 540 nanometers) area and thus reflects mostly green light, making most leaves appear green to the eye. Plant pigments absorb sunlight and use it as a source of energy in carbon dioxide fixation (the incorporation of carbon dioxide into an energy-rich organic product). The important light pigments include chlorophyll and carotenoids. Photosynthesis (the process by which plants utilize sunlight to make food) is described in detail in chapter 5.

Photoperiodism
Response to the duration and timing of day and night.

Daily Duration of Light   The length of day has an effect on two plant processes—time of flowering and plant maturity. This light-induced response is called *photoperiodism,* and plants that flower under only certain day-length conditions are called *photoperiodic.* Four photoperiodic responses in plants are a basis for classifying horticultural plants.

1. *Short-day plants* (or *long-night plants*). Short-day plants will not flower under continuous light. They require a photoperiod of less than a certain critical value within a

Table 3-4 Selected Long-Day Plants	
Plant	**Scientific Name**
Baby's breath	*Gypsophila paniculata*
Spider plant	*Chlorophytum comosum*
Sedum	*Sedum spectabile*
Evening primrose	*Oenothera* spp.
Bentgrass	*Agrostis palustris*
Fuchsia	*Fuchsia x hybrida*
Rex begonia	*Begonia rex*

Table 3-5 Selected Day-Neutral Plants	
Plant	**Scientific Name**
Bluegrass	*Poa annua*
Corn	*Zea mays*
Cucumber	*Cucumis sativus*
Pea	*Pisum sativum*
English holly	*Ilex aquifolium*
Tomato	*Lycopersicon esculentum*
Kidney bean	*Phaseolus vulgaris*

24-hour daily cycle. For example, strawberry *(Fragaria x ananasia)* requires 10 hours of light or less, and violet *(Viola papilionacea)* requires 11 hours. Poinsettia *(Euphorbia pulcherrima)* requires 12.5 hours of daylight, and cocklebur *(Xanthium strumarium)* requires about 16 hours or less of light. When planted in the field, short-day plants flower in early spring or fall.

2. *Long-day plants* (or *short-night plants*). Long-day plants are plants that flower only when light periods are longer than a certain critical length (table 3-4). These plants flower mainly in summer and include annuals such as henbane *(Hyoscyamus niger)*, which requires more than 10 hours of light, and spinach *(Spinacia oleracea),* which requires 13 hours of light. Baby's breath *(Glysophila paniculata)* requires 16 hours or more of daylight in order to flower.

3. *Day-neutral plants.* Day-neutral plants are not responsive to photoperiod and flower according to the developmental stage. Plants in this category include tomato, corn, and cucumber (table 3-5).

4. *Intermediate-day plants.* Certain grasses such as Indian grass do not flower if the days are too short or too long. These plants are said to have two critical photoperiods and are categorized as intermediate-day plants.

The Role of Darkness in Photoperiodism   Photoperiodic plants in actuality track or measure the duration of darkness or dark period rather than duration of light. Thus, short-day plants (or long-night plants) flower only if they receive continuous darkness for equal to or more than a critical value (figure 3-10). If the dark period is interrupted by light of sufficient intensity for even a minute, flowering will not be induced. Similarly, a long-day plant (or short-night plant) will not flower if the critical duration of darkness is exceeded. However, if a long-night period is interrupted by light, flowering will be induced. Interrupting the long night with such a short period of lighting is called *flash lighting*. The responses of short-day plants and long-day plants to light interruption are opposite in the two categories of plants.

The most sensitive part of the dark period regarding its response to light interruption appears to be the middle of the period of exposure. The effect diminishes before or after the midperiod. Further, the photoperiodic response can be very precise in that a deviation of even less than 30 minutes from the critical value of required exposure can cause failure to produce an induction of flowering. In henbane, for example, a photoperiod of 10 hours, 20 minutes, induces flowering, whereas a photoperiod of 10 hours does not. Further, environmental factors such as temperature can modify the photoperiodic behavior of a plant. For example, flowering in henbane is induced by exposure to 11.5 hours of light at 28.5°C, but it takes only 8.5 hours of exposure to light to induce flowering at 15.5°C.

The photoperiodic response varies among species with respect to the number of cycles of day-night treatment needed to induce flowering. Some species require only one exposure to the appropriate photoperiod to be induced to flower, whereas others require several days or even weeks (as in spinach) of exposure to the critical day-night cycle to induce flowering. Further, the stage in development (age) affects the way the photoperiod treatment is administered. Some plants respond as seedlings, but others need to attain a certain age.

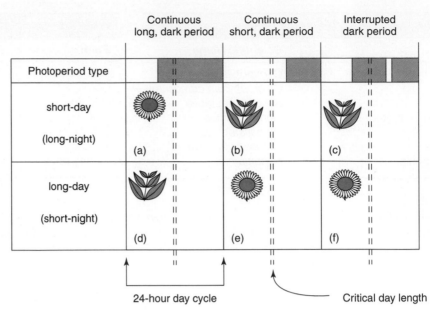

Photoperiod type	Continuous long, dark period	Continuous short, dark period	Interrupted dark period
short-day (long-night)	(a)	(b)	(c)
long-day (short-night)	(d)	(e)	(f)

24-hour day cycle   Critical day length

FIGURE 3-10 Photoperiodic response in flowering species. Light interruption of darkness affects short- and long-day plants differently.

Growers manipulate the photoperiod requirements of certain seasonal and high-income greenhouse (11.1) plants to produce plants in a timely fashion. Short-day plants such as poinsettia, chrysanthemum, and Christmas cactus are in high demand during specific times of the year. Growers start these plants under long-day conditions and then finish them under appropriate photoperiods. The required photoperiod is provided by covering the plants with a black cloth between 5 P.M. and 8 A.M. (11.5.2). The photoperiod may be prolonged during the natural short days by artificial lighting. This extended day length keeps the plants vegetative.

> **Phytochrome**
> A reversible photoreceptive protein pigment of green plants.

The Role of Phytochromes in Photoperiodism *Phytochromes* are light-sensitive, blue-colored plant pigments that exist in two photoreversible forms. That is, one form, P_r (which absorbs red light), is interconvertible with the other, P_{fr} (which absorbs far-red light). The P_{fr} form is biologically active and can induce responses such as seed germination when seeds are exposed briefly to the wavelength. The interconversions between the forms of phytochrome are summarized in figure 3-11. A molecule of P_r is converted to P_{fr} when it absorbs a photon of 660-nanometer light. This reaction is instantaneous. Similarly, when a molecule of P_{fr} absorbs a photon of 730-nanometer light it reconverts to the P_r form. These reactions are called photoconversion reactions. The ratio of P_{fr} to P_r (or P730 to P660) decreases during the growing season as the days become shorter and nights longer. At a critical level, flowering is induced in short-day plants. In long-day plants, as the day length increases and the nights become shorter during the early growing season, the P730 to P660 ratio increases. When a critical level is reached, flowering in long-day plants is initiated.

Greenhouse Effect

The earth is warmed to a limited extent by the solar radiation that strikes it directly. As indicated earlier, the earth's atmosphere filters out most of the short-wave radiation. After striking the earth, some of the solar radiation is reradiated into the atmosphere by the earth. Much of this secondary radiation is long wave (infrared). Much (85 percent) of the solar radiation is absorbed by the water vapor in the atmosphere, while the remainder is reflected back to earth (figure 3-12). The reflection back to earth is due to the accumulation of gases (especially carbon dioxide, methane, and nitrous oxide—commonly called *greenhouse gases*), which prevents the heat from escaping. Consequently, there is a global rise in temperature. This phenomenon is primarily responsible for warming (indirectly) the earth and is called the *greenhouse effect*.

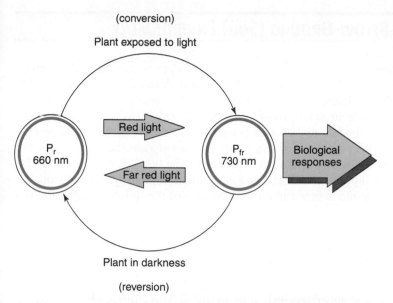

(conversion)

Plant exposed to light

P$_r$
660 nm

Red light

Far red light

P$_{fr}$
730 nm

Biological
responses

Plant in darkness

(reversion)

FIGURE 3-11 The role of phytochromes in the photoperiodic response in plants. Dark reversion of far red phytochrome to red phytochrome has so far been detected in dicots but not monocots.

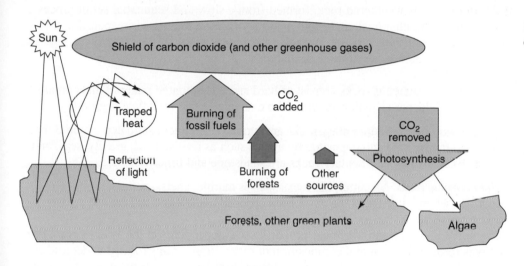

Sun

Shield of carbon dioxide (and other greenhouse gases)

Trapped
heat

Burning of
fossil fuels

CO_2
added

CO_2
removed

Photosynthesis

Reflection
of light

Burning of
forests

Other
sources

Forests, other green plants

Algae

FIGURE 3-12 The greenhouse effect of carbon dioxide.

3.2.2 BIOTIC FACTORS

Climate influences plant diseases and insect pests. For an epidemic to occur, there must be a susceptible host, pathogen, and favorable environment, called the *disease triangle* (6.6.1). Disease will not occur unless all three factors are present. However, disease can occur to varying degrees. Local weather conditions may favor the development of certain pathogens or predispose plants to diseases by lowering their resistance. Many insects have a short life span. Changes in the climate can adversely affect their population and effectiveness at any stage of their life cycle. Certain insects proliferate in specific seasons and occur in low populations in others. For example, the San Jose scale *(Aspidiotus perniciosus)* is a problem in the warm temperatures of summer but not in cold climates.

The above-ground environment contains organisms that are microscopic and largely pathogenic to plants. Plant diseases and pests are described later in this book (chapter 6). Other large-bodied organisms that occur in the environment can help or harm plants. Birds help in seed dispersal and bees and butterflies in flower pollination. On the other hand, rodents harm plants in the field in a variety of ways. Seeds may be eaten before they have a chance to germinate. Mature produce may be eaten before harvest; plant stems and foliage may be eaten by herbivores such as deer and rabbits.

3.3: Below-Ground (Soil) Environment

The *soil* is the primary medium for crop growth, although modern technology allows crops to be grown in other media (chapter 10). The role of climate in determining crop adaptation was discussed earlier in this chapter but was limited to the above-ground environment. Climate plays a significant role in determining the types of soils in which crops may be grown. This role comes from the fact that climate is a primary factor in the dynamic process of soil formation called *weathering,* the process by which *parent material* (the rocks from which soils are formed) is broken down into small particles. The type of soil formed affects the kind of vegetation it can support, which in turn further impacts on the process of soil formation by influencing the organic matter and nutrient content of the soil. Soil formation is a continuous process.

3.3.1 SOIL

Definition and Formation

> **Soil**
> The solid portion of the earth's crust consisting of mineral, air, water, and organic matter in which plants grow.

Although some people casually and erroneously refer to soil as "dirt," dirt is what you sweep off of the floor. Soil is weathered rock, formed from a slow and sequential set of processes that involve physical, chemical, and biological factors. Five factors are responsible for soil formation: *parent material, climate, organisms (organic matter), topography (relief),* and *time.* Parent rock materials are of three basic types:

1. *Igneous rocks.* Igneous rocks consist of hard and consolidated material that contains quartz and feldspar as major minerals. An example of this rock type is granite.

2. *Sedimentary rocks.* Sedimentary rocks are formed from unconsolidated material that has been transported and deposited by agents such as rivers, wind, gravity, seas, and lakes. Examples of sedimentary rocks are sandstone and limestone.

3. *Metamorphic rocks.* Metamorphic rocks such as marble, gneiss, and slate may be igneous or sedimentary rocks that have been changed under extreme heat and pressure.

The soil's physical and chemical properties are dependent on the type of rock from which it was formed. The mineral composition of rocks differs. Soils formed under arid conditions vary from soils formed under wet conditions. Similarly, soils formed under grass differ from those formed under forest climates. Leaching is more intense in the forest region, and organic matter accumulation is higher under grass cover.

> **Weathering**
> Chemical and physical alteration of rocks and minerals resulting in their disintegration, decomposition, and modification.

Rocks are decomposed into smaller particles by the agents of *weathering,* the process of soil formation. Physical processes involved in weathering include the breaking of large rocks into smaller particles by the action of moving water or the wind, which causes larger particles to knock or grind against other particles. Chemical reactions such as hydration, oxidation, and hydrolysis render the parent rock weak and prone to pulverization.

When soils are formed under heavy-rainfall climatic conditions where leaching is excessive, chemical weathering tends to be accelerated during times of high temperature. The soils formed are frequently deficient in plant nutrients, especially *micronutrients* (nutrients required by the plant in small amounts). Soils formed under dry conditions, on the other hand, have the opposite characteristics and are high in salts. Soils formed under grassland conditions have higher organic matter content than those formed under forest conditions. The vegetation of an area depends on the climate. Soils of arid regions are low in organic matter because of the small amounts of vegetation supported by the climate.

The characteristics of soils differ depending on the terrain under which they are formed. Slopes are vulnerable to erosion, and lowlands are predisposed to flooding and poor drainage. Soils age over time and change in physical and chemical properties. Hard rocks such as granite decompose slowly as compared to softer rocks such as limestone. It takes a longer period for soils to form from hard rocks than from soft rocks.

Role of Soil

The role of soil in horticultural crop production is to provide physical support and a reservoir of nutrients and moisture for growing plants. In terms of nutrition, soils may be described as fertile, marginal, or infertile. Soil nutrients are depleted with years of use and need to be replenished periodically. The soil may not be rich in native nutrients, but to be useful for crop production, it should at least be capable of holding water and nutrients for some time. If this condition does not exist, the grower should make provisions to supply supplemental nutrition to prevent deficiency problems. To be of any use for crop production, the soil should be deep enough to permit root development for good anchorage while supplying adequate nutrition.

Soil Profile

When a deep trench is dug to expose a vertical cross section of the soil, one usually can observe different layers called *horizons*. These layers together constitute the *soil profile*. The degree of profile development depends on the age of the soil, young soils showing less development than older ones.

A soil profile can be described in great detail, but for our purposes, a profile is considered to consist of three general sections (figure 3-13):

> **Soil Profile**
> A vertical cross-section of the soil showing the various layers or horizons that have developed over the period of soil formation.

1. *Topsoil (or A-horizon).* Topsoil is the upper layer of soil, where most plant roots are found. It is usually darker in color because of the high organic matter content. This section is sometimes called agricultural soil and is the portion disturbed during *tillage* (10.1) and other soil-preparation practices for the planting of crops. The topsoil is also the part of the soil that experiences the most leaching and weathering.

2. *Subsoil (or B- and C-horizons).* Subsoil is a transitional zone in the soil profile, a "catch zone" for particles and mineral elements that have moved down from the topsoil by water action. This horizon may accumulate clays, calcium carbonates, and mineral oxides. Organic matter content is minimal in this zone; some roots may occur.

3. *Parent material (substratum or R-horizon).* The R-horizon consists of the primary parent material from which the soil above was formed. With time, soil-forming factors will act on this material to produce more soil.

Properties of Soil

Some soils may consist almost entirely of organic matter, such as peat bogs, while others may consist almost entirely of mineral elements, such as sandy soils in deserts. A good agricul-

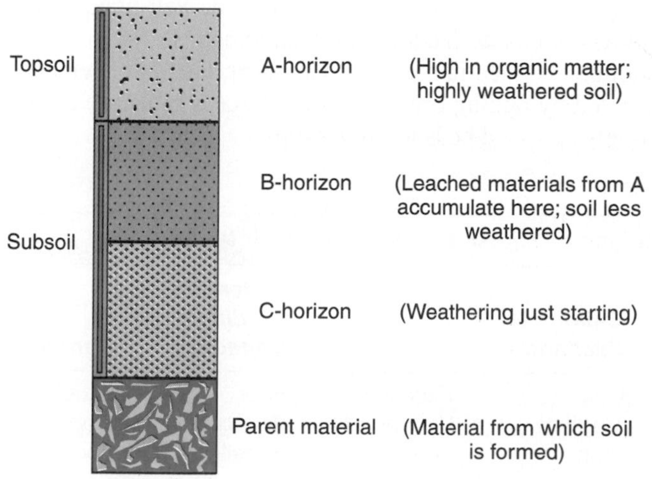

Topsoil	A-horizon	(High in organic matter; highly weathered soil)
Subsoil	B-horizon	(Leached materials from A accumulate here; soil less weathered)
	C-horizon	(Weathering just starting)
	Parent material	(Material from which soil is formed)

FIGURE 3-13 A simplified illustration of a typical soil profile.

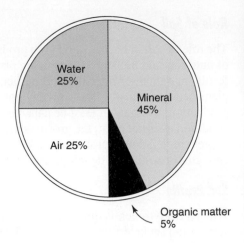

FIGURE 3-14 The composition of a typical mineral soil.

tural soil has both mineral and organic constituents. A typical soil (mineral soil) has four principal constituents—mineral, water, air, and organic matter (figure 3-14).

The soil is truly a dynamic system in which three factors (physical, chemical, and biological) interact to affect plant growth and development; soil is affected in turn by the plants. Understanding the roles of each of these factors in the soil system and their effects on plants is important in making an appropriate choice of soil for crop cultivation and knowing the best way to amend or manipulate the soil for better crop production. Aspects of these three soil properties that affect crop production are discussed next.

Physical Properties   The physical properties of interest in soil include soil texture and soil structure.

SOIL TEXTURE   Soil may be physically separated on the basis of particle size. Three basic particle size classes, called *soil separates,* are recognized; they include *sand, silt,* and *clay* (table 3-6). An agricultural soil normally contains all three soil textural classes but in varying proportions. Soil texture may be defined as the proportions (percentages) of sand, silt, and clay particles in a soil. When the three soil separates occur in equal proportions, the substance is called *loam.* A perfect loam does not occur in the field. Instead, one of the separates often predominates in the soil. Soils are therefore described as, for example, sandy loam or sandy clay loam to indicate the predominant separate or separates in the soil. A soil *textural triangle* (figure 3-15) is used to determine the textural grade of a soil.

Soil texture affects soil drainage. Clay soils do not drain well and are easily waterlogged. Clay soils are also described as heavy soils and are difficult to till. They often crack upon drying. Because root growth is hampered in heavy soils, crops whose economic parts are roots or tubers, such as sugar beet and carrot, should not be grown in clay soils. When crops are grown in sandy soils, frequent watering is necessary because these soils drain well. Well-drained soils are preferred by soil microbes and other organisms that are beneficial to plants.

Soil texture, in practice, cannot be changed in the field. However, the water-retention capacity of sandy soils can be improved by adding organic matter. Excess moisture in clay soils can be drained by using tile drains or raised beds to grow crops.

> **Soil Texture**
> The relative percentages of sand, silt, and clay in a soil.

Table 3-6   Selected Soil Properties and How They Are Affected by Soil Texture

Textural Class (Soil Separates)	Particle Size (mm)	Cation Exchange	Drainage	Water-Holding Capacity	Aeration
Sand	2.0	Low	Excellent	Poor	Excellent
Silt	0.05	Medium	Very good	Very good	Good
Clay	0.002	High	Poor	Excellent	Poor

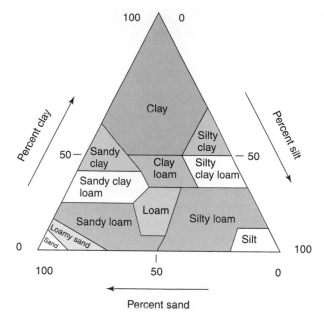

FIGURE 3-15 The USDA soil textural triangle. To use this guide, first obtain a physical analysis of the soil (i.e., the proportions of sand, silt, and clay, such as 40, 35, and 25). Locate 40 percent on the axis labeled sand and draw a line parallel to the axis labeled silt. Next, locate either the clay or silt. Locate 25 percent on the clay axis and draw a line parallel to the sand axis to intersect the previous line from the sand axis. The section of the triangle in which the intersection occurs indicates the soil textural class.

Soil texture has implications in soil fertility. Clay soils have high *cation exchange capacity (CEC)*, the ability of soil to attract and hold cations. Sandy soils have low CEC, which indicates low soil nutritional status. Sandy soils have large pore spaces and dry faster than clay soils. They also warm up much faster, making them suitable for early spring production.

SOIL STRUCTURE   Soil structure is determined by the arrangement of soil particles in the soil. The soil separates (primary particles) are arranged into secondary particles called *peds* or simply *aggregates* of different shapes and sizes.

The effects of texture on a soil are modified by its structure. Soil structure affects the pore size, water-holding capacity, infiltration rate, and permeability of water and air. A soil that drains poorly, crusts, and has poor water-holding capacity or permeability may have poor aggregation.

Peds may not be durable and thus change when the soil is disturbed. In other words, soil structure can be changed. Factors of change include raindrops, tillage, and traffic. Soils must not be tilled when wet, because the soil structure will be destroyed. The use of heavy equipment and walking over soil causes soil compaction, which impedes drainage and rooting. On the other hand, the incorporation of organic matter into the soil improves soil structure. Granular and crumb soil structures are desirable for most agricultural purposes, and they also respond to soil management. They are well aerated and have good infiltration and water-holding capacity. Compaction increases the soil *bulk density* (the weight of oven-dried soil divided by the volume of soil) and subsequently decreases the pore space, water infiltration rate, and air space.

Chemical Properties   The soil chemical properties relevant for our purposes are those that influence soil fertility and mineral nutrition of plants and soil reaction, or pH.

CATION EXCHANGE CAPACITY   Cation exchange capacity is an index of soil fertility. Many essential plant nutrients carry positively charged ions called *cations* (e.g., Ca^{2+}, Na^+, K^+, and Mg^{2+}). A fertile soil has the capacity to attract and hold these nutrients. The most efficient soil materials in cation adsorption are those with large surface areas. Clay particles called *colloids* have large surface areas that are negatively charged. Soil organic matter also has colloidal properties. The higher the capacity for attraction of cations, the higher the CEC.

Essential Soil Minerals for Plants

Most plants depend on 16 essential chemical elements for proper growth and development. Of these, three elements, namely, carbon, hydrogen, and oxygen, are obtained from the air. The soil elements essential for plant growth and development are often classified into two broad categories—*macronutrients (major nutrients)* and *micronutrients (minor elements* or

> **Soil Structure**
> The arrangement of primary soil particles into secondary particles, units, or peds.

> **Cation Exchange Capacity**
> The base-exchange capacity or measure of the total exchangeable cations a soil can hold.

Table 3-7 Soil Mineral Nutrients Essential for Plant Growth and Development

Macronutrients (Major) Primary	Secondary Nutrients	Micronutrients (Minor, Trace)
Nitrogen (N)	Calcium (Ca)	Iron (Fe)
Phosphorus (P)	Magnesium (Mg)	Manganese (Mn)
Potassium (K)	Sulfur (S)	Molybdenum (Mo)
		Copper (Cu)
		Boron (Bo)
		Zinc (Zn)
		Chlorine (Cl)

In addition to these mineral elements, plants need carbon (C), hydrogen (H), and oxygen (O); however, these essential elements are not mineral elements

trace elements). The macronutrients may be subdivided into *primary nutrients* and *secondary nutrients* (table 3-7). Primary nutrients are utilized in large amounts by plants and are often prone to deficiency in the soil. Secondary elements are used by plants in much smaller amounts than the primary elements. Micronutrients are needed in only trace or minute amounts by plants and are not frequently deficient in soils. Trace elements are especially critical in greenhouse cultivation, where artificial mixes are often used. Sandy soils and soils that experience prolonged heavy precipitation or prolonged intensive cultivation provide conditions under which micronutrient deficiency is likely.

Primary Nutrients (Macronutrients)   The three primary macronutrients are nitrogen, phosphorus, and potassium.

NITROGEN (N)   Nitrogen is one of the most widely used elements in plant nutrition. Plants absorb this element in its inorganic form as nitrate ions (NO_3^-) and occasionally as ammonium (NH_4^+). A natural cycle *(nitrogen cycle)* exists for recycling nitrogen (figure 3-16). When plants absorb nitrate ions, they become *immobilized* (mineral form is changed into organic form) by becoming part of the plant tissue. When plants die, their tissue is decomposed to release the organic form of nitrogen into inorganic ions by the process of *mineralization.* Microbes decompose dead tissue to release nitrates, and some are capable of fixating atmospheric nitrogen by the process of *nitrogen fixation.* This process, part of a *symbiotic* relationship between bacteria *(Rhizobia)* and legume roots, involves two chemical reactions: *ammonification* and *nitrification.*

Nitrogen is used in the synthesis of amino acids and proteins and is a component of chlorophyll and enzymes. It promotes vegetative growth and as a result may delay maturity. Its deficiency causes stunted growth, with leaves turning yellow *(chlorosis),* the most visible deficiency symptom. Entire leaves are chlorotic, starting with lower foliage. Nitrogen is readily lost from the soil through leaching and soil erosion and is also readily removed by plants. Deficiency can be corrected by applying organic or inorganic fertilizers.

Nitrogen is mobile in the plant. Thus, if the element is in short supply in the soil, protein nitrogen in older leaves is converted into a soluble form and translocated (3.4.5) to younger leaves, where it is most needed. The older leaves then lose color while younger leaves remain green.

PHOSPHORUS (P)   Phosphorus is absorbed primarily as orthophosphate ions (mainly $H_2PO_4^-$ and also HPO_4^{2-}). Phosphorus is found in proteins and nucleic acids (DNA and RNA) and is critical in the energy transfer process (adenosine triphosphate [ATP] and adenosine diphosphate [ADP]). Phosphorus is found to induce root proliferation and early crop maturity.

When phosphorus is deficient in the soil, leaves become dark green and plants become stunted. This deepening of color is due to an increase in nitrates in the leaves. Yield is subsequently reduced. Purplish color, especially of older leaves, or reddish-purple color on some grasses indicates phosphorus deficiency. Phosphorus is also mobile in the plant and is regu-

---

Macronutrient
A chemical element that is required in large amounts (usually greater than 1ppm) for the growth and development of plants.

Chlorosis
A condition in which a plant or a part of a plant turns greenish yellow due to poor chlorophyll development or the destruction of the chlorophyll resulting from a pathogen or mineral deficiency.

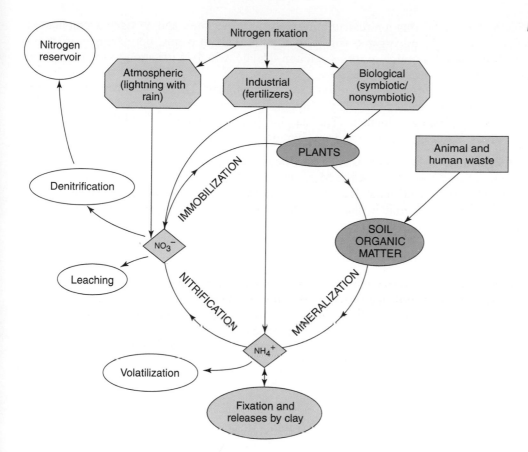

FIGURE 3-16   The nitrogen cycle.

larly recycled from older parts to younger growing parts. It has a tendency to be rendered readily unavailable (fixed) in the soil. It is most available at a pH of 5.5 to 7.0.

POTASSIUM (K)   Sandy soils may be deficient in potassium since the element is readily leached. Potassium is absorbed by plants in its ionic form (K^+). It is a catalyst for enzyme reactions and is also important in protein synthesis, translocation, storage of starch, and growth of meristematic tissue. Whereas nitrogen and phosphorus are converted into compounds for plant growth, potassium occurs in the plant tissue as a soluble inorganic salt. It is very mobile. Luxury consumption of potassium is common in plants, even though the element is required in large amounts. When potassium is deficient in the soil, the root system and stems become weak and prone to lodging; yield is reduced. Readily visible deficiency symptoms vary among species. Some plants show marginal burning of leaves *(marginal necrosis),* speckled or mottled leaves, and leaf curling. These symptoms occur in older leaves and spread upward. Potassium is removed by plants, but it is also prone to fixation and leaching, especially in soils low in organic matter and from soilless growing media.

The most abundant monovalent ion in plants is K^+, whose concentration may be equal to or more than that of nitrogen. Plants such as carnation may have as high as 9 percent potassium on a dry-weight basis. Roses have about 2 to 3 percent potassium on the average. Potassium appears to have a role in nitrogen metabolism; when deficient, plants show a high level of water-soluble nitrogen. When ammonium forms of nitrogen are used to fertilize plants in case of deficiency, the deficiency symptom is intensified and the plants become severely injured. This injury may be due to the accumulation of nitrogen (that has not been changed into protein) to excessive and toxic levels.

> **Necrosis**
> *Death of tissue associated with discoloration and dehydration of all or parts of a plant organ.*

Secondary Nutrients

CALCIUM (Ca)   Calcium is not only an essential plant nutrient but is also used in correcting soil acidity so that other soil nutrient elements can be made available to plants in appropriate amounts. It is absorbed as Ca^{2+} ions by plants. Calcium is important in cell growth and division, cell wall formation (calcium in the form of pectate), and nitrogen accumulation. The element also forms organic salts with organic acids in plants. For example, in dumb-cane *(Dieffenbachia)* calcium forms calcium oxalate, which is irritating to the tongue and throat of humans when ingested.

When deficient, plant tissue formation is incomplete. The terminal bud may cease to grow, leaving a blunt end. Deficiency symptoms for calcium are manifested frequently as defective terminal bud development. The margins of young leaves may not form, resulting in strap leaves. Roots grow poorly and are short and thickened.

MAGNESIUM (Mg)   Magnesium is released when rock minerals such as dolomite, biotite, and serpentine decompose. Absorbed as Mg^{2+} ions, magnesium is the central atom in the structure of a chlorophyll molecule. It is also essential in the formation of fats and sugars.

Magnesium is mobile in plants, and thus deficiency appears first in older leaves. Large amounts of potassium ions may interfere with its uptake due to ion antagonism created by this situation. This antagonism is prevented when the potassium to magnesium ratio in the growing medium is about 3:1 to 4:1.

SULFUR (S)   Sulfur is obtained primarily from the decomposition of metal sulfides in igneous rocks. It occurs in the soil as sulfates and sulfides, as well as in humus. It is absorbed by plants as sulfate ions (SO_4^{2-}). The unique flavors of certain vegetables such as onion and cabbage, as well as other cruciferous plants, are due to certain sulfur compounds. Sulfur is an ingredient in vitamins and amino acids. The dominant symptom of sulfur deficiency is chlorotic foliage. In addition, the stems of affected plants are weak. Sulfur is not usually added as a fertilizer element but is added indirectly when sulfate forms of other elements are applied. Sulfur is also available from air pollution.

Micronutrients (Trace Elements)   Micronutrients (trace elements or microelements) are essential elements utilized by plants in very small amounts. Low-analysis fertilizers such as 5-10-5 have trace elements as impurities. High-analysis fertilizers may be fortified with micronutrients.

BORON (Bo)   Boron is absorbed by plants as borate (BO_4^{2-}). Mobile in the plant system, it affects flowering, fruiting, cell division, water relations (translocation of sugars), and other processes in the plant. When deficient, symptoms appear at the top of the plant. Terminal buds die, producing growth described as *witches'-broom*. Lateral branches grow and form rosettes; young leaves thicken and become leathery and chlorotic.

IRON (Fe)   Though more abundant in most soils than other trace elements, iron deficiency occurs in alkaline or acidic soil. It can be absorbed through leaves or roots as Fe^{2+} ions (and also as Fe^{3+} ions to a much smaller extent since availability is reduced by being bound in plant tissue). Iron chelates can also be absorbed. Iron is a component of many enzymes and a catalyst in the synthesis of chlorophyll. Iron deficiency shows up as interveinal chlorosis of young leaves. In severe cases, leaves may become whitish, since iron plays a role in photosynthesis, as indicated previously. Iron is immobile and thus deficiency appears first in younger leaves.

MOLYBDENUM (Mo)   Vegetables, cereals, and forage grasses are among a number of species that are known to show very visible symptoms when molybdenum is deficient in the soil. This element is unavailable to plants grown under very low pH (highly acidic) conditions. In such cases, *liming* is employed as a corrective measure.

Molybdenum is involved in protein synthesis and is required by some enzymes that reduce nitrogen. The leaves of cauliflower and other cruciferous plants become narrow *(whiptail)* when the element is lacking in the soil. Plant leaves may also become pale green and roll up.

MANGANESE (Mn)   Manganese, absorbed as Mn^{2+} ions, is crucial to photosynthesis because of its role in chlorophyll synthesis. It is also important in phosphorylation, activation of enzymes, and carbohydrate metabolism. It is not mobile in plants. When deficient, interveinal chlorosis is observed in younger leaves, just as in iron deficiency.

Micronutrient
A chemical element that is required in small amounts (usually less than 1 ppm) for the growth and development of plants.

ZINC (Zn)   Zinc is an enzyme activator. It is absorbed as Zn^{2+} ions by plant roots and tends to be deficient in calcareous soils that are high in phosphorus. When deficient, plant leaves are drastically reduced in size and internodes shortened, giving a rosette appearance. Interveinal chlorosis may occur in young leaves. Kalanchoe is particularly susceptible to zinc deficiency as a greenhouse plant. In species such as peach and citrus, deficiency of zinc produces a type of chlorosis called *mottled leaf.*

COPPER (Cu)   Soils that are high in organic matter are prone to copper deficiency. Copper is important in chlorophyll synthesis and acts as a catalyst for respiration and carbohydrate and protein metabolism. Younger leaves may show interveinal chlorosis while the leaf tip remains green; with time, the leaf blade becomes necrotic. Terminal leaves and buds die, and the plant as a whole becomes stunted. Copper sulfate or copper ammonium sulfate may be administered to leaves or soil to correct deficiency problems.

CHLORINE (Cl)   Chlorine is absorbed by plants as chloride ions (Cl^-). Deficiency in the field is rare. An excessive level of chlorine is more often a problem than its absence. When deficiency occurs in the soil, plants may be stunted and appear chlorotic, with some necrosis.

3.3.2 SOIL ORGANIC MATTER

Organic matter in the soil may result from plant or animal materials. Plant residue or green manure crop incorporated into the soil by tilling the decaying plant roots is a good source of organic matter. Plant matter such as dried leaves on the surface of the soil is not considered organic matter until it is incorporated into the soil. *Erosion* of the soil depletes soil organic matter. Organic matter is important to soil productivity since it is a source of nutrients when it decomposes. It improves soil structure by binding together mineral particles into aggregates for better aeration and drainage. It helps to buffer soils against rapid changes in pH. Organic matter increases the water-holding capacity of soils and gives them their characteristic dark brown or black color. Microorganisms (e.g., bacteria, fungi, and actinomycetes) are responsible for decomposing plant parts for easier incorporation into the soil. Sugars, starches, proteins, cellulose, and hemicellulose decompose rapidly, whereas lignin, fats, and waxes are slow to decompose. Organic matter acts as a *slow-release fertilizer* (3.4.2), since its nutrients are released gradually over a particular period.

Humus is a very stable part of the soil organic matter. Much of humus is formed from two general biochemical processes. The chemicals in the plant residue undergo *decomposition* by microbial action to produce simpler products. These breakdown products undergo *synthesis,* by which the simpler products are enzymatically joined to make more complex products such as polyphenols and polyquinones. These synthetic products interact with nitrogen-containing amino compounds to produce a great portion of resistant humus. Further, the synthetic process is aided by the presence of colloidal clays. Humic particles (or humic micelles) carry a large amount of adsorbed cations (e.g., Ca^{2+}, Mg^{2+}, H^+, and Na^+) as clay micelles.

> **Soil Erosion**
> The wearing away of the land surface by geological agents such as water, wind, and ice.

3.3.3 SOIL REACTION AND NUTRIENT AVAILABILITY

A *soil test* (3.4.3) showing that adequate amounts of a nutrient are present does not indicate its availability to the plant. In addition to adequate amounts, the presence of moisture is critical, because water is the medium in which solutes are transported through the plant. Other factors that interfere with nutrient availability are soil temperature and *soil reaction,* or *pH.* Plant processes are generally slowed down by low temperatures.

Soil reaction, or pH, is a measure of the hydrogen ion concentration as an indication of the soil's degree of *acidity* or *alkalinity.* A pH of 7 is neutral. Values above 7 are considered alkaline, and values below 7 are acidic. The pH scale is logarithmic (figure 3-17), meaning that a soil pH of 5 is 10 times more acidic than a soil pH of 6 and a pH of 4 is 100 times more acidic than a pH of 6. Most horticultural crops tolerate a soil pH within the range of 4 to 8. Soil pH regulates nutrient availability. Figure 3-18 shows the relationship between pH and nutrient availability to plants. A pH of 7 ± 1 appears to be a safe range for most nutrient elements in the soil. Only iron is available at a strongly acidic pH. Conversely, iron is deficient

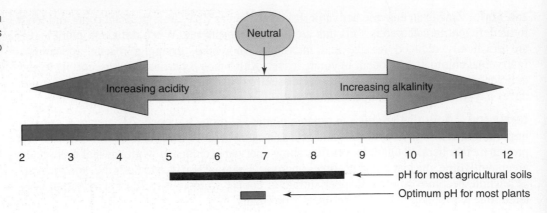

FIGURE 3-17 A representation of the pH scale. The scale is logarithmic and divided into 14 units ranging from 1 to 14.

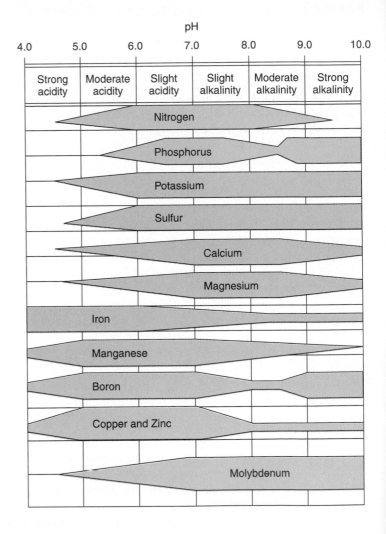

FIGURE 3-18 An illustration of the soil nutrient availability as affected by pH. The range of availability for each nutrient element depends on whether the soil is mineral, organic, or in between.

in the soil under alkaline conditions. Sensitive plants (such as bluegrass *[Poa pratensis]*) develop iron-deficiency symptoms called *iron chlorosis,* a condition in which young leaves lose their green color and become yellowish. The difference between this kind of chlorosis and that associated with nitrogen deficiency is that iron chlorosis occurs between the veins of the leaves (interveinal chlorosis) and nitrogen causes a more uniform yellowing of leaves. Soil pH affects the biotic population of soil. Fungi tend to prefer highly acidic conditions (pH of 4–5), and nitrogen-fixating bacteria *(Rhizobia)* prefer a pH range of between 6 and 8. Table 3-8 shows the pH requirements of various horticultural plants.

Table 3-8 Soil pH Requirement of Selected Plants

Acidic (4.5–5.5)	Moderately Acidic (5.5–6.5)	Near Neutral (6.5–7.5)
Azalea	Boston fern	Artichoke
Blueberry	Amaryllis	Lima bean
Bentgrass	Campanulla	Sugar beet
Cranberry	Fuchsia	Poinsettia
Camellia	Clematis	Gladiolus
Hydrangea	Gloxinia	Larkspur
Gardenia	Impatiens	Crocus
Dandelion	Grape	Dhalia
Cactus	Turnip	Apple
Strawberry	Shallot	Cabbage
Lily of the valley	Tomato	Broccoli
Laurel	Watermelon	Snapdragon
English holly	Pansy	Daphne
Potato	Rubber plant	Poppy
Trailing arbutus	Zinnia	Narcissus
Heather	Shasta daisy	Delphinium
Rhododendron	Dogwood	Pea
Gardenia	Corn	Cabbage

Scientific names of all plants are found in the appendix

Factors That Affect pH

Soil pH may rise in a soil that experiences low rainfall or is poorly drained. Salts tend to accumulate under these conditions. Soils formed on calcareous parent material have high alkalinity. Acidic soils (low pH) occur when soils are exposed to heavy rainfall and good drainage such that the bases are leached into lower depths or washed away in the runoff.

Correcting pH

Low soil pH may be corrected in practice by adding limestone ($CaCO_3$) or gypsum ($CaSO_4$) to the soil to raise the pH. The choice depends on the soil pH and other characteristics. To lower soil pH, sulfur compounds are added to the soil. Nitrogen fertilizers also tend to increase soil acidity. Liming has other added benefits: It increases the availability of phosphorus and potassium to plants, while adding calcium and magnesium to the soil. It also improves soil structure and reduces the potential toxicity of aluminum (Al) and iron (Fe), which are soluble at low pH.

3.3.4 IMPROVING SOIL FERTILITY

Soil nutrients for plant growth become depleted over years of crop production. Crops remove some nutrients, and a portion is lost through leaching and fixation. After a period of continuous farming, the nutrients in the soil must be replenished or crop productivity will decline. A variety of strategies may be adopted to improve soil fertility, including use of a fallow period, organic matter, compost, and inorganic fertilizers.

Fallow Period

A piece of land may be left uncultivated for several years to rejuvenate itself by natural processes. This tactic may be practical in areas where land is plentiful. However, most people have only one piece of land to use for gardening or other crop production activities.

Organic Matter

In this age of environmental consciousness, people are concerned about chemicals in food. *Organic farming* (chapter 21), a system of producing crops without the use of synthetic soil nutrient–enhancing chemicals, is gaining popularity. *Green manuring,* a strategy in which a field is sown with a leguminous crop for the sole purpose of plowing under, is used to improve soil fertility. Other plant residues can be treated similarly. Organic wastes from poultry, cattle, and other animal production enterprises are also used to fertilize farmland. A disadvantage with using animal waste is that the material is bulky. Further, organic fertilizers are difficult to calculate correctly and to apply uniformly.

Compost

Home gardeners may prepare *compost,* which is simply partially decomposed plant material. Composting is discussed in detail in chapter 21.

Inorganic Fertilizers

Most soil amendments are accomplished by applying inorganic fertilizers (3.4). Inorganic fertilizer amounts are easier to calculate correctly and apply uniformly, and they are less bulky.

3.4: FERTILIZERS

> **Fertilizer**
> An organic or inorganic material of natural or synthetic origin applied to the soil or plant to supply elements essential to plant growth and development.

Fertilizer sources may be *organic* or *inorganic.* Most of the fertilizers applied in horticulture are inorganic in nature.

3.4.1 ORGANIC FERTILIZERS

Organic fertilizers are derived from plant and animal residues. Sources of organic fertilizers include animal droppings (or *manure* from the barnyard or poultry house, bird droppings, and other animal wastes), dried blood, and bonemeal. *Guano* is an organic fertilizer consisting of bird droppings collected off the shores of South American islands. Other sources of organic fertilizers are cottonseed meal and liquid waste from meat- and poultry-processing factories.

Organic fertilizers are not commonly used in horticulture for several reasons, including the following:

1. They are bulky to handle, requiring large amounts of space to store the required quantity.
2. Their nutrient content is low (low analysis). Large quantities are thus required to provide appreciable amounts of the nutrients needed.
3. They are difficult to quantify and apply according to a specified rate of desired amounts of nutrient elements.
4. The nutrients they contain are released too slowly. They are released as the material decomposes, and decomposition is variable and dependent on the environmental conditions.
5. They are difficult to apply uniformly.
6. They are applied only to the soil.

3.4.2 INORGANIC FERTILIZERS

Inorganic fertilizers are most widely used to provide supplemental nutrition to horticultural plants during production in the field or in containers indoors. They are popular for reasons including the following:

1. They are easy to store.
2. They have higher analysis than organic fertilizers.
3. They can be custom formulated for specific purposes.
4. Inorganic fertilizers are easy to apply and can be applied uniformly.
5. They are available in liquid and solid forms.
6. They can be applied to both soil and leaves.
7. Accurate quantitative application is facilitated.
8. Nutrients are readily available to plants.
9. Growers can mix their own formulations accurately.

Fertilizer Formulations

Commercial fertilizers formulated to supply one nutrient element are called *straight* (or *simple* or *incomplete*) fertilizers (e.g., urea supplies only nitrogen). Some formulations supply two or three major elements and are called *compound* (or *mixed, balanced,* or *complete* if all major elements are present) fertilizers. A grower may produce homemade mixed fertilizers from straight ones.

Fertilizer Nomenclature

The major elements (nitrogen, phosphorus, and potassium) are considered in this order in describing fertilizers. Technically, phosphorus is P_2O_5 and potassium is K_2O. A label on a fertilizer bag that reads 60:0:0 means that the fertilizer supplies only nitrogen, at the rate of 60 percent by weight of the total contents of the bag, the remainder being inert fillers added during the manufacturing process to aid in fertilizer handling. Similarly, a label reading of 0:15:0 means that the fertilizer supplies only phosphorus, at 15 percent per weight. A label reading 14:14:14 indicates that this mixed fertilizer supplies 14 percent each of nitrogen, phosphorus, and potassium. Except for nitrogen, the other major nutrients are not used in the elemental form. To convert phosphorus from its elemental form to phosphate, the conversion calls for multiplication by 2.29; to achieve the reverse (convert phosphate to phosphorus) the weight of phosphate is multiplied by 0.49. Similarly, to convert potash to elemental potassium, the weight is multiplied by 0.83; to obtain the reverse, elemental potassium is multiplied by 1.20. The proportion of the elements as illustrated earlier (e.g., 15:30:15) is called the *fertilizer analysis* or *fertilizer grade* (figure 3-19). A fertilizer analysis is the minimum guaranteed analysis; an analysis of 60:0:0 means that a 100-pound bag supplies 60 pounds of nitrogen, *not* 59.9 pounds. (However, the weight could be higher than 60 pounds.) When the grade is divided by the highest denominator for the grade, the result is called the *fertilizer ratio* (e.g., 15:30:15 divided by 15 becomes 1:2:1).

Fertilizer Forms

Mixed fertilizers are obtained in the form of *granules* (granular—all components are processed together in the factory into one compound), *bulk blends* (similar to granular, but components are physically blended such that they may segregate, leading to nonuniform application), or *fluids*. Fluid fertilizers allow micronutrients and pesticides to be incorporated into one application. They may be applied through irrigation water, directly to leaves (*foliar application*).

Slow- or Controlled-Release Fertilizers

Ordinary fertilizers dissolve quickly to release their nutrients into the soil, necessitating frequent application of fertilizers. It is estimated that one month after application of the regular

> **Complete Fertilizer**
> A fertilizer formulation that supplies nitrogen, phosphorus, and potassium.

> **Fertilizer Analysis**
> The proportions of nutrient elements supplied by a fertilizer formulation.

FIGURE 3-19 Bags of selected commercial fertilizers showing fertilizer analysis.

type of nitrogen fertilizer to a lawn, little remains. In greenhouses where watering is frequent, regular fertilizers quickly leach out of the pots into the drain. Industrial processes enable fertilizers to be specially coated to release nutrients at a slow rate. For example, instead of "naked" urea, this fertilizer is coated with formaldehyde or sulfur. Such fertilizers are called *slow release* and are expensive. Sulfur-coated fertilizers may be designed for various rates of release. For example, sulfur-coated urea (SCU_{10}) releases 10 percent of the nitrogen in the first seven days, whereas SCU_{40} releases 40 percent of the nitrogen in seven days. Slow-release fertilizers are used widely in the horticultural industry in greenhouses and for lawns and golf courses.

One of the most commonly used controlled-release fertilizers is Osmocote, which is a high-analysis fertilizer (14:14:14 or 19:6:12). It is packaged in a plastic coat. The 14:14:14 analysis is formulated to release nutrients over about four months, and the 19:6:12 releases nutrients over six to nine months.

Another controlled-release fertilizer is *urea formaldehyde,* which contains 38 percent nitrogen. About 85 percent of this nutrient is available to plants over a period of six months or more. The remaining 15 percent is released so slowly that the plants may not be able to benefit from it during the growing season. This fertilizer is recommended for use under warm conditions (or in spring, summer, or early fall) because of the warm temperature needed to make nitrogen available. Urea formaldehyde should be added after pasteurization of the soil or growing medium to prevent complete release of all of the nitrogen at once.

Magnesium ammonium phosphate (Magamp) is a controlled-release fertilizer of analysis 7:40:6 plus about 12 to 14 percent magnesium. This fertilizer uses particle size to control the release of the nutrients. The finer form may release the nutrients over a period of three months, whereas the coarse form may release the nutrients over about a six-month period. The nitrogen in fertilizer is in the ammonium form; thus, when applied to media such as soilless mixes where the rate of nitrification is slow, plants may experience nitrogen deficiency.

Commercial Sources of Inorganic Fertilizers

A summary of the common sources of commercial inorganic fertilizers is provided in table 3-9.

Nitrogen   Organic sources of nitrogen include animal manures and wastes and green manures (legumes and other species grown and plowed under the soil). Commercial inorganic sources include salts of ammonia, potassium, and urea.

Table 3-9   Common Fertilizers That Supply Essential Nutrients for Plant Growth and Development

Fertilizer Element	Fertilizer	Analysis
Nitrogen (N)	Urea	46-0-0
	Ammonium sulfate	20-0-0
	Ammonium nitrate	33-0-0
	Diammonium phosphate	18-46-0
	Sodium nitrate	15-0-0
	Potassium nitrate	13-0-44
	(Ammonia gas)	(82-0-0)
Phosphorus (P)	Superphosphate	0-20-0
	Triple (treble) superphosphate	0-40-0
	Dicalcium phosphate	0-52-0
	Diammonium phosphate	18-46-0
	Monoammonium phosphate	11-48-0
Potassium (K)	Potassium chloride	0-0-60
	Potassium nitrate	13-0-44

Chemicals that provide micronutrients include magnesium sulfate (Mg), iron sulfate (Fe), borax (Bo), copper sulfate (Cu), epsom salt (Mg), sulfur (S), manganese sulfate (Mn), and calcium nitrate (Ca)

Phosphorus   Animal manures are a source of phosphorus. Commercial fertilizers are made primarily from rock phosphate (apatite). The most common fertilizer source of phosphorus is superphosphate, which contains 16 to 20 percent phosphoric acid. Other commercial forms are ammonium phosphate and triple superphosphate.

Potassium   Inorganic fertilizers are used to replenish soil supplies. Commercial fertilizers may be purchased as potassium sulfate (K_2SO_4, or sulfate of potash), potassium chloride (KCl, or muriate of potash), or potassium nitrate (KNO_3, or saltpeter). The nitrate form is the most expensive and as such is cost-effective only when applied to high-premium crops such as vegetables and orchard plants.

Calcium   Calcium is added to soil by using commercial compounds collectively called *agricultural limes*. They are available in various forms. Sources of carbonate of lime include calcite (primarily $CaCO_3$) and dolomite (primarily $CaMg[CO_3]_2$). Sources of oxide lime are called burned lime or quicklime (CaO). This form is more expensive and difficult to handle than limestone. The hydroxide form of lime is called hydrated lime ($Ca[OH]_2$). Like CaO, it is caustic and expensive.

Magnesium   Magnesium deficiency shows up as interveinal chlorosis (yellowing between veins) of older leaves. A common source of magnesium is dolomitic limestone ($CaCO_3 \cdot MgCO_3$). Others are magnesium sulfate ($MgSO_4$) and potassium magnesium sulfate ($K_2SO_4 \cdot 2MgSO_4$). Magnesium sulfate may be purchased as a foliar spray.

Sulfur   Sulfur deficiency may be corrected by adding elemental sulfur, lime sulfur, or sulfate salts of aluminum, ammonia, calcium, sodium, and others.

Boron   Boron occurs in animal manure and superphosphate and is available as sodium or calcium borate.

Iron   For foliar application, iron chelates or ferrous sulfate may be used.

Molybdenum   Foliar spraying of sodium molybdate or application to seed or soil is effective in correcting the deficiency problem.

Manganese   Like other trace elements, foliar application of manganese sulfate is commonly used to remedy any deficiency.

Zinc   Foliar application of a zinc sulfate solution or a chelate may be used to correct deficiency problems.

3.4.3 SOIL NUTRITIONAL MONITORING

The grower must regularly monitor plants in production to ensure that the best nutrition required for optimum productivity is being provided. A timely intervention can usually save the season's production from being mediocre in terms of quality and quantity. Three systems are used to monitor the nutrient status of a soil or growing medium—visual diagnosis, soil testing, and foliar analysis. These systems differ in scope, accuracy, cost, and ease of implementation. It is best to use a combination of systems in making a diagnosis.

Visual Diagnosis

As its name implies, visual diagnosis is based strictly on observation. Since it is an after-the-fact test, significant damage may already have been inflicted. The seriousness of the damage depends on the nature of the operation and the stage at which the deficiency is observed. Some symptoms are observed early in the development of the plant, whereas others manifest themselves when the plant is mature. For problems with early signals, the damage may be completely reversible; however, usually the damage is only partially reversible.

The deficiency symptoms associated with the various essential nutrients have been discussed previously in the section on nutrient elements. These symptoms range from color changes to physical deformities and death of tissue *(necrosis)*. One weakness of visual diagnosis is the lack of specificity. Often, one symptom may be unique and typical of specific nutrient deficiencies. For example, witches'-broom is typical of boron deficiency. Chlorosis (the yellowing of green leaves) is associated with deficiencies in nitrogen, sulfur, magnesium, iron, and other elements. The differences among the elements lie in the pattern of chlorosis (interveinal, veinal, or uniform), the part of the leaf (marginal or whole leaf), the age of the leaf (young or old), and whether it is the dominant symptom. Plants often become stunted in growth under conditions of poor nutrition. Deficiency symptoms are easier to spot in certain species than in others. In short, it takes experience to be able to utilize the visual diagnosis precisely. Further, deficiency associated with micronutrients is more common in greenhouse production. Other weaknesses of visual diagnosis are that pH cannot be observed, and more importantly the level of soluble salts and general nutritional status cannot be ascertained for appropriate amendments.

Soil Test

A *soil test* is a more effective and useful diagnostic evaluation than visual diagnosis for soil nutritional status. A soil test (depending on what is actually done) (1) provides a measure of soil pH (deficiency may be due to improper pH affecting the availability of the nutrient element) and (2) measures the total amount of a nutrient present in the soil; it also provides information on the proportion of the nutrient that is readily available.

A soil test can be conducted before planting, during soil preparation, or during production as the need arises. Soil testing is recommended as a routine part of a production operation. When done before planting, it allows the grower to adopt a preventive strategy rather than a curative one to nutritional problems in production. If soil pH is the problem, liming is easier and more effectively applied before planting so that it can be thoroughly mixed with the soil. A soil test helps in planning so that the needed types and amounts of fertilizers are purchased. It also helps in determining what crop production operation the soil will support naturally or what it will take to amend it for another production operation.

In a greenhouse operation (chapter 11), a fertilizer regime should be developed for the medium, type of plant, and specific production operation. Greenhouse media components react differently with plant nutrients. For example, pine bark– and peat moss–based soilless media (plant growth media containing no natural soil) (3.10) tend to experience micronutrient deficiency, especially involving iron. It is critical for a greenhouse operation to periodically monitor the pH and soluble-salt content of growing media to prevent injury to plants. Excessive salts can build up in the root medium from the application of soluble fertilizers. This high salt concentration prevents water from entering plant roots because of a drastically reduced osmotic potential. Normally, root cells have higher salt concentration than the soil solution; thus osmosis can occur and roots are able to absorb water. A signal that a soil is high in soluble salt is indicated when plants appear to be wilting on a bright day in spite of adequate moisture in the soil. Seedlings are more prone to injury from high soluble-salt levels than established plants. Salt tolerance of selected plants is presented in table 3–10.

Testing companies and institutions differ in the protocols they choose to use in their tests. For example, soluble-salt measurement depends on the amount of water used to extract the salt. The sample may be prepared by saturating the sample just enough to form a paste. Alternatively, the sample may be diluted with water (in an equal proportion by volume) to a ratio of 1:2 or 1:5 of soil to water. An *electrical conductivity* of less than 25 Mho $\times$ 10^{-5}/cm (units = milliohm/cm) (for 1:2 dilution) and less than 10 Mho $\times$ 10^{-5}/cm (for 1:5 dilution) is considered inadequate. A good level for most crops is 126–175 Mho $\times$ 10^{-5} (for 1:2 dilution) and 61–80 Mho (for 1:5 dilution) for soil-based media and 176–225 Mho $\times$ 10^{-5}/cm for soilless media. When levels reach 200 Mho $\times$ 10^{-5}/cm or more (for a 1:2 dilution), 100 Mho $\times$ 10^{-5}/cm or more (for a 1:5 dilution) for soil-based media, and over 350 Mho $\times$ 10^{-5}/cm for soilless media, plants usually will be injured. When using saturated paste, any value of 1 Mho $\times$ 10^{-5}/cm or less is considered too low. A value of 4–8 Mho $\times$ 10^{-5}/cm is adequate for most crops, and a value of 16 Mho $\times$ 10^{-5}/cm or more is excessive.

Table 3-10 Soil Salt Tolerance of Selected Plants

Relatively Salt Tolerant

Plant	Scientific Name
Bermudagrass	*Cynodon dactylon*
Beet	*Beta vulgaris*
Broccoli	*Brassica oleraceae*
Tomato	*Lycopersicon esculentum*
Cucumber	*Cucumis sativus*
Muskmelon	*Cucumis melo*
Rose	*Rosa odorata*

Relatively Salt Intolerant

Plant	Scientific Name
Bentgrass, colonial	*Agrostis tenuis*
Kentucky bluegrass	*Poa pratensis*
Strawberry	*Fragaria* spp.
Avocado	*Persea americana*
Grape	*Vitis* spp.
Carrot	*Daucus carota*
Onion	*Allium cepa*
Gardenia	*Gardenia jasminoides*
Geranium	*Pelargonium x hortorum*
Azalea	*Rhododendron* spp.
Sweet corn	*Zea mays* var. *saccharata*
Pepper	*Capsicum annuum*

For testing soil nutrients in greenhouse soils, several standard procedures have been developed, of which the *Spurway test* is the most widely used. This test is accompanied by interpretative tables.

One can conduct a soil test by purchasing a *home soil test kit*. However, for more detailed analysis, the soil sample may be sent to a commercial or university laboratory designated by the state or county, either for a fee or free of charge, depending on the particular policies of the state department of agriculture and the testing center. A good soil test starts with good *soil sampling;* the results of a test are only as good as the samples used. Samples must be representative of the field. For a small garden, about 10 to 12 samples taken to include 8 to 12 inches (20.3 to 30.5 centimeters) of the soil profile (ensuring that both the topsoil and the soil in the root zone are included), and spread over the piece of land, should be adequate. These subsamples are mixed together in a container, and from this mixture, a small sample is taken and sent to the laboratory for analysis.

A representative sample can be taken just by observing certain general guidelines. First, look at the general area to be sampled, and, if necessary, divide it into sections based on their level of uniformity. Sample these different sections separately. Avoid sampling distinctly unusual sites (e.g., waterlogged land).

Foliar Analysis

Foliar analysis is like a soil test performed on leaves. One difference between the two tests is that while a soil test provides information on nutrients available for uptake, foliar analysis provides information on nutrients taken up and accumulated in the leaf. A foliar test does not provide information on the growing medium pH or soluble-salt content, but it offers a complete analysis of all essential nutrients. The rationale of foliar analysis is that the nutrient content of the plant's tissue affects its growth. Up to a point, increasing the content of essential nutrients will result in increased growth. However, plants are able to take in large amounts of nutrients without corresponding changes in growth, a situation called *luxury consumption*. Excessive amounts of nutrients (beyond luxury consumption) can be injurious to the plant.

To conduct foliar analysis, representative samples of leaves are obtained from the plant. Species differ in which leaves are most representative of the whole plant. The age of the leaves also affects the nutrient content. Samples of foliar analysis should be obtained at intervals of four to six weeks. The results of analysis are compared with standards developed from testing a wide variety of sources. The standards for macronutrients are crop specific and vary widely; micronutrient standards are widely applicable.

3.4.4 APPLYING FERTILIZERS

The first step in a fertilizer program is to determine the type and amount of fertilizer needed. Excessive fertilization is economically wasteful and may even injure or kill the plants. A fertilizer program should take into account the cropping history of the field, the soil type, and the needs of the crop being grown. Since vegetables generally are heavy users of soil nutrients, fields that have been cropped with vegetables on a repeated basis may require fertilization.

Fertilizers may be applied before planting *(preplant)*. The recommended amount may be distributed over the area and then plowed in. Fertilizer may also be applied after germination has occurred and at various times during the growth of the crop. Whatever the time of application, fertilizers may be applied to a crop in a variety of ways.

Methods of Placement in the Field

The method of placement depends on the fertilizer form—liquid or solid (dry). Fertilizers may be applied to the soil or the plant leaves; they may be spread out or confined to a small area. Nitrogen fertilizers are available as single-element or compound forms (3.4.2). When applying nitrogen, one should consider, among other factors, the stage of development of the crop and the season. Nitrogen is needed most by almost all crops in the early stages of growth and development. When applied in cold conditions in ammonium form, the change to usable nitrate form is slow. Nitrogen fertilizers may be acidic, alkaline, or neutral in reaction.

Dry Application   Dry fertilizers are applied to the soil. They may be spread out or concentrated in bands or spots (figure 3-20). The general methods of placement of dry fertilizers are as follows:

1. *Broadcasting.* Broadcasting entails spreading the fertilizer, mechanically or manually, over the general surface area of the soil as evenly as possible. If it is done during soil preparation, a plow or disk may be used to incorporate it into the soil. Seasonal or periodic fertilizing of lawns may be accomplished by this method. It is a very speedy way of fertilizing. The disadvantage is that every part of the field is equally fertilized,

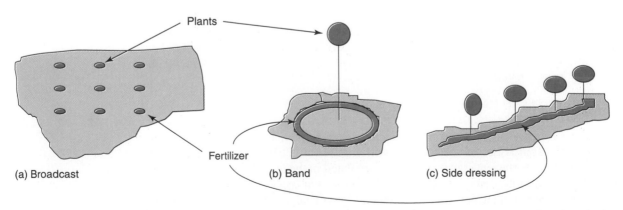

(a) Broadcast   (b) Band   (c) Side dressing

FIGURE 3-20 Methods of fertilizer placement in the field. Broadcasting (a) involves total coverage of the soil surface in the area of application, while banding (b) and sidedressing confine the fertilizer to a limited area.

and hence for widely spaced horticultural plants, such as watermelon, much of the fertilizer is not utilized by the crop.

2. *Banding*. As the name implies, banding is a method of fertilizer placement whereby the fertilizer is localized near the seed or plant. Care must be taken to place it either about 2 to 3 inches (5.1 to 7.6 centimeters) beside or the same distance below the seed or plant to prevent injury to the seedlings from the excessive presence of salts. This excess of salts is usually a concern when using especially strong nitrogen and potassium fertilizers.

3. *Sidedressing*. Because the soil is not disturbed during fertilizer application in sidedressing, the time of application is important. For example, when urea is applied on a hot summer day, the area must be irrigated to reduce the loss to the atmosphere through volatilization. Crops planted on beds such as cole crops are fertilized this way. However, care must be taken not to bring the dry fertilizer into direct contact with the plant foliage.

4. *Drill hole*. Trees have roots that penetrate deep into the soil. To make fertilizers readily available, they may be placed close to the roots by drilling holes to reach the root zone and filling the holes with the granular fertilizer at the recommended rate.

Liquid Application Liquid fertilizer applications involve applying water-soluble forms of fertilizer to either the soil or the leaves.

1. *Starter application*. In the vegetable industry (chapter 17), where seedlings are transplanted, new transplants benefit from a diluted concentration of complete fertilizer that is applied through the transplant water. Phosphorus is especially desirable in the starter solution, and a general recommendation is about 3 pounds (1.4 kilograms) of 15:30:15 fertilizer in 50 gallons (201.9 liters) of water, applied at one cup per plant.

2. *Foliar application*. Soluble fertilizers may be applied in diluted amounts to plants by spraying directly onto the leaf surface. Since only small amounts of chemicals are applied, the method is not adequate to meet the nutritional needs of plants. Supplementary soil application is usually required.

Application of Gas

Application of fertilizer as gas is possible in certain cases, such as the application of nitrogen in the form of ammonia gas under pressure. This form of nitrogen has a very high analysis. Application is accomplished by injecting the compound as pressurized gas into the soil.

3.4.5 METHODS OF APPLICATION

Fertilizers may be applied by using the bare hands or machinery or through irrigation water when water is administered by sprinklers or tubes (3.8.11). The method chosen depends on factors such as the size of the area to be fertilized, the form of fertilizer, the placement method, the crop, and the stage of application.

Dry Application

Manual For a small garden, fertilizers may be manually applied, regardless of the method of placement chosen. However, the uniformity of application may be questionable with placement methods such as broadcasting. Manually held or pushed spreaders are available to facilitate the broadcast operation and for improving uniformity of application. When applying fertilizer to trees in the landscape or potted plants using dry formulations, the operation is done by hand.

Mechanized Sometimes fertilizer can be applied at the time of planting seed. This method of application entails inserting the fertilizer along with the seed using a seed drill and is called *pop-up* application. When applied with the seed drill, the fertilizer is placed in the seed furrow. When applying fertilizer to a large area, it is done more efficiently and uniformly by using a mechanized system. Mechanized systems enable calibration of equipment to deliver fertilizer at a uniform rate over an entire area.

Liquid Application

Liquid application of fertilizer to containers on a small scale (e.g., at home) may be accomplished by dissolving the fertilizer in water and dispensing it with a calibrated container to the potted plants. A dispenser containing fertilizer may be attached to a watering hose and used for irrigation (see later section). In greenhouses and fields, liquid fertilizers are applied through irrigation water.

Foliar Application   Most fertilizer applications involve the use of major nutrients, or macronutrients, applied to the soil. The rates of application are often large and can injure vegetative tissue on direct contact. Hence, caution should be exercised in applying dry fertilizer elements by avoiding placing the elements too close to the plant. It is neither practical nor safe to apply major elements to leaves directly. Foliar spraying is often used to correct deficiency problems of trace elements. Since trace elements are required in minute amounts, it is safe to administer them by foliar application.

Fertigation   Fertigation, also called *chemigation,* entails applying fertilizer to crops through the irrigation water. This method is usually used in the greenhouse. The drip irrigation method is particularly suitable for fertigation. In effect, it is a kind of topdressing application. Critical issues of concern in fertigation include the solubility of the fertilizer and the quality of irrigation water. Water-soluble fertilizers should be used in fertigation. Hard water with excessive amounts of dissolved calcium can be problematic if the irrigation system is incapacitated through blockage of holes by calcium deposits. The advantage of this method is that in soils that drain very freely (e.g., sandy) and for fertilizer elements that are prone to leaching (e.g., nitrogen), irrigation can be controlled to deliver the right amounts of moisture and fertilizer at rates that the plant can utilize efficiently.

Small-scale fertigation systems are available for home use. They usually consist of a handheld unit containing a dry formulation of the fertilizer. The unit is connected to a watering hose that runs water through the container before sprinkling on the area to be irrigated (figure 3-21). These units are used for watering vegetable gardens, flower beds, lawns, and other plants in the landscape.

3.4.6 TIMING OF APPLICATION

Fertilizers are applied supplementarily to the nutrition plants obtain from the soil. Plants go through growth phases (4.1.1), each phase with special nutritional needs to support the growth and development activities taking place. Fertilizers are subject to a variety of environmental factors that cause them to be depleted in the soil. For example, under aerobic con-

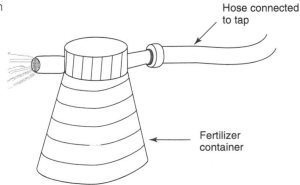

FIGURE 3-21 Applying fertilizer through a garden hose watering system.

Hose connected to tap

Fertilizer container

ditions, nitrogen is lost through *denitrification* (a process whereby bacteria convert nitrates into unusable nitrogen gas). Nitrogen fertilizers are readily leached or, in some cases, volatilized. Phosphorus and potassium are prone to fixation. Appropriate rate and type of fertilizer must therefore be chosen, and the application must be timely. Fertilizers are most beneficial to plants when they are applied as close as possible to the time of maximum need.

Fertilizers release their nutrients at various rates and over varying periods of time. Organic fertilizers release their nutrients as the material decomposes; consequently, they are not useful when immediate supplemental nutrition is required. Dry application of fertilizers requires moisture availability to enable use of the nutrients. Rain or irrigation is needed after application of dry fertilizer. On the contrary, irrigation or rain is undesirable after foliar application because it will wash the nutrients into the soil.

3.4.7 FERTILIZER RECOMMENDATIONS

After a soil test (3.4.3), fertilizer recommendations are made based on a number of factors, including the following:

1. The regional information on weather and soils.
2. The cropping history of the land.
3. The crop yield target or goal desired by the grower.
4. The crop to be fertilized (in terms of how it responds to fertilization).

For small-scale growers such as home gardeners who have no set yield and profit targets, the general goal of gardening is to produce a good, healthy, and attractive crop. Home owners and gardeners often overfertilize their gardens and lawns. For some, it is a seasonal ritual or tradition to fertilize the lawn and garden crops. Many people do not conduct soil tests on their plots and hence may not derive optimum benefits from their efforts. Since no specific yield goal is set in home gardening projects, a moderate application of fertilizer is all that may be needed. A general application may be 10 to 20 pounds per 1,000 square feet (4.5 to 9 kilograms per 92.9 square meters); a complete fertilizer (10:10:10) is often sufficient for gardens and lawns.

3.4.8 GREENHOUSE FERTILIZATION

Most greenhouse production occurs in containers (chapter 11), which restricts the amount of soil or growth medium from which a plant can obtain nutrition. Further, greenhouse media used in greenhouses for plant cultivation are often artificial and have little or no nutrition to support plants. As a result, fertilization is critical in greenhouse production. Fertilization programs should include minor or trace elements in addition to major elements. Liquid fertilizers, applied through irrigation water, are important in greenhouse fertilization. The subject is discussed further in chapter 11 (11.3.4).

3.5: SOIL ORGANISMS

The soil is teeming with life. Some soil organisms improve the plant environment for good crop growth, whereas others are pests that damage or kill plants. Soil organisms may be divided into several operational groups.

3.5.1 MICROORGANISMS

Soil *microorganisms* (or microbes) include bacteria and fungi, as well as actinomycetes. They perform very useful functions to the benefit of plants. Bacteria (6.4.2), through enzymatic digestion, decompose dead organic material. This action increases the soil's organic matter content and improves its physical properties. Microbes in the soil are also involved in nutrient cycling processes (such as the nitrogen cycle) (chapter 21).

Fungi (6.4.1) are also important decomposers. In acidic soils, fungi are crucial because they tolerate acidity. The most prominent fungi of interest are the *molds* and *mushrooms*. Microbes, although useful, can also be a menace by causing diseases that are responsible for economic loss to producers.

3.5.2 NONARTHROPOD ANIMALS

Arthropods are animals with exoskeletons and jointed legs; *nonarthropods* lack these features. Nonarthropods of horticultural interest are *nematodes* (roundworms) and *earthworms*. Nematodes are round, microscopic, and the most abundant animals in soil. Nematodes are parasitic to a wide host of horticultural plants, including tomato, carrot, turf grass, fruit trees, and ornamentals. They inhabit the roots of their hosts and cause the development of amorphous structures that resemble nodules formed by *Rhizobia*. These growths not only deform the roots but more importantly stifle the growth of plants.

Earthworms, on the other hand, have positive roles in relation to soil. They improve soil water infiltration and aeration by their earth-moving activities and increase the organic matter content of the soil through the plant materials they drag into the holes they dig. They are important in lawns, because they feed on the *thatch* (dead plant material on the soil surface) that builds up with the mowing of lawns (14.5.1). Earthworms thrive in areas that are moist and high in organic matter, avoiding dry and acidic soils.

3.5.3 ARTHROPOD ANIMALS

Examples of *arthropods* (animals with exoskeletons and jointed legs) of importance to horticulture include *termites, millipedes, centipedes, butterflies, ants, grubs,* and other insects. Termites improve drainage of the soil by the pore spaces they create. Ants and termites are also earth movers like earthworms, except that these insects can move huge quantities of material from within the soil to the surface, creating unsightly hills in lawns. Grubs are a menace in lawns because they feed on grass roots, causing dead spots to appear.

3.5.4 VERTEBRATE ANIMALS

Soil-inhabiting vertebrate animals include rodents, such as ground squirrels, mice, gophers, and rabbits. They are known for their earth-moving activities, some of which can create unsightly structures. However, the holes they burrow help in soil drainage. Rodents can ravage crops in the field, which results in economic loss for the grower.

3.6: SOIL AIR

A typical mineral soil, as previously indicated, consists of 25 percent air (3.3.1). Air is required for respiration (4.3.2) by plant roots. Waterlogged conditions cause pore spaces to be filled with water and thus force plants to respire anaerobically (respiration in the absence of oxygen) (4.3.2). Seeds require oxygen for germination (8.9). Clay soils are very susceptible to poor soil aeration, whereas sandy soils are well aerated. To improve aeration of clay soils, crops may be planted on raised beds, which helps to drain the soil pore spaces so that air can occupy them. Vegetables such as tomato and pea are susceptible to oxygen deficiency, which occurs in soils that are water saturated. These plants wilt and eventually die if the situation is not promptly corrected. Soil oxygen levels of less than about 10 to 12 percent are stressful to most plants. Plants such as water lilies have adaptive structures for respiration under water. Crops grown in water as a medium (see *hydroponics,* 11.8) require aeration to aid in root respiration.

Air is also required for decomposition of organic matter by soil bacteria (chapter 21). Under waterlogged conditions, plant materials do not decompose properly and form materials such as peat (partially decomposed organic matter) (3.10.1).

3.7: Soil Temperature

Temperature regulates the rates of all chemical reactions. Plant roots do not grow when the soil temperature is 5°C (41°F) or colder. Even though small grains can germinate at 1°C (34°F), the optimum temperature for germination is 18°C (64°F) and warmer for most crops. Soils warm up much more quickly than water. Sandy soils are said to be warm, whereas clay soils are cold. This situation results because sandy soils drain more quickly and hence retain less moisture (more air spaces) and can therefore warm up more quickly than a soil that has a high capacity for water retention.

Large bodies of water moderate the temperature responses of the soil near them (3.1). Since water heats or cools more slowly than soil, the land around large bodies of water tends to experience slow rates of temperature changes. Such a condition occurs in the Great Lakes region. Fruit trees blossom late and are able to escape the onset of killing frosts. In fall, the killing frosts are delayed, thereby extending the growing season. Vegetables and fruit trees are grown in this region along Lake Erie.

Soil temperature can be modified for crop production. Applying a mulch to a soil surface can modify its temperature; draining a soil enables it to warm up quickly. Raised beds help to improve soil drainage. Light-colored mulches reflect sunlight, whereas dark ones absorb it. Soil temperature can be lowered under light-colored mulch. Some growers use black mulches to raise soil temperature in order to grow an early crop of vegetables and melons in fall. Pale mulches are also used for weed control.

3.8: Soil Water

Soil water is critical to plant growth and development. It is the solvent in which soil nutrients are dissolved before they can be absorbed by plant roots. Once in the plant, water is the medium of transportation of solutes and is required in food manufacture (photosynthesis) (4.3.1). Plants lose large amounts of water from their surfaces by the process of transpiration (4.3.3). Soil is the primary source of water for plant use. Soil water also affects soil air and soil temperature (3.6 and 3.7) and thereby influences plant growth and development. Soil water plays a role in the loss of soil by soil *erosion*.

3.8.1 STRUCTURE AND PROPERTIES

Water consists of *polar molecules* (O^-——H^+), a property that has implications in how water behaves. The cations (e.g., positively charged ions Ca^{2+}, K^+, and Na^+) are hydrated by water molecules through the attraction of the negative oxygen end of the water molecule. Other electrostatically charged particles such as clay also attract water molecules (clay colloids are negatively charged and thus attract the positive hydrogen end of the water molecule).

Water molecules are bonded to each other by means of *hydrogen bonding*. Hydrogen bonding plays a significant role in how water is retained and moved in the soil. The two forces responsible for water retention and movement in the soil are *cohesion* and *adhesion*. Cohesion is the attraction among water molecules, and adhesion (or *adsorption*) is the attraction of water molecules to the soil's solid surfaces. Water exhibits strong *surface tension*, a phenomenon whereby, at a liquid-air interface, water molecules are attracted to each other more than to the air above. Water in this situation appears to have an elastic membrane cover. This property is significant in how water moves in the soil.

> **Surface Tension**
> The force required per unit length to separate or pull a liquid surface apart.

3.8.2 CAPILLARITY

The soil's pore spaces may be tiny (micropores) or large (macropores). Movement of water through pores is affected by the attraction of water molecules to the walls of the conduit (by adhesion or adsorption) and the surface tension of water (cohesion) (figure 3-22). The smaller the pore size, the higher the capillary rise. A formula expressing this relationship is

$$h = 2T/rdg$$

where　　h = height of capillary rise

T = surface tension

r = radius of the conduit (or tube)

d = density of the liquid

g = gravitational force

This relationship is experienced in the soil to a much less extent because of the lack of ideal conditions in the soil (conduits are not straight and smooth or air spaces occur that interrupt capillary flow). Usually, over a long period, the finer the soil texture (3.3.1), the higher the rise. Thus a sandy soil will have a low height because most of the pores are macropores rather than capillary pores or micropores (figure 3-23). On the other hand, compact clay soil has extremely small pores and will not experience as high a rise as a loam (3.3.1). It should be emphasized that capillarity does not operate only in a vertical direction but also horizontally (and, for that matter, movement in any direction is possible).

3.8.3　MEASURING SOIL MOISTURE CONTENT

Soil moisture content may be measured in one of two general ways. *Gravimetric methods* are able to measure the amounts of soil moisture directly. A soil sample is obtained and weighed before drying in the oven (at 100 to 110°C) and then reweighed. The difference in weight represents the amount of water lost during the heating. Water content is expressed in a variety of units. A practical unit used in association with irrigation (3.8.11) is *acre-foot,* representing the amount of water needed to cover an acre of land to a depth of 1 foot.

Another way of determining soil moisture content is by measuring the *moisture tension.* This procedure is used to measure how strongly water is held in the soil. An instrument called a *tensiometer* measures soil moisture tension.

3.8.4　HOW WATER ENTERS THE SOIL

Soil water comes from precipitation (e.g., rain, ice, and snow) or irrigation. Water from these sources may enter the soil at rates dependent on the *infiltration capacity* of the soil. This infiltration capacity in turn is dependent on soil texture, structure, and the presence of impervious layers in the soil, among other factors. If water is provided at a rate in excess of the infiltration capacity, water will pool, or *pond,* on the surface of the soil and eventually flow away as *surface runoff.* Ponding further decreases soil surface infiltration because it causes capping, which blocks the pores. To measure the amount of water the soil receives, a *rain gauge* may be installed at the site (figure 3-24).

FIGURE 3-22 Forces that affect the movement of water in the soil.

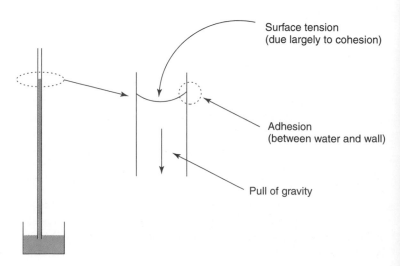

Surface tension
(due largely to cohesion)

Adhesion
(between water and wall)

Pull of gravity

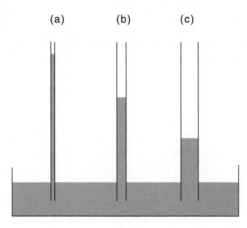

(a) (b) (c)

FIGURE 3-23 A demonstration of the capillary rise of water in the soil. Water rises to the highest level in the capillary tube (a) representing micropores and least in larger pores (b), (c).

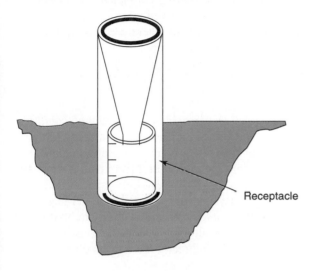

FIGURE 3-24 A diagrammatic representation of a rain gauge.

Receptacle

3.8.5 HOW WATER MOVES IN THE SOIL

When all of the pore spaces in the soil are filled with water, the soil is said to be *saturated*. The movement of water through the soil environment under a state of water saturation depends on the *hydraulic force* and *hydraulic conductivity*. Hydraulic force drives the water through the soil. This force is usually gravitational pull. Hydraulic conductivity describes the ease with which the soil pores allow water to be moved through the soil. Water moves rapidly through a sandy loam with a narrow profile and slowly through a clay loam with a wide profile (spreads due to the greater horizontal movement). Most of this rapid movement occurs through macropores.

When soil is not saturated, water movement is dependent on hydraulic force and conductivity, but most importantly on *matrix potential gradient,* the difference between the *matrix potential* of the moist areas and the drier areas into which water is moving. A high matrix potential occurs in areas of high moisture content. Further, hydraulic conductivity is higher in sand than in clay when the matrix potential is high. However, under drier conditions, water flow in clay is aided by the occurrence of higher capillarity.

The downward movement of water depends on layering, or *stratification,* of the soil. Soil water movement changes from one horizon (3.3.1) to another. Sometimes certain layers (such as clay pans [3.3.1]) are impervious. If soil water moving through a medium-textured soil encounters a layer of coarse-textured soil, downward flow becomes retarded and occurs mainly horizontally. Water flows horizontally because the large pores of the coarse soil have less attraction for water than the finer pores from where it is coming. Downward flow resumes only when moisture accumulates sufficiently and creates a high enough matrix potential gradient.

> **Hydraulic Conductivity**
> The proportionality in Darcy's Law indicating the soil's ability to transmit flowing water.

3.8.6 HOW SOIL RETAINS WATER

The goal of irrigation is to provide and retain moisture in the root zone of the plant. After a heavy irrigation or rain, the soil may receive water to fill all pore spaces and become saturated. The soil at this stage is at its *maximum retentive capacity*. Under this condition, water drains freely under the force of gravity (figure 3-25). The matrix potential at this stage is high. After a period of time, drainage ceases. The water drains out of the macropores first, leaving only the micropores to remain filled with water. This remnant water is resistant to gravitational force. The soil at this stage is said to be at *field capacity*. The water at field capacity may be depleted by evaporation from the surface of the soil or transpiration from the plant leaf surface. These two processes together are called *evapotranspiration* (discussed later in this chapter).

There comes a stage where soil moisture cannot be readily absorbed by roots. At this point the plants show signs of moisture stress and start to wilt. If this condition persists for a long time, the plants remain wilted both day and night. They will recover if water is provided soon. The status of soil moisture at this stage is called the *wilting coefficient* or *permanent wilting percentage*. The remnant water occurs only in the smallest of micropores.

Extreme drying of the soil depletes it of water to the extent where water molecules are tightly bound to the soil particles. At this stage water moves in a vapor phase. The soil moisture content at this point is called the *hygroscopic coefficient*.

> **Field (Water) Capacity**
> The amount of water remaining in a soil after the soil layer has been saturated and the free (drainable) water has been allowed to drain away.

3.8.7 TYPES OF SOIL MOISTURE

From a physical point of view, three forms of water are recognizable—*gravitational, capillary,* and *hygroscopic*. Gravitational water is temporary and not of much use to plants; it can be detrimental to plants unless rapidly drained. Similarly, hygroscopic water is not useful to plants because it is largely in vapor form. Capillary water is thus the most useful form of soil water for plant growth and development. The difference between soil moisture at field capacity and wilting coefficient is described as *available water*. Any water held at a potential of lower than −15 bars is unavailable to most plants. Generally, as soil texture decreases, its ability to hold water increases. Sand holds the least amount of water, whereas silt loam retains the most moisture. Soil organic matter contributes to soil moisture retention indirectly through its effect on soil structure. Plants grow best when soil moisture is near field capacity.

It should be mentioned that a combination of factors affect the availability of soil moisture to plants, including soil structure and texture, soil moisture suction, humidity, wind velocity, air and soil temperature, and density of plant roots. Some plants use up more soil water than others to produce an equal amount of dry matter.

3.8.8 WATER TABLE

Water drained from upper horizons of the soil accumulates below. As the accumulation continues, the soil and underlying parent material become saturated with water, creating what is

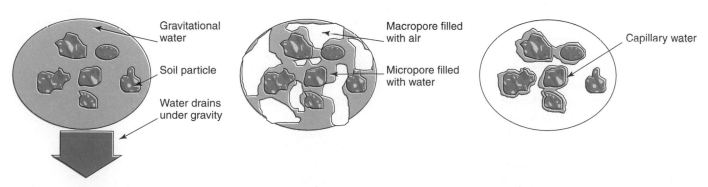

(a) Soil is water-saturated   (b) Soil at field capacity   (c) Soil at permanent wilting point

FIGURE 3-25   Drainage and retention of water in soil. (a) Water-saturated soil drains largely under gravitational force. (b) After the gravitational water has drained out of the macropores, or large pores, the soil is said to be at field capacity. (c) Capillary water is tightly bound to soil particles.

called *groundwater.* The top of the saturated zone in the soil is called the *water table.* The water table varies with seasons and is usually highest in winter. A high water table means water is close to the soil surface, a situation occurring in areas called *wetlands.* Without drainage (removal of excess water from the soil) (3.8.12), such soils cannot be used to produce horticultural plants. Most plant roots cannot thrive under waterlogged conditions.

Sometimes, water accumulates in the soil profile because the downward flow is impeded by an impervious layer (e.g., clay pan). This condition, described as a *perched water table,* develops when plants are grown in containers. A perched water table means that the pore spaces at the bottom of the pot are all filled with water. Shallow pots thus have less room for plants to grow in before they reach this induced water table (figure 3-26). To prevent the formation of a perched water table, the growing media must be properly constituted to allow complete drainage of the entire depth of the soil in the container. The soil should be porous, and drainage holes in the bottom of the pot should be open at all times.

3.8.9 HOW SOIL LOSES MOISTURE

Soil water is subject to loss by a variety of ways. Some of it (up to about 50 percent) may be lost to drainage (3.8.12) of water from the plant root zone. Percolating water is not permanently lost, since some of it can be returned to the root zone by capillary rise (3.8.2). Some of the remaining soil water is moved to the soil surface by capillarity and lost by evaporation (3.8.10). Another avenue to soil water loss is from the root zone via the process of transpiration (plants absorb water and lose it, especially through their leaves) (4.3.3).

3.8.10 EVAPOTRANSPIRATION

Loss of moisture by the process of evaporation and transpiration together produces a combined effect called *evapotranspiration.* As moisture is lost from the soil surface, water moves slowly below to restore the equilibrium. However, if the rate of evaporation is so rapid that it overwhelms the rate of replacement of water from below, the surface layer can become so dry that the continuity of water flow from below is interrupted. This dry surface layer serves to protect the lower layers and conserves moisture.

Evapotranspiration is influenced by certain factors. The rate of evaporation is directly related to the net radiation from the sun. Solar radiation is higher on a clear day than a cloudy one. Further, plant cover or shading by plant leaves reduces the effective soil surface for evaporation. This factor, called *leaf area index* (leaf area of plants per soil surface area occupied by plants measured), affects the amount of radiant energy reaching the soil surface.

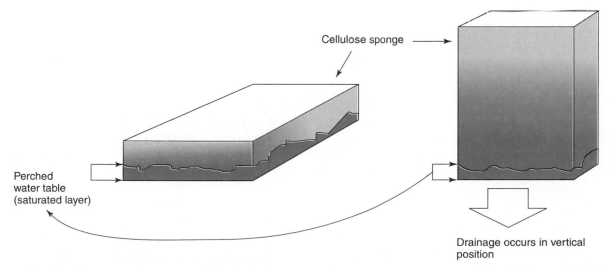

FIGURE 3-26 Demonstration of a perched water table. After drainage ceases when the piece of foam is in the horizontal position, a water-saturated layer occurs at the bottom. When stood on its shorter end, water drainage resumes because of increased space above until a new perched water table develops.

Evaporation is thus higher in a field of newly planted seedlings than in an established one with more dense leaf cover.

Evapotranspiration is also affected by atmospheric vapor pressure. When the atmospheric vapor pressure is less than that at the plant and soil surfaces, evaporation occurs. Temperature increases the vapor pressure at the leaf and soil surfaces, with little effect on the vapor pressure of the general atmosphere. With a drier atmosphere, evaporation proceeds rapidly. Similarly, wind tends to remove moisture vapor from wet surfaces. Thus, leaves and soil surfaces lose moisture more rapidly on dry, windy days. Evapotranspiration is higher in soils near field capacity (3.8.6) than in soils with low moisture content.

Evapotranspiration can be controlled by crop selection and cultural management. Application of mulch (21.1.5) can significantly reduce evaporation and conserve moisture. Cover crops or ground cover (13.12) also reduces evaporation from the soil. Tillage (3.9.2) as a method of weed control eliminates the wasteful use of water by weeds.

Plants may need supplemental moisture during cultivation. Evapotranspiration losses can be reduced when plants are irrigated because the timing of application and amount of water applied can be controlled. Watering should be done at a time of day when the water vapor gradients are low. In this regard, crops are best grown in the cool season, whenever possible. In the dry period when irrigation is needed, the soil surface should be kept only as moist as is needed by the particular plant. However, the watering schedule should allow water to penetrate to the root zone. Methods that allow water to be concentrated around the roots (drip irrigation (3.8.11) are most efficient in this regard. To further reduce surface moisture loss, subsurface irrigation (3.8.11) is even more efficient because the pipes are buried in the ground to supply water to the root zone.

3.8.11 PROVIDING SUPPLEMENTAL WATER FOR PLANTS

Water is a critical requirement in crop production. To be adequate, moisture must be supplied in the right amounts and at the right time. In certain regions, crops can be entirely rain fed, needing no supplemental moisture supply for growth and development. However, summer production of crops requires additional water for economic production. Many horticultural products, crops, and ornamentals require a good supply of water for high yield and quality. High-value crop production often relies on artificial moisture supplementation.

For a small garden operation, additional moisture may be delivered by hand carrying it in a variety of receptacles such as buckets and watering cans. A garden hose may also be used for watering garden crops. A continuous moisture supply for a given period is provided by a variety of methods described next, some of which are adaptable to large- and others to small-scale productions.

Methods of Irrigation

The method of irrigation used in a particular situation depends on the following factors:

1. *Crop.* The method used to irrigate depends on the type of crop or plant and its water needs, size, and way it is being cultured. Because plants that creep on the ground cannot lie in water, the method of irrigation should not permit water to pond on the soil surface. When growing trees that are spaced widely, it is efficient to supply water to the trees individually rather than watering the large spaces in between plants. Irrigating tall trees from above is difficult. Overhead systems of irrigation are unsuitable for tree irrigation. Certain plants prefer or need large volumes of water to perform well, and the irrigation system should be capable of providing this volume.

2. *Source of water.* Different sources of irrigation water vary in cost and availability. Certain methods of irrigation (flood) require the entire soil surface to be covered with water. This method is not practical when the source of water is the domestic water supply. It is more suited to water from rivers or other large bodies of water. Tap water is adequate for watering lawns and residential and urban landscape plants.

3. *Soil type.* Soils differ in water infiltration rate. If the soil is sandy, the water infiltration rate is high, making the area unsuited to flood irrigation. Much of the water is wasted near the supply source because of deep and rapid infiltration. Water moves slowly over the surface.

4. *Slope of the land.* Certain irrigation methods require water to be moved by gravity and thus are adaptable to fields in which the ground slopes.

5. *Rainfall regime of the region.* In areas of erratic rainfall and low annual totals, moisture must be administered efficiently. A system such as the drip system places less demand on the scarce water supply.

6. *Crop rotation.* If crops rotated in the production system require different land preparation and other cultural practices, a method that is flexible and adaptable should be selected. Not all plants can tolerate flooding. The differences in spacing requirements necessitate a flexible system that can be adjusted readily.

7. *Surface of the land.* Installation of an irrigation system is affected by the nature of the terrain. Certain methods such as flooding require grading and leveling of the land. Where pumps are needed to lift water, the relief of the area determines the kind of pump needed to move water from the source to the field.

8. *Cost.* Irrigation methods differ in cost in terms of initial installation and maintenance.

The three general categories of methods for applying supplemental moisture to plants are sprinkler irrigation, drip or trickle irrigation, and gravity irrigation.

Sprinkler Irrigation   Sprinkler irrigation is a method of mimicking the rain by supplying moisture from above. The equipment may be as simple as a *lawn* or *garden sprinkler* (16.1.2) or as elaborate as a large, self-propelled *center pivot* system used on large commercial farms (figure 3-27). The sprinkler system may consist of portable or fixed pipes. Equipment costs can be high. Sprinklers are desirable where soils have high infiltration rates or uneven surfaces. Sprinklers are also used where irrigation is not the primary source of moisture for crop growth. In this case, a portable system may be transported to the area when needed.

Sprinkler irrigation has the effect of modifying the microclimate of plants, since water has a high specific heat. Sprinkler irrigation is used in the horticultural industry for frost protection in the production of vegetables and fruits in winter (3.2.1). Sprinkler irrigation is discussed further in chapter 16 (16.2).

Drip or Trickle Irrigation   In methods of irrigation besides drip irrigation, water is applied perhaps to areas where it is not needed. *Drip irrigation* is a spot application system for watering plants in which minute amounts of water are applied almost continuously to plants throughout

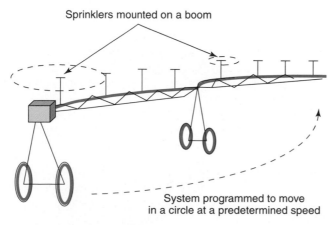

Sprinklers mounted on a boom

System programmed to move
in a circle at a predetermined speed

FIGURE 3-27   A center pivot irrigation system.

the growing period (figure 3-28). It is economical to use and also highly desirable in areas where the water supply is limited. Plastic pipes with emitters are used to deliver the water. Drip irrigation is especially desirable for high-value crops that require uniform soil moisture for good development. For example, when tuber crops such as potatoes are grown under conditions where the soil experiences fluctuations in moisture levels, tuber formation is irregular, leading to low market value. The drip irrigation system is discussed further in chapter 16 (16.2).

Gravity Irrigation   In sprinkler and drip irrigation, a pump may be required for appropriate pressure to move water through pipes. Gravity systems depend on the slope of the land to move water. Initial land preparation includes leveling of the land (in *flood irrigation*) or digging of ditches (in *furrow irrigation*). These methods are not suitable for sandy soils or soils that have high infiltration rate, especially if water is to be moved over long distances. Flood irrigation is commonly used in orchards (figure 3-29). Furrow irrigation is a version of flood irrigation in which the surface flow is limited to channels between ridges (figure 3-30). Vegetables may be irrigated in this way.

3.8.12   SOIL DRAINAGE

Plant roots do not tolerate excessive moisture. *Drainage* is the method by which excess moisture (gravitational) is removed from the soil to enable the soil to be adequately aerated. An area of land can be drained by one of two systems—surface drainage or subsurface drainage.

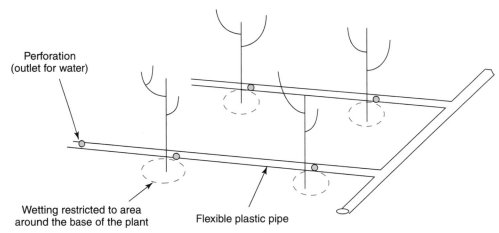

FIGURE 3-28   Area of soil wet under drip irrigation.

FIGURE 3-29   Area of soil wet under flood irrigation.

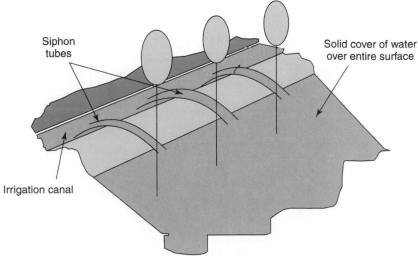

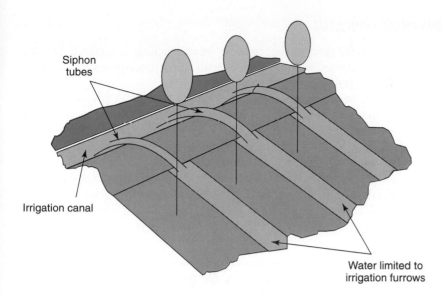

FIGURE 3-30 Area of soil wet under furrow irrigation.

Siphon tubes

Irrigation canal

Water limited to irrigation furrows

1. *Surface drainage.* Surface drainage uses open ditches to remove surface running water before it infiltrates the soil. The land is prepared such that it slopes very gently toward the ditches. Large amounts of water can be drained by surface irrigation in a short period. These ditches require maintenance to keep them open; open drains occupy land that could otherwise have been cropped and also interrupt the continuity of the land.

2. *Subsurface drainage.* In subsurface, or underground, drainage systems, channels are provided in the zone of maximum water accumulation to drain the excess water. The system must be placed below the plow depth. In one system, that most commonly used, the channels consist of perforated plastic pipes laid at predetermined intervals across the field. Water enters the pipes through the perforations and is moved to an outlet ditch. Other less common systems are the mole drainage and the clay-tile drainage systems. Installation of the mole system is less expensive than clay tiles, which are very costly. Outlets of the subsurface systems of drainage must be protected from clogging by sediments, dead animals, bird nests, and other obstructions.

Advantages of Drainage

Drainage is beneficial to plants and horticultural operations in several ways.

1. *Strong structural foundation.* Greenhouses and other structures constructed for horticultural use are more stable when located on well-drained soils. Greenhouses typically use large amounts of water for a variety of activities—washing, irrigation, and the like. The site of the facility should be well drained to handle the excess water.

2. *Aeration.* By draining the root zone, plant roots grow in an environment with a good balance of air and moisture. Aerobic bacteria are able to function properly to decompose organic matter to release nutrients for plant use.

3. *Warm soil.* Drainage of excess water decreases the specific heat of soil, allowing it to warm up more quickly. In clay soils and poorly drained areas (which are described as cold soils), plants should be grown in raised beds. Raised beds improve drainage and free up pore spaces for air to warm the soil.

4. *Timely field planting.* Seeds require warm soil to germinate. Seeds can be planted at the usual time in the season if cold soils are drained to make them warmer.

3.8.13 ENVIRONMENTAL CONCERNS

Land is irrigated mostly in drought-prone areas. In these areas, leaching is minimal, leading to the accumulation of salts in irrigated soil. Excessive fertilization predisposes soil to salinity. Crops have different salt tolerance levels.

The use of fertilizers and irrigation water, while desirable for high economic production, has both short- and long-term environmental consequences. Groundwater and surface waters are polluted as a result of this crop production activity. Excess water from irrigation, either as runoff or drainage, ends up in the groundwater or rivers. This water is loaded with salts and as such causes the salt concentration of rivers to rise.

Phosphates are readily fixed in the soil (adsorbed to soil particles). Phosphate pollution occurs in surface water as a result of soil erosion, which moves soil into surface waters. Nitrates, on the other hand, are readily leached and hence end up in the groundwater. Nitrates may accumulate in some leafy vegetables beyond normal levels. Nitrate toxicity is a problem in ruminants.

3.9: FIELD GROWING MEDIA

Most horticultural plants are grown in solid media, in the field, in containers outside or inside of the house, or in controlled-environment greenhouses. Soil properties (3.3.1) have implications in the way they are managed for optimum productivity. Most plant production or display occurs in the field in real soil (the product of weathering of rocks). Depending on the soil properties and the needs of the plant to be cultivated, soils are physically and sometimes chemically manipulated to provide the best medium for growth. Growing plants indoors either at home or in the greenhouse requires the growth medium to be placed in a container. This situation poses new challenges in managing the growth medium, which should drain freely while holding adequate moisture and nutrients for plant use. Potting media frequently consist either entirely or partly of artificial or *soilless* components.

Soils are disturbed to varying degrees to prepare them for crop production. *Tillage* is the term used for the manual or mechanical manipulation of soil to prepare it for use in crop cultivation. Conventional soil preparation *(conventional tillage)* involves turning over all of the soil in an area of land. Other conservative systems such as the *minimum tillage (zero tillage* or *no till)* disturb only the spots where seeds or plants are to be located.

3.9.1 PURPOSES OF TILLAGE

How soil is tilled depends on its intended use.

1. *Seedbed preparation.* A seedbed provides an environment in which a seed can germinate and grow. It is loose, well drained, deep enough, makes good contact with the seed, retains adequate moisture, and is free of weeds. If the seeds to be sown are tiny, such as those of carrot and lettuce, the *tilth* (fineness of tillage result) should be fine. A rough finish is sufficient for large seeds and transplants.

2. *Level land.* Land leveling may be required while preparing the land to make it amenable to a chosen method of irrigation.

3. *Weed control.* Weeds are a menace to crop production and compete with crops for plant environmental growth factors. They must be controlled before planting and during the growth of the crop by appropriate tillage methods.

4. *Incorporation of organic matter and soil amendments.* Green manures, crop residues, fertilizers, and other chemicals may be added to the soil by plowing them under or mixing them in at the time of preparation.

5. *Improved physical properties of soil.* Compaction of the field impedes drainage, rooting, and general crop growth. A *pan* caused by traffic or tillage may be broken up by tillage methods.

Pan
A layer in the soil that is highly compacted or very high in clay content.

---

6. *Erosion control.* The soil surface after tillage may be such that it impedes runoff. Stubble may be incorporated into the topsoil or ridges constructed to curb erosion. On a slope, the ridges or direction of plowing should be across the slope to impede surface runoff.

3.9.2 TYPES OF TILLAGE

In terms of the depth to which a soil is tilled and the purpose of tillage, there are two general classes of tillage. Each requires different implements.

Primary Tillage

In primary tillage, the soil is tilled to a depth of about 6 to 14 inches (15 to 36 centimeters). The topsoil is turned over, burying the vegetation and other debris. Primary tillage may be used to incorporate fertilizer and also to aerate the soil. The end product is a rough soil surface (clods). Because of the depth of plowing, heavy machinery and implements are used in primary tillage operations. The end product is meant to be transitional and is not ready for planting seeds or seedlings.

Secondary Tillage

Primary tillage is usually followed by a secondary tillage operation to break up the clods to produce a finer tilth for a good seedbed. Another purpose of secondary tillage is to control weeds. The implements used work up to about half the depth (2 to 6 inches or 5 to 15 centimeters) of primary tillage implements.

3.9.3 TILLAGE SYSTEMS

The term *tillage system* refers to the nature and sequence of tillage operations adopted in preparing a seedbed for planting. Each system has specific objectives relating to the condition of the seedbed and soil conservation strategies emphasized.

Conventional Tillage

The goal of conventional tillage is to obtain a clean field in which all weeds and stubble are plowed under the soil. The soil is deeply plowed, requiring both primary and secondary tillage operations and benefiting from both methods. Three general steps are involved in conventional tillage:

1. A preliminary clearing of the land to remove excessive amounts of plant debris. This step depends on the kind of vegetative cover. Shrubs and trees usually must be removed. Grasses can be plowed under the soil without difficulty, unless the plants are very tall.

2. Primary tillage is then conducted to turn the soil over to bury the remaining debris. This activity is called plowing and involves the use of heavy implements.

3. Secondary tillage completes the tillage by providing a fine seedbed. The operation requires several passes of various light implements.

In conventional tillage, the soil is completely exposed to environmental factors. Further, because several trips are made over the land by the various tillage equipment, the soil is prone to compaction (3.3.1), a condition that impedes soil drainage. Deep tillage is thus required periodically under conventional tillage to break the hardpan (3.3.1) formed as a result of soil compaction. Conventional tillage is expensive.

Conservation Tillage

The goal of conservation tillage is to conserve the soil in terms of soil loss to erosion and moisture loss. It may entail producing a fine tilth that breaks capillarity and reduces moisture loss from evaporation, or plant matter may be left on the surface for mulching and impeding soil erosion.

Minimum Tillage

The goal of minimum tillage (also called zero tillage or no-till) is to disturb the soil as little as possible during the entire soil preparation and crop growth periods. Only the spot where the seed or plant will be located is disturbed. Minimized soil disturbance requires special equipment and also relies on chemicals to control weeds.

Listing

Listing is a tillage operation in which an implement called a *lister* is used to create beds or form ridges on which seeds or plants are sown. Sometimes, in well-drained soil, seeds are sown in the furrows between the ridges instead of on the ridges. The soil in the ridge is warmer and has a quick infiltration rate. Ridges also help to prevent soil erosion.

Cultivation

Cultivation is a tillage operation performed as needed during the crop production cycle, any time between seed germination and crop harvest. The primary goal of cultivation is to control weeds. However, soil is also cultivated to improve infiltration and aeration.

3.9.4 TILLAGE IMPLEMENTS

Implements used in tillage may be as simple as a handheld hoe or as complex as a tractor-powered implement. The power of the tractor depends on the size of the implement and the conditions under which tillage is being performed. Tillage implements may be grouped into two categories, primary and secondary tillage implements.

Primary Tillage Implements

Primary tillage implements are operated at lower depths (deeper) to turn over large amounts of soil. They are used to plow the land, as indicated previously.

1. *Moldboard plow.* Moldboard plows vary in design. They can be adjusted to completely invert the soil in order to bury all of the crop residue on the surface. They may be rigged to have a single moldboard or a set of several. These plows are generally set to plow at a depth of 12 to 24 inches (30 to 60 centimeters). Moldboard plows work best when the soil is moist.

2. *Disk plow.* The disk plow is adaptable to difficult situations such as tilling of virgin soil; hard, dry soil; and heavy soil. It turns over the soil but buries the weeds and other plant matter only partially. The plow consists of a set of concave disks and varies widely in design and size. A multipurpose implement, it can be used to provide a satisfactory soil bed for planting if the land being plowed already has a good tilth.

3. *Chisel plow.* The chisel plow is equipped with chisels (sizes and shapes vary) and is useful for breaking up hardpans that impede soil drainage and root development.

4. *Powered rotary tiller.* For shallow tillage (about 6 inches), the powered rotary tiller is useful for incorporating plant residue into the soil and preparing fine seedbeds. Smaller versions (rototillers) are used for small-scale operations such as garden and lawn cultivation.

5. *Lister or bedder.* The lister is essentially a plow equipped with two moldboards. The V-shaped arrangement causes the plow to throw soils sideways to form ridges on beds.

Secondary Tillage Implements

Secondary tillage implements are used after primary tillage implements to further pulverize the soil for the fine tilth required in seedbed preparation. They are operated at shallower depths (up to 6 inches) than primary tillage implements. These implements are used to level and firm the soil.

1. *Disk harrow.* Rows of disk blades are arranged such that the clods created after primary tillage are repeatedly hit and pulverized in one pass of the implement over the soil.

2. *Other harrows.* Harrows are fitted with prongs (called teeth) for stirring the soil. Teeth types vary and may include spikes, tines, and springs (thus called spring-tooth harrow).

3. *Field cultivator.* Cultivators are also varied in construction, but they are similar in basic design to chisel plows, only lighter.

The grower should select the appropriate implement for each operation, bearing in mind that certain secondary implements may be adjusted for use as primary implements (e.g., cultivators and disk harrows are multipurpose implements).

3.10: POTTING MEDIA

Soils, as they occur in nature, consist of mineral elements (3.3.1). They are dense and bulky. Plants are not always grown outdoors in fields. Some are grown indoors and require containers to hold the soil. Because of the bulky nature of natural soil, scientists have developed methods for synthesizing growing media for a variety of purposes. The ingredients in such mixes may be natural or artificial. The goal of such creations is to use proportions of these ingredients in mixes such that the results mimic the environment that a natural soil would provide for a seed or plant. In fact, since humans are in control, they are able to manipulate the proportions of ingredients to create a wide variety of growing conditions not available in nature. These mixes are sometimes called *soilless mixes* because they consist of materials that are not true soil ingredients. However, certain mixes contain real or true soil material.

3.10.1 FORMULATING A MIX

Properties of a Good Mix

Potting mixes are formulated according to need. Certain mixes are constituted for germination and others for growth. Various plants prefer various characteristics in mixtures. Difference notwithstanding, all mixtures should have certain basic physical, chemical, and biological properties.

1. Physically, the mix should drain freely (good infiltration) and be well aerated.

2. The materials used should have high particle stability (i.e., it should not decompose rapidly) and be easily wetted.

3. It should have good moisture-holding capacity and good bulk density.

4. For use in automated pot-filling machines, the mix should flow easily.

5. Chemically, the material used should not produce any toxins. For example, sawdust or chipping from treated lumber should be washed before use. Natural toxins in plants usually break down when plant materials decompose properly.

6. The mix should have good CEC (about 50 to 100 mEq/100 g of soil) and buffer capacity. The pH should be about 5.5 to 6.0 or according to the need of the plant to be grown.

7. Nutrient element needs for proper growth and development should be provided in a balanced amount. Certain mixes include special fertilizer conditioners.

8. Artificial soil mixes are usually pasteurized to kill pathogens. However, beneficial microbes should be present in the soil.

Materials

Materials used in formulating a mix include those of organic and inorganic origin.

Materials of Plant Origin In natural soils, plant roots and other plant remains, along with decaying animal material, provide the organic matter content. In artificial mixes, plant materials are incorporated to fulfill the role of organic matter. The common sources of organic matter in soilless mixes are peat moss, wood by-products, and bark.

PEAT MOSS *Peat moss,* or simply *peat,* is an organic material composed of partially decomposed plant matter that has been preserved under water. It is recovered from underneath bogs and swamps, for example, and has high water-holding capacity. It also supplies some nutrients, especially nitrogen. The types of peat differ according to plant material acidity and degree of decomposition. The best and most widely used is *sphagnum peat.*

Sphagnum peat consists of partially decomposed plant material (*Sphagnum moss*) that is dehydrated. Its varying degrees of decomposition form the basis for material grading. Grade H1 occurs at the top of the bog and is the most decomposed. Grade H10 occurs at the bottom of the pile and is least decomposed. The highest-quality horticultural peat is obtained from layers H3 through H5. This peat has high moisture- and nutrient-holding capacity. It is recovered from acidic bogs. The grades used in horticulture have a pH of between 4.0 and 4.5. Sphagnum moss is lightweight and relatively disease free. It inhibits the growth of the organisms responsible for damping-off disease in seedlings. It holds nutrients well and can be pasteurized without being destroyed.

Apart from sphagnum peat, other peats differ according to the plant material from which they are formed. These types include sedge peat, hypnum peat, and reed peat, which are not commonly used in greenhouse operations. They are more decomposed and have a pH range of between 4.0 and 7.5 or even higher.

WOOD BY-PRODUCTS Wood shavings and sawdust may be used in soilless mixes. The source of wood products is important since certain woods, such as redwood, contain high levels of manganese, an element injurious to young plants. Before use, such problem material should be leached (watered heavily for several hours) to remove this element and other toxins that may exist. Not all toxins can be leached out of the plant material. Sawdust from cedar and walnut should be avoided because leaching is not effective in removing the toxins they contain. Sawdust may be acidic or alkaline in reaction. Including sawdust in a growing medium often creates temporary nitrogen deficiency because the bacteria that decompose sawdust first utilize some of the existing nitrogen. Thus, some nitrogen fertilizer should be included in the mix when sawdust is used as an ingredient.

In areas of certain crop production, by-products of the industries may be good sources of material for soil mixes. Usable plant by-products include peanut shells, bagash (from crushing sugarcane), straw, and corncobs.

BARK Bark (2.7) from hardwoods such as oak and maple or softwoods such as conifers may be used in a growing mix. Bark consists primarily of lignin (2.7), which decomposes slowly. Hardwood bark tends to be higher in nitrogen, phosphorus, and potassium and lower in calcium content than softwood bark. Softwood bark, on the other hand, is higher in micronutrients, especially manganese. Just like wood by-products, bark from certain species may contain toxins. Growth inhibitors have been found in the barks of walnut, cherry, cedar, and white pine. Loblolly pine (*Pinus taeda*) and slash pine (*Pinus caribaea*) barks do not have growth inhibitors and thus are the most widely used in artificial soil mixes.

Materials of Mineral (Rock) Origin   A variety of inorganic materials (containing no carbon) are used in soil mixes.

SAND   *Sand* is a heavy ingredient in growing mixes (3.3.1). Easy to pasteurize, its role in the mix is to improve drainage and infiltration; it does not hold moisture. Sand does not supply any nutrients to the mix. Silica sands with particle sizes between 0.5 and 2.0 millimeters are desirable. If larger particles are used, the mix may settle and become compacted, thus reducing infiltration. Sand may be obtained from the riverbed or mined from white mountain sand deposits. The latter source produces particles that have flat sides (called sharp sand). This angular property gives preference to this kind of sand because it does not settle or pack down in a mix as does sand with rounded particles such as that from riverbeds. Packing down reduces the pore spaces in the soil.

PERLITE   *Perlite* is a light rock material of volcanic origin. It is essentially heat-expanded aluminum silicate rock. The volcanic ore is heated to extreme temperatures of about 1,800°F (982°C) to cause the rock particles to expand to produce the white product used in mixes. Its role in a mix is to improve aeration and drainage. It can be pasteurized. If this ingredient is required in a mix, the horticultural grade should be selected since it has larger particle size and is thus more effective.

Perlite is neutral in reaction and provides almost no nutrients to the mix (except for small amounts of sodium and aluminum). A disadvantage of the use of perlite is its low weight, which makes it float when the medium is watered. Further, during mixing, it produces dust, which can be eliminated by wetting the material lightly before use.

VERMICULITE   *Vermiculite* is heat-expanded mica. This mineral is heated at temperatures of about 1,400°F (760°C) to produce the folded structure associated with the material. It is very lightweight and has minerals (magnesium and potassium) for enriching the mix, as well as good water-holding capacity. Neutral in reaction (pH), it is available in grades according to sizes. Grade 1 includes the largest particles, and grades 4 and 5 are fine in texture. The most commonly used grades are 2 through 4. Its fineness, incidentally, makes it prone to being compressed easily in the mix. To reduce this potential, a mix including vermiculite should not be pressed down hard.

ROCK WOOL   Rock wool is produced from a mixture of basalt, coke, and limestone heated at extremely high temperatures of about 2,700 to 2,900°F (1,482 to 1,600°C). Rock wool fibers are molded into different shapes (e.g., cubes or slabs) and sizes for a variety of uses in horticultural production in the greenhouse. Rock wool supplies small amounts of micronutrients (calcium, magnesium, sulfur, iron, zinc, and copper). Since this material is not biodegradable, its use in media is problematic from an environmental perspective.

Other Additives for Mixes   In addition to the four inorganic materials described previously, several materials are added to improve the quality of the general mix from the standpoint of nutrition, soil reaction, and stability. These materials include the following:

1. *Fertilizer.* Because most of the component materials of a soilless mix do not supply plant nutrition, it is important to include some fertilizer to provide at least a starter nutrition for plants when these materials are used in a soil mix. Slow-release fertilizer (3.4.2) is often included in a mix for this purpose. Fluoride-containing fertilizers (such as superphosphates) should be avoided, since the fluoride is injurious to certain greenhouse-cultivated foliage plants such as chlorophytum.

2. *Limestone.* Limestone, or calcium carbonate, is an ingredient in soil mixes whose principal role is to correct the pH of the mix (3.3.3). It is added in the form of a powder.

3. *Wetting agent.* To transport commercial soilless mixes at reduced cost, manufacturers prefer to use dry ingredients because of their lighter weight. However, dry peat moss repels water. To correct this problem, manufacturers of soil mixes include a wetting agent in their products. Wetting agents are easiest to apply in granular form. Liquid formulations of these agents are also available.

4. *Polystyrene pieces.* The major use of the inert material polystyrene is to lighten the weight of the mix. It has no nutritional value. When watered, particles of this ingredient are found floating on the surface of the water that pools above the soil in pots that contain this material.

3.10.2 FACTORS TO CONSIDER IN CHOOSING MATERIALS FOR A MIX

A large variety of ready-made mixes can be purchased from reputable suppliers. However, if a grower decides to prepare a homemade mix, the following factors should be considered.

Quality of Ingredients

Since mixes vary in physical and chemical characteristics, growers usually adapt specific production practices to soil mixes (e.g., watering and fertilizing regimes). Changes in the quality of the ingredients alter the physical and chemical properties of the mix, making it less responsive to the production practice in use.

Availability

If one is operating in a peanut-growing region, peanut shells could be used in formulating mixes. Establishing sources of regular and ready supplies of raw materials is important to prevent interruption in the production cycle of an enterprise.

Cost

Materials vary in cost. Inorganic components can be especially expensive. Substitutes can be made, as previously indicated, if one understands the role of ingredients in the mix.

Use

The mix prepared should meet the needs of the operation in terms of plant requirements and any automation available. Certain mixes are not conducive for use in potting machines.

Ease of Preparation

Pasteurization is critical to the formulation of mixes. Certain ingredients are difficult to pasteurize. If chemicals (i.e., fertilizers) are to be added, care should be taken to avoid hazards from incorrect calculations and mixing.

3.10.3 CONSTITUTING MIXES

All of the ingredients previously described do not have to be included in each mix. Mixes are formulated for specific purposes, based on the needs of the plant. Certain mixes have broad application. Examples of such mixes are

1. 1:1 of sphagnum moss and vermiculite
2. 1:1:1 of sand, sphagnum moss, and a wood product (e.g., bark or wood shavings)

The appropriate amounts of lime and fertilizer should be added to these mixes. These ingredients are included in these proportions on the basis of volume rather than weight, since they all vary in bulk density. It is important that the mix be disease free; therefore, commercial products are often sterilized by either *heat (steam) pasteurization* or chemical treatment (e.g., methyl bromide).

Standard Mixes

Certain growing mixes have been developed and popularized over the years. These media form the standard, which may be modified for specific purposes. The advantages of these standard mixes include first the fact that they have been thoroughly researched and proven to be successful and second that using standard mixes removes the guesswork from the for-

Table 3-11 Selected Standard Mixes

1. Cornell Peat-Lite Mix A

Materials	Quantity Per Cubic Yard	Quantity Per Cubic Meter
Sphagnum peat moss	0.5 yd^3	0.5 m^3
Vermiculite	0.5 yd^3	0.5 m^3
Ground dolomitic limestone	5 lb	3 kg
Single superphosphate	1–2 lb	0.6–1.2 kg
Calcium of potassium nitrate	1 lb	0.6 kg
Fritted trace elements	2 oz	74 g
Wetting agent	3 oz	111 g

2. Cornell Peat-Lite Mix B

Substitute horticultural perlite for vermiculite

3. The University of California Mix

Materials	Quantity Per Cubic Yard (lb)	Quantity Per Cubic Meter (kg)
Hoof and horn or blood meal (13%)	2.5	1.47
Potassium nitrate	0.25	0.15
Potassium sulfate	0.25	0.15
Single superphosphate	2.5	1.47
Dolomitic lime	7.5	4.42
Calcium carbonate	2.5	1.47

The University of California Mix, as constituted above, should be used fresh; if storage is needed, the mix should exclude the hoof and horn or blood meal

mulation of mixes. They can be reproduced and thus are useful in quantitative studies where uniformity of research materials is needed. These standard mixes are predictable in their effects. Two of the commonly utilized standard mixes are described in table 3-11. Soil mixing can be done on a large scale using commercial-batch soil mixers or continuous media mixing systems.

Roles of Ingredients in a Mix

Each soilless mix ingredient has a specific role to play in the mix. The property of a mix and the ingredient responsible are as follows:

1. *Good water retention.* Soilless mixes are utilized mainly in plants cultured in containers. The volume of soil is usually limited. The mix materials should be able to retain moisture so that the medium does not dry out too quickly. Organic materials are used as ingredients for increasing the water retention of the medium.

2. *Good drainage and aeration.* The mix should be well drained to reduce the danger of overwatering, which can cause a perched water table (3.8.8). Most plant roots are intolerant of anaerobic conditions. Materials such as perlite and sand are used for good aeration. Good soil drainage guards against the buildup of harmful salts.

3. *Proper pH.* Improper soil reaction interferes with nutrient availability for plants (3.3.3). Certain organic materials used in mixes produce acidity or alkalinity in the medium. The use of lime is necessary to correct soil acidity.

4. *Fertility.* Most soil mix ingredients do not contribute appreciably to the fertility of the growing medium. Additional materials such as fertilizers are needed to provide the required nutrition for proper plant growth and development.

Advantages of Soilless Mixes

The major advantages of soilless mixes include the following:

1. *Uniformity of mix.* The physical and chemical properties of a mix are uniform throughout the mix, a condition not found in field soil. The homogeneity of the medium makes it possible for plants to grow and develop uniformly, provided other growth factors are also consistent.

2. *Ease of handling.* Mixes are lightweight and easy to transport. Ingredients are easy to scoop and move around during manual preparation.

3. *Versatility.* Mixes can be made to order (custom mixed) for specific needs. They can be used to amend soils in the field by mixing into flower beds, lawns, or garden soil. They are convenient to use in container plants.

4. *Sterility.* Mixes, initially at least, are free from diseases and pests. Seedling germination is less prone to diseases such as damping-off (8.9).

5. *Good drainage and moisture retention.* A mix can be custom-made to provide the appropriate degree of drainage and moisture retention.

6. *Convenience of use.* Mixes are ready to use when purchased.

Disadvantages of Soilless Mixes

Disadvantages of soilless mixes include the following:

1. *Light in weight.* Some mixes are very light, especially when dry. When used in potted plant production, they are easily toppled over by even a gentle wind or push.

2. *Limited nutritional supply.* Mixes that incorporate fertilizers provide nutrition for a limited period of time. The component ingredients of mixes are generally void of any appreciable amount of plant growth nutrients. Micronutrients are especially lacking and should be supplemented with an appropriate fertilizer program when growing a crop.

3. *Lack of field correspondence.* Mixes are constituted to provide minimal problems to germinating seeds and rooting plants. As such, the physical conditions of mixes are different than field soil conditions. Roots grow rapidly and ball up in pots. During transplanting into the field, care should be taken to create good root contact with the soil. Some plants may not establish quickly if root contact with the soil is poor.

4. *Cost.* Some mixes are expensive (but worth the investment).

3.11: SOIL STERILIZATION

3.11.1 FIELD STERILIZATION

Sterilizing field soil can be accomplished by harnessing the energy of the sun. *Solar pasteurization,* or *soil solarization,* is a method of pasteurizing the soil by using solar energy. The goal of pasteurization is to rid the soil of harmful bacteria, fungi, nematodes, and weeds. The

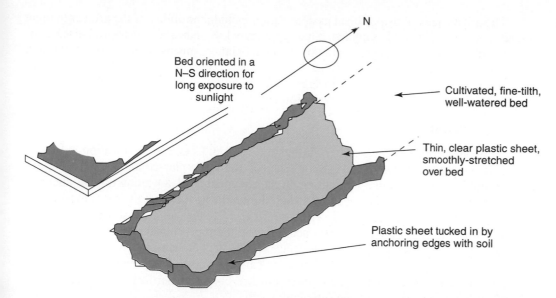

FIGURE 3-31 Solarization of soil.

N

Bed oriented in a
N–S direction for
long exposure to
sunlight

Cultivated, fine-tilth,
well-watered bed

Thin, clear plastic sheet,
smoothly-stretched
over bed

Plastic sheet tucked in by
anchoring edges with soil

area to be treated is first cleared of grass and weeds. The soil is then cultivated, raked, and watered uniformly to a depth of about 12 inches (30 centimeters). A clear plastic is stretched tightly over the area and tucked under a border of soil (figure 3-31). The plastic cover should be left in place for about four to eight weeks for effective solarization. Black plastic reflects heat, even though it controls weeds much better. Beds with north-south orientation receive sunlight most of the day, and the effect is better. Field temperature under solarization can exceed 120°F.

Plant diseases known to be controlled by this treatment include fusarium wilt, verticilium wilt, pinkroot, and southern blight. Weeds effectively controlled include cheeseweed *(Malva neglecta),* annual bluegrass *(Poa annua),* and pigweed *(Amaranthus* spp.). Plants grown in recently solar-pasteurized soil grow faster and larger than those in nonpasteurized soil. The soluble nutrient levels in soil with reference to nitrogen, calcium, and magnesium have been found to increase after solarization. Solarization is used by organic gardeners.

3.11.2 STERILIZATION OF GREENHOUSE SOIL

Growing media for indoor use can be sterilized by a variety of methods. The goals are identical to sterilization of field soil.

Steam Pasteurization

Steam pasteurization is a nontoxic method of sterilizing soil by which steam is passed through a pile of soil. Soil to be steamed should be prepared by mixing thoroughly and readied for planting. Fertilizers and lime should be mixed in before sterilizing. (Certain fertilizers, such as Osmocote, cannot be sterilized and should be added after the process is complete.) The soil to be steamed is covered with a tarpaulin or some other appropriate material. The soil should not be too dry or too wet; either condition requires additional heating time to properly sterilize the soil. The hot steam may be introduced via buried pipes with perforations. Alternatively, in surface steaming, a porous canvas hose may be laid on top of the evenly graded topsoil and covered. The buried pipe method is more effective when ground beds are to be sterilized.

Steam aerated by introducing air into the steam flow is preferred to using steam alone. At a temperature of 60°C (140°F), the aerated steam is administered for 30 minutes. In the pure steam method, the soil temperatures are higher (84°C, or 180°F). Overheating is detrimental to the soil because it can kill beneficial organisms and cause buildup of toxic substances, especially in soils that have a high organic matter content. Heating results in a high soluble-salt level in the soil. Further, since beneficial organisms are killed, soilborne diseases that occur after pasteurized soil has cooled tend to spread quickly.

The advantages of using steam pasteurization include the ability of the grower to plant immediately after the soil has cooled. Although it may kill undesirable pests and disease organisms, if done at the proper temperature, the beneficial organisms are not harmed. Further, applying steam is relatively inexpensive. Soil drainage and aeration have been known to improve after steam pasteurization due to aggregation of particles. Steam sterilization is also applicable to containers, tools, and other equipment.

Chemical Sterilization

Soil can be sterilized by using certain fumigants (7.10.1). Chemical sterilization is less effective than steam sterilization and is toxic. The soil cannot be planted immediately after the process, unlike after steaming. Depending on the chemical used, a waiting time of about 24 hours to several weeks may be needed. Further, during the application, the room must be vacated to prevent worker poisoning. To effectively permeate the soil, it must not be cold but a desirable warm temperature that differs according to the type of chemical used. Chemicals used include *chloropicrin* (or tear gas). Because of the effect on humans, precautions must be taken to control the gas, such as watering the area after application to provide a water seal or providing the treatment under airtight cover. Further, chemical sterilization should be used only when the greenhouse is vacant (no plants).

Other chemicals used in sterilization include formalin and vapam. Vapam is not toxic to humans but must not be used in their presence. Treated growing media are ready for use only after several weeks. Some plants are sensitive to the residual effects of certain chemical treatments. For example, carnations are damaged when grown in media treated with Dowfume MC-2 (a mixture of methyl bromide and 2 percent chloropicrin). This chemical is extremely toxic to humans and must be used with great caution.

3.12: LIQUID MEDIA

As indicated earlier, soil is the most popular medium for the cultivation of crops. Soil, or any solid medium, provides physical support for plants to grow. Today, plants are cultivated on a large scale under controlled-environment conditions indoors. A variety of media have been developed to facilitate such an undertaking. Perhaps the most revolutionary is growing plants in liquid media (water fortified with nutrients). The science and methods of water culture, or *hydroponics,* are presented in detail in chapter 11.

One of the earliest and still widely used liquid media in plant culture is *Hoagland's nutrient solution,* an all-nitrate solution. Various *nutriculture systems* have their own unique liquid media recipes. Crop needs and cultural conditions call for special nutrient systems to be used for the successful production of specific crops.

SUMMARY

Plants require appropriate temperature, moisture, light, and air for good growth and development. These conditions occur both above and below ground. In addition, plants need a medium in which to grow, the common one being soil, even though modern technology enables plants to be grown in soilless media. Air is required for respiration (for energy); water is needed for germination, photosynthesis, and translocation of solutes within the plant, among other functions. Light is needed for photosynthesis and also affects reproductive activities in some species through photoperiodism. Light affects plants through its intensity, duration, and quality. Long-day plants require more than 12 hours of daylight to flower. Short-day and day-neutral plants have varying daylight requirements. Temperature is the intensity factor of heat energy and regulates the rate of chemical reaction. Soil has layers called horizons that make up a soil profile. The topsoil is of most importance to horticulture. Soils vary in texture (particle size distribution) and structure (particle arrangement), as well as in chemical characteristics such as soil reaction (pH) and nutrient level. The nutrients used by plants in large amounts, called major nutrients, or macronutrients, are nitrogen, phosphorus, and potassium. Those required in small or minute amounts are called micronutrients, minor nutrients, or trace elements and include copper, molybdenum, zinc, chlorine, manganese, magnesium, and calcium. Nutrient elements can be artificially supplied by fertilizers. The soil is home to a large number of organisms, some of which are beneficial to plants.

Soil is the primary medium in which plants are grown. Plants may be grown directly in the soil in the field or in soil placed in containers. Sometimes the medium used for potting may be entirely or partially nonsoil or artificial. Before use, the soil has to be prepared. This preparation is called tillage. Primary tillage (rough) precedes secondary tillage (finer tilth). Several concepts are important in tillage. In conventional tillage, the field or plot is deeply plowed using primary or secondary operations before use. In minimum tillage, only the spot where the seed is to be placed is disturbed. Conservation tillage includes practices that conserve soil moisture and prevent erosion. Different implements, including plows, harrows, listers, and tillers, are available for various tillage operations. Potting soil frequently includes artificial ingredients such as perlite and vermiculite. Organic materials used include peat moss, sphagnum moss, and wood products; inorganic ingredients include sand and limestone.

REFERENCES AND SUGGESTED READING

Boodley, J. W. 1998. The commercial greenhouse, 2d ed. Albany, N.Y.: Delmar.

Bunt, A. C. 1976. Modern potting composts: A manual on the preparation and use of growing media for pot plants. University Park: Pennsylvania State University Press.

Janick, J. 1986. Horticultural science, 4th ed. San Francisco: W. H. Freeman.

Kramer, J. 1974. Plants under lights. New York: Simon & Schuster.

Maracher, H. 1986. Mineral nutrition in higher plants. New York: Academic Press.

McMahon, R. W. 1992. An introduction to greenhouse production. Columbus Ohio Agricultural Education Curriculum Materials Services.

Nelson, P. V. 1985. Greenhouse operation and management, 3d ed. Reston, Va.: Reston Publishing Co.

Rice, L. W., and R. P. Rice, Jr. 1993. Practical horticulture. Englewood Cliffs, N.J.: Prentice-Hall.

Singer, M. J., and D. N. Munns. 1987. Soils: An introduction. New York: Macmillan.

PRACTICAL EXPERIENCE

1. Plant 16 pots of a selected plant. Group pots into sets of four. Select a source of nitrogen fertilizer and four different levels (0, 1, 2, 3). The amount chosen depends on the size of the pot and the type of plant. Your instructor will give you guidance. Apply one level to each set of pots. Observe the differences in plant growth at the end of the course by measuring a selected number of plant characteristics.

2. Repeat exercise 1 using other nutrients (phosphorus and potassium) and different plant species. You may increase the number of pots per set so that you can measure plant weight and other characteristics at periodic intervals. These intermediate measurements are destructive, requiring the samples to be uprooted or cut and dried. You may also take measurements at periodic intervals without destructive sampling.

3. Collect samples of soil from various areas—a garden, a flower bed, and a lawn. Test for soil reaction (pH) and other macroelements by using a simple garden soil testing kit.

4. Place a set of seedlings of corn and one set of beans under bright light. Place similar sets of plants in darkness. Observe the differences in plant growth and development after three weeks (or more) and explain the results.

5. Visit a farm equipment rental company or dealership to see the various available tillage implements and learn about their functions.

6. Obtain samples of the various components of soilless mixes—vermiculite, perlite, peat moss, and sphagnum moss—and create your own mixes. Test the physical properties of the mixes, including weight per unit volume, texture, and drainage.

OUTCOMES ASSESSMENT

PART A

Please answer true (T) or false (F) for the following statements.

1. T F   Plant roots require air.
2. T F   Smog is a combination of smoke and fog.
3. T F   Short-day plants flower only when day length is less than 14 hours.
4. T F   Soil horizons make up a soil profile.
5. T F   Soil texture is concerned with the arrangement of particles in a soil.
6. T F   Silt has smaller particle size than clay.
7. T F   A typical mineral soil consists of 10 percent organic matter.
8. T F   A fertilizer grade of 10-15-25 consists of 25 percent phosphorus.
9. T F   Sodium is a major plant nutrient.
10. T F   A soil pH of 8 is alkaline.
11. T F   A fertilizer of analysis 0-20-0 is a complete fertilizer.
12. T F   Manual or mechanical manipulation of soil to prepare it for planting is called tillage.
13. T F   Primary tillage precedes secondary tillage.
14. T F   A moldboard plow is a secondary tillage implement.
15. T F   Peat moss is an organic material.
16. T F   Perlite is volcanic in origin.
17. T F   Tillage may control soil erosion.

PART B

Please answer the following questions.

1. List the four abiotic environmental factors required for plant growth.

   _____   _____

   _____   _____

2. List four major environmental pollutants.

   _____   _____

   _____   _____

3. Etiolation occurs when plants grow_____.
4. Plant response to duration of day length is called _____.
5. Equal proportions of sand, silt, and clay constitute a _____.
6. Yellowing of leaves due to nitrogen deficiency is called _____.
7. _____ is a method of reducing soil acidity.
8. The hardened layer of soil caused by compaction is called_____.
9. Provide the alternate names for minimum tillage.

   _____

10. _____ is the fineness of tillage results.

11. List three each of the following types of implements:
    a. primary tillage _____ _____ _____
    b. secondary tillage _____ _____ _____
12. Potting media that consist of nonsoil material are generally called _____.
13. Give two advantages and disadvantages of soilless mixes.

_____ _____

_____ _____

Part C

1. Compare and contrast peat moss and sphagnum moss.
2. Describe the physical characteristics of vermiculite.
3. Distinguish between climate and weather.
4. State and explain the Hopkins bioclimatic law.
5. Describe how ozone is formed.
6. What is the role of soil nitrogen in plant nutrition?

CHAPTER 4

Plant Physiology

PURPOSE

The purpose of this chapter is to discuss the primary plant physiological processes and to show how they affect the growth and development of horticultural plants. The discussion includes how an understanding of these processes enables scientists and growers to manipulate them for the higher quality and productivity of plants.

EXPECTED OUTCOMES

After studying this chapter, the student should be able to

1. Describe the generalized pattern of growth in organisms (sigmoid curve).
2. Describe the generalized phases of plant growth.
3. Discuss vegetative growth and development in plants and how growth patterns are used as a basis for the classification of plants.
4. Describe reproductive growth and development in plants.
5. Describe the role of environmental factors on plant growth and development.
6. Describe specific growth processes—photosynthesis, respiration, transpiration, translocation, and absorption—and their roles in plant growth and development.
7. Discuss specific ways in which growers may manipulate physiological processes for increased plant productivity and quality.

KEY TERMS

Absorption
Accessory pigments
Aerobic respiration
Amino acids
Anaerobic respiration
Apical dominance
C_3 plants
C_4 plants

Calvin cycle
Cohesion-tension theory
Crassulacean acid metabolism
Dedifferentiation
Differentiation
Fatty acids
Fermentation

Growth regulators
Hydrolysis
Light-dependent reaction
Light-independent reaction
Monosaccharides
Nucleic acids
Parthenocarpy
Phloem loading

Phloem stream
Phospholipids
Photoperiodism
Photorespiration
Photosynthesis
Photosystem
Physiological maturity

Polysaccharides
Pressure flow
Primary structure
Quaternary structure
Secondary structure
Self-incompatibility

Senescence
Tertiary structure
Transpiration
Transpiration stream
Vernalization
Water potential gradient

OVERVIEW

The genotype of an organism specifies its course of development within a given environment. A tall plant will grow tall, first because it has the *genes* for tallness, and second because it is provided with the appropriate *environment* to support the expression of the tallness trait. In growing to become tall, certain specific physiological activities must occur to provide the materials and energy required to translate genetic information into a physical appearance or phenotype. In other words, it is through physiological processes that genes are expressed.

Physiological processes allow the embryo in a seed to develop into a mature plant. As a plant grows and develops, it does so because of the roles of physiological processes such as *photosynthesis, respiration, transpiration, translocation,* and *absorption.* A variety of external and internal factors affect physiological processes. Light and hormones are notable factors that affect the growth of plants. By understanding how these factors and the processes themselves function, scientists (and eventually growers) are able to manipulate plants to their advantage by altering certain environmental conditions. Some processes may be slowed down and others speeded up. The major physiological processes that affect plant growth and development are discussed in this chapter.

4.1: GROWTH AND DEVELOPMENT

Growth and development involve three basic activities:

1. Cellular division
2. Enlargement
3. Differentiation

4.1.1 GROWTH

Growth is an irreversible phenomenon that occurs in a living organism, resulting in an increase in its overall size or the size of its parts. Growth is accompanied by energy-dependent metabolic processes (4.3.2). As such, whereas producing a leaf or root is growth, an increase in size due to swelling from water absorption is not.

> **Growth**
> A progressive and irreversible increase in size and volume through natural development.

Cellular Division and Enlargement

The processes of cell division and enlargement are frequently associated. Cellular enlargement often precedes cellular division, as found in meristematic cells. However, the two processes are regulated independently. Cell enlargement does not induce cell division, and the latter is not always preceded by the former. For cells to enlarge, the cell walls (which are rigid [2.1.2]) must first be "loosened." This loosening occurs when certain acids are secreted into the cell walls (cell wall acidification). These acids in turn activate certain pH-dependent enzymes that act on cellulose (4.2.1) in the walls. This cell-loosening process is influenced by a plant hormone called *auxin* (4.6). In addition to loosening of the cell wall, cellular enlargement requires *positive turgor pressure,* which results as a cell takes in water by *osmosis.* Because enlargement is required for growth and turgor is required for enlargement, the water status of a plant is critical

> **Osmosis**
> The diffusion of water or other solvents through a differentially permeable membrane from a region of higher concentration to a region of lower concentration.

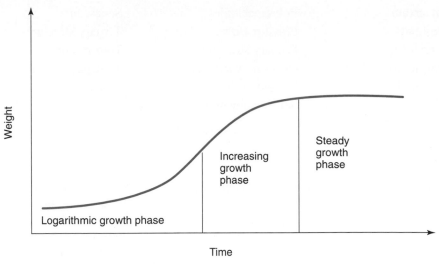

FIGURE 4-1 A sigmoid growth curve.

to its growth and development. Cellular enlargement usually occurs in only one direction. Expansion occurs where the cell wall is most elastic. Cell wall elasticity is dependent on the orientation of the cellulose *microfibrils* in the wall. Where longitudinal cellulose microfibrils occur, the cell wall enlarges in a transverse fashion. On the other hand, transverse deposition of microfibrils in the cell walls produces longitudinal expansion. Random cellulose microfibrils produce equal expansion in all directions.

Growth produces an increase in dry matter when the plant is actively *photosynthesizing* (or manufacturing food) (4.3.1). It is influenced by genetic, physiological, and environmental factors. Some plant cultivars are bred to be tall and large bodied, whereas miniature cultivars are bred for other purposes. For example, tomato cultivars used in salads produce small and near bite-size fruits; those used in canning are much larger in size. Plants can be manipulated by the application of chemicals to change their size by either enhancing or inhibiting growth. By changing the environment of the plant (e.g., soil, water, nutrients, temperature, photoperiod, and light intensity), a plant's growth can be enhanced or hindered. These environmental factors of plant growth and development are discussed in chapter 3.

Growth in an organism follows a certain general pattern described by a *sigmoid curve* (figure 4-1). The pattern and corresponding developmental stages as they occur in plants are as follows:

1. *Logarithmic growth phase.* The logarithmic growth phase is characterized by an increasing growth rate and includes seed germination and vegetative plant growth periods.

2. *Decreasing growth phase.* During the decreasing growth phase, growth slows down. This stage includes flowering, fruiting, and seed filling.

3. *Steady growth phase.* Growth rate either declines or stops during the steady growth phase. This phase is associated with age, or the plant's maturity.

The characteristics of the sigmoid growth pattern vary among species and plant parts. In fruit growth, food materials are translocated from one part of the plant to another. In species such as apple, orange, pear, tomato, and strawberry, fruit growth follows the simple sigmoid growth curve (with variations in characteristics). However, in certain species, including stone fruits such as plum, peach, and cherry, fruit growth follows a *double sigmoid curve* pattern whereby the single pattern is repeated (figure 4-2). During the plateau of the first sigmoid curve, the fruit size barely changes; most of the activities involve seed development. In stone fruits, the hardening of the endocarp (pit) occurs during the second phase of fruit development.

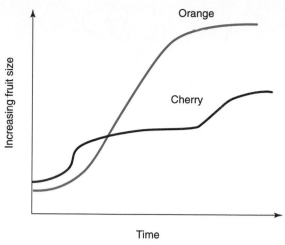

FIGURE 4-2 The sigmoid and double sigmoid growth curves. Stone fruits are characterized by the double sigmoid curve growth pattern. The characteristics of either curve differ from one species to another.

Development

Cellular Differentiation and Dedifferentiation   *Differentiation* is the process by which meristematic cells (2.2.2) (genetically identical) diverge in development to meet the requirements for a variety of functions. The specialization of cells occurs as a result of differential activation of the cell's genome (total number of genes) (5.2.1), resulting in the structure and function of previously genetically identical cells becoming different.

Differentiated cells can revert to meristematic status; when this happens they are said to be *dedifferentiated*. Cells dedifferentiate when the plant structural pattern is disrupted by, for example, wounding. To repair the damage, the cells in the damaged regions first become "deprogrammed" so that they can be reprogrammed to form the appropriate types of cells needed for the repair.

> **Differentiation**
> *The process of change by which an unspecialized cell becomes specialized.*

Polarity

Plants and their parts exhibit strong directional differences that are described as polarity. Classic examples of such directional growth and development are the existence of abaxial and adaxial (2.4.2) parts of a leaf and the shoot and root ends of a full plant that are very different in structure and function (figure 4-3). *Polarity* may be influenced by environmental factors such as gravity and light or it may be genetic in origin. In angiosperms, polarity appears to be fixed and difficult to alter. Thus, a stem cutting will always produce roots at the basal end and shoots at the apical end (figure 4-4). Polar transport of growth hormones (such as auxin) is implicated in this phenomenon.

4.1.2   THE ROLE OF SIGNALS IN GROWTH AND DEVELOPMENT

Plant growth and development are regulated by complex signals. Plant hormones have been shown to stimulate differentiation of procambium. In tissue culture (9.12), the addition of a hormone such as auxin stimulates leaf formation in callus (a mass of undifferentiated tissue). In plants, removing the shoot apex stimulates lateral bud growth (15.3). Genetic control similar to that found in animals has been seen in maize studies. Other known control signals are *positional control* (due to the position of a cell in plant tissue), *biophysical control* (due to physical pressure generated by growing organs), and *electrical currents* (generated by growing plants).

FIGURE 4-3   Polarity in plants.

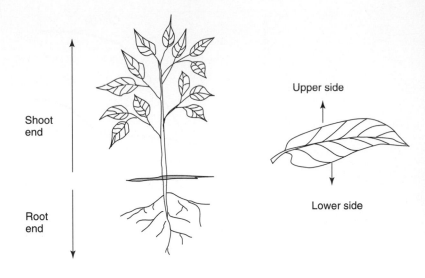

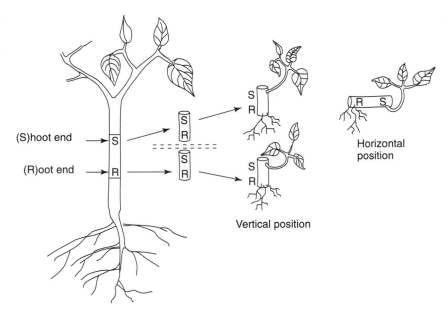

4.2: ORGANIC COMPOUNDS OF PLANT CELLS

Major and minor inorganic chemicals obtained by plants from the soil are discussed in chapter 3 (3.3.1). These inorganic elements are utilized by the plant to synthesize the organic components of cells. Water is one of the most abundant inorganic compounds in plant cells. Carbon, hydrogen, and oxygen occur in all organic molecules. Sulfur and phosphorus occur in very few organic molecules and, even then, only in small amounts. The major classes of cellular organic constituents are *carbohydrates, lipids, proteins,* and *nucleic acids.* Of these, proteins and nucleic acids are relatively large molecules and hence are called *macromolecules.*

4.2.1 CARBOHYDRATES

Carbohydrates are the most abundant organic molecules in nature. In plants, they are the principal components of cell walls. They are also the primary energy-storage molecules in most organisms. Carbohydrates are made up of three elements—carbon (C), hydrogen (H), and oxygen (O)—usually in the ratio of 1:2:1 of carbon to hydrogen to oxygen. The general molecular formula of carbohydrates is thus $(CH_2O)_n$. The three principal kinds of carbohy-

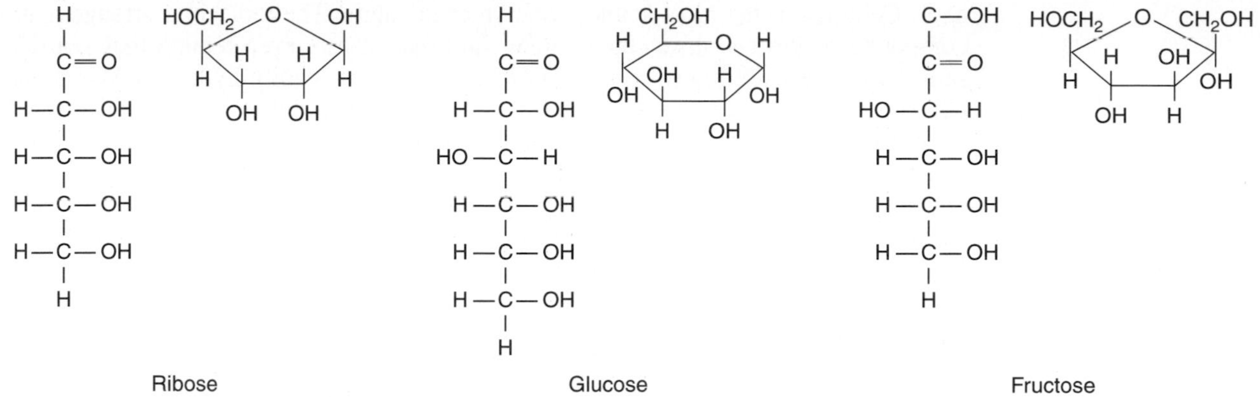

FIGURE 4-5 Selected monosaccharides of physiological importance. Ribose is a pentose (five-carbon sugar), while glucose and sucrose are hexoses (six-carbon sugars). They are represented in both open-chain and closed-ring molecular forms.

drates—*monosaccharides, disaccharides,* and *polysaccharides*—are classified on the basis of the number of sugar molecules they contain.

Monosaccharides

Monosaccharides ("one sugar"), or *simple sugars,* consist of a chain of subunits with the basic structure $(CH_2O)_n$. Simple sugars are thus the building blocks of carbohydrates. Sugar names have the suffix *-ose* and a prefix that indicates the number of carbon atoms each sugar contains. For example, a *pentose sugar* has five carbons, and a *hexose sugar* has six carbons (figure 4-5). Hexoses and pentoses are the most important simple sugars in plants. They occur as cell wall constituents and are important in energy aspects of cellular function. Glucose, a hexose, is used in the synthesis of other complex molecules such as starch (see polysaccharides). It is a major product of photosynthesis (4.3.1). The sweet taste of ripened fruits is due in part to the glucose and fructose sugars, as well as hexose sugars. Fructose can be converted to glucose and hence perform the functions of glucose. Ribose sugar is a constituent of deoxyribonucleic acid (DNA) (5.2.4) and ribonucleic acid (RNA) (5.2.4).

Disaccharides

Disaccharides consist of two monosaccharides joined by the process of *condensation* (removal of a water molecule by an enzyme-catalyzed process). Disaccharides can be broken down into component molecules by *hydrolysis* (addition of a water molecule to each linkage). *Sucrose* (or common table sugar from sugarcane or sugar beet) is a disaccharide consisting of glucose and fructose. Whereas the sugar transported in animal systems is commonly glucose, sucrose is the form in which sugars are most often transported in plants.

> **Hydrolysis**
> The breakdown of complex molecules to simpler ones resulting from the union of water with the compound.

Polysaccharides

A *polymer* is a large molecule consisting of identical or similar molecular subunits called *monomers*. These monomers can polymerize into long chains. When three or more sugar molecules polymerize, the product is a polysaccharide. In most plants, accumulated sugars are stored (in seed, leaves, stems, and roots [9.11]) in the form of a polysaccharide called *starch*. Starch is made up of two different polysaccharides—*amylose* and *amylopectin*. Amylose consists of glucose molecules linked in a nonbranching pattern (1,4-linkages), whereas amylopectin consists of molecules linked in a branching pattern (1,6-linkages). The figures in the linkage patterns refer to the position of carbons in the rings involved in the linkage. In certain species, such as temperate grasses, the commonly stored polysaccharides in leaves and stems are polymers of fructose called *fructans*.

Polysaccharides are important structural compounds. The most important in plants is cellulose, a polymer of β-glucose monomers (instead of α-glucose, as in starch) in 1,4-linkages. Cellulose is the most abundant polymer in nature. The molecular arrangement in this compound makes cellulose chains more rigid than starch, even though they consist of the same monomers. Cellulose is resistant to enzymes that readily hydrolyze starch and other polysaccharides. The biological functions of starch and cellulose are thus different. In fact, once incorporated into the cell wall as a constituent, cellulose cannot be utilized as a source of energy by plants. In ruminant animals such as cattle, it takes the action of microbes in the digestive tract to make cellulose an energy source. Other important polysaccharides in plants are *pectin* (polymer of *galacturonic acid,* a six-carbon sugar containing an acid group) and *hemicellulose* (composed of a complex mixture of sugars).

4.2.2 LIPIDS

Lipids are a group of fats and fatlike substances. They differ in two major ways from carbohydrates. Lipids are not water soluble, unlike most carbohydrates (except large polymers such as starch and cellulose). Also, lipids structurally contain a significantly larger number of C-H bonds and as a result release a significantly larger amount of energy in oxidation than other organic compounds. Lipids can be classified as follows.

Fats and Oils

Fats and *oils* have a similar chemical structure, consisting of three fatty acids linked to a *glycerol* (a three-carbon alcohol) molecule (figure 4-6). This structural arrangement is the origin of the term *triglyceride*. Fats and oils are storage forms of lipids called *triglycerides*. Whereas fats and triglycerides are solid at room temperature, oils remain liquid. Cells synthesize fats from sugars. Plants, as previously indicated, store excess food as starch. To a limited extent, fat is stored as droplets within the chloroplasts of some species of plants such as citrus.

Triglycerides differ in nature by the length of their fatty acid chains and whether all of the carbon atoms are linked to as many hydrogen atoms as possible (and thus called *saturated*) or some of the carbon atoms are double bonded to hydrogen (and thus called *unsaturated*), thereby allowing such carbon atoms to be bonded with other atoms. Triglycerides containing unsaturated fatty acids tend to behave like oils (being fluid or liquid) at room temperature. Unsaturated fatty acids are found in plants such as corn and peanut, both of which are sources of edible oil.

Phospholipids

Phospholipids differ from triglycerides in that only two of the three fatty acids are attached to glycerol, the third one being attached to a phosphate group. The presence of the phosphate at the end of the phospholipid molecule makes this end of the fatty acid molecule water soluble (the other end remains water insoluble). Phospholipids are important cellular membrane constituents.

FIGURE 4-6 A typical fat molecular structure.

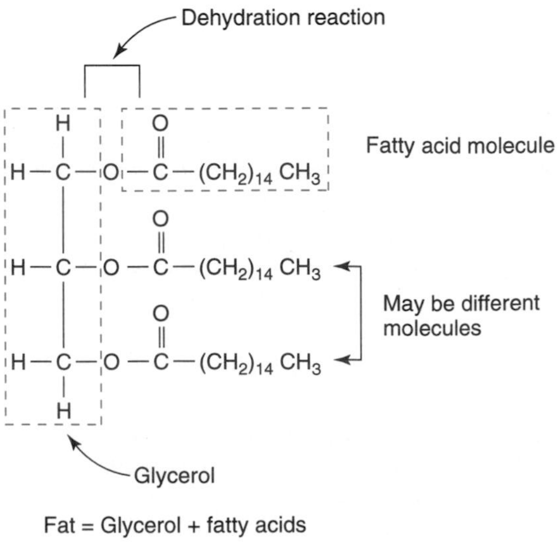

Fat = Glycerol + fatty acids

Cutin, Suberin, and Waxes

Cutin, suberin, and *waxes* make up a group of lipids that are insoluble and create a structural barrier layer in plants. Cutin and suberin form the structural matrix within which waxes are embedded. Waxes consist of fatty acids combined with a long-chain alcohol. The cutin-wax complex forms a water-repellent layer that prevents loss of water from the protected area. The outer walls of epidermal cells have a protective layer called a cuticle that consists of cutin embedded with wax (2.2.2). Suberin occurs in significant amounts in the walls of cork cells in tree bark (2.7.).

4.2.3 PROTEINS

Proteins are structurally more complex than carbohydrates and lipids. They are composed of building blocks called *amino acids,* which are nitrogen-containing molecules. Amino acids polymerize by linkage of *peptide bonds* (a bond formed between two amino acids) to produce chains called *polypeptides* (5.2.4). The 20 commonly occurring amino acids are used to form all proteins by bonding in a variety of sequences. The basic structure is the same for all amino acids—an *amino group,* a *carboxyl group,* and a *hydrogen atom* bonded to a *central carbon atom.* What distinguishes amino acids is the unique *R group.* The R-group may be *polar* or *nonpolar* (according to its tendency to dissolve in a polar solvent such as water), polar groups being more soluble in water than nonpolar ones. An amino acid may have a *net charge* (either positive or negative) or be *neutral* (no net charge). Of the 20 commonly occurring amino acids, 15 are neutral, 3 are basic, and 2 are acidic (table 4-1).

Proteins have four levels of molecular organization. The *primary structure* refers to the basic sequence of amino acids in the polypeptide chain. These primary chains may coil or twist to assume the *secondary structure,* the most common form being the *alpha-helix* (5.2.4). A chain may fold back on itself to form globular structures (as found in most enzymes). This level of organization produces the *tertiary structure* of proteins. The fourth level of organization, called the *quaternary structure,* occurs when two or more polypeptides group together to form a complex protein.

Enzymes

Enzymes are large, complex, globular proteins that act as *catalysts* in biochemical reactions. Catalysts accelerate the rate of chemical reactions by lowering the energy required for activation. That is, reactions can be sped up while occurring at relatively low temperatures. In the process of the reaction, these substances remain unaltered and are hence reusable. An enzyme has an *active site* to which the *substrate* (the substance acted on by the enzyme) attaches to form an enzyme-substrate complex. A substrate might be a compound such as glucose or adenosine triphosphate (ATP). These substances are changed into new products at the end of the reaction. An example of an enzyme reaction is illustrated in figure 4-7.

> **Enzyme**
> *A complex protein that speeds up a chemical reaction without being used up in the process.*

About 2,000 enzymes are known to occur in nature. Each enzyme catalyzes a specific reaction. Sometimes an enzyme-catalyzed reaction requires the presence of a third substance in order to proceed. These additional substances are called *cofactors* and may be organic or inorganic. Inorganic cofactors are also called *activators* and are usually metallic ions required in trace amounts in plant nutrition, such as iron, magnesium, and zinc (3.3.1). Organic cofactors (e.g., nicotinamide adenine dinucleotide [NAD^+] and nicotinamide adenine dinucleotide phosphate [$NADP^+$]) are called *coenzymes.* These substances accept atoms that are removed by the enzyme during the reaction. Enzymes that remove hydrogen atoms from substrates are called *dehydrogenases.*

4.2.4 NUCLEIC ACIDS

Nucleic acids are chemicals involved in hereditary aspects of cellular life. They are polymers of nucleotides that consist of a *phosphate group,* a *five-carbon sugar,* and a *nitrogenous base.* The two major types of nucleic acids are *DNA* and *RNA.* The subject of nucleic acids is discussed further in chapter 5, which deals with genetics.

Table 4-1   Classification of the 20 Common Amino Acids

A.   Neutral Amino Acids

1.   With Nonpolar R Groups

Alanine
Isoleucine
Leucine
Methionine
Phenylalanine
Proline
Tryptophan
Valine

2.   With Polar R Groups

Asparagine
Cysteine
Glycine
Glutamine
Serine
Threonine
Tyrosine

B.   Basic Amino Acids (with net positive charge)

Arginine
Histidine
Lysine

C.   Acidic Amino Acids (with net negative charge)

Aspartic acid
Glutamic acid

In addition to these 20 amino acids, two other important, modified amino acids (i.e., cystine and hydroxyproline) occur; these are formed by modification of cysteine and proline, respectively

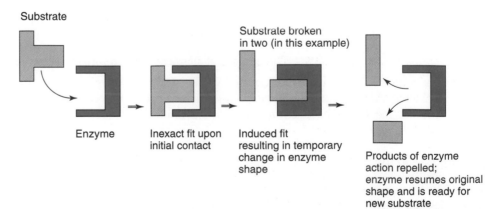

FIGURE 4-7   The induced-fit hypothesis of enzyme reaction. The active site or the enzyme molecule binds to the substrate, and in the process a closer fit between the two is induced.

4.3: PLANT GROWTH PROCESSES

Plant growth processes provide the raw materials and the energy required for building new tissues and nurturing them to maturity. The major processes are discussed in the following sections.

4.3.1 PHOTOSYNTHESIS

Photosynthesis accounts for more than 90 percent of the dry matter yield of horticultural plants and is the ultimate source of food and fossil fuel. Photosynthesis is the single most important chemical reaction in nature. It impacts the environment significantly through its effects on the oxygen content of the air. This major physiological process is important not only because of its tremendous impact on a variety of functions in nature but also because an understanding of the process enables scientists to maximize its rate for higher crop productivity.

> **Photosynthesis**
> *The process by which plants convert light to chemical energy.*

Photosynthesis is a reaction occurring in green plants whereby plants utilize water and the energy of sunlight to fix inorganic carbon dioxide in the form of organic compounds, releasing oxygen in the process. In other words, the sun's energy is transformed by plants through photosynthetic processes into chemical energy usable by other living organisms. The importance of this process is more readily apparent when we understand that plants are ultimately the source of all food. Plants may be used *directly* as food (e.g., vegetables, fruits, grains, nuts, and tubers) or may be used by animals and then *indirectly* become available through animal products (e.g., poultry, fish, meat, and dairy products). Lest we limit the importance of plants to food, it should be made clear that plants are also sources of materials for fuel, clothing, and medicines. They are utilized widely in the beautification of the landscape and performance of other functional roles. [chapters 10 and 12].

The general chemical reaction of photosynthesis is:

$$6CO_2 + 12H_2O \xrightarrow[\text{light energy}]{\text{green plant}} C_6H_{12}O_6 + 6O_2 + 6H_2O$$

This reaction occurs in the chloroplasts (2.1.2), using chlorophyll as an enzyme. Carbon dioxide comes from the air and water from the soil.

Phases of Photosynthesis

Light-Dependent Reactions

LIGHT The nature of the electromagnetic spectrum is discussed in chapter 3 (3.2.1). Figure 3-9 shows that visible light is only a small portion of the vast electromagnetic spectrum. Only certain wavelengths of light are involved in photosynthesis. These specific wavelengths depend on the *absorption spectrum* (the range of wavelengths of light absorbed) of the various pigments involved in photosynthesis. A pigment may absorb a broad range of wavelengths, but certain ones are more effective than others in performing specific functions. The *action spectrum* of a pigment describes the relative effectiveness of different wavelengths of light for a specific light-dependent process such as flowering or photosynthesis.

PIGMENTS INVOLVED IN PHOTOSYNTHESIS Three types of pigments, *chlorophylls, carotenoids,* and *phycobilins,* are known to be involved in photosynthesis. A variety of chlorophylls occur in nature. Chlorophyll is a large molecule that is water insoluble. All photosynthetic eukaryotes contain chlorophyll *a* (figure 4-8). In addition, vascular plants and bryophytes, among others, require chlorophyll *b* for photosynthesis. Chlorophyll *b* broadens the range of wavelengths that can be used in photosynthesis and is hence called an accessory pigment. Chlorophyll *a* occurs at about two times the concentration of chlorophyll *b*. Other varieties of chlorophyll are *c* and *d*. Both chlorophylls *a* and *b* have absorption peaks in the blue and red regions of the visible light spectrum. Absorption peaks for chlorophyll *a* occur at 430 and 662 nanometers, and absorption peaks for chlorophyll *b* occur at 453 and 642 nanometers.

Carotenoids are a group of water-insoluble (but fat-soluble) accessory pigments that occur in the chloroplasts of all photosynthesizing cells. These pigments are red, orange, or yellow

FIGURE 4-8 The structure of a chlorophyll. The difference between chlorophyll *a* and chlorophyll *b* lies in the R group, being —CH₃ in chlorophyll *a* and —CHO in chlorophyll *b*.

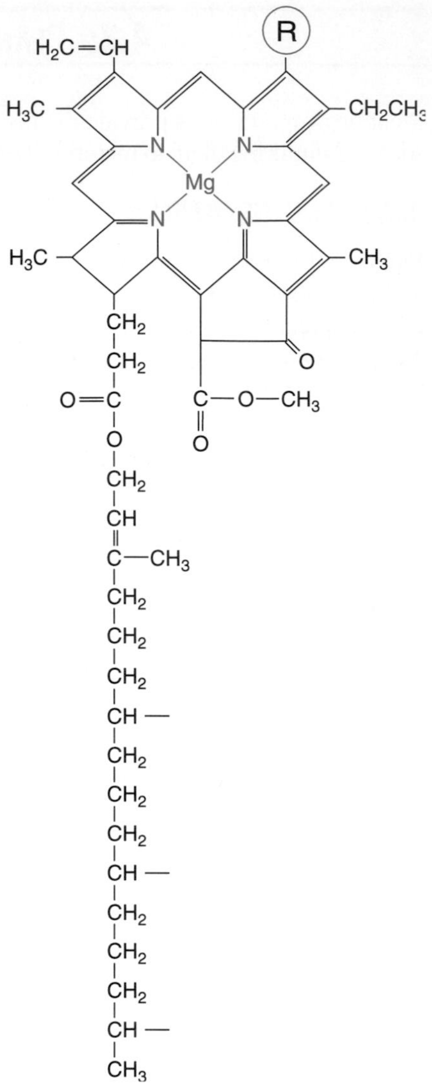

in color. However, the expression of these colors is masked by the overwhelming presence of green-colored chlorophyll. The other colors are expressed when chlorophyll breaks down, as occurs seasonally in autumn, revealing the dazzling colors associated with the season. There are two types of carotenoids—*carotenes,* which are nonoxygen-containing pigments, and *xanthophylls,* which contain oxygen in their molecular structure. The beta-carotene found in plants is the primary source of vitamin A needed by humans and other animals.

Phycobilins, water-soluble accessory pigments, are not found in higher plants.

> **Photosystem**
> A collective term for a specific functional aggregation of photosynthetic units.

PHOTOSYSTEMS The light-absorbing photosynthetic pigments are packaged in the *thylakoid membranes* (saclike membranes of pigment-containing structures within chloroplasts) (2.1.2) in discrete units called *photosystems.* There are two such discrete units, called *photosystem I (PS I)* and *photosystem II (PS II).* Although all of the pigments in a photosystem are capable of absorbing light, only a pair of chlorophyll molecules can utilize the light energy in the photochemical reaction. This pair of molecules forms the *reaction center chlorophyll;* the other pigments are called *antenna pigments.* The light absorbed is transferred in a relay fashion from one pigment to another until it reaches the reaction center molecules. In PS I, the reaction center chlorophyll is a special molecule of chlorophyll *a,* with optimal absorption peak at 700 nanometers; it is thus called P_{700}. Photosystem II is called P_{680} for similar reasons. Even though PS I can work independently, it works together with PS II. Further, PS II is more effective than PS I in terms of light harvesting.

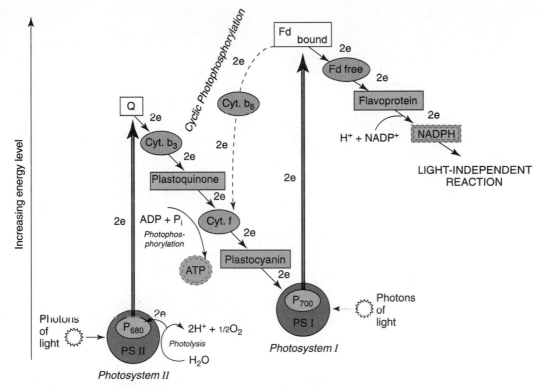

FIGURE 4-9 A summary of the light reaction of photosynthesis.

When the photosynthetic pigments absorb light, electrons are excited and temporarily boosted to a higher energy level. As these high-energy electrons return to a lower energy level, the energy may be lost as heat, reemitted (by fluorescence), or utilized in the formation of a chemical bond. In photosynthesis, the energy is used to form a chemical bond. The executed electrons in PS II are transferred by P_{680} to a molecule called *plastoquinone,* leaving P_{680} electron deficient. By a light-dependent oxidative splitting of water molecules (called *photolysis*), the electrons are regained (figure 4-9). The high-energy electrons from plastoquinone are passed down along a proton gradient in an electron transport chain to PS I. During this transport, adenosine triphosphate (ATP) is formed from adenosine diphosphate (ADP) and phosphorus by the process of *photophosphorylation.*

Light energy simultaneously excites electrons of a P_{700} molecule, which releases electrons to an acceptor molecule called *bound ferredoxin.* As the energized electrons pass downhill along a proton gradient, they are transferred to the coenzyme $NADP^+$, which is then reduced to NADPH. The electrons that moved from P_{700} are replaced by those from P_{680}. This unidirectional flow of electrons from water to $NADP^+$ is called noncyclic electron flow, and hence the photophosphorylation is called *noncyclic photophosphorylation.* Noncyclic photophosphorylation can occur when PS I works independently of PS II.

The enzyme-catalyzed process of photolysis depends on manganese and chloride, thus underscoring the importance of these elements in plant nutrition. The result of the light-dependent phase of photosynthesis is the oxidation of water (the only substance with a net loss of electrons) to release oxygen and the reduction of $NADP^+$ (the only substance with a net gain in electrons) to NADPH. It is this release of molecular oxygen that sustains animal and plant life on earth.

Light-Independent Reactions   The light-independent reactions stage is also collectively called the *dark reaction* of photosynthesis. At this stage, carbon dioxide is reduced to carbohydrate, a process called *carbon dioxide fixation.* This stage does not depend directly but rather indirectly on light, since the ATP and NADPH required are produced by the light-dependent reactions. Further, the two chemicals do not accumulate in the cell but are used up as fast as they are produced. The fixation of carbon dioxide (CO_2) comes to a halt soon after the light supply is terminated.

> **Carbon Dioxide Fixation**
> *A cyclical series of reactions in which carbon dioxide is reduced to carbohydrate.*

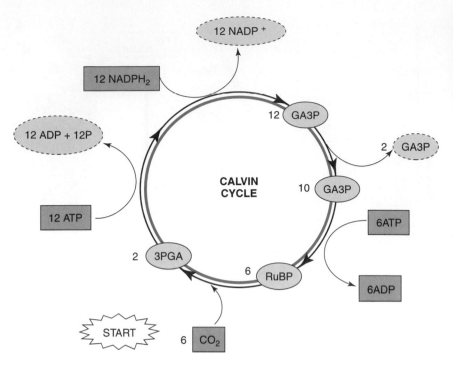

FIGURE 4-10 The Calvin cycle or C_3 pathway of carbon dioxide fixation.

Carbon dioxide fixation occurs by one of two major pathways, which are distinguished by the first product formed. These are the three-carbon (C_3) pathway and the four-carbon (C_4) pathway.

1. *The Calvin cycle.* Named after its discoverer, the first product of the *Calvin cycle* is a three-carbon compound. Thus this pathway is also called the *C_3 pathway*. Plants that photosynthesize by this pathway are called *C_3 plants*. Carbon dioxide enters the cycle and becomes covalently bonded to a five-carbon sugar with two phosphate groups called *ribulose 1,5-bisphosphate (RuBP)* (figure 4-10). This process (fixation) is catalyzed by the enzyme *RuBP carboxylase,* also commonly called *rubisco.* Because this enzyme is abundant in chloroplasts, rubisco is said to be the most abundant protein in nature. The overall process can be summarized by the following equation:

$$6CO_2 + 12NADPH + 12H^+ + 18ATP \rightarrow 1 \text{ glucose} + 12NADP^+ + 18ADP + 18P_i + 6H_2O$$

The intermediate product is *glyceraldehyde 3-phosphate.*

Glucose is indicated in the preceding summary equation, but in practice, photosynthesizing cells generate a minimal amount of this sugar. Most of the fixed carbon dioxide is either converted to sucrose (which is the principal form in which sugar is transported in plants) or stored in the form of starch.

2. *The four-carbon pathway.* Many plants are known to be able to fix carbon dioxide by a pathway whose first product is a four-carbon substance. This pathway is also called the *C_4 pathway,* and plants that photosynthesize by this pathway are called *C_4 plants.* First, a CO_2 molecule is bonded to *phosphoenol pyruvate (PEP),* a three-carbon acceptor compound, resulting in the production of *oxaloacetate.* The reaction is catalyzed by the enzyme *PEP carboxylase* (figure 4-11).

C_4 plants are less efficient than C_3 plants in terms of the energy requirements in fixing CO_2. To fix one molecule of CO_2, C_4 plants need five ATPs, whereas C_3 plants need only three. However, C_4 plants have higher photosynthetic rates than C_3 plants and also are able to continue photosynthesizing under conditions such as high temperatures and high light intensity when C_3 plants cannot (figure 4-12). Generally, C_4 plants are adapted to tropical conditions. Select examples of both categories of plants are presented in table 4-2. Under hot,

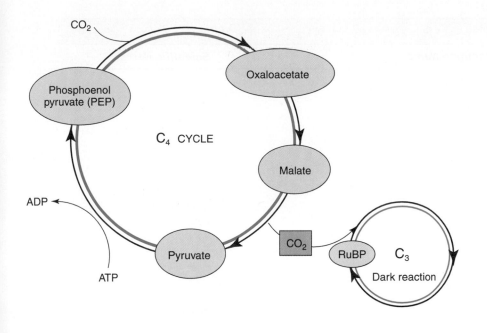

FIGURE 4-11 The C_4 pathway of carbon dioxide fixation.

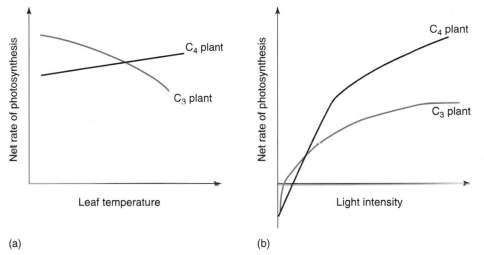

(a)

(b)

FIGURE 4-12 The relative rates of photosynthesis in C_3 and C_4 plants as influenced by (a) temperature and (b) light intensity.

sunny skies, C_3 plants undergo a process called *photorespiration* (light-dependent production of glycolic acid in chloroplasts and its subsequent oxidation in peroxisomes [4.3.2]). C_4 plants use CO_2 more efficiently and hence are able to function at only partially closed stomata, as occurs on hot, sunny days.

C_3 and C_4 plants are different structurally. The bundle sheath cells of C_3 leaves have small chloroplasts. Photosynthesis occurs only in the mesophyll cells. However, the bundle sheath cells of C_4 plants are large and contain large chloroplasts. These chloroplasts exhibit the Calvin cycle, while the mesophyll cells exhibit the C_4 pathway. All plants known to use the C_4 pathway are flowering plants. In lawns, where C_4 species such as crabgrass (*Digitaria sanguinalis*) occur among C_3 species such as Kentucky bluegrass (*Poa pratensis*), the crabgrass grows rapidly in summer and tends to suppress the fine-leafed species with its broad leaves.

3. *The Crassulacean acid metabolism.* The Crassulacean acid metabolism (CAM) is a photosynthetic pathway that allows certain plants to fix carbon dioxide in the dark by the activity of PEP carboxylase. Because the stomata of leaves are closed during the hot day, CAM plants depend on CO_2 that accumulates in the leaf during the nighttime. This reaction provides *malic acid,* which accumulates in the vacuoles of cells. During the next light period, the malic acid is decarboxylated. The resulting CO_2 is transferred to the Calvin cycle within the same cell.

Table 4-2 Selected Examples of C_3, C_4, and CAM Plants

Common Name	Scientific Name
C_3 Plants	
Kentucky bluegrass	*Poa pratensis*
Creeping bentgrass	*Agrostis tenuis*
Sunflower	*Helianthus annuus*
Scotch pine	*Pinus sylvestris*
Tobacco	*Nicotiana tabacum*
Peanut	*Arachis hypogaea*
Spinach	*Spinacia oleracea*
Soybean	*Glycine max*
Rice	*Oryza sativa*
Wheat	*Triticum aestivum*
Rye	*Secale cereale*
Oats	*Avena sativa*
C_4 Plants	
Crabgrass	*Digitaria sanguinalis*
Corn	*Zea mays*
Bermudagrass	*Cynodon dactylon*
Sugarcane	*Saccharum officinale*
Sorghum	*Sorghum vulgare*
Pigweed	*Amaranthus*
Euphorbia	*Euphorbia* spp.
Millet	*Pennisetum glaucum*
Sedge	*Carex* spp.
CAM Plants	
Wax plant	*Hoya carnosa*
Snake plant	*Sansivieria zeylanica*
Maternity plant	*Kalanchoe diagremontiana*
Pineapple	*Ananas comosus*
Spanish moss	*Tillandsia usneoides*
Jade plant	*Crassula argentea*
Ice plant	*Mesembryanthemum* spp.
Century plant	*Agave americana*
Cacti (many spp.)	

Most CAM plants inhabit environments in which moisture stress and intense light prevail. Many are succulents such as members of the cactus (Cactaceae), stone crop (Crassulaceae), and orchid (Orchidaceae) families. Houseplants with CAM include wax plant (*Hoya carmosa*) and snake plant (*Sansevieria zeylanica*).

CAM plants are relatively slower growing than C_3 or C_4 plants under favorable conditions. They grow more slowly because plants, by nature, tend to conserve moisture and in so doing close the stomata most of the day, thus limiting the CO_2 intake needed for fixation.

Environmental Factors Affecting Photosynthesis

The rate of photosynthesis is affected by a number of factors, including light intensity, carbon dioxide concentration, temperature, water availability, photoperiod, and growth and development.

Light Intensity   Light is important for the production of ATP and NADPH. Thus, at low light intensities, these products are not produced in adequate amounts. However, when light intensity is extreme, other factors such as CO_2 may be limited, causing the rate of photosynthesis to decline.

Maximum photosynthesis occurs near noon. For most houseplants and trees, such as oak, the photosynthetic rate is not affected by an increase in light intensity. Many houseplants are shade loving, and other plants such as turf grasses, corn, and some fruit trees are sun loving. For example, bermudagrass (a tropical grass) responds to changes in light intensity. The horticultural practice of pruning trees permits light to penetrate the canopy so that more leaves can receive direct sunlight for increased photosynthesis. Plants grown in high light intensities tend to have broader and thicker leaves than those grown in lower light intensities. The quality of leaf vegetables is thus influenced by light intensity.

Carbon Dioxide Concentration   Rapid photosynthesis can deplete cells of carbon dioxide. An increase in carbon dioxide concentration is beneficial to C_3 plants, since they have a high *CO_2 compensation point* (the equilibrium concentration of carbon dioxide at which the amount evolved in respiration is equal to the amount fixed by photosynthesis). C_4 plants have a lower CO_2 compensation point than C_3 plants because the former are more efficient in trapping CO_2. Some commercial growers enrich the greenhouse atmosphere with an artificial source of carbon dioxide (from liquid carbon dioxide or by burning methane or propane) to grow plants such as orchid (*Cattleya* spp.), carnation (*Dianthus caryophyllus*), and rose (*Rosa* spp.). This method of providing supplementary greenhouse carbon dioxide is called *carbon dioxide fertilization* (11.3.5). This practice may be necessary during the winter when carbon dioxide levels in airtight greenhouses may be lowered to a degree where photosynthesis could be limited on sunny days. However, unless the purpose of providing additional carbon dioxide is to increase productivity, the carbon dioxide concentration in winter may be readily restored by frequent ventilation of the facility rather than adopting carbon dioxide fertilization.

Temperature   Photosynthetic rate is decreased in cold temperatures because the fixation stage is temperature sensitive. However, under conditions in which light is a limiting factor (low light conditions) the effect of temperature on photosynthesis is minimal. Generally, if light is adequate, the photosynthetic rate is found to approximately double the rate in plants in temperate areas for each 10°C (50°F) rise in temperature. The quality (sugar content) of certain fruits such as cantaloupe is reduced when they are grown under conditions in which the photosynthetic rate is reduced but respiration is high because of high temperatures. C_3 plants grow poorly at high temperatures (e.g., lawn grasses that follow the C_3 pathway perform poorly in summer, whereas C_4 plants that are weeds, such as crabgrass, thrive).

Water Availability   When plants grow under conditions of moisture stress due to low soil moisture or dry winds that accelerate transpiration, enzymatic activities associated with photosynthesis in the plants slow down. Stomata close under moisture stress, reducing carbon dioxide availability and consequently decreasing photosynthetic rate.

Photoperiod   The duration of day length (photoperiod) affects photosynthesis in a directly proportional way. Generally, plants that are exposed to long periods of light photosynthesize for a longer time and as a result tend to grow faster. In the winter season when sunlight is less direct and of shorter duration, growing plants indoors is more successful if additional lighting at appropriate intensity is provided to extend the period of natural light.

Growth and Development   The general plant growth and development needs also influence the rate of photosynthesis. The photosynthetic rate is lower in a young expanding leaf than in a fully expanded one. On the other hand, as plant leaves begin *senescence* (an aging process involving degradation of proteins), the photosynthetic rate in mature leaves declines and eventually ceases in certain species.

> **Senescence**
> The breakdown of cell components and membranes that leads to the death of a cell.

Other factors affect the rate of photosynthesis. One such factor is nutrition (deficiency of nitrogen and magnesium, which are both required by chlorophyll, can decrease the rate). Environmental pollutants such as ozone and sulfur damage horticultural plants by causing loss of chlorophyll in leaves and thus reducing the available photosynthetic surface.

4.3.2 RESPIRATION

Respiration may occur in an environment that is oxygen rich or oxygen deficient.

Aerobic Respiration

Aerobic respiration accomplishes the reverse of photosynthesis by using oxygen from the air to metabolize organic molecules into carbon dioxide and water to release stored energy in the form of ATP. In fact, the primary purpose of respiration is this energy-production function. *Polysaccharides* (4.2.1) are stored in different forms by different organisms; for example, bananas store it as sucrose, whereas onions store it as fructose. The general reaction that occurs in mitochondria in the cytoplasm is as follows:

$$C_6H_{12}O_6 + 6O_2 \rightarrow 6CO_2 + 6H_2O + \text{energy}$$

Respiration consists of three distinct stages. The first reaction, *glycolysis,* occurs in the cytoplasm, where glucose is broken down into pyruvic acid. In the first part, ATP is converted to ADP; in the second part, ADP is converted to ATP. Pyruvic acid from glycolysis enters the mitochondria to complete the next two phases, which are the *Krebs cycle (tricarboxylic acid cycle* or *citric acid cycle)* and the *electron transport chain* (requires oxygen). This three-phase process is called *aerobic respiration,* since it requires oxygen. When oxygen is limited, another form of respiration, called *anaerobic respiration* (or *fermentation*), occurs, the end product being alcohol.

Glycolysis   Glycolysis ("sugar splitting") occurs in the cytoplasm of the cell. It involves the breaking down of glucose into pyruvic acid (figure 4-13). In a series of reactions, two three-carbon sugar phosphates are produced from one six-carbon glucose molecule. The sugar phos-

> **Glycolysis**
> *The initial phase of all types of respiration in which glucose is converted to pyruvic acid without involving free oxygen.*

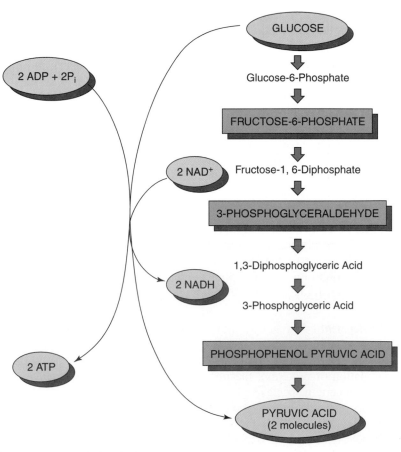

FIGURE 4-13   A summary of glycolysis.

phates are then converted to pyruvic acid (a pyruvate), accompanied by the production of ATP and the reduction of NAD^+ to NADH. The overall equation for glycolysis is as follows:

$$\text{glucose} + 2NAD^+ + 2ADP + 2P_i \rightarrow 2 \text{ pyruvic acid} + 2NADH + 2H^+ + 2ATP + 2H_2O$$

In effect, one glucose molecule is converted into two molecules of pyruvic acid. The formation of ATP by the enzymatic transfer of a phosphate group from a metabolic intermediate to ADP is called *substrate-level phosphorylation.*

Krebs Cycle   The Krebs cycle occurs in the mitochondrion following the entry of pyruvic acid from glycolysis. The series of enzyme-catalyzed reactions involved in this cycle constitute what is called *oxidative decarboxylation* (figure 4-14). The reactions may be summarized by the following equation:

$$\text{oxaloacetic acid} + \text{acetyl CoA} + ADP + P_i + 3NAD^+ + FAD \rightarrow \text{oxaloacetic acid} +$$
$$2CO_2 + CoA + ATP + 3NADH + 3H^+ + FADH_2$$

Electron Transport Chain   Electrons from reduced coenzymes formed during glycolysis and the Krebs cycle are passed on from one carrier to another in a downward fashion to oxygen. The energy released is used to form ATP from ADP, a process called *oxidative phosphorylation.* This stage of respiration is known as the electron transport chain (figure 4-15). For each pair of electrons transferred from NADH to oxygen, three molecules of ATP are produced. Similarly, for each pair of electrons transferred from $FADH_2$ to oxygen, two molecules of ATP are produced.

Energy Yield in Respiration   The energy yield from the respiration of one molecule of glucose is presented in table 4-3. The net yield of 36 ATPs from aerobic respiration of one molecule of glucose is common to most organisms.

Anaerobic Respiration

Aerobic respiration results in the complete oxidation of pyruvic acid to CO_2 and water. When oxygen is absent, pyruvic acid (or pyruvate) is not the end product of glycolysis. Instead, pyruvate is broken down to *ethyl alcohol* (ethanol) and CO_2 in most plant cells (figure 4-16).

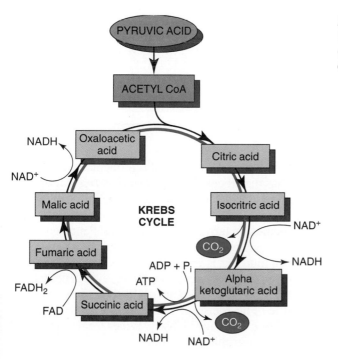

FIGURE 4-14 A summary of the Krebs cycle. The cycle always begins with acetyl CoA, which is its only real substrate.

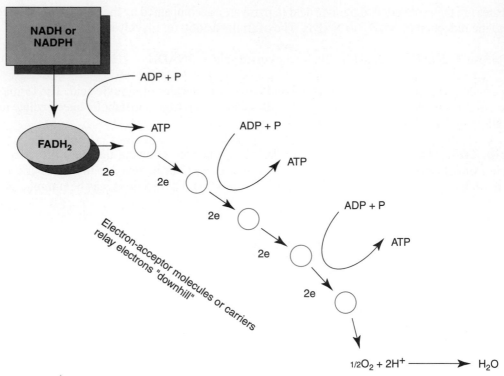

FIGURE 4-15 Electron transport involves the flow of electrons in an energetically downhill fashion, resulting in energy release for the formation of ATP.

Table 4-3 Energy Produced from Aerobic Respiration

Metabolic Reaction	Coenzyme Type Produced	ATP Yield Coenzyme	Total
Glycolysis			
	(Direct)	2	2
	2 NADH	2	4
Oxydative decarboxylation of pyruvate to acetyl CoA	2 NADH	36	
Krebs cycle	(From GTP)	2	2
	6 NADH	3	18
	2 FADH$_2$	2	4
Net yield of ATP			36

Fermentation
The metabolic breakdown of an organic molecule in the absence of oxygen or with low levels of oxygen to produce end products such as ethanol and lactic acid.

This anaerobic process is called anaerobic respiration, or alcoholic fermentation. In some bacteria, the end product is lactic acid, and thus the process is called *lactate fermentation*. The anaerobic respiratory pathway is very inefficient. In aerobic respiration, the initial energy of 686 kilocalories (kcal) per mole of glucose yields 263 kcal (39 percent) at the end of the process, which is conserved in 36 ATP molecules. In fermentation, only 2 ATP molecules are produced, representing about 2 percent of the available energy in a molecule of glucose.

When horticultural plants are grown in mud (air is limited), they are forced to respire anaerobically. Similarly, houseplants die when they are overwatered or grown in containers with poor drainage because they are unable to respire aerobically; instead they resort to anaerobic respiration, which yields very little or no energy at all. Fermentation, however, is a very important process utilized in the alcoholic beverage industry. Wine (from grapes) and apple cider are a few products that depend on anaerobic respiration.

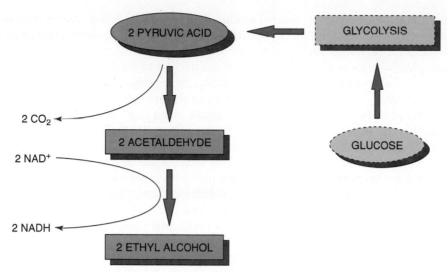

FIGURE 4-16 A summary of the process of fermentation.

Respiration is the source of energy for all life processes. Since it depends on products of photosynthesis (the two processes work in opposite directions), it is critical that a desirable relationship between them be maintained for proper growth and development of plants. If food is broken down faster than it is manufactured, plant growth will be severely hampered. This imbalance may cause the eventual death of certain plants when they are grown in the shade. Whereas light is required for photosynthesis, it is not required for respiration, and hence the latter proceeds even in shade. Fortunately, photosynthesis generally occurs at a higher rate than respiration, such that there are excess photosynthates for growth or production of fruits and seed through storage. It is estimated that a photosynthetic rate of about 8 to 10 times higher than the respiration rate is required for good production of vegetables.

To decrease respiration of carbohydrates, the temperature may be lowered to slow the reaction. However, this action also slows the photosynthetic rate. A warm temperature, moderate intensity of light, and adequate supply of water are desired for maximizing the photosynthetic rate. This condition occurs on warm, bright days with cool night temperatures of about 5°C (41°F) colder than day temperatures. Respiration is reduced during the cool period of the night, while adequate duration and intensity of light exist during the day for photosynthesis.

Respiration does not occur only when plants are growing. Freshly harvested plant produce should be stored appropriately to prevent degradation from respiration. For some plants, such as apple, proper storage may mean the provision of cold storage at about 0°C (32°F); other plants, such as banana and flowers, require less-cold storage conditions (about 10°C [50°F]). Anaerobic respiration may be detrimental to most plants, but it may be utilized to reduce respiration rate during storage of certain vegetable crops and fruits.

4.3.3 TRANSPIRATION

The other two major plant growth processes—transpiration and translocation—are associated with the movement of water in plants. The movement of water in the soil is discussed in chapter 3 (3.8.5).

Species such as corn (*Zea mays*) lose large amounts of water, whereas others, such as tomato (*Lycopersicon esculentum*), lose only moderate amounts of water during cultivation. Other plants such as cowpea (*Vigna sinensis*) are relatively highly resistant to water loss by transpiration. Transpiration occurs through the stomata (2.4.2), the pores through which the much-needed CO_2 for photosynthesis passes. Transpiration is regulated by closure of the stomata, an event that excludes CO_2 and reduces the rate of photosynthesis. However, respiration produces some CO_2 that can be trapped and used by plants after the stomata have closed.

The stomata open or close according to changes in turgor pressure in the guard cells (2.4.2). Water in most cases is the primary factor that controls stomatal movements. However, other factors in the environment also affect stomatal movements. Generally, an increase in CO_2

> **Transpiration**
> The loss of water from plant surfaces by evaporation and diffusion.

concentration in the leaf causes stomatal closure in most species. Some species are more sensitive than others to the effects of CO_2. Similarly, the stomata of most species open in light and close in the dark. An exception to this feature is plants with CAM pathways of photosynthesis. Photosynthesis uses up CO_2 and thus decreases its concentration in the leaf. Evidence suggests that light quality (wavelength) affects stomatal movements. Blue and red light have been shown to stimulate stomatal opening. The effect of temperature on stomatal movements is minimal, except when excessively high temperatures (more than 30°C [86°F]) prevail, as occurs at midday. However, an increase in temperature increases the rate of respiration, which produces CO_2 and thereby increases the CO_2 concentration in the leaf. This increase in CO_2 may be part of the reason stomata close when the temperature increases.

Apart from environmental factors, many plants have been known to accumulate high levels of abscisic acid (4.6), a plant hormone. This hormone accumulates in plants under conditions of moisture stress and causes stomata to close.

Transpiration is accelerated by several environmental factors. The rate of transpiration is doubled with a more than 10°C (50°F) rise in temperature. Transpiration is slower in environments of high humidity. Also, air currents may accelerate the transpiration rate by preventing water vapor from accumulating on the leaf surface.

Certain plants are adapted to dry environments. These plants (called *xerophytes* [2.4.2]) have special anatomical and physiological modifications that make them able to reduce transpiration losses. These species, which include a large number of succulents, are able to store and retain large amounts of water. Physiologically, the plants photosynthesize by the CAM pathway and close their stomata at night.

4.3.4 HOW WATER MOVES IN PLANTS

Water moves in plants via the conducting elements of the xylem (2.2.2). It moves along a water potential gradient from soil to root, root to stem, stem to leaf, and leaf to air forming a continuum of water movement. The trend is for water to move from the region of highest water potential to the region of lowest water potential, which is how water moves from the soil to the air. Transpiration is implicated in this water movement, because it causes a water gradient to form between the leaves and the soil solution on the root surface. This gradient may also form as a result of the use of water in the leaves. Loss of moisture in the leaves causes water to move out of the xylem and into the mesophyll area, where it is depleted. The loss of water at the top of the xylem vessels causes water to be pulled up. This movement is possible because of the strong cohesive bond among water molecules. The movement of water up the xylem according to this mechanism is explained by the *cohesion-tension theory*. It is the cohesiveness of water molecules that allows water to withstand tension. Water is withheld against gravity by capillarity (3.8.6). The rise is aided by the strong adhesion of water molecules to the walls of the capillary vessel.

Capillary flow is obstructed when air bubbles interrupt the continuity of the water column, called *embolism*. This event is preceded by *cavitation,* the rupture of the water column. Once the tracheary elements (2.2.2) have become embolized, they are unable to conduct water. In the cut flower industry, the stems of flowers are cut under water to prevent embolism (chapter 22).

Water enters the plant from the soil via the root hairs, which provide a large surface area for absorption. Once inside the root hairs, water moves through the cortex and into the tracheary elements. There are three possible pathways by which this movement occurs, depending on the differentiation that has occurred in the root (e.g., presence of endodermis, exodermis, or a transcellular pathway suberin). Water may move from cell to cell, passing from vacuole to vacuole. Sometimes water may move via the *apoplastic pathway,* through the cell wall, or the *symplastic pathway,* from protoplast to protoplast through the pores in the plasmodesmata (minute cytoplasmic threads that extend through openings in the cell wall and connect the protoplast of adjacent living cells).

Root pressure plays a role in water movement, especially at night when transpiration occurs to a negligible degree or not at all. Ions build up in the xylem to a high concentration and initiate osmosis, so that water enters the vascular tissue through the neighboring cells. This pressure is called root pressure and is implicated in another event, *guttation,* whereby

> **Root Pressure**
> The development of positive hydrostatic pressure in the xylem followed by osmotic uptake of water.

droplets of water form at the tips of the leaves of certain species (e.g., lady's mantle [*Achemilla vulgaris*]) in the early morning. These drops do not result from condensation of water vapor in the surrounding air but rather are formed as a result of root pressure forcing water out of hydathodes (2.2.2). Root pressure as a water movement mechanism is least significant during the daytime when water moves through plants at peak rates. Further, some plants such as pine (conifers) do not develop root pressure.

The absorption and movement of water in plants occurs via the xylem vessels. Inorganic nutrients are also transported through these vessels. Solutes are moved against a concentration gradient and require an *active transport* mechanism that is energy dependent and mediated by carrier protein. Some amount of exchange between xylem and phloem fluids occurs such that inorganic salts are transported along with sucrose and some photosynthetic products are transferred to the xylem and recirculated in the transpiration stream.

4.3.5 TRANSLOCATION

Translocation is the long-distance transport of organic solutes through the plant. Photosynthetic products are moved out of the leaves and into the assimilate stream from the *sources* (especially leaves but also storage tissue [9.11]) to where they are used or stored (*sinks*). The primary translocation source (leaves) is located above the primary sink (root). However, the movement of organic solutes is not unidirectional or fixed.

During vegetative growth, assimilates are distributed from leaves to growing parts in upward and downward directions. However, when the plant enters the reproductive phase of growth, developing fruits require large amounts of assimilates and hence there is redistribution so that most of the flow from neighboring sources and even from distant ones are redirected to the fruits. Movement in the assimilate stream occurs via the phloem vessels as *sap* (a fluid consisting mainly of sugar and nitrogenous substances) (figure 4-17).

Phloem transport is believed to occur by the mechanism of *pressure flow*. This hypothesis suggests that assimilates are moved from translocation sources to sinks along a gradient of hydrostatic pressure (turgor pressure) of osmotic origin. Sugar is asserted to be transported in the phloem from adjacent cells in the leaf by an energy-dependent active process called *phloem loading*. The effect of this process is a decrease in water potential in the phloem sieve tube, which in turn causes water entering the leaf in the transportation stream to move into the sieve tube under osmotic pressure. The water then acts as a vehicle for the passive transport of the sugars to sinks, where they are unloaded, or removed, for use or storage. The water is recirculated in the transpiration stream because of the increased water potential or the sink resulting from the phloem unloading.

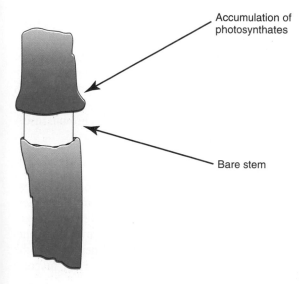

Accumulation of photosynthates

Bare stem

FIGURE 4-17 Accumulation of carbohydrates and sugars above the position of girdling or stem that interrupts the continuity of the phloem.

4.4: DEVELOPMENTAL STAGES OF GROWTH

A growing plant seedling goes through a number of developmental changes:

embryonic → juvenile → transitional → maturity → senescence

4.4.1 EMBRYONIC STAGE

The seed consists of an embryo or miniature plant. Until the conditions for germination are right, the embryo remains dormant. *Seed dormancy* is said to occur when a viable seed fails to germinate under favorable environmental conditions (8.6). This biological mechanism is especially advantageous when plants are growing in the wild. It ensures that seeds will germinate only when adequate moisture and other necessary environmental conditions exist to sustain growth after germination. The condition may be physical (*physical dormancy*) or physiological (*physiological dormancy*) in origin. A frequent source of physical dormancy is the presence of an impervious seed coat that does not permit water imbibition for germination. In modern cultivation, such seeds may be mechanically scratched (a process called scarification) to make the seed coat permeable to water. Persistence of seed dormancy may cause undesirable delay in germination, resulting in nonuniform seed germination and consequently an incomplete stand and irregular maturity in the field that leads to unnecessary delays in harvesting. Leguminous species are most plagued by mechanical seed dormancy due to an impervious seed coat.

Dormancy may be due to physiological causes stemming from chemicals in various parts of the seed or the fruit that inhibit germination. Certain desert species require a good soaking rain in order to leach away a germination inhibitor before seeds will germinate. Consequently, adequate moisture is ensured for the young seedlings to become established and be able to fend for themselves before a dry spell occurs. Seeds seldom germinate while they are in the fruit. However, once excised and washed, the mature seeds promptly germinate. In tomato, the chemical *coumarin* is implicated in this inhibitory condition. Abscisic acid is another chemical known to impose seed dormancy in plants (4.6).

In some species, seed maturity lags behind fruit ripening. In the American holly (*Ilex opaca*), an *after-ripening* process is required to bring the embryo in the seed to maturity, without which it will not be able to germinate. Seed dormancy in certain situations can be broken through a cold temperature treatment (*cold stratification*). In these species, seeds germinate after a brief storage at temperatures above freezing (e.g., 5°C [41°F]).

Buds can also become dormant under certain environmental conditions. *Bud dormancy* is a period of quiescence during which growth is temporarily suspended. Bermudagrass becomes dormant and changes color to a dull brown with the onset of cold temperature. It resumes active growth in spring. Both flower and vegetative buds may experience dormancy. Buds on the same plant have different dormancy characteristics. Just as seed dormancy protects seeds from damage due to adverse weather, woody plants in temperate climates are protected by internally imposed dormancy. Such dormancy ensures that new growth does not occur until the danger of damage from adverse temperatures is minimized. However, the requirements for breaking dormancy vary between the two structures (vegetative and reproductive). In plants such as peach and cherry, the period of cold treatment required to break dormancy is shorter for flower buds than for vegetative buds. If a peach cultivar with a long *low chill requirement* is grown, insufficient chilling can cause some fruiting to occur in the absence of vegetative growth.

4.4.2 JUVENILE STAGE

Juvenility starts after germination and is the stage during which the plant undergoes vegetative growth without any reproductive activities. The period of juvenility is variable among species, being several weeks in some and several years in others. Certain species display telltale characteristics of this stage. For example, in the English ivy (*Hedera helix*), immature (juvenile) plants trail or climb and have three to five deeply lobed palmate alternative leaves. When mature, the plants produce upright shoots with nonlobed ovate opposite leaves. In fruit trees, juvenile

Dormancy
The failure of seeds, bulbs, buds, or tubers to grow due to internal factors or unfavorable environment.

Stratification
The practice of exposing imbibed seed usually to cold temperatures for a period of time prior to germination in order to break dormancy.

branches appear on older ones as vertical shoots. These shoots, called *suckers,* or *water sprouts,* may occur at the base of mature tree trunks (chapter 15). Another trait of juvenility is the retention of leaves or juvenile plant parts through the winter, as occurs in oak trees. In *Acacia melnoxylon,* juvenile leaves have compound bipinnate forms, but mature plants have simple leaves.

4.4.3 TRANSITIONAL STAGE

The stage of transition occurs between juvenility and maturity. Plants at this stage may display characteristic features of both stages simultaneously. In the transitional stage, mature plants may revert to juvenility with changes in the environment.

4.4.4 MATURITY OR REPRODUCTIVE STAGE

The stage of maturity is characterized by reproductive activities (flowering and fruiting). As mentioned previously, plants such as English ivy have a specific adult (or mature) leaf form that is different from the juvenile form. It should be mentioned that the attainment of maturity does not mean activities (such as flowering) occur automatically. The appropriate conditions for flowering and fruiting must prevail for such activities to take place. Though the plant in this phase is relatively stable, the application of growth regulators can induce juvenility in mature tissue (9.12). Other activities in mature plants are aging and senescence (4.5.2).

Physiological maturity is the stage at which the plant has attained maximum dry weight. At this stage, it will not benefit from additional growth inputs. To realize the potential yield of a crop that has been well cultivated, it should be harvested at the right time, using the appropriate methods, and then stored under optimal conditions. Since some horticultural crop products are highly perishable (such as fresh fruits, vegetables, and cut flowers), if they are to be harvested for sale at a distant location from the farm, temporary storage is a critical consideration. If crops are not harvested at the right time, their quality may not be acceptable to consumers. A fresh produce may be too fibrous or not succulent enough or it may taste bitter if not harvested at the proper time. In some crops, such as green beans, consumers prefer a tender product and hence growers harvest beans before they are fully mature.

4.5: PHASES IN THE PLANT LIFE CYCLE

The pattern of growth takes a flowering plant through two distinct phases in its life cycle: (1) *vegetative growth* and (2) *reproductive growth.* Plants do not grow continuously but have periods in which they are dormant or in a resting phase.

4.5.1 VEGETATIVE PLANT GROWTH AND DEVELOPMENT

The vegetative phase of growth is characterized by an increase in the number and size of leaves, branches, and other characteristics of the shoot. Shoot growth follows certain patterns that may be used as a basis for classifying plants.

Shoot Elongation

In certain plants, most shoot elongation ceases after a period of time, and the terminal part is capped by a flower bud or a cluster of flower buds. Plants with this habit of growth are said to be *determinate.* These plants are also described as bush types and are usually able to stand erect in cultivation. They flower and set seed within a limited period. The pods mature and can all be harvested at the same time. In other plant types, shoot elongation continues indefinitely, such that flower buds arise laterally on the stem and continue to do so for as long as the shoot elongates. Plants with this growth habit are said to be *indeterminate.* Mature pods and flowers occur simultaneously on such plants. Trees behave in this way, as do vines. Some indeterminate vegetables require support (staking) in cultivation. Certain species have growth habits between these two extremes. Breeders sometimes breed for bush or erect plant types in certain crops to adapt them for mechanized culture.

Duration of Plant Life

A complete life cycle in flowering plants is the period it takes from seed to seed (i.e., germination to seed maturity). On the basis of life cycle, there are three main classes of flowering plants. These classes are discussed in detail in chapter 1.

1. *Annuals.* Annual plants complete their life cycles in one growing season (1.3.1). They are herbaceous and have no dormancy during the growing season (the dormant stage being the seed). Garden crops are mostly annuals, and so are many bedding plants (e.g., lettuce and petunia). Climatic factors may cause species such as impatiens and tomato to behave like annuals, although they are not true annuals.

2. *Biennials.* Biennial is a less common plant growth pattern found in plants such as evening primrose (*Primula* spp.), sugar beet, carrot, cabbage, and celery (*Apium graveolens*). These herbaceous plants complete their life cycles in two growing seasons. They grow vegetatively in the first season, remain dormant through the winter months, resume active vegetative and reproductive activities in the second season, and finally die. Plants such as carrot and sugar beet are grown for their roots (the storage organs), which are harvested at the end of the first growing season. These plants are often grown as annuals.

3. *Perennials.* Perennial plants live for more than two seasons and may be herbaceous or woody. In climatic zones where frost occurs, herbaceous perennials may lose their vegetative shoots, leaving only the underground structures to go through the winter. Woody perennials, on the other hand, maintain both above- and below-ground parts indefinitely. In flowering perennials, the flowering cycles are repeated over and over, year after year. Examples include rhubarb, asparagus, bulbs, tomato, eggplant, fruit trees, and ornamental trees and shrubs. As indicated previously, tomato and eggplant are frequently cultivated as annuals in temperate zones.

4. *Monocarp.* Monocarpic plants are a kind of perennial in the sense that they live for many years. However, they flower only once in a lifetime, after which they die. An example is the century plant.

4.5.2 REPRODUCTIVE GROWTH AND DEVELOPMENT

Reproductive activities occur in phases.

Flower Induction

Most agricultural plants are self-inductive for flowering. However, some need a cold temperature treatment to overcome the resting period, or dormancy. The chemical reaction of flower induction is the first indication that the plant has attained maturity. In certain species, the process is known to be dependent on environmental factors, the most common being temperature and light.

> **Vernalization**
> A cold temperature treatment required by certain species in order to induce flowering.

Vernalization   *Vernalization* is the cold temperature induction of flowering required in a wide variety of plants. The necessary degree of coldness and duration of exposure to induce flowering vary from species to species, but the required temperature is usually between 0 and 10°C (32 and 50°F). Although some plants such as sugar beet and kohlrabi can be cold sensitized as seed, most plants respond to the cold treatment after attaining a certain amount of vegetative growth. Some plants that need it are not treated because they are cultivated not for flower or seed but for other parts such as roots (in carrot and sugar beet), buds (brussels sprout), stems (celery), and leaves (cabbage). Apple, cherry, and pear (*Pyrus communis*) require vernalization, as do winter annuals such as wheat, barley, oats, and rye. Flowers such as foxglove (*Digitalis* spp.), tulip, crocus (*Crocus* spp.), narcissus (*Narcissus* spp.), and hyacinth (*Hyacinthus* spp.) need cold treatment, which may be administered to the bulbs, making them flower in warmer climates (at least for that growing season). However, they must be vernalized again to flower in subsequent years. In some of these bulbs, the cold treatment is needed to promote flower development after induction but not for induction itself. Sometimes vernalization helps plants such as pea and spinach to flower early, but it is not a requirement for flowering.

In onion, the bulbs are the commercial products harvested. Cold storage (near freezing) is used to preserve onion sets during the winter. This condition vernalizes the sets, which will flower and produce seed if planted in the spring. To obtain bulbs (no flowering), the sets should not be vernalized. Fortunately for growers, a phenomenon of *devernalization* occurs, in which exposure to warm temperatures of above 27°C (80°F) for two to three weeks before planting will reverse the effect of vernalization. Onion producers are therefore able to store their sets and devernalize them for bulb production during the planting season.

Flowering in certain species is affected by a phenomenon called *thermal periodicity,* in which the degree of flowering is affected by alternating warm and cool temperatures during production. For example, tomato plants in the greenhouse can be manipulated for higher productivity by providing a certain cycle of temperature. Plants are exposed to a warm temperature of 27°C (80°F) during the day and cooler night temperatures of about 17 to 20°C (63 to 68°F). This treatment causes increased fruit production over and above what occurs at either temperature alone.

Photoperiodism   The effect of photoperiodism is discussed in detail in chapter 3. As explained there, photoperiod is a phenomenon whereby day length controls certain plant processes. Further, it is actually the length of darkness, rather than light, that controls flowering, making the terminology a misnomer. Plants such as *Xanthium* require only one long-night (short-day) treatment to induce flowering, whereas others such as poinsettia (*Euphorbia pulcherrima*), chrysanthemum (*Chrysanthemum* spp.), and kalanchoe (*Kalanchoe* spp.) require several days (three to four). Although photoperiodism and vernalization both influence flowering, they seem to do so by different mechanisms and are not interchangeable. That is, a plant that needs both treatments will not flower if only one is provided.

Other Factors   Apart from duration of light, its intensity also affects flowering. Under the controlled environment of a greenhouse, light is provided in adequate amounts regarding intensity and duration. When potted flowering plants are purchased for the home, flowering may be poor or lacking because of the low light intensity in most homes. Flowering has been known to be stimulated under conditions of stress from drought or crowding. On the other hand, excessive moisture during the period of flower initiation in philodendrons causes a disproportionate number of seeds to be vegetative. Generally, woody plants tend to flower more copiously in spring if the preceding summer and fall were dry than if these seasons were wet.

Flower Initiation and Development

After being appropriately induced to flower, flowers are initiated from vegetative meristems that change into flowering meristems. This change is an irreversible process. The meristems differentiate into the flower parts. Proper temperature conditions are required for success, since high temperatures can cause flower abortion. The duration of the developmental process varies from one species to another. Time of flower initiation is important in horticulture. Flower primordia are laid down for a few to many months before flowering in many perennial species. For example, in the crocus, flowers in spring are produced from buds initiated in the previous summer. Flower primordia are initiated under a short photoperiod of August to September in plants such as June-bearing cultivars of strawberry.

Flowers can be chemically induced by externally applying certain growth regulators such as auxins (4.6). Commercial production of certain crops such as pineapple is aided by artificial flower induction. Other growth regulators are used to control the number of flowers set on a plant.

Pollination

Pollination is the transfer of pollen grains (male gametes) from the anther to the stigma of the flower. If the pollen deposited is from the anther of the recipient flower, or from another flower on the same plant, the pollination process is called *self-pollination* (8.1). Nut trees are mostly self-pollinated. Sometimes pollen from different sources is transferred to a flower, as is the case in most fruit trees. This type of pollination mechanism is called *cross-pollination* (8.1). Some species may be self-pollinated but have a fair capacity for outcrossing. When different cultivars are planted close together, there is a good chance for outcrossing. Wind and insects are agents of cross-pollination. Wind is particularly important for pollinating plants with tiny flowers such as grasses. Many fruit crops and vegetables are pollinated by insects that are attracted to the bright colors and nectaries of the flowers. In commercial production, growers of certain crops (such as orchard crops and strawberry) deliberately introduce hives of domestically raised bees to aid pollination for a good harvest. Without pollination, flowers will drop off, thus reducing crop yield.

Flowers that are specifically adapted to pollination by bees, wasps, and flies have showy and brightly colored petals. These flowers are usually blue or yellow in color. The nectary is located at the base of the corolla tube and has special structures that provide for convenient landing by bees. Bee flowers include rosemary (*Rosemary officinalis*), larkspur, lupines, cactus (*Echinocereus*), foxglove (*Digitalis purpurea*), California poppy (*Eschscholzia californica*), and orchids of the genus *Ophys*.

Flowers specifically adapted to pollination by moths and butterflies are typically white or pale in color so that they are visible to nocturnal moths. These flowers also have strong fragrance and a sweet, penetrating odor that is emitted after sunset. The nectary of such flowers is located at the base of a long and slender corolla tube. Such nectaries can be reached only by the long sucking mouth parts found in moths and butterflies. Examples of flowers associated with moths and butterflies include the yellow-flowered species of evening primrose (*Oenothera*), pink-flowered *Amarylis belladonna,* and tobacco (*Nicotiana* spp.).

Birds are associated with flowers that produce copious, thick nectar and have very colorful petals (especially red and yellow). These flowers are generally odorless. Examples are bird-of-paradise (*Strelitzia reginae*), columbine (*Aquilegia canadensis*), poinsettia (*Euphorbia pulcherima*), banana, fuchsia, passion flower, eucalyptus, and hibiscus.

Bats pollinate certain flowering species. Similar to "bird flowers," "bat flowers" are generally large and strong, with dull colors. Most of these flowers open at night, when nocturnal bats operate. Some flowers hang down on long stalks below the foliage of the plant. Bat flowers have strong fruitlike or musty scents. Examples are banana, mango, and organ-pipe cactus (*Stenocereus thurberi*).

Fertilization

Fertilization
The union of two gametes to form a zygote.

Fertilization is the union of a sperm from the pollen with the egg of the ovary to form a *zygote.* For this union to occur, the pollen must grow rapidly down the style to unite with the egg in the ovary (8.1). In some plants, pollen from a flower is unable to fertilize the eggs of the same flower, a condition called *self-incompatibility.* These plants, of necessity, must then be cross-pollinated. Most fruit trees are self-incompatible (and thus self-sterile), which may be due to the inability of pollen to germinate on the stigma or unsynchronized maturity of male and female parts of the flower such that the pollen is shed after the stigma has ceased to be receptive or before it becomes receptive (*protandry* and *protogyny,* respectively). Some incompatibility mechanisms are genetic in origin (figure 4-18). Plants such as date palms and willows are compelled to cross-pollinate because of a condition called *dioecy,* in which plants are either male or female. This condition arises because staminate and carpellate flowers occur on different plants. In monoecious plants such as corn and oak, both staminate and carpellate flowers are found on the same plant (2.8). The stigma should be mature and ready physically and chemically in terms of the presence of fluids in the right concentration and nutrient content (especially certain trace elements such as calcium and cobalt) to stimulate pollen germination.

Fruit Formation and Development

Parthenocarpy
The development of fruit in the absence of fertilization.

Fertilization normally precedes fruit formation. However, under certain conditions, the unfertilized egg develops into a fruit, the result being a seedless fruit. This event is called *parthenocarpy* (5.15). Since seedlessness is desired in certain fruits, horticulturalists sometimes deliberately apply certain growth regulators to plants to induce parthenocarpic fruits.

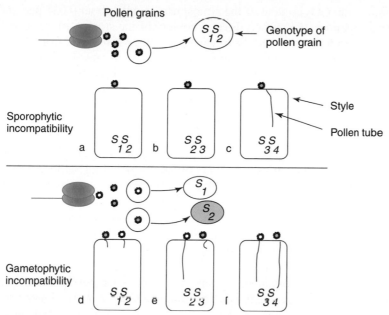

FIGURE 4-18 Sporophytic and gametophytic incompatibility. Notice that in sporophytic incompatibility the incompatible pollen is completely inhibited, while germination is arrested in gametophytic incompatibility.

In tomato, high temperatures have been known to induce seedlessness. Seeds appear to play a significant role in the growth and development of certain fruits. As such, if the embryo dies or pollination is not complete, some eggs will not be fertilized. Fruits formed as a result of this event are malformed, as is observed in cucumber and apple. In the case of stone fruits (e.g., cherry, peach, and apricot), the death of an embryo results in fruit drop. Each plant has an optimal number of fruits it can support in relation to vegetative matter (leaves). In commercial production, growers of orchard crops undertake what is called *fruit thinning*, whereby the number of fruits set is reduced artificially. Fruit set is adversely affected by improper temperature (high or low), low light intensity, and inadequate moisture and nutrition. As fruits develop, they enlarge and become filled with soluble solids.

Fruit Ripening

Fruit ripening starts after enlargement ceases. The results of ripening are usually a change in color (breakdown of chlorophyll to reveal other pigments), softening of the fruits (due to breakdown of pectic substances that strengthen cell walls), and change in flavor (sour to sweet). Ripening is a physiological event that signifies the end of fruit maturation and the onset of senescence. This event is associated with a sudden and marked increase in the rate of respiration of a fruit and the concomitant evolution of carbon dioxide. Levels of enzymes such as hydrolases, synthetases, and oxidases increase. Color change of the fruit exocarp is one event that is visible to the grower in many crops. The color change depends on the crop and cultivar. In banana, raw (unripe) fruit is green and while ripening goes through shades of color changes to yellow. The green chlorophyll is broken down to reveal the hitherto masked yellow color of carotenoids. The acceptable degree of ripening depends on the crop, the consumer (or use), and the marketing system. As ripening progresses, fruits become more susceptible to fungal attack and rot due to the activities of enzymes. Fruits become softer. Ethylene is associated with ripening; it initiates and also accelerates the process (4.5.2).

Ethylene gas can be biosynthesized in the plant from methionine, an amino acid, with the aid of indoleacetic acid (IAA). In fact, high amounts of endogenous IAA can trigger the production of ethylene in large quantities. Wounding of a plant can stimulate the production of ethylene. Ethylene gas is used in the banana production industry to hasten ripening. Bananas are harvested and shipped green from production centers and are induced to ripen at their destination.

By volume, the CO_2 content of the atmosphere is only about 0.035 percent. To reduce the rate of fruit ripening, the carbon dioxide content of the storage room atmosphere is increased to about 2 to 5 percent, and the oxygen percentage is reduced from 20 percent to between 5 and 10 percent. This environment results in a reduced rate of respiration. Respiration rate can also be slowed by reducing the storage room temperature to about 5°C (41°F). Ripening fruits respire at a higher rate.

Senescence

Senescence precedes death, the final stage in the life cycle of a plant. This phase may occur naturally or be accelerated by environmental conditions including pathogenic attack. During senescence, cells and tissues deteriorate. The effect of senescence is physically visible. In this state of decline, yield progressively decreases and the plant becomes weak. The whole plant eventually dies, as in annuals; however, in deciduous perennials, the leaves drop in the fall season and the rest of the plant remains alive.

Senescence is a complex process that is not clearly understood. It occurs in patterns that appear to be associated with the life cycle of plants (annual, biennial, and perennial). In terms of the ultimate end of living organisms, plants may experience *partial senescence* (in which certain plant organs age and eventually die) or *complete senescence* (in which the whole plant ages and eventually dies). In annual plants, death occurs swiftly and suddenly, and senescence is complete. It occurs after maturity in fruits; dramatically, a whole field of annuals can deteriorate and die in concert in a short period. In biennials, the top portions of the plants wither after the first season and the bottom part remains dormant in the winter. Deciduous perennial plants shed their leaves in fall and resume active growth with a new flush of leaves in spring. Whereas senescence in animals is terminal, growers can employ the horticultural method of pruning (chapter 16) and nutritional supplementation to revitalize an aged plant.

4.6: PLANT HORMONES

Growth Regulator
A natural or synthetic compound that in low concentrations controls growth responses in plants.

Plant *hormones* are organic molecules produced in small amounts in one (or several) parts of the plant and then transported to other parts called *target sites,* where they regulate plant growth and development (figure 4-19). Because of this physiological role, plant hormones are also called *plant growth regulators,* a broad term that includes and is often associated with synthetic chemicals that have effects similar to hormones. Plant hormones are classified on the basis of their origin, *natural* or *synthetic.* Unlike animal hormones, which have specificity in site of production and target site, plant hormones tend to be more general with respect to both source and target. There are five basic groups of natural plant hormones: auxins, gibberellins, cytokinins, ethylene, and abscisic acid.

4.6.1 AUXINS

Auxins are produced in meristematic tissue such as root tips, shoot tips, apical buds, young leaves, and flowers. Their major functions include regulation of cell division and expansion, stem elongation, leaf expansion and abscission, fruit development, and branching of the stem. Auxins move very slowly and in a polar (unidirectional) fashion in plants. In shoots, the po-

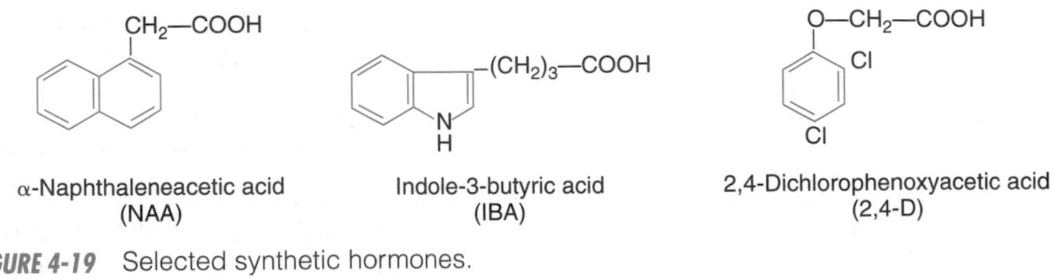

α-Naphthaleneacetic acid
(NAA)

Indole-3-butyric acid
(IBA)

2,4-Dichlorophenoxyacetic acid
(2,4-D)

FIGURE 4-19 Selected synthetic hormones.

larity is *basipetal* (toward the base of the stem and leaves), whereas it is *acropetal* (toward the tip) in roots. Auxin movement is not through defined channels such as sieve tubes (phloem) and vessels (xylem) but through phloem parenchyma cells and those bordering vascular tissues. This hormone is believed to move by an energy-dependent diffusion mechanism. The only known naturally occurring auxin is indole-3-acetic acid (IAA). Synthetic hormones that are auxins include 2,4-dichlorophenoxyacetic acid (2,4-D), which is actually a herbicide for controlling broadleaf weeds such as dandelion in lawns (α-naphthalcneacetic acid [NAA]) and indole-3-butyric acid [IBA]). Other uses of auxins in horticulture are as follows:

1. As rooting hormones to induce rooting (adventitious) in cuttings (4.6), especially when propagating woody plants.
2. To prevent fruit drop (control of abscission) in fruits trees (e.g., citrus) shortly before harvest.
3. To increase blossom and fruit set in tomato.
4. For fruit thinning to reduce excessive fruiting and thus produce larger fruits.
5. For defoliation before harvesting.
6. To prevent sprouting of stored produce, for example, in potato; when applied to certain tree trunks, basal sprouts are suppressed.

The concentration of auxin can be manipulated in horticultural plants in cultivation. Auxins are produced in relatively higher concentrations in the terminal buds than in other parts. This localized high concentration suppresses the growth of lateral buds located below the terminal bud. When terminal buds are removed (e.g., by a horticultural operation such as pruning or pinching), lateral buds are induced to grow because of the abolition of *apical dominance*. This technique makes a plant fuller in shape and more attractive (figure 4-20). *Phototropism,* the bending of the growing point of a plant toward light, is attributed to the effect of auxins. Light causes auxin to be redistributed from the lit area to the dark side, where it causes cell elongation, leading to curvature (figure 4-21).

> **Apical Dominance**
> The regulatory control of the terminal bud of a shoot in suppressing the development of lateral buds below it.

4.6.2 GIBBERELLINS

Gibberellins are produced in the shoot apex and occur also in embryos and cotyledons of immature seeds and roots. They occur in seed, flowers, germinating seed, and developing flowers. The highest concentration occurs in immature seeds. This class of hormones promotes cell division, stem elongation, seed germination (by breaking dormancy), flowering, and fruit development. In carrot (*Daucus carota*) and cabbage (*Brassica oleracea* var. *Capitata*), among others, exposure to long days or cold is required to induce flowering (*bolting*). Application of gibberellic acid eliminates the need for these environmental treatments. Gibberellins are noted for their ability to overcome dwarfism in plants, allowing compact plants to develop to normal heights. Gibberellic acid is used to induce seedlessness in grapes; the size of seedless grapes is also increased through the application of this hormone. An example is gibberellic acid (GA_1), one of the numerous (more than 70) closely related terpenoid compounds that occur naturally.

4.6.3 CYTOKININS

Cytokinins are hormones that stimulate cell division and lateral bud development. They have been isolated mainly from actively dividing tissue. Cytokinins occur in embryonic or meristematic organs. Examples of natural cytokinins are isopentenyl adenine (IPA) and zeatin (Z). Zeatin, isolated from the kernels of corn, is the most active naturally occurring cytokinin. Kinetin was first isolated from yeast. Benzyl adenine (BA) is also a commonly used cytokinin. Cytokinins interact with auxins to affect various plant functions. A high cytokinin-auxin ratio (i.e., low amounts of auxin, especially IAA) promotes lateral bud development because of reduced apical dominance. Relatively high amounts of auxin induce root formation in callus. The principal role of cytokinins in plant physiology is the promotion of cell division. They are important in tissue culture work and are more effective when IAA is also added (9.12). The effect of cytokinins when used this way is to cause cells to remain meristematic (undifferentiated) in culture, producing large amounts of callus tissue.

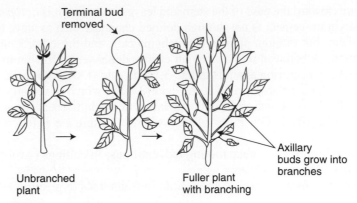

FIGURE 4-20   The effect of abolishing apical dominance in plants.

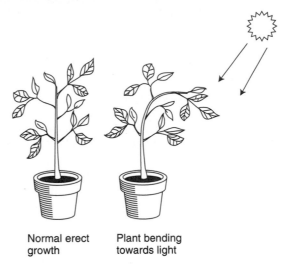

Normal erect
growth

Plant bending
towards light

FIGURE 4-21   Phototropism displayed by a plant located near a window.

4.6.4 ETHYLENE

Many fruits, such as tomato and apple, exhibit a significant increase in cellular respiration during ripening, consuming large amounts of oxygen. This phase of development is called *climateric,* and thus such fruits are called *climateric fruits.* In *nonclimateric* fruits, such as grape and citrus, fruit ripening proceeds at a gradual pace. *Ethylene* is a gas found in the tissues of ripening fruits and stem nodes. It promotes fruit ripening and leaf abscission. In the horticultural industry, cthylene is used to aid in uniform ripening of apple, pineapple, and banana, and in changing the rind color of fruits (as in orange and grapefruit from green to yellow and tomato from green to uniform red). On the other hand, ripening apples produce this gas in large quantities, which tends to shorten the storage life of fruits. Storage life can be prolonged by removing the gas with activated charcoal, for example. Ethaphon is used commercially to induce ripening. Ethylene in the growing environment may cause accelerated senescence of flowers and leaf abscission. Commercially, ethylene is used to promote fruit loosening in grape, blueberry, and blackberry to facilitate mechanical harvesting. Carnations close and rosebuds open prematurely in the presence of ethylene. In cucumber and pumpkin, ethaphon spray can increase female flowers (disproportionately) and thereby increase fruit set. Ethylene is also implicated in the regulation of sex expression and promotion of femaleness in cucurbits (cucumber and squash). Male flowers are associated with high gibberellic acid levels but can be changed to females by the application of ethylene.

4.6.5 ABSCISIC ACID

Abscisic acid (ABA) is a natural hormone that acts as an inhibitor of growth, promotes fruit and leaf abscision, counteracts the breaking of dormancy, and causes the stomata of leaves to close under moisture stress. It has an antagonistic relationship with gibberellins and other growth-stimulating hormones; for example, ABA-induced seed dormancy may be reversed by applying gibberellins. Commercial application of ABA is limited partly by its high cost and unavailability.

Plant hormones may also be classified based on their effect on plant growth as *stimulants* or *retardants*. Cytokinins and gibberellins have a stimulating effect on growth and development, whereas ABA inhibits growth. Alfalfa is known to produce the alcohol *triacontanol*, which stimulates growth. Naturally occurring inhibitors include benzoic acid, coumarin, and cinnamic acid. A number of synthetic growth retardants are used in producing certain horticultural plants. Their effect is mainly a slowing down of cell division and elongation. As such, instead of a plant growing tall, with long internodes, it becomes short (dwarf), compact, fuller, and aesthetically more pleasing. Examples of these commercial growth retardants include the following:

1. Daminozide (marketed under trade names such as Alar and B-nine): plants that respond to it include poinsettia, azalea, petunia, and chrysanthemum.

2. Chlormequat (CCC, cycocel): retards plant height in poinsettia, azalea, and geraniums.

3. Ancymidol (A-Rest): effective in reducing height in bulbs, such as Easter lily and tulip, as well as chrysanthemum and poinsettia.

4. Paclobutrazol (Bonzi): used to reduce plant height in bedding plants including impatiens, pansy, petunia, and snapdragon.

5. Maleic hydrazide: used to prevent sprouting of onions and potatoes.

4.7: NONPATHOGENIC (PHYSIOLOGICAL) PLANT DISORDERS

Horticultural plants are plagued by numerous diseases and pests, some of which can completely kill the affected plants. Generally, if a grower observes certain precautions and adopts sound cultural practices, the chance of experiencing a disease problem in plants is reduced drastically. Some diseases are endemic in certain areas. As such, plants that are susceptible to such diseases should not be grown there. If they must be grown in such areas, resistant cultivars should be used; otherwise, disease control measures such as spraying with pesticides will be required for successful production. Other cultural observances such as using high-quality, clean seeds; weed control; right timing of planting; phytosanitation; and preplanting seed treatment will minimize the occurrence of diseases in a production enterprise.

However, many other disorders are *nonpathogenic* (not caused by pathogens) in origin. They are caused by improper or inadequate plant growth environmental conditions—pertaining to light, moisture, temperature, nutrients, and air—as well as improper cultural operation, involving compaction of soil and pesticide application, for example. Since these disorders are nonpathogenic, their effect is localized and usually within complete control of the grower. Further, since plant production under uncontrolled environmental conditions is subject to the uncertainties of the weather, certain disorders are unpredictable and sometimes difficult to prevent.

This section is devoted to nonpathogenic disorders, many of which are weather related. Chapter 3 contains related information. Plants are not affected equally by adverse weather conditions. An important caution to observe when inspecting plants for disorders is not to hastily attribute every disorder to parasitic or pathogenic causes and exercise caution before initiating pest-control measures.

4.7.1 WEATHER-RELATED PLANT DISORDERS

The following are different categories of weather factors and how they inflict damage on horticultural plants when they prevail in adverse levels. It should be emphasized that these factors often interact or interplay in producing an effect. The role of these factors in plant growth has been discussed previously.

Temperature

Extreme Cold Plants in temperate zones may suffer one of two kinds of injury from extremely cold temperatures. Similar to frostbite in humans, plants may suffer from *frost damage* when temperatures suddenly drop below seasonable levels. Affected plants may show signs of wilting overnight. When this cold strikes during the blooming period, the plant may lose most or all of its flowers. Frost damage occurs more frequently in younger tissues, and herbaceous species are more susceptible than woody ones.

A much more severe cold damage called *winter kill* occurs when plants are subjected to prolonged periods of freezing temperatures. Under such conditions, branches may die back (tips wilt); when roots are severely impaired, however, the plant may die. In evergreens, such as pine, extreme cold may cause the foliage to "burn" (turn brown).

Extreme Heat Microclimates, both natural and man-made, occur in the landscape in places such as underneath trees and the eaves of homes. Brick structures absorb heat during the day and radiate it at night. This property is advantageous during the cold months, because radiated heat protects plants in the vicinity from frost. However, in hot months, these same walls, especially those that face south, can create extremely hot microclimates, thereby injuring plants within their spheres of influence. Similarly, the hot asphalt of parking lots, concrete, and some pavements can radiate intense heat that damages plants. Heat-sensitive plants may show marginal scorching of leaves.

Moisture

Excess Moisture Excess moisture overwhelms the pore spaces in the soil leading to waterlogged conditions. Poorly drained soils create anaerobic conditions that lead to root death (root rot), if they persist for an extended period. Plants vary in their response to poor drainage. Root rot eventually results in plant death through stages, starting with stunted growth and yellowing and wilting. Excess moisture received after a period of drought might cause tubers and roots of root crops, as well as the walls of fruits such as tomato, to crack.

Excess Dryness (Drought) Lack of moisture usually is expressed as wilting of plant leaves.

Low Humidity The tips and margins of leaves of tropical plants brown (tip burn) under conditions of low humidity. This browning is caused by rapid transpiration, which overwhelms the rate at which water is moved through the leaf to the ends. As a result, water fails to reach the edges of the leaf, leading to drying and browning.

Light

Intense Light Strong and direct sunlight may scorch certain plants. Potted plants placed in south-facing windows receive direct sunlight unless the presence of a tree in the direction of the sun's rays filters the light. Intense light also causes the foliage of certain plants to bleach and look pale and sickly.

Low Light Inadequate sunlight induces etiolated growth (spindly) and yellowing of leaves. Plant vigor is reduced, and leaves drop prematurely.

Nutrients

Nutrient Deficiency Generally, an inadequate supply of any of the major plant nutrients, especially nitrogen, causes plants to be stunted in growth and leaves to yellow (chlorosis). Deciduous plant leaves may prematurely senesce and defoliate. In addition to yellowing, lack of potassium shows up later as marginal leaf burns of leaves; young and expanding leaves show purple discoloration when phosphorus is lacking in the soil. Calcium deficiency in tomato shows up as *blossom end rot*.

Nutrient Excess Excess acidic soils may cause excess availability of trace elements (e.g., iron and aluminum), which can lead to toxicity in certain plants.

4.7.2 HUMAN-RELATED ACTIVITIES

Air Pollution

Industrialized and heavily populated areas often experience excessive amounts of chemical pollutants in the air. These toxic gases damage horticultural plants. Acute amounts of sulfur dioxide cause chlorosis and browning of leaves and sometimes necrosis (cell destruction and death). Fluoride injury has been recorded in sensitive plants such as ponderosa pine as reddish-brown bands that appear between necrotic and green tissue. Ozone is a major pollutant that is produced primarily from the photochemical action of sunlight on automobile emission. It can cause chlorosis and necrosis in a wide variety of plants.

Pesticide Application

Improper application of pesticides may cause collateral damage to cultivated plants. Applying sprays on a windy day may cause the chemicals to drift onto desirable plants, resulting in deformed leaves, discoloration, and in some cases death of tissue and possibly the entire plant. Herbicide damage appears suddenly and may last through the cropping season. Often, the symptom is bleaching; in severe cases, it may be followed by leaf drop. Unlike the effect of herbicides, collateral damage from insecticides shows up as browning of the foliage.

Fertilizer Application

Chemical fertilizers are frequently applied to houseplants or outdoor plants in production. Eagerness for good yield may lead some growers to overfertilize their plants, resulting in a buildup of excessive fertilizer in the soil. High amounts of salts create *sodic* soil conditions. A higher salt-soil concentration than root fluids can cause dehydration of roots. Instead of the roots absorbing soil moisture, they become depleted of moisture. Plant growth is inhibited under such conditions, and plants wilt (as they would under drought conditions) and eventually die.

SUMMARY

The variety of activities that have been described to occur at various phases in plant growth and development are the results of certain growth processes. These processes provide the raw materials and the energy required for building new tissues and nurturing them to maturity. The major processes include photosynthesis, respiration, transpiration, and translocation. Photosynthesis is the process by which green plants manufacture food from water and nutrients absorbed from the soil and light energy. Photosynthates are translocated to other parts of the plant where, through the process of respiration, the energy locked up in the food is released for use by the plants. Gaseous exchange between the plant and its environment occurs through pores in the leaves called stomata. Plants lose moisture by the process of transpiration. Plant development and growth occur in phases described by a sigmoid curve—a rapid logarithmic growth phase, followed by a decreasing growth phase, and then a steady growth phase. Plants have two general phases in their life cycles—a vegetative phase and a reproductive phase.

Based on the duration of the life cycle, plants can be categorized into annuals, biennials, perennials, and monocarps. In flowering plants, a reproductive phase follows a vegetative phase. Flowers are produced and eventually become pollinated and fertilized to produce seed and fruit. When plants are subject to adverse conditions in the environment, they develop a variety of physiological disorders such as wilting, drying, cracking, and abnormal growth.

REFERENCES AND SUGGESTED READING

Galston, A. W. 1980. Life of the green plant. 3d ed. Englewood Cliffs, N.J.: Prentice-Hall.

Nicklel, L. G. 1982. Plant growth regulators: Agricultural uses. New York: Springer-Verlag.

Noggle, G., and G. F. Fritz. 1983. Introductory plant physiology, 2d ed. Englewood Cliffs, N.J.: Prentice-Hall.

Wareing, P. F., and I. D. J. Phillips. 1985. Growth and differentiation in plants. New York: Pergamon Press.

More, R., and W. D. Clark. 1995. Botany: Form and function. Dubuque, Iowa: Wm. C. Brown Publishers.

Raven, P. H., R. F. Evert, and S. E. Eichhorn. 1992. Biology of plants, 5th ed. New York: Worth Publishers.

PRACTICAL EXPERIENCE

1. *Translocation.* Obtain a young tree (about a year old) and girdle the midsection by carefully removing the bark and making sure to scrape away the phloem layer completely. Maintain the plant under proper growth conditions in which it will be able to photosynthesize adequately. After some time, the upper edge of the girdle should begin to swell from accumulation of photosynthates being translocated down to other parts of the plant from the leaf.

2. *Growth regulators.* Cytokinins and gibberellins stimulate plant growth, and abscisic acid is an inhibitor. Commercial growth regulators are available. Plant geraniums in pots and divide the pots into two groups, each containing plants of equal size. To one set apply a growth regulator (e.g., Cycocel) and leave the other as a control. You may also apply various concentrations of the hormone. Observe the changes in growth after a period by comparing hormone-treated plants with controls.

3. *Apical dominance (pinching).* Obtain young poinsettia plants. Pinch off the apical buds in some plants and leave others unpinched. After a period, observe the changes in the plants' branching.

4. *Fruit ripening* Obtain two bunches of green bananas. Place one bunch in the open air in the laboratory and tie up another bunch in an airtight plastic bag. The latter will trap the ethylene gas in the bag and accelerate ripening.

OUTCOMES ASSESSMENT

PART A

Please answer true (T) or false (F) for the following statements.

1. T F Photosynthesis accounts for more than 90 percent of dry matter yield of horticultural plants.

2. T F C_3 plants photorespire.
3. T F   Most plants are C_3 plants.
4. T F   Aerobic respiration requires carbon dioxide.
5. T F   Transpiration occurs through the xylem vessels.
6. T F   2,4-D is an auxin.
7. T F   Indole-3-acetic acid (IAA) occurs naturally in plants.
8. T F   Fertilizers may be applied by foliar application.
9. T F   Annual plants complete their life cycles in two growing seasons.
10. T F   Juvenility may be induced in mature tissue.

PART B

Please answer the following questions.

1. Complete the following formula:
   $6CO_2 + 6H_2O \rightarrow$ _____
2. Give two examples each of C_3 and C_4 horticultural plants.
   _____   _____

   _____   _____

3. The process of the loss of moisture through above-ground plant parts is called
   _____.
4. The horticultural practice of inducing lateral branching by removing the terminal bud of the plant is called_____.
5. The cold temperature treatment given to certain plants to induce flowering is called
   _____.
6. List the three generalized growth phases in organisms.
   _____   _____   _____
7. Growth involves three basic activities. List them.
   _____   _____   _____
8. Juvenile shoots that occur at the base of the trunk of a mature tree are called
   _____.
9. Name three plants that require cold treatment to flower.
   _____   _____   _____
10. The union of two gametes to form a zygote is called _____.
11. Give three specific examples of the application of growth regulators in horticulture.

   _____

   _____

   _____

PART C

1. Distinguish between C_3 and C_4 plants.
2. Distinguish between photosynthesis and respiration.
3. What is photorespiration?
4. Discuss the role of ethylene in horticulture.
5. Describe the process of senescence.
6. Discuss the process of ripening.
7. Describe the process of translocation.

CHAPTER 5

Plant Genetics and Improvement

PURPOSE

This chapter is designed to review basic genetic principles and concepts and how they are applied in the breeding of new horticultural plants. The first part, dealing with genetic principles, should not be studied as an independent unit but as an integral part of this chapter, which is devoted to applications. The breeding methods and other applications discussed in the latter part have their foundation in the principles and concepts discussed in the first part.

EXPECTED OUTCOMES

After studying this chapter, the student should be able to

1. Discuss the importance of genetics in horticultural plant improvement.
2. Explain the genetic basis of biological variation.
3. Describe how plant breeding objectives are developed.
4. Describe and discuss the steps in a simple plant breeding program.
5. Discuss the use of molecular biotechnological tools in plant improvement.
6. Discuss specific practical applications of classical genetics and molecular biology in plant improvement.

KEY TERMS

Artificial selection	DNA mutations	Gamete
Backcrossing	DNA transcription	Gene
Center of origin	DNA translation	Gene penetrance
Central dogma of biology	Dominant trait (allele)	Gene pool
Clone	Double helix	Genetic linkage
Complementary	Emasculation	Genetic map
Crossing over	Epistasis	Genetic marker
Cross-pollination	Extrachromosomal	Genotype
Dihybrid cross	inheritance	Germ plasm bank
DNA	Fertilization	Haploid

Heritability	Law of segregation	Phenotype
Heterogeneous	Meiosis	Polynucleotide chain
Heterosis	Meristems	Protein synthesis
Heterozygous	Messenger RNA	Quantitative trait
Homologous chromosome	Mitosis	Recessive trait (allele)
Hybrid	Mutagenesis	Segregate
Hybrid vigor	Mutation breeding	Self-pollination
Inbred	Natural selection	Simply inherited trait
Inbreeding depression	Nature	Somatic cell
Incomplete dominance	Nucleoside	Variable gene expression
Law of independent assortment	Nucleotide	Zygote
	Nurture	

OVERVIEW

Plant breeding or improvement is a science and an art. Genetics is the underlying science of plant breeding. In fact, breeders are sometimes referred to as applied geneticists. They try to nudge nature to the advantage of humans by manipulating plants to perform according to their schedule and needs. They manipulate the nature (heredity) of plants and thereby create new types that are adapted to new environments (nurture) and produce higher-quality products that are disease resistant. Every year, new and improved flower and vegetable garden cultivars are released by plant breeders for use by growers. These are products of calculated and deliberate manipulation of plants by scientists who understand the genetics and environment of those plants.

As agents of heritable change, plant breeders are sometimes also described as applied evolutionists. This chapter is devoted to describing how these scientists operate, highlighting the role genetics plays in their endeavors. This is not to say that plant breeding is an exact science; it is also an art, as already stated. Experience (breeder's eye) is a valuable asset in breeding. Conventional or classical methods of breeding depend more on this artistic component. However, thanks to advances in science and technology, new and more effective methods of plant improvement are now available to breeders. Instead of manipulating plants at the whole-plant level, breeders are able to manipulate plants directly at the deoxyribonucleic acid (DNA) level *(molecular biotechnology),* thereby expanding the degree to which they are able to affect the course of nature. It should be emphasized that conventional and modern tools are used side by side for best results. As such, discussions in this chapter include both types of techniques.

MODULE 1 REVIEW OF SELECTED PRINCIPLES AND CONCEPTS OF GENETICS

Living things have a hereditary mechanism that ensures that parental genes that condition their traits are transmitted to their offspring. Heredity has both a chemical and a physical basis. Natural laws govern the transmission of hereditary characteristics from one generation to another, making it possible to predict to a varying extent, and depending on the trait in question, the outcomes of a cross between two parents. By understanding the laws of genetics, scientists are able to manipulate plants to create new and more desirable variants.

5.1.1 MENDELISM

Gregor Mendel is credited with providing the scientific community with the first accurate interpretation of heredity. His work on the garden pea *(Pisum sativum)* culminated in the discovery of two laws that govern the transmission of hereditary characteristics from one generation to the next. Mendel observed contrasting expression of a number of traits in the pea (e.g., green or yellow cotyledons, round or wrinkled seed coat, yellow or green pod color, and white or purple flower color). Unaided by humans, pea plants perpetually produce their own kind; that is, white-flowered plants always produce white-flowered offspring. Mendel crossed *(hybridized)* plants with contrasting traits and observed the nature of their progeny with respect to specific traits. For example, a cross between a round-seeded parent and a wrinkled-seeded parent always produced a round-seeded offspring. The first generation of the hybrid is called the *first filial generation (F_1)*. Since the F_1 in the seed shape cross yielded a round-seeded offspring, the round seed shape trait is said to be *dominant* to the wrinkled seed shape, which is described as a *recessive* trait (figure 5-1).

When the F_1 generation was self-fertilized, or *selfed* (i.e., pollen grains from a flower were deposited on its own stigma), the unexpressed trait (recessive) reappeared in the *second filial generation (F_2)*, along with the dominant trait in a ratio of 3:1 of dominant (round seed) to recessive (wrinkled seed). The heterozygote and homozygote dominant genotype produced indistinguishable *phenotypes* (figure 5-2). Therefore, in the F_2 generation, the two traits sorted out, or *segregated,* into two different classes. Each trait is conditioned by a distinct factor, or *gene,* which has alternative expressions or forms called *alleles* (e.g., *A, a*). The uppercase letter is used to represent the dominant allele and the lowercase the recessive allele. The pattern of segregation is the same, regardless of the number of genes considered. In the case of two contrasting traits considered simultaneously (a *dihybrid cross*), Mendel observed a segregational ratio of 9:3:3:1 of round/yellow to wrinkled/yellow to round/green to wrinkled/green (figure 5-3). However, when taken one trait at a time, the 3:1 ratio of either round to wrinkled or green to yellow was observed, indicating that two 3:1 ratios had been superimposed. In effect, Mendel discovered that the hereditary factors or genes occurred in a pool that permits all possible combinations to occur during *gamete* (pollen grains and eggs in plants) formation. These pollen grains combine at random with the eggs to form *zygotes*. Mendel's profound discoveries are summarized into what are known as Mendel's laws of heredity:

> *Law I (law of segregation).* The pair of hereditary factors (genes) separate (segregate) during gamete formation such that only one from each pair ends up in each gamete.

> *Law II (law of independent assortment).* Pairs of genes controlling different traits separate independently of each other during gamete formation and combine randomly to form zygotes.

Allele
One of a pair of genes at the same locus on homologous chromosomes.

Zygote
The product (a protoplast) of the union of two gametes.

FIGURE 5-1 Dominance and recessiveness.

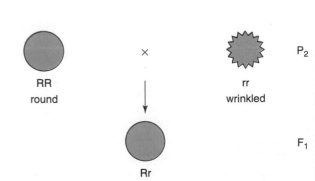

5.1.2 GENOTYPE AND PHENOTYPE

The totality of all of the genes an individual possesses (its genetic makeup) is called its *geno-type*. In the true sense of the word, an individual's genotype is difficult—if not impossible—to ascertain. The term is often used to refer to a fraction of the genes of an individual as it pertains to a specified trait or traits. For example, if the color gene is represented by *P* (for purple color) and purple color is dominant to white color *(p)*, then an individual, with respect to this gene, could have a genotype of *PP, Pp,* or *pp* (figure 5-4). If other traits are included, the genotype could be *PPSs* (dihybrid), *PPSsGg (trihybrid),* and so forth.

The expression of the genotype in an environment is the *phenotype.* In other words, a phenotype is the product of an interaction between a gene and its environment. Classical genetic procedures may be used to infer the genotype of a plant, with respect to a trait, from the phenotype expressed. The *testcross* is used to determine whether a genotype consists of identical alleles *(homozygous)* or different alleles *(heterozygous).* It is accomplished by crossing the plant (genotype) in question with one known to possess the double recessive allele combination for the trait in question. If the unknown genotype is heterozygous, a segregation ratio of 3:1 (phenotypic ratio) is expected. If homozygous, only one phenotypic class is observed (figure 5-5). Genotypes can be described at the biochemical level when, for example, certain patterns are observed on a gel after *electrophoresis* (5.16); they may also be described at the morphological level (e.g., in the leaf and flower) in which the genotype is being expressed.

> **Testcross**
> *The mating between an individual of unknown genotype for a certain trait and an individual that is homozygous recessive for the same trait.*

5.1.3 HOMOZYGOSITY AND HETEROZYGOSITY

As previously discussed, a *locus* (site on a chromosome where a gene is located) is said to be homozygous when identical alleles occur. Plants that have a capacity to perpetually produce an identical expression of a trait are said to be homozygous with respect to the trait. Symbolically,

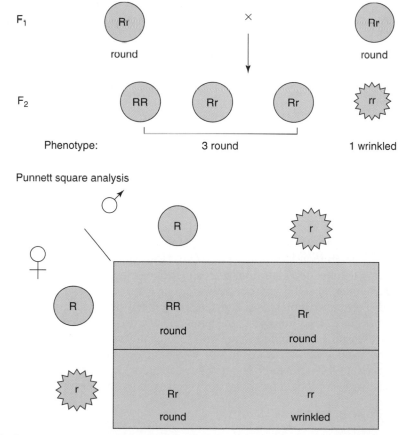

> **Punnett Square Analysis**
> *A largely illustrative grid used by geneticists to predict the probability (or frequency) of the genetic and phenotypic constitution of an F₂ generation, given the genotypes of the parents in a cross.*

FIGURE 5-2 Segregation in a monohybrid cross with dominance; an example of the 3:1 phenotypic ratio.

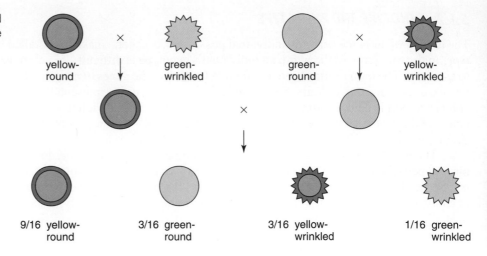

FIGURE 5-3 Segregation in a dihybrid cross with dominance; an example of the 9:3:3:1 phenotypic ratio.

yellow-round × green-wrinkled

green-round × yellow-wrinkled

9/16 yellow-round

3/16 green-round

3/16 yellow-wrinkled

1/16 green-wrinkled

Punnett square analysis

	1/4 RY	1/4 Ry	1/4 rY	1/4 ry
1/4 RY	1/16 RRYY	1/16 RRYy	1/16 RrYY	1/16 RrYy
1/4 Ry	1/16 RRYy	1/16 RRyy	1/16 RrYy	1/16 Rryy
1/4 rY	1/16 RrYY	1/16 RrYy	1/16 rrYY	1/16 rrYy
1/4 ry	1/16 RrYy	1/16 Rryy	1/16 rrYy	1/16 rryy

FIGURE 5-4 Specifying genotypes in a cross.

PP × pp
(purple)   (white)

Pp

(a) Monohybrid cross

PpSs × PpSs

(b) Dihybrid cross

PpSsGg × PpSsGg

(c) Trihybrid cross

in a *diploid* organism (two alleles of a gene present at each locus), identical letters (e.g., *DD, cc,* or *NN*) are used to describe the genotype. When plants with two contrasting traits are crossed, (*DD* × *dd*) the F_1 generation is *heterozygous* (*Dd*). However, as Mendel observed, repeated selfing produces a tendency toward homozygosity, because heterozygosity is reduced by 50 percent with each round of selfing (figure 5-6).

Species that are *self-pollinated* (or self-fertilized) (5.9.1) tend to be homozygous at most loci. As such, they *breed true* from seed, meaning one can use harvested seed from a plant to produce the next season's crop and have the same results in terms of genetic purity of the harvest. On the other hand, *cross-pollinated* plants (which can be fertilized by pollen from sources other than the plant) (5.9.2) tend to be heterozygous at many loci. They do not breed true from seed. In this case, growers should purchase fresh seed each season for planting.

5.1.4 PARTIAL DOMINANCE

Partial or *incomplete dominance* is a condition in which the F_1 hybrid expression of a trait is unlike the expression of the phenotype of the parent with the dominant allele, as would be ex-

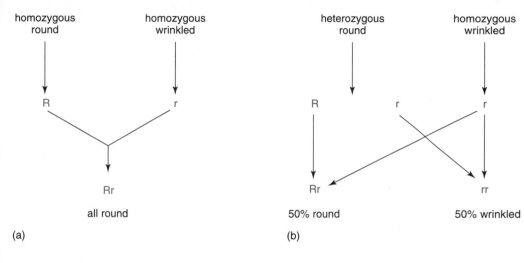

FIGURE 5-5 A testcross finding unknown genotypes.

RR × rr	Rr × rr
homozygous round / homozygous wrinkled	heterozygous round / homozygous wrinkled
R r	R r / r r
Rr	Rr / rr
all round	50% round / 50% wrinkled
(a)	(b)

FIGURE 5-6 The effect of repeated selfing. Heterozygosity is reduced by half with each round of selfing but never completely exhausted in practice.

pected in the case of *complete dominance* (e.g., $DD = Dd > dd$). The F_1 phenotype in this case is an intermediate expression of the parental phenotypes. An example of this gene interaction is exhibited in a cross between red- and white-flowered plants of *Camallia japonica* or the four-o'clock plant *(Mirabilis jalapa)* (figure 5-7). The F_1 generation is a light shade of red (or pink), whereas the F_2 generation produces a segregation ratio of 1:2:1 of pure red to light red (or pink) to pure white.

5.1.5 EPISTASIS (MULTIGENIC INHERITANCE)

An observed trait may be the result of the combined effects of several genes, each controlling one of a series of steps in the chemical processes that produced it. In other words, more than one gene affects the same biochemical pathway. One gene (dominant or recessive) may mask (or suppress) the expression of another. In snapdragon, for example, the gene *R* is masked by another *niv.* The *R* gene is responsible for red color, and *Niv* (and *niv*) genes determine the formation of the basic flower pigment precursor upon which other color genes act to produce var-

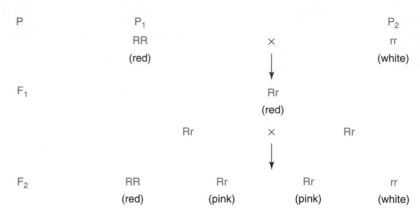

P	P₁		P₂

$$P \qquad\qquad P_1 \qquad\qquad\qquad\qquad\qquad\qquad P_2$$

RR × rr
(red) (white)

F₁ Rr
(red)

Rr × Rr

F₂ RR Rr Rr rr
(red) (pink) (pink) (white)

Phenotypic ratio: 1 red : 2 pink : 1 white

FIGURE 5-7 Partial dominance.

ious colors. The recessive gene *(niv)* blocks the synthesis of this precursor; as a result, $R{-}/niv$ *niv* and *rr/niv niv* genotypes produce white flowers (no pigment precursor development). This is an example of the phenomenon of *epistasis,* in which genes interact in a nonreciprocal fashion. Such an interaction causes one gene to interfere with, or prevent the expression of, another. Other types of epistatic interactions occur in nature. In this case, *niv* is epistatic (suppresses another) to *r* and *R* genes (the *r* and *R* being *hypostatic* [suppressed by another] to *niv*). Another example is found in tomato, in which a cross between two yellow-fruited genotypes *(rrA^tA^t × RRa^ta^t)* produces red fruits *(RrA^ta^t)*. The dominant genes are both required to produce *lycopen* (the red pigment in red tomato). Many of these epistatic events associated with color expression have been identified. The phenomenon of epistasis is observed as a result of different types of gene interaction other than the one described previously.

5.1.6 GENETIC LINKAGE

Genetic Recombination
The production, by crossing over, of chromosomes with gene combinations that differ from those in the original chromosomes.

Genes are located in a linear order on a chromosome and, by virtue of this structural arrangement, are physically associated with each other, or said to be linked (a condition called *genetic linkage*). The consequence of linkage is that genes do not assort independently as Mendel's second law states (5.1.1). Instead, pairs of genes located on different homologous chromosomes exhibit independent assortment. Linkage creates a tendency of linked genes to be inherited together as a block. Certain genes are more tightly linked than others (i.e., they are physically closer together). Due to the process of *crossing over* (5.2.3), linkages can be broken. This breakage occurs with higher frequency between genes that are widely spaced than those that are closely spaced. The amount (percentage) of *recombination* is thus a function of the distance between genes. Breeders sometimes manipulate plants to break undesirable linkages. A geneticist may make strategic crosses and, by examining their progenies, obtain mathematical estimates of the distance between genes. By arranging the genes according to the percentage recombination, a gene order, called a *genetic map,* or *chromosome map,* can be produced (figure 5-8). A common method of mapping is accomplished by using the *three-point testcross.* Breeders use these maps in decision making related to breeding strategies and methodologies (5.7.1).

Linkage, to a breeder, may be desirable or undesirable. It is desirable when two traits of interest are tightly linked. This condition means that improving one trait automatically improves the other. Another desirable association is between a *qualitative trait* (simply inherited; controlled by one or a few genes) that is readily detectable or visible and a *quantitative trait* (controlled by many genes) (5.1.8). This situation is often vigorously sought by breeders. Since it is difficult to select for quantitative traits directly because of a large environmental influence, breeders use the associated qualitative trait as a *marker* (5.17) to indirectly, effectively, and efficiently select for the quantitative trait.

Undesirable linkage occurs when only one of the two linked traits is desirable. For example, in peach, small fruit size is linked with smooth skin. Peaches have coarse skin, and nectarines have smooth skin. The two are crossed in breeding programs aimed at producing

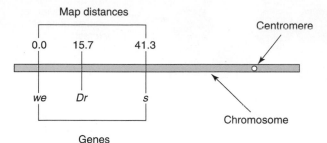

Map distances

0.0   15.7   41.3

Centromere

we   Dr   s

Chromosome

Genes

FIGURE 5-8 A hypothetical genetic map. Genetic maps are obtained by calculating the amount of genetic recombination occurring between genes or known markers. The units of measurements are recombination frequencies. Physical mapping is accomplished by assigning markers from the genetic map to loci on the chromosome. The distances on a physical map are units of physical length expressed as the number of nucleotides separating markers.

bigger nectarine fruits (large-fruited peach × small-fruited nectarines). To be able to distinguish nectarines from small peaches (both have smooth skin), the linkage needs to be broken. To accomplish breakage, a large segregating population needs to be produced to increase the chances of finding the desired recombinant. Another kind of undesirable linkage is one in which increasing the expression of one trait diminishes the expression of the other. Protein and oil contents of seed display this negative association.

5.1.7 PENETRANCE AND EXPRESSIVITY

It has already been said that the environment in which a gene occurs influences how it is expressed. The source of this environmental effect could be as close and intimate as the immediate cellular environment or as remote as the general plant environment. In plants, breeders transfer genes from one genetic background to another in their improvement work through *hybridization* (5.9.2). Sometimes they encounter a situation in which the gene may be successfully transferred, but the desired effect is not observed. In one case, a disease-resistance gene, for example, may offer resistance in one plant but fail to do the same in another plant from the same species. This phenomenon is described as *variable gene penetrance* (figure 5-9). Sometimes changes in the plant environment may cause the same plant to produce different phenotypes or degrees of expression of the trait under these different conditions. For example, the hibiscus plant normally produces single flowers (a flower has one set of petals). A double-flowered mutant (additional petals added to the primary set) has been developed. However, the number of petals of the double flower is influenced by temperature. When grown at between 1.5 and 10°C (35 and 50°F), the double-flower characteristic is lost or diminished; flowers produce fewer petals. This gene interaction is called *variable gene expressivity* (figure 5-10).

5.1.8 QUANTITATIVE TRAITS

Mendel studied seven traits that are described as *simply inherited* (qualitative) traits. These traits are controlled by one or a few genes and expressed such that phenotypes can be readily categorized into distinct, nonoverlapping groups (e.g., purple versus white flowers or round versus wrinkled seed shape). Certain traits, called quantitative traits, however, are controlled by many genes, each contributing small effects toward the total expression of the trait. This inheritance pattern is also described as *polygenic* (and the genes sometimes called *modifiers, polygenes,* or *multiple factors*). They cannot be classified in a discrete fashion and are often measured rather than counted. Quantitative traits are therefore sometimes called *metrical traits*. The expression of quantitative traits is usually more influenced by the environment than are simply inherited traits. Quantitative traits such as yield, weight, and size are influenced by the environment such that a good growing environment (e.g., good moisture, fertility, and temperature) results in larger and juicier fruits than those from a droughty environment.

Even though the genetics of quantitative traits is an extension of that of Mendelian traits (qualitative traits), the former are not amenable to the techniques of the latter. Instead, a variety of mathematical and statistical techniques are employed to estimate the number of genes and their average contribution to the observed phenotype. The larger the difference in the number of genes between the two parents in a cross with respect to the trait of interest, the larger the number of F_2 genotypes. In Mendelian cases involving just one gene difference, three genotypes were

> **Quantitative Trait**
> A trait that is controlled by a group of genes at different loci with each allelic pair having a specific quantitative effect on the expression of the trait.

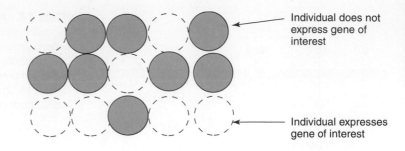

FIGURE 5-9 Penetrance of a gene in a population.

Individual does not express gene of interest

Individual expresses gene of interest

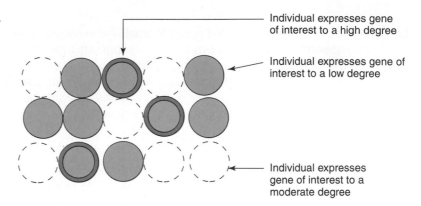

FIGURE 5-10 Variable expressivity of a gene in a population.

Individual expresses gene of interest to a high degree

Individual expresses gene of interest to a low degree

Individual expresses gene of interest to a moderate degree

observed in the F_2 generation (see figure 5-2), of which two were duplicates of the parental genotypes. If the number of gene differences increases to n, the number of individual genotypes in the F_2 generation increases exponentially from 3 to 3^n. The number of either parent (parental genotype), however, reduces from 1/4 to $(1/4)^n$ of the population. The genetic characteristics of quantitative traits have implications in their manipulation by breeding techniques. Simply inherited traits are relatively easier to manipulate, and quantitative traits are much more difficult to breed for. Large numbers of segregants are required to increase the chance of finding the desired recombinant (with the desired combination of parental genes). Since many traits of economic importance are quantitatively inherited, breeders are constantly searching for effective and efficient breeding methods. Molecular techniques are utilized for best results (5.15).

5.1.9 MUTATION

Genes are expected to remain the same (and nature ensures that this is so) through the existence of elaborate natural machinery for repairing the DNA (5.2.4) as it undergoes *replication, transcription,* and *translation* processes. (5.2.4) Unfortunately, permanent alterations in the DNA, called *mutations,* occur when the appropriate repair mechanism fails. Mutations are heritable, and their consequences depend on the function or functions the mutated gene controls. Mutations may arise *spontaneously* (unaided by humans) or deliberately *(induced).* Breeders sometimes resort to induced mutations to develop new horticultural plants. When mutations occur in recessive forms, an individual can harbor a number of deleterious ones without adverse consequences, as long as they do not occur in the double recessive state. Mutations are the ultimate source of biological variation. The subject is further discussed under applications of genetics in breeding (5.10).

5.2: CELLULAR AND SUBCELLULAR BASES OF HEREDITY

5.2.1 THE CELL

The *cell* is the fundamental unit of organization of life (2.1.1). This unit is endowed with the complete complement of genetic information to code for the materials required for the formation of the entire organism from which it was derived, a capacity called *totipotency.* Thus, in theory, it is possible to remove a living cell from any part of an organism and expect it to regenerate into the full organism, provided the appropriate nurturing conditions are supplied. This, in practice, is easier said than done. Totipotency is a dominant principle in *tissue culture* (5.19, chapter 9).

Eukaryotes have various levels of organization (chapter 2). These organisms are comprised of units that are progressively organized into more complex forms until ultimately the whole organism is formed. The simplest forms of life are *unicellular* (consisting of one cell), whereas complex forms are *multicellular* (composed of many cells). Most cells can live as independent entities performing all life-sustaining functions. In algae, for example, the independence of existence is a natural function, whereas cells from some eukaryotes can be sustained only on artificial media under laboratory conditions.

Totipotency
The notion that every cell in an organism has the same genes and thus the same genetic potential to make all cells other cell types.

5.2.2 CHROMOSOMES

When a cell is in its resting (nondividing stage), a densely staining region of its nucleus, called *chromatin,* can be identified upon staining and microscopic examination. This region consists of DNA and protein materials. The chromatin region is made up of a discrete number of threadlike structures called *chromosomes.* Chromosomes consist of DNA (5.2.4) and histones (basic proteins). The DNA-histone complex is the basic structure of a chromosome, accounting for about 60 to 90 percent of the chromatin mass. Each species has a characteristic number, or set, of chromosomes. The number in a set is called the *haploid (n)* number. Examples of chromosome numbers *(n)* of selected species are carrot (9), garden pea (7), lettuce (9), onion (8), redwood (11), and rose (7). *Sex cells* (pollen grains and eggs) contain the haploid number of chromosomes, whereas body cells, or *somatic cells,* contain 2*n*. Chromosomes within a cell differ in size.

In animals, certain chromosomes *(sex chromosomes)* are responsible for sex determination in the organism. The remainder are called *autosomes.* There are two kinds of chromatin. The chromosome strand stains differentially, the densely staining region being made up of *heterochromatin,* whereas the less densely staining region contains *euchromatin* (figure 5-11). The heterochromatic regions consist of *repetitive DNA,* and the euchromatin consists of *nonrepetitive* or *single copy DNA* and is the coding part of the chromosome. As previously stated, genes lie on chromosomes in a linear order. Most chromosomes have a structure called a *centromere,* which may be located in the center *(metacentric chromosome),* off center *(submetacentric* or *acrocentric),* or at the tip *(telocentric)* (figure 5-12). This structure plays a significant role in cell division (5.2.3).

5.2.3 CELL DIVISION

Growth and reproduction in plants depend on two *cell division* processes:

Cell Division
The reproduction of cells.

1. *Mitosis.* Mitosis is a cell division process that occurs in somatic cells. In this conservative process, a cell simply replicates or duplicates itself by doubling up its chromosomes and dividing into two, such that the daughter cells are identical to each other and to the parent cell in terms of genic content. The steps that lead to the perpetuation of chromosome constitution of a cell are described in figure 5-13. The cell division cycle has distinguishable stages during which certain activities occur (figure 5-14). The duration of each stage is variable, depending on the organism, type of cell, temperature, and other factors. Mitosis produces new cells for growth and maintenance of the plant. These cells are undifferentiated but are candidates for *differentiation* (assignment of specific function). When a plant is bruised, new cells are produced by this process to heal the wound. The capacity to divide is often limited to cells concentrated in specific regions of plants called *meristems* (2.2.2). Undifferentiated cells are also called *meristematic cells.* A common place to find actively dividing cells is the root tip. Cells cease to divide once their walls acquire certain kinds of strengthening material such as lignin (the cell becoming lignified). Mitosis occurs when cells proliferate under tissue culture conditions.

2. *Meiosis.* Meiosis occurs only in specialized tissues in the reproductive parts (flowers) of plants (i.e., meiosis does not occur in nonflowering plants). The purpose of meiosis is to

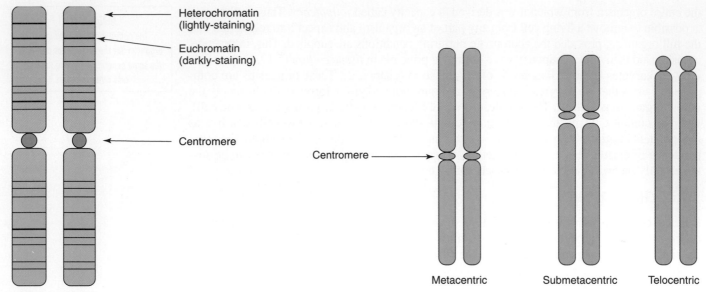

FIGURE 5-11 Variable staining of chromosomes.

FIGURE 5-12 Centromere location on chromosomes.

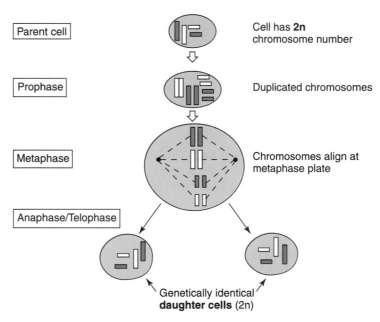

FIGURE 5-13 A summary of mitosis.

reduce the somatic number by half in order to produce gametes (germ cells) for fertilization. Unlike mitosis (monoparental), meiosis is necessitated by biparentalism. In *fertilization,* two parental genomes combine to form a zygote. Without a mechanism to reduce chromosome number to the basic number before fertilization, the chromosome number would be cumulative with each subsequent fertilization (e.g., $2n + 2n = 4n$ and $4n + 2n = 6n$). This process entails two cell division stages (figure 5-15). Most significantly, the daughter cells are *not* identical to each other or the parent, as is the case in mitosis. Instead of two cells, four unidentical daughter cells are produced that develop into *gametophytes* that produce *gametes* (figure 5-16). During meosis, a random phenomenon called crossing over, in which pairs of homologous chromosomes exchange parts be-

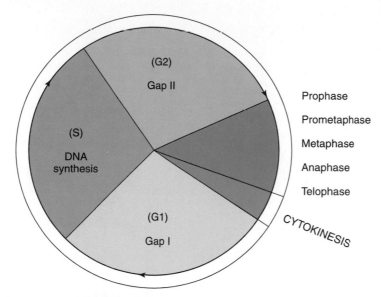

FIGURE 5-14 A summary of the cell cycle.

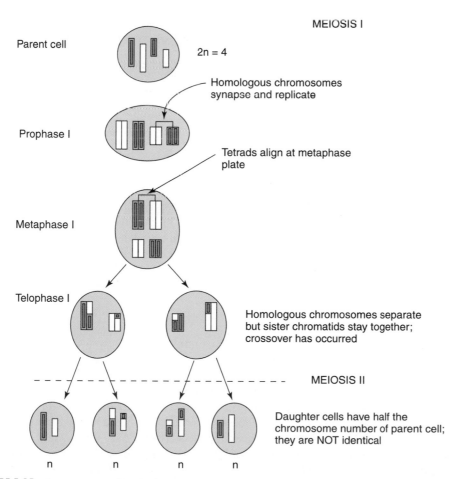

FIGURE 5-15 A summary of meiosis.

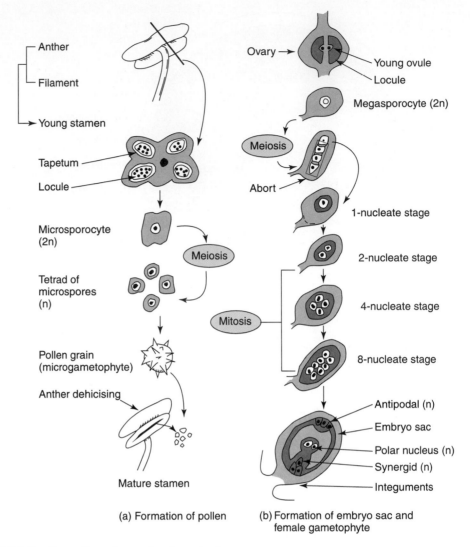

Anther
Filament
Young stamen
Tapetum
Locule
Microsporocyte (2n)
Tetrad of microspores (n)
Pollen grain (microgametophyte)
Anther dehicising
Mature stamen

Meiosis

(a) Formation of pollen

Ovary
Young ovule
Locule
Megasporocyte (2n)
Meiosis
Abort
1-nucleate stage
2-nucleate stage
4-nucleate stage
8-nucleate stage
Mitosis
Antipodal (n)
Embryo sac
Polar nucleus (n)
Synergid (n)
Integuments

(b) Formation of embryo sac and female gametophyte

FIGURE 5-16 Formation of gametes in a plant.

tween adjacent strands, occurs. This event occurs predominantly in meiosis and to a limited extent in mitosis. Meiosis increases diversity among related individuals in a species through two processes—*chromosome segregation* and *gene recombination.* It is meiotic crossing over that makes each individual unique (i.e., no two individuals are genetically identical, except *clones* and identical twins). This phenomenon generates genetic variation that is critical to plant improvement. Through crossing over, alleles in parental chromosomes are reshuffled to produce new types called *recombinants.*

5.2.4 DNA: ITS STRUCTURE AND FUNCTION

Structure

> **DNA**
> Deoxyribonucleic Acid (DNA) is the genetic material organisms inherit from their parents.

Deoxyribonucleic acid is the genetic material of living organisms. It consists of *nitrogenous bases, sugar,* and *phosphate.* There are four bases: *adenine (A), cytosine (C), guanine (G),* and *thymine (T).* Adenine and guanine are called *purines,* and cytosine and thymine are *pyrimidines.* The letters *A, C, G,* and *T* are the genetic alphabets (figure 5-17). The sugar is a *pentos* (five-carbon ring) and is of the *deoxyribose* variety. The sugar and base link up to form a *nucleoside,* which then combines with a phosphate to form a *nucleotide* (figure 5-18). Nucleotides link up to produce a chain called a *polynucleotide,* in which the sugar and phosphate form a

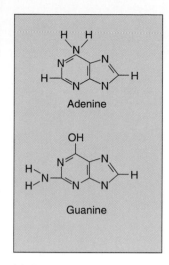

(a) Purines

FIGURE 5-17 The four nitrogenous bases of DNA.

FIGURE 5-18 A nucleotide.

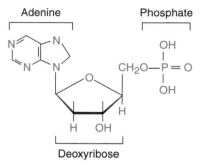

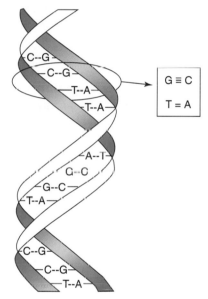

FIGURE 5-19 A DNA double helix.

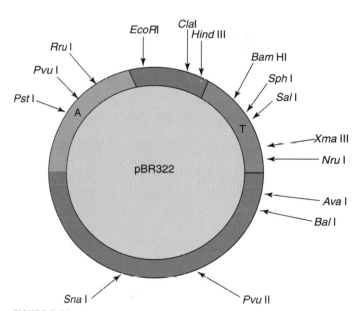

FIGURE 5-20 A restriction map of a plasmid.

backbone from which the bases extend. Two polynucleotide chains pair up in an antiparallel (running in opposite sequence) and complementary fashion. That is, *A* always pairs with *T* (by a double hydrogen bond), and *G* always pairs with *C* (by a triple bond). The pair of chains then winds or coils up into a *double helix* (figure 5-19). The double helix is right-handed (turns run clockwise along the axis of the helix). This description of DNA structure constitutes the *B form* (there are other forms or models of DNA—A, C, D, and Z) and is what is believed to occur in living cells.

Whereas nuclear DNA exists as linear and double-stranded molecules, DNA of prokaryotes (e.g., bacteria) and organelles in eukaryotes (e.g., chloroplasts) exists as a circular structure called a *plasmid* (figure 5-20). Plasmids are important in *genetic engineering* (5.15) protocols. The DNA duplex in a resting cell is supercoiled (twisted around its axis). Before replication, the DNA is unwound to separate the strands. The double helix of DNA can be disrupted by heating, which breaks the hydrogen bonds, an event called *denaturation*.

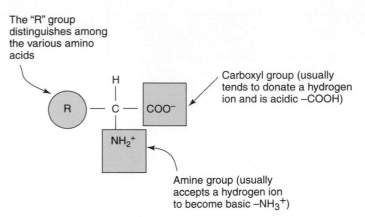

The "R" group distinguishes among the various amino acids

H

R — C — COO⁻

Carboxyl group (usually tends to donate a hydrogen ion and is acidic –COOH)

NH₂⁺

Amine group (usually accepts a hydrogen ion to become basic –NH₃⁺)

FIGURE 5-21 Basic amino acid structure.

Function

> **The Genetic Code**
> The set of rules giving the correspondence between mRNA and amino acids in protein.

The sequence of bases in the polynucleotide chain holds the key to DNA function. The sequence is critical because it represents the *genetic code* for the synthesis of the corresponding *amino acids* that constitute *proteins* (or *enzymes*). DNA does not code for adult traits directly, there being no genes for adult traits as such. Instead, genes code for various developmental processes. The variety of protein products in a cell undertake catalytic and structural activities that eventually result in adult phenotype.

Protein Synthesis   There are about 20 commonly occurring amino acids (figure 5-21). The structures of selected amino acids are presented in figure 5-22. According to the prescribed sequence (based on the genetic code), amino acids are joined together by *peptide bonds* to form polypetide chains (figure 5-23). The genetic code is a *triplet code*. Three adjacent bases form a code for an amino acid. Each trinucleotide sequence is called a *codon* (figure 5-24). The genetic code is read from a fixed starting point of the DNA strand.

The information in the DNA is decoded by the process of *protein synthesis*. The genetic message occurs in the sequence of nitrogenous bases. The sites of protein synthesis are outside of the nucleus on the ribosomes in the cytoplasm. Copies of the nuclear DNA must first be made and transported to the ribosomes to serve as *templates*. Another nucleic acid, called *ribonucleic acid (RNA),* is responsible for this genetic transport. The RNA differs from DNA in the types of sugar. RNA has *ribose,* whereas DNA has *deoxyribose.* Also, RNA uses *uracil (U)* as a base in place of thymine in DNA. The DNA double helix first unwinds (temporarily) to allow the transcription of one of its strands. This process is called *RNA synthesis* and results in a copy of nuclear genetic information called the *messenger RNA (mRNA).* The mRNA is transported to a ribosome by another type of RNA, *transfer RNA (tRNA),* which has a unique clover-leaf structure with two recognition sites. One site recognizes a specific amino acid and picks it up, and the other identifies a specific codon on the mRNA template. Once this tRNA "adaptor" has been properly charged with an amino acid for which its *anticodon* (complementary codon) is appropriate, it proceeds to locate the correct codon on the template with which the anticodon is compatible and attaches to it temporarily (figure 5-25). The amino acids carried by adjacent tRNAs are then peptide bonded to a growing chain until the template is completely translated. This translation process is facilitated by ribosomes and consists of three steps—*initiation, elongation,* and *termination.* The translated mRNA template disintegrates afterward.

> **Messenger RNA (mRNA)**
> The ribonucleic acid that encodes genetic information from DNA and transports it to ribosomes for translation into amino acids.

Each gene codes for one protein (according to the *one gene–one protein theory*), but certain proteins comprise more than one polypeptide chain, making it necessary to modify the previous theory to *one gene–one polypeptide.* All genes do not code for proteins, and, further, all genes in a cell are not actively transcribing mRNA all of the time. Genes are regulated such that they are selectively turned on or off as needed. A cell in the root and one in the leaf have different sets of genes turned on, since they have different roles in the plant. In other words, cells in the adult plant eventually become specialized (differentiated). However, it is possible to rejuvenate mature cells and cause them to become meristematic (5.18).

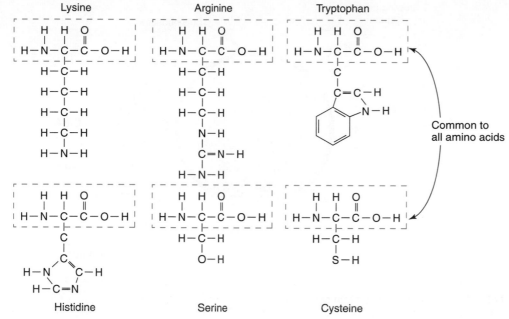

FIGURE 5-22 Structure of selected amino acids.

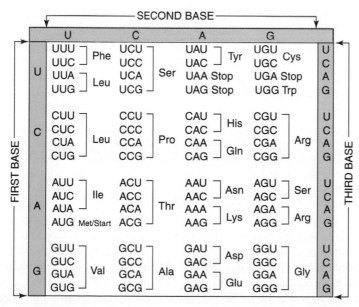

FIGURE 5-23 A polypeptide.

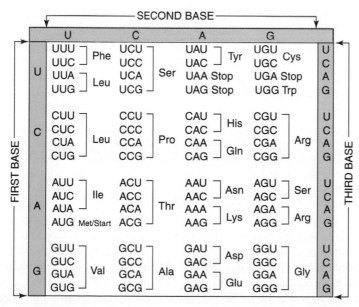

FIGURE 5-24 The triplet genetic codes.

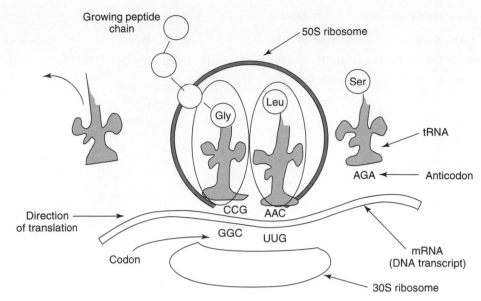

FIGURE 5-25 A summary of protein synthesis.

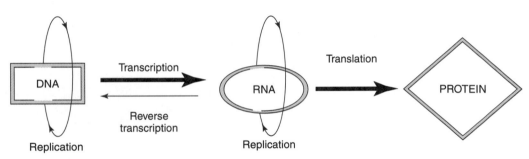

FIGURE 5-26 The central dogma of molecular biology.

Protein has several levels of organizational complexity. The polypeptide chain, its basic structure, is called the *primary structure* of protein. It folds into conformations, the first level of folding producing the *secondary structure*. The second level of folding produces three-dimensional structures that form the *tertiary structure* of protein. Certain proteins consist of aggregates of more than one polypeptide *(multimeric)*. This level of organization is the *quaternary structure* of proteins.

5.3: CENTRAL DOGMA OF MOLECULAR BIOLOGY

Complementary DNA (cDNA)
DNA that is made by reverse-transcribing mRNA into its DNA complement.

The genetic information of the DNA is changed into biological material principally through proteins, according to the *central dogma of molecular biology*. The dogma states that genetic information flow is generally unidirectional from DNA to proteins, except in special cases (figure 5-26). This flow, mediated by transcription (copying of the DNA template by synthesizing the RNA molecule) and translation (synthesis of a polypeptide using the genetic information encoded in an mRNA molecule) and preceded by replication (process of DNA synthesis), can now be reversed in vitro (in the test tube) by scientists. Thus, once a protein is known, the nucleotide sequence in the prescribing DNA strand can be determined and synthesized (the product called *complementary DNA,* or *cDNA*) (5.16).

5.4: EXTRACHROMOSOMAL INHERITANCE

The cell has a number of organelles (2.1.2), but most of the DNA in eukaryotes is found in the chromosomes in the nucleus. Chloroplasts and mitochondria have their own DNA. *Extrachromosomal* or *cytoplasmic DNA* (DNA occurring outside of the nucleus) is not subject to the Mendelian mode of inheritance. Since the male gamete (pollen grain in plants) has virtually no cytoplasm, an offspring always inherits the maternal cytoplasm and along with it any unique hereditary factors carried by the cytoplasmic DNA. Sterility genes have been found in the cytoplasm of plants, including corn (5.9.2). In some horticultural foliage plants, leaf patterns (variegation) are attributed to chloroplasts.

MODULE 2 APPLICATION OF GENETICS: CONVENTIONAL BREEDING

Plant improvement, or breeding, depends on an understanding of genetic principles and concepts, some of which have been discussed in module 1 of this chapter. Plant breeding involves the genetic manipulation of plants at various levels of organization (chapter 2 overview). Manipulation at the whole-plant level is described as *conventional* or *classical breeding.* In this type of plant manipulation, an observed phenotype is assumed to have a genetic basis. (In practice, this assumption is usually tested and confirmed before breeding.) In other words, a flower is purple because a certain gene (or genes) makes it so. It should be recalled that a gene is expressed in an environment, not a vacuum. Breeders also need to know how the gene is inherited.

Classical manipulation entails the assembling of desired genes into a new genetic matrix through crossing or hybridization. By using this technique, the breeder in essence restructures the genomes of the two parental plants by *indirect* means. Unconventional methods of breeding have a more direct approach to restructuring the genome. The DNA of the individual is *directly* manipulated after identifying and isolating the gene or genes responsible for conditioning specific traits. No crossing is involved, and therefore the technology is able to traverse sexual barriers to unite genes between parents that will not be successfully mated. Molecular biotechnology, or *genetic engineering,* as this level of manipulation is called, is discussed in module 3.

5.5: BREEDING SYSTEMS OF FLOWERING PLANTS

Flowering plants may be divided into broad categories according to how they breed, by self-pollination or cross-pollination.

5.5.1 SELF-POLLINATION

Self-pollination is the mating system in which pollen grains from a plant are deposited on the stigma of a flower of the same plant (5.8). Self-pollinated species tend to have a narrow genetic base since very little alien genetic material is incorporated into the original genome. They are true breeding and can remain pure without any special maintenance strategy. Spontaneous mutation (5.1.9) may introduce genetic variation into a true-breeding cultivar. Even though self-pollinated, some degree of outcrossing can occur in such species, ranging from negligible (less than 1 percent) to significant (more than 4 percent). The amount of outcrossing depends on the species and the environmental conditions. Self-pollinated species are naturally inbred but do not suffer the consequences associated with *inbreeding* (5.9.2). Various mechanisms enforce self-pollination in plants. *Cleistogamy* is a condition in which self-pollination occurs before the flower bud opens.

5.5.2 CROSS-POLLINATION

Cross-pollination is a system of mating in which a stigma of a flower is open to receive pollen from other sources (more than 40 percent of the total) than itself (5.8). Cultivars of such species are best described as populations of plants. They do not breed true from seed and are heterozygous and heterogeneous (5.1.3). Instead of inbreeding, cross-pollinated species constantly experience genetic recombination, resulting in the creation of variability. When they are selfed, they experience a loss of vigor *(inbreeding depression)* (5.9.2). Natural mechanisms such as *dioecy* (the occurrence of separate female and male plants in one species) and *self-incompatibility* (the inability of flowers to be fertilized by their own pollen) occur in some species to enforce cross-pollination.

5.6: Biological Variation

As stated previously in this chapter, no two individuals are genetically identical, except for clones or identical twins. This variation is due to the phenomenon or crossing over that occurs during meiosis (5.2.3). Crossing over produces the genetic variation seen in progeny following a cross. Plant breeders depend largely on this source of variation for plant improvement. To make biological variation readily available to scientists, expeditions are sometimes undertaken to regions that are described as *centers of origin* of specific plants to collect variant types of these plants for storage in a germ plasm bank. Sometimes the type of variation needed does not occur naturally or has not yet been discovered. The breeder's last opportunity to obtain this variability is to artificially induce it by using *mutagens* (or mutation-causing factors). Mutation, hence, is the ultimate source of variation (5.10).

> **Center of Origin**
> *A geographical area in which a species is believed to have evolved through natural selection from its ancestors.*

5.6.1 SOURCES OF VARIATION

Without variation, breeding or plant improvement is impossible. The plant breeder has several sources from which to draw variability for a breeding program. These sources are listed and discussed in the following sections.

Adapted Local Cultivars

A plant may have a genotype that includes all desirable genes for yield, quality, disease resistance, and the like, but if it is not adapted to its environment of culture, it will not perform optimally. Breeding programs are hence regionally based; that is, crops are bred in the general region in which they will be grown. You do not breed a plant in the Midwest for use in the Southwest region. To avoid the need to adapt a cultivar to a region of cultivation, it is best to use locally adapted genetic variability (if possible) for breeding. These cultivars include *land races* (unimproved local varieties) and existing cultivars being used (or any older ones not being used).

Recombination

Crossing among diverse individuals creates an opportunity for new types (recombinants) to arise because the practice causes genes to be organized in a new genetic matrix. Hybridization with subsequent recombination also offers opportunity for desirable traits found in different individuals to be assembled in one individual. Breeders therefore frequently employ hybridization to create variation when improving self-pollinated species.

Breeders' Seed

Breeders' seed is derived from recombination and other variation-inducing techniques. Plant breeders usually retain samples of materials that do not make it as commercial cultivars but have certain desirable qualities. These seeds can be incorporated into future breeding programs. They may also have breeding lines that are known sources of specific genes.

Introductions

New genotypes may be imported from other regions or countries. Plant breeders in different parts of the world are actively involved in crop improvements for their regions. They have different breeding objectives because of differences in the environment and consumer preferences. Sometimes breeding objectives may be international in scope. A disease may be prevalent in several regions of the world. Scientists in one region of the world may succeed in breeding for resistance to a disease that is a problem in another region. Cultivars may be imported and utilized in breeding programs in areas where resistance is being sought.

Wild Plant Resources

The process of domestication often results in the "breeding out" of cultivars' undesirable natural protective strategies such as prolonged dormancy, thick seed coat, bitterness, numerous small fruits, and indeterminate growth habit (5.7). These attributes protect the plant in the wild from pests and the uncertainties of the environment. Modern cultivation practices and consumer preferences do not make such natural protection necessary. Plant breeders have discovered that wild relatives are a good source of desirable genes to restore a protective capacity to modern cultivars. Crop expeditions are periodically organized to collect samples of such plants from regions of greatest diversity. These areas are called centers of origin of crops. Vavilov, the great botanist, described such centers of origin of important plants (table 5-1), including China, South Asia, Central Asia, Asia Minor, the Mediterranean, Abyssinia, Central America, and South America. The first six are classified as Old World and the last two New World centers. Some of the important crops associated with these areas are presented in table 5-1. Harlan proposed an alternative to Vavilov's theories (figure 5-27). Sometimes incorporation of materials from wild relatives by the breeding methods of *wide crossing* (5.12) is problematic due to incompatibility. To succeed, a variety of biotechnological methods must be employed (5.15).

> **Harlan's "Noncenters" Theory**
> *States that agriculture did not originate in definite centers but rather many plant species originated at the same time over wide geographic areas.*

Germ Plasm Banks

To facilitate the collection, maintenance, and distribution of plant germ plasm resources for use by scientists, certain organizations and institutions have established *germ plasm banks* for specific crops according to their mandates for operation. The International Agricultural Research Centers usually hold large accessions of the crops they are mandated to research (table 5-2).

Mutation

It has already been stated that mutation is the ultimate source of genetic variation. It can arise spontaneously or be induced (5.10). Because spontaneous mutants occur too infrequently to make them a reliable source of variability for plant improvement, scientists use artificial methods to induce mutations. For plants propagated vegetatively, the benefits of recombination through hybridization are not attainable. New variability must be induced artificially.

5.7: CONVENTIONAL BREEDING

Plant breeding is the science and art of nudging nature (with respect to plants) to the advantage of humans. In the process of *domestication,* plants are introduced from the wild and cultivated under human supervision. While in the wild, plants need to fend for themselves and thus develop a wide variety of natural means of protection. For example, some plants have thorns (on the leaves, stems, or branches), and others store toxic chemicals in various parts to ward off predators. To ensure survival, plants in the wild have mechanisms for dispersal of seeds or other propagules. For example, pods on some pod-bearing plants shatter their seeds when mature and dry to disperse them afar to colonize new grounds. Other plants in the wild have numerous small fruits that are less juicy and fleshy. The large numbers ensure that some

> **Domestication**
> *The process of introducing plants from the wild into cultivation under human supervision.*

Table 5-1 Selected Horticultural Plants Associated with Major Centers of Origin of Cultivated Plants as Described by Vavilov

1. **Chinese Center**
   Crops: Bamboo, radish, eggplant, cucumber, peach, apricot, walnut, and some citrus

2A. **Indian Center (India and Burma)**
   Crops: Gourd, orange, other citrus fruits, mango, and black pepper

2B. **Indo-Malayan Center (Indochina, Malaysia, Java, Borneo, Sumatra, and Philippines)**
   Crops: Giant bamboo, ginger, banana, coconut, nutmeg, and breadfruit

3. **Central Asiatic Center (Afghanistan, Northwest India, Adjacent Soviet Provinces)**
   Crops: Garden pea, mustard, garlic, carrot, onion, basil, pear, almond, and apple

4. **Near-Eastern Center (Asia Minor, Iran, Transcaucasis, and Turkmenistan Highlands)**
   Crops: Poppy, cabbage, lettuce, fig, pomegranate, cherry, hazelnut, and cantaloupe

5. **Mediterranean Center**
   Crops: Beet, parsley, leek, chive, celery, parsnip, rhubarb, thyme, lavender, peppermint, sage, and rosemary

6. **Abyssinian Center (Ethiopia and Somaliland)**
   Crops: Okra and garden cress

7. **South Mexican and Central American Center (South Mexico, Guatemala, El Salvador, Honduras, Nicaragua, and Costa Rica)**
   Crops: Corn, common bean, lima bean, sweet potato, pepper, papaya, cashew, and cherry tomato

8A. **South American Center (Peru, Ecuador, and Bolivia)**
   Crops: Potato, tomato, pumpkin, marigold, and guava

8B. **Chiloe Center (South Chile)**
   Crops: White potato and strawberry

8C. **Brazilian-Paraguayan Center**
   Crops: Passion fruit, pineapple, and Brazil nut

---

of the fruits will be distributed to other parts of the area and thereby provide an opportunity for the species to continue. Some desert plants germinate only after a heavy rain that washes away a germination-inhibitory substance. The requirement of a soaking rain in the desert also ensures that, once germinated, the seedling will have adequate moisture to sustain it through the initial critical period of growth before being able to use its own roots to forage for nutrients and moisture.

Once out of the wild, plants gradually shed their natural protective features, not voluntarily but because humans, in the domestication process, find these traits a nuisance at best. In fact, humans deliberately select against some of these natural protective traits through breeding programs. Humans prefer sweeter and juicier fruits, so breeders select against the bitter chemicals some plants use for protection in the wild. To be juicier, fruits often have to be bigger, so selection is directed against numerous small fruits in favor of fewer and bigger fruits. Instead of waiting endlessly for a soaking rain to induce germination, modern crop producers want seeds to germinate within several days and with predictability and uniformity. As such, seed dormancy is often eliminated through breeding. Another survival mechanism in the wild is a prolonged fruiting period of plants. In the wild, some plants simultaneously carry flowers, immature pods, and mature pods. Productivity of the crop is thus extended over a long period, ensuring that, in the event of an adverse weather condition during the crop growing period, the plant has the opportunity to produce new sets of flowers. However, in modern crop production, the method of cultivation of many crops and the economics of pro-

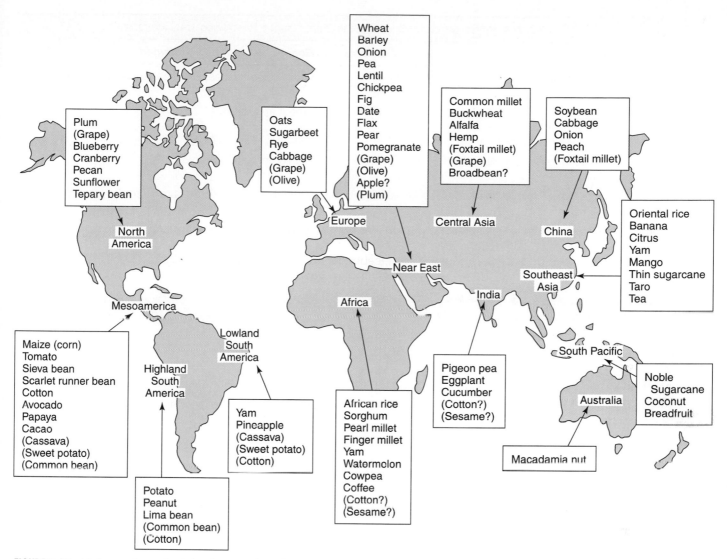

FIGURE 5-27 Major centers of origin of crops described by Harlan recognizes less definite centers. He proposed the "noncenters" theory of origin of crops.

duction demand that crops mature at the same time for mechanized harvesting. The negative side of domestication is that, as natural protection is removed from domesticated plants, they become increasingly vulnerable to pests. Therefore, to produce a good yield, cultivated crops need to be artificially protected from pests by spraying chemicals. Chemicals not only pollute the environment but also contaminate horticultural produce, thus posing a health risk to consumers. Plant breeders seek to create plants that are more productive in terms of yield and high quality. In floriculture, aesthetic value and variety are important.

5.7.1 THE UNDERLYING CONCEPTS

An understanding of the following equation will help in appreciating the strategies adopted by plant breeders:

$$P = G + E$$

where P refers to phenotype, G to genotype (the total genetic constitution), and E to the total environment (within which the genotype is expressed).

Genes, as previously explained, develop in a given environment to produce their effect. That is, as the formula suggests, what you see (phenotype) is dictated by *nature* (the kinds of

Table 5-2 The International Agricultural Research Centers and the Plant Genetic Resources They Maintain

Center and Location	Acronym	Genetic Resources Maintained
Centro Internacional de Agricultura Tropical: Cali, Colombia	CIAT	Grain legumes, cassava, rice, and forage grasses
Centro Internacional de Majoramiento de Mais y Trigo (International Maize and Wheat Improvement Center): El Batan, Mexico	CIMMYT	Wheat and maize
Centro Internacional de la Papa (International Potato Center): Lima, Peru	CIP	Potato and sweet potato
International Board for Plant Genetic Resources: Rome, Italy	IBPGR	Coordinates genetic resource centers (gene banks)
International Center for Agricultural Research of Dry Areas: Allepo, Syria	ICARDA	Wheat, barley, fava bean, lentil, chickpea, pigeon pea, and forage grasses
International Council for Research in Agroforestry: Nairobi, Kenya	ICRAF	
International Crops Research Institute for the Semi-Arid Tropics: Hyderabad, India	ICRISAT	Sorghum, millet, chickpea, pigeon pea, and peanut
International Food Policy Research Institute: Washington, D.C.	IFPRI	
International Irrigation Management Institute: Colombo, Sri Lanka	IIMI	
International Institute of Tropical Agriculture: Ibadan, Nigeria	IITA	Rice, potato, sweet potato, peanut, cowpea (blackeye pea), soybean, maize, and cassava
International Livestock Center for Africa: Addis Ababa, Ethiopia	ILCA	
International Laboratory for Research on Animal Diseases: Nairobi, Kenya	ILRAD	
International Network for the Improvement of Banana and Plantain: Montferrier-sur-lez, France	INIBAP	Banana and plantain
International Rice Research Institute: Los Banos, Philippines	IRRI	Rice
International Service for National Agricultural Research: The Hague, Netherlands	ISNAR	
West Africa Rice Development Association Bouake, Cote d'Ivoire	WARDA	Rice

genes) and *nurture* (the environment in which the genes are expressed). If you do not like what you see and desire to change it, you may alter the genetic constitution, change the environment, or both. Several factors influence the choice of the course of action taken. It is important to note that a change in *G* is *permanent*, whereas a change in *E* is *temporary*. For example, you can increase the yield *(P)* of tomatoes by planting a new, higher-yielding cultivar *(G)* or by applying fertilizers *(E)* to your old cultivar. A high-yielding cultivar has bred into it genes that promote high yield (has yield genes, you might say). It will always have the potential to be a high yielder, unless the genotype is altered (spontaneously or deliberately). On the other hand, old cultivars will produce much higher yields if fertilizers and other growth-enhancing factors are provided. In practice, however, a high-yielding cultivar also requires a good environment (including fertilizers), as all plants do. The critical difference is that whereas the old cultivar under fertilization may produce say 10 pounds (4.5 kilograms) per unit area, your new, improved cultivar could produce say 25 pounds (11.25 kilograms) per unit area under similar conditions. The other point to note is that there are often several ways to change *G* in order to change *P*. Using the tomato example, you can increase *P* by planting a cultivar that produces 50 small fruits or one that produces 20 bigger fruits. Small size may be preferred by consumers in one situation, whereas bigger fruits may be preferred in another. Therefore, how *G* is changed should take other factors into account.

Phenotype can be any outcome desired (relating to, for example, height, color, shape, size, number of plant parts, disease resistance, response to fertilizer, drought resistance, and so forth). The role of the environment is always critical in breeding efforts. Since genes may not be expressed in certain genetic backgrounds or in certain environments, breeding efforts tend to be on a regional (or even narrower environmental restriction) basis. The highest yielding cultivar in one region may be the lowest yielding in another. Certain cultivars have wider adaptation than others and are able to perform well across several locations. Another reason breeding programs tend to be regionalized is because what may be a problem in one area may not be a priority problem in another for the same crop. Also, it does not make much sense to establish a breeding program for a crop in an area in which it cannot be grown. Therefore, breeding programs are often associated with the areas of production of the plants being improved.

Plant breeders focus attention on genetic alteration (permanent, heritable changes) of plants. *Agronomists* concentrate more on determining the proper environmental conditions for optimal plant performance. There are limits to manipulating either *G* or *E*. After a certain level of environmental input, further provision of production enhancers becomes cost-ineffective (diminishing returns on investment). In fact, yield or performance might decline because excessive inputs might become detrimental to plant growth and development. For example, an overfertilized plant may grow to be too leafy, very tall, succulent, prone to lodging and insect attack, and late maturing and sometimes even have poor produce quality.

In the case of genetic manipulation, a similar situation is encountered where rapid gains are made initially, after which gains from further genetic improvement slow down or level off. Sometimes limitations to genetic improvement are imposed by the technology and techniques of the day. Modern technology has enabled circumvention of some of the obstacles to genetic manipulation. Another obstacle to genetic manipulation is genetic linkage, the physical association of genes on a chromosome (5.1.6). A tight linkage is difficult to break, but sometimes some good and bad traits are linked such that improving the good trait simultaneously enhances the bad one. Also, some traits may be associated or correlated in a negative way so that as one trait increases in expression the other simultaneously decreases. The latter situation is difficult when increased expression of both traits is desired for good plant growth or produce quality.

5.7.2 DETERMINING BREEDING OBJECTIVES

Even though plant improvement is an age-old activity, modern plant breeding depends on a good understanding of the principles of genetics and allied sciences, as has already been stated. In addition to special techniques and methodologies (which are based on scientific principles), the breeder needs to understand the environment in which the work is being undertaken, as well as the needs of producers and consumers. If producers need a short determinate cultivar to facilitate production, the breeder has to address this need. If consumers prefer sweeter fruits, efforts should be focused in that direction by making taste a breeding objective. Producers and consumers do not have to dictate breeding objectives all of the time. Breeders can also take the initiative to develop new cultivars and introduce them to producers and consumers. By following the trends in the horticultural industry, breeders can organize breeding programs to meet the current and anticipated needs of producers. Breeders and engineers who design crop production equipment should work together in developing practical equipment. Equipment manufacturers focus on making the work of the producer more efficient and easier. However, machines have to be designed to be gentle on plants or breeders need to breed tougher plants to withstand the action of those machines. In this age of high environmental awareness, there is an ever-increasing pressure to limit the use of pesticides in crop production, meaning that crops must have some natural means of defense against pests. Unfortunately, some of these means of natural protection were removed through breeding programs to meet certain other objectives.

Sample Breeding Objectives of Selected Horticultural Plants

The following examples are designed to show breeding objectives in a few horticultural plants. Common breeding objectives are *yield, quality,* and *disease resistance.* In addition to these general goals, certain crops have unique breeding needs to make them perform optimally or meet consumer approval. The following are examples of breeding goals pursued by breeders working to improve the selected crops:

1. *Onion (Allium cepa* L.). Bulb quality, a general goal in onion improvement, is determined by size, shape, color, firmness, soluble solids, pungency, and dormancy. Phenotypic correlations indicate a relationship among large size, softness, low pungency, and poor storage. That is, smaller onions are firmer, more pungent, and more durable in storage. Bulbs with high carbohydrate concentration (soluble solids) are preferred by industries that produce onion chips and powdered onion for seasoning. Consumers differ in their preference for bulb shape, color, and pungency. Dormancy in onions is critical so that they may be stored for a long time without sprouting.

    The other major breeding goal for onion is disease and insect resistance. Onions are plagued by a host of diseases, resistance to some of which has been identified and successfully incorporated into certain commercial cultivars. Important diseases include *Fusarium* basal rot *(Fusarium oxysporum),* pink rot *(Pyrenochaeta terrestris),* white rot *(Sclerotium cepivorum),* and downy mildew *(Peronospora destructor).* Sources of resistance to some of these diseases (e.g., downy mildew, white rot, and neck rot) have been identified. Commercial cultivars with *Fusarium* basal rot and pink rot resistance have been developed. Insect pests of economic importance are thrips *(Thrips tabaci)* and onion maggot *(Delia antiqua).*

    Introgression of disease resistance via wide crosses is promising for some of these pest problems. However, sterility of progeny of such crosses hampers progress.

2. *Cabbage (Brassica oleracea* L.). Cruciferous plants contain certain compounds *(glucosinolates)* whose hydrolytic products are harmful to humans and livestock. Upon rupturing, glucosinolates in cells become hydrolyzed by myrosinase to produce a variety of products, including isothiocyanates (mustard oils), thiocyanates, and nitriles. Glucosinolates have been linked with a certain goiter *("cabbage goiter")* due to its antithyroid properties. The products of hydrolysis of glucosinolates are involved in the production of the pungent and bitter flavor associated with *Brassica* vegetables. Breeding for low-level glucosinolates is a goal in cabbage improvement.

    Insect and disease resistance goals in cabbage include breeding for resistance to cabbage worm *(Artogeia rapae* L.), cabbage looper *(Trichoplusia* spp.), and diamondback moth *(Plutella xylostella* L.). This pest trio is described as the *lepidopterous complex.* Sources of resistance to these insects have been identified. Important pathogenic diseases include powdery mildew *(Erysiphe cruciferarum* Opiz), *Fusarium* yellow *(F. oxysporum* Schlechtend), and downy mildew *(Peronospora parasitica).*

3. *Cucumber (Cucumis sativus).* Hybrids are important to cucumber cultivars and thus efforts are devoted to the development of inbred lines. In terms of disease and insect resistance, genetic resistance to viral, fungal, and bacterial diseases including angular leaf spot, bacterial wilt, anthracnose, downy mildew, and fusarium wilt has been identified and incorporated into certain cultivars. Fruit rot caused by *Rhizoctonia solani* has received attention and needs further research into techniques of innoculation and discovering additional sources of resistance.

    Resistance to pickle worm *(Diaphania nitidalis)* needs further attention. Similarly, resistance to cucumber beetles *(Diabrotica undecimpunitata* and *Acalymma vittatum),* aphids *(Aphis gossypii),* leaf miners *(Lyriomyza sativae),* and nematodes is being sought. Quality attributes include stem color, presence or absence of spines, fruit firmness, and fruit shape. Consumer preference differs from one culture to another. Bitter principles (due to *cucurbitacins)* are being bred out of cultivars.

5.7.3 IMPORTANCE OF VARIATION

Without heritable variation, it is not possible to effect changes by breeding. This fact cannot be overemphasized. If you desire to make a current cultivar resistant to a particular disease, you assume that somewhere in the *gene pool* for the crop, there is a gene (or genes) that conditions the disease. If not, no matter how hard you try, you cannot accomplish your goal. However, before you give up, you should realize, as previously stated, that while recombination is the major source of biological variation, the ultimate source of variation is mutation. So, when the desired variation does not occur naturally, it may be possible to induce it artificially (*induced mutagenesis,* or *mutation breeding*) (5.10). However, this method of breeding is risky and has limited success. Most breeding programs depend on existing natural variation. Such variation is available for major crops in existing germ plasm banks. Various researchers may have limited amounts of variation in their immediate possession for the plant of interest, but administrators of germ plasm banks routinely sponsor germ plasm collection expeditions to collect wild relatives and other natural plant variations. These varieties are kept in cold storage and maintained for distribution to scientists upon request.

5.7.4 IMPORTANCE OF HERITABILITY OF A TRAIT

So far, we have learned that the plant breeder should be familiar with the botany of the plant, the environment in which breeding will be undertaken, and whether the variation for the trait in question exists; he or she should also have a breeding objective. Before actual breeding can be initiated, however, one crucial issue remains to be determined. Is the variation for the trait heritable or due to environmental causes? Plants receiving different nurturing (e.g., different amounts of light, moisture, and fertility) may have the same genotype but show variations that are purely environmental. The fact is, unless the variation available for the trait is heritable, the breeder is merely spinning his or her wheels, since no progress can be made in selective breeding without the desired genetics. The key to making progress in breeding is being able to select from among a large population of plants the right ones (those with the desired combination of genes) at each step in the usually long breeding program. This art is called the breeder's eye, but the science component of breeding is increasingly diminishing the guesswork by helping breeders to identify recombinant plants more accurately and readily.

As discussed earlier in this chapter, the number of genes that control a trait influences the degree to which the environment impacts its expression. Quantitatively inherited traits are more difficult to manipulate and are hence handled differently. Breeders adopt a statistical procedure to determine the proportion of the phenotypic variation that is attributable to genetic causes. The relative importance of genetic and environmental factors provides an index called *heritability of a trait (H^2)*. In this form, the index is also called *broad sense heritability,* and it measures the degree to which *phenotypic variance (V_P)* is due to variation in genetic factors as pertains to a specific population and specific environment. It does not measure the proportion of the total phenotype attributed to genetic factors but rather estimates the proportion of observed variation in the phenotype that is attributed to genetic factors vis-à-vis environmental factors. Mathematically, phenotypic variance is expressed as

$$V_P = V_G + V_E + V_{GE}$$

where V_G refers to genetic variance, V_E to environmental variance, and V_{GE} to the interaction between the two. The interaction is usually negligible and therefore frequently omitted. Broad-sense heritability is therefore simplified to

$$H^2 = V_G/V_P$$

H^2 values range between 0.0 and 1.0 (0 to 100 percent). The closer the value to 1.0, the lesser the impact of the environmental conditions on the phenotypic variance in the population. H^2 estimates are not very accurate. A more accurate estimate is provided by the *narrow sense heritability (h^2)*. Plant breeders rely on this estimate in making decisions in plant selection,

> **Heritability**
> The degree of phenotypic expression of a trait that is under genetic control.

since h^2 measures this *potential response* to selection. That is, a high h^2 indicates that breeding for the trait is likely to be successful since the environment will not mask its phenotypic expression. The increased accuracy derives from the partitioning of the *genetic variance* (V_G) into its components:

$$V_G = V_A + V_D + V_I$$

where V_A refers to *additive variance* (the basis of continuous variation), V_D to *dominance variance* (deviation from additivity as a result of phenotypic expression of a heterozygote not being exactly midway between the two homozygote parents), and V_I to *interactive variance* (due to epistasis at two or more loci). Mathematically,

$$h^2 = V_A/V_P$$
$$= V_A/V_G + V_E$$

or $V_A/V_A + V_D + V_E$, since V_I is negligible.

The strategy adopted to improve a cultivar depends on the nature of the improvement desired. If a plant cultivar is highly productive but needs improvement in only one aspect (e.g., disease resistance or protein quality), a different approach is required than when a genotype is being manipulated for complex traits such as yield.

5.7.5 CROSSING (HYBRIDIZATION)

Crossing is one of the most common operations in a breeding program involving flowering plants. This process is required to transfer the desired gene or genes from the parent (source) to the recipient. Certain hereditary factors may occur in the cytoplasm. As previously indicated, pollen grains are practically void of cytoplasmic material. Therefore, when a breeding program is designed to improve a high-yielding, high-quality cultivar in only one specific way (such as resistance to a particular disease), the breeder wants to keep the genes of the desired cultivar as intact as possible while incorporating only the desired gene or genes from the donor. In this case, the established cultivar is used as the maternal parent so that the desirable genes in the cytoplasm are not lost.

When crossing self-pollinated plants, it is critical that the recipient (female plant) be pollinated at the right stage (in the bud stage). A fully mature (fully open) flower will have already been self-pollinated and is of no use. Breeders of self-pollinated species need to perform a procedure called *emasculation*, whereby one plant is turned into a female by removing all of the male parts (anthers) (figure 5-28). This process is often delicate and painstaking. Once emasculated, the pollen from the other parent (male) may be deposited on the stigma in a variety of ways, depending on the breeder's preference and experience with the plant, as well as the floral characteristics. To identify a cross, the breeder should mark the flower by tying a tag around the base of the pedicel.

The success of crossing is variable and depends on the operator's experience, the species, and the environmental conditions. How certain can the breeder be that the cross was truly a success? Breeders sometimes use *genetic markers* to ensure the fidelity of a cross. A genetic marker is a gene with readily identifiable expression associated with an outcome that is more difficult to observe. When the marker is observed, the other event is assumed to have also occurred. In a cross, if a plant with purple flowers (dominant trait) is used as a male and the female parent has white flowers (recessive), then the F_1, to be a successful cross, must have purple flowers. Any seed harvested as a putative cross but producing white flowers must have come from self-pollination.

5.7.6 SELECTION

Selection is a critical activity in a breeding program, because it determines the efficiency of the program. It is based on a breeder's ability to discriminate among often large segregating populations to identify and select recombinants or individuals that have the desired combination of genes from the parents used in the cross. If the trait being improved is simply in-

---

Emasculation
The removal of anthers from a bud or flower before pollen is shed.

Genetic Marker
An allele tracked during a genetic study.

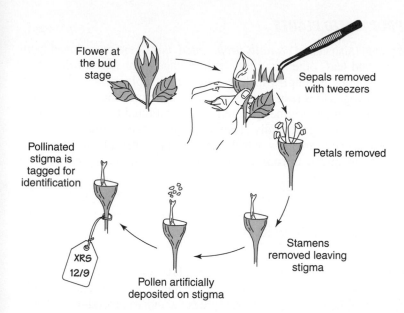

Flower at the bud stage

Sepals removed with tweezers

Petals removed

Stamens removed leaving stigma

Pollen artificially deposited on stigma

Pollinated stigma is tagged for identification

XRS 12/9

FIGURE 5-28 Steps in making a cross. These steps for emasculation are not necessary if techniques such as male sterility are involved in the crossing programs.

herited, selection for it is often easy. On the other hand, if the trait is quantitatively inherited, the breeder often needs to select a large sample to advance to the next generation to ensure that the desired recombinants are included. For increased selection efficiency, breeders rely on genetic markers that are associated with the trait being improved.

5.8: How Breeding Is Done

The science of plant breeding has developed to the extent that general methodologies are available for a wide variety of breeding applications. There are general guidelines for all breeding work and some for specific problems. One of the bases for classifying breeding methods is the type of natural mating process *(mating system)*.

5.8.1 SELF-POLLINATED SPECIES

Some plants are self-pollinated (i.e., the flower uses its own pollen for pollination) (5.5.1). Many annual vegetables and flowering species are self-pollinated. Because they use their own pollen, these plants tend to be highly inbred and pure, in the sense that their progeny is very much like the parent. The seeds from these plants produce uniform plants. Since such plants have a tendency to remain pure, breeding methods employed to improve them must be geared toward maintaining this characteristic. Examples are pea, lima bean, and chickpea.

5.8.2 CROSS-POLLINATED SPECIES

Cross-pollination is the process of transferring pollen grains from one plant and depositing them on the stigma of a different one (5.5.2). Cross-pollinated (also called *outcrossed* or *open-pollinated*) plants are heterozygous by nature. They produce progeny that are *heterogeneous* (nonuniform). Heterozygosity confers an advantage on cross-pollinated plants because it gives them vigor. Cross-pollinated plants are more difficult to manage in production and are not valuable in markets that require uniformity of product. A field may have plants that vary widely not only in physical appearance but also in physiological responses and product quality. Some plants may mature earlier than others in the population, making it difficult to harvest a field mechanically. On the other hand, in the case of environmental adversity, there is a built-in insurance against total loss, since the plant population consists of individuals with a wide range of adaptations and resistance. Examples of cross-pollinated horticultural plants are corn, cabbage, and spinach.

5.8.3 *VEGETATIVELY PROPAGATED PLANTS*

Vegetatively propagated plants (also called *clonally propagated plants*) are reproduced from plant parts (e.g., root, stem, and leaf) other than the seed. Foliage plants, or nonflowering plants, are propagated vegetatively. Examples include fruit trees (apple, pear, and peach), bulbs (tulip and daffodil), and vegetables (asparagus). A unique characteristic of such plants is that, notwithstanding how they were originally genetically constituted (i.e., homozygous or heterozygous), they *always* produce genetically identical products called clones. This outcome results because in the absence of meiosis, which creates variation, replicates from somatic material are genetically identical. Certain species can be propagated by either seed (sexual) or vegetative (asexual) methods.

5.9: BREEDING METHODS

As previously indicated, the breeding system of the species plays a major role in deciding which method of breeding is used. A number of standard breeding procedures can be modified to suit particular situations. The following are examples of standard procedures.

5.9.1 *SELF-POLLINATED SPECIES*

Mass Selection

Mass selection is the oldest of all breeding methods and the easiest to conduct; it is based on phenotype. Selection is conducted by growers unknowingly when they select and save seeds from the most desirable fruit for planting the next season. When conducted more formally, plants are selected based on appearance (phenotype) according to breeding objectives and advanced from one generation to the next as a *bulk* (figure 5-29). The cultivar released from mass selection is a heterogeneous mixture of types that may individually be homozygous. The method is suitable for breeding *horizontal resistance* to diseases in plants (chapter 6).

Pedigree Selection

The *pedigree selection* method of plant breeding is conducted after creating variability by crossing two parents. The final product is descended from a single plant. Records are kept such that the cultivar can be traced to its original parents (figure 5-30). The pedigree breed-

FIGURE 5-29 Summary of a bulk breeding method.

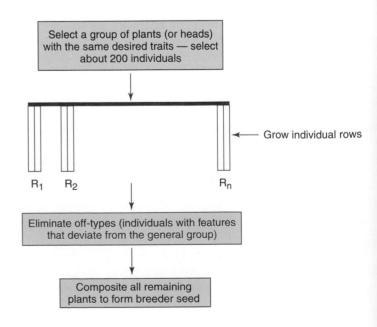

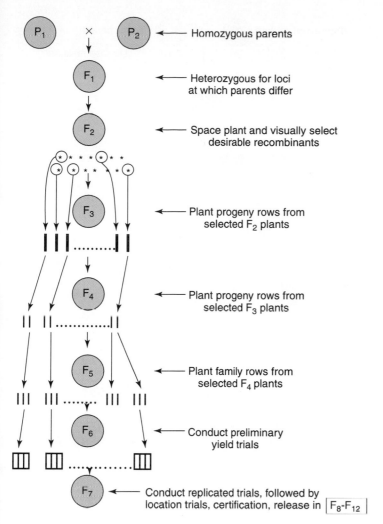

FIGURE 5-30 Summary of a pedigree breeding method.

P₁ × P₂ ← Homozygous parents

F₁ ← Heterozygous for loci at which parents differ

F₂ ← Space plant and visually select desirable recombinants

F₃ ← Plant progeny rows from selected F₂ plants

F₄ ← Plant progeny rows from selected F₃ plants

F₅ ← Plant family rows from selected F₄ plants

F₆ ← Conduct preliminary yield trials

F₇ ← Conduct replicated trials, followed by location trials, certification, release in F_8-F_{12}

ing method is used to breed self-pollinated species. Self-pollinated species are true breeding since they only use pollen from the same flower or plant.

Backcrossing

Backcrossing is used when the current cultivar is highly desirable, except for a deficiency in one or a few specific aspects (e.g., lack of resistance to a particular disease). In this case, the breeder's strategy is to keep the genotype of the desirable cultivar as intact as possible while adding on the genes from elsewhere for disease resistance (or whatever trait would be in question). This method is employed to improve self-pollinated species (figure 5-31).

Other methods of breeding self-pollinated species include bulk and hybrid (5.9.2). Those previously described may be modified as the occasion demands. Some of these methods are applicable to breeding cross-pollinated species.

5.9.2 CROSS-POLLINATED SPECIES

Mass selection, *recurrent selection, synthetics,* and other methods of breeding may be applied to cross-pollinated species. However, one of the most noted methods of breeding this category of plants is *hybrid breeding.* To understand and use this method effectively, several principles and concepts should be understood.

Hybrid Vigor and Inbreeding Depression

Hybrid vigor, or *heterosis,* is the phenomenon of the expression of a trait by the product of a cross, over and above that of the parent. Hybrid vigor occurs when a cross between two parents

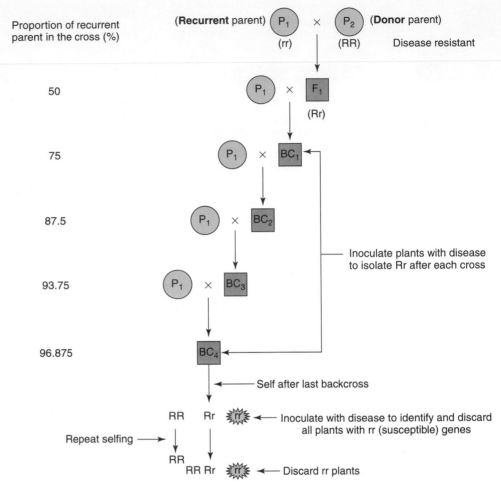

Proportion of recurrent parent in the cross (%)	
	(Recurrent parent) P₁ × P₂ **(Donor** parent)
	(rr) (RR) Disease resistant

Proportion of recurrent parent in the cross (%)

(Recurrent parent) P₁ × P₂ **(Donor** parent)

(rr)   (RR)   Disease resistant

50   P₁ × F₁

(Rr)

75   P₁ × BC₁

87.5   P₁ × BC₂

Inoculate plants with disease to isolate Rr after each cross

93.75   P₁ × BC₃

96.875   BC₄

Self after last backcross

RR   Rr   rr ← Inoculate with disease to identify and discard all plants with rr (susceptible) genes

Repeat selfing →

RR

RR Rr   rr ← Discard rr plants

FIGURE 5-31   Summary of a backcross breeding method.

Hybrid Vigor
An increase in vigor or growth of a hybrid progeny in relation to the average of the parents.

produces an expression of a specific trait over and above its expression in either parent. For example, if the parent P_1 yields 2 kilograms and parent P_2 yields 4 kilograms, a cross of $P_1 \times P_2$ might yield 5 kilograms, thus exhibiting hybrid vigor. By crossing diverse parents, the F_1 hybrid is heterozygous at many loci, giving it a broad genetic base. Hybrid vigor occurs when genetically divergent parents (unrelated) are crossed and increases as the two parents increase in divergence (within limits). Heterosis is greatest in the F_1 generation and declines progressively with repeated selfing. Selfing, as previously discussed, has the genetic consequence of making loci homozygous. Heterozygosity is reduced by 50 percent from the previous generation (5.1.3).

Inbreeding is the crossing of parents that are related, the extreme mating system being selfing or self-pollination. It has the opposite effect of heterosis, resulting in a diminished capacity of expression of plant characteristics by the hybrid. This loss of vigor is described as inbreeding depression. With repeated selfing, the genotypes eventually become homozygous at all loci, at which stage the plant is said to be highly inbred, or pure.

Species that are naturally self-pollinating do not suffer the consequences of inbreeding. On the contrary, outcrossed species lose vigor when inbred. However, when two unrelated inbred lines of a cross-pollinated species are crossed, hybrid vigor results. Heterosis is exploited in breeding horticultural plants, the observed hybrid vigor usually being greater in cross-pollinated plants.

The genetic basis of heterosis (and inbreeding depression) has not been completely elucidated. The *dominance theory* explains the loss of vigor as a consequence of the manifestation of the genetic load in an individual. Whereas deleterious recessive alleles remain in the heterozygous state, their expression is masked by the dominant alleles. Upon selfing, these recessive alleles become homozygous, unleashing their lethal effects and consequently causing a re-

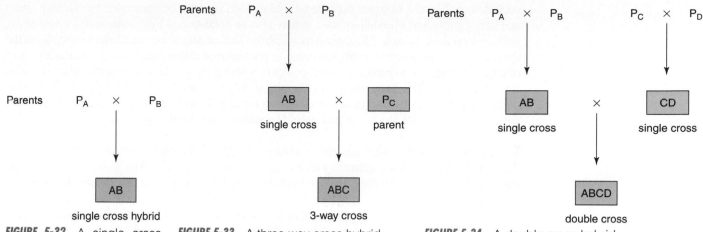

FIGURE 5-32 A single cross hybrid.

FIGURE 5-33 A three-way cross hybrid.

FIGURE 5-34 A double cross hybrid.

duction in vigor. After cycles of inbreeding, vigor is dramatically restored when two divergent inbreds are crossed. The divergence between the inbreds makes them complement each other in a cross, since the resulting hybrid will have a combination of all dominant alleles.

$$AAbbCC\ (P_1) \times aaBBcc\ (P_2) = AaBbCc\ (F_1)$$

Whereas P_1 has the benefit of two dominant alleles and P_2 only one, F_1 benefits from all three dominant alleles. If the number of dominant alleles is responsible for the increased vigor of F_1 generation, then it follows that if one is able to develop an individual that is homozygous dominant at all loci, the resultant hybrid would be at least as vigorous, or perhaps more, than the heterozygote hybrid (i.e., *AABBCC* is superior to *AaBbCc*). Such homozygous types are difficult to obtain because of undesirable linkages between deleterious and beneficial alleles.

The *overdominance* theory attributes hybrid vigor to heterozygosity itself. Heterozygosity is believed to have advantages over homozygosity (i.e., *AA, aa* < *Aa*). Physiologically, both dominant and recessive alleles contribute to the expression of the trait. Further, the two different alleles could interact, as in the case of incomplete dominance.

Types of Hybrids

There are three basic types of hybrids, with several modifications. The *single cross,* involving two diverse inbred lines, is the most commonly produced hybrid (figure 5-32). A two-parent cross produces the greatest amount of heterosis along with high yields. Products of single crosses have maximum uniformity for seed maturity and other important plant growth and productivity characteristics. The limitation of this cross is that the hybrid crop is produced as a single inbred line (as female parent) that is usually lower in vigor (than other female parents in other crosses) and produces lower seed yields. A single cross may be modified by replacing the female with a less-inbred line (less expensive to produce) that is more productive. To accomplish this cross, two closely related inbred lines are first crossed and the resulting F_1 generation crossed with the other parent (as male).

The second type of cross is the *three-way cross* (figure 5-33). It involves three divergent inbred lines. A single cross is first made between two lines, and the F_1 crossed to a third parent. Similar to the single cross, a modified three-way cross can be made. The third type of cross is the *double cross,* which involves four unrelated inbred lines. The double cross is a product of crossing two single crosses (figure 5-34).

> **Hybrid**
> The offspring of two parents that differ in one or more inherited traits.

Using Male Sterility in Hybrid Breeding

What Is Male Sterility?   Male sterility is a condition in certain plants manifested in a variety of phenotypes including pollen abortion, lack of stamens, and malformed stamens. Plants with such defects become obligate outbreeders. Male sterility may be controlled by *nuclear genes* and hence subject to Mendelian laws. It may also be conditioned by *cytoplasmic factors* or *cytoplasmic-genetic factors*. The degree to which the abnormalities are manifested depends on the species and environmental conditions such as photoperiod and temperature. Male sterility is of practical application in breeding plant species in which it occurs. It is used to eliminate the need for emasculation. The three categories of male sterility described have pros and cons. To be used in breeding, there must be a means of restoring fertility. To use any of these genetic-based systems, the sterility and fertility genes should be strategically bred into parents.

Genetic Male Sterility   Genetic male sterility is under the control of nuclear genes in which the alleles for sterility are generally recessive to alleles for fertility. The genotypes of the various conditions and how they are used are described in figure 5-35. The male sterile plants *(msms)* are maintained by crossing with male fertile plants *(Msms* or *MsMs)*. The problem with using this system in breeding is that it is very difficult to eliminate the undesired male fertile plants from the male sterile ones in the female population, before either harvesting or sorting harvested seed.

Cytoplasmic-Genetic Male Sterility   Three factors are involved in cytoplasmic-genetic male sterility—the normal cytoplasm *(N)*, the male sterile cytoplasm *(S)*, and the fertility restorer *(Rf, rf)*. The restorer genes are nuclear in origin. The genotypes of male fertile and male sterile plants are shown in figure 5-36. In using this system, one has to bear in mind that the cytoplasm of the hybrids is entirely maternal in origin.

Male Sterility in Hybrid Breeding   The use of male sterility in breeding (using corn as an example) is summarized in figure 5-37. It should be noted that the particular method adopted depends on the species. The male-sterile lines are designated as *A lines,* the maintainer lines are called *B lines,* and the fertility restorer lines are called *R lines.*

Hybrids are created on the principle that heterozygosity confers vigor *(hybrid vigor)* on a plant. They grow faster and have greater vigor and frequently yield very well. They are uniform in the expression of their traits. Hybrid varieties are commonly used in growing certain crops such as corn. Hybrid seeds are difficult to produce and hence cost more than nonhybrid seeds. Since selfing reduces heterozygosity, the advantage of heterozygosity is maximum in the F_1 seed and decreases rapidly with selfing. The grower should not save seed from the current year's crop for planting in the next. A fresh supply of hybrid seeds should be purchased annually from reputable seed growers.

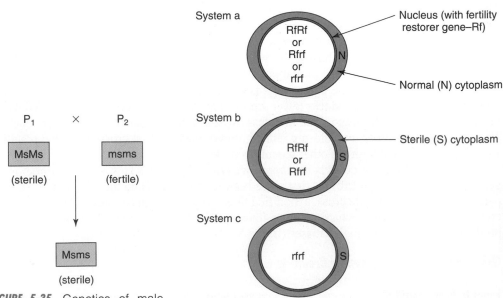

P₁ × P₂

MsMs   msms
(sterile)   (fertile)

Msms
(sterile)

FIGURE 5-35 Genetics of male sterility in plants.

System a
RfRf or Rfrf or rfrf
Nucleus (with fertility restorer gene–Rf)
Normal (N) cytoplasm

System b
RfRf or Rfrf
Sterile (S) cytoplasm

System c
rfrf

FIGURE 5-36 Cytoplasmic male sterility in plants.

Examples of Hybrid Programs

Commercially produced hybrid seed is available for both self-pollinated and cross-pollinated species. Ornamentals with hybrid seeds include petunia, pansy, impatiens, zinnia, snapdragon, begonia, cyclamen, marigold, and African violet. Vegetable crops with commercial hybrids include tomato, sweet corn, onion, pepper, and others.

In onions, large *general combining ability* and *specific combining ability* (discussed next) for yield and its components, bulb maturity, firmness, and storage ability have been reported. Both cytoplasmic male sterility type S and T (CMS-S and CMS-T) occur in onions. However, CMS-T is more difficult to use successfully because of its complex inheritance and the common occurrence of the restorer alleles. To improve the seed yield of hybrids, many of the U.S. hybrids are produced by three-way crosses.

Cytoplasmic male sterility occurs in chives and is used to aid hybridization programs. Chives are prone to severe inbreeding depression. Male lines used in F_1 hybrids should not exceed I_3 (third-generation inbreds). Fertility restoration is not an important consideration in chive hybrid programs since the commercial or harvested product is the vegetative part (leaves).

Most calabrese cultivars of broccoli are F_1 hybrids. Hybrids produce uniform produce and mature early. In brussels sprout breeding, uniformity of product is important, and consequently F_1 hybrids have been preferred over three-way and double crosses. However, a high yield of uniform products has been achieved by using near isogenic lines in three-way and double crosses.

Cabbage commercial hybrid seed production makes use of the occurrence of self-incompatibility. Single cross hybrids produce uniform products and are easier to produce. To

A: single cross

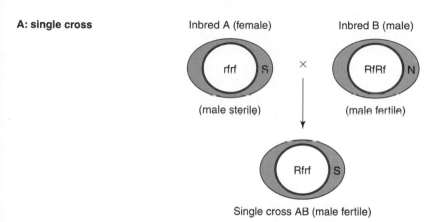

B: double cross

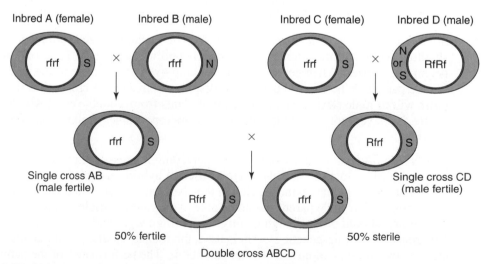

FIGURE 5-37 Using cytoplasmic male sterility in corn hybrid breeding.

use double crosses, four homozygous incompatible inbred lines must be used as follows: $(S_1S_1 \times S_2S_2) \times (S_3S_3 \times S_4S_4)$. *Topcrossing* is the most common method of hybrid production of cabbage in the United States. A single self-incompatible parent is used as female, and a good open-pollinated cultivar serves as the pollen source.

Cauliflower hybrid breeding faces a major problem of "pollution" of the product as a result of the selfing of parental plants (or sib mating within parental lines). This event produces a significant amount of off types when hybrid seeds are planted. The undesired matings occur as a result of an ineffective or weak self-incompatibility system. Further, the incidence of off types increases when time of flowering between parents to be crossed is nonsynchronous.

Most commercially produced cucumbers in the United States are hybrids. Since cucumber is cross-pollinated (and also self-incompatible), the production of inbred lines and identification of best hybrid combination constitutes a major aspect of hybrid breeding in cucumber. Inbreeding depression is not a significant problem in producing inbred lines. Corn is one of the crops in which hybrids have been used for a long period of time.

Conditions for Economic Production of Hybrids

Conditions for a viable commercial hybrid seed enterprise are both biological and economic. If hybrid seed cannot be produced reliably and economically, it may not be worthwhile. The following conditions are necessary to increase the chance of success.

1. *Hybrid vigor (heterosis).* Hybrid seed production exploits hybrid vigor. The parents used in a cross should be tested to find out how compatible they are. All parents are not capable of producing good hybrids. Inbred lines used in crosses are evaluated on the basis of their *combining ability,* or productivity, in a cross. A set of inbred lines are crossed in all possible combinations and evaluated on the basis of *general combining ability (GCA)* or *specific combining ability (SCA)* (the performance of parents in hybrids). Inbred lines with high GCA are productive in a wide variety of combinations, whereas those with a high SCA are productive with only specific cross combinations. Hybrid vigor is usually lower in self-pollinated species, in which *additive gene action* predominates, and higher in cross-pollinated species, in which *dominance gene action* is predominant (5.7.4).

2. *Elimination of fertile pollen from the female parent.* Parents in a cross are carefully selected to combine for best results. If a female parent is male fertile, it might contribute pollen to reduce the effectiveness of the male parent and result in reduced productivity of the cross. Male infertility in female plants may be accomplished by manually removing male reproductive parts in the flower, a process called emasculation (5.7.5). Male sterility controlled by genetic factors has been identified in many species and is of practical application in breeding of those species (5.9.2).

3. *Adequate pollination and fertility restoration.* The method utilized for breeding should produce sufficient pollen to completely pollinate all plants. This condition becomes important when genetic factors are employed to eliminate fertile pollen from female parents. Whereas 100 percent of offspring from a single cross and a three-way cross will be male sterile, only 50 percent of plants from a double cross will be male fertile. However, pollen shed is usually adequate for complete pollination of plants in the field.

4. *Availability and maintenance of parents.* Inbreeding in cross-pollinated species is accompanied by loss of vigor. The parents used in hybrid seed production are inbred lines that should be maintained such that they perform at a desired level for an adequate period. The male parent should be able to provide a sufficient amount of pollen. The female should produce high-quality seed of high vigor and good germination. Where a cytoplasmic-genetic system is used, the female should be able to retain male sterility over a range of environmental conditions. The performance of the parents in hybrids should be high.

5. *Efficient pollen transport.* There are two primary agents of natural pollination—insects and wind. Pollination may also be accomplished manually. For commercial production, pollination by hand is tedious, laborious, and expensive; it is used in producing hybrid tomatoes. Natural methods of pollination are most economical, and, as such, hybrid seed production is most economical in plants that are naturally cross-pollinated by wind. Insect pollination is less reliable because in fields in which chemicals are used, their production could be reduced by poisoning. Further, insects may prefer the male sterile flowers more than male fertile ones, thus resulting in smaller amounts of viable pollen being transported.

6. *High economic returns on investment.* Hybrid seed production is expensive and hence uneconomical in plants with low economic returns. In high-premium crops such as tomato, hybrid seed production by manual pollination is cost-efficient.

5.10: MUTATION BREEDING

5.10.1 MUTATIONS

A change in the sequence of genomic DNA is called a mutation. Isolation of mutants has been the conventional method of gene analysis. The products of mutation are called mutants, and the agents of mutation are called mutagens.

Mutations may arise naturally, or spontaneously. *Spontaneous mutations* arise as a result of normal cellular operations or interactions of cells with their environments. Normal cellular activities, such as DNA replication, provide opportunities of alteration in the sequence of nucleotides that, as described earlier, is responsible for producing a specific sequence of amino acids, which in turn leads to the production of a specific protein. A change in the nucleotide sequence results in a change in the protein. Changes in the nucleotide sequence can arise in a variety of ways. A base pair (A-T or G-C) can be altered, the resulting mutation being called a *point mutation.* One pyrimidine may be substituted for the other and, similarly, one purine for the other. This substitution is called a *transition* and results in the exchange of A-T for G-C and vice versa. In another example, a purine may be replaced by a pyrimidine (A-T → T-A) or vice versa (C-G → G-C) a substitution called a *transversion.* It should be mentioned that natural mechanisms exist in the cells to correct (or attempt to correct) these sequence alterations as they occur. Mutations arise when corrective measures fail.

Mutations may be induced by using a variety of *mutagens* that may be classified as either *chemical mutagens* or *physical mutagens.* Certain chemical mutagens such as nitrous acid can oxidatively deaminate cytosine to uracil. Mutations may arise when *base analogs* are incorporated into DNA. For example, when DNA replicates in the presence of bromouracil (BrdU), the latter is incorporated into the former in place of thymine. This occurrence may cause A to be replaced by G and consequently A-T base pairs to be replaced by G-C base pairs. Certain chemicals may interfere with the spindle apparatus and result in abnormal chromosome separation and distribution into daughter cells.

Mutations may be caused by physical agents such as X rays, gamma rays (both *ionizing radiations*), and ultraviolet (UV) *(nonionizing radiation)* light. Plant materials or whole plants may be subjected to irradiation in *acute* (one-time application of high dose) or *chronic* (split doses over a period) application. Physical mutagens often cause chromosomes to physically break. In the process of reconstruction of fragmented chromosomes, the linear order of genes may be changed, resulting in structural aberrations classified as *deletions, duplications, inversions,* and *translocations.* The structural changes in the chromosomes pose problems during cell division, resulting in consequences of reduced fertility or sterility. Translocations are common in plants of the genera *Oenothera* and *Paeonia.* Unlike conventional breeding methods, one cannot predict the outcome of exposing organisms to mutagens; the breeder can only hope that one of the random changes that these mutagens cause will be what is desired.

Apart from point mutations and chromosomal structural aberrations, variations in the number of chromosomes can be induced or arise spontaneously. These variations may be one

> **Mutation**
> A heritable change in an individual resulting from a change in the nucleotide sequence or the number of chromosomes.

Table 5-3 The Occurrence of Aneuploidy in Plants

Chromosome Number	Term	Number of Copies of Problem Chromosome
$2n$	Diploid number (normal)	2 (normal)
$2n - 1$	Monosomy	1 (one missing)
$2n + 1$	Trisomy	3 (one extra)
$2n + 2$	Tetrasomy	4 (two extra)
$2n + 3$	Pentasomy	5 (three extra)

of two types: (1) *aneuploidy* (variations in which the somatic complements are irregular multiples of the basic haploid number that characterizes the species) and (2) *euploidy* (variations in which the somatic complements are exact multiples of the basic haploid number that characterizes the species). Aneuploidy is known to be caused by the phenomenon of *nondisjunction* (failure of duplicate chromosomes to separate during cell division), as well as *anaphase lag* (certain chromosomes do not migrate to the poles to be included in the telophase nuclei). These events result in meiotic or mitotic products that may be deficient (e.g., $n - 1$) or have extra (e.g., $n + 1$) chromosomes (table 5-3). Consequences or benefits of fewer or additional chromosomes depend on the genetic constitution of the chromosome in question. Ornamentals such as spring beauty *(Claytonia virginica)* and garden hyacinth *(Hycinthus orientalis)* have viable aneuploids. Trisomics have been used in locating genes in particular linkage maps of tomato.

5.10.2 EXAMPLES OF APPLICATIONS

Mutation breeding is used in developing new variability for breeding programs. In bitter gourd *(Momordica* spp.), gamma radiation was used to induce a new variety from a local cultivar. In okra *(Abelmoschus* spp.), a mutant carrying resistance to yellow vein mosaic virus and tolerance to fruit borer using ethyl methane sulfonate has been reported. Ethyl methane sulfonate was used to generate potentially useful variability in lettuce.

5.11: POLYPLOIDY

> **Polyploidy**
> A condition in which a cell nucleus has more than two complete sets of chromosomes for the species.

Euploids of all kinds are collectively called *polyploids.* Polyploidy is very important in plant breeding not only because it can be induced but also because many important horticultural plants are natural polyploids (e.g., banana, cultivated strawberry, tart cherry, and blueberry). Table 5-4 describes commonly occurring multiples of chromosome sets in plants. A haploid contains one genome, or set of chromosomes; a diploid has two. Where more than one genome is involved, the sets of chromosomes may be similar. Such organisms are called *autopolyploids* (e.g., AAA and AAAA). However, when dissimilar genomes are involved, the organisms are called *allopolyploids* (e.g., AAB and AABB). Of necessity, allopolyploids arise through hybridization. When two species are crossed, the F_1 generation is sterile. However, by doubling the chromosome number of the F_1, the resulting allotetraploid can be viable and productive.

A common method of inducing polyploidy is use of an alkaloid *(colchicine)* obtained from the autumn crocus *(Colchicum autumnale).* This chemical interferes with the spindle apparatus, destabilizing it such that chromosomes do not separate after duplication, leading to doubling.

The *basic number of chromosomes* in a cell is symbolized by n, which in diploid organisms $(2n)$ represents one-half the number of chromosomes (or the number in gametes). However, a polyploid consists of several sets of chromosomes. Half of this number (n) is not equivalent to the half, as in the case of a diploid. To avoid this confusion when discussing the genetics of polyploids, the basic number is represented by x. For example, if the organism has

Table 5-4 The Occurrence of Polyploidy in Plants

Chromosome Number	Symbol	Term	Number of Sets of Chromosomes
2n	AA	Diploidy (normal)	
Euploidy		(Multiples of n)	
Autopolyploidy			Identical sets
3n	AAA	Triploidy	3
4n	AAAA	Tetraploidy	4
5n	AAAAA	Pentaploidy	5
6n	AAAAAA	Hexaploidy	6
Allopolyploidy			Unidentical sets
3n	AAB	Allotriploidy	3
4n	AABB	Allotetraploidy	4
5n	AABBC	Allopentaploidy	5
6n	AABBCC	Allohexaploidy	6

Unlike aneuploidy, euploidy involves whole chromosome set abnormalities

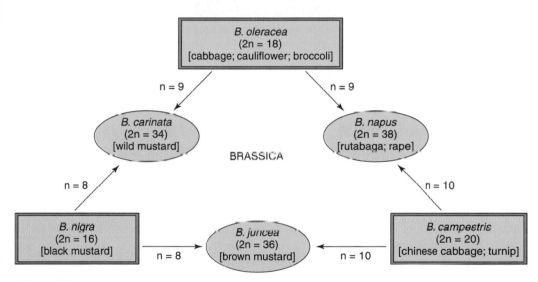

FIGURE 5-38 Amphidiploidy in *Brassica*.

a genome consisting of six sets of seven chromosomes (a total of 42 chromosomes), $x = 7$. The somatic cells contain 42 chromosomes, which is equal to $2n$. The gamete (e.g., pollen) contains one-half (or n) the number of chromosomes, or 21.

Polyploidy is utilized in plant breeding for a number of purposes. It is associated with *gigas* phenotype, whereby polyploids are larger than their diploid counterparts. This has application in the breeding of certain horticultural plants; for example, polyploid flowers are larger and more showy and their fruits are usually more fleshy and succulent. Polyploidy is used in specific ways in breeding. The product of wide crosses (5.12) is frequently infertile because of meiotic complications. Doubling of the chromosomes of the F_1 sterile hybrid often results in fertile plants that are called *amphidiploids* and also in the creation of a new species *(instant speciation)*. The sterility induced by polyploidy is also taken advantage of in the development of seedless fruits; for example, in the watermelon, a cross between a tetraploid and a diploid yields triploids with small and underdeveloped seeds. Apart from these specific uses, polyploidy results in the creation of new variability for use by plant breeders. Certain variability has arisen by natural means; for example, rape (*Brassica napus* var. *olerifera*) is an amphidiploid of *B. campestris* and *B. oleraceae*. Similarly, rutabaga (*B. napus* var. *rapifera*) is a cross between turnip and kale (figure 5-38).

The problem with polyploidy is reduced fertility or sterility, resulting from meiotic problems: Odd-number genomes are usually sterile. Sterile triploids are vegetatively propagated. Natural autopolyploids are not common. The potato *(Solanum tuberosum)* and banana *(Musa paradisiaca)* are autotriploids. Other natural autoploids are the hyacinth *(Hycinthus orientalis)* and the apple *(Malus pumila).* High fertility in polyploids is possible when pairing between homologues results in bivalents and anaphase segregation is normal. High levels of ploidy beyond autotetraploidy result in abnormalities, increased infertility, dwarfing, and plants with reduced vigor.

5.12: WIDE CROSSING

The preferred choice of parents for a breeding program are those from the same species, since such genotypes usually combine freely without infertility problems or any hindrance to genetic recombination. At times the sources of genes required to accomplish a breeding objective are located in unrelated species. In such instances, breeders may have to resort to *wide crosses* involving different species *(interspecies cross)* or different genuses *(intergeneric crosses).* Interspecific hybridization is difficult, and even more so intergeneric cross, partly because they frequently result in sterile offspring *(hybrid sterility).* Fertility may be restored by ploidy manipulation. Sterility may result from distances between parental genomes or between the cytoplasm of one parent and the genome of another. Sterility caused by structural differences between the two chromosome sets is called *chromosomal hybrid sterility.* If sterility is induced by the presence of specific gene complexes, it is called *genetic hybrid sterility.* Apart from sterility, wide crosses may also be plagued by *hybrid weakness,* or *inviability,* in which the resulting F_1 generation is too weak to be useful to the breeder. When the F_1 generation is vigorous and fertile, the next generation (F_2) may yield sterile or weak products, the condition referred to as *hybrid breakdown.*

5.12.1 EXAMPLES OF APPLICATION

Wide crossing has been used to broaden the genetic base of broccoli and to select superior types from segregating populations. Interspecific hybridization between *Apium graveolens* L. and *A. prostratum,* a wild species resistant to *Septoria* and immune to leaf miners, has been reported. Parsley and celery have been successfully crossed. The products from this cross include new forms of leaf celery, which is high in vitamin C, carotene, and essential oils, among others.

Vegetatively propagated species have *fixed genotypes* since there is no opportunity for recombination to occur through meiosis. They produce products that are uniform (clones) year after year. Clonally or vegetatively propagated species are improved by using other techniques such as *mutation breeding* (5.10) or *tissue culture* (chapter 9). Mutation breeding is the ultimate procedure to adopt when the desired variation does not occur in nature. That is, if the gene does not occur naturally, it may be induced artificially. To this end, mutation breeding is applicable to both sexually and vegetatively propagated plants.

5.13: DURATION OF A BREEDING PROGRAM

Plant breeding is a form of *organic evolution.* The major differences between breeding and evolution are that humans do the selecting *(artificial selection)* and the process is much quicker (several years as compared to estimated millions of years by *natural selection).* Even so, completion of a plant breeding program may require between five and ten years. Breeders are therefore constantly looking for ways to shorten the duration of their breeding programs and to reduce the cost of breeding. Some breeding efforts may cost millions of dollars. A way to speed up breeding programs is to discover and adopt methods that increase selection efficiency, such as using genetic markers for early generation selection.

5.14: Limitations of Conventional Breeding Methods

Conventional breeding methods have certain limitations, including the following:

1. Long duration of programs.
2. Limited to crossing within species and occasionally between species. If plants are not genetically compatible, they cannot be hybridized.
3. Lower selection efficiency.
4. Large segregating populations to handle and hence more space required.

5.15: Parthenocarpy

Certain plants have the ability to develop fruits without pollination and are said to be *parthenocarpic.* The fruits of such plants are seedless. *Parthenocarpy* should be distinguished from *apomixis,* which is the development of seed without fertilization. Parthenocarpy is desirable in certain fruits in which seedlessness is preferred by consumers, such as cucumber, grape, and watermelon. Fruit set is ovary enlargement, followed by further development to full size. Subsequently, fruit development is influenced by seed formation. The inhibitory effect of fruit set and development is bypassed by parthenocarpy, resulting in early and more regular fruit production patterns. Synthetic auxins may be used to induce parthenocarpy in fruit plants such as tomato. In monoecious plants such as cucumber and squash, constant femaleness should be maintained in the plant bearing the fruit. To promote the development of such pistillate flowers (female flowers), plants may be treated with ethylene or cytokinin. Morphactin (chlorfluorenol) may be used to induce artificial parthenocarpic fruit set in some plants such as cucumber. Greenhouse-grown slicing cucumber and pickling cucumber are commonly parthenocarpic.

> **Parthenocarpy**
> The development of fruit in the absence of fertilization.

Module 3 Application of Genetics: Molecular Biotechnology

5.16: Underlying Principles

The cutting-edge *recombinant DNA technology* (also called *gene cloning* or *molecular cloning*) is a battery of experimental protocols employed to transfer DNA (genetic information) from one organism to another. The organisms do not have to be related for the transfer to be successful. Further, there is no need for hybridization to occur for gene transfer. Consequently, DNA may in theory be transferred from tomato to oak. To understand the protocols, one needs to understand the structure and function of DNA (5.2.4). Further contributions from a variety of scientific disciplines (including biology, biochemistry, genetics, and microbiology) continue to furnish critical information in areas such as nucleic acid enzymology, molecular genetics of bacterial virus, and plasmids, which form the knowledge base of *genetic engineering,* another term for the recombinant DNA technology.

The successful application of this technology in plant improvements includes certain essential steps. Each step may involve the use of several biotechnological procedures, such as enzymatic restriction of DNA, tissue culture, molecular markers, sequencing, and vectors. Some of these procedures are described in the following sections.

> **Genetic Engineering**
> Synchronous with recombinant DNA technology, it is artificial manipulation or transfer of genes from one organism to another.

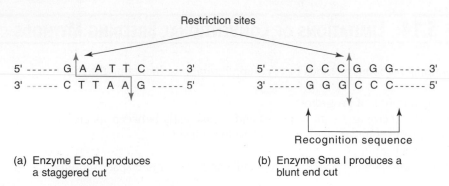

(a) Enzyme EcoRI produces a staggered cut

(b) Enzyme Sma I produces a blunt end cut

FIGURE 5-39 Selected restriction endonucleases and their characteristics.

5.16.1 RESTRICTION ENDONUCLEASES

To transfer a piece of DNA to another material, it must first be excised from its source. This excision is accomplished by using enzymes to cleave source DNA to remove the target fragment. The enzymes used in this fashion are called *restriction endonucleases.* These molecular "scissors" are isolated from bacteria, in which they serve as a defense mechanism against foreign DNA by digesting them to disable the intruding organism. They are designed to cut at predetermined sites *(restriction sites)* on a DNA strand. Further, the signal to cut depends on the enzyme identifying a specific sequence of nucleotides *(recognition sequence)* and binding to the region. Each recognition sequence is unique to each restriction enzyme and usually a few bases long (figure 5-39). Each sequence consists of nucleotides arranged in a palindromic fashion.

5.16.2 CLONING VECTORS

After cleavage, the piece of DNA *(restriction fragment)* must be transported in a "biological vehicle" to the recipient cell. The vehicles used, called *cloning vectors,* are circular extrachromosomal DNA molecules (plasmids) that occur in bacteria (5.2.4). Plasmids are doubled-stranded DNA and capable of self-replication. They vary in size and hence the size of restriction fragment that can be used for cloning. Those used for cloning large fragments are called *cosmids.* Plasmids are usually genetically engineered by scientists, who insert specific nucleotide sequences such as recognition sequence for a restriction endonuclease and a selection system (such as resistance-to-antibiotics genes).

5.16.3 ELECTROPHORESIS

Electrophoresis
A technique for separating a mixture of charged compounds in an electric field on a matrix.

One of the most widely used techniques in biotechnology is *electrophoresis,* the separation of charged biological molecules in an electric field. This versatile biochemical technique is used to detect genetic variation. The mixture to be separated is first prepared in an appropriate buffer. Separation occurs in a solid support medium (a *gel*) that may be prepared from one of several materials such as starch, agarose, or polyacrylamide. Appropriate amounts of the sample are deposited in wells (or delivered via strips of absorbent material). A power pack is used to supply direct current at an appropriate charge. The charged molecules migrate from the well in the direction opposite to their charge and at speeds proportional to their size, charge, and the characteristics of the support medium (figure 5-40).

After separation, the support medium is removed from the electrophoresis unit for observation. In certain procedures, the gel may have to be chemically stained to reveal the relative positions of the migrated molecules. In other cases, the fragments may be viewed under special light (ultraviolet) or photographed under X rays to reveal the pattern of migration (figure 5-41).

5.17: GENETIC ENGINEERING OF PLANTS

Like conventional breeding, it is easier to genetically engineer plants for traits that are simply inherited (controlled by one or a few genes) or involve a small cluster of genes.

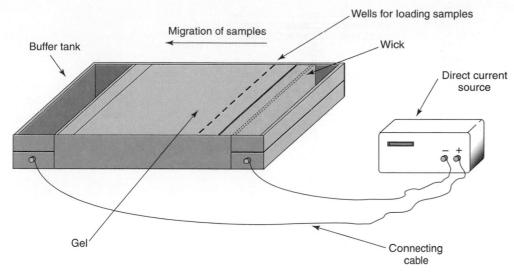

FIGURE 5-40 An electrophoresis unit.

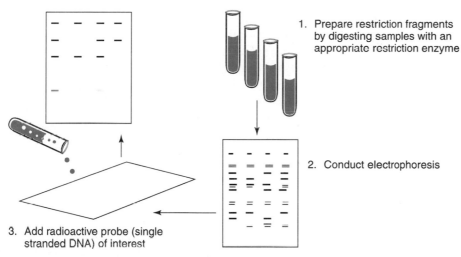

4. Autoradiography results in a photographic film imprinted with the bands complementary to the probe

1. Prepare restriction fragments by digesting samples with an appropriate restriction enzyme

2. Conduct electrophoresis

3. Add radioactive probe (single stranded DNA) of interest

FIGURE 5-41 The results of electrophoresis.

Candidates for manipulation include traits such as resistance to herbicides, resistance to viral infection, tolerance of environmental stress, improved nutritional quality of seed protein, changed flower pigmentation, and delay of senescence. The steps involved are as follows.

5.17.1 GENE IDENTIFICATION

The critical first step in genetic engineering is the identification and isolation of the specific gene or genes responsible for the trait of interest. The plant with the desired trait must first be identified, which is often easy. The more difficult part is associating this trait with a specific gene product in the cell, a lengthy process involving the comparative analysis of the genetic profiles *(genetic fingerprint)* of individuals with and without the expression of the desired trait. The unique band is extracted and further analyzed. When protein electrophoresis is conducted, the piece of DNA responsible must be synthesized backward from the protein (figure 5-42), against the central dogma of molecular biology (5.3). The synthetic gene is called a complementary DNA (cDNA).

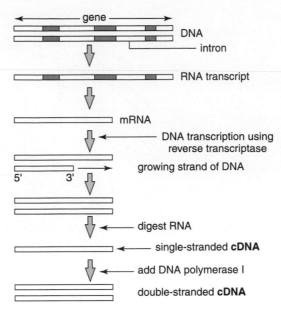

FIGURE 5-42 Summary of the steps in creating a complimentary DNA (cDNA).

5.17.2 CLONING

Once isolated, the gene is packaged in a carrier. Because the piece of DNA by itself is not capable of replication, it must become part of an existing replication machinery. First, the DNA is inserted into a plasmid DNA obtained from a bacterium. This plasmid, called a *vector,* is a double-stranded molecule capable of replication. Upon receiving the piece of foreign DNA, the plasmid becomes a *recombinant plasmid* (figure 5-43). The next step is to place the plasmid back into a bacterium (the host cell), which is accomplished by a process called *transformation* and may be done by one of several methods (e.g., *calcium chloride transformation* or electrically [*electroporation*]). Once inside the host, the bacterium is said to be transformed and the gene cloned. The recombinant plasmid will be able to replicate clonally to produce more of the alien DNA. Vectors are hence also called *cloning vehicles.*

5.17.3 TRANSFORMING PLANTS

The ultimate goal of cloning is not to transform a bacterium but to transfer the cloned gene into a plant—in this case to become part of the plant (host) genome (figure 5-44). There are several methods for transferring the cloned gene into a plant, with varying degrees of difficulty and success. Plants infected with the crown gall bacterium (*Agrobacterium tumifaciens*) develop tumors as a result of the transfer and incorporation of bacterial DNA into the host genome. Scientists capitalize on this natural means of DNA transfer by using this bacterium as a vector. First, the bacterium is disarmed so that it is unable to cause tumors in the host. It is then transformed with the cloned gene. Next, the transformed crown gall bacterium is allowed to infect pieces of tissue or material from the host plants under tissue culture conditions. Successfully infected cells are *regenerated* into full plants using appropriate tissue culture systems. Hopefully the cloned gene is then expressed in the host system. Plants that receive an alien gene from a different genus are called *transgenics.*

Another method of transferring cloned genes into host plants is to bombard the cells of the recipient plant with tungsten particles coated with the transformed plasmids. The instrument used is called a *particle gun,* and the procedure is conducted under tissue culture conditions (figure 5-45). By using this method, it is hoped that at least one of the cells in the host tissue will be penetrated by the alien DNA and thus transformed. The procedures described are only general and simplified; in practice, they are more elaborate and involved, and have built-in checks to ensure that the alien DNA is transferred at each step.

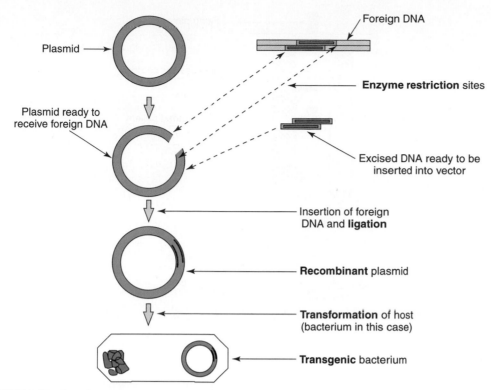

FIGURE 5-43 Creating a recombinant plasmid.

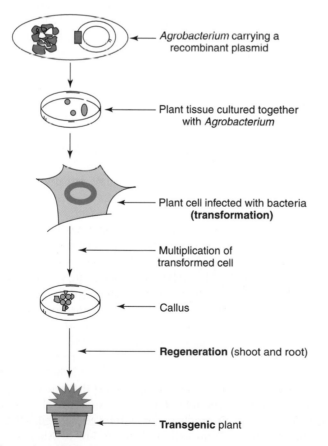

FIGURE 5-44 Transformation of plants.

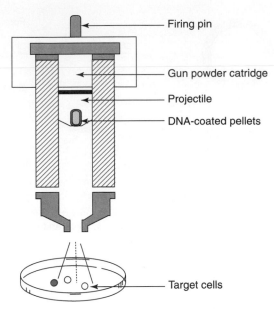

Firing pin

Gun powder catridge

Projectile

DNA-coated pellets

Target cells

FIGURE 5-45 A particle gun used in transformation.

Examples of Applications

Plant transformation has been accomplished in melon, using *Agrobacterium tumefaciens* via the particle gun technique. In celery and eggplant, stable kanamycin-resistant *transgenic* plants have been produced through transformation of celery by *A. tumefaciens*. Tomato biotechnology has yielded outstanding success. The insect control protein of *Bacillus thurringiensis* protein has been transferred successfully into tomato. The resulting transgenic plants are resistant to insects such as tomato fruitworm *(Heliothus zea)* and tobacco hornworm *(Manduca sexta)*.

Tomato has also been engineered for increased fruit quality. Called *Flavr Savr tomato* by its developers at *Calgene,* a biotechnology company based in California, this tomato has been genetically engineered to vine ripen before harvest. Tomatoes bruise easily during transportation and are hence harvested while hard and green. The fruits are then induced to ripen at the destination before selling by exposing them to ethylene gas in the storehouse (chapter 21). However, tomatoes handled this way are less tasty. The challenge to scientists was therefore to slow down or inhibit fruit softening while other ripening processes continued. Fruits could then be harvested vine ripened and have better taste while remaining firm and tough enough to resist damage during transportation.

Fruit softening is caused by the enzymatic breakdown of pectins in the cell wall by an enzyme called *polygalacturonase (PG)*. Scientists used the *antisense technology* to accomplish their objective. A copy of the PG gene was synthesized in the reverse order (complementary) and cloned. It was then used to transform tomatoes by *Agrobacterium-mediated transformation,* which places two complementary copies of the PG gene in one plant. The plant then becomes *transgenic* (having received DNA from another genus). The cell transcribes both genes, one into mRNA (normal) and the other into mRNA (antisense). The complementary sequence then anneals to the normal sequence (just as a Velcro fastener works), thus preventing the transcription of the PG enzyme. This selective disabling of genes is the rationale behind antisense technology. The expression of the alien antisense gene did not affect the ethylene production and lycopene accumulation, both of which are associated with fruit ripening. The genetic marker (5.17) used in this transformation process is the gene for kanamycin resistance *(kan-r)*. This gene is transferred into the host along with the antisense gene, which is a source of criticism by those who suggest that bacteria feeding on the transgenic tomato plant debris might eventually develop resistance to the antibiotic. Others criticize biotechnologically engineered products for unforeseen dangers that might spring unwelcome surprises in the future. Transgenic plants that are resistant to herbicides such as glyphosate, bromoxynil, and bialaphos have been developed in tomato.

5.18: MOLECULAR MARKERS

Apart from transferring alien DNA, molecular biotechnology is used in other beneficial ways, such as accelerating breeding programs through the use of *molecular markers*. Genetic markers are heritable characteristics associated with certain traits; when a marker is observed, it means that the associated trait is also present. It provides a diagnostic tool for predicting or confirming the presence of a trait. Plant breeders use selection as a tool for discriminating among existing variations. Morphological differences are conditioned by genetic variations at the molecular level. Instead of waiting for a morphological trait to develop to maturity (which might take years in some cases), the breeder can instead predict the adult phenotype by examining the phenotype at the molecular level. This procedure can drastically reduce the duration of breeding projects. Molecular markers include *isozymes* (protein markers) and those generated from *restriction fragment length polymorphism (RFLP)* (figure 5-46) and *polymerase chain reaction (PCR)* (figure 5-47). Both RFLP and PCR technologies produce DNA markers. A unique capability of the PCR technique is its capacity for amplification or increase of DNA given even a minuscule amount of the source. Starting with only a pinhead-sized amount of DNA, PCR techniques may be used to increase it (exponentially) a millionfold.

Another application of biotechnology in horticulture is in the area of molecular markers for facilitating plant breeding. Isozymes of acid phosphatase were used to identify sibs in F_1 hybrids of *B. oleraceae* (cabbage). Isozymes have been used successfully for testing seed purity in this crop. In celery, isozymes and storage protein markers have been employed in variety identification and pedigree analysis. These markers have been used in hybrid identification and outcrossing determination. The gene for tomato root knot resistance *(Mi)* is reported to be linked with isozyme marker *Asp1* (acid phosphatase). Similarly, resistance to race 3 of *Fusarium* I3 is linked with *Got2*.

In cabbage, RFLP-based genetic maps of *B. campestris, B. oleraceae,* and *B. napus* have been constructed to tag agronomic traits and to study chromosome organization within and between these three species. Two dominant quantitative trait loci (QTLs) for resistance to race 2 of *Plasmodiaphora brassicae* (clubroot disease) have been mapped in relationship with several RFLP markers. Random amplified polymorphic DNA (RAPD) technique is used to certify hybrids of canola and in nondestructive early identification of clubroot resistance in segregating

<table>
<tr><td>

Polymerase Chain Reaction (PCR)
A technique for amplifying a target DNA sequence.

</td></tr>
</table>

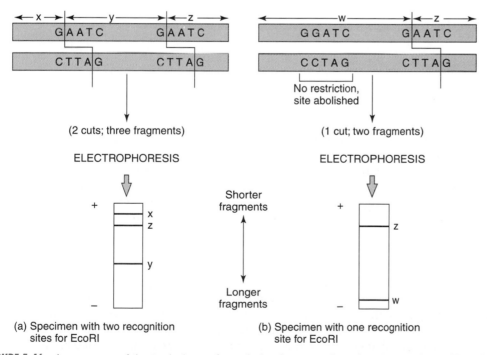

(a) Specimen with two recognition sites for EcoRI

(b) Specimen with one recognition site for EcoRI

FIGURE 5-46 A summary of the technique of restriction fragment length polymorphism (RFLP).

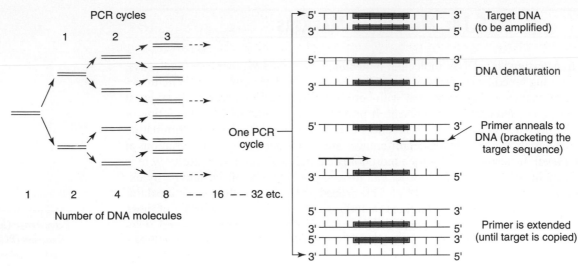

FIGURE 5-47 A summary of the technique of polymerase chain reaction (PCR).

populations. Insect resistance in tomato, as measured by the presence of 2-tridecanone (2-DT), is reported to be associated with four RFLP loci. Further, five RFLP loci are known to be associated with soluble solids of tomato fruits. RFLPs are used to map QTLs in sweet corn and to study its diversity and phylogeny.

5.19: TISSUE CULTURE

Tissue culture is another aspect of biotechnology that can be used to manipulate plants. The underlying principle in tissue culture is that each cell is endowed with the full genetic complement of its particular species to enable scientists to generate a complete plant from one cell. Each cell is therefore said to be *totipotent*. Clones can be produced through tissue culture as a means of plant propagation. For this reason, tissue culture is also called *micropropagation*. This technique is discussed in more detail under plant propagation (chapter 9).

SUMMARY

Plant improvement involves the manipulation of the genome of the plant with a specific goal in mind. A good understanding of genetics helps breeders to be effective and efficient in their endeavors. DNA, hereditary material, is located mostly in the nuclei of plants in linear structures called chromosomes. Genes, the factors that condition traits, are pieces of DNA arranged in a linear order in chromosomes. Genes—not traits—are inherited by the offspring from their parents. The transmission of genes in this fashion is governed by certain laws that were discovered by Mendel: gene pairs segregate during cell division so that only one from each pair ends up in each gamete, and genes assort independently during meiosis. Some genes (dominant) mask the expression of others (recessive) when they occur at the same locus.

Meiosis is the major source of biological variation, resulting from the event of crossing over. However, the ultimate source of variation is mutation. The expression of genes is subject to the environment in which they occur. Certain traits are governed by one or a few genes (simply inherited), whereas others are governed by many genes with small effects (quantitative traits).

The normal pattern of genetic information flow is from DNA to protein, not the reverse (central dogma of molecular biology). However, scientists are now able to isolate a protein and work backward to synthesize, in vitro, the corresponding gene (complementary DNA). DNA may undergo changes (mutations) to produce new variants in the population. Breeders sometimes deliberately induce mutations to create new variations.

The underlying principle in plant breeding is that the phenotype (what is observed) is a product of the interaction between the genotype (genetic makeup) of the individual and its environment. Phenotype can therefore be changed by manipulating either the genotype (plant breeding) or its environment (agronomy). Before starting a breeding program, breeders should first ascertain whether adequate heritable variation is available. Then they should have clear goals or objectives. The specific methods used depend on the mating system of the plant (self- or cross-pollinated or vegetatively propagated). Genetic changes may be made in flowering plants through crossing (hybridization). In nonflowering plants, mutation breeding may be conducted for plant improvement.

Molecular biotechnological procedures are being utilized by plant breeders to accelerate their breeding programs through the use of molecular markers for effective selection. Further, breeders are able to import genes from outside of the genus with which they are working and thereby develop transgenic plants. Problems that once were impossible to solve with conventional breeding are being tackled with amazing success using molecular biotechnological tools.

It should be emphasized that conventional and molecular biotechnological methods of breeding are complementary tools. Geneticists and breeders work together to improve horticultural plants.

REFERENCES AND SUGGESTED READING

Campbell, N., L. Mitchell, and J. Reece. 1997. Biology: Concepts and connections, 2d ed. New York: Benjamin/Cummings.

Klug, W. S., and M. R. Cummings. 1996. Essentials of genetics. Englewood Cliffs, N.J.: Prentice-Hall.

Wallace, R. A. 1997. Biology: The world of life, 7th ed. New York: Benjamin/Cummings.

PRACTICAL EXPERIENCE

LABORATORY

---

1. Select one (or more) vegetable and one ornamental flowering crop and assemble seeds of as many variant types as possible. For this activity, sources of seed include the market, farms, research stations, other universities, and germ plasm banks. Place the collection in test tubes in a rack and label clearly.

2. Plant samples of selected specimens from activity 1 and describe the differences in morphology.

FIELD WORK

Attempt to cross two different cultivars of tomato, pepper, or an ornamental flowering plant.

OUTCOMES ASSESSMENT

PART A

Please answer true (T) or false (F) for the following statements.

1. T F   DNA is genetic material.
2. T F   Pollen grains combine randomly with eggs to form zygotes.

3.   T F   Recombination is the ultimate source of biological variation.
4.   T F   Without heritable variation, it is impossible to conduct a breeding program.
5.   T F   Meiosis occurs in somatic cells.
6.   T F   Pollen grains are products of meiosis.
7.   T F   The environment in which a gene occurs affects its expression.
8.   T F   The cell is the fundamental unit of organization of life.
9.   T F   The totality of all of the genes an individual has is its phenotype.
10. T F   A true-breeding plant is genetically heterozygous.
11. T F   Permanent alteration in the DNA is called domestication.

PART B

Please answer the following questions.

1. The meiotic event responsible for creating variation is called _____.
2. A plant whose genome has incorporated an alien piece of DNA from another genus is called a _____.
3. The process of making a flower female by removing all male organs is called

   _____.
4. Variation that can be passed on from one generation to another is called

   _____.
5. Tissue culture is also called _____.
6. List some possible breeding objectives in horticulture.

   _____
   _____

PART C

1. Distinguish between *genotype* and *phenotype*.
2. Describe the steps in a molecular biotechnological procedure for producing a transgenic cultivar.
3. Describe the backcrossing method for plant improvement.
4. Phenotype = genotype + environment. Discuss.
5. Describe how hybrids are developed.
6. Define the term *biotechnology*.

PART 2

PROTECTING HORTICULTURAL PLANTS

CHAPTER 6

Biological Enemies of Horticultural Plants

PURPOSE

The purpose of this chapter is to present and discuss the various classes of pathogens and parasitic factors that cause disorders in horticultural plants.

EXPECTED OUTCOMES

After studying this chapter, the student should be able to

1. Discuss the economic effects of pests in horticulture.
2. List the categories of organisms that are pests to plants.
3. Distinguish among insect pests on the basis of life cycle.
4. Distinguish among insect pests on the basis of feeding habits.

KEY TERMS

Adaptation	Hemiparasites	Obligate parasites
Competition	Marginal conditions	Parasites
Epiphytes	Metamorphosis	Saprophytes
Facultative parasites	Necrosis	Symbiosis
Heat therapy		

OVERVIEW

Pests are organisms that are harmful to plants. They include organisms ranging from tiny microorganisms (or microbes) to large animals and other plants. They damage above-ground plant parts and below-ground parts, the latter kind of damage being more difficult to detect. The activities of pests may cause physical injury to plants. They may also cause disorders in plants, making them unable to grow and develop properly. In some cases, pests do not come into contact with desirable plants but, just by being in close vicinity, compete for growth factors and deprive the desirable plants of adequate nutrition. Certain pests only cause blemishes on plants, making them unsightly and unattractive without physically damaging or upsetting the physiology of the plant. Whatever the form of pests associated with cultivated plants, they need to be controlled when their presence threatens to cause economic loss to the grower.

Pests of cultivated plants include weeds (other plants), insects, nematodes, and mites. Those that cause diseases are bacteria, fungi, and mycoplasma-like organisms. These organisms can be divided into two broad classes: plant pests and animal pests. The life cycles of these organisms are important in their control. Control measures are usually targeted at the stages in their life cycles when they are most vulnerable. In terms of plant diseases, fungal problems predominate.

MODULE 1 PLANTS AS PESTS

6.1: WEEDS

> **Weed**
> A plant that is growing where it is not wanted.

A *weed* may be broadly defined as a plant out of place. A corn plant in a bed of roses is as much a weed as a rose plant in a cornfield, no matter how attractive it may look. In theory, any plant can be considered a weed at some point. However, the term is commonly used to refer to certain undesirable plants, often with no economic use to growers. Such plants vary widely in characteristics and include algae, mosses, ferns, and flowering plants. Weeds tend to be more aggressive than cultivated plants and have characteristics that enable them to survive under very harsh conditions. They can thrive on marginal soils and have seeds that have longevity, dormancy, and other allied survival characteristics associated with life in the wild.

6.1.1 ECONOMIC IMPORTANCE

Weeds are undesirable in cultivated fields or the landscape—or for that matter anywhere. They should be controlled because they usually result in economic losses. The following are some of the significant ways in which weeds show their undesirable characteristics.

1. They compete with cultivated plants for growth factors (light, moisture, air, and nutrients [Chapter 3]), thus diminishing the performance of crop plants. Weeds often are better competitors than cultivated plants. They are more vigorous in growth and produce many seeds for effective dispersal. They have long viability and efficient dormancy mechanisms.

2. Weeds may harbor pests of cultivated plants and other dangerous animals. For example, chickweed harbors whitefly, red spider mites, and cucumber mosaic virus. Charlock may harbor clubroot organisms, which attack brassica crops. Groundsel may harbor nematodes. Snakes and rodents hide among weeds.

3. They are an eyesore in the landscape, diminishing the aesthetic value of lawns and other ornamental displays.

4. When they arise, weeds are removed at a cost. Mechanical removal of weeds by using a hoe or uprooting is tedious. Herbicides (7.10) are commonly used in large operations to control weeds—at a tremendous cost, not to mention the adverse environmental impact.

5. Weeds increase the costs of crop production. Additional equipment and machinery are needed to control weeds. When weed infestation is high and weeds are allowed to flower, the harvested grain may be infested with weeds. These unwanted weeds are cleaned at additional cost to the operation.

6. They reduce the quality and market value of horticultural products. Weeds compete with crops for nutrients and cause desired crops to grow and develop improperly. Yield is decreased, and along with it income. Seeds infested with weeds command a low market price. Impurities may be eliminated, adding to production costs.

7. They may be poisonous plants that can harm people. Pollen from weeds causes allergies in humans. Plant species such as ragwort and buttercup are toxic to animals. If the fruits of weeds such as black nightshade contaminate the desired crop seed and are eaten by accident, humans may be poisoned.

8. Weeds may clog drains, waterways, and other water bodies such as ponds. A pond or lake infested with weeds soon becomes shallow, thus reducing its effective use (e.g., for irrigation). Waterways choked with weeds encourage flooding.

6.1.2 BIOLOGY, CLASSIFICATION, AND DISTRIBUTION

Weed species have wild characteristics for adapting to marginal conditions and have effective mechanisms for self-perpetuation in the environment. Many weed seeds can persist in the soil for long periods of time in a dormant state until appropriate conditions for germination and growth occur. Many weeds such as groundsel establish rapidly and thereby can suppress crop seeds that germinate later. Some weeds germinate during particular seasons (e.g., orache in spring), whereas others such as annual meadow grass germinate throughout the year.

Soil moisture, pH, and nutritional status affect the distribution of weeds (3.3.3). Some weed species adapt to acidic soils, and others prefer poorly drained soils. Weeds, like other plants, are adapted to various environments. Some are tropical, whereas others prefer temperate conditions. Within the same environment, certain weeds are more of a problem in fields of cultivated crops than others. Weeds also differ in growth pattern and means of propagation.

Some weed species have underground vegetative structures (e.g., swollen roots and rhizomes) used for propagation (9.11). These weeds include bindweed, dandelion, bracken, and plantain. Unfortunately, some cultural practices designed to control weeds in actuality help them to establish more rapidly. Cultivation cuts these underground structures into pieces and spreads them around. Grasses and other species with rhizomes are among the most difficult weeds to control.

Types of Weeds Based on Life Cycle

Weeds may be classified into three groups based upon their life cycles (1.2):

1. *Annuals.* Annual weeds complete their life cycles in one year. *Summer annuals* germinate in spring and grow through summer. They produce seed and then die in fall or winter. Examples of summer annual weeds are lamb's-quarter, cocklebur, foxtail, and crabgrass, which infest lawns and vegetable fields. *Winter annuals* germinate in fall, live through winter, and produce seed in spring. The seeds remain dormant in summer, germinating in fall. Examples include chickweed, shepherd's purse, and hairy chess, which appear in plots of winter vegetables and fall nursery plants. Annual weeds are relatively easy to control, in spite of the fact that their short life cycle means they produce seeds more often.

2. *Biennials.* A few biennial weeds occur in crop cultivation and the landscape. Biennial weeds germinate in the spring of one year and remain vegetative until the next spring, when they flower. Weeds with biennial habits include hogweed and ragwort.

3. *Perennials.* Perennial weeds are very difficult to eradicate once established. They may be started from seed initially, but once mature, they also propagate by vegetative means. Many noxious perennial weeds are grasses (e.g., bermudagrass, nutgrass, and quackgrass). Nongrass perennial weeds include plantain and dandelion. In fact, bermudagrass, guineagrass, barnyardgrass, johnsongrass, goosegrass, and purple nutsedge are among the most difficult weeds to control.

6.1.3 CONTROLLING WEEDS

Since weeds may cause economic loss, they should be controlled when they occur. When weeds appear after the crop is mature or ripened, controlling them may not be economical.

However, one of the preventive measures against weeds is to exclude weed seeds from crop seeds at harvest. Planting impure seeds is a means of spreading weeds. Poor-quality compost and unsterilized soil are ways by which weed seeds can be introduced into the greenhouse or garden. Poor-quality sod often contains weed seedlings. Further, improperly laid sod with gaps between strips creates opportunities for weeds to rapidly infest a lawn.

A plant's life cycle or growth pattern influences the control strategy employed in weed control. Winter annuals are best controlled while still in the seedling stage during the fall and early spring. Similarly, summer annuals should be controlled soon after germination. Biennials are vegetative in the first year of growth; they should be controlled in that year. Perennials are very difficult to control and hence should not be allowed to become established. Once established, they have a persistent root system, making them difficult to control. Perennials can be controlled effectively during the period of rapid growth before flowering or during the regrowth period after cutting. Flowering species should be prevented from setting seed.

Biological, chemical, and cultural control of weeds and other pests is discussed in chapter 7. Annual weeds may be effectively controlled by mulching (21.1.5). However, mulching is ineffective in controlling perennial weeds. Controlling any type of weed is easier when the plants are in the seedling stages. Application of herbicides (7.10) during the seedling stage is often effective in controlling weeds; however, it is ineffective when weeds are mature. Annuals and biennials also respond less favorably to chemical control when plants are entering the reproductive phase. In perennials, chemical control is effective at the bud stage, just before flowering.

6.1.4 WEEDS AS INDICATORS OF SOIL FERTILITY

Weeds may be plants out of place, but they arise where conditions are most favorable for their existence. The type of weed species found on a piece of land is often a fairly good indicator of the soil characteristics (especially fertility, pH, and type). Examples of such telltale signs are as follows:

1. An area of land on which a good population of, for example, goosegrass, thistles, chickweed, and yarrow are found usually indicates that the soil is fertile and nutritionally balanced.

2. When dandelion, poppy, bramble, shepherd's purse, bulbous buttercup, and stinging nettle occur in dense populations, the soil is likely to be light and dry.

3. Sedge, buttercup, primrose, thistle, dock, comfrey, and cuckooflower are found in wet soils.

4. Acidic soils support acid-loving plants such as cinquefoil, cornflower, pansy, daisy, foxglove, and black bindweed.

5. White mustard, bellflower, wild carrot, goat's beard, pennycress, and horseshoe vetch are found in alkaline soils.

6. Clay and heavy soils hold moisture and favor crops such as plantain, goosegrass, annual meadowgrass, and creeping buttercup.

It should be emphasized that large populations of mixtures of several of the associated species listed must occur for the diagnosis to be reliable.

6.2: PARASITIC PLANTS

Parasite
An organism that lives on or in another species and derives part or all of its nourishment from the living host.

There are more than 3,000 species of parasitic angiosperms. These plants have little or no chlorophyll and thus are incapable of photosynthesizing to meet their nutritional needs. Parasitic plants connect themselves to their host plants' water- and food-conducting tissues through specialized rootlike projections called *haustoria* (singular: *haustorium*). The dodder (*Cuscuta salina*), a parasite with bright-yellow or orange-colored strands, is a member of the family Convolvulaceae (morning glory family). It is a stem parasite. Others in this category are *Loranthes* and *Arienthobium*. Root parasites include the broomrapes (family Orobanchaceae). The Indian pipe (*Monotropa uniflora*) is also a root parasite that obtains its food from the host through the fungal hyphae it inserts into the host. Similarly, the world's largest flower, *Rafflesia arnoldii*, and others in the genus are root parasites in the grape family (Vitaceae).

Plants in five families have the capacity to trap and use insects to supplement their nutritional needs. These plants are described as carnivorous and include the Venus's-flytrap, pitcher plant, and sundew. Plants that live on the surfaces of other plants (not symbiotically but parasitically) are classifiable into three types:

1. *Epiphytes.* Epiphytes are plants that are *autotrophic* (photosynthesizing) but depend on others for support and some of their water and nutrient needs. These plants absorb water and minerals from the surfaces of their host plants and from the atmosphere. They are particularly adapted to the humid tropical forest; an example is Spanish moss.

> **Epiphyte**
> A plant that grows on another plant without deriving nutrition from the host.

2. *Hemiparasites.* Hemiparasites are equipped to photosynthesize but depend on the host species for water and nutrients. For example, mistletoe (*Phoradendron* spp.) is parasitic to many broadleaf trees including oak, birch, and huckleberry. Certain cultivars of these species are resistant to the parasite. Like mistletoe, the Indian warrior plant also produces haustoria. These plants are green plants and can therefore photosynthesize to some degree.

3. *True parasites.* Plants that are true parasites lack photosynthetic structures and are completely dependent on the host for all nutritional needs. An example is the dodder (*Cuscuta* spp.), which is parasitic to some vegetables and woody perennials.

6.3: SELECTED COMMON WEEDS

The following are examples of common weeds that occur in the vegetable garden or landscape. A weed may be more of a problem in one part of the country than another because of adaptation.

1. *Canada thistle (Cirsium arvense).* Canada thistle is more of a problem in the northern states of the United States. It has well-branched, deep roots. A perennial that reproduces both sexually and asexually (by rhizomes), it has disk flowers that may be white, lavender, or rose-purple.

2. *Hedge bindweed (Convolvulus sepium).* Bindweed, a perennial weed that reproduces by seed or rhizomes, is a twining weed with white or pinkish flowers. It occurs in different places, including cultivated fields, especially in the eastern half of the United States.

3. *Field bindweed (Convolvulus arvensis).* Field bindweed differs from the hedge variety by being more widespread and having extensive and deep root systems. It is also a perennial and reproduces by seed and rhizomes. It has twining habits and may also spread on the ground; its white or pink flowers are bell shaped.

4. *Dandelion (Taraxacum officinale).* Dandelion is widespread, occurring nearly everywhere in the United States. It is a perennial with a deep taproot system, and the crown is branched. Dandelion produces flower heads that are golden yellow and easily dispersed by the wind. This weed is particularly common in lawns.

5. *Common milkweed (Asclepias syriaca)*. Common milkweed is a broadleaf weed with milky sap. It has sweet-smelling, pink-white flowers. This perennial has a well-branched root system with long rhizomes and reproduces by seed or rhizomes. It is more of problem in fields in the eastern half of the United States.

6. *Common lamb's-quarter (Chenopodium album)*. Lamb's-quarter is an annual weed that reproduces by seed. It has a taproot system and green flowers that are borne in spikes in a clustered panicle at the end of the branch and leaf axil. This weed is problematic, especially in the eastern half of the United States.

7. *Common cocklebur (Xanthium pensylvanicum)*. Cocklebur is a monoecious annual weed. Its stem is hairy and taproot system strong and branched. It may attain 90 centimeters in height under certain conditions and reproduces by seed.

8. *Redroot pigweed (Amaranthus retroflexus)*. Redroot pigweed has a shallow taproot system and green flowers borne in dense panicles at the end of the stem and branches. Under certain conditions, it may grow as tall as 200 centimeters.

9. *Large crabgrass (Digitaria sanguinalis)*. An annual grassy weed, large crabgrass is a problem in most parts of the United States. It is found in lawns and cultivated areas and has a dense, fibrous root system.

10. *Smooth crabgrass (Digitaria ischaemum)*. Crabgrass is an annual and a common lawn weed in most parts of the United States.

11. *Barnyardgrass (Echnochloa crus-galli)*. An annual grassy weed, barnyardgrass has a bunch form, as opposed to crabgrass, which has a spreading form (14.2.2). It prefers a relatively moist growing environment.

12. *Quackgrass (Agropyron repens)*. Quackgrass is a perennial weed that reproduces by seed and rhizomes. It has an extensive, fibrous root system. It is not a problem in many of the southern states.

13. *Johnsongrass (Sorghum halepense)*. Johnsongrass is a perennial weed that reproduces by seed and rhizomes. It is a greater problem in the southern states.

SUMMARY

Weeds are plants out of place. They may be annual, biennial, or perennial in life cycle. They compete with garden and landscape plants for growth factors. They may also harbor diseases and pests. Weeds are more adapted than cultivated plants and hence more competitive. The preponderance of certain weed species in a location often indicates soil fertility status. Weeds are spread through use of impure seeds, lack of phytosanitation, and other factors. Grass weeds are especially difficult to eradicate. Certain plants live parasitically on others.

OUTCOMES ASSESSMENT

PART A

Please answer true (T) or false (F) for the following statements.

1. T F A rose plant in a cornfield is a weed.
2. T F Weeds tend to be less aggressive than cultivated plants.
3. T F Dandelion is an annual weed.
4. T F Certain weeds are associated with certain cultivated plant species in the field.

PART B

Please answer the following questions.

1. List four weed species that grow naturally in acidic soils.

    _____ _____ _____ _____

2. List three each for the following:
    a. annual weeds _____ _____ _____
    b. perennial weeds _____ _____ _____

PART C

1. Why should weeds be controlled?
2. Describe the characteristics that make weeds more adapted and competitive than cultivated plants under marginal and harsh environmental conditions.

MODULE 2 ANIMAL PESTS OF PLANTS

6.4: INSECTS

Insects are classified under the phylum Arthropoda (characterized by jointed legs, exoskeleton, segmentation, and bilateral symmetry). They belong to the class Insecta (true insects) and along with species in the class Arachnida (spiders and mites) are the sources of most plant pests.

6.4.1 ECONOMIC IMPORTANCE

An estimated 80 percent of known animal life consists of insects. There are more than 800,000 species of insects known to humans, out of which less than 1 percent are classified as pests. Insects are not only abundant in numbers and diversity but also are widely distributed and adapted. Horticultural plants are attacked by a wide variety of insect pests. Some of them cause gradual and progressive damage, whereas others, such as swarms of locusts, can strip a vast acreage of green plants of every leaf in a matter of only a few days. Many modern plant species are grown from improved seeds and are able to resist a number of insect pests. However, modern cultivation often depends on the use of *insecticides* (7.2) to obtain economic yields. Insects inflict damage on plants, and some are carriers (vectors) of disease-causing organisms that are harmful to animals including humans.

Fortunately, certain insects are beneficial to the economic production of certain horticultural plants. For example, orchard crops depend on insects for pollination. In fact, in the production of crops such as strawberry, growers often deliberately introduce hives of bees into their fields for effective pollination and good yield.

6.4.2 IMPORTANT INSECT ORDERS

Insects are distinguished from other insectlike animals such as mites and spiders by having three pairs of jointed legs (the others have four pairs). Insects also have wings and antennae. Taxonomically, there are four major insect orders of economic importance in horticultural production:

1. *Lepidoptera.* Only the larva stage (6.4.3) of Lepidoptera causes economic damage to plants. Adults feed mainly on plant fruits. Of this order, several important families cause severe damage to plants. The family Gelechiidae includes stem borers, which are larvae that bore through shoots, stems, and fruits. The family Pyralidae includes stem borers and leaf rollers, whereas Noctuidae (owls and moths) consists of a large variety of leaf eaters, stem and fruit borers, armyworms, and cutworms.

2. *Coleoptera.* Coleoptera is the order for beetles, which are biting and chewing insects. Important families are Curculionidae (weevil proper), Chrysomalidae (leaf beetles) and Bruchidae (pulse beetles), which attack pods of legumes. The family Coccinellidae includes the beneficial lady bird beetle. The family Scarabaeidae includes white grubs, which inhabit the soil and feed on roots and tubers.

3. *Hymenoptera.* The family Formicidae (ants) of the Hymenoptera order consist of leaf-cutting ants. Bees belong to this order but are considered beneficial insects rather than pests.

4. *Diptera.* The adults of the order Diptera (flies) are not pests. The larvae are the pests of this order. The family Tephritidae (fruit flies) includes important pests of ripening fruits.

Other important orders are Thysanoptera (thrips) and Orthoptera (grasshoppers and crickets). The order Homoptera, which includes families such as Cicadellidae (leaf hoppers), Aphididae (aphids), and Pseudococcidae (mealybugs), is also important.

6.4.3 CLASSIFICATION OF INSECT PESTS

Based on Life Cycle

Like all living things, insects have a life cycle that consists of various stages of growth and development (starting from egg to adult). At one or more of these stages, insects may be capable of causing damage to plants. The processes of change that insects pass through is called *metamorphosis,* which is a basis for classifying insects into one of four classes.

> **Metamorphosis**
> Transformation of larva into an adult.

1. *No metamorphosis.* Insects that do not metamorphose are hatched as miniature adults, meaning they look like small versions of their adult forms (figure 6-1). Insect orders such as Thysanura and Collembola exhibit this type of metamorphosis.

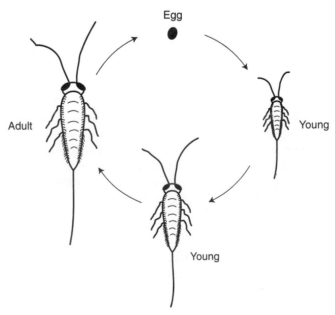

Egg

Young

Young

Adult

FIGURE 6-1 Life cycle of an insect with no metamorphosis.

2. *Gradual metamorphosis.* In gradual metamorphosis, eggs hatch into young insects called *nymphs* that do not have all of the adult characteristics. As they mature, nymphs exhibit adult characteristics (figure 6-2). Common orders of insects in this class are Homoptera, Hemiptera, Orthoptera, and Isoptera.

3. *Incomplete metamorphosis.* Insects exhibiting incomplete metamorphosis change shape gradually through the maturation process (figure 6-3). Insects in the equivalent of the nymph stage are described as *naiads*. An example of this order is Odonata (which includes, for example, dragonfly).

4. *Complete metamorphosis.* A complete metamorphosis consists of four distinct stages, in none of which the insect looks like the adult (figure 6-4). The egg hatches into a *larva* that feeds on plant foliage and other parts. The larva passes through a dormant stage called *pupa.* The mature adult emerges from the pupa as a beautiful butterfly, which is harmless to plants. Examples of insects in this class are Lepidoptera (butterflies) and Hymenoptera (bees). Bees and butterflies are important in aiding crop pollination. The larvae differ from one order to another. In Diptera (flies), the larva is called a *maggot*

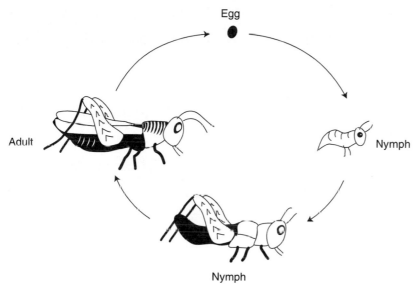

FIGURE 6-2 Life cycle of an insect with gradual metamorphosis.

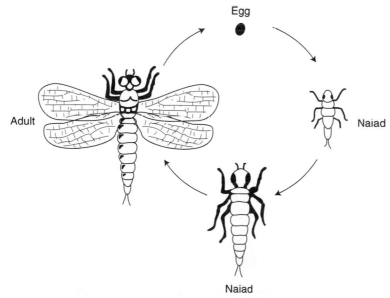

FIGURE 6-3 Life cycle of an insect with incomplete metamorphosis.

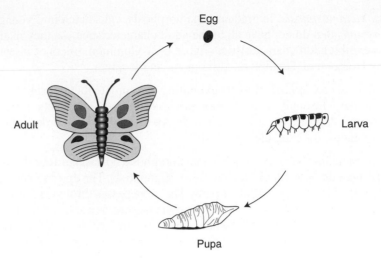

FIGURE 6-4 Life cycle of an insect with complete metamorphosis.

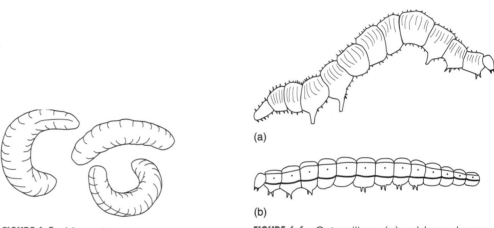

FIGURE 6-5 Maggots.

FIGURE 6-6 Caterpillars: (a) cabbage looper and (b) corn earworm.

(figure 6-5). The larvae of Lepidoptera (butterflies and moths) are called *caterpillars* (figure 6-6). Certain larvae tunnel through leaves, as occurs in the Diptera orders; they are called *miners*. Larvae of moths and beetles are able to tunnel through stems and other tissue and are called *borers*.

Classification Based on Feeding Habit

Insect pests can also be classified based on how they inflict damage on the plant through their eating habits. On this basis, there are two categories of insects, chewing insects and sucking and piercing insects.

Chewing Insects Chewing insects, as their name implies, chew plant parts (e.g., petals, leaves, stems, fruits, and flowers) during feeding. They have chewing mouthparts. Their damage is not limited to the above-ground parts of plants but also includes roots. This group includes larvae such as caterpillars and grubs, as well as adults such as grasshoppers, beetles, and boring insects. The tissue of the plant is destroyed in the process of feeding, the damage caused being more serious as the insect matures. Damage from chewing insects is easy to identify (figure 6-7). Symptoms include the following:

> *Defoliating.* Insects such as leaf beetles, caterpillars, cutworms, and grasshoppers devastate the foliage of plants by chewing portions of leaves and, in severe cases,

FIGURE 6-7 Typical damage caused by a chewing insect pest such as a cabbage worm. (*Source:* Photo provided courtesy of Oklahoma Cooperative Extension Service, Oklahoma State University)

stripping the plant completely of leaves. Defoliation causes a reduction in photosynthetic area, thereby reducing plant vigor and productivity.

Boring. The chewing mouthparts of borers are used to bore channels into succulent tissue—stems, fruits, tubers, and seeds. Examples of pests that bore into plant parts are corn borers and white grubs.

Leaf mining. Unlike borers, which bore directly into the tissue, leaf miners only tunnel between epidermal layers.

Root feeding. Root feeders damage plants from below in the soil by chewing roots and other underground structures. Examples are white grubs and wireworms.

Sucking and Piercing Insects   Sucking and piercing insects puncture the plant part on which they feed in order to suck out plant fluids. Examples of such insects include aphids, scales, mealybugs, thrips, and leaf hoppers. Insects in this group are often small in size (figure 6-8). Even in their adult stage, they can be microscopic, making them difficult to readily detect and control.

Their effect is often recognized as curling up or puckering of leaves (leaf distortion) or bleaching of leaves. Sucking and piercing insects also damage fruits. They are often found on the undersides of leaves. Certain sucking insects may inject toxins into the plant in the feeding process. Another characteristic of damage due to sucking insects (also found with some chewing insects) is an abnormal growth called a *gall* (figure 6-9).

Identifying the kind of insect is a critical first step in their control. Insects, like other living things, have a preference for feeding time. Some avoid the bright daylight, feeding only when it is dark. As such, some insect pests hide on the undersides of leaves or even retreat and hide in the soil at the base of the plant during the day. Being aware of such habits not only helps in identifying insect pests but also aids in their control. Mealybugs and scales often dwell in colonies and at some stage may secrete a waxy layer over themselves for additional protection; that layer is difficult to penetrate by many insecticides.

6.4.4 CLOSE RELATIVES OF INSECTS

Mites

Mites are insectlike organisms. They belong to the order Acrina, of which the family Tetranychidae (spiders and mites) is very important in horticultural plant production. They have four pairs of legs (not three) and have no antennae or wings. They are among the most widely distributed of pests, affecting a variety of plant species including ornamentals, fruits, vegetables, and field crops. They occur in the field and greenhouse (11.1). Mites are sucking insects that reproduce rapidly and frequently. Where infestation is high, they form fine webs

(a)

(b)

FIGURE 6-8 Sucking insects: (a) aphids and (b) mealybugs. (*Source:* Photos provided courtesy of Oklahoma Cooperative Extension Service, Oklahoma State University)

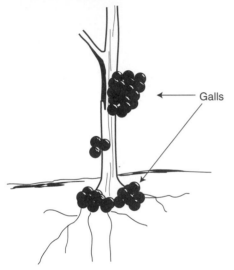

← Galls

FIGURE 6-9 Galls can develop on both the above- and below-ground parts of plants.

FIGURE 6-10 Typical spider mite damage. (*Source:* Photo provided courtesy of Oklahoma Cooperative Extension Service, Oklahoma State University)

on plant parts (figure 6-10). Spider mites (*Bryobia praetiosa, Tetranychus urticae,* and *Panomychus ulmi*) are economically important horticultural pests.

Spiders

Spiders are closely related to true insects. They belong to the class of arthropods (jointed legs) called Arachnida, which also includes mites.

Millipedes and Centipedes

Millipedes (Diplopoda) and *centipedes* (Chilopoda) occur most commonly in damp areas such as beneath stones, piles of leaves, and logs. They rarely damage plants directly and prey on small insects and spiders.

6.4.5 STORAGE PESTS

Insects inflict considerable damage on stored food. A variety of weevils, moths, beetles, and flies are known postharvest pests that damage stored products. Examples include the sawtoothed grain beetle, dried fruit beetle, raisin moth, and *Drosophila.*

FIGURE 6-11 A Colorado potato beetle. (*Source:* Photo provided courtesy of Oklahoma Cooperative Extension Service, Oklahoma State University)

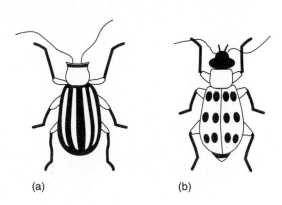

(a) (b)

FIGURE 6-12 Cucumber beetles: (a) striped and (b) spotted.

6.4.6 SOME IMPORTANT GENERAL INSECTS AND MITES

Horticultural plants are plagued by numerous insects and mites. The following are some of the important types:

1. *Colorado potato beetle (Leptinotarsa decimilneata).* Commonly called the potato beetle, this insect also feeds on tomato, eggplant, and others. Both the larvae and adults are pests. They occur throughout the country, wherever the host plants are found (figure 6-11).

2. *Cucumber beetles (Diabrotica undecimpunctata [12-spotted beetles]; Acalymma trivittata [stripped beetle]).* Cucumber beetles feed on leaves, stems, and fruits. The larvae attack plant roots. Two species are very important—the 12-spotted and stripped beetles (figure 6-12).

3. *Cutworms.* Cutworms are pests of vegetable garden crops. They are nocturnal and thus hide under rocks and debris during the day. They especially damage recently transplanted seedlings by cutting them off at or below soil level (figure 6-13).

4. *Cabbage looper (Trichoplusia ni).* The cabbage looper feeds on cruciferous garden plants (cabbage family) and many other garden species. It is one of the most common garden caterpillars and is identified by a characteristic loop made by its body as it moves forward. The larvae make holes in the plant parts as they feed, but sometimes the plant may be completely devoured.

5. *Corn earworm (Heliothus zea).* Sometimes called tomato fruitworm, corn earworm prefers to feed on corn and tomato fruits. The larvae damage corn by eating the succulent kernels. In the feeding process, the larvae leave behind large quantities of feces. In tomato, the earworm bore through the succulent fruits (figure 6-14).

6. *Squash vine borer (Melittia calabaza).* The squash vine borer attacks the vines of pumpkin, squash, cucumber, and melon, among others. They tunnel through the vines of these plants, resulting in wilting and eventual death.

7. *Leaf miners.* Leaf miner damage is visible as tunnels between leaf surfaces (figure 6-15). These tunnels are caused by larvae of flies such as the serpentine leaf miner (*Liriomyza* spp.). When a leaf is heavily infested, it yellows and eventually drops. Leaf miners attack a variety of garden plants such as tomato, pepper, eggplant, and especially lettuce and melon.

8. *Spider mites (Tetranychus spp.).* Mites are piercing or sucking insects that live on the undersides of leaves. They vary in color from reddish-yellow to light green. They attack a variety of garden crops, especially tomato and cucurbit. After feeding, the

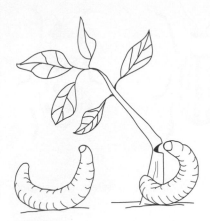

FIGURE 6-13 Cutworms chewing a young shoot.

FIGURE 6-14 Corn earworm damage. (*Source:* Photo provided courtesy of Oklahoma Cooperative Extension Service, Oklahoma State University)

FIGURE 6-15 Leaf miner damage on spinach. (*Source:* Photo provided courtesy of Oklahoma Cooperative Extension Service, Oklahoma State University)

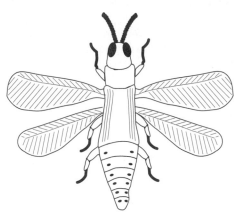

FIGURE 6-16 Thrips.

leaves appear bronzed or discolored. Severely attacked plants whither and drop. Spider mites have the capacity to produce 6 to 10 generations each year. They devastate ornamentals, fruits, nuts, vegetables, and numerous other species.

9. *Thrips.* Sucking insects, thrips are very common in the garden. Plants attacked have distorted leaves. Onion thrips (*Thrips tabaci*) and western flower thrips (*Frankliniella occidentalis*) are two of the most common thrips in the garden and landscape (figure 6-16).

10. *Aphids.* Also called plant lice, aphids are common on young leaves and growing tips of stems. One of the most common of the aphid species is the green peach aphid (*Myzus persicae*), which attacks many garden crops. Aphid populations are effectively held in check where natural enemies such as lacewings occur. This sucking insect affects many ornamentals, orchard crops, and vegetables. They are important because they are able to transmit a number of viruses that cause devastation to horticultural crops, examples being cucumber mosaic, bean mosaic, and lettuce mosaic. Aphids cause plants to grow with less vigor.

11. *Ants.* Ants are common in the garden and the landscape. One of the most common is the Argentine ant (*Iridomyrmex humilis*). It feeds on honeydew secreted by aphids. The leaf cutter ant (*Acromyrmex* spp.) cuts pieces of leaves for food.

6.4.7 CONTROL

Insects are commonly controlled by the use of chemicals. However, they may be controlled by cultural, legislative, and biological methods. Control methods are discussed in detail in chapter 7.

6.5: DISEASES

Cultivated plants are usually more susceptible than their wild relatives to diseases. This susceptibility is due in part to the fact that many different cultivars have uniform genetic backgrounds as a result of the activities of plant breeders (5.7) using the same breeding stock in developing new cultivars. Further, modern cultural practices allow plants to be grown in dense populations (as in monoculture), which facilitates the spread of disease.

Horticultural plants are attacked not only by insects but also by a large number of microorganisms that cause infectious diseases. These parasitic *pathogens* (disease-causing organisms) can be placed into four categories—*fungi, bacteria, viruses,* and *mycoplasma-like organisms.* Like all living things, these organisms have their own life cycles. Since they are parasitic, they need a susceptible host on which to survive.

Plant diseases may be classified on the basis of the causal organism, such as fungi, bacteria, viruses, and microplasma-like organisms.

> **Pathogen**
> *An organism that causes disease.*

6.5.1 FUNGI

An estimated 75 percent of all seed plant species live in some form of association with fungi in their roots, which is known as *mychorrizae.* This association is called *mutualism,* where both host and fungus benefit, and is similar to *symbiosis,* a mutually beneficial plant-bacteria association found in legumes only. Mychorrhizal fungi are known to be more efficient than plant roots in absorbing phosphorus. They are also known to be essential for the growth and development of forest trees and herbaceous species. Acid rain (3.2.1) destroys michorrhizae.

The reproductive structures from which fungi develop are called *spores.* Spores occur in a tremendous variety of shapes, sizes, and colors. A few fungi have no spore stage. When spores germinate, they produce *hyphae,* which grow and branch to form the fungus body called *mycelium.* Certain spores are visible as mold growth on the leaf surface. Fungal diseases such as *Helminthosporium* leaf spots, rusts, powdery mildew, and cercospora leaf spots are examples of this mold growth form. Other fungi occur as tiny dark fruiting bodies that are embedded in the tissue of the diseased plant. Examples of fungi are *Septoria* and *Ascochyta.*

Fungal spores are transported in a variety of ways— by wind, water, birds, insects, spiders, slugs, and mites. Some of them have a protective covering that enables them to survive adverse environmental conditions. In order to infect, hyphae of fungi gain access to the host through wounds (caused by equipment, pests, hail, ice, and so forth) or natural openings such as stomata (2.4.2). Sometimes the pathogen penetrates the epidermal layer by direct action on that layer.

Most of the infectious plant diseases are attributed to fungi. Fungi are either unicellular or multicellular (mostly) plants that lack chlorophyll. More than 250,000 species of fungi have been described, of which about 22,000 are known to cause plant disease. Some of them can live only on dead tissue (*saprophytic*), such as organic matter, while others live on living tissue (*parasitic*). Some of them are restricted (*obligatory*) to one host type (dead or living), while others have flexibility (*facultative*). Those that feed on dead and decaying matter are beneficial to plants because they aid in the decomposition of organic matter or compost to release nutrients for plant use. In the lawn, they aid in the decomposition of thatch (accumulated dead grass on the surface of the soil) (14.5.1).

All fungi are not pathogenic. Many are useful to humans and plants. *Penicillin* (from *Penicillium*) is one of the most important antibacterial drugs. Mushrooms (figure 6-17) used for food are fungi; fermented beverages and foods (e.g., bread, wine, cheese, and beer) depend on fungi (yeast) in their production.

> **Saprophyte**
> *An organism that derives its nutrients from the dead body or the nonliving products of another plant or animal.*

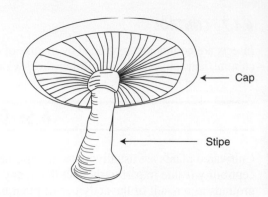

FIGURE 6-17 A typical mushroom.

Cap

Stipe

Table 6-1 Selected Fungal Diseases of Horticultural Plants

Common Name	Pathogen	Example of Plant (Host)
Damping-off	*Pythium* spp. *Rhizoctonia* spp.	Common problem of seedlings in the nursery
Downy mildew	*Plasmopara viticola*	Grape
Rust	*Puccinia striiformis* *P. graminis*	Turfgrasses
Late blight	*Pythium infestans*	Tomato and potato
Powdery mildew	*Erysiphe polygoni* *E. graminis* *E. cicoracearum*	Many different plants Cereal Cucurbits
Brown rot	*Monilinia fructicola*	Peach, plum, almond, and other stone fruits
Dutch elm	*Ceratocystis ulmi*	Elm tree
Fusarium wilt	*Fusarium oxysporum*	Pea, tomato, and watermelon

Even though most plant diseases are caused by fungi, they are usually relatively easy to control. Methods of disease control are described later in this chapter. Table 6-1 shows some important plant fungal diseases, their symptoms, causal organisms, hosts, and methods of control.

6.5.2 BACTERIA

Bacteria are unicellular organisms that may occur in one of three basic shapes—spherical, rodlike, or spiral. They are ubiquitous in the environment. Bacteria are also classified according to *Gram reaction* (positive or negative for violet or pink-red reaction to Gram's stain).

Bacteria cause very few known economically important diseases in horticultural plants. However, the diseases they cause are difficult to control. Pathogenic bacteria have rodlike shapes and prefer a pH of about 6.5 to 7.5 (3.3.3). Most of them are intolerant of high temperatures and will die if exposed to temperatures of 125°F (51.7°C) for about 10 minutes. The pathogenic bacteria that attack plants are mostly facultative parasites. They enter the host through wounds or natural pores. These unicellular organisms are not always pathogenic, some, like fungi, being beneficial to humans and plants. The bacterium *Escherichia coli* (commonly called *E. coli*) occurs in the intestinal tract of humans. Bacteria are involved in the decomposition processes in septic tanks and other sewage systems and compost heaps (21.2). Bacteria-plant associations (symbioses) involving legumes and *Rhizobia* species result in the fixation of nitrogen in the host plant's roots. Bacteria multiply rapidly by simple fission (mother cell divides in half).

Pathogenic bacteria and their hosts, symptoms, and control measures of diseases are described in table 6-2. Bacteria cause rots, cankers, spots, wilts, and blights. Bacterial diseases

Table 6-2 Selected Bacterial Diseases of Horticultural Plants

Common Name	Pathogen	Plant (Host) Attacked
Bacterial wilt	*Erwinia tracheiphila*	Cucumber and other cucurbits
Bacterial soft rot	*Erwinia carotovora*	Vegetables, fruits, and tubers
Crown gall	*Agrobacterium tumifasciens*	Tree fruits and woody ornamentals
Bacterial canker	*Pseudomonas syringae*	Pitted fruits: cherry, peach, and plum
Common blight	*Xathomonas phaseoli*	Field bean, lima bean, and snap bean

Table 6-3 Selected Viral Diseases of Horticultural Plants; Viruses Are Given Descriptive Names Based on the Diseases They Cause

Tobacco mosaic virus (TMV)
Tomato ring spot virus (TomRSV)
Tomato spotted wilt virus (TSWV)
Potato leaf roll virus (PLRV)
Potato virus Y (PVY)
Cucumber mosaic virus (CMV)
Bean common mosaic virus (BCMV)
Curly top virus of sugar beet (CTV)
Prunus necrotic ring spot virus (PNRV) (affects most stone fruits)

can be reduced through the observance of phytosanitation, but chemicals are not very effective against them. Bacterial disease–resistant cultivars are available for many species. Bacteria can overwinter in or on plant materials.

6.5.3 VIRUSES

Technically, *viruses* cannot be described as animals. They generally consist of a core of RNA or DNA (5.2.4) encased in protein or lipoprotein. They cannot be cultured in vitro. They act like living organisms only when found inside of a living cell of a plant or animal. Viruses are not like bacteria and fungi in that they are not capable of digestion and respiration. They are obligate parasites and operate by infecting a host cell and taking over its hereditary machinery for the production of viral DNA (or RNA). The consequence is an altered host metabolism. Viruses are microscopic and can be transmitted from infected plants to healthy ones through the feeding action of insects (*vectors*) such as aphids, leaf hoppers, and mites.

Viral diseases plague horticultural plants but to a lesser extent than fungi (table 6-3). However, once infested, viral diseases are virtually impossible to treat. The disease may not kill the plant but instead stunt growth. The vectors may be controlled to curtail the spread of the disease. Exposure of infested plants to high temperatures (38°C or 100.4°F) for 20 to 30 days (*heat therapy*) has been used to inactivate certain viruses. Actively growing tips of plants are virus free and provide another effective means of producing disease-free plants from infected plants (see propagation of plants by *tissue culture* [9.12]). Viruses are often described according to the species they infect (host specificity), (e.g., lettuce mosaic virus, maize dwarf mosaic virus, and sugar beet curly top virus). They are also described according to serological properties, particle morphology, mode or modes of transmission, morphology, and nucleic acid type (DNA or RNA).

Common symptoms of viral infection are yellowing of leaves, loss of vigor, poor growth, and stunting. Viruses are primarily systemic in the host plant's vascular fluids and hence transmitted readily by sucking insects. They commonly remain indefinitely in biennial or perennial hosts for as long as the host remains alive (except tobacco mosaic virus [TMV], which can thrive on dead plant tissue). To control viruses, one may use resistant cultivars, control the vectors, or limit their transmission in vegetative plants or their parts. Some viruses are transmitted by seed. Sometimes infected plants may show no symptoms (symptomless).

6.5.4 MYCOPLASMA-LIKE ORGANISMS

Mycoplasmas, which occur in the phloem of plants, are parasites that are intermediate in size between viruses and bacteria. Like viruses, mycoplasmas are technically not animals. Some of the symptoms associated with mycoplasma-like infections include yellowing, stunting, wilting, and distortions, as in viral infections. Aster yellows have been identified in, for example, carrot, strawberry, lettuce, phlox, and tomato.

6.6: OTHER PESTS OF HORTICULTURAL PLANTS

Apart from insects, fungi, bacteria, viruses, and mycoplasma-like organisms, horticultural plants are plagued by other animal species. They may not cause diseases but can nonetheless cause economic loss to growers. The major ones include small animals, birds, nematodes, and snails and slugs.

6.6.1 SMALL ANIMALS

Rabbits, mice, gophers, bats, and moles are among the rodents and other small animals that plague cultivated plants. They may attack plants from above or below the ground. Moles, for example, damage lawns from underground. They may eat seeds when planted, causing incomplete plant stand in the field, or they may eat mature fruits and seeds. These mammals, especially the ones that bore holes in the ground, may improve drainage.

6.6.2 BIRDS

Birds cause damage similar to that effected by small animals. Fruits and cereal crops are particularly prone to bird damage. On the other hand, birds feed on insects at all stages in their life cycles thereby playing a beneficial role in controlling some insect pests.

6.6.3 NEMATODES

Nematodes are unsegmented roundworms. They are microscopic plant parasites that attack above- and below-ground parts of plants, including vegetables, fruit trees, ornamentals, and foliage plants. However, most that are parasitic on plants are soilborne. The most common nematodes include root knot (*Meloidogyne* spp.) and cyst (*Heterodera* spp.). They are spread by equipment, water, shoes, and other means. On the roots, they cause irregularly shaped knots that interrupt nutrient flow in the plant. This amorphous growth is distinguishable from the well-defined and usually round shape of *root nodules* produced in legume-*Rhizobia* symbiosis (figure 6-18). The plant may be discolored (abnormal yellowing) and stunted in growth. Tuber crops may have disfigured and unsightly skin.

Many nematodes are inactive at temperatures below 10°C (50°F). A field infested with nematodes can be managed by cultural operations to lower the population of the parasites. When the populations are high, fumigation is most effective, even though it is not possible to irradicate the pest from the soil. Nematodes are especially difficult to control in dormant form (cysts or eggs).

6.6.4 SNAILS AND SLUGS

Nucturnal creatures, snails and slugs are similar in many respects, one distinguishing feature being the lack of a shell in slugs (figure 6-19). The presence of these mollusks is betrayed by the slimy trail they leave behind as they move on the soil surface. They feed on leaves and young plant stems. Because of their nocturnal habits, they hide during the daytime in cool and moist places such as under rock, debris, and mulch. Slugs cause more damage than snails to horticultural plants. The most common garden snails are the brown garden snail (*Helix aspersa*) and the decollate snail (*Rumina decollata*). The former has a globular spiral shell and the latter a cone-shaped spiral shell. The most common slugs in the home garden are the spot-

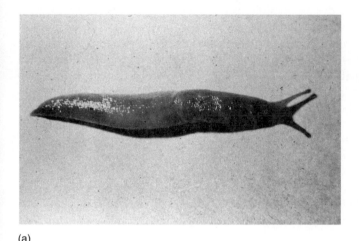

FIGURE 6-18 Nematode damage as distinguished from root nodules formed as a result of *Rhizobia* infection.

Root nodules
by *Rhizobia*

Root nematode attack

(a)
(b)

FIGURE 6-19 (a) Slug and (b) snail. (*Source:* Photos provided courtesy of Oklahoma Cooperative Extension Service, Oklahoma State University)

ted garden slug (*Limax maximus*) and the tawny garden slug (*Limax flavus*). The tawny slug leaves a trail of yellow slime. The gray garden slug (*Agricolimax reticulatus*) is small and leaves a clear slime trail.

6.7: How Disease Occurs

6.7.1 DISEASE FACTORS

Every organism has a life cycle during which it reproduces, develops, grows, and dies. Disease conditions involve the interaction between the life cycles of the host and the pathogen. To understand the nature of disease, it is important to distinguish between *parasitism* and *pathogenicity*. An organism that lives in or on another for the purpose of deriving food is called a parasite, and the relationship between the two individuals (parasite and host) is called parasitism. A parasite deprives its host of food, making the latter less vigorous and less productive. This one-sided relationship may not go beyond depriving the host of nutrients. In certain cases, the relationship is mutually beneficial (symbiotic), as occurs in the root nodules of legumes where *Rhizobia* reside. The bacteria derive nutrients from the legumes and in return fix atmospheric nitrogen into usable form for the plant. Sometimes the activities of the parasites coupled with the reaction of the host result in abnormal physiological activities in the host and physical degeneration of cells and tissue. This condition describes the state of pathogenicity, and the organism associated with it is called a *pathogen*.

Pathogens differ in the plant types, parts, tissues, and organs they can successfully attack and grow on. Obligate parasites tend to be *host specific* (limited to one species), whereas

> **Pathogenicity**
> The capability of a pathogen to cause disease.

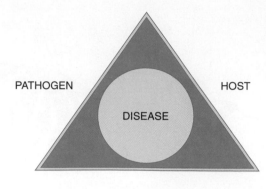

FIGURE 6-20 The disease triangle.

PATHOGEN

HOST

DISEASE

ENVIRONMENT

nonobligate parasites can produce disease symptoms on a variety of plants. The variety of plants a pathogen can grow on is called its *host range*. For disease to develop, three ingredients must be present: (1) a pathogen (causal agent), (2) a susceptible host (plant on which the pathogen can grow), and (3) a favorable environment. This grouping constitutes the *disease triangle* (figure 6-20). Disease will not occur if one of these components is not present. Each component can occur to a variable extent, which affects the degree of disease development. For example, the pathogen may be present in small or large numbers, be a highly virulent or less virulent race, and be in active or dormant stage. The host may be genetically resistant to the pathogen, or it may be too young or too old. The environment may be too dry or too wet, too cold or too warm, or very favorable.

Disease development occurs in stages. The chain of events leading to the development and perpetuation of the disease is called the *disease cycle*. The primary events in a disease cycle include inoculation, penetration, infection, dissemination, and overwintering or oversummering.

Inoculation

The coming into contact of host and pathogen is called *inoculation*. The pathogen (or any part of it) that can initiate infection is called the *inoculum*. Examples are spores and sclerotia (in fungi) or the whole organism in bacteria, viruses, and mycoplasmas. Since an inoculum may consist of one or more individual pathogens, the unit of inoculum of any pathogen is called a *propagule*. The inoculum that causes the original infection is called the *primary inoculum*, and the subsequent infection is called the *primary infection*. Primary inoculum produces secondary inoculum, which produces secondary infection. The inoculum may be harbored by plant debris, soil, or planting material (e.g., seeds and tubers). It is transmitted to the host either passively, by wind, water, or insects, or actively, by *vectors* (e.g., insects and mites).

Penetration

Whereas all pathogens in their vegetative state are capable of initiating infection immediately, fungal spores and seeds of parasitic higher organisms must first germinate. Germination requires good conditions. Before penetration, recognition between the host and the pathogen must occur. The recognition process may trigger a defense reaction in the host to resist penetration. If successful, disease development will be curtailed.

Penetration may occur through natural openings (e.g., stomata), wounds, or intact membrane boundaries (figure 6-21). Fungi frequently penetrate the host directly. Bacteria and viruses require existing wounds or those caused by vectors to gain entry into the host.

Infection

Infection
The establishment of a parasite within a host.

The infection stage in disease development involves contact, growth, and multiplication of pathogens in host tissues. The symptoms of successful infection may include necrosis and discoloration of tissue. Symptoms may not always develop immediately; certain infections

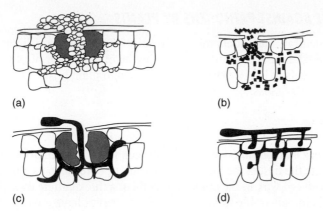

FIGURE 6-21 Penetration of pathogens. (a) Bacteria penetrating through a natural opening like a stoma. (b) Bacteria penetrating a leaf through an injured part. (c) Fungus penetrating through a natural opening. (d) Fungus penetrating through an intact, undamaged leaf surface.

remain *latent* until conditions are more favorable. The time interval between inoculation and the onset of disease symptoms is called the *incubation period.* During infection, the tissue of the host may or may not be killed. The success of infection depends on favorable environmental conditions.

After penetration, pathogens invade the host cells or tissue. Invasion relates to the spread of infection. Many infections caused by fungi, bacteria, and nematodes are *local* (limited to one or a few cells and involving small areas on the plant). On the other hand, infections caused by mycoplasmas and all natural viral infections are *systemic* (spreads to or invades all susceptible cells and tissue throughout the plant). Certain fungal diseases (e.g., downy mildew, rusts, and smuts) are systemic in the way they invade the host. Part of the infection process is growth and reproduction, which leads to colonization of the host cells or tissue. With reproduction, new cells are invaded by the pathogen.

Responses to Invasion by Pathogens   When plants respond to an invasion by a pathogen, a disease is said to have developed. There are three general responses to invasion:

1.   The affected host may experience excessive abnormal growth such as galls (as in crown gall) or curling (e.g., peach leaf curl). This growth is called *overdevelopment of tissue.*

2.   Instead of excessive growth, the plant may be stunted (as occurs with most viral infections) or plant organs may be only partially developed. This response is described as *underdevelopment of tissue.*

3.   To contain the infection, the cells around the point of attack may die so that the organism cannot spread any further. This response of tissue death is called *necrosis (hypersensitive reaction).* Diseases such as leaf spots, blights, cankers (e.g., anthracnose), and decays (e.g., soft rot) are symptoms of diseases involving tissue death.

Dissemination

Agents of dissemination are air, water, insects, other animals, and humans. Fungal spores are usually spread by air currents. Bacteria and nematodes are often disseminated by water. Viral infections are transmitted by vectors. Pathogens are carried on farm tools and equipment and on the feet or bodies of animals.

Overwintering or Oversummering

Annual plants die after a season, infection or no infection. The pathogen must therefore be able to survive in the absence of its host. Similarly, pathogens of perennial plants must survive during the period of dormancy while adverse weather prevails. For example, nematodes overwinter as eggs in plant debris. Some fungi inhabit the soil.

6.7.2 DEFENSE AGAINST PATHOGENS BY PLANTS

Plants have structural and biochemical means of combating an invasion by pathogens. Structural and biochemical defense mechanisms involve the use of physical barriers or the secretion of chemicals to block entry of pathogens into plant hosts. Plants differ in the types of mechanisms they possess.

Structural Defense

Preexisting Defense Mechanisms　　When a pathogen lands on the surface of the plant and attempts to penetrate the tissue, the plant's first means of defense is its epidermis (2.2.2). Some plants have deposits of wax in the epidermal cells or a thick cuticle over the epidermal layer, either of which is impenetrable to pathogens. Some leaves have a waxy, water-repellent surface or a pubescent surface that makes it difficult for water to settle to provide the moisture certain pathogens such as fungi and bacteria require to grow or multiply. The presence of such a tough material severely limits direct penetration of pathogens, thus reducing infection because the remaining avenues for entry are the natural openings (stomata).

Induced Defense Structures　　If the pathogen gains entrance into the tissue, the presence of the alien matter may induce the formation of a cork layer for containment of the invasion. In some cases, especially in stone fruits, an abscission layer may be induced to form in active leaves. This layer causes a permanent break between the site of infection and the rest of the leaf, consequently forestalling the advancement of the infection. In some plants, an injury (e.g., resulting from penetration) induces gums to be formed around the point of infection, thus limiting its spread. Another induced structure is tissue necrosis, a hypersensitive reaction to invasion. A cell that is penetrated by the hyphae of fungi, for example, soon experiences degeneration of the nucleus, followed ultimately by death. Death of tissue means the fungus has no access to nutrients, leading to the eventual death of an obligate parasite.

Biochemical Defense

Plants under attack by pathogens may secrete certain metabolites to ward off attack.

Preexisting Mechanisms　　Certain plants under normal conditions exude chemicals onto the plant surface (stem, root, or leaves). Tomato, for example, is known to release *fungitoxic exudates* in its leaves, which in high enough concentrations inhibit the germination of the fungus *Botrytis*. The scales of red onions are known to exude red pigments (protocatechaic acid and catechol) that make this variety resistant to onion smudge fungus; the white-scaled variety remains susceptible.

In another scenario, a pathogen may fail to attach to and infect a plant if it is unable to recognize the plant as one of its hosts. For example, in a viral attack, no infection may occur if the viral nucleic acid is not compatible with the host nucleic acid. Some pathogens require certain essential nutrients for proper growth and development that, when absent, make the host less susceptible, or more resistant, to the pathogen. In bacteria soft rot (*Erwinia carotovora* var. *atroseptica*), the disease is less pronounced in potato cultivars that are low in reducing sugars than in cultivars that are high in reducing sugars.

Induced Mechanisms　　When a plant is injured, the cells in the area secrete certain chemicals to initiate healing. These chemicals include a variety of phenolic compounds and others such as chlorogenic acids and caffeic acids. These chemicals are already present in the plant. However, an injured plant may secrete a special group of compounds called *phytoalexins*, which do not occur naturally in healthy plants. Phytoalexins are produced in large amounts when a plant is stimulated by injury. Most of the known phytoalexins are induced by fungal infections. Examples are *phaseolin* (from bean), *pisatin* (from pea), *capsidol* (from pepper), and *rishitin* (from potato). In certain plants, detoxification of toxins from pathogens (e.g., pyricularin) is known to occur.

Resistance to certain pathogens, including fungi, bacteria, viruses, and insects, has been induced (*induced resistance*) in many plants by treatment with biotic agents. In tobacco, an infection of a hypersensitive cultivar with TMV induces a nonspecific resistance to TMV, fungi (*Phytophthora*), and aphids, among others.

6.7.3 GENETICS OF DISEASE RESISTANCE

True resistance is disease resistance genetically controlled by one, a few, or many genes. There are two kinds of true resistance—horizontal and vertical.

Horizontal Resistance

Horizontal resistance is also called *polygenic* or *multigenic resistance* because it is controlled by numerous genes. Each gene contributes toward total resistance and hence is said to have *minor gene resistance*. Horizontal resistance is controlled largely by environment and thus may vary under different environmental conditions. Generally, it does not protect plants from becoming infected. Instead, it slows down the spread of disease and the development of epidemics in the field. Every plant has some degree of horizontal resistance.

> **Resistance**
> The ability of an organism to exclude or overcome, partially or completely, the effect of a pathogen or other damaging factors.

Vertical Resistance

Vertical resistance, also called *qualitative, specific,* or *differential resistance,* is usually controlled by one or a few genes (*monogenic* or *oligogenic resistance*). Plants with this kind of resistance may be resistant to certain races of a pathogen while remaining susceptible to others. Because the resistance gene plays a major role in the expression of resistance (instead of contributing in small amounts toward the overall resistance), this kind of resistance is also called *major gene resistance.* When a pathogen attacks a plant with vertical resistance to the disease, the plant usually reacts in a hypersensitive way, thus preventing the pathogen from becoming established.

Plants with vertical resistance usually show complete resistance to a particular pathogen under a wide variety of environmental conditions. However, it takes only one or a few mutations in the pathogen (new race) to make the previously resistant plant susceptible to the disease. This kind of resistance is hence not *durable* (horizontal resistance is *durable resistance*). The ideal genetic system for disease resistance is both horizontal and vertical resistance in one cultivar.

6.7.4 APPARENT RESISTANCE

Certain plants known to be susceptible may not be adversely affected by pathogens under certain conditions. Two kinds of apparent resistance are known—escape and tolerance.

Escape

Plants may not show symptoms of disease because one of the three critical ingredients in the disease triangle may be missing (e.g., required temperature may not be present or moisture may be inadequate). In some plants, the young tissue may be susceptible to pathogens and an older one may not. For example, powdery mildew (caused by *Phytium*), bacterial, and viral infections affect younger tissue more severely than older tissue. By manipulating the cultural environment (e.g., spacing, planting date, pH, fertility, and moisture), susceptible plants may escape diseases.

Tolerance

Plants are said to be tolerant when they are able to grow, develop, and be fairly productive even when infected with a pathogen. Most plant viral infections produce this kind of resistance, in which the pathogen (virus) does not kill the host but causes reduced productivity or performance.

Table 6-4 The Genetics of Disease Resistance in Plants

Virulent or Avirulent Genes in the Pathogen	Resistance of Susceptibility Genes in the Plants	
	R (resistant dominant)	r (susceptible recessive)
A (avirulent dominant)	AR (−)	Ar (+)
a (virulent recessive)	aR (+)	ar (+)

Where − is incompatible (resistant) reaction (no infection) and + is compatible (susceptible) reaction (infection develops).

AR is resistant because the plant (host) has a certain gene for resistance (R) against which the pathogen has no specific virulence (A) gene. This does not mean other virulence genes do not occur.

Ar is susceptible due to lack of genes for resistance in the host and hence susceptible to other virulence genes from the pathogen. aR host has the resistance gene, but the pathogen has a virulence gene that can attack it.

ar is susceptible because the plant is susceptible and the pathogen is virulent.

6.7.5 GENETIC BASIS OF DISEASE INCIDENCE

> **Virulence**
> The degree of pathogenicity of a pathogen.

As previously discussed, a virulent pathogen, a susceptible host, and an appropriate environment are all required for disease to occur. The virulence of a pathogen and the resistance of a host have genetic bases. For each gene that confers resistance in the host, there is a corresponding gene in the pathogen that confers virulence to the pathogen and vice versa, called the *gene-for-gene concept* of genetics of disease resistance and susceptibility. Generally (exceptions occur) in the host (e.g., plants), the genes for resistance are dominant (R) and the genes for susceptibility or lack of resistance are recessive (r). However, in the pathogen, the genes for avirulence (inability to infect) are usually dominant (A) and those for virulence are recessive (a). Therefore, when two cultivars, one carrying the gene R (for resistance to a certain pathogen) and the other carrying the gene r (for susceptibility) are inoculated with two races of the pathogen of interest, one of which carries the gene A (for avirulence against R) and the other race carrying the gene r (for virulence against R), the progeny will have the genotypes summarized in table 6-4.

SUMMARY

Insects are a major class of horticultural pests both indoors and outdoors. The economically important insect orders that affect plants are Lepidoptera, Coleoptera, Hymenoptera, and Diptera. Some insects chew, whereas others suck, during feeding. Some insects attack stored products. Important insect pests include aphids, fruit flies, and corn earworms. Spider mites (not true insects) are common pests of horticultural importance. Plant diseases are caused by fungi, bacteria, viruses, and mycoplasma-like organisms. Most infectious plant diseases are caused by fungi. Bacterial diseases of horticultural plants are few. Other animal pests include birds, rodents, and nematodes.

REFERENCES AND SUGGESTED READING

Agnos, G. N. 1988. Plant pathology. New York: Academic Press.

Bohmont, B. L. 1997. The standard pesticide user's guide. Englewood Cliffs, N.J.: Prentice-Hall.

Cravens, R. H. 1977. Pests and diseases. Alexandria, Va.: Time-Life.

Dixon, G. R. 1981. Vegetable crop diseases. Westport, Conn.: AVI Publishing.

Klingman, G. C., F. M. Ashton, and L. J. Noordhoff. 1982. Weed science: Principles and practices, 2d ed. New York: John Wiley & Sons.

Prone, P. 1978. Diseases and pests of ornamental plants, 5th ed. New York: John Wiley & Sons.

Ware, G. W. 1988. Complete guide to pest control, 2d ed. Fresno, Calif.: Thomson Publications.

OUTCOMES ASSESSMENT

PART A

Please answer true (T) or false (F) for the following statements.

1. T F All fungi are pathogenic.
2. T F Viruses are obligate parasites.
3. T F Green peach is a kind of aphid.
4. T F A mite is an insect.
5. T F Most plant diseases of economic importance are caused by fungi.
6. T F Butterflies belong to the order Hymenoptera.
7. T F A saprophyte lives on living tissue.
8. T F Necrosis refers to tissue death in plants.

PART B

Please answer the following questions.

1. List the four important insect orders that are pests of horticultural plants.

   _____ _____ _____ _____

2. Beetles belong to the insect order _____.
3. Insect pests may be classified as sucking or _____.
4. Bacteria cause diseases such as cankers, _____, and _____.
5. Necrosis is _____.

PART C
---

1. Describe the life cycle of an insect with complete metamorphosis.
2. What are the signs that a chewing insect pest has infested a plant?

CHAPTER 7

Principles and Methods of Disease and Pest Control

PURPOSE

This chapter is designed to classify the methods of disease and pest control and discuss the rationale behind their use, their effectiveness, and the environmental consequences of their use.

EXPECTED OUTCOMES

After studying this chapter, the student should be able to:

1. Discuss the general principles of pest control.
2. Discuss the rationale behind each of the four control strategies.
3. Classify pesticides.
4. Classify insecticides.
5. Classify herbicides.
6. Discuss the strategies for the safe and effective use of herbicides.
7. Describe the equipment used in the application of pesticides.
8. Describe the pros and cons of each pest-control strategy.

KEY TERMS

Active ingredients	Contact poisons (action)	Integrated pest management
Aerial applications	Dusts	Label
Aerosols	Eelworms	Lethal dose (LD_{50})
Alternative host	Emulsifiable concentrates	Miticides
Antagonism	Fogs	Molluscides
Band application	Formulation	Mollusks
Chlorinated hydrocarbons	Herbicides	Nematicides
Condensation	Insecticides	Nonselective herbicide

Organophosphates	Preplant application	Selective herbicide
Parasitism	Pyrethroids	Sterilization
Pesticides	Pyrethrums	Stomach poisons (action)
Phytoallexins	Repellents	Systemic poisons (action)
Phytosanitation	Rodenticides	Toxicity
Postemergence application	Scale insects	Wettable powders
Preemergence application		

Overview

In chapter 5 we learned that the expressed phenotype depends on the genotype (the kinds of genes) and the environment in which the genes are expressed ($P = G + E$). The environment *(E)* should not be limited to the growth factors (light, moisture, temperature, and nutrients), even though these are the essential components. *Biological competitors* in the general environment may compete with useful plants for these growth substances to the detriment of the latter or destroy tissues and interrupt physical and developmental functions of cultivated plants. In terms of the environment, some of these competitors are native, or endemic, to particular areas. Others are imported by a variety of modes.

Diseases and pests must be controlled because they cause economic loss to a horticultural operation. The loss may come as a result of

1. Increased cost of production (additional inputs)
2. Decreased yield
3. Decreased quality

For the home growers or those who cultivate plants as a hobby and not for sale, losses to horticultural plants may come in more subtle ways. There may be emotional drain from the disappointment of a ruined crop or blemishes on plants in the landscape that reduce their aesthetic value.

Diseases and pests must be controlled safely and economically. A cost-effective control calls for a good understanding of the nature of the disease (pathogen), the environment, and the plant species (host).

From the formula $P = G + E$, the effective control of pests can be handled by either changing the crop growing environment or improving on the nature of the plant (genetic constitution) to include disease resistance genes.

Module 1 Principles of Pest Control

The purpose of pest control is to minimize or completely eliminate the economic loss to a horticultural operation through reduced productivity, reduced product quality, or reduced aesthetics.

7.1: Control Strategies

7.1.1 PRINCIPLES OF CONTROL

Four basic principles are involved in pest control—exclusion, eradication, protection, and resistance.

Exclusion

Exclusion involves activities that prevent the pathogen from being introduced into a given area in the first place. If introduced, the conditions should prevent a pathogen from becoming established. Excluding the pathogen on a large scale often involves the enactment of government policies that make it illegal to import or export certain plant materials. This legislative control (or quarantine) is discussed in detail in module 2 of this chapter.

Eradication

If a pathogen succeeds in entering and becoming established in an area to some extent, measures may be undertaken to curtail its spread while reducing the current population until the pathogen is eventually eliminated completely from the area.

Protection

Protection entails the isolation of the host from the pathogen. Such isolation is usually accomplished by applying a chemical to the host. Physical methods of protection are also used.

Resistance

A form of protection of genetic origin is resistance, whereby a plant or host is equipped with disease-resisting genes. Resistance breeding (5.9) is undertaken by breeders to incorporate these genes into new cultivars through planned crosses and the use of other plant improvement strategies. The result of resistance breeding is a plant armed with natural means of defense, thus eliminating the need to use chemicals.

7.1.2 PREVENTING PEST ATTACK

Pest control is an additional production cost that can be eliminated or reduced by adopting certain preventive strategies. Some of these strategies are described as follows and include the observance of good cultural practices.

1. *Certain environmental conditions predispose plants to diseases.* Chapter 6 stated that one of the three factors that must be present for disease to develop is the proper environment (6.7.1). This environment includes proper temperature, light, and humidity. Warm and humid conditions often invite diseases; aeration is needed to reduce the creation of this kind of microclimate around plants. Plants should be properly spaced and humidity controlled (e.g., by watering at the right time of day to allow excess moisture to evaporate).

2. *Select and use adapted cultivars.* Plants have climatic conditions under which they grow and perform best (3.2.1). When grown in the wrong regions, plants are unable to develop properly and are more likely to succumb to diseases.

3. *Use pest-resistant cultivars.* If certain pests are prevalent in the production area, it is best to use resistant cultivars, if available, in any production enterprise. For soilborne diseases, tree seedlings grafted onto an appropriate stock may be desired (19.3).

4. *Plant at the best time.* Seasonal planting may prevent exposure to unfavorable climatic conditions. This approach applies mostly to annuals (1.3.1), which complete their life cycles in one growing season. Sometimes short-duration cultivars may be selected to successfully grow a crop in a short window of opportunity where conditions reduce pest incidence.

5. *Provide adequate nutrition.* Strong and healthy plants resist diseases better than malnourished ones. Soil testing (3.4.3) reveals the nutritional status of the soil so that fertilizer amounts (3.4) can be amended for adequate plant nutrition.

6. *Observe good sanitation.* Because plant remains left on the soil surface can harbor pathogens, debris should be removed or buried in the soil. Diseased plant parts can spread the problem to healthy plant parts. Similarly, wounds provide easy entry to disease organisms (6.7.1). As such, the horticultural operation of pruning (15.1) should be undertaken as needed to remove diseased plant parts, following up with proper wound dressing. Tools should be cleaned and disinfected periodically.

7. *Remove weeds.* Weeds compete for nutrients and also harbor diseases and other pests (6.1). Since weeds are volunteer plants in a cultivated plot, they are usually found in areas in which they are capable of performing under the prevailing conditions. They are thus more competitive than the cultivated crops, which require the grower's care.

8. *Use quality seeds or seedlings (or appropriate planting material).* Poor-quality seeds may have a high proportion of weed seeds and may also carry seedborne diseases. Obtain all planting materials from reputable nurseries or growers.

9. *Prepare the soil or growing medium properly for planting.* Depending on the tillage operation desired, weeds must be controlled by either killing them with chemicals (7.2) or plowing them under the soil. In greenhouse culture (11.1), the growing medium should be sterilized to eliminate pests. Garden soils can also be sterilized by the method of solarization (3.11). A well-prepared soil or growing medium should drain freely to avoid waterlogged conditions, which prevent good plant development.

These general strategies of disease prevention are applicable to all production types. In addition, some production enterprises may use unique strategies that help to reduce pest incidence.

7.1.3 DESIGNING CONTROL STRATEGIES

Effective control of diseases and pests should take into account the pathogen (e.g., bacterium), the host (species), and the environment (6.7.1). A good strategy should be effective, inexpensive, and safe (in terms of both application and residual consequences for consumers and the environment). From the perspective of the pathogen, the strategy should exploit the vulnerability of the pathogen by administering the control at the stage in the life cycle at which it is most vulnerable. The strategy should also consider the stage in the life cycle at which the organism is destructive to horticultural plants. Control measures should be effected *before* the destructive stage. Certain organisms are destructive at more than one phase in their life cycles. For example, an insect may lay unsightly eggs on the flowers or leaves of ornamental plants or fruits. When the eggs hatch, the larvae may be destructive and the adult harmless. What may be a problem or undesirable in one case may have no economic consequence in another. Eggs on flowers may be unsightly and undesirable, but for a plant whose economic part is the seed, blemishes on the pods may not affect the quality of the seeds they carry. A control strategy should also consider the feeding habits of the organism, since insects may either suck or chew plant parts.

From the perspective of the host plant, control strategies should consider inherent genetic capacity. Some diseases affect young plants and others older plants. Certain cultivars have disease resistance genes and are able to resist infection (6.7.3). To be effective, a control strategy should consider the environment in several ways. Weather factors such as temperature, precipitation, and winds limit the effectiveness of the control measure. Rainfall after pesticide application may wash away the chemicals. In this age of environmental awareness, there is a call for reduction in pesticide use, since residues end up in groundwater as pollutants.

Cultural practices should be considered in adopting a strategy for disease control. For example, by changing crop spacing, pruning, adopting crop rotation, weeding, and taking other cultural measures, disease incidence can be effectively controlled. Finally, the cost of the control measure should be considered.

7.1.4 PRINCIPLES AND METHODS OF CONTROL

Controlling Insect Pests

Based on the nature of the agents employed, there are six general methods of control.

1. *Biological control.* The principles involved in the biological control of pests are geared toward favoring organisms (natural enemies) that are antagonistic to the pest or pathogen, and improving the resistance of the host.

2. *Cultural control.* Cultural control employs the principles of protection and eradication by helping plants avoid contact with the pest or pathogen and reducing the population of or eradicating the causal organism in the area. Cultural methods depend on certain actions of the grower.

3. *Regulatory or legislative control.* Regulatory control involves the intervention of government with laws aimed at excluding the pathogen or pest from a given geographic area.

4. *Chemical control.* The chemical control of pests involves protecting plants from the pathogen or pest, curing an infection when it occurs, and destroying the pest if the attack is in progress.

5. *Mechanical control.* Insects can be controlled by mechanical methods that employ devices to prevent them from making contact with the plants or lure and entrap the insects.

6. *Integrated pest management (IPM).* The method of integrated pest management, as its name suggests, entails the use of a combination of the other general methods of pest control in a comprehensive approach to disease and pest control. However, efforts are made to minimize the use of chemicals.

The preceding methods are discussed in detail in the various modules in this chapter.

Controlling Diseases

The principles of disease control are exclusion, eradication, protection, and resistance.

1. *Exclusion.* The principle of exclusion entails the use of a method such as regulation to prevent the introduction of the pathogen into an area where it does not currently exist.

2. *Eradication.* When disease incidence occurs to a limited extent or is restricted in distribution, it is feasible to completely eliminate the pathogen from the area.

3. *Protection.* Plants can usually be protected from pathogens by applying a chemical that prevents the pathogen from infecting the host.

4. *Resistance.* Plant breeding programs aim at providing a level of resistance (not total immunity) to a disease in plants. Whereas certain plants may be susceptible to a disease, others may be able to resist it, depending on a variety of factors such as the age of the plant, the environment, and the aggressiveness of the pathogen.

These four basic principles apply to controlling insects. Similarly, some methods of insect control (e.g., cultural, regulatory, and chemical) are applicable to diseases. In controlling diseases and insect pests, four general strategies may be adopted, depending on whether the attack is yet to occur or is already in progress. A single strategy may involve the use of one or more principles and methods of pest control. These strategies are summarized in table 7-1.

7.2: Classification of Pesticides

Pesticides are chemicals designed to kill pests. They are frequently very toxic to humans and thus should be used judiciously and with care. Pesticides differ not only in chemical composition but also in killing action, toxicity, residual effect, specificity, species destroyed, cost, and effectiveness, among others.

Based on the type of organisms on which they are used, there are two broad categories of pesticides in horticulture:

Pesticide
A substance or mixture of substances used to control undesirable plants and animals.

1. *Pesticides used to control unwanted plants.* Pesticides used to control unwanted plants are called *herbicides.* Plants that are pests are generally called *weeds.* Weeds

Table 7-1 Strategies and Methods of Pest Control

Strategy 1:	**Exclude Pathogen from Host**	
Methods:	Quarantine (7.8)	
	Crop inspection	
	Crop isolation	
	Use of pathogen-free planting materials (9.12)	
Strategy 2:	**Reduce or Eliminate Pathogen's Inoculum**	
Methods:	Cultural	Crop rotation (7.7.10
		Host eradication (7.7.4)
		Improved sanitation (7.7.2)
		Improved crop growth environment
		Soil drainage (3.8.12)
		Aeration of soil (3.6)
		Proper soil pH (3.3.3)
		Proper soil nutrition (3.3.3)
		Remove weeds (7.9.6; 7.11)
	Physical	Heat treatment
		Solarization (3.1)
		Sterilization (7.9.6)
		Traps-polyethylene sticky sheets (7.9.1)
		Mulches (7.7.5)
	Chemical	Seed treatment (8.8)
		Soil fumigation (7.10.2)
	Biological	Trap crops (7.6.8)
		Antagonistic plants (repellants) (7.6.7)
Strategy 3:	**Improve Host Resistance**	
Methods:	Cultural	Improved crop growth environment
		Nutrition, moisture, drainage
	Biological	Genetic resistance (plant breeding)
		Resistant cultivars (7.7.3)
Strategy 4:	**Protect Host Directly**	
	Biological	Use natural antagonists (7.6)
	Chemical	Use pesticides
		Seed treatment
		Spray plants
	Physical	Use protective aids (e.g., tree guard)

may be defined as plants out of place. The different types of herbicides are based on killing action, specificity, active ingredient, and other characteristics as described in detail in module 4 of this chapter.

2. *Pesticides used to control nonplant pests of plants.* Horticultural plants are attacked by a wide variety of pests that may be grouped on the basis of animal class. The major pesticide groups are described in the following sections. They are readily identified because the prefix denotes the class of animals (the suffix *-cide* is common to all).

7.2.1 INSECTICIDES

Insecticides are pesticides designed to control insects. They are the most widely used pesticide for killing animal pests. Module 3 of this chapter is devoted to a detailed discussion of insecticides.

> **Insecticides**
> *Pesticides used to control unwanted insects.*

7.2.2 FUNGICIDES

Fungicides are pesticides designed to control fungal pathogens. There are two basic kinds:

1. *Protective fungicides.* Unlike the protection offered by other kinds of pesticides, protective fungicides offer protection only to the part of the plant surface that is covered by the chemical. It is critical therefore to apply fungicides in a uniform and even manner over the entire surface to be protected. One of the oldest and still widely used nonsystemic fungicides is Captan, which is applied as a protective spray or dust in vegetables, fruits, seed treatments, and ornamentals.

2. *Systemic fungicides.* Systemic fungicides penetrate the plant tissue and circulate through all parts of the plant to combat infection. A relatively new group of fungicides, they are more efficient in controlling pests. An example is the benzimidazole (e.g., benomyl), which is effective against *Botrytis, Sclerotinia,* and others.

In terms of chemistry, fungicides may be classified as *organic* or *inorganic.*

> **Phytotoxicity**
> The immediate (acute) or continuous low (chronic) impact of a chemical on a plant or its part.

1. *Organic fungicides.* Organic fungicides are more selective and pose less environmental danger. The newer types are especially readily biodegradable and less *phytotoxic* (damaging to plant tissue). The most widely known class of organic fungicides is the dithiocarbamates, which include old and still useful fungicides such as thiram, maneb, zineb, and mancozeb. Thiram is used in apple and peach orchards, turf, and vegetable gardens. Other classes of organic fungicides are the substituted aromatics, thiazoles, triazines, and dicarboximides.

2. *Inorganic fungicides.* The core elements in organic fungicides are sulfur, copper, and mercury. Sulfur is available in one of several formulations (7.10.3): powder, colloidal sulfur, or wettable powder. When applied, it may kill by direct contact at high environmental temperatures (above 70°F [21°C]) by fumigant action. Sulfur is used in controlling powdery mildew. One of the most popular copper formulations is *Bordeaux mixture,* the oldest fungicide (consisting of $CuSO_4$ and hydrated lime), which is effective against downy mildew. It also repels insects such as flea beetles and leaf hoppers. Inorganic copper fungicides are not water soluble.

7.2.3 NEMATICIDES

Nematicides are chemicals designed to penetrate the relatively impermeable cuticle of nematodes. They are generally applied by professionals by injecting fumigants of halogenated hydrocarbons under pressure into the soil.

7.2.4 RODENTICIDES

Rodenticides are pesticides designed to kill rodents. Rodents are most effectively controlled by poisoning. The most widely used class of rodenticides are the coumarins. They must be ingested repeatedly to kill the pest. Thus, they are safe in case of accidental ingestion. An example is Warfam, which was developed by the University of Wisconsin. Coumarins are anticoagulants.

7.2.5 MOLLUSCIDES

Chemicals that are designed to kill mollusks are called *molluscides.* These chemicals are usually formulated as baits; an example is methiocarb, which is very effective against snails and slugs in ornamental plantings. Metaldehydes are one of the oldest and most successful molluscides.

7.2.6 MITICIDES

Miticides are pesticides designed to kill mites.

7.2.7 AVIACIDES

Aviacides are pesticides designed to kill birds. They are commonly included in grain and used as bait. Strychnine is an aviacide.

7.3: GROWTH REGULATORS IN PEST CONTROL

Plant growth regulators (4.6) are used to manipulate plant height, promote rooting, and reduce fruiting, among other uses. A high concentration of certain plant hormones can reduce infection by some pathogens. This effect has been observed in tomato with respect to *Fusarium* and in potato with respect to *Phytophthora*. Viral and mycoplasma infections are known to cause reduced vigor and stunting in plants. However, an application of gibberellic acid spray overcomes stunting and axillary bud suppression. Sour cherry yellows is a common viral infection of cherries that is commercially controlled by the application of gibberellic acid on a significant scale.

7.4: CHOOSING A PESTICIDE

7.4.1 STEPS IN THE DECISION-MAKING PROCESS OF PEST CONTROL

The following are general steps that may be followed in the development and implementation of a pest-control strategy (figure 7-1). Depending on whether the pest problem is new, as well as the experience of the person making the decisions, some of the steps may be skipped.

1. *Detection.* A pest-control program always starts with a problem. The pest must first be detected. The presence of a pest may be detected by visual observation of the organism or the damage it causes. While certain pests can be identified from a distance, some pests have hiding places (e.g., the underside of the leaf, under stones, or under debris) and require the grower to make an effort to search at close quarters. The key to successful pest control is early detection. It is advisable, therefore, that the grower routinely inspect the plants and look for pests and disease organisms known to be associated with the production operation and those prevalent in the area.

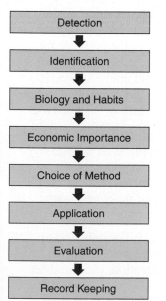

FIGURE 7-1 Steps in the decision-making process for pesticide application.

2. *Identification.* When a problem has been observed, it is important to make a positive identification of the insect or pathogen. Without knowing the organism involved, no sound control measure can be developed. Trial and error is wasteful. If the grower does not have the expertise to positively identify an insect or disease organism, a sample of an infected plant or the organism itself should be collected and sent to an appropriate identification center, such as the department of agriculture at a land grant institution or the U.S. Department of Agriculture (USDA) Extension Office. County extension agents should be consulted before contacting the national office.

3. *Biology and habits.* An organism has a life cycle. In insects that undergo a complete metamorphosis (6.4.3), there is a dramatic change from one stage of development to another. Each of these stages has its unique characteristics and habits. One stage may be more vulnerable than another to a particular control measure. The most vulnerable stage should be targeted for controlling the organism. It is important to know the habits of the organism in order to plan a control strategy. Certain insects hide on the undersides of leaves. Others inhabit the soil, and still others live on plant materials. Some insects are nocturnal in feeding habit, and others feed during the daytime. It has already been stated that certain insects have chewing mouthparts and others have piercing mouthparts. Some organisms have or secrete a protective covering, whereas others do not. These and other biological characteristics and habits are important in designing effective pest-control strategies.

4. *Economic importance.* It is economically wasteful of resources and time if it costs more to control a pest than the returns expected from the enterprise without protection. In other words, if the pest incidence does not pose an economic threat, the grower should ignore the pest. Certain pests can completely wipe out a production operation and must therefore be controlled immediately at the sign of their presence.

5. *Choice of method of control.* The most effective, economic, safe, and environmentally sound method of control should be selected after identifying the pest and assessing the potential damage. It may be found that a combination of methods rather than one particular method may be most effective (7.5). Some methods are easier to apply than others. Further, some methods may require the use of special equipment or the hiring of professional applicators.

6. *Application.* If chemicals are to be used, they must be applied at the correct rate. Premixed chemicals are available for purchase in certain cases. Otherwise, the user must follow the instructions provided with the chemical to mix the correct rate. Timeliness of application is critical to the success of a pest-control method. To eradicate a pest, it is important to know its life cycle. Certain applications may destroy the adults without damaging the eggs. By knowing when eggs hatch, an appropriate schedule can be developed to implement repeated application of the pesticide for more complete control. The environmental conditions under which application of a pesticide occurs is critical to its effectiveness. Pesticides should not be applied if rainfall is expected soon after the application. Further, a calm day is required to contain chemicals applied as sprays and dusts (7.10.3) within the area of intended use.

7. *Evaluation.* The effectiveness of an application should be evaluated within a reasonable period after application to determine whether a repeat application is necessary. Evaluating the impact of pest control on the total operation is important. Controlling pests is expected to significantly increase productivity and returns on investment. If this is not the case, the grower should review the operation and make necessary changes.

8. *Record keeping.* Keeping records of one's operations is critical. The only way to make alternative choices is to have data for comparison. Such a record should include the type of pesticide, rate of application, cost of application, yield, and net returns.

7.4.2 ENVIRONMENTAL AND SAFETY CONCERNS

To avoid indiscriminate use of pesticides and to protect the environment, laws and guidelines that govern the use of pesticides are enforced. These laws vary from place to place. Pesticide manufacturers and governmental agencies conduct extensive tests on pesticides before they are approved for use. That a pesticide is approved for use does not mean it may be used in any situation desired. Regional and local factors such as climatic factors, soil characteristics, and agricultural production may preclude the use of approved chemicals in certain situations. It is imperative that only legally approved chemicals (with respect to the particular area) be used. Such information is available through the local extension service. Local nurseries and vendors often carry only state-approved pesticides. For the inexperienced grower, it pays to seek expert advice on the correct pesticide to use. This kind of information is usually only a phone call away and free of charge.

In a competitive industry, a variety of pesticides abound for the same problem. Some are more effective than others; some are safer than others. Pesticides also vary in their ease of use and *formulation* (7.10.3). Some pesticide formulations are suited for outdoor use only. In fact, the utmost care should be taken when using pesticides in the home. When you purchase a pesticide, be sure—at the very least—to read the label and understand the recommendations for its safe use. You may also let the seller know whether the problem is indoors or outdoors.

In sum, the chemical selected should be

1. Legally approved for use in the area.
2. Effective against the pest.
3. Appropriate for the conditions under which it will be used.
4. Accompanied by detailed instructions about its proper use.
5. Relatively safe (to humans, the environment, and the economic parts of the plant). Certain chemicals can be applied only by certified personnel.
6. One that the grower can apply safely (based on available equipment, location of a problem, and other factors).

7.4.3 IMPORTANCE OF PESTICIDE LABELS

A *label,* the piece of paper (or other suitable material) affixed by the manufacturer to the container of a product, provides certain specific information about the product (figure 7-2). The information is not arbitrary and must meet specific governmental guidelines. A label is much

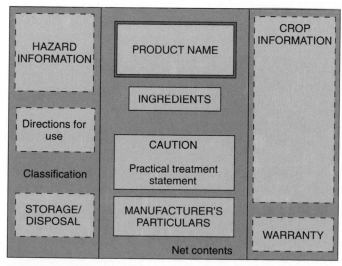

FIGURE 7-2 A typical pesticide label.

more than an advertisement. Several categories of information are provided on a label, including the following:

1. Name of the product, which may include a trademark name and chemical name.
2. Company name, address, and logo (where applicable).
3. Type of pesticide (e.g., fungicide or insecticide).
4. Product chemical analysis and characteristics: active ingredients and proportions (common and/or chemical names of ingredients) and formulation of substances (e.g., dust, emulsion, and wettable powder).
5. Pests it controls.
6. Directions for proper use and any restrictions.
7. Hazard statements (appearing as *caution, warning, danger,* or *poison*).
8. Storage and disposal directions.
9. Governmental administrative stipulations (e.g., Environmental Protection Agency [EPA] approval and EPA number).
10. Net content.

Typically, the trade name of the chemical, the type of chemical, and the company name are most visible. The hazard term is also quite conspicuous. Much of the other information is often in fine print. However, the user must endeavor to *read and follow all directions* very carefully. Understanding the warning signs of toxicity should be a high priority, since pesticides are generally very toxic to humans. As more research information becomes available on various aspects of the chemical product and also its impact on the environment, the governmental regulatory agencies may require additional information to be included on the product label.

7.4.4 PESTICIDE TOXICITY

> **Toxicity**
> The relative capacity of a substance to be poisonous to a living organism.

It is wise to treat all chemicals and especially pesticides (designed to kill) as poisonous until information to the contrary is clearly available. A measure of the danger associated with pesticides is the *hazard* rating, which is a function of *toxicity* and *exposure*. Toxicity is a measure of the degree to which a chemical is poisonous to the organism. A pesticide that is very toxic as a concentrate may not be hazardous when formulated as a granule (7.10.3). On the other hand, a pesticide of low toxicity may become very hazardous when used at high concentrations. Apart from formulation, the frequency of use and the experience of the operator may increase or decrease the hazard level.

> **LD$_{50}$**
> The milligrams of toxicant per kilogram body weight of an organism that is capable of killing 50 percent of the organisms under the test conditions.

The standard way of measuring toxicity is by determining the *lethal dose (LD$_{50}$)* of the chemical. LD$_{50}$ represents the dose sufficient to kill 50 percent of laboratory test animals, usually rats. Since the test is performed on mammals, the LD$_{50}$ is sometimes described as *mammalian toxicity*. The higher the LD$_{50}$ value, the less poisonous the chemical. LD$_{50}$ is measured in units of milligrams of substance per kilogram of animal body weight. It follows, therefore, that children (low total body weight) could be killed by a dose that would only make an adult very sick. The LD$_{50}$ values for selected garden chemicals are presented in table 7-2. Since a pesticide may be ingested, inhaled, or absorbed through the skin, toxicity is sometimes broken down into three kinds—*oral toxicity, toxicity on inhalation,* and *dermal toxicity.*

Chemicals gain access to humans through ingestion, touching, or inhaling. Some chemicals are highly corrosive and burn the skin upon contact. Chemicals that produce fumes or are formulated as dusts or powders are easily inhaled. Certain chemicals have strong odors that alert the user to their potential danger if inhaled.

7.4.5 USING PESTICIDES SAFELY

A first rule to using pesticides safely is to treat them as health hazards. After all, they are designed to *kill* pests. They are to be used as the last resort for controlling pests. If pesticides are necessary, the user should

Table 7-2 LD$_{50}$ Values (Oral) of Selected Pesticides

Pesticide	LD$_{50}$
Fungicides	
Captan	9,000–15,000
Maneb	6,750–7,500
Thiram	780
Zineb	8,000
PCNB	1,550–2,000
Insecticides	
Carbaryl	500–850
Dursban	97–279
Malathion	1,000–1,375
Pyrethrum	820–1,870
Rotenone	50–75
Herbicides	
DCPA	3,000
EPCT	1,600
Simazine	5,000
Oxyzalin	10,000

Source: Extracted (and modified) from extension bulletin B-751, Farm Science Series, Michigan State University, University Cooperative Extension Service

1. Choose the correct one. Look for safer or less toxic alternatives of pesticides.

2. Purchase or mix only the quantities needed. Do not store excess chemicals, since it poses a serious health hazard, especially in homes where children live. Leftover chemicals also create disposal problems.

3. Read the label and act accordingly. Be very familiar with the manufacturer's directions for safe use. Note and observe all warnings. Use *only* in the way prescribed by the manufacturer. Follow directions for use. Be sure to use the correct concentration.

4. Wear protective clothing and avoid contact with the skin. Wear gloves or at least wash your hands thoroughly and immediately after using any chemical. Protect your eyes and cover your nose and mouth with a mask to prevent inhalation or ingestion of chemicals.

5. Do not eat food, drink, or chew anything while handling chemicals, and do not eat afterward until you have washed your hands with soap.

6. Apply chemicals under the best conditions possible. Do not spray on a windy day. If a light wind prevails, do not spray into it, proceeding such that the wind is behind you as you move along. If rain (or irrigation) occurs after application, much of the chemical will be lost to the ground. When applying chemicals outdoors, it pays to listen to the weather report to know the best time for application. When applying chemicals indoors, avoid spraying onto cooking utensils and food. All such items should be covered before spraying. Children, especially, should leave the house for a period of time if extensive spraying is to be done. Moving a diseased plant outside of the house rather than spraying it indoors is recommended. Chemicals should be applied in conditions of adequate ventilation. A closed area such as a greenhouse should be adequately ventilated before people return to work in the area.

7. Apply with extreme care. Certain pesticides are injurious to both pests and humans and will kill indiscriminately. Plants can be damaged through accidental splashing or drifting during a spraying operation. The operator can be injured through carelessness. Premixed chemicals are less concentrated and safer to handle. Concentrated chemicals should be handled with extra care.

8. Know what to do in case of an accident. The label should indicate proper actions in case of a spill, ingestion, or inhalation. Water and a detergent should be readily available to wash any body part that comes in direct contact with the chemical. Cleaning certain spills requires more than water.

9. Clean all applicators thoroughly after use and store them in a safe place.

10. Store chemicals as directed by the manufacturer. A cool, dry place is often required for chemical storage. Keep *all* chemicals out of the reach of children and pets. It is best to store unused chemicals in their original containers for ease in recognizing the chemicals and avoiding accidents through misidentification.

11. Be very careful of using pesticides near the time of produce harvest. Pesticide poisoning may occur if produce is harvested before the pesticide effect wears off. Where applicable, produce grown with pesticides should be washed before being eaten or fed to animals.

7.4.6 METHODS OF PESTICIDE APPLICATION

Pesticides may be applied to plants, products, or the growing medium, according to need. The general ways in which pesticides are used are as follows:

1. *Foliar application.* Pesticides may be applied to plant foliage in the form of a liquid or dust (powder).

2. *Soil treatment.* A soil may be fumigated (by treating with volatile chemicals) to control nematodes and other soilborne diseases. Sometimes various formulations including drenches, granules, and dusts may be applied.

3. *Seed treatment.* Planting materials (e.g., seeds, bulbs, corms, and tubers) may be treated with a pesticide to control soilborne diseases that cause seed decay or damping-off of young seedlings.

4. *Treatment of wounds.* When large branches of trees are pruned (15.5), or cut, the wounds are large and heal slowly. Therefore, the wound must be protected from infection and rotting of the stem. First, a 10 to 20 percent household bleach is used to sterilize the surface of the wound. Ethyl alcohol (70 percent) may also be used. The wound is then painted with a tree wound dressing, which may consist of, for example, a mixture of lanolin, rosin, and gum in a ratio of 10:2:2. Alternatively, asphalt-varnish tree paint may be used.

5. *Control of postharvest pests.* Fruits may be dipped in dilute solutions of fungitoxic chemicals to protect them from rotting in storage.

7.5: INTEGRATED PEST MANAGEMENT

> **Integrated Pest Management (IPM)**
> An approach to pest control which attempts to use all the best management methods available to keep pest populations below the economic and/or aesthetic injury level, with least damage to life and the environment.

Integrated pest management (IPM) is a pest-control strategy whose goal is not to eradicate but to manage a pest such that its population is maintained below that which can cause economic loss to a production enterprise or aesthetic injury. In this strategy, human health and the general environment are paramount considerations. By nature, IPM depends on a broad and interdisciplinary approach to pest control, incorporating various aspects of the basic control methods (cultural, biological, legislative, and chemical) (figure 7-3).

7.5.1 GOALS OF IPM

The goals of IPM may be summarized as follows:

1. *Improved control of pests.* Methods of pest control should be reviewed regularly so that the best strategy is always used. As scientific knowledge abounds and technology

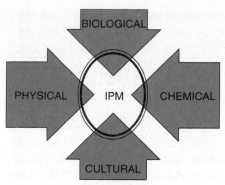

FIGURE 7-3 The components of an integrated pest management (IPM) system.

advances, new and improved alternative measures will become available. Strategies should draw on the strengths of all of the basic control methods in a truly interdisciplinary fashion to develop the best control package. Improvements in control should consider the fact that nature has built-in means of controlling population growth by the presence of natural enemies of organisms in the environment. Preference should be given to natural methods of control over the use of chemicals.

2. *Pesticide management.* Controlling pests should be a planned activity such that pesticides are used judiciously. Pesticides should be used so that natural enemies of plant pests are not destroyed in the process. Further, pesticides should be contained and applied only when absolutely necessary. Care must be taken to minimize the side effects of chemical application.

3. *Economic protection of plants.* As previously stated in this module, the mere sight of a pest does not necessarily mean that it constitutes a threat to economic production. A control measure should be enforced only when its necessity has been determined. When needed, IPM ensures that a control package includes the bare minimum for effectiveness. By using minimal quantities of pesticides (through reduced rate of application and reduced frequency of application), the cost of protection can be reduced significantly.

4. *Reduction of potential hazards.* A paramount objective of IPM is the responsible use of pesticides. When pesticides are deemed necessary, they should be used at the lowest effective level to prevent adverse environmental impact. IPM strategies are aimed at maintaining ecological stability and protecting the health of the grower (or user), the consumer, and the general environment. Hazardous chemicals should be prevented from entering the *food chain* (the sequence of transfer of food from producers through several levels of consumers in the biological community). Excess pesticides and other agricultural chemicals seep into the groundwater and also pollute the general environment, posing serious health problems to humans and wildlife.

7.5.2 DECISION-MAKING PROCESS IN DEVELOPING AN IPM PROGRAM

An IPM program is developed by following the basic principles for a conventional pest-control program, but with certain modifications. It is important to remember that an IPM program is broad based and interdisciplinary, including strategies involved with the basic methods of control (7.1). The general decision-making steps are as follows:

1. Identify the pest and the beneficial organisms in the area of interest. IPM uses natural enemies to control pests. By knowing these two groups of organisms, one can make the appropriate choices of control method that will selectively kill the pest while protecting the beneficial insects.

2. Know the biology of the organisms involved and how the environment influences them.

3. Select an appropriate cultural practice that will be detrimental to the pest while favoring the desirable organisms. Cultural control is usually considered first because the method of plant production plays a role in the pest problems encountered in cultivation. Cultural practices include selecting cultivars that are resistant to pests in the production area, preparing the soil properly, planting at the right time, providing adequate nutrition and water, and so forth (7.7).

4. Develop a pest monitoring schedule for the production enterprise to record the kinds and populations of various organisms. This step requires some expertise to be successful.

5. Determine tolerable threshold levels of pest populations. Since one of the goals of IPM is economic pest control, it is important to know what level of pest population constitutes an economic threat. When this threshold is reached, an appropriate control strategy should be implemented. The strategy should take into account the pest plant species, the stage of plant development, the economic product, and the general environment. Whereas a threshold may be deemed an economic threat when it occurs in the early stages of plant growth, the same threshold may not be economically threatening when the plant is approaching maturity.

6. If an economic threat is deemed eminent, decide on a definite course of action to forestall the danger. Sometimes the best decision is to wait a while. If intervention is delayed, the grower must increase monitoring of the population dynamics of the pest to avoid any surprise disaster.

7. Evaluate and follow up on the IPM program to make appropriate strategic adjustments for increased effectiveness and efficiency of control.

7.5.3 TOOLS FOR AN IPM PROGRAM

Tools used in implementing an IPM program reflect the basic methods of pest control. Using these tools in an informed and responsible manner results in a successful pest-control program with minimal adverse environmental impact.

1. *Cultural tools.* Cultural tools include proper soil preparation, proper time of planting, and use of resistant cultivars (7.7).

2. *Biological tools.* Biological tools include use of natural enemies of the pests and pheromones as sex traps (7.5).

3. *Chemical tools.* Chemical tools include the use of pesticides and improved methods of application (7.10).

4. *Legislative tools.* Legislative tools include the use of laws to restrict the transport of plant materials that may be contaminated (7.8).

SUMMARY

Diseases and pests can be controlled by implementing certain preventive measures such as observing phytosanitation, using quality disease-resistant seeds, and planting adapted cultivars. If diseases and pests become a problem, they may be controlled by one of four strategies: biological, cultural, chemical, or governmental controls. A fifth strategy, integrated pest management, uses a combination of all four strategies. Pesticides that control weeds are called herbicides. Nonplant pests may be controlled by one of several specifically designed pesticides including insecticides, fungicides, nematicides, miticides, rodenticides, and molluscides. Pesticides are toxic to humans and should be handled with care. Pesticide labels should be read very carefully and the directions followed meticulously. In pest control, the first step is to identify the specific pest and determine its potential for damage as well as the damage already caused.

PRACTICAL EXPERIENCE

Collect five labels (or obtain containers) from pesticides and compare and contrast their characteristics, using the guide provided in section 7.4.3.

OUTCOMES ASSESSMENT

PART A

Please answer true (T) or false (F) for the following statements.

1. T F Control measures should be used after the destructive stage of the pest.
2. T F The higher the LD_{50}, the higher the toxicity of a chemical.
3. T F Nematicides destroy mites.
4. T F The first step in the successful control of pests is to identify the causal organism.
5. T F Weeds are plants out of place.

PART B

Please answer the following questions.

1. List the four general methods of pest control.

   _____   _____

   _____   _____

2. Give two factors to be considered in the design of a pest-control strategy.

   _____   _____

3. EPA is an abbreviation for _____.
4. The standard way of measuring the toxicity of a chemical is by_____.
5. Fungicides destroy_____.
6. Chemicals used for controlling insects are classified as _____.
7. IPM is an abbreviation for _____.

PART C

1. Discuss the importance of a pesticide label.
2. Discuss the concept of IPM.
3. Describe how diseases and pests can be controlled by adopting preventive measures.
4. List and discuss any four cautions to observe in the safe use of pesticides.
5. Why is it important to identify the pest and the damage before implementing a control measure?

7.6: RATIONALE OF BIOLOGICAL CONTROL

Nature has built into its dynamics a variety of mechanisms whereby it maintains a state of equilibrium. Whenever a destabilizing force comes into effect, this natural balance shifts in the direction of the force. Unaided by humans, nature effectively maintains this desirable state of balance, or equilibrium. Unfortunately, when humans interface with nature, they tend to nudge it to their advantage and in the process often wind up destroying the delicate balance, which may lead to serious environmental consequences.

Every organism has its natural enemies. Unless a disaster or a drastic change occurs in their living conditions, the danger of extinction of organisms is minimized, in part because one organism does not dominate nature through overpopulation. This natural means of mutual control makes *biological control* the oldest method of pest control. Biological control is the control of diseases and pests by the direct activities of living organisms or the indirect activities of their products.

Industrialization and technical advancement have been major contributors to the destabilization of natural balance. Such advancements changed the lifestyles of people and caused consumers to be more demanding in terms of product quality. Subsistent agriculture was gradually replaced by mechanized farming of large tracts of land. Instead of allowing the farmland to lay idle (fallow) for a period of time to rejuvenate, the same tract of land was repeatedly farmed, predisposing it to depletion of plant growth nutrients. This practice ushered in the era of artificial soil amendments with fertilizers (chapter 3). Instead of *mixed cropping,* which is closer to what occurs in nature, *monocropping* which encourages the buildup of pests associated with one particular species is common today. To curb this disproportionate increase in the population of one pest, farmers use more chemicals (pesticides) to control pests. As previously indicated, an unfortunate aspect of chemical use is that these toxins often kill indiscriminately, depleting the population of both pests and desirable organisms.

> **Biological Control**
> *The use of other organisms to control populations of pathogens.*

7.7: STRATEGIES OF BIOLOGICAL CONTROL

A variety of strategies may be adopted to control diseases and pests biologically. In fact, biological control involves the exploitation of natural defense mechanisms and managing and controlling them to increase their effectiveness. The natural systems exploited in biological control are discussed in the following sections.

7.7.1 STRUCTURAL

Some species have characteristics that condition resistance to certain pests and disease-causing organisms. Certain species, for example, have hairs (pubescents) on their leaf surfaces and other parts that interfere with oviposition in insects. In this way, the multiplication of insects is impeded, thus hindering their spread and devastation to the plant. Other plants have genetically conditioned structural features such as a thick cuticle that sucking and chewing insects have difficulty penetrating.

7.7.2 CHEMICALS

Certain chemicals extracted from plants have insecticidal action. Common ones include the widely known *pyrethrum* extracted from plants in the chrysanthemum family, *rotenone,* and *nicotine* (rotenone being more common). Other species, including the neem tree, mamey, and basil, contain a chemical that repels insect pests or hinders their growth and development into adults.

7.7.3 PHYTOALLEXINS

In nature, certain plants exude toxins from their roots into the soil. These toxins prevent the growth of other species in the immediate vicinity. The species hence maintains a kind of territorial boundary similar to that which occurs in the animal kingdom.

7.7.4 PARASITISM

The Japanese beetle *(Tiphia),* for example, is attacked by the larvae of a beetle, while the adult alfalfa weevil is a host for the eggs of the stingless wasp *(Microstomus aethipoides),* which hatches inside the weevil, eventually destroying it. Cyst nematodes (Heterodera and Globodera) are parasitized by certain fungi (e.g., *Catenaria auxilianis),* and the root-knot nematode *(Meloidogyne* spp.) is parasitized by the fungus *Dactylella oviparasitica.* Similarly, bacteriophages are viruses that destroy bacteria. These viral parasites occur in the environment.

7.7.5 PREY-PREDATOR RELATIONSHIPS

Birds may prey on insects and rodents. Snakes also prey on rodents that destroy horticultural plants. Carabid beetles (ground beetles) prey on aphids, caterpillars, slugs, and others. Lacewings *(Chrysopa)* prey on aphids, spiders prey on flying insects, and social wasps prey on caterpillars.

7.7.6 ANTAGONISM

As described previously in this chapter, nature has built-in mechanisms for maintaining balance so that no single organism dominates. If an organism is introduced into a new environment where its antagonizing organism is not present, the organism can multiply rapidly and pose a great economic threat to vulnerable cultivated crops in the area. In olive orchards, for example, the olive parlatoria scale, an economic pest, can be controlled effectively by biological means if its antagonistic organism, the parasitic wasp, is introduced into the environment. Antagonistic plants that exude toxins against nematodes are known to occur in nature.

> **Antagonism**
> The phenomenon of one organism producing toxic metabolic products that kill, injure, or inhibit the growth of some other organism in close proximity.

7.7.7 REPELLENTS

Some plant species exude strong scents that are repulsive to certain insects. Onion, garlic, and leek have been known to repel aphids, and mint repels cabbage butterflies and flea beetles. Horseradish repels potato bugs, and sage repels cabbage pests and carrot flies. Marigolds repel root nematodes. By planting the appropriate combinations of plants in a particular area, the grower can gain some degree of crop protection from a specific pest.

7.7.8 ALTERNATIVE HOST (TRAP PLANTS)

Pests have preference for the plant species they attack. If two hosts are available, one may be preferentially attacked. Slugs prefer lettuce to chrysanthemums, and, as such, a good crop of the latter can be produced in the field by planting lettuce among them as "decoy" plants, or trap plants. Similarly, nematodes may be controlled by planting certain species that prevent the development of larvae into adults. This practice has the effect of decreasing the population of nematodes in the soil. *Clotalaria* plants are used to trap the larvae of root-knot nematodes *(Meloidegyne* spp.).

7.7.9 BIOCONTROL

In the storage of horticultural produce, *biocontrol* is employed in the postharvest control of diseases in stone fruits such as peach and plum. This control is effected by treating fruits with a suspension of the bacterium *Bacillus subtilis,* which is found to delay brown rot caused by the fungus *Monilinia fruticola*. Biocontrol measures involving other bacteria and fungi exist. Bacteria are involved in the frost damage of certain frost-sensitive plants by aiding in the formation of ice (called *ice nucleation*). Ice-nucleated active bacteria (e.g., *Pseudomonas syringae*) are replaced by applying non-ice-nucleated bacteria, which reduces bacteria-mediated frost injury.

7.7.10 MICROBIAL SPRAYS (BIOPESTICIDES)

Scientists have identified and cultured natural enemies of certain horticultural plants. An infected field is sprayed with large populations of laboratory-cultured microbes. For example, aerial application of spores of the fungus *Collectotrichum gloesporiodes* has been successfully used to control the northern jointvetch in rice fields. Also, the fungus *Talaromyces flavius,* when applied to the soil, is effective in controlling soilborne diseases such as wilts of potato (potato wilt) and eggplant (verticilium wilt). Through breeding efforts, more aggressive and effective strains of these microbes are being developed, as in the case of plant cultivars. Another commercially available microbial spray is the *Bacillus thuringiensis* spray, which is effective against caterpillars or cutworms, corn borers, cabbage worms, and others. The effect on caterpillars starts upon ingestion of the bacteria.

Advantages and Disadvantages of Biological Control

Advantages   The advantages of biological control include the following:

1.   Pesticides are harmful to the environment and are hazardous to humans and wildlife. Biological control uses organisms already present in the environment.
2.   Seeds of improved cultivars (resistant cultivars) are cheaper to use than spraying against pests with chemicals.
3.   Biological control is safer to apply than chemicals.

Disadvantages   The major disadvantages of biological control include the following:

1.   Availability and application are limited to relatively few crop species.
2.   Handling of organisms is less convenient than chemicals, often requiring special care.

Other Examples of Biological Control

A variety of beetles have been identified as predators of pests of cultivated crops:

1.   The European seven-spotted lady beetle *(Coccinella septempunctata)* preys on aphids.
2.   The ladybug *(Crystalaemus montrocizieri)* destroys mealybugs.
3.   The larvae of the Japanese beetle feed on the larvae of other beetles.

In addition to beetles, the bacterium *Bacillus thuringiensis* is known to infest and kill a variety of insects, including the larvae of butterflies, moths, and corn borers, while being harmless to plants. Other examples are presented in table 7-3. Figure 7-4 presents examples of various organisms in action in effecting biological control. These exhibits represent only a selected few examples.

7.8: CULTURAL CONTROL

A variety of strategies are employed to implement cultural control of diseases and pests in plants.

7.8.1 CROP ROTATION

As indicated in the introduction to this chapter, monoculture and repeated cultivation of one species on the same area of land encourages the buildup of the diseases and pests that plague

Table 7-3 Selected Examples of Biological Control of Horticultural Pests

Biological Agent	Some Pests Controlled
Ladybug	Aphid
Bacillus thuringiensis	Colorado potato beetle and caterpillar
Bacillus popillise	Japanese beetle
Green lacewing	Aphid and mealybug
Parasitic wasp	Tomato hornworm and cabbage looper
Nedalia beetle	Citrus cottony scale
Tilleteopars	Powdery mildew
Talaromyces flavius	Potato wilt and verticillium wilt

In addition to these organisms, the use of resistant cultivars, plants with repellent scents, and crop rotation are other nonchemical methods of pest control

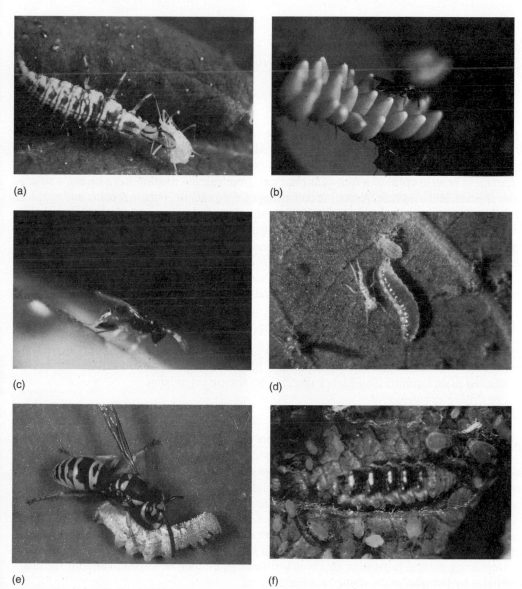

(a)

(b)

(c)

(d)

(e)

(f)

FIGURE 7-4 Selected examples of biological control. (a) A lacewing attacking prey. (b) *Tetrastichus gallerucae* attacking elm leaf beetle eggs. (c) Female *Aphytis* piercing scale insect with ovipositor. (d) Predaceous midge larva eating aphid. (e) Yellow jacket attacking a caterpillar. (f) Flower fly larva eating aphids. (*Source:* Photos provided courtesy of Oklahoma Cooperative Extension Service, Oklahoma State University)

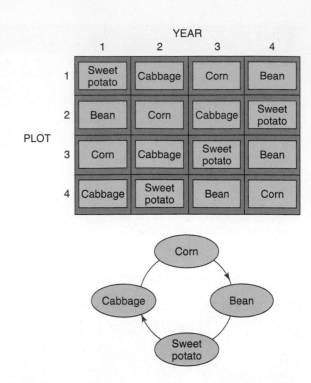

FIGURE 7-5 A crop rotation cycle.

the cultivated species. Crop rotation is a strategy whereby no one species is perpetually planted on the same plot of land (figure 7-5). Additionally, a species is not followed by its relative. Instead, species with different soil requirements or use are rotated in a definite cycle (e.g., corn to tomato to bean to corn, or a four-year rotation). Note that the rotation has a cereal, a solanaceous species, and a legume. A rotation consisting of, for example, potato, tomato, and eggplant (all solanaceous species) certainly violates the rule of not following a species with its relative. Rotation of crops is effective in reducing the populations of certain soilborne diseases such as tomato wilts. The causal organisms of such diseases need the host plant in order to thrive and consequently cannot persist in the soil if the host is absent for about two to three years. Rotations are particularly effective in controlling diseases and pests whose causal organisms do not travel long distances (such as nematodes, weevils, certain wilts, and phytophthora).

7.8.2 SANITATION

Disease-causing organisms and insects remain in the field if infected plant debris is left on the ground. Sometimes infected plant remains have to be incinerated to kill the pathogens. Uninfected plant remains may harbor insects and disease organisms.

7.8.3 USE OF RESISTANT CULTIVARS

Plant breeders genetically manipulate the genotypes of plants to the advantage of humans (5.7). Through scientific inquiry, some of the protective strategies of plants in the wild have been discovered and studied. Some species resist certain diseases and pests because they have genes that condition such characteristics. Through breeding, scientists are able to transfer the ability to resist diseases and pests from the wild into cultivated species and from cultivar to cultivar within the species. Sometimes resistance is even transferred across species and genus boundaries (see biotechnology [5.16]). Resistant cultivars exist for most of the major pests of horticultural plants. As indicated in chapter 5, a great deal of the efforts of plant breeding go into replacing natural protection that was lost through deliberate selection against the characteristics by humans in the domestication process.

7.8.4 HOST ERADICATION

Host eradication is often a drastic preemptive method of pest control that involves the elimination of all susceptible hosts when a pathogen is known to have been introduced into the production area. This tactic is used to forestall an eminent disease epidemic. For example, all citrus trees in a production area where a pathogen has been introduced may be completely eradicated. On a small scale, host eradication is conducted in nurseries and greenhouses by rouging (removing off types) infected plants. Certain pathogens require two alternative hosts to complete their life cycles. In such a case, the less economically important host should be eliminated to interrupt the developmental cycle of the pathogen. For example, *Cronartium ribicola* requires pine and currant plants to complete its life cycle.

7.8.5 MULCHING

Plastic mulching has the capacity to trap heat, which causes the soil temperature to increase. The high temperature destroys some soil pathogens, including *Verticilium*.

7.9: LEGISLATIVE CONTROL

Various regulatory or legislative restrictions are placed on the movement of live plants and produce from one place to another. Controlling the spread of plant pests through laws is called *plant quarantine*. The purpose of such laws it to prevent the importation and spread of pathogens and insect pests into areas where they do not already occur. In the United States, the Plant Quarantine Act of 1912 established the laws that govern such restricted movement of materials. International and local (with and between states) restrictions are designed to curb the spread of diseases and pests that are associated with specific plants. Enforcement is especially strict with regard to plants that are of high economic importance. Plant breeding efforts may succeed in developing resistant cultivars to certain pests and diseases. However, the production enterprise can be wiped out with the introduction of a new strain of the pathogen from another country or region. When imports are detained in quarantine procedures, the inspectors observe the materials over a period of time (at least for the duration of the incubation period of the insect or pathogen), after which plants are released, treated, or destroyed, depending on the results during the detention.

> **Plant Quarantine**
> *The use of legislation to control the import and export of plants or plant materials to prevent the spread of plant pests and diseases.*

Quarantines are not foolproof since their success depends on the experience of the inspectors. Further, pathogens may escape detection if they exist in less conspicuous stages in their life cycles, such as eggs or spores. Where plants are treated, certain latent infections may exist in seeds and other plant materials even after a period of growing plants in the field.

7.10: MECHANICAL AND PHYSICAL CONTROL

Pests may be controlled by a variety of mechanical or physical methods.

7.10.1 TRAPS

A number of mechanical control measures may be adopted to control pests in the greenhouse and field. A fly catcher is a strip of paper or polyethylene (commonly yellow in color) that has been coated on both surfaces with a sticky substance. Insects that land on the paper get stuck to it and eventually die (figure 7-6). Larger mechanical traps are used to catch rodents. Certain lights are designed to attract insects, which then become trapped by other devices installed for that purpose.

7.10.2 HANDPICKING

In a small garden, caterpillars and other large bugs may be handpicked (not necessarily with bare hands) and destroyed.

FIGURE 7-6 Polyethylene trap.

7.10.3 BARRIERS

Rodents can be kept out of a garden by fencing it. A band of sticky paper (similar to a fly catcher) wrapped around the base of a tree prevents crawling insects on the ground from climbing up the tree.

7.10.4 TILLAGE

As a pest-control measure, tillage (cultivation) (3.9.1) is used to remove weeds from the area.

7.10.5 MULCHING

Mulching can be a method of mechanical control because a material such as plastic prevents weeds from growing through the material.

7.10.6 HEAT TREATMENT

In the greenhouse, soil and other growing media are routinely sterilized before use. Depending on the temperature, sterilization may kill nematodes and water mold (at 50°C [122°F]). To kill bacteria, fungi, and worms, the temperature should be about 72°C (162°F). Weed seeds and some bacteria and viruses are killed at temperatures of about 82°C (180°F). When soil is oven sterilized, beneficial microbes (e.g., bacteria involved in the nitrogen cycle) may be killed, and toxic levels of salts (e.g., that of magnesium) may occur.

Hot water is used in the greenhouse to clean certain dormant planting materials (e.g., seeds and bulbs) to remove pathogens. Heat treatment by hot air is used to dry cut surfaces of vegetative plant materials such as tubers to accelerate healing and thus prevent rot. For example, sweet potato may be air dried at 28 to 32°C (82 to 90°F) for about two weeks. Drying of grains and nuts is required before long-term storage. Fruits such as grapes and plums can be dried to produce raisins and prunes, respectively. The latter products can be stored for long periods without decay.

7.10.7 COLD TREATMENT

Most postharvest protection of fresh produce is achieved through cold storage to maintain quality. Cold storage does not kill pathogens but slows their activity.

7.10.8 RADIATION

Exposure of harvested products to the appropriate dose of radiation (e.g., gamma radiation) is known to prolong shelf life. Similarly, it is known that certain pathogenic fungi (e.g., *Botrytis* and *Alternaria*) produce spores only under conditions in which the light received contains ultraviolet (UV) radiation (3.2.1). Some greenhouses are glazed (11.2.4) with UV-absorbing material so that radiation with a wavelength below 390 nanometers is not received within the greenhouse. Vegetables may be produced without infection by these pathogens.

SUMMARY

Every organism has natural enemies. Biological control exploits natural defense mechanisms, managing and controlling them to increase their efficiency. Some plants have structural features that confer upon them resistance to particular pests. Certain plants contain chemicals such as pyrethrums that repel pests. Parasitism and prey-predator relationships occur in nature. Plant breeders are able to breed resistance to pests and diseases into cultivars. Diseases and pests can be controlled by adopting crop rotation practices. Some microbial sprays are available for use against a number of pests. Governments enact legislation to restrict the movement of live biological material from one place to another to limit the spread of contagious diseases. Such quarantine laws differ from place to place. Sometimes pests have to be physically or mechanically removed by, for example, trapping or handpicking.

OUTCOMES ASSESSMENT

PART A

Please answer true (T) or false (F) for the following statements.

1. T F Every organism has natural enemies.
2. T F The Japanese beetle is scientifically called *Tiphia*.
3. T F Biological control uses organisms that already exist in the environment.
4. T F Rotenone is an organic insecticide.
5. T F Onion, garlic, and leek are known to repel aphids.

PART B

Please answer the following questions.

1. The chemicals obtained from chrysanthemum that have insecticidal properties are called _____.
2. Name two organisms that are used in biological pest control.

    _____   _____
3. Carabid beetles prey on _____.
4. Lacewings prey on_____.
5. Horseradish repels _____.
6. Give an example of a bacterial spray._____
7. Controlling the spread of diseases and pests through laws is called _____.

1. Define and explain the term *biological control*.
2. Describe a specific example of biological pest control.
3. Describe how structural characteristics of plants act as agents of biological control.
4. Describe how phytoallexins work.
5. Develop a four-year crop rotation planting program.

MODULE 3 CHEMICAL CONTROL OF PLANT PESTS: INSECTICIDES

7.11: INSECTICIDES AND THEIR USE

Insecticides are chemicals used to control insect pests. They are classified in several standard ways.

7.11.1 CLASSIFICATION BASED ON KILLING ACTION

Chemicals used to control insect pests vary in the way they kill, which thus provides a basis for classifying insecticides. This method of classification is outmoded. Since insects differ in morphology and feeding habits, for example, it is critical that the insecticide attack the pest where it is most vulnerable. The various action modes under which insecticides may be classified are as follows:

1. *Contact action.* Insecticides that kill by contact action are also called *contact poisons*. They are effective when sprayed directly onto the pests or when the pests come into contact with poisons as they move on plant parts that have been sprayed. Once in contact with the pest, contact poisons attack the respiratory and nervous systems, with lethal consequences. Most insects succumb to contact poisons (e.g., malathion). Insects that hide on the undersides of leaves are hard to hit directly by contact poisons.

2. *Stomach action. Stomach poisons* must be ingested by the pest to be effective. As such, chewing insects (e.g., grasshoppers, beetles, and caterpillars) are effectively controlled by this class of poisons. Once ingested, the poison (e.g., rotenone) is absorbed through the digestive tract.

3. *Systemic action. Systemic insecticides* permeate the entire plant so that any insects that suck or chew are exposed to the poisons. They may be applied as foliar sprays or directly to the soil to be absorbed by roots. Insects cannot hide from this chemical since once they feed (whether by sucking or chewing) they ingest the toxin. The caution to observe with systemic poisons is that when applied to food crops, the produce must not be eaten until the toxin (e.g., orthene) has broken down to a safe level.

4. *Fumigation. Fumigants* are volatile chemicals that enter the target pest through its respiratory system. They are effective when used in closed systems such as storage houses and greenhouses (11.1). The soil can also be fumigated to control soilborne diseases such as root-knot nematodes. Though gaseous, fumigants have contact action. The fine particles settle on the body of the insect before entering through the pores.

5. *Repellent action.* Most insecticides are designed to kill pests. However, some chemicals, called *repellants* (e.g., Bordeaux mixture), repel insects (e.g., leaf hopper and potato flea beetle) from plants without any killing action.

6. *Attractant action.* Females of many insect species secrete certain chemicals called *pheromones* that attract male partners. Scientists have successfully synthesized these chemicals for use in luring male insects to traps, where they are caught and destroyed. The Japanese beetle moth and gypsy moth are easily baited by the use of pheromones. By baiting males and destroying them, most females are left unfertilized, thus reducing the population of the insects.

7. *Suffocation.* Scale insects are widely controlled by spraying oils that plug the breathing holes in their bodies and suffocate them.

7.11.2 CLASSIFICATION BASED ON CHEMISTRY OF ACTIVE INGREDIENT

Modern classification of insecticides is based on chemical composition, since many modern insecticides have both contact and stomach actions. The two broad classes of insecticides are based on the chemistry of the *active ingredient* (the compound responsible for the killing action).

> **Active Ingredient (a.i.)**
> The amount of actual pesticide in a formulation that is toxic or inhibiting to the pest.

Inorganic Compounds (Inorganics)

Insecticides made up of inorganic compounds or minerals are becoming increasingly less common. They are usually designed to kill by stomach action and include compounds such as arsenic (lead arsenate or calcium arsenate), sulfur, and fluorine.

Organic Compounds (Organics)

Organic insecticides may be *natural* or *synthetic*.

Natural (Botanicals) Botanicals are products from plants that have insecticidal effects. Many plant species produce organic compounds that are toxic to insect pests that feed on them. Plant organic substances are usually safe and nontoxic to humans. An example of a botanical is *pyrethrum,* which is obtained from chrysanthemum. Another organic substance is *nicotine,* which is obtained from tobacco plants and is an addictive substance found in cigarettes. Other botanicals are rotenone, ryania, and sabadilla. Many organic compounds act as stomach or contact poisons.

Synthetic Organic Compounds Synthetic organic chemicals are artificially compounded and are effective against a wide variety of insects and pests. On the basis of the active ingredients in chemicals, several classes of synthetic organic insecticides are identified:

1. *Organochlorines (chlorinated hydrocarbons).* Organochlorines are most readily associated with dicarbo-, dihydrotetrachloride (DDT), which is one of the earliest and most successful to be developed. It has wide application, being effective against horticultural field pests, mosquitoes, fleas, and flies. It has a long residual effect, working long after initial application. Organochlorines are not readily biodegradable, which contributes to their rapid buildup in the environment in the soil and water, as well as in the tissues of plants and animals and their products. They are thus not only toxic to humans through their action on direct contact but also through the ingestion of food that has been contaminated, such as dairy and meat products (i.e., through a cow eating contaminated feed) and fish from contaminated waters. Consequently, DDT (as well as its close relatives) has been banned in many parts of the world. Other organochlorines are chlordane, lindane, methoxychlor, heptachlor, and aldin.

2. *Organophosphates.* Unlike chlorinated hydrocarbons, organophosphates (or organic phosphates) have shorter residual action (breakdown within 30 days) and are more readily biodegradable. As such, they do not build up in the environment. However, certain types (e.g., parathion) are extremely toxic to humans. Organophosphates are generally effective insecticides. Malathion is less toxic and widely used as a horticultural spray. Other organophosphates include diazinon, phorate, dameton, and chlorpyrifos.

3. *Carbamates.* Carbamates are relatively safer to use than those previously described. They have somewhat low mammalian toxicity and short residual action (breakdown within seven days). They are effective against sucking and chewing insects. One of the earliest and most successful was carbaryl (trade named Sevin). Others are carbofuran, aldicarb, and propoxur.

4. *Pyrethroids.* Pyrethroids are synthetic equivalents of natural pyrethrins found in species such as chrysanthemum. They are less toxic to humans and effective against a broad spectrum of insects.

Fumigants   Fumigants act in the gaseous state and are best used in closed environments (e.g., as storage pesticides) or injected into the soil. One of the most common types is methyl bromide, an odorless and colorless gas that is highly toxic to humans but is used widely to fumigate stored vegetables, seeds, fruits, and grains. It is also used to chemically sterilize soil mixes for use in greenhouses. Malathion is one of the most widely used fumigants of stored grain.

Spray Oils   Spray oils are obtained by specially distilling and refining crude oils. They are used to combat scale insects and mites in orchard plants and ornamentals. A common form is called dormant spray.

Biologicals (Microbial Insecticides)   Biologicals are commercially produced pathogens (e.g., bacteria, fungi, and viruses) that are applied to the foliage of plants to prey on specific insect pests. For example, commercial preparations of the bacterium *Bacillus thuringiensis* are applied to foliage to control several species of *Lepidoptera* (caterpillars).

7.11.3 FORMULATIONS OF INSECTICIDES

Formulation
The form in which a pesticide is offered for sale.

The chemical that actually controls the target pest *(active ingredient)* is not marketed or utilized directly but mixed with an inert ingredient to create what is called a *formulation*. Although some formulations are ready to use, others require diluting with a solvent or water before use. The two general types of formulations are dry and liquid.

Dry Formulations

Dusts   Dusts are chemicals formulated as powders and applied as such without mixing or diluting. They usually contain low concentrations of the active ingredient or ingredients (about 1 to 10 percent) mixed with fine-powdered inert material (e.g., chalk, clay, or talc). Dusts are applied by using simple equipment called *dusters*. They are easy to apply but leave unsightly residue on foliage. Further, when applied in even the slightest wind condition, *drift* (blowing away in the wind) may be a problem. Drifting of a pesticide onto plants not intended to be sprayed may result in collateral damage.

Drift
The movement by air of pesticide particles outside the intended target area during or shortly after application.

Wettable Powders   Wettable powders are concentrated chemicals formulated as dusts or powders that require dilution before use. Wettable powders are usually formulated to a high concentration of active ingredient (about 50 percent or greater). The addition of water decreases drift, but since the powders, even in solution, tend to settle, care should be taken to stir the mixture frequently so that the chemical is applied uniformly and at the desired rate. To increase the effectiveness of pesticides that require mixing with water before use, they are mixed with surfactants (agents that help pesticides to stick or spread better by lowering surface tension). This mixing is necessary because the plant surface naturally repels water to a varying extent. Wettable powders require constant agitation during use.

Granules   Sometimes insecticides are formulated as coarse particles called granules. Granules are applied to the soil in the same way as granular fertilizer formulations are applied. They may require dissolution in water before roots can absorb the chemicals. Some may have to be incorporated into the soil. Systemic herbicides may be formulated this way.

Other granules are designed to act like fumigants and thus need no water to initiate their effects. Granules are ready to use and pose little danger to the user. There is no danger of drift, and application requires only simple equipment such as a spreader.

Pellets   Pellets differ from granules in that the former consists of particles that are uniform in size and of specific weight. Unlike granules, pellets can be applied by precision applicators.

Baits   When an active ingredient is mixed with food or some other substance that attracts pests, the formulation is called a bait. Pests are attracted to baits and die when they ingest the poisons. Baits are usually low in active ingredient (less than 5 percent). They are commonly used to control indoor pests including mice and cockroaches.

Liquid Formulations

Aerosols   Aerosols contain one or more active ingredients in a solvent. Household chemicals are frequently formulated as aerosols. These insecticides are contained in pressurized cans and are ready for use. They are very convenient to use but still require adherence to the safety measures that apply to all insecticides. The insecticides are propelled by special gases (propellants) including fluorocarbons (e.g., freon), isobutane, and isopropane. Aerosols may be formulated for use as smoke or fog in special generators under enclosed conditions (e.g., warehouses or greenhouses). The advantage of this formulation is that the entire space is filled with the pesticide. However, because aerosols are difficult to confine to the target, everything in the area is exposed to the pesticide. Injury due to inhalation is possible if aerosols are used without proper protection.

Emulsifiable Concentrates   Emulsifiable concentrates consist of an active ingredient mixed in a petroleum solvent and an *emulsifier*. The emulsifier is an *adjuvant* that allows the formulation to be mixed with water. Emulsifiable concentrates are used widely in horticultural applications. They are adaptable to a variety of methods of application and equipment, including mist blowers, aerial applicators, and portable sprayers. Emulsifiable concentrates are desirable for several reasons, including the fact that, unlike wettable powders, they do not separate out in solution and hence do not require frequent stirring in the tank as dusts. They are mixed with water to the required concentration and leave very little residue on plants and fruits. However, mixing errors may occur, leading to a high potential for *phytotoxicity* (plant damage from chemicals).

> **Emulsifier**
> A surface-active agent that facilitates the suspension of minute droplets of one liquid in another to form a stable emulsion.

Flowables   Flowables are suspensions of active ingredients. They are easy to use but may leave some residue on plants.

Fumigants   Fumigants are formulations that produce gases during application. They may be liquids or solids. Fumigants are best applied in closed environments such as granaries, warehouses, and greenhouses. They are used in controlling some soilborne pests. A major advantage of fumigants is their ability to invade any space in the areas of application (such as cracks and crevices). A single application is usually effective in controlling the pest. The disadvantages include the need for special equipment, limitation to use in enclosed areas, and a high potential for human respiratory injury.

Solutions   Sometimes, by including special additives in the formulation, the active ingredient may become soluble in water. Solutions have the advantage of leaving no residue on surfaces and requiring no agitation during use.

7.11.4 APPLICATION EQUIPMENT

The type of equipment used in insecticide application depends on several factors, including the area to be treated, the kinds of plants to be treated, and the formulation. Some equipment is mechanically operated, and others are motorized. Although some are handheld, others require the use of tractors and other means of transportation. Common insecticide applicators are described in the following sections.

Small-Scale Applicators

Pressurized Cans   Aerosols come ready to use in pressurized cans (figure 7-7). Special equipment is not needed. The pesticide is released by simply pressing down on the nozzle to deliver a fine, misty spray.

Compressed-Air Tank Spray   A compressed-air tank is a much larger version of the handheld aerosol can, but the principle of operation is the same (figure 7-8). The tank is partially filled with the correctly prepared chemical solution. The remainder of the space is occupied by air, which is compressed in a variety of ways depending on the design of the equipment. A handle may be used to mechanically pump air into the tank. The tank is attached with a flexible tube fitted at the tip with a nozzle for delivering fine sprays in patterns according to its design. The flexible tube enables the operator to spray hard-to-reach places. The design of the compressed-air tank sprayer may allow the equipment to be carried as a knapsack on the back or hung on the shoulder in a sling.

Atomizer Sprayer   An atomizer sprayer is a simple handheld implement with a plunger that is pushed to draw air into the tube for dispersing the chemical solution in a fine spray at each stroke (figure 7-9).

Dusters   Pesticides formulated as dust are applied with the aid of dusters that may have plungers similar to those of atomizers. Some dusters have a squeezable bulb (figure 7-10).

Large-Scale Applicators

Motorized Ground Applicators   Motorized ground applicators work on the same principle as those for small-scale applications, except that certain functions that are manually performed

FIGURE 7-7   Pressurized can used for spraying aerosols.

FIGURE 7-8   Compressed-air tank sprayer.

in small applicators are automated in large applicators. The designs and sizes of the equipment vary:

1. *Portable (usually on the back of the operator).* Some compressed tanks are small enough to be carried around (figure 7-11). They are fitted with small motors so that the operator needs only control the delivery tube and direct the spray at desired targets.
2. *Tractor mounted.* The power take-off of the tractor may be used to provide the source of power for operating large sprayers, which are drawn or attached to the rear of tractors.
3. *Truck-mounted.* Some trucks are equipped with tanks and other devices for spraying or spreading (granular) pesticides, such as the fan jet applicator for orchards.

Aerial Applicators   Where very large acreages must be treated, using airplanes or helicopters equipped with sprayers may be most economical (figure 7-12). Some commercial companies specialize in aerial applications.

7.11.5 EQUIPMENT CALIBRATION

Chemicals used to control insects are very toxic to both pests and humans and as such should be used very carefully, only if needed, and also in the minimum strength or concentration needed to be effective. Overapplication is not only wasteful but may damage the plants and produce being protected and may be injurious to operators and consumers of the produce.

Insecticide manufacturers provide adequate instructions with their products for their correct and safe use. The recommended rate of application should be adopted. Premixed insecticides are available for easy and ready application and are especially recommended for the novice.

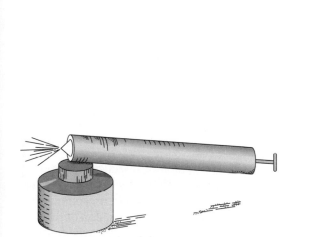

FIGURE 7-9   An atomizer sprayer.

FIGURE 7-10   A bulb duster.

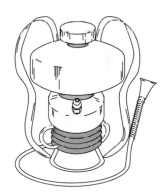

FIGURE 7-11   A portable motorized back sprayer.

FIGURE 7-12   Aerial application of pesticides.

Small applicators are usually not calibrated beyond that for which the manufacturer designed them, and as such the operator should be careful to deliver just the right quantity of the correctly mixed chemical. When liquid chemicals are used, a good application is one that covers the leaf surface to a point where dripping is about to occur. Dusts should similarly cover the leaf surface uniformly in one round of application. The amount applied depends on the distance between the leaf and the applicator.

Large applicators need calibration beyond the manufacturer's settings. Factors to take into account in calibration include the rate of application; the speed of the tractor, truck, or airplane; the nozzle type at the end of the sprayer; and the distance of the nozzle from the plants.

7.11.6 STRATEGIES FOR EFFECTIVE AND SAFE APPLICATION

Identify Pest

Treatment is most successful if the disease is identified in its early stages. In horticulture, then, the grower should visit the field or inspect the plants regularly. Certain insects are known to be perpetual pests at certain times in the growing season and should be anticipated. In insect control, the grower should determine whether the problem is caused by chewing or sucking insects. Chewing insects chew away the leaves especially, leaving holes, a network of veins, whitish patches, or partly eaten leaves. Stomach and contact poisons are effective for their control. It is more difficult to detect the presence of sucking insects. Plants are deprived of nutrients and may appear weak, grow less vigorously, or show rolled leaf edges. Sometimes, as in the case of aphids, these sucking pests can be found by turning over the leaf to examine the underside. Systemic and contact insecticides are effective against sucking insects.

Determine Economic Damage Potential

Certain insects are unable to inflict enough damage to cause the grower any significant economic loss. Since chemicals are not only expensive but hazardous to health, they should be applied only when the grower has determined that the potential loss is significant enough to warrant their use.

Insect Biology

Knowing the biology of the insect enables the timely application of insecticides for effectiveness. Insects must be controlled before they reach the stage where they cause devastation to the crop and before they have a chance to multiply. Insect pests are more susceptible when they are active than when they are in the egg or pupa stages. If, upon examination of the field or garden, only dormant stages are observed, chemical control should be delayed until the eggs hatch or adults emerge. Immediate control measures are required when mixed stages occur. In such a situation, a follow-up application should be made as appropriate (with respect to the life cycle of the insect) to coincide with the hatching of eggs.

7.11.7 HOUSEPLANT PESTS AND THEIR CONTROL

Common houseplant pests and their means of control are presented in table 7-4.

7.11.8 GARDEN PESTS AND THEIR CONTROL

A selected number of garden pests and their means of control are summarized in table 7-5.

7.11.9 LANDSCAPE PESTS AND THEIR CONTROL

A selected number of landscape pests and their means of control are presented in table 7-6.

Table 7-4 Suggestions for Control of Selected Houseplant Pests

Pest	Plants Attacked	Control
Mite	African violet, begonia, cyclamen, gloxinia, palm, geranium, and English ivy	Spray with dicofol
Mealybug	Begonia, African violet, gardenia, palm, dracaena, and gloxinia	Use malathion, diazinon, or orthene or remove by hand
Aphid	Gloxinia and begonia	Use malathion or remove by hand
Whitefly	Geranium, coleus, and begonia	Use malathion, rotenone, or orthene
Scale	Philodendron, azalea, citrus, fern, and palm	Remove by hand or use nicotine sulfate
Fungus gnat	Philodendron, fuchsia, and fern	Use nicotine sulfate

It is recommended, whenever possible, to use pesticides on plants on which they are registered and to follow the manufacturer's directions

Table 7-5 Suggestions for Control of Selected Home Garden Pests

Pest	Plants Attacked	Control
Aphid	Cole crops (cabbage, broccoli, cauliflower, and brussels sprout)	Use diazinon or malathion
Mite	Bean (dry, lima, and snap) and tomato	Use malathion, diazinon, or kelthane
Caterpillar	Cole crops	Use Sevin, malathion, or thiodan
Cutworm or white grub	Cole crops	Use Sevin or diazinon
Earworm or fruitworm	Corn and tomato	Use Sevin or diazinon
Damping-off	Eggplant, pepper, cucumber, muskmelon, pumpkin, and tomato	Use captan or thiram
Downy mildew	Cole crops and beans	Use zineb or maneb
Powdery mildew	Bean, pumpkin, and squash	Use sulfur or benomyl
Wilt	Tomato, eggplant, and pepper	No chemical control
Cercospora leaf spot	Beet and carrot	Use zineb or maneb

It is recommended, whenever possible, to use pesticides on plants on which they are registered and to follow the manufacturer's directions

SUMMARY

Chemicals used to control insect pests are called insecticides. On the basis of killing action, they are classified as contact poisons, stomach poisons, systemic insecticides, fumigants, repellents, attractants, and those that kill by suffocation. Active ingredients of insecticides may be organic or inorganic. Organic insecticides may be created from natural or synthetic compounds. Synthetic organic compounds have several classes: chlorinated hydrocarbons (e.g., DDT), organophosphates (e.g., malathion), carbamates (e.g., Sevin), pyrethroids (synthetic pyrethrins), and others. Insecticides may be formulated as dusts, wettable powders, emulsifiable concentrates, granules, or aerosols. Insecticide applicators vary in size from handheld to aerial sprayers.

Table 7-6 Common Landscape Pests and Control

Pest	Description and Suggested Control
Lawn Pests	
Ants	Inhabit the soil in nests and destroy vegetation, leaving denuded spots in the lawn; their mounds are unsightly Control: Drench nests with pesticides (e.g., diazinon, Sevin, and malathion)
Chigger	Also called red bug, it is the larval stage of a small mite; it sucks sap from stems and causes stunted growth Control: Apply insecticide dusts and sprays (e.g., durban, Sevin, and diazinon)
Billbug	Also called a snout beetle, it damages the roots and crowns of grasses Control: Apply diazinon or carbaryl
Armyworm	Feeds on stems and leaves Control: Apply diazinon or carbaryl
Sod webworm	Also called a grass moth; as it feeds, it spins threads that bind soil and leaves into tubelike structures on the soil surface Control: Spray carbaryl, aspon, or diazinon
Japanese beetle	Destructive in both larval and adult stages; destructive also to trees, fruits, and other ornamentals Control: Apply diazinon
Leaf bug	Damaged leaf shows yellow dots initially and eventually yellows and dies Control: Apply malathion or diazinon
General Landscape Pests (Annuals, Perennials, Trees, and Shrubs)	
Aphid	Sucking insect found on the underside of leaves; causes puckering or curling of leaves; it secretes honeydew that attracts other pests (e.g., flies, mites, and ants) Conrol: Apply orthene, diazinon, or malathion
Caterpillar	Larva of many insect pests (e.g., cankerworm, gypsy moth, eastern tent caterpillar, webworm, California orange dog, and sawfly) feed on the foliage of landscape plants Control: Use diazinon, Sevin, or orthene
Borer	Larvae of certain insects (e.g., peach twig borer and other wood borers) damage flowering fruit trees and other ornamental trees Control: Apply dimethoate, bendiocarb, or lindane
Beetle	Adult and larva may inflict damage to plant foliage Control: Apply methoxychlor or carbaryl
Mite	Presence characterized by discolored patches on leaf as a result of feeding (sucking) from underneath the leaf; affected plants are less vigorous and may eventually brown and die Control: Apply dicofol, malathion, or kelthane
Gall	May occur on stems, branches, twigs, or leaves; leaf galls usually cause less damage, being primarily unsightly and reducing aesthetic value Control: Apply carbaryl, diazinon, or malathion
Scale	Scale insects overwinter as eggs or young. They may be armored or unarmored according to scale characteristics. They suck plant juice by using their piercing mouthparts. Control: Apply dormant oil, acephate, or carbaryl

---

PART A

Please answer true (T) or false (F) for the following statements.

1. T F Stomach poisons kill by contact action.
2. T F Fumigants attack the respiratory systems of victims.
3. T F DDT is a natural organic insecticide.
4. T F Organophosphates have shorter residual action than chlorinated hydrocarbons.
5. T F Household chemicals (pesticides) are often formulated as aerosols.
6. T F A duster is an implement for applying wettable powders.

PART B

Please answer the following questions.

1. Classify insecticides on the basis of killing action.

   _____

2. Classify insecticides on the basis of active ingredients.

   _____

3. Name three insects that may be controlled by stomach poisons.

   _____   _____   _____

4. Give an example of a microbial insecticide.

   _____

5. List four different kinds of small-scale insecticide applicators.

   _____   _____
   _____   _____

PART C

---

1. Describe how systemic insecticides work.
2. Discuss the success of DDT in pest control.
3. What is an advantage of emulsifiable concentrates over wettable powders?

MODULE 4 CHEMICAL CONTROL OF PLANT PESTS: HERBICIDES

OVERVIEW

Herbicides are chemicals used to control weeds. Chemicals are the method of choice in killing weeds in large-scale plant production operations. Although their use facilitates plant production, the collateral damage to the environment and health hazard they pose to humans often detract from their role in agricultural production. Herbicides are also convenient for use in controlling weeds in the landscape and along railroads and highways. They work by interfering with the metabolic processes of the plant. The challenge in their design and application is to minimize damage to cultivated plants while killing unwanted plants. Indiscriminate use of herbicides should be avoided. Certain chemicals are restricted for use by professionals (licensed or certified applicators). As with all toxic chemicals, strict adherence to the directions for their safe use minimizes the danger to the health of humans and cultivated plants.

7.12: Classification of Herbicides

Herbicides may be grouped in one of several ways—by selectivity, how they kill (mode of action—by contact or systemic [translocation]), timing of application, and chemistry.

7.12.1 SELECTIVITY

On the basis of selectivity, there are two types of herbicides, selective and nonselective.

Selective Herbicides

True selectivity is achieved when a herbicide applied at the proper dose and timing is effective against only certain species of plants but not against others. Selective herbicides are designed to kill only certain plants without harming others. Generally, they are designed to discriminate between broadleaf and narrowleaf (grasses) morphologies. Selective herbicides are the most widely used because most situations require only certain plants to be killed but not others. For example, in lawn (grass), broadleaves of any kind are not desired. Broadleaf weeds such as dandelions and wild mustards can be safely eliminated by spraying the lawn with a selective herbicide such as (2,4-dichlorophenoxy acetic acid), which kills only broadleaf plants. Certain chemicals can be manipulated to be selective by changing the concentration at which they are applied. At a high rate of application, a particular herbicide may kill a certain species but fail to do so at a lower concentration. However, even at lower concentrations, these chemicals remain toxic and should be handled with care. As such, by spraying older cultivated plants with low application rates, younger weeds may be controlled without harming the desired crop plants.

Nonselective Herbicides

Nonselective herbicides literally kill all plants exposed to them—weeds and crops alike. These nondiscriminating herbicides are used to control weeds in areas where no plant growth is desired, such as driveways, parking lots, and along railroad tracks. Examples of nonselective herbicides are Roundup and atrazine. Nonselective herbicides may be made selective through manipulation of the concentration or rate of application. For example, using atrazine at low concentrations decreases its killing action to certain plant types.

7.12.2 CONTACT VERSUS TRANSLOCATED

Some chemicals use two modes of action. *Contact herbicides* kill by direct contact with plants and are very effective against annual weeds. To be most effective, the application must completely cover the plant parts. *Translocated herbicides* are absorbed through either roots or leaves. Those applied to the soil have residual action and thus are most suitable for controlling perennial weeds. Complete coverage is not necessary when using these chemicals.

7.12.3 TIMING OF APPLICATION

Regarding crop (or weed) growth cycle, three stages are important for herbicide application.

Selectivity
The ability of a pesticide to kill some pests and not others without injuring related plants or animals.

Systemic Pesticide (translocated)
One that is absorbed and moved from the site of uptake to other parts of the plant.

Preplant

Preplant herbicides are applied to the soil before planting the crop. Depending on the kind, it may or may not require incorporation into the soil to be effective. Preplant applications are made at low rates or concentrations and have the advantage of damaging weeds when they are in the most vulnerable seedling stage.

Preemergence

Like preplant herbicides, preemergence herbicides are applied after planting the crop, either before crops or weeds emerge or after crop emergence but before weed emergence. These herbicides kill only germinating seedlings and not established plants. Whenever soil is disturbed, weeds arise. A newly planted ground cover may be sprayed with a preemergence herbicide to suppress weeds that may have been stirred up.

Postemergence

Herbicide application after cultivated plants have emerged is described as postemergence treatment. In several situations, such as occurs in orchards, the grower has no choice but to adopt postemergence application. However, the grower may choose to apply a herbicide before weeds emerge (preemergence). It is important, therefore, that emergence always be in reference to either the weed or the crop plant.

7.12.4 CHEMISTRY

Herbicides may be classified according to their chemical nature as either organic or inorganic.

Organic Herbicides

The various classes of organic herbicides include organic arsenicals and phenoxy herbicides.

Organic Arsenicals   Organic arsenicals are translocated herbicides and thus are effective against plant species with underground structures (e.g., rhizomes and tubers), as occurs in nutsedges and johnsongrass. They are relatively less toxic than inorganic chemicals and are salts of arsenic and arsenic acid derivatives.

Phenoxy Herbicides   Phenoxy herbicides are also referred to as hormone weed killers. One of the most common phenoxy herbicides is 2,4-D. Another is 2,4,5-trichlorophenoxy acetic acid, which is used in the control of woody perennials and is associated with the Agent Orange episode in Vietnam, where it was used to defoliate large forest areas. The latter has been banned by the EPA.

Diphenyl Ethers   An example is Fusilade.

Substituted Amide   These herbicides are readily biodegradable by plants and in the soil. An example is Diphenamid.

Substituted Ureas   Selective preemergence herbicides, substituted ureas have strong residual effects in the soil. An example is Siduron.

Carbamates   This class of herbicides is formulated generally for preemergence application. An example is EPTC.

Triazines   An example of this class of herbicides is Simazine. It is used in driveways and around patios.

Aliphatic Acids   An example is Dalapon. It is used to control grasses.

Arylaliphatic Acid   An example is DCPA.

Substituted Nitriles   These herbicides are fast acting and also have broad action. An example is Dichlobenial.

Bipyridyliums   Examples are diquat and paraquat. These are contact herbicides.

Inorganic Herbicides

Inorganic herbicides have great residual effects and thus are strictly regulated by the EPA. They are not recommended for use around the house.

7.13: FORMULATIONS

Herbicides are formulated to be applied as either *liquids* or *granules*.

7.13.1 LIQUIDS

Liquid formulations are applied as either wettable powders or water-dispersible granules in water. Most herbicides are applied as sprays, making the sprayer the most important implement in herbicide application. Sprayers come in a variety of designs and may be hand or power operated (7.10.4). Sprayers may be mounted on trucks or tractors. Sprayer application may also be at low volume (high herbicide concentration delivered in small amounts per unit area) or high volume (low concentration of a herbicide applied in large amounts per unit area).

7.13.2 GRANULES

When granules are used, they may be applied at low or high rates. When applying at low rates (small amounts of granules), a carrier material such as sand may be mixed with the granules to increase the bulk for more effective and uniform application. Granular formulations are more expensive than others because of the bulk and shipping costs. Further, they do not provide uniform application. However, the equipment for application is less expensive and there is no need to haul large amounts of water to the field during application.

7.14: METHODS OF APPLICATION

Depending on the areas to be treated and the distribution of vegetation, herbicides may be applied in several ways, including broadcast, band, and spot application.

7.14.1 BROADCAST APPLICATION

When the entire area is to be treated, the herbicide may be broadcast without fear of damaging other plants. Liquids and granules may be broadcast. Aerial application by aircraft is a form of broadcasting.

7.14.2 BAND APPLICATION

In orchards and vineyards, the paths between rows may be readily cultivated, but the plant canopy may not permit the equipment to get close enough to clear all of the weeds from around the stem. Herbicides may be applied by band application to control the narrow strips of weeds around plants.

7.14.3 SPOT APPLICATION

Weeds that break through gaps in the driveway or walkway and masses of weeds concentrated in a small or hard-to-reach area are often spot treated (figure 7-13).

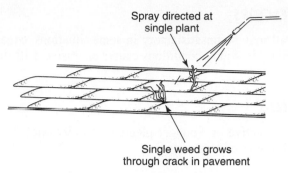

Spray directed at single plant

Single weed grows through crack in pavement

FIGURE 7-13 Spot application of pesticides.

7.15: FACTORS INFLUENCING HERBICIDE EFFECTIVENESS

Herbicides are applied directly to plants or the soil. As such, plant and soil conditions, coupled with the conditions in the general environment in which application occurs, influence the effectiveness of herbicides. Although some of these factors are environmental, the effectiveness of herbicide application depends largely on the operator. Some sources of error in application are discussed in the following sections.

7.15.1 WEED IDENTIFICATION AND ASSESSMENT OF INFESTATION

Since a weed is simply a plant out of place, a volunteer corn plant in an orchard is a weed. The presence of a few plants of corn in a tomato field does not warrant chemical control. The corn (the weed) may be cut down or uprooted by hand. The weeds to be controlled must be correctly identified for the correct treatment to be prescribed.

7.15.2 HERBICIDE SELECTION

Since most herbicides are selective, it is important that the correct plant species be identified so that the appropriate chemical is used.

7.15.3 TIME OF APPLICATION

Certain weeds are extremely difficult to eradicate. However, their spread may be slowed if the weeds are controlled before they flower so that they do not bear seed. Also, since herbicides may be applied as preplant, preemergence, or postemergence herbicides, it is important that they be applied at the correct time in relation to the emergence of the crop or weed. If one is depending on rainfall to wash the herbicide into the soil, the operator should follow the weather forecast to ensure that rain will fall within about a week after application or make provisions for irrigation. Not all herbicides require rainfall (water) after application to be effective.

7.15.4 CORRECT RATE OF APPLICATION (EQUIPMENT CALIBRATION)

The recommended rate of application should be adopted in a weed-control program. The equipment used should be properly calibrated to deliver the desired rate. If the chemical is overdiluted, the application will be a waste of time and money since the weed will not be controlled. On the other hand, too high a rate of application may injure desired crops and also waste money.

7.15.5 GOOD GROUND COVERAGE

If the ground is not completely covered during an application, weeds may germinate and survive at untreated spots. This result may necessitate an additional spot application or manual removal of weeds, increasing production costs.

7.15.6 WEATHER FACTORS

Although little rainfall may be unsatisfactory in some situations, excessive rain may wash granules away. Spraying in windy conditions causes excessive drift that may damage crop plants.

7.15.7 AGE OF WEEDS

Herbicides are more effective on younger plants than older ones. Weed-control measures should be effected before weeds are mature.

7.15.8 SOIL CHARACTERISTICS

Herbicides such as Dacthal are readily absorbed by the soil organic matter, making them less effective on soils that are high in organic matter. Clay soils also absorb certain herbicides. In such situations, a higher rate of application may be necessary for better results.

7.16: INDOOR WEED CONTROL

Weeds should not become a problem indoors, especially in the home. Most houseplants are planted in pots or suitable containers. Weeds may occasionally arise because of the source of the growing medium. Such unwanted plants should be uprooted by hand. There is absolutely no need to use chemicals to control weeds in the home. In greenhouses without concrete floors, weeds could become a problem if neglected. Weeds could also arise through cracks in the concrete. Such weeds may be spot treated with chemicals. Certain herbicides are approved for use in greenhouses.

7.17: SUGGESTED HERBICIDES FOR THE LANDSCAPE

Common herbicides used in ornamental plant culture are presented in table 7-7.

Table 7-7 Suggested Herbicides for the Landscape

Problem Weeds	Suggested Herbicides
Lawn	
Annual grass weeds	Trifluralin
	Dacthal (DCPA) - apply as preemergent
Annual broadleaf weeds	Bromoxynil
Perennial and other weeds	Glyphosate - as spot treatment; use for non-selective control or as preemergent for new lawns
General control of broadleaf weeds	2,4-D amine
Flower beds	
Annual weeds	Dacthal
Perennial weeds	Eptam (EPTC)
Around shrubs and trees	
Annual weeds	Dacthal, Bensulide
Perennial grass weeds	Eptham, Glyphosate

7.18: Suggested Herbicides for the Home Garden

Weed control in the home garden can be accomplished without chemicals. Annual weeds (e.g., purselane, pigweed, and crabgrass) and perennial weeds (e.g., bindweed and quackgrass) can be controlled by, hoeing, mulching, cultivating, or pulling by hand. Chemicals such as Dacthal may be used if necessary.

Summary

Herbicides are chemicals used to control weeds. They may be classified according to selectivity (as selective herbicides, which kill only certain plants, or nonselective herbicides), timing of application (as preplant, preemergence, or postmergence herbicides), mode of action (as either contact or translocated), and chemistry (as either organic or inorganic herbicides). Herbicides may be formulated as liquids or granules. They are applied by either broadcast, or band application. Sometimes they are applied as a spot application to control highly localized weed infestation. The effectiveness of herbicide application depends on several factors, including plant species, type of herbicide, time of application, weather factors, age of weeds, and soil factors.

Practical Experience

Calibrate a sprayer.

Outcomes Assessment

Part A

Please answer true (T) or false (F) for the following statements.

1. T F A herbicide that kills indiscriminately is said to be nonselective.
2. T F Aerial application is accomplished by broadcasting.
3. T F Roundup is a selective herbicide.
4. T F Most herbicides are applied as sprays.
5. T F Nonselective herbicides are the most widely used herbicides.
6. T F Preemergence herbicides kill only germinating seedlings and not established plants.
7. T F Low-volume herbicide application entails the administration of low concentrations of herbicides in large amounts per unit area.

Part B

Please answer the following questions.

1. Selective herbicides can distinguish between _____ and _____ plants.
2. Classify herbicides on the basis of time of application.

    _____

    _____

    _____

3. Herbicides are formulated as liquids or _____.
4. Equipment _____ ensures the correct rate of application.

1. Describe three specific situations in which spot application of herbicides is appropriate.
2. Describe how one specific factor influences herbicide effectiveness.

MODULE 5 GREENHOUSE PEST CONTROL

Greenhouses are enclosed structures and hence have environmental conditions that are different from the general open environment. Further, the greenhouse environment can be controlled, and thus the pest incidence can be controlled to some extent. However, even with best efforts, certain pests occur in greenhouses.

7.19: COMMON GREENHOUSE INSECT PESTS

Some of the most common and economically important greenhouse pests are described in the following sections.

7.19.1 APHIDS

The most common aphid found in greenhouses is the green peach aphid *(Myzus persicae)* (6.4.6). They may be winged or wingless. Whereas the winged green peach aphids are brown in color, the wingless ones are yellowish-green or pink. Aphids attack a wide variety of greenhouse plants. Younger leaves that are attacked become distorted, and older leaves show chlorotic patches.

7.19.2 FUNGUS GNATS

Important species of the fungus gnat are *Bradysia* spp. and *Seiara* spp. Gnats are gray-colored, long-legged flies. They live on the soil. The stage in their life cycle that constitutes a pest problem is the larva stage. Gnat larvae are white worms with black heads. They prefer soils that are rich in organic matter, feeding on soil fungi and decaying organic matter. However, they can also feed on underground storage organs and the roots of young seedlings.

7.19.3 LEAF MINERS

Leaf miners are insects that in the larval stage tunnel between the outer layers of leaves. Their activity blemishes leaves severely. Chrysanthemums are particularly susceptible to leaf miner attack; the chrysanthemum leaf miner is called *Phytomyza atriconis*. Other species exist.

7.19.4 MEALYBUGS

Mealybugs (*Pseudococcus* spp.) are oval-shaped piercing insects that secrete a waxy covering over their bodies, causing them to appear white. This waxy layer forms a protective covering that makes pesticidal control of bugs difficult. To be most effective, mealybugs are controlled by spraying the nymphs (6.4.6). The economic damage they cause to plants is similar to aphid attack. Mealybugs also secrete honeydew, which attracts black mold to grow on the plant.

7.19.5 MITES

Mites (6.4.6) belong to the spider or scorpion family (Arachnida). They are very tiny in size and develop well under conditions of high humidity and low temperatures of about 16°C (60°F). A common species is the *Steneotarsonemus pallidus* (Cyclamen mite). However, the most important mite in greenhouse production is the two-spotted mite (or red spider)

(Tetranychus urticae). Mites usually hide on the undersides of leaves. The red spiders may spin unsightly webs over leaves and flowers.

7.19.6 SCALE INSECTS

Scale insects are similar to mealybugs. Some of them also secrete honeydew and thus cause black mold growth to appear, as in mealybug attack. They may be armored and have rubbery outer coatings or be without such a coating and armor.

7.19.7 SLUGS AND SNAILS

Slugs and snails are mollusks (which include shell animals such as oysters) (6.6.4). These pests chew tender seedlings and leaves. They are nocturnal in feeding habit and thus hide during the day under stones, leaves, and other objects; they prefer very damp environments. Slugs and snails gain access to the greenhouse area through growing media and attachment to plants and containers.

7.19.8 THRIPS

Thrips are very tiny insects (6.4.6). They feed on a wide variety of greenhouse plants, usually congregating on buds, petals, or leaf axils. A gentle tap on the hiding place causes aphids to become dislodged. During feeding, aphids scrape the surface of the leaf, resulting in whitish streaks that may eventually turn brown. They excrete brown droplets that eventually turn black.

7.19.9 WHITEFLIES

Whiteflies are also tiny insects. The greenhouse whitefly *(Trialeurodes vaporariorum)* is covered with a white, waxy powder. Greenhouse plants most affected by whiteflies include petunia, poinsettia, ageratum, chrysanthemum, and tomato. Since they are attracted to the color yellow, sticky strips (7.9.1) are usually suspended over benches to trap these flies. Whiteflies also secrete honeydew.

7.19.10 CATERPILLARS

Caterpillars, or the worm stage of some moths, are a menace to greenhouse production. Examples are corn earworms (which attack buds and succulent parts of plants such as chrysanthemum), European corn borers (bore through stems), cutworms (attack shoots), and beet armyworms (attack plants such as geranium, chrysanthemum, and carnation).

7.20: COMMON GREENHOUSE DISEASES

7.20.1 VIRUSES

Viruses, as previously indicated (6.5.3), cause stunting of affected plants and discoloration of leaves in the form of streaks, rings, or blocks. They are seldom transmitted by seed, exceptions including tomato ring spot and tobacco ring spot, which affect geraniums. Other greenhouse viruses include carnation mottle, carnation mosaic, chrysanthemum stunt, and chrysanthemum mosaic.

7.20.2 BACTERIA

Few greenhouse bacterial diseases exist. Major diseases include bacterial blight of geranium *(Xanthomonas pelargonium)*; bacterial leaf spot of geranium and English ivy *(Xanthomonas hederae)*; bacterial wilt of carnation *(Pseudomonas caryophylli)*; and crown gall of rose, chrysanthemum, and geranium *(Agrobacterium tumifasciens).*

7.20.3 FUNGI

Major greenhouse fungal diseases include the following:

1. *Powdery mildew.* Mildew occurs under conditions of high humidity. Plants affected by powdery mildew have a whitish, powdery growth on plant parts. Some plants (e.g., rose) are susceptible at an early stage and thus become very distorted. In other species, such as zinnia and dahlia, powdery mildew occurs on older plant parts. In the latter scenario, the economic damage is blemishing, which makes plants less aesthetically desirable and thus not usable as cut flowers. Sulfur may be applied for both prevention and control, along with monitoring the humidity and temperature of the greenhouse to prevent high humidity levels.

2. *Botrytis blight. Botrytis* blight is known to affect numerous species of plants. The species *Botrytis cinerea* (common gray mold) causes rots and blights of many greenhouse plants, including carnation, chrysanthemum, rose, azalea, and geranium. Depending on the species, the stem, leaf, flower, or other tissue may be affected. Observing greenhouse sanitation reduces the incidence of *Botrytis.* Ample ventilation is required, as is preventing irrigation water from splashing on plants.

3. *Root rot.* Three important causal agents of root and basal rots are *Rhizoctonia, Phythium,* and *Thielaviopus.* These fungi are soilborne and transmitted by mechanical means such as splashing of water during irrigation and contamination of tillage tools and containers. They are controlled by soil pasteurization and sterilization of tools, containers, and greenhouse bench tops. Root media should be well drained. Observance of good sanitation is necessary to control this pest.

4. *Damping-off.* When damping-off occurs before germination of seeds, the seeds tend to rot in the soil. Postemergence infection causes young seedlings to topple and eventually die. Preemergence damping-off is caused by *Phythium,* and *Rhizoctonia* causes damping-off of seedlings. Since the fungi are soilborne, damping-off is controlled largely by planting seeds in a pasteurized soil or medium that is well drained. Also, care should be taken when watering to prevent splashing.

5. *Verticillium wilt. Verticillium* wilt is caused by a fungus that inhabits the soil. It affects a wide variety of plants including rose, geranium, begonia, and chrysanthemum. The symptoms vary from one species to another and depend also on the stage of plant development. Some plants may not show any symptoms until they reach the reproductive stage, at which time the flower buds wilt. Generally, affected plants show wilting and yellowing of leaf margins, starting from the lower and older leaves. Once infected through the soil, the fungus grows upward in the plant through the xylem tissue. Thus, cuttings (chapter 9) from infected plants also spread the disease. Soil pasteurization is effective in controlling the disease.

6. *Nematodes.* Nematodes (eelworms) are soilborne organisms (6.6.3), one of the widely known species being the root-knot nematode (*Meloidogyne* spp.). Infected plants have knotted roots and amorphous growth of the roots. The knots in the roots interrupt vascular flow, and affected plants soon experience stunted growth. Soil pasteurization and general aeration help to control this pest.

    Certain nematodes cause leaf spots and eventually leaf drop. The spots start on the lower sides of leaves as light-colored brown spots that eventually turn black. Leaf nematodes require plant materials to survive in the soil. Thus, good sanitation including removal of plant debris helps to control these organisms. Spraying the affected plant foliage with appropriate pesticides (e.g., parathion or diazinon) is an effective control measure.

7.21: CONTROL METHODS

The common methods of pest control in the greenhouse are described in the following sections.

7.21.1 PESTICIDE SPRAY

Most greenhouse pest problems are controlled by spraying appropriate chemicals. The pesticides may be emulsifiable concentrates or wettable powders (7.10.3). Certain pesticide formulations are approved for use in greenhouses.

7.21.2 AEROSOL

Aerosols are usually applied in the greenhouse when immediate killing of pests is desired, since very little residue is left on the plant, and, even then, only the upper surfaces of leaves show residue. Aerosols should be applied to dry leaves on a calm day to prevent the material from being drawn out of the greenhouse through openings. The greenhouse must usually be kept closed overnight after an aerosol application.

7.21.3 DUST

Dusts are not commonly used to control pests in the greenhouse.

7.21.4 FOG

Fogs are applied by using fogging equipment. Fogs are oil based and usually prepared to 10 percent strength of the regular insecticide or fungicide. The fogging equipment heats up the pesticide, which breaks down into a white fog that spreads throughout the facility. However, leaks in the greenhouse may cause uneven distribution of the gas.

7.21.5 SMOKE

Unlike fogs, dusts, and sprays, smokes do not require special equipment for application. Instead, a combustible formulation of the pesticide packaged in containers is placed in the center isle of the greenhouse and ignited. Smokes are generally not as phytotoxic to foliage as other gas applications. When in use, all vents and doors must be closed.

7.21.6 APPLICATION TO ROOT MEDIA

Soilborne diseases and pests may be controlled by drenching the soil with pesticides. Granules or powder formulations may be applied to the soil surface and washed down in irrigation water.

7.22: CONTROL STRATEGIES

To be effective, the timing of application of the pesticide is critical. Three factors should be considered regarding the intervals between pesticide applications:

1. The residual life of the pesticide
2. The life cycle of the pest
3. The killing action of the pesticide

It is important to know the life cycle of the insect to be controlled and the developmental stage at which it is most susceptible to the pesticide. For example, if the insect has a seven-day life cycle and the pesticide to be applied (e.g., smokes and aerosols) is not effective against eggs, the pesticide should be applied at six-day intervals. The rationale is that the first application will kill most of the adults. Since aerosols and smoke leave no residue, the eggs will hatch on schedule. However, before they develop to adult stage, when they can lay a new batch of eggs, the next round of treatment kills that population, along with any that survived the first treatment. The survivors of the first round of treatment have a chance to lay eggs, which will be missed by the second application. However, the eggs will hatch within six days, in time for the third application. This third application is usually adequate in completely eradicating the pest within 12 days. Usually, as temperatures increase, the life cycle of the pest increases in length. Spraying intervals of between five and seven days are generally effective in controlling many greenhouse insect and mite attacks.

When controlling mites, it should be kept in mind that mites are known to develop resistance to pesticides rapidly. They are also known to occur as a heterogeneous mixture in a given population. A three-miticide cycle of control is recommended. First, one miticide should be selected for use at one time until the mites develop resistance to it. A second miticide should then be selected and used repeatedly until resistance against it has also been developed, followed by use of the third pesticide. When resistance to the third pesticide has been developed, the first should be reintroduced, since by this time the population would have lost its resistance to the first pesticide. The three-miticide cycle is then repeated.

Certain pesticides are specially formulated or approved for greenhouse application. Pesticides are also registered for use on certain plants. States may have preferred chemicals for use under various circumstances. It is important to check with appropriate local authorities such as the cooperative extension service to find out which pesticide is best for use on a particular occasion.

7.23: PREVENTING GREENHOUSE DISEASES

The greenhouse is an enclosure in which plant growth factors are under artificial control. The conditions are often ideal for plant production and similarly favor greenhouse pests associated with the specific production. Diseases can be prevented by adopting several cultural practices: environmental control; strict observance of sanitation; use of clean, healthy plant materials; and use of sterilized soil.

7.23.1 ENVIRONMENTAL CONTROL

One problem associated with environmental control that is a frequent source of diseases is condensation. This problem is caused by the combination of temperature and moisture content of the greenhouse atmosphere. During the daytime, the sun causes air inside of the greenhouse to warm. Warm air holds more moisture than cold air. Thus, during the night when it is cooler, the warm air is progressively cooled until it reaches the dew point. At this stage, water starts to condense on the surfaces of plants and greenhouse structures. Drops of water on plant leaf surfaces provide the condition needed for spores of pathogens to germinate and thus promote the incidence of diseases such as *Botrytis* blight caused by the gray mold fungus.

To reduce condensation, the greenhouse should be equipped with exhaust fans and ventilation that will circulate air by bringing in fresh, cooler air periodically. If the vents are closed, still air can be avoided by using a horizontal airflow system to move the air through the greenhouse and reduce the incidence of cold spots, which cause condensation of moisture.

Greenhouse structures, including floors, have surfaces that can retain water over a period of time. High humidity levels favor diseases such as mildews. When plants are watered in the morning, the moisture on solid surfaces has enough time to evaporate. With appropriate ventilation, excessive humidity is eliminated. Excessive moisture in rooting media can be avoided by using well-drained media and also watering only when needed.

7.23.2 SANITATION

Weeds are known to harbor insects and disease-causing organisms, apart from being unsightly. They may be mechanically removed or controlled by using approved herbicides.

Debris remaining after a crop harvest should be promptly removed and disposed of. Plant material from horticultural practices such as pinching (15.1) and disbudding (11.7.4) should be gathered during the operation and discarded outside of the greenhouse. Decaying plant materials provide a fertile medium in which disease organisms thrive.

Greenhouse containers—flats, pans, and pots—should be sterilized before reuse. Clay pots may be steam sterilized. Pots can be chemically sterilized by soaking containers a in formaldehyde solution (1 part formalin [40 percent formaldehyde] to 100 parts water) for 30 minutes, followed by rinsing and air drying. Apart from containers, all greenhouse tools should be cleaned and sterilized after use. First, the dirt should be scraped off and the tools washed with water; the tools should then be dipped in household bleach (sodium hypochlorite) at 1 part bleach (5.25 percent sodium hypochlorite) to 9 parts water. Watering hoses should not be left on the greenhouse floor such that the ends touch the floor; they should be hung with the ends turned upward.

Greenhouse benches should be sterilized periodically. All workstations should be cleaned and wiped with disinfectant or bleach after use. Wooden benches and flats may be painted or dipped in a 2 percent solution of copper naphthenate. Soil and growing media used in the greenhouse should be sterilized. If bulk root media are purchased, they should be stored such that they will not become contaminated by unsterilized soil or water.

Restricted access to parts of the greenhouse where preparation takes place (e.g., media and seeds) should be enforced to prohibit the general public from entering those areas. Visitors and customers come from a variety of places and may carry infected soil on the soles of their shoes. It is important to routinely clean and disinfect the floor at frequent intervals.

SUMMARY

Because of their enclosed nature, greenhouses have specific pest problems. The most common insect pests include aphids, fungus gnats, leaf miners, mealybugs, mites, scale insects, slugs, snails, thrips, whiteflies, and caterpillars. Common diseases include a variety of viral and bacterial problems and fungal diseases (e.g., powdery mildew, *Botrytis* blight, root rot, damping-off, *Verticillium* wilt, and nematodes). Because of the enclosed condition, pesticide formulations for greenhouses such as smokes, fogs, and aerosols are suitable for use.

When controlling pests, it is important to know the life cycle of the organism, the residual life of the chemical to be used, and the pesticide's mode of action. The environment in the greenhouse can be controlled and monitored to reduce pest incidence. Further, observance of good sanitation and hygiene, as well as sterilizing growing media, tools, and other greenhouse structures, reduces the incidence of disease.

REFERENCES AND SUGGESTED READING

Bohmont, B. L. 1997. The standard pesticide user's guide, 4th ed. Englewood Cliffs, N.J.: Prentice-Hall.

Boodley, J. W. 1996. The commercial greenhouse, 2d ed. Albany, N.Y.: Delmar.

Cravens, R. H. 1977. Pests and diseases. Alexandria, Va.: Time-Life.

Dixon, G. R. 1981. Vegetable crop diseases. Westport, Conn.: AVI Publishing.

Klingman, G. C., F. M. Ashton, and L. J. Noordhoff. 1982. Weed science: Principles and practices, 2d ed. New York: John Wiley & Sons.

Nelson, P. V. 1985. Greenhouse operation and management, 2d ed. Retson: Retson Publishing.

Prone, P. 1978. Diseases and pests of ornamental plants, 5th ed. New York: John Wiley & Sons.

Outcomes Assessment

Part A

Please answer true (T) or false (F) for the following statements.

1. T F The green peach aphid is the most common species of aphid in the greenhouse.
2. T F Chrysanthemum is particularly susceptible to leaf miner attack.
3. T F Mealybugs are chewing insects.
4. T F Fogs are oil-based pesticides.
5. T F Mites prefer a dry environment.
6. T F Thrips occur on plant roots.
7. T F Powdery mildew is a bacterial disease.
8. T F Aerosols are best applied on a windy day.

Part B

Please answer the following questions.

1. List
    a. four important greenhouse insect pests.

    _____   _____   _____   _____

    b. four important greenhouse diseases.

    _____   _____   _____   _____

2. How do slugs and snails gain access to the greenhouse?

    _____

3. Greenhouse tools may be sterilized by dipping into a solution of _____.
4. Describe the best conditions under which mites develop in the greenhouse.

    _____

    _____

5. Why are mealybugs difficult to control?

    _____

    _____

Part C

---

1. Discuss the importance of timing in controlling greenhouse pests.
2. Describe the recommended strategy for controlling mites.
3. Describe two specific ways by which disease incidence in the greenhouse may be reduced.

PART 3

PROPAGATING HORTICULTURAL PLANTS

CHAPTER 8

Sexual Propagation

PURPOSE

This chapter discusses the principles of plant propagation and describes the various methods of propagation, including the advantages of each and their best time of use.

EXPECTED OUTCOMES

After studying this chapter, the student should be able to

1. Describe how seeds are commercially produced.
2. List and discuss the factors that affect the use of seeds in plant propagation.
3. Describe how seed germination can be improved.
4. List and discuss the environmental conditions for seed germination.

KEY TERMS

Breeders' seed	Germination test	Registered seed
Cavity seeding trays	Germ plasm	Scarification
Certified seed	Hardening off	Seed dormancy
Clones	Hypogeous germination	Stratification
Enhanced seed	Longevity	Tetrazolium test
Epigeous germination	Percent germination	Transplanting
Fertilization	Percent pure live seed	Transplanting shock
Flat	Plug production	True to type
Fluid drilling	Primed seed	Viability
Foundation seed	Propagation	

OVERVIEW

Plant propagation is simply the reproduction or duplication of a plant from a source (mother plant). The ultimate objective of propagation is to produce more plants exactly like the parent. Flowering plants produce seeds that can be used for propagation. Seeds vary in size and other characteristics, which influence how they can be used for propagation. Certain plants do not produce seed and hence can be propagated by using parts of the plant other than seed (chapter 9). In some species, either seed or vegetative parts can be used for propagation. There are advantages and disadvantages to either method of propagation. This chapter explores the wide variety of methods of propagating horticultural plants by seed.

There are two major purposes of propagation:

1. *Preservation of germplasm.* Scientists on a crop expedition collect samples of plant materials intended for use in research endeavors. Such materials are often present in only small quantities and must be increased. Sometimes scientists discover novelties that arise spontaneously in nature or in a research project. These rare finds must be preserved, and, to utilize them in any research, they must be increased. In developing a new cultivar, plant breeders create the new genotype through a variety of breeding methodologies. The prototype must first be preserved and then multiplied for use by growers. The ultimate goal of preservation, therefore, is to ensure that the original characteristics of the plant are maintained.

2. *Crop production.* Growing crops or plants involves propagation using an appropriate material (e.g., seed, cutting, and others). The goal in propagation in commercial production is to increase or replicate the source material. In this instance, the product of propagation could be thousands of plants or seeds.

The success of propagation, whether to produce a small or large quantity, is judged by how closely the daughter plants resemble the parent. In some cases, such as in research, exact copies are required. In other cases, as long as the products express the major desirable traits of the parent or source material, the propagation is deemed successful.

The outcome of propagation depends on the genetics of the plant (genotype) as well as the mating system (chapter 5). Some parent plants have identical alleles at each locus and are said to be homozygous. If they reproduce by means of seed, employing the self-fertilization mating system, these plants will reproduce true to type by seed. On the other hand, the strength of some plants lies in the fact that they are heterozygous, as is the case in hybrids (5.9.2). Such plants do not reproduce true to type from seed. Many horticultural plants, including fruits and ornamentals, are heterozygous. The only way of perpetuating the desirable qualities in the progeny is to use means of propagation other than use of the seed. Using seed always requires that the genotype be subjected to reorganization in a new genetic matrix through the process of meiosis (except in the case of apomixis [5.15]). Since meiotic products are not alike, the resulting plants are different from each other. To circumvent meiosis, some plants may be regenerated from vegetative tissue (e.g., leaf, stem, or root). When this route is followed, the products of propagation are genetically identical *(clones).*

Flowering plants are capable of producing seed that can be used for their propagation. A seed houses a miniature plant that is nurtured to life when appropriate conditions are provided. Growing plants from seed is more convenient than using vegetative methods. Seed planting is easier to mechanize, and seeds are also less bulky and easier to store than other plant parts. The seed industry has played a significant role in the growth of the horticultural industry.

8.1: SEED FORMATION

Sexual propagation starts with seed formation. The process of seed formation follows *fertilization,* the union of gametes (formed by the process of meiosis [5.2.3]), which occurs after *pollination* (deposition of pollen on the stigma of the flower). The processes involved in the reproduction of seed plants are described in figure 8-1. The two processes were introduced in chapter 2 (2.8).

Anthers contain pollen grains (2.8). When anthers dehisce (shed their contents), pollen grains are released and ready to be transported to the stigma (2.8) by various vectors or agents

Pollination
The transfer of pollen from an anther to a stigma.

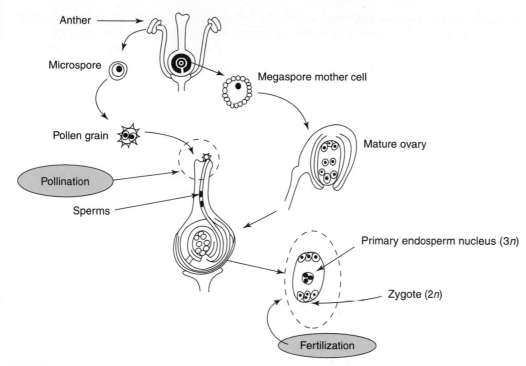

FIGURE 8-1 Reproduction in flowering plants involves pollination and fertilization.

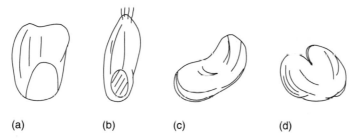

FIGURE 8-2 A sample of seed types. (a) Corn and (b) rice caryopsis; (c) bean seed and (d) seed of clover.

of pollination (2.8). The pollen grain germinates on the stigma and produces a tube *(pollen tube)* that carries two *sperm nuclei (n)* down the style into the *embryo sac.* The sperm nuclei are produced by the mitotic division of the *generative cell.* In the embryo sac, one sperm fertilizes the egg to form a *zygote (2n).* Fertilization thus restores the diploid number of the cell. The other sperm fuses with the large *central cell* of the embryo sac to produce a *triploid (3n)* cell, a process called *triple fusion.* The process whereby both a zygote and a cell with 3n nuclei are formed within the embryo sac is called *double fertilization.*

The next step after fertilization is the development of the ovule (containing the zygote and the 3n central cell) into a seed. The 3n cell divides repeatedly and develops into the *endosperm,* a nutrient-rich mass of cells that provides nourishment to the embryo until it develops to a stage where it becomes self-supporting (seedling stage).

The mature ovule is the *seed* (figure 8-2). The outer tissue of the ovules (the inner and outer *integuments*) fuse and lose their water to become the *seed coat,* or *testa* (figure 2-57). The endosperm contains large amounts of starch that serve as food reserves to be broken down into glucose for use during seed germination. In monocots, the outermost layer of the endosperm is called the *aleurone layer* and is the site of protein storage (figure 2-58). In seeds such as sunflower, mustard, pine, and fir, the endosperm stores a large amount of fats and oils.

The embryo is variable in size in terms of the volume of the seed it occupies. It occupies a large volume in species such as pea and oak, but occupies a relatively small volume in

> **Fertilization**
> *The union of gametes, an egg and a sperm, to form a zygote.*

> **Seed**
> *The mature ovule of a flowering plant.*

cereals. In species such as holly and orchid, the embryo remains relatively undifferentiated. However, in grass, the embryo is highly differentiated into structures including the *shoot apex, scutellum* (single cotyledon), and *coleoptile* (a cylindrical protective leaf). The *root apex* is protected by the *coleorrhiza* (figure 8-2, 2-57).

8.2: SEED PRODUCTION AND CERTIFICATION PROCESS

Plant breeders are engaged in plant improvement (5.7). Old cultivars are changed by importing new genes that condition improved characteristics, making new cultivars more disease resistant, higher yielding, more beautiful, and better in other ways according to breeding objectives. Once the breeder has completed the breeding program and tested the new material extensively, the seed is then released by the researchers to special producers for propagation.

Before the seed becomes available to the ordinary grower, it goes through stages of increase and certification. According to the stages, there are four classes of seed (figure 8-3):

1. *Breeder seed.* A small amount of seed is developed and released by a breeder as the source of foundation seed.

2. *Foundation seed.* Breeder seed is increased under supervision of agricultural research stations and monitored for genetic purity and identity.

3. *Registered seed.* Foundation seed is distributed to certified seed growers to be further increased for distribution.

4. *Certified seed.* The progeny of registered seed is sold to farmers. During the process of increase, certifying agencies in the state or region of production monitor the activity to ensure that the product meets standards set for the crop. Before sale, the certified seed grower is required to perform certain basic analyses *(seed quality analyses)* of the seed as described in the following section.

> **Certified Seed**
> *The progeny of registered seed that is maintained at a satisfactory genetic identity and purity, and approved and certified by an official certifying agency.*

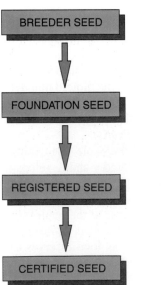

Small quantity of original seed produced by plant breeders through a carefully designed and conducted breeding program and according to specific objectives. Controlled by breeder/institution or company.

An increase of the breeder seed by Agricultural Experimental Stations or reputable growers under supervision of researchers.

An increase of foundation seed by registered growers.

An increase of registered seed (or foundation seed) that is inspected and approved by certifying agency for sale to farmers and other growers.

FIGURE 8-3 Commercial classes of seed. The differences depend on the quantities of seed handled and the person responsible for seed maintenance.

8.3: SEED QUALITY ANALYSIS

Certified seed producers are required to declare their *seed analysis* on the seed packet or tag (figure 8-4). To do so, a seed test must be conducted. The common seed tests available in the seed industry include the germination test, cold test, tetrazolium test, and purity test.

8.3.1 GERMINATION TEST

A germination test involves growing a number of seeds (usually in germination paper or sand) under moist and warm conditions. The number of seeds that germinate to produce healthy, normal seedlings is recorded, along with the number of ungerminated seeds.

8.3.2 COLD TEST

A cold test is conducted by subjecting seeds to a period of cold temperature (10°C or 50°F) before germination under warm conditions. The goal of this test is to find out how seeds will germinate and grow under the cold, wet conditions often found in the field.

8.3.3 TETRAZOLIUM TEST

The tetrazolium test depends on the use of a chemical called *tetrazolium chloride* (2,3,5-triphenyltetrazolium chloride). It is a rapid test in which seeds are soaked in water and then dissected longitudinally. One set of halves is placed in petri dishes containing 0.1 percent tetrazolium solution; the other set is left untreated. Living tissue in the seed is stained, whereas the dead tissue remains the same color. Percentage germination is given by the proportion of stained seeds. This test only indicates that seeds are respiring and does not equate the respiring seed with germination.

8.3.4 PURITY TEST

Seed purity analysis provides information on the physical condition of the seed and the presence of unwanted material (weed seed and inert material other than desired seed). It provides information on the proportion of the total seeds being offered for sale that are the desired seeds.

The various tests provide specific information, which is listed on the seed envelope or tag. The information includes the following:

1. *Percent germination.* Percent germination is determined using a simple ratio:

$$\frac{\text{number of healthy normal seedlings}}{\text{total number of seeds tested}} \times 100$$

The higher the percentage, the better.

FIGURE 8-4 A typical seed tag.

2. *Percent pure seed.* Percent pure seed measures the proportion of desired seed in the lot.

3. *Percent of other crop seed.* When harvesting equipment is not cleaned properly after harvesting one kind of seed and before harvesting another, remnant seed from the previous harvest may contaminate the latter harvest. This contamination may also arise as a result of crop weeds (unwanted plants [6.1]) being allowed to mature and being harvested along with the desired crop without rouging. Contamination of this kind is common in situations in which impure seed is used to establish a crop.

4. *Percent inert material.* Inert materials such as stones, pieces of wood, and the like may be present in seed packaged for sale because of improper cleaning of seed before packaging. Sometimes the harvester blades may be set too low such that they not only cut plant material but also scoop up some soil and rocks. Inert material may also come from seed processing. For example, if corn seed is being produced, pieces of the cob may be included if the corn is shelled improperly or is not dried to the proper moisture content before shelling. In seed purity tests, weeds may be classified as *common, restricted* (allowed in minimal amounts), or *noxious* (most undesirable and hardest to control).

5. *Percent weed seed.* Weed species are associated with certain crops. Seeds offered for sale should be free from noxious weed seeds. Weed seed contamination occurs when the seed producer does not implement proper weed control during production.

6. *Percent pure live seed.* From the preceding information, one can calculate percent pure live seed (the percentage of desired cultivar seed that will germinate) by the formula

$$\text{percent live seed (PLS)} = (\% \text{ germination} \times \% \text{ purity})/100$$

Seed should be purchased from a reputable grower or shop to ensure the highest quality. Further, the grower should purchase fresh seed each season if possible and purchase only as much as is needed to avoid the need to store remnant seed. If seed is to be stored for planting the next season, cold storage is often required. Even when high-quality seeds are purchased for planting in the current season, they can deteriorate before harvesting if there is a delay in planting, unless the seed is stored properly.

Mechanical Seed Damage

Seed suffers mechanical damage during harvesting, threshing, and handling. It is critical to harvest seed crops when they have dried to the right moisture content. If too dry, seeds are prone to splitting or cracking. Also, if the mechanical harvester is not set properly, excessive force may be delivered to the plant during harvesting, which contributes to physical damage. Cracked seeds may not be a critical problem for some markets (e.g., for processing into flour or meal). However, cracked seeds can significantly reduce the market value of the produce in other markets.

Mechanical damage may be determined by a simple laboratory procedure as follows:

1. Obtain 20 to 50 seeds (randomly selected) from the lot.
2. Place a sample in a petri dish containing a 1 percent sodium hypochlorite solution. (If household bleach is used, it should be noted that it contains 5.25 percent sodium hypochlorite.)
3. Allow seeds to soak for 15 minutes.
4. Count the number of swollen seeds.
5. Count the number of swollen seeds after soaking for 30 minutes.
6. Calculate percent cracked seed as follows:

$$\frac{\text{number of swollen seed}}{\text{total number of seed}} \times 100$$

8.4: Seed Viability and Longevity

Seeds may look healthy but fail to germinate when planted. Two other seed qualities that are essential but not displayed on the seed envelope are viability and longevity. *Viability* is a measure of the proportion of seeds in a lot that are capable of germinating. *Longevity* is a measure of how long seeds remain viable. Viability is measured using germination and tetrazolium tests. Species in the pumpkin family (Cucurbitaceae) (e.g., cucumber, squash, and cantaloupe) remain viable for a long time (sometimes for several years), whereas species in the lily family (Liliaceae), including onion and leek, lose viability only about two to three months after harvesting. Seeds of pine, hemlock, and spruce remain viable with proper storage for several years. On the contrary, seeds of maple, elm, and willow have a very short period of viability lasting only a few weeks.

Seed longevity depends on the species and the conditions at harvest and during storage. Seeds should be stored only when they have attained the appropriate moisture content (usually less than 15 percent). Seed dried to about 7 percent moisture and under low relative humidity can be stored for a long time in a refrigerator (0 to 4°C or 32 to 39°F). Very low relative humidity of about 5 percent allows seeds to be stored for a long time while retaining the highest level of viability. However, a relative humidity of 50 to 65 percent maintains the viability of most seeds for about one year. To increase longevity, seeds may be stored *cryogenically* in liquid nitrogen at −192°C (−313.6°F) for years.

> **Seed Viability**
> The proportion of seed in a lot that is capable of germinating.

> **Seed Longevity**
> A measure of how long seed remains viable.

8.5: Purchasing Seed

Select seed cultivars that are adapted to your locality. Extension agents can assist you in making the right choice. Select fresh seeds if possible, and have them ready to plant on time. Seeds should be purchased from a reputable grower or store. The date of harvest is usually printed on the seed packet. Mail-order purchases are available through a variety of outlets. Seed companies produce seed catalogs on an annual basis. Some small seed companies may provide seed at a lower cost to growers. However, more-established companies provide a variety of information on the seed packet to guide the grower with limited knowledge about growing plants.

8.6: Seed Dormancy

All seeds do not germinate, even when optimal conditions are provided. A physiological or structural adaptive mechanism called *dormancy* imposes further restrictions on the requirements for germination. Seeds germinate only after the dormancy is overcome or broken. Dormancy is desired in the wild, where plants depend entirely on nature for survival. It prevents germination in the face of adverse weather, which will kill the vulnerable seedlings after emergence. *Structural dormancy* is imposed via the seed coat (*seed coat dormancy*). Hard seed coats are impermeable to the much-needed moisture that is critical for germination. The seed coat may be softened before planting by one of several methods, such as *scarification,* a method of mechanically scratching the seed coat (by, for example, tumbling seeds in a drum containing coarse material) (8.7.1). Seeds may also be scarified by soaking them in concentrated sulfuric acid or household bleach for a period (8.7.2).

Physiological dormancy (embryo dormancy) occurs when the embryo requires a special treatment to induce it to start active growth. A cold temperature application (called *stratification*) of about 1 to 7°C (34 to 45°F) is commonly required to break the dormancy (8.7.1).

A number of chemicals in plants inhibit germination of seeds while they are still embedded in the pulp of the fruit (e.g., in tomato and strawberry). In some species, such as

> **Seed Dormancy**
> The failure of viable seed to germinate under adequate environmental conditions.

Pinus and *Ranunculus*, the fruits are shed before the embryo fully matures. Such *physiologically immature* seeds must undergo certain enzymatic and biochemical changes to attain maturity. These changes are collectively called *after ripening*. Immature embryos cannot germinate. Of necessity, such fruits are stored for a period of time to allow embryos to mature completely.

Seeds of ancient origins have been reported to germinate after the hard seed coat was weakened. The sacred lotus *(Nelumbo nucifera)* is reported to have germinated after 2,000 years, and the arctic lupine *(Lupinus articus)* germinated in 48 hours after being dormant for 10,000 years.

> **After Ripening**
> *A period after harvesting required by certain fruits to complete embryo maturity.*

8.7: IMPROVING GERMINATION CAPACITY OF SEEDS

Ideally, seeds should germinate within the reasonably expected period that is characteristic of the species, provided the right conditions for germination are present. Germination of seeds should be quite predictable by the grower, especially in commercial operations where timing of harvest is essential to obtain premium prices for horticultural produce. Furthermore, in commercial operations, uniformity of germination and maturity are critical, especially where any aspect of production is mechanized. When machines are used to harvest a crop, they are unable to distinguish between ripe and unripe fruits. Additional labor costs are incurred to manually sort out the harvest to remove immature fruits before marketing.

Some horticultural practices initiate germination processes in a seed to a safe stage and then discontinue it (see *primed* seed in 8.7.2). These seeds germinate quickly. Because of dormancy, seeds extracted from mature pods or fruits fail to germinate if planted immediately without a rest period. Through breeding activities, dormancy has been bred out of some cultivated species, since after a species becomes domesticated, the responsibility of caring for and protecting plants is transferred from nature to humans. In some plants, such as legumes, seeds germinate in the pod while still on the plant. At the other extreme, some seeds need special treatment to germinate.

Structural dormancy is imposed by seed testa or other protective covering, which prohibits the entry of moisture and air to start the physiological process of germination (8.6). Species with hard, impervious seed coverings occur in trees such as pine and in some legumes. To help such seeds germinate, certain methods (mechanical and chemical) may be employed to loosen the seed covering or initiate physiological processes for germination. These processes are described in the following sections.

8.7.1 PHYSICAL

Mechanical

Mechanical processes include a variety of methods adopted to scratch the surface of the seed to loosen the covering, which is called scarification (8.6). Seeds may be mechanically rubbed between sheets of abrasive tissue (such as sand paper). They may also be scratched by tumbling them in a rotating drum lined with a coarse material (such as emery cloth). Some people also mix abrasive material such as gravel with seeds in the drum. Mechanical bruising of seed must be done carefully to avoid damaging the embryo.

Temperature

Scientists have observed that forests that have experienced a fire produce new seedling growth soon after a rainfall, whereas forests that have not burned do not have such new growth. The inference is that the heat (high temperature) weakens the seed covering enough to enable it to imbibe moisture. In horticultural applications, some seeds may be placed in boiling water (with heat turned off) for about 24 hours. Some seeds require cold temperature treatment (stratification) to germinate (8.6).

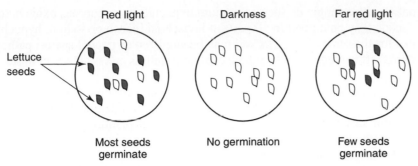

FIGURE 8-5 The role of phytochrome in seed germination. Red light stimulates germination while darkness inhibits it. Far red light is only mildly stimulatory of germination of lettuce seed.

Light

Light is required by many weed species and some small seeds such as lettuce *(Lactuca sativa)* before germination can occur. Soil tillage exposes buried seeds to light. Small seeds should be sown in loose soil and to a shallow depth to expose them to light.

The light requirement to stimulate germination must be of a certain quality (wavelength) (3.2.1). Exposure of lettuce to red light (about 660 nanometers) induces germination, but far red light (730 nanometers) inhibits it. It was discovered that if seed received red light after exposure to far red light, germination occurred. In fact, germination occurred as long as the last treatment before sowing was red light (figure 8-5).

8.7.2 CHEMICAL

Acid Treatment

Seed dormancy due to a hard seed covering may be overcome by soaking the seed in concentrated sulfuric acid for a period ranging from a few minutes to several hours (8.6). Concentrated acid is very injurious and must be handled with great caution. It is best to test a small sample first to find the best duration of soaking before treating the whole lot.

Leaching with Water

Leaching may be used where dormancy is due to chemical factors (such as coumarin). As previously mentioned, the seed of some fleshy fruits, such as strawberry and tomato, will not germinate in the fruit because of the presence of chemical germination inhibitors. Some desert plant seeds germinate only after a heavy downpour of rain that is able to wash away the inhibitors. By germinating only after a heavy rain, the seedlings are assured adequate moisture for survival and development until their roots are developed enough to absorb moisture.

Primed Seeds

Primed or enhanced seeds are seeds that have been soaked in a solution (e.g., ethyl alcohol or potassium chloride [KCl]) to give the germination process a head start by activating the enzymes and hormones. Seeds must be primed carefully to ensure success. Primed seeds germinate early and uniformly.

> **Primed Seed**
> Seed soaked in a specific solution to initiate physiological processes for quicker germination.

8.8: SEED TREATMENT

Unlike soilless mixes that are sterile, field soil harbors a wide variety of pathogens that are harmful to seeds. Some seeds benefit from seed treatment (8.12) in which appropriate pesticides are applied to seeds to protect against soilborne diseases prevalent in that soil. Seeds may be coated with a fungicide such as calcium hypochlorite or cuprous oxide before planting. Also, seeds may be dipped in 10 percent household bleach or sodium hypochlorite for five minutes for the same effect. The soil may be sterilized or treated against pathogens that cause damping-off in seedlings. Greenhouse soil and soilless mixes are easier to sterilize than field soil.

Direct seeding is extremely difficult when seeds are very tiny (8.11). To overcome this problem, tiny seeds may be *pelleted,* by coating with clay, for example. Bedding plant seeds may be treated in this way before planting. Pelleting makes planting easier and spacing more precise. New methods of preparing seeds for planting are being investigated. For uniform germination and establishment of a crop, planting of pregerminated seeds is being investigated. Recently, sprouted seeds have been suspended in a protective gel or drilled with some water, a method called *fluid drilling.*

8.9: ENVIRONMENTAL CONDITIONS FOR SEED GERMINATION

Germination involves physiological and biochemical processes. The below-ground environmental conditions (chapter 3) must be adequate for seeds to germinate. The critical factors are as follows.

8.9.1 MOISTURE

Seeds must imbibe moisture to a certain degree for the germination process to be initiated. Moisture is needed to initiate the enzymatic breakdown of food reserves. The critical degree of imbibition differs among species. For example, whereas soybean needs to imbibe about 50 percent of its weight before germination, sorghum requires only about one-third of its weight. Horticultural plants that germinate best under conditions of high moisture include beet, celery, and lettuce. When raising seedlings in flats (8.12), a glass plate may be used to cover the flat to prevent moisture loss through evaporation (figure 8-6). Similarly, the flat may be placed in a plastic bag to accomplish the same purpose (figure 8-7). Although moisture is critical to germination, excessive moisture encourages rotting and other diseases.

8.9.2 TEMPERATURE

Temperature regulates seed germination. In seeds that require cold temperature treatment to break dormancy, abscisic acid or other inhibitors are broken down under low temperatures (as obtained in winter). When warm spring temperatures arise, levels of endogenous gibberellins increase, resulting in germination. The rates of biochemical reactions are controlled

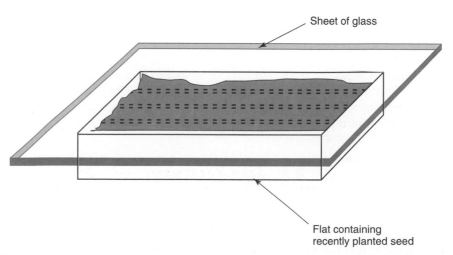

Sheet of glass

Flat containing
recently planted seed

FIGURE 8-6 Seed germination in a flat covered with a sheet of glass to retain humidity.

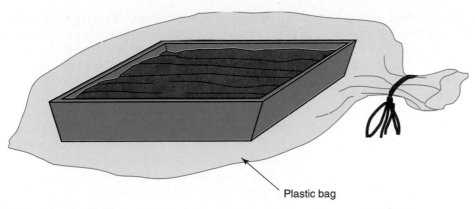

FIGURE 8-7 Seed germination in a flat enclosed in a plastic sack to retain humidity.

by temperature. When soils are too cold, growth processes slow down or cease altogether. Generally, a warm seedbed is desirable for seed germination, but warm season crops (such as bean and squash) do better at warmer temperatures (15 to 25°C or 59 to 77°F) and cool season crops (such as cole crops) do well at cooler temperatures (less than 10°C or 50°F).

8.9.3 LIGHT

Most seeds do not need light to germinate. A number of horticultural species including some herbaceous garden flowers, some vegetables (e.g., lettuce and celery), and grasses require light to germinate. The species are therefore planted shallowly in the soil. Light inhibits germination in some species (such as onion), which must be planted deeper in the soil or covered with a dark cloth or other material in the nursery. Geraniums require darkness to germinate.

8.9.4 AIR

In most species, germination is an aerobic process. The low oxygen concentration of soil air is inhibitory to most species. The seedbed must be well drained for good aeration.

8.9.5 DISEASE FREE

As indicated previously, soilless media are sterile, but field soil may contain pathogens that can overwhelm a developing embryo or young seedling. One of the most common diseases of seedlings is *damping-off,* a fungal attack caused especially by *Pythium ultimum* and *Rhizoctonia solani* (7.19). These fungi are active at warm temperatures (20 to 30°C or 68 to 86°F) and thus are less of a problem when germination occurs under cooler conditions.

It is important to note that these factors work together for good germination. The conditions must occur in a good balance. If, for example, moisture is present in adequate amounts but temperature is excessive, seeds might be heated in the soil and rot.

8.10: SEED GERMINATION AND EMERGENCE

Seed germination is a complex process involving metabolic, respiratory, and hormonal activities. The dry seed first imbibes water to initiate enzymatic breakdown of stored metabolites. These metabolites provide the source of nutrition for the developing embryo and are located in the cotyledons of dicots and endosperms of monocots (1.6.1). As these stored foods (proteins, fats, and oils) are metabolized, respiration (4.3.2) occurs to synthesize chemical energy, or adenosine triphosphate (ATP). Also, deoxyribonucleic acid (DNA) and ribonucleic acid (RNA) (5.2.4) are synthesized, the RNA being required for the production of certain hydrolytic enzymes such as amylases, proteases, and lipases (4.2). The net result of these biochemical and enzymatic processes is the production of new cells and formation of new tissue, leading to growth and development of the embryo into a seedling.

> **Seed Germination**
> A sequence of events in a viable seed starting with water imbibition and leading to embryo growth and development.

The two basic modes of seedling emergence are associated with what parts of the seed emerge first from underground. In certain species, the cotyledons (seed leaves) emerge above the ground, preceded by a characteristic arching (hook) of the *epicotyl* (a stemlike axis). This mode of shoot emergence is called *epigeous (epigeal) germination* (figure 8-8) and is associated with dicots. However, in certain dicots such as the pea *(Pisum sativum)*, the cotyledons remain underground. The cotyledons, once above ground, gradually change from a creamish color to green. In squash *(Cucurbita maxima)*, the cotyledons are important in photosynthesis. However, in castor bean and garden bean, these seed leaves do not contribute significantly to photosynthesis. The cotyledons gradually shrivel in size and eventually drop when the seedling becomes completely autotrophic. These are not true leaves and are called *seed leaves.*

In other species such as the grasses, the endosperm remains underground so that the plumule emerges first. This mode of seedling emergence is associated with monocots and is called *hypogeous (hypogeal) germination* (figure 8-9). The cotyledon remains underground and eventually decomposes. In this mode of germination, the *hypocotyl* (rather than the epicotyl) elongates.

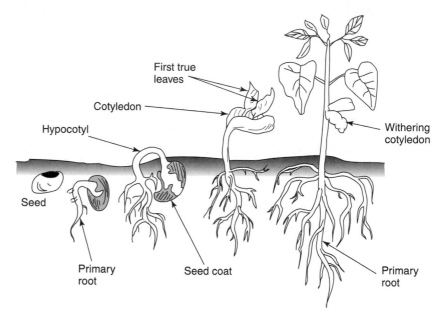

FIGURE 8-8 Epigeous or epigeal germination of bean seed. Cotyledons emerge above the ground.

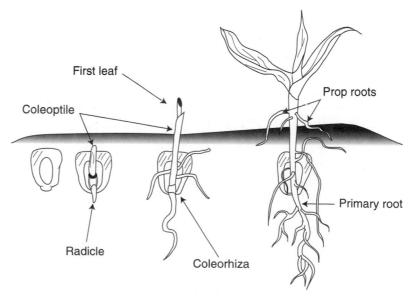

FIGURE 8-9 Hypogeous or hypogeal germination of corn seed. Cotyledons remain underground.

There are two methods of establishing or propagating a plant by seed—direct and indirect seeding.

8.11.1 DIRECT SEEDING

Direct seeding is a one-step planting method in which seeds are placed permanently in the spots in the field where they will germinate, grow, and go through the entire reproductive cycle. Table 8-1 provides a selected listing of crops and ornamentals that can be seeded directly. Direct seeding entails certain advantages and disadvantages.

Advantages

1. *Convenient to use.* Seeds are easy to handle, transport, and store. They can be stored longer than vegetative material (9.12). Further, seeds do not require any special provisions in transit. Vegetative materials may require special packaging and sometimes environmental control in transit. At the destination, seeds need only be stored in a cool, dry place while awaiting planting. Vegetative materials may require refrigeration or moist conditions to keep from drying.

2. *Readily adaptable to mechanization.* Seed planters (handheld or motorized) are available for planting a wide variety of seeds (figure 8-10).

Table 8-1 Selected Plants That Are Direct Seeded

Plant	Scientific Name
Pea	*Pisum sativum*
Corn	*Zea mays*
Melon	*Cucumis melo*
Bean	*Phaseolus vugaris*
Beet	*Beta vulgaris*
Lettuce	*Lactuca sativa*
Carrot	*Daucus carota*
Lima bean	*Phaseolus limensis*

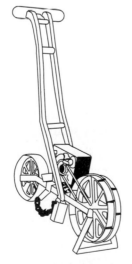

FIGURE 8-10 A mechanical seed planter. It can be adjusted to plant a variety of seed sizes at varying spacing.

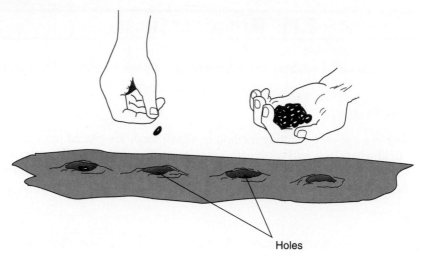

FIGURE 8-11 Sowing seeds by hand.

3. *Easy to plant.* Seeds can be broadcast or planted by spot placement, in which a hole is dug for each seed (or several seeds) (figure 8-11). It is not important to orient the seed in any particular way.

4. *Lack of bulk.* A small packet of seeds can be used to plant a large area. It is easy to order seeds by mail.

Disadvantages

1. *Species with tiny seeds are difficult to seed directly.* Tiny seeds require a much finer tilth of soil and should not be planted deep in the soil.

2. *Even with 100 percent germination, 100 percent stand is not assured.* A seed may be likely to fail in germination because, for example, it was not planted at the right depth or did not have good contact with the soil to imbibe water.

8.11.2 INDIRECT SEEDING

Indirect seeding is a two-stage process of establishing plants in the field. Instead of placing seeds where they will grow to maturity, they are started in a *nursery*, where they are grown to seedling stage (8.12). Healthy seedlings are then *transplanted* to their permanent locations in the field. Home gardeners may produce their own seedlings. However, the horticultural industry offers professionally grown seedlings for variety of crops. Table 8-2 presents a list of some of the crops and ornamental plants that are commonly seeded indirectly.

Transplanting
The process of relocating plants usually from a nursery to the field.

Advantages

1. *Good establishment.* Only healthy seedlings are transplanted, ensuring a 100 percent initial establishment. With direct seeding, some seeds may not germinate, even if the highest quality of seed is used.

2. *Early maturity.* Establishing a crop by seedlings has been known to hasten maturity.

3. *Shortened field growing period.* By using a greenhouse (or other facility) to raise seedlings, the grower can have a head start on the season. During adverse conditions (e.g., cold temperature), seedlings may be raised indoors for several weeks while the ground conditions remain too cold for seeds to germinate. When good weather arrives, these seedlings are transplanted into the field for an early crop.

Disadvantages

1. *Nursery care is an additional production activity that increases production costs.* Nursery care requires the provision of artificial sources of heat (in some cases), light, and space.

Table 8-2 Selected Plants That Are Not Direct Seeded

Plant	Scientific Name
Ageratum	*Ageratum houstonianum*
Alyssum, sweet	*Lobularia maritima*
Begonia	*Begonia x semperflorens-cultorum*
Coleus	*Coleus blumei*
Dusty miller	*Centaurea gymnocarpa*
Geranium	*Pelargonium x hortorum*
Impatiens	*Impatiens* spp.
Marigold	*Tagetes* spp.
Pansy	*Viola tricolor*
Phlox	*Phlox drummondii*
Snapdragon	*Antirrhinum majus*
Verbena	*Verbena x hybrida*
Zinnia	*Zinnia elegans*
Tomato	*Lycopersicon esculentum*
Cabbage	*Brassica oleracea*
Eggplant	*Solanum melongena*
Pepper	*Capsicum annuum*
Parsley	*Petroselinum crispum*
Cucumber	*Cucumis sativus*
Onion	*Allium cepa*

2. *Seedlings are not as readily amenable as seeds to mechanized planting.* Mechanized planters for seedlings have been developed for some crops, but vegetative planting materials are more delicate than seeds. In some cases, semimechanized planting operations, in which people feed seedlings into the planter manually as the tractor moves along, are used.

3. *Seedlings are bulky to handle.* Seedlings are raised in containers that need to be transported to the field.

4. *Seedlings should be planted promptly.* Seedlings must be transplanted before they are too old. Seedlings may become too big for their containers and develop root problems (pot-bound [8.13]).

5. *More immediate postplanting care is needed.* Transplanting shock (8.12) is a problem when indirect seeding is used. If seedlings are not properly prepared for transplanting, they will take a longer time to become established. Transplanting must be done at a certain time of day and may require a starter application of fertilizer for good and rapid establishment. Newly transplanted seedlings must be watered immediately and more frequently until well established.

8.12: SEED NURSERY ACTIVITIES

When raising seedlings at home, a section of the garden may be reserved for this purpose. A raised bed should have a very fine tilth. Seedlings may be grown in beds or in a variety of containers as described in the following section. Seedlings may be raised indoors in the basement of a home or some other convenient place. Special units may be purchased for this purpose. These units vary in size, design, and versatility; some are equipped to control temperature, humidity, and light.

8.12.1 CONTAINERS

Containers are discussed more fully in chapter 10.

Flats

A horticultural *flat* was originally a wooden box measuring, for example, 24 × 18 × 3 inches (61 × 45.7 × 7.6 centimeters) (*L* × *W* × *H*) (figure 8-12). Plastic flats are more durable and easier to manage and thus more popular today. Styrofoam flats are also used. Provision must be made for drainage by either drilling holes in the bottom boards or by leaving spaces between the pieces of wood. To prevent losing the soil through the openings in the bottom of the box, a sheet of paper (e.g., newsprint) may be used to line the bottom before filling with the appropriate mix of planting media. The flat should be filled and leveled off to about 1 inch (2.54 centimeters) below the top of the box.

Cavity Seeding Trays

Cavity seeding trays are plastic containers that may be obtained as individual cells (cells vary in size) but often come in sets (2, 4, 6, and so on) (figure 8-13). Each cell has a drainage hole in the bottom. The advantage of this container system is that it facilitates transplanting because the roots of individual plants are confined to a cell. Commercial producers of bedding plants employ a large-scale mechanized system called *plug production,* which entails the sowing of individual seeds into individual plastic cells by specially designed machines. The sown seeds are automatically covered with the appropriate soil mix and then transferred to a nursery where, under conditions of appropriate moisture, temperature, and light, the seeds are nursed to the size ready for sale.

Seedlings are sold in units called *packs.* Home owners and other growers who do not wish to raise their own seedlings find these packs a very convenient source of planting materials. Since seedlings have individual cells in the pack, they are easier to transplant.

Peat Pots

Peat pots are made of compressed sphagnum peat moss and newspaper fiber. They are biodegradable, and, as such, when used as material for containers for seed germination, the containers are planted directly along with the seedling (figure 8-14). Peat pots are usually treated with a fungicide and nitrogen fertilizer to aid in germination.

8.12.2 TIMING OF SEEDING

Horticulture, as previously indicated, is a science, art, and business. Timeliness of operation is crucial for success, especially if production is geared toward seasonal or holiday market. If you want to produce pumpkins for Halloween, you must know the date for the event, the

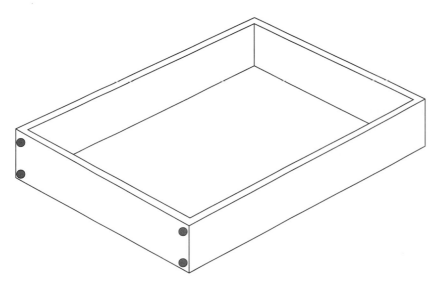

FIGURE 8-12   A typical wooden flat; more durable plastic flats are popular.

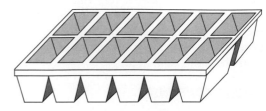

FIGURE 8-13   A cavity seeding tray.

FIGURE 8-14 Peat pots are biodegradable.

duration of growth for pumpkins, and when you should sell for maximum profit (definitely before the event); you can then decide when to plant the crop. Some horticultural products have year-round markets.

8.12.3 SEED TREATMENT

Most ornamental seeds are not pretreated by the seed production company. If properly produced, no pretreatment may be necessary for seeds. However, pretreatment with a fungicide is helpful in protecting against soilborne diseases (7.2.2).

8.12.4 SOWING SEEDS

The seed packet displays a variety of information, including the name of the cultivars and directions for production (sowing, spacing, care, and maturity). When in doubt, always consult the instructions on seed packets.

The soil surface should be of fine tilth. Instead of broadcasting over the entire soil surface, seeds should be planted in rows. This strategy helps in transplanting the seedlings. A number of shallow grooves should be made by placing a straightedge on top of the soil and drawing a line with a finger or stick (figure 8-15). If the seeds are not too small, they can be evenly distributed in the grooves. With care, the seeds can also be distributed fairly evenly in the grooves by tilting the opened packet and gently tapping it as it is moved along the groove. A simple handheld seed planter, a folded paper, or even the seed envelope itself may be used to drill seeds (figure 8-16).

Most seeds must be covered lightly after sowing. This covering may be part of the germination medium or other material such as shredded sphagnum moss. Fine seeds (e.g., petunia and begonia) are best left uncovered to be washed down with water during watering. Since a flat may contain rows of different plant species or cultivars, it is a good practice to always label each different item with a plastic (or wooden) label that indicates the plant name, cultivar, and date of planting. Some seed companies enclose labels in their packages. Some growers simply invert the empty packet over a stake that marks the rows sown to a particular cultivar. However, sometimes the paper rots under repeated watering or may be blown away by wind. Use indelible or water-insoluble ink in writing the label.

8.12.5 AFTER-SOWING CARE

Mechanized sowing by mechanical seeders (plug seed production) has revolutionized the bedding plant industry. However, sowing certain popular seeds such as zinnia and marigold is problematic because of anatomical features that interfere with the process.

Moisture

Seeds should be watered immediately after sowing, using a fine mist of a fine-spray nozzle to apply water to avoid washing seeds away or burying them too deeply. Excessive moisture is undesirable. Where possible, the container may be watered from the bottom (*subirrigation*) by placing the flat in water. If cavity seeding trays are used, the water may be added to the bottom of the holding tray. The flat may be covered with a pane of glass or plastic sheet to retain moisture in the soil. It is important to keep the soil surface moist throughout the germination period until seedlings emerge. A seed will die if allowed to dry after imbibing water or initiating sprouting.

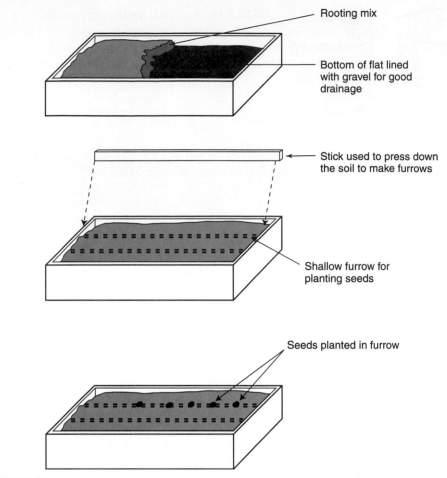

Rooting mix

Bottom of flat lined
with gravel for good
drainage

Stick used to press down
the soil to make furrows

Shallow furrow for
planting seeds

Seeds planted in furrow

FIGURE 8-15 Steps in sowing seeds in a flat.

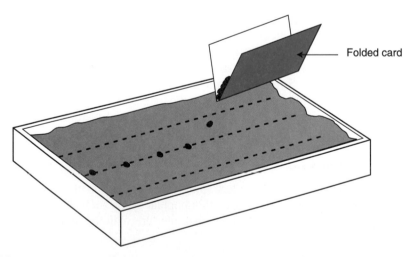

Folded card

FIGURE 8-16 A folded card may be used to facilitate the sowing of small seeds.

Temperature

The soil should be kept warm for good germination by most seeds. Place the flat in a warm place at 20 to 26°C (68 to 79°F), depending on the species. Even though 70°F (21°C) may be adequate, a warmer temperature of about 75°F (24°C) will cause germination to occur more rapidly. Certain seeds such as pansy and snapdragon prefer a relatively cooler germination temperature of about 65°F (18°C).

Temperature fluctuations in the seedbed are caused by evaporation and irrigation water. Evaporation of moisture may cool the soil surface water by about 5 to 10°F (1 to 3°C). When cold water (45 to 50°F [7 to 10°C]) is used in watering, it can cause the soil temperature to drop. A cold soil takes a long time to regain its heat. To stabilize soil temperature, some growers use *hotbed cables*. Some of the small-scale propagation boxes have heating coils. In commercial greenhouses, *biotherm,* or *bioenergy, systems* or hot water tubes are used to provide plants growing in containers with heat from the bottom.

Fertilizing

Fertilizers are not required until the seeds germinate. The stored food in the cotyledon is usually adequate for the period of germination. High fertility in the germination medium results in weak seedlings. The soil medium should have a good pH (6.0 to 6.8). Phosphorus is required for strong root development; consequently, the element should be available as soon as possible to ensure seedling survival.

8.12.6 AFTER-GERMINATION CARE

Fertilizing

Once the seeds germinate, the seedlings should be provided with a low-level application of a complete fertilizer analysis such as 20-20-20 at the rate of 1 ounce (28.35 grams) per 3 gallons (11.4 liters) of water. Fertilizer should be applied no more often than weekly to avoid the development of weak seedlings. Weak seedlings are difficult to transplant and less likely to survive the operation.

Hardening Off

Hardening off is a horticultural practice whereby seedlings are prepared for transplanting to the field. The danger transplants face in the field is the drastic change in the growing environment from a more controlled one to a harsh one. If unprepared, transplants are prone to *transplanting shock* (stress that may be suffered by transplanted seedlings). This preparative process is called hardening off and is usually accomplished by manipulating the temperature and moisture levels. Seedlings are gradually exposed to slightly cooler temperatures (13 to 15°C or 55 to 59°F) and/or reduced moisture.

> **Hardening Off**
> *The process of preparing seedlings for transplanting by gradually withholding water, nutrients, and decreasing temperature.*

8.13: TRANSPLANTING

8.13.1 PRETRANSPLANTING STORAGE

If seedlings cannot be transplanted when they are ready, they should be stored at a cool temperature to slow their growth (e.g., in a refrigerated environment at about 40°F [5.5°C]). They should be brought into the open for at least a day before transplanting.

> **Transplanting Shock**
> *A temporary setback in growth suffered by fresh transplants due to adverse conditions.*

Seedlings may be transplanted from the nursery bed directly to the field. If the grower is raising seedlings for sale, the seedlings are transplanted into containers such as peat pots or cavity seedling trays and nursed to marketable size. In the second case, seedlings are transplanted at a younger age and then must be transplanted a second time by the gardener or grower. Whatever the situation, seedlings should be transplanted after at least the first true leaves have developed (figure 8-17).

Seedlings are fragile and must be handled with care. Leaves may be damaged without great repercussion; new ones form in a short while, unless the terminal bud is damaged. An injured stem is a much more serious situation, and, as such, seedlings should be handled by their leaves (not stem). Instead of pulling seedlings from the soil, a trowel or hand fork may be used to dig them up. Watering plants before transplanting makes the operation easier. Seedlings should have some soil around the roots to aid in reestablishment. For transplanting, a tool called a *dibble* may be used (figure 8-18). However, a trowel is just as effective.

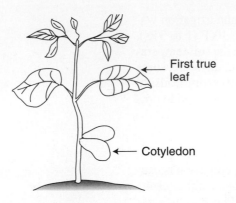

FIGURE 8-17 Seedlings should be transplanted after true leaves have developed.

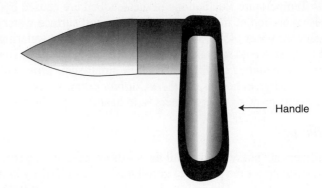

FIGURE 8-18 A dibble may be used to make holes for transplanting small seedlings.

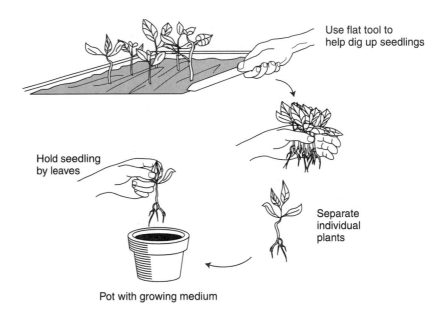

FIGURE 8-19 Steps in transplanting seedlings raised in a flat.

Seedlings are placed in holes in the receiving medium to a depth of at least that used in the nursery. Once in place, soil is added and gently patted down around the base of the stem (figure 8-19). After transplanting, the plants should be watered.

When transplanting from pots or cells, all one needs to do is position the stem between a pair of fingers and then invert the pot onto the palm (figure 8-20). A gentle tap at the bottom of the pot or tapping the edge of the pot against the edge of a table aids in removing the plant from the pot. If the roots are *pot-bound* (coiled over themselves at the bottom of the pot) due to limited space, they should be loosened before transplanting. An advantage with peat pots is that, since the pot is biodegradable, there is no need to remove the seedlings from the pot before transplanting (figure 8-21).

Transplanting into the field should be undertaken at a good time of day. Freshly transplanted seedlings need time to establish and resume normal activities. The roots will not immediately conduct moisture as they did in pots, and hence the plants are in danger of moisture stress. Transplanting at high noon magnifies this danger. Seedlings should be transplanted in the late afternoon so that they have ample time to recover before the next day's noon temperature.

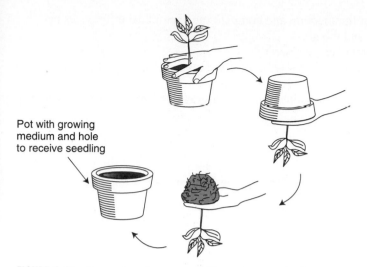

Pot with growing medium and hole to receive seedling

FIGURE 8-20 Removing larger seedlings from pots for transplanting or repotting.

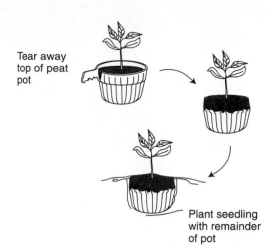

Tear away top of peat pot

Plant seedling with remainder of pot

FIGURE 8-21 Transplanting seedlings raised in peat pots. Since peat is biodegradable, there is no need to remove peat pots before transplanting.

SUMMARY

Flowering plants produce seed from which they can be propagated. Seeds are products of the meiotic process (except apomictic seeds, or seeds produced without fertilization) and hence produce offspring that are not clonal or identical to the mother plant. A seed contains the miniature plant (the embryo), which is protected by the cotyledon(s). Commercial seed production goes through four stages (breeders' seed, foundation seed, registered seed, and certified seed) before it is made available to growers. Seeds vary in viability and longevity. Some seeds fail to germinate even under optimal conditions because of either physiological or structural barriers that induce dormancy. Under such conditions, seeds may be scarified or stratified to break the dormancy. Other chemical treatments may be applied to enhance seed germination. Seeds require moisture, air, temperature, light, and a disease-free environment for germination. Species differ in their requirements for germination. Some plants can be propagated by direct seeding, whereas others are transplanted after nursing. Direct seeding is generally easier to undertake, especially on a large scale. Transplanting ensures good establishment and hastens crop maturity. Seedlings can be raised indoors while waiting for good weather to plant in the field.

REFERENCES AND SUGGESTED READING

Copeland, L. O. 1976. Principles of seed science and technology. Minneapolis: Burgess.

Hartmann, H. T., and D. E. Kester. 1983. Plant propagation: Principles and practices, 4th ed. Englewood Cliffs, N.J.: Prentice-Hall.

Mayer, A. M., and A. Poliakoff-Mayber. 1975. The germination of seeds, 2d ed. Oxford: Pergamon Press.

Stefferud, A., ed. 1961. Seeds: USDA yearbook of agriculture. Washington, D.C.: U.S. Department of Agriculture.

PRACTICAL EXPERIENCE

1. Plant seeds of a legume and a grass in a soilless medium.
   a. Observe the pattern of germination between the two plant types.

b.   Wash away the potting medium and compare the anatomical differences between the legume and the grass.
2.   Test the germination percentage of a seed
   a.   Purchased from various sources (e.g., grocery, nursery, and supermarket).
   b.   Produced in different seasons (1, 2, 3 years).
   c.   Stored under different conditions (e.g., cold storage, freezer, and room temperature).

OUTCOMES ASSESSMENT

PART A

Please answer true (T) or false (F) for the following statements.

1.   T F   The seed covering of dicots is called a pericarp.
2.   T F   Longevity is a measure of how long a seed remains viable.
3.   T F   Damping-off is a disease caused by a fungus.
4.   T F   Pea has hypogeous germination.
5.   T F   Indirect seeding does not require transplanting.
6.   T F   A caryopsis is the seed of a grass.

PART B

Please answer the following questions.

1.   List the four classes of seed.
   _____   _____

   _____   _____.
2.   The percentage of seed from a lot that will germinate is called _____.
3.   The percent pure live seed is _____.
4.   _____ is the mechanical scratching of seed to improve germination.
5.   List the environmental factors for seed germination.
   _____

   _____
6.   _____ germination is exhibited by grasses.
7.   When roots of plants grown in a container coil over themselves in a ball, they are said to be _____.
8.   What are the advantages of indirect seeding?
   _____

   _____

PART C
---

1.   What are the major purposes of propagation?
2.   Discuss the importance of seed dormancy to plants and growers.
3.   Describe the method and purpose of priming seed.
4.   Describe the practice and importance of hardening off.

CHAPTER 9

Asexual Propagation

PURPOSE

This chapter discusses the methods of propagating horticultural plants by using materials other than seed.

EXPECTED OUTCOMES

After studying this chapter, the student should be able to

1. Distinguish between sexual and asexual methods of propagation.
2. Discuss the advantages and disadvantages of asexual propagation.
3. Describe propagation by cutting.
4. Describe propagation by budding.
5. Describe propagation by grafting.
6. Describe propagation by layering.
7. Describe propagation by underground plant parts.
8. Describe and discuss micropropagation.

KEY TERMS

Air layering
Bark graft
Bridge grafting
Chip budding
Cleft graft
Clone
Corms
Crowns
Dormant stock
Graft
Grafting wax
Graft junction
Herbaceous

Inarching
Laminate bulb
Mauling rootstocks
Mound layering
Nontunicate bulb
Patch budding
Rootstock
Runners
Scion
Semihardwood
Separation
Simple layering
Slips

Softwood
Stem tubers
Stolons
Stool layering
Suckers
T-budding
Tip layering
Topgrafting
Topworking
Tuberous root
Tunicate bulb
Whip or tongue graft

OVERVIEW

When plants are propagated by any material other than seed, they are said to be *asexually,* or *vegetatively, propagated.* Since the sexual mode of reproduction is the principal means by which biological variation is generated (through recombination [5.1]), propagation that circumvents the sexual mechanism (asexual, or vegetative, propagation) results in no genetic changes. The progeny or offspring are genetically identical (*clones*). Therefore, instead of meiosis, mitosis (5.2.3) is the mechanism governing asexual propagation. Species that are heterozygous and hence fail to reproduce true to type may be propagated vegetatively or clonally to preserve the genotype. Trees or woody perennials benefit from this method of propagation, since it drastically shortens the time of breeding of such plants.

The vegetative tissues commonly used in asexual propagation are the stem, leaf, and root. Modern technology allows full-fledged plants to be regenerated from tissues and even single cells (*micropropagation* [9.16]). A common phenomenon that occurs in some species enables vegetative propagation to be conducted using seeds. These seeds are different in that they are produced without a sexual union between male and female gametes. Instead, certain elements in the sexual apparatus (such as the nucellar tissue that has not undergone meiotic division) spontaneously develop into seed. Plants that are capable of this kind of reproduction are said to be apomictic, and the phenomenon that produces such seeds in these plants is called *apomixis.* In citrus, a sexual embryo and nucellar embryo can be produced side by side. It should be mentioned that since an individual's appearance (phenotype) is determined by its genes (genotype) but influenced by the environment, differences in the environment can cause clones to perform differently.

Asexual propagation has certain advantages and disadvantages.

> **Clone**
> *An individual or a group of individuals that develop asexually from cells or tissues of a single parent individual.*

9.1: ADVANTAGES

1. *Plants are uniform.* The offspring are genetically identical (clones) and are the same in appearance as the parental source (true offspring). Uniformity (homogeneity) of produce quality is critical to the success of certain production enterprises where the market demands uniform products.

2. *Quick establishment of plants.* Only strong and healthy seedlings are transplanted, and thus a good stand is always obtainable. Asexual propagation is generally known to hasten crop maturity. In geranium, for example, asexual propagation causes plants to flower about three weeks sooner than if propagated from seed.

3. *Only means of propagation in certain species.* Certain plant species such as banana and grape produce seedless or nonviable fruits. Other species including foliage plants do not produce seed at all. On such occasions, the grower has no choice but to use asexual means of propagation to produce these crops.

4. *Seedborne diseases avoided.* Seedborne diseases are not a problem when plants are asexually propagated.

5. *Less expensive.* For certain species, asexual propagation is less expensive than propagation by seed.

6. *Heterozygous material may be propagated without genetic alteration.* Certain plants can be propagated either sexually or asexually. If through plant breeding activities a hybrid (5.7) is developed with high heterosis (hybrid vigor or hybrid superiority over parents [5.7]), this heterozygous advantage is "locked up" and can be expressed indefinitely (until further genetic alteration occurs). In seed propagation, the hybrid must be reconstituted each time for planting because each sexual cycle changes the genetic constitution of the offspring (5.2.3).

9.2: DISADVANTAGES

1. *Systemic viral infection can spread to all plants, making planting material taken from infected plants a carrier of the infection.* This viral infection may be eliminated by
    a. Starting with a new, disease-free seedling.
    b. Using heat treatment (where an infected clone is held at 37 to 38°C [98.6 to 100.4°F] for about three to four weeks).
    c. Using tissue culture (9.16) of terminal growing points that, even in infected plants, are usually disease free.
2. *Planting materials are bulky.* Unlike seed, which can be stored in a small seed envelope, products of cuttings (9.5) and seedlings in containers are bulky to handle and transport.

3. *Storage of asexual material is cumbersome and usually short term.* Seeds are dormant propagating materials, whereas asexual materials such as seedlings are actively growing and thus need special care to sustain growth processes until transplanted. If they are not transplanted on time, they need to be held under certain conditions to slow growth. Overgrown seedlings have a lower success rate in the field.

4. *All plants are genetically identical and thus subject to the same hazard to the same degree.* Because of extreme uniformity among plants, a disease infection can wipe out entire populations of plants.

5. *Mechanized propagation in some cases is not practical.* Automation of asexual propagation methods is often problematic.

9.3: ADVENTITIOUS ROOTING IN ASEXUAL PROPAGATION

True roots (*seminal roots*) are produced only by seed. When vegetative parts of the plant are used for propagation, it is critical that roots be induced on these materials. The roots are non-seminal and called *adventitious roots* (2.5.2). They originate from root initials that are formed adjacent to vascular tissue. Some plants are easier to root than others. In fact, plants such as rubber tree and oak have not been successfully rooted adventitiously. Rooting ability is a function of plant characteristics and the environment.

9.4: APOMIXIS

Apomixis is a form of asexual reproduction through which seeds are produced without fertilization (i.e., no fusion of gametes). The common dandelion, a known weed in lawns, is capable of reproducing by sexual and ordinary vegetative means, as well as apomictically. Wild blackberry can also reproduce apomictically. This phenomenon should be distinguished from parthenocarpy (5.15), which is the development of fruits from unfertilized eggs. These fruits (parthenocarpic fruits) are seedless (e.g., navel orange, banana, certain grapes, and fig). However, all seedless fruits are not parthenocarpic; for example, seedlessness may be artificially induced in tomato by spraying flowers with a dilute hormone.

Apomixis, however, involves normal structures such as ovaries. For example, an embryo of a seed may develop from a $2n$ nutritive cell or other diploid cell in an ovule. Because usually the embryo of a seed develops from a zygote, plants raised from apomictic seeds are genetically identical to the parent plant (clones).

> **Apomixis**
> A form of sexual (vegetative) reproduction through which seeds are produced without fertilization, the seeds being entirely of maternal origin.

9.5: TYPES OF CUTTINGS

> **Cuttings**
> A detached vegetative material from a plant used to produce a new plant.

Cuttings are pieces of vegetative material obtained from any of the three primary plant organs—stem, leaf, or root (2.3). These tissues are nursed under appropriate conditions to develop into full-fledged plants. Cuttings are by far the most commonly used asexual propagation method in the horticultural industry. A variety of cuttings may be grouped on the basis of several factors, including parts of the plant used, age of the plant part, and succulence of the tissues. It is possible to obtain more than one kind of cutting from a single plant. However, certain materials are more successful or easier to use for establishing certain species than others.

9.5.1 STEM-TIP (TERMINAL) CUTTINGS

In *stem-tip cutting* the tip of the stem is cut (or in some cases snapped with the fingers) and used to produce a seedling. The piece cut is about 3 inches long and has leaves (figure 9-1). Terminal cuttings may be obtained from herbaceous, softwood, semisoftwood, and hardwood plants. The cut may be made at either the node or internode of the stem. Stem-tip cuttings produce seedlings much more rapidly than stem-section cutting.

9.5.2 STEM-SECTION CUTTINGS

Stem-section cutting (or simply *stem cutting*) is the asexual method of propagation whereby pieces of stem material containing at least one bud are used for planting. The difference lies in the type of wood from which the cuttings are made.

Softwood Cuttings

Softwood cuttings are made from soft tissues (nonlignified) of shrubs or deciduous trees. They are taken from the new growth of the current season (spring growth). The parent source should be actively growing and have leaves (figure 9-2). Examples of plants that can be propagated by this method are rose, forsythia, plum, dogwood, and lilac. Because of the presence of leaves, these cuttings must be maintained under high humidity during root induction to avoid desiccation.

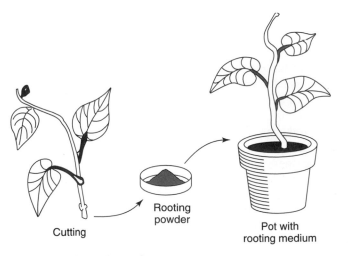

Cutting

Rooting powder

Pot with rooting medium

FIGURE 9-1 Stem tip cutting.

FIGURE 9-2 Softwood cuttings are made from young, leafy shoots from spring growth.

Semihardwood Cuttings

As the name implies, semihardwood cuttings are derived from tissues that are more woody than softwood cuttings. Semihardwood cuttings are made from the spring growth of trees and shrubs, but they are more mature than softwood. This material may be obtained around midsummer. The cuttings should have some leaves on the top parts and be handled in the same manner as softwood cuttings (figure 9-3). Plants such as azalea, rhododendron, holly, and other broad-leaved evergreen ornamentals are propagated by this method.

Deciduous Hardwood Cuttings

Deciduous hardwood cuttings are made from plant parts that are more hardened and woody. They are taken before the plants produce a flush of spring growth, in early spring or late winter. The materials are thus obtained from the previous summer's growth. These cuttings do not have leaves and are about 6 to 12 inches (15 to 30 centimeters) long (figure 9-4). They are best rooted in a well-drained, sandy medium. The cuttings are inserted vertically into the rooting medium. Some fruits (e.g., grape, fig, and currant), deciduous shrubs (e.g., rose, forsythia, and honeysuckle), and deciduous trees (e.g., willow) yield hardwood that is used in their propagation.

> **Hardwood Cutting**
> *A cutting derived from mature or woody stem material.*

Conifer Cuttings

Conifers, narrow-leaved evergreens, are propagated by hardwood cuttings obtained from plants in early winter. These cuttings should have needles on the upper part (figure 9-5). Conifer cuttings root slowly, sometimes requiring months to produce adequate rooting. The preferable rooting environment is cool and humid. Sometimes a cold frame (11.2.2) with high light intensity can be used to accelerate the rooting process. Examples of plants in this category are juniper, spruce, hemlock, yew, and pine.

Herbaceous Cuttings

Herbaceous stem cuttings are also considered softwood cuttings. Numerous potted succulent greenhouse plants are propagated by herbaceous cuttings (figure 9-6). Geranium, carnation,

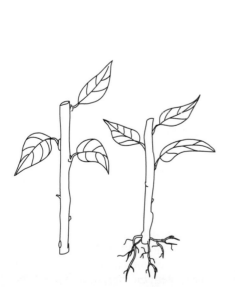

FIGURE 9-3 Semihardwood cuttings. Broadleaf evergreens are commonly propagated in this way.

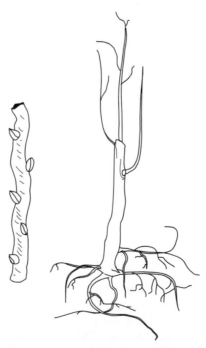

FIGURE 9-4 Deciduous hardwood cuttings are obtained from leafless plants in late winter or early spring.

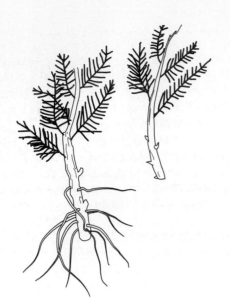

FIGURE 9-5 Narrow-leaf evergreen or conifer cuttings.

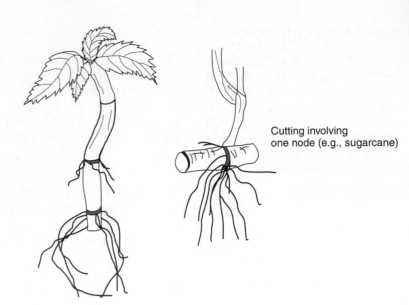

Cutting involving one node (e.g., sugarcane)

FIGURE 9-6 Herbaceous stem cuttings may be rooted vertically or horizontally.

chrysanthemum, coleus, ivy, spider plant, and lanta are propagated by herbaceous stem cuttings. Many of these species root easily from cuttings. In species that are intolerant of high moisture levels, cuttings can be rooted in a flat covered with a wire frame with a sheet of white polyethylene. The cuttings are watered lightly once or twice daily. The covering maintains adequate humidity.

9.5.3 LEAF CUTTINGS

Full or Partial Leaf Cuttings

> **Leaf Cutting**
> *A whole leaf or part of one that is detached and used to raise a new plant.*

Leaf cuttings are herbaceous cuttings that involve the use of either a piece of or an entire leaf. Popular species propagated in this way include begonia, gloxinia, peperomia, echeveria, crassula, sansivieria, and African violet. In plants such as African violet, the leaf is picked with a leaf stalk attached. The leaf stalk may be dipped in a rooting hormone mixture before inserting it into a rooting medium in a pot (figure 9-7). The leaf may also be rooted in a container of water before transplanting it into soil (figure 9-8). Gesneriads are commonly propagated by leaf cuttings.

Leaf-Vein Cuttings

The leaf-vein cuttings method may be used to raise plantlets by cutting through the veins at various points. The leaf is then placed face down so that the cut parts touch the propagation medium (figure 9-9) Plantlets grow from these cut points.

Leaf-Bud Cuttings

Some plants, including rhododendron, may be propagated from leaf-bud cuttings obtained from soft- or hardwood (figure 9-10). A leaf-bud cutting consists of a short piece of stem with an attached leaf and a bud in the leaf axil. High humidity and bottom heat are usually required for rooting to occur. This method of leaf propagation is useful where the source of cuttings is limited. Instead of using a long piece of stem as a single stem cutting, each bud on the stem can be removed and nursed to become a single plant.

Leaf petiole is dipped in rooting powder

Leaf is rooted in a rooting medium

Rooted leaf cutting ready for replanting

FIGURE 9-7 Propagation by leaf cutting. The petiole may be dipped in a rooting hormone to aid rooting.

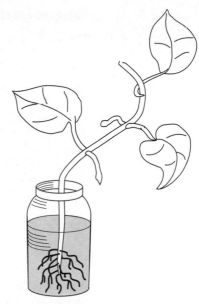

FIGURE 9-8 Rooting in water.

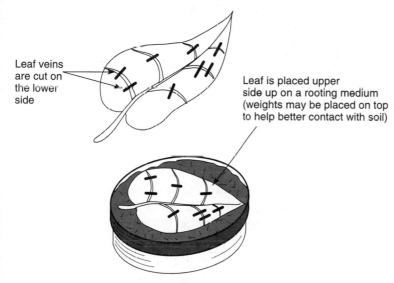

Leaf veins are cut on the lower side

Leaf is placed upper side up on a rooting medium (weights may be placed on top to help better contact with soil)

FIGURE 9-9 Propagation by leaf-vein cutting.

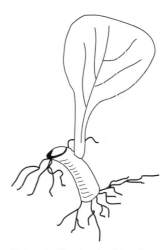

FIGURE 9-10 Propagation by leaf-bud cutting.

9.6: FACTORS AFFECTING ROOTING OF CUTTINGS

9.6.1 PLANT FACTORS

Nutritional Status of the Plant

Cuttings obtained from healthy, well-nourished plants that have been growing in the sun and have accumulated carbohydrates are more successful in rooting than those obtained from malnourished plants. Plants low in nitrogen make better cuttings. The leaves on cuttings act as sources of carbohydrates, and the buds are sources of auxins.

Hormone Level

The auxin level in the plant affects the rooting capacity of the cutting. Nursery growers routinely dip cuttings in commercially prepared rooting hormones (4.6) to stimulate rooting. This practice is especially important when handling hard-to-root species. Rooting hormones may be formulated as a powder (dust) or liquid (solution). To avoid contamination in case any of the cuttings are diseased, they should not be dipped into the same container of powder or solution. For powder application, a puffer duster (squeezable bottle) may be used; a spray bottle may be used for liquid applications to spray the ends of the cuttings. Cuttings should never be dipped into the original container of the hormone. The amount needed should be poured out or removed and any leftover amount discarded.

Rooting hormones may be mixed with fungicides to protect the cut surface from rotting under humid conditions. Fungicides are especially helpful in species that require a long period of time to root. A suggested ratio of hormone to fungicide (e.g., Captan) is 9:1 by weight.

Juvenility

Cuttings from mature woody and fruiting stock plants are less successful than those from newer shoot growths. The juvenile stage is thus more conducive to rooting. Cuttings are commonly made from one-year-old plant parts.

Position of Plant Part

The position of plant parts used as cuttings plays a role in their capacity to root. Lateral shoot cuttings root better than terminal shoot cuttings.

9.6.2 ENVIRONMENTAL FACTORS

Time of Year Cutting Is Made

Whereas cuttings may be taken any time of the year for herbaceous and perennial plants, hardwood cuttings root better when the materials are obtained in late winter when plants are dormant. Similarly, softwood cuttings from deciduous woody plants are best taken in spring; semihardwood cuttings root best when obtained in midsummer.

Darkness

Darkness promotes etiolation (3.2.1) in plants. Although this condition indicates improper development, materials obtained from etiolated plants root better than plants exposed to full light. A practical application of this condition is the keeping of basal plant parts of shoots to be used for cuttings in darkness for a period of time. This technique may be applied in species that are very difficult to root.

Light

Even though etiolated plant materials root better than normal ones, rooting cuttings in full light under a mist makes them root more quickly. However, when light intensity is excessive, plants may be in jeopardy of moisture stress. During such periods, shading is required. Similarly, in winter, supplemental lighting is required.

Temperature

Bottom heat in cutting beds warms the soil to induce quicker rooting. It is important to have adequate rooting before the shoot starts to grow vigorously. The bottom heat is usually about $10°F$ ($6°C$) higher than the air temperature, which should be maintained between 65 and $75°F$ ($18°$ and $27°C$).

Moisture

Cuttings do not have roots and hence are unable to absorb moisture. However, the exposed plant parts are subject to evaporation. To reduce moisture loss, cuttings are generally maintained under conditions of high moisture by misting them. Initially, misting may be required almost continuously. As time goes by, the misting schedule is modified, being less frequent and less intense. In some greenhouses, a mist is replaced by a *fog* (very fine spray of water under intense pressure).

Nutrition

No fertilization is needed for unrooted cuttings since they can take up very little of it. Light fertilization may be applied through the mist. Once rooted, a complete fertilizer application of 20:20:20, for example, at a rate of 1 pound (0.45 kilogram) per 100 gallons (378.5 liters) of water may be applied.

Rooting Medium

Certain species, including coleus, African violet, and *Philodendron scandens,* can root in water alone. Most cuttings are rooted in solid media that must be sterilized, freely draining, and of good moisture-holding capacity. Sand is well drained but poor in moisture-holding capacity. If sand is used, the rooting should be conducted under shady conditions and should be hand watered. Vermiculite and perlite make good propagation media. The medium for propagation must be very loose to enable rooted cuttings to be readily removed from the rooting medium and planted with little loss of roots.

In place of the conventional rooting media, a variety of preformed, lightweight materials are widely used (figure 9-11). These media include rockwool media, compressed peat pellets, and other artificial materials marketed under various brand names such as Oasis root cube, Oasis wedge, and Horticubes. Sometimes cuttings are rooted directly in the finish pots. Species such as geranium and poinsettia are propagated in this way.

Sanitation

Cuttings have exposed surfaces and hence are prone to disease attack. The propagating medium must be sterilized before use. Steam sterilization is effective in controlling most soil-borne diseases.

FIGURE 9-11 Rooting stem-tip cutting in an Oasis wedge.

Cuttings are ready to be transplanted when a mass of roots has formed. The rooted cutting should be lifted gently with little pulling. It should be planted no deeper than it was in the propagating medium. These materials are transplanted like any other seedling (8.11.2).

SUMMARY

Cuttings may be obtained from the stem, leaf, or root for propagating certain plants. Once taken, the pieces may be directly planted in some cases, but in other cases must be rooted before transplanting into pots or the field. Cuttings may be obtained from softwood or hardwood. The application of a rooting hormone may be required in certain cases for rooting to occur or to hasten rooting.

OUTCOMES ASSESSMENT

PART A

Please answer true (T) or false (F) for the following statements.

1. T F Cutting is the most common method of vegetative propagation in the horticultural industry.
2. T F Softwood cutting is obtained from the new growth of the spring season.
3. T F Conifers are propagated by softwood cuttings.
4. T F It is not possible to root cuttings without the aid of rooting hormones.
5. T F Hardwood cuttings have more wood than semihardwood cuttings.
6. T F Hardwood and softwood cuttings can be obtained from the same plant.
7. T F Some cuttings can be rooted in water.

PART B

Please answer the following questions.

1. Give three examples each of plants that may be propagated by
   a. softwood cuttings _____ _____ _____
   b. semihardwood cuttings_____ _____ _____
   c. deciduous hardwood cuttings_____ _____ _____
   d. herbaceous cuttings_____ _____ _____
2. Give two examples each of plants that may be propagated by
   a. stem cutting_____ _____
   b. leaf cutting_____ _____

PART C

1. Describe how plants are rooted in water.
2. Explain why products of cuttings are genetically uniform.
3. Discuss the advantages and disadvantages of vegetative propagation.

9.8: NATURE OF GRAFTING

Grafting is an asexual propagation method in which parts of two different plants are joined so that they continue their growth as one plant. To accomplish this, one of the two plants serves as the bottom part, which is in contact with the soil and is called the *rootstock* (or simply *stock*), and the other as the top part, or *scion*. In this plant union, the plant material used as the scion is being propagated and is usually the only one allowed to grow. In effect, the scion becomes the new shoot of the plant union, and the stock serves as the root, conducting nutrients across the graft junction and into the shoot. There are two basic strategies for bringing about the union between the two plant parts in a graft:

1. *Detached scion grafting.* Detached scion grafting is the strategy in which the scion is detached; only the stock remains rooted (figure 9-12). This technique is the most commonly used for grafting.

2. *Approach grafting.* In this strategy, two plants are united at a predetermined and prepared site (figure 9-13). That is, the scion and the stock both remain an integral part of

> **Grafting**
> *A technique of uniting two plants so they grow as one.*

> **Rootstock**
> *The bottom part of a graft that is in contact with the soil and not allowed to produce side shoots.*

> **Scion**
> *The plant part that is the top part of a graft and grows to become the desired shoot.*

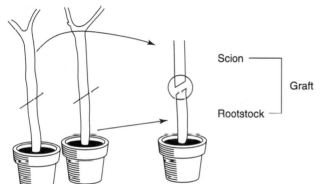

FIGURE 9-12 The basic elements in a graft union: a rootstock and scion.

Scion —
Graft
Rootstock —

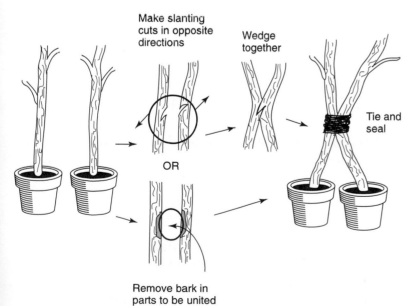

FIGURE 9-13 Strategies in making an approach graft.

Make slanting cuts in opposite directions

Wedge together

Tie and seal

OR

Remove bark in parts to be united

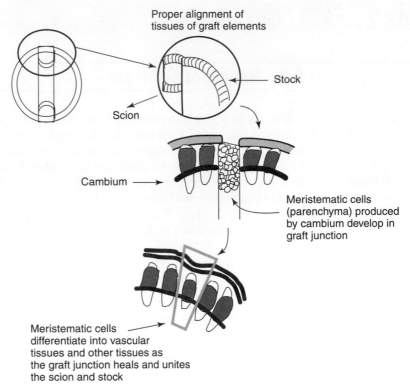

FIGURE 9-14   Steps in the healing of a graft union.

the respective parent plant. This procedure is used when detached scion grafting is not feasible. The produce harvested from the plant reflects only the characteristics of the scion parent and not both stock and scion. The scion serves only as ground support.

9.8.1 HEALING OF A GRAFT JUNCTION

A graft is successful if healing is complete and vascular transport restored between the two parts (scion and stock). It is critical that the tissues in the two parts be correctly aligned—xylem for xylem and phloem for phloem. Healing starts with the production of callus (undifferentiated cells) by mitosis (5.2.3) and occurs in the cambium region of the two parts. Next, some of the cells differentiate to form new cambium tissue to join the old in the two parts (figure 9-14). This stage is followed by further differentiation of cells to form vascular tissue, which completes the repair of the cuts and allows for uninterrupted vascular transport.

9.9: WHEN TO USE GRAFTING

Grafting may be used for relatively simpler and more routine propagation of plants and for more complicated and specialized repair of damaged plants.

9.9.1 GENERAL USES

The routine uses of the technique of grafting include the following:

1.   *To propagate plants whose cuttings are difficult to root.* Grafting provides an opportunity for clones to be produced without rooting the cuttings in a propagating medium. The cutting is "rooted" on another plant.

2.   *To provide disease resistance to a susceptible but desirable cultivar.* In this case, the desirable cultivar is grafted onto a cultivar that is able to resist soilborne diseases and pests.

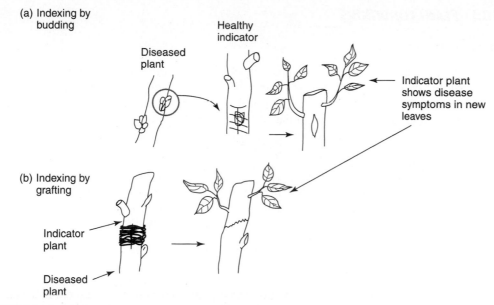

(a) Indexing by budding

Diseased plant

Healthy indicator

Indicator plant shows disease symptoms in new leaves

(b) Indexing by grafting

Indicator plant

Diseased plant

FIGURE 9-15 Virus indexing using (a) budding and (b) grafting.

3. *To rapidly increase the number of a desirable cultivar.* The plant does not have to go through a reproductive cycle to produce seed to propagation. Plant parts can be taken and rooted to produce new plants.

9.9.2 SPECIALIZED USES

Grafting may be used to repair damaged plants, invigorate them, or change their form.

1. *To change plant size and vigor.* Special rootstocks with the capacity to dwarf scion cultivars exist. Called *mauling rootstocks,* they have a dwarfing effect on the plant.

2. *To repair damaged plant (established tree) parts.* Grafting may be used to repair a damaged trunk, for example, from girdling caused by rodent attacks. A damaged root may also be repaired by grafting.

3. *To change plant form.* One or several cultivars may be grafted onto several limbs of a single plant. Such grafting may be done to change the plant form and improve aesthetics.

4. *Virus indexing.* Plant pathologists may use grafting (or budding, discussed in module 3) in a special technique to find out whether a plant is free of virus. This technique is called virus indexing. Any of the grafting or budding methods may be used in this evaluation. In a simple budding (9.12) procedure, a bud is obtained from a diseased plant and budded onto susceptible or sensitive species called *indicators.* The indicator plant is decapitated above the budding site so that the remaining stem of the indicator plant has about two to three buds (figure 9-15). As these sensitive buds grow, they exhibit symptoms characteristic of whatever virus (or mycoplasma) has infected the plant from which the bud was obtained.

> **Virus Indexing**
> *A procedure used to determine whether a given plant is infected by a virus.*

9.10: GENERAL CONDITIONS FOR SUCCESS

Grafting success is influenced by plant conditions, prevailing environment, and experience of the operator. Grafting is a skill, and thus some operators are able to perform it more successfully than others.

9.10.1 PLANT CONDITIONS

Compatibility

The two plants to be united must be compatible (*graft compatibility*). Even though this union is physical, the two plants should be as closely related genetically as possible for success. It is easiest to stay within species and graft apples onto apples. However, in certain cases, interspecies grafting is successful, as in the case of some almonds and plums, which are successful as scions on peach rootstocks.

Diameter of Parts

The stock diameter must be equal to or larger than the scion diameter. The scion is usually no larger than the size of a regular pencil, but some methods of grafting use larger stocks so that several scions can be grafted onto one stock. Scions are usually derived from healthy one-year-old plants.

Physiological State

Grafting is usually done using dormant plants. These plants have no leaves (except in the case of evergreens). In some cases, the rootstock may be actively growing, but the scion should not be growing.

Alignment of Tissues

Since grafting is a physical union that depends on healing of the cut surfaces (wounds) through mitotic division, the cambium tissues of both parts must be properly aligned. They must make contact over as wide an area as possible. If the tissues are not aligned properly, the graft will fail. The graft junction may be tied to keep the alignment in place throughout the healing period.

9.10.2 ENVIRONMENT

The worst environmental enemy of a new graft is desiccation. Therefore, a newly made graft should be waterproofed. After tying, *grafting wax* may be applied over the entire surface. Some operators use plastic or rubber ties instead of wax. Either way, the purpose is to prevent desiccation from occurring at the graft junction.

9.10.3 THE OPERATOR

In addition to all of the mentioned factors, the operator should always use a sharp knife and make sharp, clean cuts to ensure good contact of tissue. A more experienced operator is likely to have greater success than a novice at grafting.

9.11: METHODS OF GRAFTING

Grafting may be accomplished by one of several methods, depending on the species, the age and size of the plant, the problem to be corrected, and the purpose. These methods also differ in difficulty and the skill required for success. Although some methods are for general purpose use, others are used to solve specific and specialized problems. Notwithstanding the method, the principles are the same. There are two basic methods—one in which the scion is removed from its source and transferred onto another plant and a second in which no detaching is done before the formation of the graft union.

9.11.1 GENERAL PURPOSE METHODS

Detached-Scion Grafting

Common methods of detached-scion grafting are described in the following sections.

Whip or Tongue Grafting   The steps involved in the whip or tongue method of grafting are illustrated in figure 9-16. This method is suited for plant materials that are about 1/4 to 1/2 inch (0.64 to 1.3 centimeters) in diameter. The scion and stock should be as close in diameter as possible to provide maximum contact between the cambia of both parts. This method is best employed in the winter. The cuts in both scion and stock are made in the internode region. All cuts should be sharp and clean and the angles as close as possible in shape in both parts. The scion and stock should fit as tightly as possible. It is critical that the buds on the scion point upward. After properly aligning the two parts, the graft junction is tied or taped. A wax coating may be applied over the tie. Fruit trees are grafted by this root grafting method. After the graft is tied and waxed, the plants are stored under conditions that encourage callus formation (7.2 to 10°C or 45 to 50°F) for about three to four weeks. Top growth should not be allowed to occur before healing of the graft. Storing the grafted plants at 0 to 4.4°C (32 to 40°F) discourages top growth. New growth occurs in spring as the temperature begins to rise. *Grafting machines* are available and used widely in grafting grapes.

Cleft Grafting   As the name indicates, in cleft grafting the scions are inserted into a cleft in the stock (figure 9-17). Unlike the whip or tongue method in which both plant parts are usually of the same diameter, the stock is considerably wider in cleft grafting, thus allowing multiple grafts to be made on one stock. After cutting the stock at right angles, a meat cleaver or some similar knife is hammered into the center to produce a wedge cut. A wedge is inserted into the shallow split to keep it open for the scions to be inserted. Care should be taken to avoid tearing the bark away. This method is particularly suited for *topgrafting,* or *topworking*, the grafting strategy in which several scions are grafted onto one large scion to change fruiting cultivars in a fruiting tree. No tying is necessary in this method since the split in the stock has enough pressure from both sides to hold the scions together. A wax coating is essential, especially when the parts are not tied after grafting. Cleft grafting is done in late winter or early spring.

Bark Grafting   Bark grafting is applicable to species whose bark separates easily from the wood. Since the bark must be separated, this method is employed during the early to middle spring. This method is also used for topgrafting or topworking trees such as broadleaf evergreens, including citrus and olive. One unique characteristic is that instead of tying, the scion is nailed to the stock (figure 9-18).

Approach Grafting

Approach grafting may be tried when other general detached-scion methods have failed. At least one of the two plants involved should be in a container (although both can be in containers).

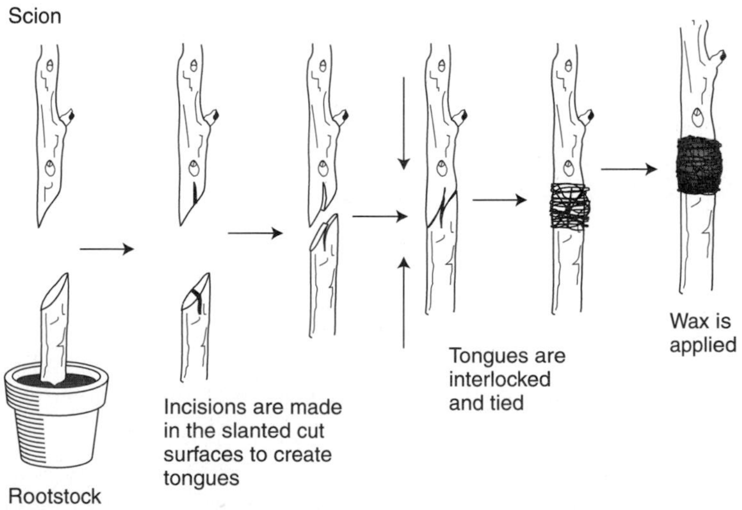

Whip
A young tree seedling (about one year old) with a slender single stem (less than ½ inch wide).

Topgrafting
A method of grafting in which the cultivar of a tree is changed by grafting the main scaffold branches or the stem using a new cultivar.

Scion

Rootstock

Incisions are made in the slanted cut surfaces to create tongues

Tongues are interlocked and tied

Wax is applied

FIGURE 9-16   Steps in the method of whip or tongue grafting.

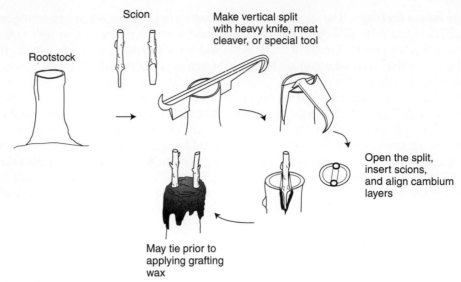

Scion

Rootstock

Make vertical split with heavy knife, meat cleaver, or special tool

Open the split, insert scions, and align cambium layers

May tie prior to applying grafting wax

FIGURE 9-17 Steps in the method of cleft grafting.

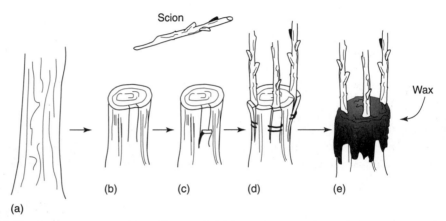

Scion

Wax

(a)

(b)

(c)

(d)

(e)

FIGURE 9-18 Steps in bark grafting. The stock (a) is cut horizontally and slots prepared by making two parallel vertical cuts (b). The bark is peeled back (c) so the scion can be inserted and nailed in place (d). The grafting process is completed by waxing.

Smooth cuts in opposite directions but identical in size, shape, and depth and occurring at the same height are made on the plants. Sometimes a tongue cut may be used in this method. The prepared parts are appropriately united, after which the graft junction is tied and sealed to keep the union in place for healing. After healing is complete, the top of one plant and the base of the other are cut so that only one stem remains.

9.11.2 SPECIALIZED METHODS

Side Grafting

In side grafting, an angled cut is made into the stock. A scion is prepared and fitted into the cut by bending the scion gently backward to open up the cut. The scion is secured in place by tying (figure 9-19). This technique may be used to improve the shape and symmetry of a plant. The scion and the shoot from the stock parent are both allowed to grow and develop.

Bridge Grafting

The problem that calls for bridge grafting occurs in temperate or colder regions. When rodents are prevented from reaching food on the ground by the presence of snow around the

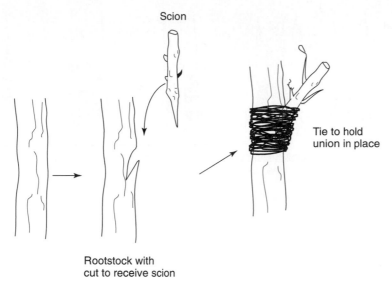

Scion

Rootstock with
cut to receive scion

Tie to hold
union in place

FIGURE 9-19 Steps in side grafting.

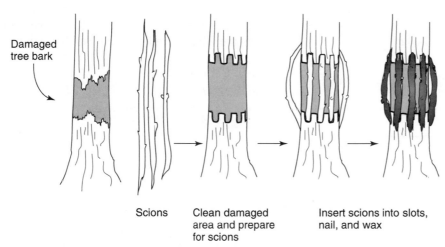

Damaged
tree bark

Scions

Clean damaged
area and prepare
for scions

Insert scions into slots,
nail, and wax

FIGURE 9-20 Steps in bridge grafting. This method is similar to bark grafting except that it requires both ends of the scion to be grafted.

base of the tree, they may turn to the tree and gnaw away a portion of the bark in a ring fashion, called *girdling* a tree. Since this activity interrupts the transport of food via the phloem, the roots may be starved to death, eventually affecting the whole tree. The barkless ring may be bridged to reconnect the interrupted bark by using several pieces of scion material to form bridges across the ring (figure 9-20). It is critical that the twigs used as scion be properly oriented (tips facing up the tree and bottom parts facing down). The scions are then nailed in place.

> **Girdling**
> *The interruption of phloem transport by removing a ring of bark from the stem.*

Inarching

Inarching is similar to bridge grafting except that the girdling occurs at the base of the tree and the bridging is done by using growing seedlings. The seedlings are grown in a circle around the trunk. The tips are cut off and the stumps prepared to be attached to the trunk (figure 9-21). Nailing is required to hold the seedlings in place. This radical grafting may be used to save a very valuable tree (specimen tree [13.8]).

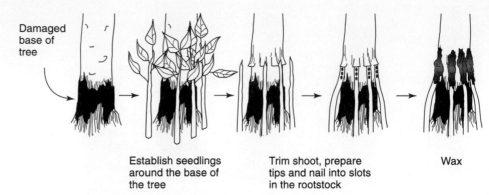

Damaged base of tree

Establish seedlings around the base of the tree

Trim shoot, prepare tips and nail into slots in the rootstock

Wax

FIGURE 9-21 Steps in inarching. This method is similar to bark grafting except that undetached scions are used and the scion are inserted from below the graft junction.

SUMMARY

A graft is a union between two different plants such that they grow as one plant. The bottom part of the graft is called the rootstock and the top a scion. The union heals by mitotic processes. The scion is the part of the union usually encouraged to grow and develop to maturity. To be successful, the two plant parts to be grafted should be compatible, of the same diameter (or the rootstock larger), and in the proper physiological state where they are dormant or the rootstock is actively growing. Success of a graft also depends on the skill of the operator. The common methods of grafting are whip or tongue, cleft, and bark grafting.

OUTCOMES ASSESSMENT

PART A

Please answer true (T) or false (F) for the following statements.

1. T F   The top part of a graft is called the stock.
2. T F   A graft junction is healed through the process of meiosis.
3. T F   In a graft, the stock should be at least as large as the top.
4. T F   It is possible to graft more than one plant on a single rootstock.
5. T F   Scions are usually derived from two- to three-year-old plant parts.

PART B

Please answer the following questions.

1. The special rootstock that has the capacity to dwarf a scion is called a _____.
2. The method of grafting used to repair damaged tree trunks is called _____.
3. List two general uses of grafting.

    _____ _____

4. List two specialized uses of grafting.

    _____ _____

5. List four plants that can be grafted.

    _____ _____

    _____ _____

6. What is the most common method of grafting fruit trees?

    _____

PART C

1. What is the purpose of grafting wax in a grafting operation?
2. Describe a grafting union.
3. Compare and contrast whip or tongue and cleft methods of grafting.
4. Why is it important for the two parts in a graft to be properly aligned?
5. Describe the roles of each part of a graft.

MODULE 3 BUDDING

9.12: TYPES OF BUDDING

Budding may be described as psuedografting and is sometimes called *bud grafting*. A major difference between budding and grafting is that budding uses a single bud as the scion, whereas grafting uses a piece of plant material consisting of several buds. Budding is less involved and much easier to accomplish than grafting. Since a bud is inserted into an opening in the bark of a stem or branch, it is important that budding be done when the rootstock is actively growing (i.e., in spring, late summer, or fall).

> **Budding**
> *A form of grafting in which the scion consists of a single vegetative bud.*

9.12.1 T-BUDDING

T-budding gets its name from the shape of the cut made in the bark of the stock in which the bud (scion) is inserted. The stock should be actively growing and young (one to two years old) so that the bark can be easily separated from the stem. T-budding is widely used to propagate fruit trees (e.g., apple, pear, peach, and citrus) and roses. Both the bud and the stock are encouraged to grow, but after the bud has attained a good size, the stock is cut off above the site of budding. The scion and rootstock must be compatible.

Buds are obtained from *bud sticks,* which are small pieces of shoot. The shoot material should be the current season's shoot and also be vigorously growing. When collecting bud sticks, one should be careful to collect only vegetative (not fruiting) shoots, which have buds that are slender in shape and are more pointed. Vegetative shoots to be used for bud sticks are collected on the same day budding is to be done. They are wrapped in waterproof paper or placed in plastic bags to prevent desiccation. A good stock is about one to two years old and is considered ready when its bark peels off easily without tearing. Buds on the middle section of the bud stick are usually best for budding since they are mature.

After making the T-shaped cut, the bark is opened up with the end of a knife (figure 9-22). A bud is extracted from the stick in the shape of a shield and includes a small piece of wood. The bud shield is inserted into the opening and pushed down. When in place, it is tied with rubber bands. A successful operation shows a take in about three weeks.

9.12.2 PATCH BUDDING

Plants that are difficult to bud by the T-method (e.g., pecan and walnut) because of a thick bark may be patch budded. *Patch budding* involves removing a piece of bark (figure 9-23). A patch of equal size containing a bud is obtained and fitted into the stock plant. The patch is tied in place, ensuring that the bud is exposed. It is also important that the plants be actively growing to ensure that the bud can be more readily extracted from the parent. The best time for patch budding is mid to late summer. As in other methods, it is necessary to remove all sprouts occurring below the budded area.

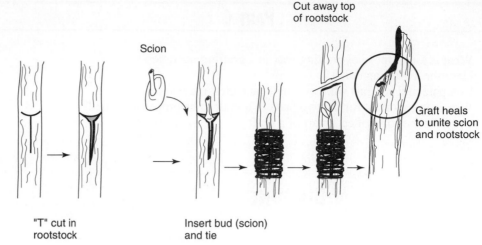

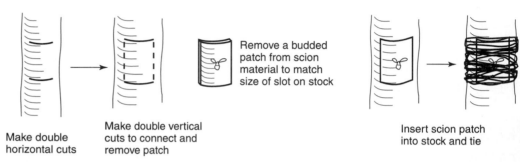

FIGURE 9-22 The method of T-budding.

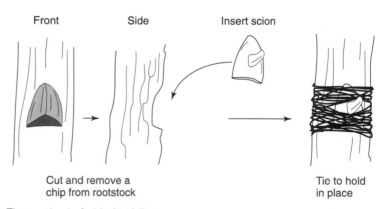

FIGURE 9-23 Steps in the patch budding method of propagation.

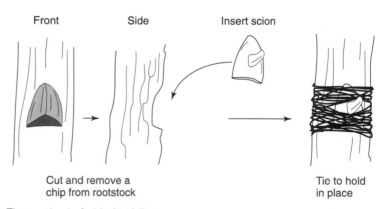

Front   Side   Insert scion

Cut and remove a
chip from rootstock

Tie to hold
in place

FIGURE 9-24 The method of chip budding.

9.12.3 CHIP BUDDING

Chip budding differs from T-budding in several ways. Chip budding involves a cut that includes a larger chip of wood—in fact, more wood than bark. The method is applicable to a dormant stock since the bark does not need to be separated from the wood as in T-budding. Therefore, chip budding can be done in summer or fall. The removed chip is replaced by a bud that is cut to fit the hole in the stock (figure 9-24). It is important, as always, to align the cambia in both bud (scion) and stock. After inserting the bud, it is taped with rubber tape or other similar waterproof material. When the bud takes and grows, the stock is cut above the bud.

SUMMARY

Budding is similar to grafting in that part of two different plants are united. The difference is that in budding the scion is a single bud rather than a piece of twig with several buds as in grafting. Budding is a relatively easier procedure than grafting. It is done while the plant that is designated as stock is still actively growing, since the bark needs to be opened up for the bud to be inserted. Vegetative (not fruiting) bud sticks are used and are collected on the same day the operation is to be performed. The method of T-budding is commonly used to propagate fruit trees. Other methods include patch and chip budding. Chip budding can be done in summer or fall when the plant is not actively growing since it does not require the bark to be opened up but only a chip to be cut from the wood.

OUTCOMES ASSESSMENT

PART A

Please answer true (T) or false (F) for the following statements.

1. T F Budding is done when the stock is actively growing.
2. T F Fruit trees are commonly T-budded.
3. T F Fruiting bud sticks are collected for use in budding.
4. T F Bud sticks are best collected on the same day as budding.
5. T F Pecans and walnuts are easy to T-bud.
6. T F Chip budding involves the removal of a piece of wood.
7. T F Buds take (or are successful) within about three weeks after T-budding.
8. T F Vegetative buds tend to be slender and more pointed in shape.

PART B

Please answer the following questions.

1. Name three specific plants that are T-budded.

   _____ _____ _____

2. The best times for T-budding are in_____.
3. _____ involves a cut that includes a relatively larger piece of wood.
4. The best times to patch bud are _____.

PART C

1. Compare and contrast budding and grafting.
2. Describe how T-budding is done.
3. Distinguish between T-budding and chip budding.
4. Describe a specific situation in which budding is not practical.

9.13: TYPES OF LAYERING

Layering
A method of vegetative propagation of plants usually with flexible limbs (shrubs, vines) in which roots are generated on the limb before being severed for planting as an independent plant.

Layering may be described as modified cutting. The plant part to be cut is rooted before it is completely cut away from the parent plant. Roots or stems may be propagated by layering. The part of the plant that is eventually cut off to be grown independently is called the *layer*. Layering is accomplished in a variety of ways that may occur naturally or with the help of humans.

9.13.1 NATURAL LAYERING

Layering occurs naturally in certain species because their anatomy permits portions of the plant to come into contact with soil at some point.

Tip Layering

Species such as boysenberry and black raspberry (cane fruits) have been known to propagate naturally by *tip layering*. When the tips of the current season's long canes bend down and touch the soil, they turn around to grow upward once again. At the point of contact with the soil, roots start to develop, provided the portion is adequately covered with soil (figure 9-25). The layer may then be severed from the parent and dug up for replanting as an independent plant.

Runners

Natural layering involving roots occurs in horticultural species such as the strawberry. As the plant grows, it produces *runners,* or *stolons* (above-ground creeping stems) (2.3.3), in various directions. When the nodes on these structures come into contact with the soil, roots develop, and eventually new plants arise at these nodes; these new plants may be harvested by cutting and digging out for replanting (figure 9-26).

Suckers

Suckers are adventitious shoots produced by species including spirea, red raspberry, and blackberry. These shoots arise from the horizontal roots produced by these plants (figure 9-27). The result of this habit is that a single original plant may produce several new plants clustered together. Each of these adventitious shoots may be harvested for replanting.

Crowns

The *crown* is the root-shoot junction. In species such as shasta daisy and African violet, the crown grows larger with time and produces lateral shoots from the underground parts of older

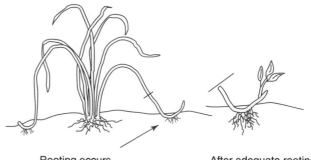

Rooting occurs where cane touches ground

After adequate rooting new shoot is detached for replanting

FIGURE 9-25 Propagation by tip layering.

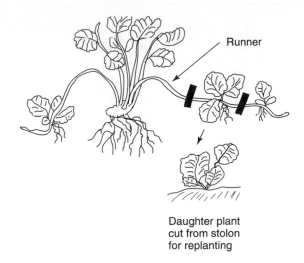

FIGURE 9-26 Propagation by using runners.

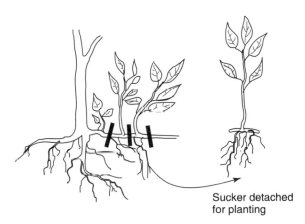

FIGURE 9-27 Using suckers for propagation.

stems and roots. These new shoots may be harvested by *crown division*, which entails cutting the crowns into pieces such that each piece has roots and a shoot (figure 9-28). These new shoots may then be replanted.

9.13.2 HUMAN-AIDED LAYERING

Simple Layering

The *simple layering* method of propagation is easier to employ in species that produce long, flexible shoots that arise from the plant at ground level because it requires the part to be layered to bend to touch the ground (figure 9-29). The selected stem (one-year-old stem preferred) is girdled or nicked about halfway through the portion that will be in contact with the soil. Nicking or girdling the stem causes auxins and carbohydrates to accumulate in the area of the stem for quick rooting. A shallow hole (4 to 6 inches or 10.2 to 15.2 centimeters) is dug at an appropriate spot. The stem is gently curved such that the nicked part of the stem is positioned erectly, aided by another peg. The hole is then filled with soil, mulched, and watered regularly. In some cases, an additional weight (e.g., a large stone) may be placed on top of the mulch. When adequately rooted, the layer is cut from the parent and dug up for replanting. Foliage plants including *Philodendron* and *Dieffenbachia* and other species such as climbing roses are propagated by simple layering.

Rooted shoot from
crown division

FIGURE 9-28 Using materials from crown division for propagation.

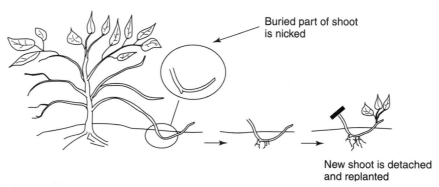

Buried part of shoot
is nicked

New shoot is detached
and replanted

FIGURE 9-29 Simple layering.

Serpentine Layering

Serpentine layering is sometimes called *compound layering* because several layers can be obtained from one shoot that is anchored to the ground. The flexible shoot is anchored to the soil at various sites rather than buried along the entire length (figure 9-30).

Trench Layering

In *trench layering,* the midsection of the flexible stem is buried in the soil after nicking in several places (figure 9-31). This type of layering causes several seedlings to develop. Species including rose and rhododendron can be propagated by trench layering.

Mound Layering

Mound layering is also called *stool layering.* It is accomplished by first cutting back the mother plant close to the ground, most often in late winter. This pruning causes new shoots to grow in spring. At the onset of new growth, soil is heaped around the base of these shoots to form a stool bed (figure 9-32). The size of the mound is increased as the shoots grow bigger. It is critical to keep the mound of soil moist continually. Roots will develop from the base of these shoots. The individual shoots are harvested by removing the soil and cutting them off from the mother stump for replanting. Species such as rose, apple, and currant are commonly propagated by this method.

Air Layering

In the methods of layering previously described, the plant parts are brought into contact with the soil to initiate rooting and a new shoot. In *air layering,* the soil is brought to the part of

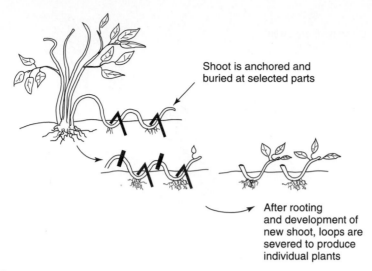

Shoot is anchored and
buried at selected parts

After rooting
and development of
new shoot, loops are
severed to produce
individual plants

FIGURE 9-30 Serpentine layering.

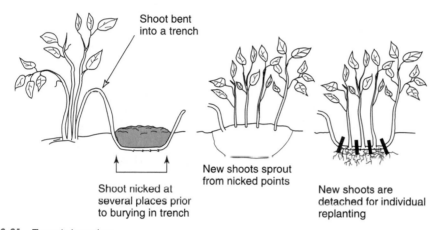

Shoot bent
into a trench

New shoots sprout
from nicked points

New shoots are
detached for individual
replanting

Shoot nicked at
several places prior
to burying in trench

FIGURE 9-31 Trench layering.

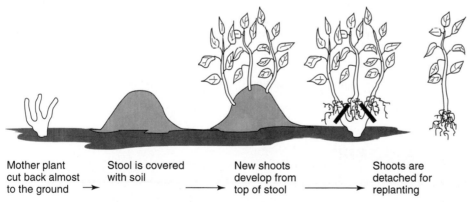

Mother plant
cut back almost
to the ground

Stool is covered
with soil

New shoots
develop from
top of stool

Shoots are
detached for
replanting

FIGURE 9-32 Mound layering.

the plant where a layer is to be produced. This method, which enables plants with stiff, hard-to-bend stems to be layered, is more involved than others. However, the principles are the same. The selected stem is girdled at a desired section (about 6 to 12 inches or 15.2 to 30.5 centimeters from the tip) by removing about an inch-wide section of the bark, including the cambium layer. Girdling stimulates root formation. This open area is treated with a rooting hormone (an auxin, such as IBA or Hormodin #3). Finally, a ball of sphagnum moss is spread

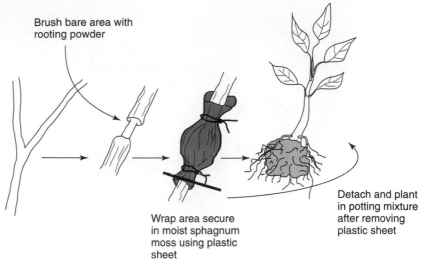

Brush bare area with rooting powder

Wrap area secure in moist sphagnum moss using plastic sheet

Detach and plant in potting mixture after removing plastic sheet

FIGURE 9-33 Air layering.

around the area and completely wrapped in polyethylene plastic or aluminum foil (figure 9-33). The ends of the polyethylene wrapping are tied with tape (e.g., electrician's tape). The wrapping retains moisture in the moist sphagnum moss. After a good amount of roots have formed, the layer is detached from the parent plant by cutting below the wrap. The layer is then replanted. Species including *Dieffenbachia,* litchi (*Litchi chinensis*), and the Indian rubber plant (*Ficus elastica*) may be propagated by this method.

SUMMARY

In effect, layering is a modified form of cutting in which the plant part is cut after it has been either naturally or artificially induced to root while still attached to the mother plant. When ample rooting has occurred, the part (the layer) is cut away from the parent and planted as an independent plant. This procedure may be accomplished in a variety of ways. In air layering, a nick in the branch is wrapped in soil to induce rooting. The nicked part of the branch may be arched down and buried under a mound of soil, called mound layering. Layering occurs without the aid of humans in certain plants such as the cane fruits (e.g., raspberry) and rhizomatous plants (e.g., strawberry). In these species, roots develop at the point of contact between the soil and flexible branches (or creeping branches as in strawberry). The tips or portions of these structures with the new adventitious roots may be detached and planted as new, independent plants.

OUTCOMES ASSESSMENT

PART A

Please answer true (T) or false (F) for the following statements.

1.   T F   Strawberry can produce layers naturally.
2.   T F   Cane fruits frequently propagate naturally by layering.
3.   T F   Aerial roots may be induced on certain plants.
4.   T F   Apples are commonly propagated by mound layering.

Please answer the following questions.

1. List two methods of natural propagation by layering.
    _____   _____

2. List two methods of artificial propagation by layering.
    _____   _____

3. Give one example each of plants that may be propagated by the methods listed in questions 1 and 2.
    _____   _____

4. Mound layering is also called _____ layering.

PART C

1. Distinguish between layering and cutting as methods of vegetatively propagating plants.
2. Describe how air layering is accomplished.

MODULE 5 SPECIALIZED UNDERGROUND STRUCTURES

9.14: TYPES OF SPECIALIZED UNDERGROUND STRUCTURES

Many herbaceous species that die back at the end of the growing season have underground food storage organs that survive the dormant winter period. These organs are also vegetative propagation structures that produce new shoots in the growing season. The variety of underground storage organs may be grouped into two classes based on how they are propagated: plants propagated by *separation* and plants propagated by *division*.

9.14.1 PLANTS PROPAGATED BY SEPARATION

Separation is the breaking away of daughter structures from the parent structure to be used to establish new plants. Two specialized underground structures—*bulbs* and *corms*—produce such materials.

> **Separation**
> A method of propagation in which underground structures of plants are divided not by cutting but by breaking along natural lines between segments.

Bulbs

Structurally, a bulb is an underground organ that consists predominantly of fleshy leaf scales growing on a stem tissue (basal plate). The scales wrap around a growing point or primordium to form a tight ball. Lateral *bulblets,* or miniature bulbs, originate in the axils of some of these scales and when developed (*offsets*) may be separated from the mother bulb to be planted independently as new plants (figure 9-34). The two types of bulbs are tunicate and scaly bulbs.

Tunicate Bulbs   The onion is an example of a *tunicate bulb* (also called *laminate bulb*). This type of bulb consists of concentric layers of tightly arranged scales, the outermost layer being a dry, membranous protective layer (tunic) (figure 2-18). Other examples are daffodil, tulip, and hyacinth.

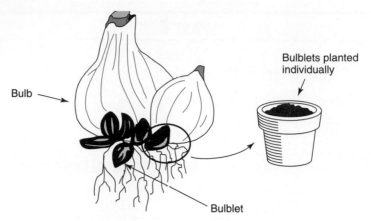

FIGURE 9-34   Propagation by using bulblets.

Scaly Bulbs   *Scaly,* or *nontunicate,* bulbs lack an outer dry protective membrane (figure 2.19). They are more delicate and require special handling to prevent drying and damage. The scales are not tight but loose and can be removed individually from the bulb. The lily is a nontunicate bulb. The daughter bulb or bulblets develop at the base of the scales of the mother bulb.

When the foliage of the plant dies back, the bulb resumes a dormant state. Bulbs may be dug up, separated, cleaned of soil, and then stored at a cool temperature to keep them dormant. The ability of a bulb to flower depends on its size. If it is too small, it may have to be grown for as long as several years before it will reach flowering size. Small bulb size may be due to premature removal of the foliage of the plant. The bulb should be harvested after the top has turned brown by natural processes.

Corms

Whereas the bulb consists predominantly of modified leaves, the corm is a modified stem. Food is stored in this compact stem, which has nodes and very short internodes and is wrapped up in dry, scaly leaves. When a corm sprouts into a new shoot, the old corm becomes exhausted of its stored food and is destroyed as a new corm forms above it. Several small corms, or *cormels,* arise at the base of the new corm (figure 9-35). These cormels may be separated from the mother corm at maturity (die back) and used to propagate new plants. Plants that produce corms include crocus and gladiolus.

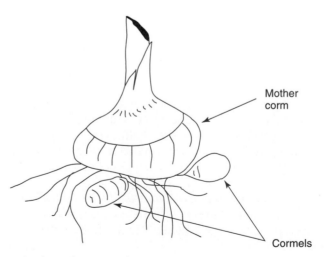

FIGURE 9-35   Propagation by using cormels.

9.14.2 PLANTS PROPAGATED BY DIVISION

In plants propagated by separation, complete and individual miniature plants are naturally produced and attached to the mother plants. These plants can be separated by simply breaking the bunch apart or cutting along natural boundaries. In division, no such clearly defined and individually packaged miniature plants exist. Rather, the large mass of mother plants is strategically divided by cutting into pieces so that each piece consists of certain basic structures to permit development into a new plant. When using the division method for propagation, it is advisable to treat the cut surfaces of materials with fungicides to prevent rotting when planted. Various underground storage structures found in plants may be used for propagation.

> **Division**
> *A method of propagation in which underground stems are cut into pieces and replanted.*

Rhizomes

Rhizomes are underground stems that grow horizontally. These features vary in diameter from one species to another, some being slender and others thick. Similar to stolons (above-ground horizontally growing stems) in strawberry, these rhizomes have nodes that produce adventitious roots that support shoots at these junctions. Rhizomes are also used for propagation by cutting dormant ones into pieces at the internodes (figure 9-36). Examples of plants propagated by rhizomes are ginger, banana, Kentucky bluegrass, and iris.

Stem Tubers

Certain plants store food underground in modified stems. These swollen ends of stems do not have nodes but rather buds (or eyes), each of which can be nurtured to produce a new plant. To produce a new plant, the tuber is divided into sections so that each bud has a good amount of flesh or stored food (figure 9-37). Once divided, the cut surface should be allowed to dry before planting. An example of a stem tuber is found in the edible Irish potato.

Tuberous Roots

Whereas the Irish potato produces a swollen stem, the sweet potato produces a swollen root. When a tuberous root is buried in the soil, it produces a number of adventitious shoots called *slips* (figure 9-38). These slips can be detached and planted individually to produce new plants. In other tuberous roots, propagation is accomplished by dividing the crown (*crown division*) or cluster of roots of dominant plants. Another example of a tuberous root is the dahlia.

9.14.3 SUCKERS

Suckers are adventitious shoots that arise from roots (2.3.3). In raspberry, the shoots are harvested from horizontal roots and used to propagate the plant.

FIGURE 9-36 Rhizomes may be divided and used for propagation.

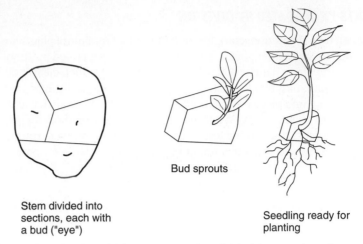

Bud sprouts

Stem divided into
sections, each with
a bud ("eye")

Seedling ready for
planting

FIGURE 9-37   Stem tubers such as Irish potato may be divided into sections for propagation.

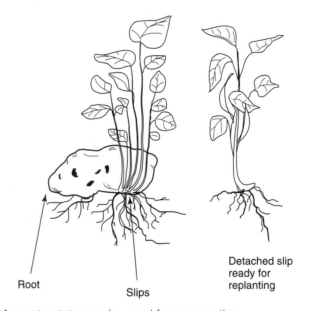

Root

Slips

Detached slip
ready for
replanting

FIGURE 9-38   Slips of sweet potato may be used for propagation.

9.14.4   OFFSHOOTS

The variety of adventitious shoots (suckers, crown division, slips, and offsets) that arise from the stems or roots of plants are called *offshoots*. The pineapple plant is quite versatile in terms of parts that may be used to propagate the plant. The crown on top of the fruit, slips arising from axillary buds at the base, and suckers that originate from the lower part of the stem may all be used for propagation.

Offshoot
The generic term for adventitious shoots that arise from various parts of the plant.

SUMMARY

Apart from roots, certain species develop a variety of underground swollen structures that are, in some cases, the economic, edible, or usable part of the plant. These modifications may be in the roots or stem. Bulbs such as onion, daffodil, and tulip have modified leaves that are scaly; corms such as gladiolus and crocus have modified stems that have assumed a globelike shape. In some plants, including ginger and banana, modified underground stems grow horizontally and are called rhizomes; in the Irish potato, the underground stem is swollen. However, in the sweet potato, the root is swollen into a tuberous root that is edible. The modified underground structures are used in propagating the respective plants.

PART A

Please answer true (T) or false (F) for the following statements.

1. T F   Irish potato is an underground root.
2. T F   Onion is a tunicate bulb.
3. T F   Corms are modified stems.
4. T F   Bulbs are modified leaves.
5. T F   A tunicate bulb is also called a laminate bulb.
6. T F   The ability of a bulb to flower depends on the size of bulb used for planting.

PART B
---

1. Scaly or _____ bulbs lack an outer protective scaly leaf.
2. The small corms that develop from the mother corm are called _____.
3. Potatoes may be propagated by plantlets with adventitious roots called _____ that develop from the tuber.
4. Bulbs and corms are propagated by the method of_____ .

PART C
---

1. Why do scaly bulbs require more careful handling?
2. Describe how a bulb is propagated.

MODULE 6 MICROPROPAGATION (TISSUE CULTURE)

9.15: THE TECHNIQUE

Micropropagation (or *tissue culture*) is a technique by which tissue obtained from a plant (mature or immature) is cultured in an artificial medium. This tissue first changes into a mass of undifferentiated (2.2.2) cells called *callus,* from which differentiation (2.2.2) into shoot and roots or embryos may later occur (figure 9-39). Molecular biotechnological procedures often incorporate tissue culture as one of their critical methods (5.16). Micropropagation may be used in herbaceous (e.g., strawberry, gladiolus, tobacco, carnation, and gloxinia) and woody plants (e.g., apple, rose, kalmia, and rhododendron).

Tissue culture requires a completely sterile environment to be successful. Propagation can be initiated with any part of the plant: leaf, stem, root, pollen grain, embryo, and others. These excised plant parts used to initiate propagation by tissue culture are called *explants*. Both immature and mature plant tissue may be used. Explants are aseptically prepared by *surface sterilization* before being placed in a sterile medium (figure 9-40). The medium is fortified with all of the nutrients required for plant growth and development, including mineral salts (major and minor elements), sugar, vitamins, and growth regulators (auxins and cytokinins). One of the most commonly used tissue culture media is the Murashige and Skoog medium. By

> **Micropropagation**
> The technique of producing new plants from single cells, tissue, or small pieces of vegetative material.

> **Callus**
> A mass of undifferentiated cells that can be induced or arise naturally as a result of wounding.

> **Explant**
> Generic term for the living vegetative plant material extracted for tissue culturing on an appropriate medium.

FIGURE 9-39 Callus formation in tissue culture.

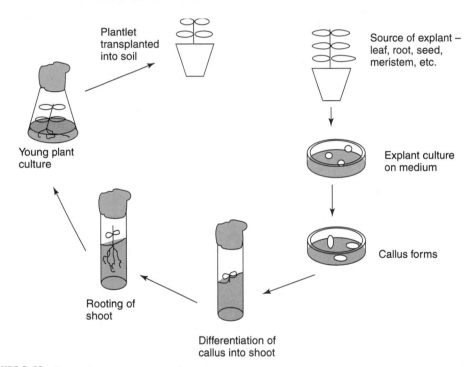

FIGURE 9-40 Steps in micropropagation.

modifying the composition of the medium, with respect to growth regulators, one can cause it to sustain normal plant growth or induce the explant to develop callus tissue, the result of repeated mitotic cell division.

Sometimes scientists are able to manipulate callus tissue through the introduction of chemicals into the growth medium that induce heritable variations (mutations) (5.10). Such variations occasionally arise spontaneously even in normal media culture and are called *somaclonal variations*. Some of these variants have agronomic value and are utilized in crop improvement by plant breeders.

The technique is used widely in crop improvement and other biotechnological procedures (5.16). It can be used to replicate asexual plant materials clonally to produce numerous plantlets for planting the crop. This practice is undertaken in the laboratory and is space saving. Viral infections are noted for crippling plants without killing them. Further, it is known that new tissue that develops in an infected plant is usually virus free. This healthy tissue may be extracted and tissue cultured to produce numerous virus-free plants, an effective way of purifying an infected parental stock.

> **Somaclonal Variation**
> Heritable variation that arises spontaneously as callus forms on a tissue culture medium.

9.16: APPLICATIONS

9.16.1 ARTIFICIAL SEEDS

The concept of *artificial seeds* is being pursued by several biotechnology companies. It entails the mass production of clones of embryos through tissue culture. These embryos are then coated with growth nutrients in a biodegradable carrier as protective layer. This covering may also carry pesticides, fertilizers, and even nitrogen-fixing bacteria. Once packaged, the artificial seeds may be planted like natural ones. The technology is currently in its infancy, and hence artificial seeds are more expensive than natural ones.

9.16.2 SHOOT MERISTEM CULTURE

Shoot meristem culture, or *mericloning,* is another innovative micropropagation technique. It is used to facilitate improvement in certain crops such as orchard crops, which require about seven years from planting to first blooming. In orchards, it appears that a certain seed-fungi association is a normal prerequisite for germination to occur. A quick way of obtaining planting material is through tissue culture of shoot meristems in a medium fortified with certain hormones and other substances to perform the role of fungi. After about a month of tissue culture, adventitious roots begin to develop. Before this stage, the tissue is separated into small pieces, each of which is capable of developing into a new plant. By this process, numerous genetically identical plants are produced in a short time.

9.16.3 EMBRYO RESCUE

Sometimes in plant improvement programs, crosses are made between genetically divergent plants. Crossing plants within the same species usually has no genetic consequences. The cross is able to develop normally to produce seed and subsequently a plant upon planting the seed. On certain occasions, the breeder would have to import desirable genes from a source different from the species of the plant being improved. Such gene transfers, when performed by the conventional method of crossing, usually have genetic consequences. The embryo may fail to develop normally. When this happens, it is possible to remove the developing embryo and nurture it in an artificial medium under tissue culture conditions.

9.16.4 APPLICATIONS IN SELECTED CROPS

Practical applications of tissue culture in horticulture have been widely reported. In garlic, in vitro cultivation has been used to produce quality planting material that is free from viral and bacterial diseases. In leek, explants from the stem, leaves, flower head, and basal plate have been micropropagated. Onion meristem tip culture was used to rid vegetatively propagated shallots of viral infection. Similarly, meristem culture has been used to develop clones of self-incompatible inbreds for use in F_1 hybrid seed production in broccoli. Anther culture is used to develop instant homozygous inbred lines for hybridization. Following an interspecific cross between broccoli and cauliflower to transfer clubroot resistance for *Brassica napus* to *B. calabrese,* embryo rescue was used to regenerate surviving interspecific embryo cabbage.

Mass production in vitro of virus-free and disease-resistant *Brassica* plants occurs. *Embryo* and *ovary rescue* techniques are frequently employed in the rescue of nonviable interspecific hybrids. *Protoplast fusion* was successfully employed in creating hybrid plants of red cabbage and radish. The somatic hybrids were male sterile but female fertile. Carrot is easy to manipulate in cell culture. A very creative use of biotechnology is exemplified by work done with carrots whereby scientists have produced synthetic seed by encapsulating asexual somatic embryos with polyethylene oxide or sodium oxide. Somaclonal variants with resistance to downy mildew and lettuce mosaic virus have been reported in lettuce.

Tissue culture techniques can be utilized to create what may be described as in vitro hybrids. Cells obtained from different plants are prepared by removing their cell walls using enzymatic procedures to leave the protoplasts. These naked protoplasts are induced to fuse

> **Protoplast Fusion**
> *The techniques of in vitro union of two protoplasts into somatic hybrids.*

through the use of chemicals such as diluted polyethylene glycol (antifreeze used in automobiles). This technology is called *protoplast fusion,* and the resulting plants are called *somatic hybrids.* Through further manipulation, the nucleus of one of the two protoplasts may be destroyed so that only one of the two cytoplasms fuse. The product then is called a *cybrid.*

REFERENCES AND SUGGESTED READING

Boodley, J. W. 1998. The commercial greenhouse. Albany, N.Y.: Delmar.

Hartmann, H. T., and D. E. Kester. 1983. Plant propagation, 4th ed. Englewood Cliffs, N.J.: Prentice-Hall.

Kyte, L. 1983. Plants from test tubes: An introduction to micropropagation. Portland, Oreg: Timber Press.

McMillan, B. P. 1979. Plant propagation. New York: Simon & Schuster.

Thorpe, T. A., ed. 1981. Plant tissue culture: Methods and applications in agriculture. New York: Academic Press.

PART 4

GROWING PLANTS INDOORS

CHAPTER 10

Growing Houseplants

PURPOSE

This chapter discusses the environmental factors required for growing plants indoors and how to choose, grow, care for, and use houseplants.

EXPECTED OUTCOMES

After studying this chapter, the student should be able to

1. Distinguish between the practices involved in growing plants in the house and under controlled environments (greenhouses).
2. List and discuss the factors that one should consider in growing houseplants.
3. List 10 common houseplants.
4. Discuss the correct ways of watering plants.
5. Discuss the correct ways of feeding plants.
6. Describe how plants are repotted.

KEY TERMS

Azalea pots
Bulb pots
Controlled environment
Florescent lights
Foliar sprays
Garden room

Hygrometer
Incandescent lights
Phototropism
Plantscaping
Potting media

Pot-bound
Rose pots
Slow-release fertilizer
Soluble fertilizer
Standard pots

OVERVIEW

Plants can be successfully grown indoors in a *controlled environment* (chapter 11) in which all of the required plant growth factors are supplied in appropriate amounts. Growing plants indoors at home is an activity that enables home owners to enjoy plants year-round. However, the home environment is not as controlled as a greenhouse environment because the home is shared by humans, and sometimes pets, whose needs take precedence over the needs of plants. Home conditions are generally less than ideal for general plant growth, especially with respect to light. For this reason, not all plants can be successfully grown indoors.

Growing plants indoors necessitates that plants be grown in containers, which restricts the growing environment for plant roots. Choice of proper growth medium is critical to the success of growing potted plants. Also, proper choice of plants is essential in growing indoor plants. After choosing the right plants and the appropriate medium, the grower must employ good management practices to maintain the plants in good health. A green thumb is not hereditary but an acquired trait obtained through experience and knowledge of basic horticultural principles.

In this chapter, a list of indoor plants is provided, along with the principles of indoor plant culture. The reader will learn how to constitute a good growth medium and is provided with guidelines for caring for houseplants. Further, use of plants indoors either for purely decorative purposes or to serve certain functions is discussed.

10.1: FACTORS THAT INFLUENCE THE CHOICE OF HOUSEPLANTS

The variety in outdoor horticultural plants is tremendous. Similarly, many ornamentals are adapted to indoor environments, although they are fewer in number than field plants. The choice of which plants to grow depends on several factors.

10.1.1 PERSONAL PREFERENCE

Some people are cactus enthusiasts, and others love roses. For some people, it is love at first sight—they decide what to grow when they see a plant growing somewhere. Such people often visit a nursery, ask to be shown some possibilities, and then buy what most appeals to them. People are more likely to invest time and resources in growing and caring for plants that they like, than in plants with less appeal. If one decides to be a collector of a certain kind of plants, variety and diversity occur in some species such as *Peperomia* and *Ficus.* Peperomias vary in shape, size, texture, and color. In *Ficus,* one can find trees, shrubs, and creeping and trailing plants.

10.1.2 GROWING CONDITIONS

Some houseplants can do well in nearly any part of the room. Others may require special conditions that only an avid flower enthusiast can afford the time and patience to provide. Many people just want a plant that will grow without intensive care (table 10-1). As will be discussed, houseplants need good care to grow properly.

10.1.3 ROOM DECOR

Interiorscaping
The use of ornamental plants for functional and aesthetic purposes.

Plants are used to enhance the room decor. They come in different shapes, sizes, colors, and textures. For best results, when different kinds of plants are used in interior decoration (*interiorscaping* or *plantscaping*), they should not only complement the room furnishings, but also relate well to each other and blend to create a pleasant environment.

Table 10-1 Selected Easy-to-Grow Plants

Plant	Scientific Name
Rubber plant	*Ficus elastica*
Spider plant	*Chlorophytum comosum*
Snake plant	*Sansevieria trifasciata*
Philodendron	*Philodendron* spp.
Medicine plant	*Aloe vera*
Pothos	*Scindapus aureus*
Aspidistra	*Aspidistra* spp.

10.1.4 PLANT CHARACTERISTICS

Plant characteristics include the following:

1. *General attractiveness.* The plant should be aesthetically pleasing to behold. The foliage and/or flowers should be attractive.

2. *Appearance at maturity.* People seldom purchase fully grown plants for the home. Instead, indoor plants grow and change (e.g., in shape, size, and height). A young plant may not be as appealing as when it is much older.

3. *Growth cycle.* Some plants are only attractive when they flower and may be unattractive in the vegetative state. Some foliage plants are attractive as such and become even more so when they flower.

4. *Growth (maturity) rate.* Houseplants do not grow or attain maturity at the same rate. Although some plants, such as annuals, grow rapidly, others, such as palms, take several years to attain a good size that is aesthetically pleasing.

10.2: USING PLANTS IN THE HOME

The key to the successful use of plants in the home is creativity and experimentation. Plants are living things and, like people, need regular (in some cases daily) attention. Ornamentals in the home are meant to be enjoyed, so the use of horticultural plants should not be on a scale such that it becomes a chore instead of a joy. Indoor use of plants may involve one, a few, or even a whole room full of plants (*garden room*).

10.2.1 LOCATING PLANTS

Locating plants in the room depends on several factors, described in the following sections.

Architecture of the Room

Some homes have high ceilings and can accommodate tall plants including larger tropical plants such as *Dracaenas*. Some homes are designed with skylights, which can provide the additional lighting needed by some plants. Leaf form and plant shape should complement the architectural style of the room. If the architecture is big and bold, big and bold plants should be selected. Traditional interiors usually require plants with delicate foliage, such as ferns and grape ivy (*Rhoicissus rhomboidea*). Existing features such as fireplaces and mantlepieces should be utilized in the design of plant displays. Plants with distinctive foliage such as the rubber plant (*Ficus elastica*) and Swiss cheese plant (*Monstera deliciosa*) fit in well with the straight lines of contemporary architecture.

Space

Large plants do better in large rooms. Large plants make small rooms appear too crowded. For example, a fully grown creeping fig (*Ficus benjamina*) is out of place in a cottage drawing room. Similarly, a small plant in a large room has virtually no impact.

Level

Plants may be placed on the floor or on pieces of furniture (such as on tabletops or bookshelves). Elevated positions are suited to small plants (figure 10-1). The top shelves of bookshelves or other high levels are suitable for plants that have trails or long vines. Small plants should be placed on tabletops.

FIGURE 10-1 A potted plant displayed on top of a piece of furniture.

Color of Walls and Upholstery

Plants should be placed against a background that will bring out their colors. Plants with strong foliage forms are effective against walls with patterns, provided the motifs in the pattern and leaf size contrast sufficiently. The leaves of the umbrella plant (*Schefflera phylla*) are effective against a background of small-patterned wallpaper. Where the background consists of bold, abstract designs, it can be balanced with a display of plants with delicate foliage such as asparagus fern.

10.2.2 SPECIFIC USES OF INDOOR PLANTS

Plants may be used to perform certain functional roles in the room.

Fill in Gaps

Plants are often placed in areas too awkward for a piece of furniture, such as a corner.

Brighten Up an Area

Flowering plants in bloom can brighten up the room. A variety of dull spots occur in a room (e.g., empty walls, unused fireplace, stairwell, and corners). Plants with distinct leaves, such as the silhouette plant (*Dracaena marginata*), climbers (e.g., trained on poles), or other trailing and cascading plants mounted on wall brackets may be displayed against an empty wall. Because some dull spots such as corners usually have poor conditions for plant growth, plants adapted to such conditions (e.g., *Aspidisaenas, Sansevieria,* and *Philodendrons*) should be chosen.

Cover Up Sharp Edges

Potted plants can be positioned to cover the edges of walls or architectural features. For example, climbers such as *Fatshedera lizei* can be used in the stairwell.

Create Room Dividers

Instead of using wooden structures, for example, appropriate plants may be arranged to form a wall (figure 10-2). Where dividers are used, plants such as *Philodendron* or *Hedera* may be trained to grow over these physical structures. Trailing or climbing plants need to be monitored and pruned or trained to keep them within desired boundaries. Plants may be positioned to climb up structures or cascade down them. Sometimes smaller displays such as a *terrarium* (chapter 25), bottle garden, or a small group of plants may serve the purpose of separating one area of the room from another.

FIGURE 10-2 Room dividers created with living plants.

Window Displays

Plants in windows enhance the room decor (10.2.5). Since light enters the house primarily through windows, the selection of plants to use should consider the position of the windows, the plant sensitivity to light, and the plant size. While south-facing windows receive sunlight year-round, north-facing windows receive the least amount of light, especially during the winter. North-facing windows favor foliage plants such as aspidistra and sansevieria. Desert cacti do well in unshaded south-facing windows.

Fragrance

Certain plants exude sweet scents that freshen the indoor atmosphere. For example, scented pelargonium has a pleasant fragrance.

Direct Traffic

Plants can be strategically arranged to steer people away from certain parts of the room and to prevent people from using certain spaces as pathways.

Cover Up Undesirable View

A *window garden* (10.2.5) may be planted to block the view to unattractive areas on the outside. Plants may be arranged to hide unsightly parts of a room.

Environmental Quality (Air Quality Control)

Certain plants are known to improve the air quality indoor areas by absorbing contaminants in the air. Contaminants include fumes from cleaning solvents, radon, secondhand smoke, and furniture and carpeting and ozone from copying machines. Some of the most effective plants for this purpose are gerbera daisy, chrysanthemum, golden pothos, and *Spathiphyllum*.

10.2.3 IMPROVING THE DISPLAY OF HOUSEPLANTS

Apart from selecting appropriate plants and arranging them effectively in the room, there are several specific ways in which the display can be enhanced:

1. Use a spotlight to draw attention to conversation pieces or specimen plants (unique plants that invite conversation). Backlighting enhances the display of some species such as Boston fern (*Nephrolepsis*). Certain plants may be displayed in less-than-ideal conditions with appropriate lighting.

2. Use decorative containers to hold the plants (double potting). The plastic pot from the nursery may be placed in a very attractive container to enhance the display (figure 10-3).

> **Double Potting**
> A method of enhancing the display of potted plants by placing the potted plant in a more decorative pot.

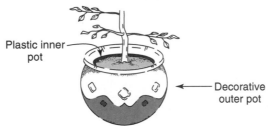

Plastic inner pot

Decorative outer pot

(a)

(b)

FIGURE 10-3 (a) Displaying a plant potted in a plastic pot in a more attractive decorated pot. (b) Examples of decorative containers.

Other containers include wicker baskets, brass saucepans, and in some cases patterned containers. When using patterned pots, the color of the foliage and flowers should blend well with the pot color and pattern. Growers do not often plant directly into decorative containers. Instead, they are used as outer coverings to hide the ordinary flowerpot. Containers may be clay or china.

3. Group plants (10.2.4). Instead of scattering plants throughout a room, a number of plants of the same type can be grouped together (massed). Compact plants may be grouped on a stand, on a pebble tray, or on a table as a centerpiece. Small specimens may also be effectively displayed in tiny, unusual containers (e.g., egg cup). Colorful seasonal plants such as tulip, hyacinth, azalea, and geranium can be massed on a windowsill.

4. Use hanging baskets (10.2.7). By themselves, hanging baskets can have very attractive holders. Plants grown in hanging baskets offer some of the most attractive displays.

5. Use ornamental paper. Wallpaper and plants can be used together to provide an effective display, with the paper as the background.

6. Use plant support. Plants may be displayed on pedestals, wooden tables, glass-topped, wrought-iron tables, and other such specially designed supports (wooden or metal jardinieres, tiered plant stands, and aspidistra stands). Other pieces of furniture in the house can be adapted as flower stands (e.g., corner cupboards and washstands). Plants may be grown and displayed on a plant trolley.

10.2.4 GROUPING PLANTS

A large plant can be effectively displayed alone. Smaller specimens do better when grouped. Grouping can be accomplished by arranging individual potted plants together (e.g., on a gravel tray) or by planting a mixture of plants in large troughs (figure 10-4). A wide variety of contain-

FIGURE 10-4 Planting different species of plants in one container.

FIGURE 10-5 When different species or types of plants are grown in a single container they may be selected and arranged to create an overall shape.

ers are available for use. They vary in type of material (e.g., plastic, wrought iron, wood, and clay), shape, size, and decorative appearance. After choosing a container, the next task it to choose the right combination of plants. Plants should be grouped according to their need (e.g., sun loving, partial light loving, and moisture loving). Certain plants such as sansevieria are adapted to less-than-ideal conditions and hence can be utilized in a variety of groups. The combination of plants should also consider the plant size, color, form, and texture. Plants can be grouped to create an overall shape (figure 10-5). Further, one may include a flowering plant in an arrangement to give it some color. It may be necessary to prune periodically to maintain a good balance in the display. Plants should be repotted, or replanted, when the container becomes too small for them. During replanting, the original set of plants may be retained or new ones included.

10.2.5 GROWING PLANTS IN THE WINDOW

Displaying plants in windows is very popular because windows (especially south-facing windows) are the source of most of the natural light entering the house. Although plants in south-facing windows are prone to scorching due to excessive light, those in north-facing windows may not receive enough light. Windows experience temperature fluctuation, some of which is due to either cold or warm drafts from air conditioners or radiators located beneath windows. Nonetheless, with good care, one can raise healthy, attractive plants on a windowsill or near a window. They can range from single potted plants (figure 10-6) to an elaborate plant window (figure 10-7). Plants in the window display do not have to be displayed on the sill and can instead be hung in hanging baskets. Shelves may also be constructed in windows so that tiers of potted plants may be arranged. A window may be modified to create a container (such as a terrarium) in which plants can be grown (figure 10-8). After arranging the pots, the base of the container may be filled with moss to hide the pots. Lighting may be installed, as well as automatic mist spraying and temperature control units. Plants for display in well-lit windows are presented in table 10-2.

10.2.6 GARDEN ROOMS

Garden rooms are usually extensions of the main part of the house designed to be sunny. A large variety and number of plants are housed in the room, but some space is reserved for large furniture. Ideally, the garden room is adjacent to the living room. The structure may consist of panes of glass or some other durable and transparent material. Designs vary widely, with some home owners installing pools in their garden rooms. The floor of the room may be made of wood, ceramic tile, or some other kind of material.

Plants may be grown in pots or ground beds or hung in baskets. A large variety of plants are grown successfully by strategically placing plants in the locations where they receive the best available conditions. For example, sunlight-loving (or light-loving) plants should be lo-

FIGURE 10-6 A single plant displayed on a windowsill.

FIGURE 10-7 A window garden.

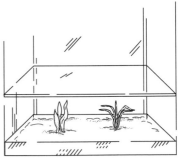

FIGURE 10-8 A window terrarium is a unique way of displaying plants.

Table 10-2 Plants That Grow Well Under Well-Lit (Window) Conditions (Full Sun)

Plant	Scientific Name
Bougainvillea	*Bougainvillea spectablis*
Medicine plant	*Aloe vera*
Amaryllis	*Hippeastrum* spp.
Bird-of-paradise	*Strelitzia reginae*
Lipstick vine	*Aeschynanthus lobbianus*
Rubber plant	*Ficus elastica*
Coleus	*Coleus blumei*
Hen and chickens	*Escheveria peacockii*
Fucshia	*Fuchsia x hybrida*
Gardenia	*Gardenia jasminoides*

cated near windows. It should be remembered that a garden room is meant to provide a comfortable environment for people before plants. As such, plants that prefer high humidity should be avoided, since such an environment will make it uncomfortable for humans to use the garden room. Plant species suitable for greenhouse production, including *Acacias, Musa enseta* (dwarf banana), *Eucalyptus,* dwarf conifers, potted roses, garden annuals, cacti, and some *bonsai* (chapter 26), can be raised in a garden room.

10.2.7 HANGING BASKETS

Hanging baskets provide another avenue for displaying houseplants (figure 10-9). An advantage of using hanging baskets is that it allows plants to be displayed in very awkward places such as over doorways and suspended from ceilings, patios, and walls. In this way, plants can be grown at eye level. Hanging basket containers can be made of wood, wire, ceramic, or plastic. For wall attachments, the container is halved (i.e., flat on one side) so that it can be fixed to the wall. Suspended baskets are usually round. The container may also be solid sided or made of wire. In the latter case, a lining of plastic (less attractive) or moss is needed before the planting medium is placed. The advantage of wire baskets is that plants can be planted on both the inside and outside of the container to cover it up completely (figure 10-10).

Like potted plants, it is critical that hanging basket containers drain properly. However, hanging baskets dry out much more quickly than potted plants on the floor or tabletop because they are exposed to warmer temperatures (since warm air rises) at the level at which they are suspended (about 3°C or 5°F warmer) and airflow around them is much greater. Hanging baskets should be watered more frequently than regular potted plants. Wire basket designs dry out more quickly than solid-sided container designs. To add to the decor, some home owners replace the wires attached to the pots for suspension with decorative chains or fabric support. Hanging baskets need attention similar to that for potted plants. Some species prefer sunny conditions, whereas others prefer shade (table 10-3). Plants should be fertilized as needed.

> **Hanging Baskets**
> *A potted plant grown and displayed usually by freely suspending it or attaching it to a wall.*

FIGURE 10-9 Plants with vines and other hanging structures can be effectively displayed in hanging baskets.

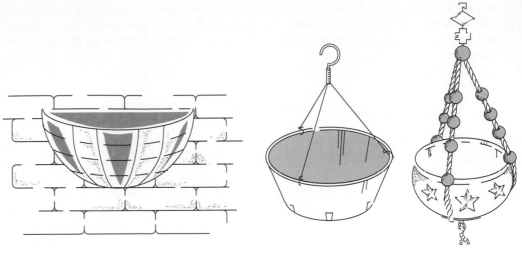

FIGURE 10-10 Wire baskets may be wall mounted or hung like a hanging basket.

Table 10-3 Selected Plants for Hanging Baskets

Plant	Scientific Name
Begonia	*Begonia* spp.
Spider plant	*Chlorophytum* spp.
Aparagus fern	*Asparagus* spp.
English ivy	*Hedera helix*
Boston fern	*Nephrolepis exaltata*
Coleus	*Coleus blumei*
Wandering Jew	*Zebrina pendula*
Lipstick vine	*Aeschynanthus* spp.
Swedish ivy	*Plectranthus australis*
Pothos	*Scindapus aureus*
Heat leaf philodendron	*Philodendron scandens*
Christmas cactus	*Zygocactus truncatus; Schulmbegergia hybrids*

10.3: CARING FOR HOUSEPLANTS

Houseplants need all of the growth factors that are obtained in the outside environment—good *soil, air, water, light,* and *nutrients.* Houseplants, however, differ in the quality and quantity of each factor required for optimal growth.

Caring for houseplants starts with bringing home healthy plants. Commercial nurseries grow plants under controlled conditions year-round. These conditions are adjusted to suit the needs of plants. However, at home, even though home owners adjust the house temperature as the seasons change, these adjustments are designed for the comfort of people, not plants. Rooms are often evenly heated or cooled. It is very easy for the home owner who is not a houseplant enthusiast to forget about the special needs of plants in the home.

10.3.1 BRINGING PLANTS HOME SAFELY

In spring or summer, temperatures in nurseries and homes are not likely to be significantly different. More significant differences are likely to occur between the home and the nursery during the cold period (early fall to early spring). Buying plants for use in the home requires the most attention and care during this period to reduce shock to plants. Plants purchased in the cold season should be transported with some insulation. The car should be heated before plants are moved into it from the nursery. For long-distance transportation, plants may be placed in cardboard boxes and wrapped in several layers of paper. Some plants are more delicate than others.

10.3.2 MONITORING LIGHT

A photographic light meter is used to determine light intensity. Light is a critical requirement for plant growth and development. The average light intensity in a house is about 55 lux, compared with more than 130,000 lux outside on a bright sunny day. A room is not uniformly lit. While plants are in the care of commercial nurseries, light conditions are maintained at optimal or near-optimal conditions, which usually means at higher intensities than would be found at home. For example, tropical foliage plants generally prefer high light intensities (above 10,000 lux and in some cases even above 30,000 lux).

> **Lux**
> *The metric unit expressing the illumination falling on all points on a surface measuring one meter square, each point being one meter away from a standard light source of one candle; 1 lux = 0.093 footcandles.*

Chloroplasts in leaves are known to orient themselves differently to suit high- and low-intensity light conditions. The problem with houseplants arises when they are transferred from the greenhouse (high intensity) to the home (low intensity). This change is drastic for many plants, which immediately begin to readjust to adapt to the home environment. Plants readjust differently, with variable consequences in terms of their aesthetic value. Some plant species such as *Ficus benjamina* and *Coleus hybridus* adjust to low light levels by losing chlorophyll and subsequently dropping their leaves. These plants develop new leaves that are much thinner and have chloroplasts that are uniformly distributed throughout the lamina, obviously for better interception of light. Other species including palms and lilies respond to low light levels by changing color from green to yellowish, but without abscission. By remaining attached to the plant, these sickly leaves reduce the aesthetic value of the plant. Some nursery owners, anticipating the eventual transfer of plants to homes by customers, may put plants through a weaning period to acclimatize and prepare them for the home environment.

A variety of sunlight intensities are experienced in the home, depending on the season, architectural design of the house, and other landscape activities around the house or even nearby houses (figure 10-11):

1. *Direct sunlight (full light).* A house that has no large trees in its immediate vicinity or structures that may block sunlight can receive 100 percent sunlight for parts of the day through windows that face east, west, southeast, and southwest.

2. *Indirect sunlight (filtered bright light).* When trees obstruct the direct sunlight, it enters the house after going through the leaves. Some windows have decorative curtains and blinds that also filter direct sunlight, making only a portion of its intensity (about 50 to 75 percent) available indoors.

3. *Bright light.* Bright light occurs near the areas where sunlight directly or indirectly enters the house. Such areas are less bright (about 25 percent) than the primary source.

4. *Medium light.* North-facing windows do not receive direct sunlight. Even windows facing east or west do not receive direct sunlight if heavily obstructed. The lighting condition near these windows is of medium intensity (about 10 percent).

5. *Poor light.* As one moves away from windows into the center of the room, natural light intensity diminishes drastically (to about 5 percent). Room corners that are not near light sources are usually dimly lit. Poor lighting conditions can be contributed to by the interior design of the room. Dark colors absorb light, whereas lighter colors reflect it. A white wall reflects light to a plant placed in front of it. Furniture arrangement can create dark sections in the room.

Fortunately for home gardeners, most indoor foliage plants do not require high light intensities. However, when growing flowering plants such as azalea, chrysanthemum, geranium, and poinsettia indoors, high light intensity is desired for flower initiation. Plants such as cacti prefer higher light intensities to compensate for reduced photosynthetic surface from

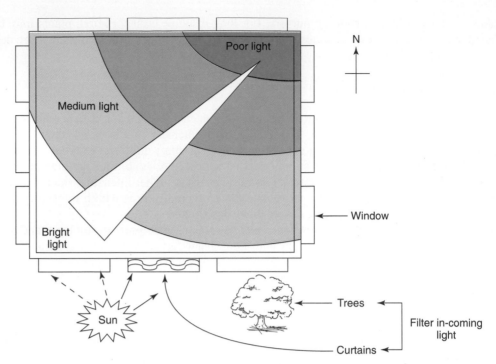

FIGURE 10-11 Natural light received in a room is influenced by factors such as the number and kinds of windows, the window treatment, the presence of trees near the windows, and the orientation of the house.

thickened leaves. Also fortunate for home gardeners is that many plants are quite tolerant of imperfect lighting conditions, at least for a period. Therefore, plants can be shifted around in a room without adverse consequences, as long as efforts are made to periodically provide the proper lighting conditions. The best approach, however, is to purchase plants that suit the lighting conditions one has or can provide, because shade-loving plants such as fern and aglaonema are less tolerant of improper lighting conditions (table 10-4).

When plants at home experience improper lighting conditions, they may grow spindly, have poor coloration (look pale) and small leaves, and exhibit *phototropism* (growth toward light) (chapter 4). The shape and aesthetic value of plants can be drastically diminished when, in search of more light, plants turn their leaves in the direction of the source.

In philodendron (*Monstera deliciosa*), plants grown under dim light have unsplit leaves, whereas those grown under bright light have split leaves (figure 10-12). Generally, plants with colored leaves (such as the red color of coleus) require more intense light to reach the chloroplasts masked by the red pigment. Similarly, variegated plants require more light to compensate for the lack of chlorophyll in certain parts. When grown in dim light, they do not variegate but show solid green color.

> **Phototropism**
> The response of a plant to nonuniform illumination, usually resulting in bending toward the strongest light.

10.3.3 SUPPLEMENTARY LIGHTING (ARTIFICIAL LIGHTING)

Plants generally need 12 to 16 hours of light per day for proper growth and development. Plants in the house receive a varying duration of light depending on how long people stay at home. Lights are turned on and off as people come into and leave the house. As such, plants may receive long periods of exposure to light on some days and little on others. Artificial lighting may be used in large houses for several reasons.

Decorative

A home owner may desire to draw attention to specimen plants. These showcase plants may have unique and very attractive features and may be placed under a spotlight to emphasize their beauty. Lights and plants can be placed strategically to enhance the decor of a room.

Table 10-4 Selected Plants Adapted to Various Light Conditions Indoors

Plants Adapted to Direct Natural Light (Place in South-Facing Window)

Norfolk Island pine (*Araucaria heterophylla*)
Croton (*Codiaeum* spp.)
Florist's chrysanthemum (*Chrysanthemum x morifolium*)
Florist's cyclamen (*Cyclamen persicum*)
Poinsettia (*Euphorbia pulchernima*)
Dutch hyacinth (*Hyacinthus orientalis*)
English ivy (*Hedera helix*)
Christmas cactus (*Schulmbergia hybrids. Zygocactus truncatus*)
Azalea (*Rhododendron* spp.)

Plants Adapted to Low Light (Place in North-Facing Windows)

Dumbcane (*Dieffenbachia* spp.)
Lipstick plant (*Aeschynanthus* spp.)
Iron plant (*Aspidistra elatior*)
Staghorn fern (*Platycerium* spp.)
Snake plant (*Sansevicria trifasciata*)
Pothos (*Scindapsus aureum*)
Heart philodendron (*Philodendron scandens*)
Chinese evergreen (*Aglaonema commutatum*)
Peacock plant (*Calathea* spp.)
Corn plant (*Dracaena fragrans* 'Massangeana')

Plants Adapted to Medium Light (Place in East-Facing Windows)

Zebra plant (*Aphelandra squarrosa*)
Spider plant (*Chlorophytum comosum*)
Gold-dust plant (*Dracaena surculosa*)
Weeping fig (*Ficus benjamina*)
Indian rubber tree (*Ficus elastica*)
Baby's tears (*Helxine soleirolii*)
African violet (*Saintpaulia ionantha*)
Wax plant (*Hoya carnosa*)
Sentry palm (*Howea* spp.)

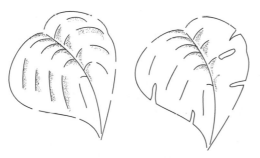

Split leaf of a plant
grown under bright light

FIGURE 10-12 An effect of intense light on philodendron leaf development.

Physiological

Plants need light to grow and develop. No plant can survive in darkness. As such, in dimly lit parts of the room, as well as in winter when natural light is least available, artificial light may be used to supplement natural light for plants to grow properly.

10.3.4 SOURCES OF ARTIFICIAL LIGHT

When additional lighting is required in a room, the type used is influenced by how it fits into the general decor of the room. When needed in a garden room, the styling may be compromised, but in the living room, styling of the light source is an important consideration for most people. There are three general sources of artificial light for indoor use—incandescent, mercury vapor, and fluorescent lights.

Incandescent Lights

Incandescent lights (11.3.2) are commonly used in homes. The light they emit is high in orange-red and low in blue-violet wavelengths (chapter 3). Even though they produce adequate light, the major disadvantage of this source of light is the tremendous amount of heat generated in the process of providing light. Only about 30 percent of the energy from an incandescent bulb is in the form of light, the remainder being given off as heat. When used for supplemental lighting, incandescent bulbs should be placed at a safe distance (depending on the power rating) to prevent scorching the plant. However, when placed too far away, incandescent light does little to help the plant, since most of its energy is heat and not light. This light source is hence largely decorative (e.g., spotlights). Floodlight models of incandescent lights are available and are more efficient. They also come in a variety of appealing styles.

Mercury Vapor Lights

Mercury vapor lights are more useful than incandescent lights in terms of providing light. They also emit less heat. However, they operate at high power (minimum of 250 watts) and are expensive.

Fluorescent Lights

Fluorescent lights are the most efficient of all sources and most recommended for houseplants. They (11.3.2) are very energy efficient, cost less than the other types to operate, emit little heat, and can be placed close to plants without scorching them. They are available in a variety of colors, which adds to their decorative use at home.

Fluorescent tubes are also designed to emit different qualities of light. The spectrum of light usable by plants includes the violet-blue and red wavelength. Thus, it is important to take note of the spectrum on the label. Daylight fluorescent tubes provide mostly blue light and little red light. They are suitable for foliage plants. The best fluorescent tube lights for plants are those that provide a reddish hue, especially if flowering species are being grown. For extra light, the wide-spectrum light may be used. Extra light is desired by plants such as orchids, cacti, and pelargoniums. This requirement may be satisfied by using the very high output (VHO) fluorescent tubes. Cool white light, though poor in orange-red quality, provides excellent conditions for foliage plants to develop rich colors, branch more, and have a slow rate of stem elongation, resulting in fuller and more attractive plants. Unlike incandescent lights, which burn out suddenly, fluorescent lights age and lose intensity slowly. They have to be replaced after about four months of use.

Skylight

A skylight is not a light fixture but an architectural design strategy to allow more natural light to reach the interior of a room through the roof (figure 10-13). For best results, the shell covering the opening in the roof should be constructed out of a material with *translucent* (not transparent) glazing. Translucent material allows the incoming solar radiation to be better distributed over a larger area without hot spots.

Caution: Even though light is very important for plant growth, it is better to provide too little than too much light. The danger of overexposure to light is greatest in summer. Note that when you place a plant in a window, only one-half of it, at best, receives full sunlight. Intense light may bleach or scorch the foliage of plants. Glass in a window is a filter of light pre-

FIGURE 10-13 A skylight provides an additional source of natural light in a room.

venting most of the ultraviolet rays from reaching the plant. When growing sensitive plants, one should be aware of their needs. As already indicated, variegated plants (e.g., *Hedera helix*) cease to variegate but instead produce dark-green leaves under light intensity lower than optimum. Other light-related disorders are discussed in chapter 3.

10.3.5 TEMPERATURE

Houseplants generally prefer temperatures of between 18 and 24°C (64.4 and 75.2°F) for good growth and development. This condition often prevails in the average home in temperate climates. For most foliage ornamentals, a room night temperature of 21°C (70°F) is satisfactory, while growth is stalled at temperatures of 15°C (59°F). Flowering houseplants do well at 15°C (59°F) night temperatures. Even though plants may tolerate less-than-optimum temperatures above or below (10 to 30°C [50 to 86°F]), the danger to houseplants lies in the fluctuations in temperature. Night outdoor temperature may drop below freezing (0°C or 32°F) whereas indoor temperature may be 28°C (82°F) or higher. A change in temperature of more than 20°C (36°F) is detrimental to houseplants. As such, plants that are positioned close to windows or on windowsills run the risk of exposure to drastic temperature changes (warm inside and freezing outside) and may die as a result. Summer temperatures of 26 to 32°C (78.8 to 89.6°F) are tolerable for most indoor plants, provided the humidity is maintained at a high level. It is advisable to invest in a thermometer, preferably a maximum-minimum type (which helps to determine the temperature fluctuations in the room) (figure 10-14).

Some caution in the care of houseplants can reduce the risk of loss to adverse temperature:

1. Keep humidity high in winter by periodically spraying plants with mist sprayers to keep microclimates humid.

2. Do not place plants in the path of drafts (from air conditioners or heaters). Some plants may benefit from the additional warmth from a radiator as long as it is not direct and the humidity level is high.

3. Windows and doors close to plants should be airtight to prevent unsuspected cold drafts in winter.

FIGURE 10-14   A minimum-maximum thermometer.

4. If plants must be placed close to windows, they should be protected by having a storm window installed to prevent cold chills at night.

5. When curtains are drawn at night during winter, be sure that the plant on the windowsill is inside of the curtain.

6. If you have flowers in the kitchen, avoid placing them near sources of heat (e.g., stoves and refrigerator tops).

7. Plants directly facing doorways are subject to crosscurrent air.

10.3.6 HUMIDITY

Humidity and temperature work together. Humidity, the relative amount of water vapor in the air, is measured by using a *hygrometer,* an instrument recommended for homes with plants (figure 10-15). A high humidity level is uncomfortable for humans. Generally, a relative humidity (RH) of 60 percent is satisfactory for most houseplants. Many plants experience stress when RH is below 40 percent. Under such conditions, drying of leaf tips occurs in plants such as palms. Indoor RH is seasonal, being lower in winter than at other times. Plants with thick leaves are often able to tolerate low levels of humidity in the air, unlike those with thinner leaves.

10.3.7 PROVIDING SUPPLEMENTARY HUMIDITY FOR PLANTS

In winter, the use of heaters tends to cause indoor air to be dry. Dry air encourages excessive evaporation from surfaces. Humidity indoors can be increased by several methods:

FIGURE 10-15   A hygrometer.

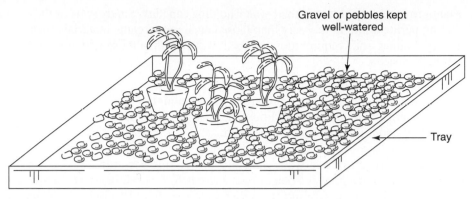

FIGURE 10-16 A pebble tray may be used to keep the plants' microenvironment humid.

1. *General humidification.* Provide additional humidity for the whole room by using a domestic humidifier.

2. *Localized humidification*
   a. *Mist spraying.* Periodically, plants can be misted with water, but this practice provides only short-term effects.
   b. *Pebble tray.* Potted plants can be placed on pebbles in a tray (figure 10-16) and the tray watered (but not above the pebbles). This approach provides a more continuous humid microclimate. Pebble trays may be used for one pot or a group of several pots. Alternatively, a pot may be set on a wooden block placed in a saucer and treated like the pebble tray setup.

3. *Enclosed chamber.* For plants that are very sensitive to low humidity, enclosed glass or plastic chambers or bottles may be used to hold potted plants or grow such plants (as is done in a terrarium [Chapter 25]).

4. *Move plants.* Certain parts of the home have higher humidity levels than others (e.g., bathroom and kitchen). Plants may be moved close to these humid sections during the winter months.

Caution: Relative humidity and temperature work together. If the temperature is high, RH will be low. On the contrary, when RH is high and the temperature in the house drops, water will bead on the leaves. These droplets of water provide a humid environment for maintenance of disease organisms.

10.3.8 WATER

Water plays a very important role in plant nutrition. One of the most common problems in houseplant culture is overwatering. Plant species differ in their moisture requirements. Several factors determine the water needs of houseplants:

1. The room condition (which varies with the season) determines how plants must be watered. Warm or hot environments cause plants to lose more water than cooler environments.

2. Actively growing plants use more water than dormant plants.

3. Plants with thin leaves and larger leaf surfaces transpire more and thus need more water than other types of plants.

4. The container material plays a role in the moisture needs of plants. Plastic, styrofoam, and glazed containers retain more moisture in the growing medium than clay pots or unglazed containers. Plants grown in unglazed pots require frequent watering.

5. Plant growth media differ in their water-holding capacity. Sandy soils or those containing perlite drain more freely than those containing organic materials such as peat. Freely draining soils require more frequent watering than those with good water-holding capacity.

6. The size of the container in relation to plant size is also critical. When a large plant is grown in a small pot that can hold only a small amount of water at a time, more frequent watering is required. In addition, the roots of large plants in small pots become pot-bound, requiring repotting of the plant.

The greatest danger of overwatering is due to watering according to a set schedule. It is safest to always determine that a plant needs water before providing it. If a week after watering the soil is still reasonably moist, further watering should be delayed. Symptoms of overwatering are wilting (in the presence of abundant moisture), yellowing of leaves, and rotting; symptoms of lack of moisture include drooping of leaves and wilting. Although these symptoms are stress alerts, it is best to avoid them. Plants cannot be revived after a certain stage of wilting, especially in the case of plants with thick leaves. Sometimes, when the plants are revived, portions of the leaves (the edges) may be permanently scorched, leading to disfigured leaves and decreased aesthetic appeal. Physiologically, alternate between drought and adequate moisture conditions (just like repeated freezing and thawing of food) offsets developmental processes in the plant including reproductive processes and may cause tissue death.

Many different types of equipment are available for measuring soil moisture. A moisture meter may be purchased by the avid gardener. For most people, a simple moisture indicator that changes color based on the dryness of the soil is satisfactory. There are other ways in which soil moisture can be determined without using instruments, although one of them is not soil surface dryness. Most plants are overwatered because growers look at the soil surface without knowing what is going on beneath it and decide that watering is required. It is best to stick a finger into the potting medium to a depth of about 1 inch or more (or use a piece of stick) to determine the stickiness or moisture level of the soil.

10.3.9 HOW MUCH WATER TO PROVIDE

Plants have different water needs. It is a waste of resources and a danger to plants to supply more water than is needed. For a particular plant, the amounts applied may be varied depending on the growth phase. In terms of quantity applied, plants may be watered in these general ways:

1. *Plentifully (or liberally).* In watering liberally or plentifully, the potting medium is kept constantly moist. When water is needed, plants may be drenched with water until the medium can hold no more. At this stage, excess water collects in the drip tray or saucer. This excess water should be discarded. Drenching can be done from the top or by placing the pot in a container of water and allowing it to soak up water until it can take in no more. The pot is then removed from the water.

2. *Moderately.* When water is applied in moderate amounts, only a small amount of excess water drains into the saucer. If a grower is using the soaking method, only a little water should be added to the container at a time. Water is added continually until the surface of the soil is moist. The plant is watered again when the soil feels slightly dry.

3. *Sparingly.* Watering sparingly keeps the growing medium only partially moist. Water never drains out of the pot. The plant is rewatered when most of the soil is dry.

Caution: These three general watering regimes are not alternatives. The method chosen for a situation depends on the specific needs of the plant, the conditions under which it is growing, and its growth phase.

10.3.10 ROLE OF PLANT GROWTH CYCLE IN WATER NEEDS

The plant growth cycle is often overlooked in the management of houseplants. Deciduous plants have a visible and predictable alteration between active growth (spring to fall) and rest period (winter), because they shed their leaves in the cold season. Without leaves, it is not difficult to guess that the plant does not need as much nutrition (water and minerals) as it does when it has leaves. Many bulbs and corms also have periods of rest during which their above-ground portions die back. However, many plants also have rest periods in their biological clocks that are less obvious. These species do not exhibit any dramatic signals to prompt the grower to make the necessary adjustments in management practices. These evergreen species, like most foliage houseplants, retain their foliage year-round. Some horticulturalists recommend that many indoor plants be forced to rest in winter when daylight is reduced. Induced resting may be accomplished by reducing the amount of water supplied and discontinuing fertilizer application. The best approach, however, is to consult the growing instructions supplied with plants or seeds purchased from a nursery.

10.3.11 METHODS OF APPLYING WATER

Water may be applied to plants by using any convenient container such as a cup. However, it is advisable and most convenient to use a watering can to water houseplants. Very inexpensive, lightweight plastic watering cans with long and thin spouts may be purchased from supermarkets. A long spout enables watering without spilling and splashing onto leaves. It also increases the maneuverability of the operator so that plants positioned in hard-to-reach places are readily watered. For homes with glasshouses, garden rooms, atriums, or a large collection of plants, a watering hose with an on/off control switch at the nozzle may be more convenient. There are two basic methods of watering houseplants:

1. *Water soil directly from above.* In this method, care is taken not to splash water on the leaves (figure 10-17). The soil level in the pot should permit a good amount of water to collect on top without spilling over.

2. *Water soil from below.* Potted plants may be placed in a saucer or container into which water is poured (figure 10-18). Water is then slowly absorbed through the drainage holes in the bottom of the pot.

As much as possible, plants should be watered carefully to avoid wetting the plant leaves. If the water is hard, it leaves unsightly marks on the leaves. Also, if fertilizers are applied through the irrigation water, marks from the salts are left on leaves if splashing occurs. Water on leaves may also create humid conditions in which disease-causing organisms thrive.

FIGURE 10-17 Watering a potted plant from above.

FIGURE 10-18 Watering a potted plant from below.

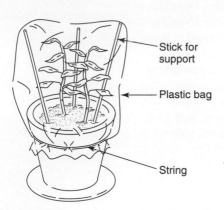

FIGURE 10-19 A potted plant enclosed in a plastic bag to maintain high humidity.

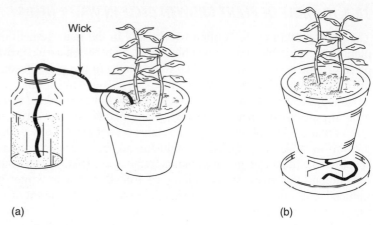

FIGURE 10-20 Watering potted plants by using a wick (a) inserted into the pot from above from a feeder bottle and (b) inserted through the drainage hole in the bottom of the pot.

Some plants, such as bromeliads, can tolerate water on the foliage. When watering plants, lukewarm water should be used. Tap water in some places can be very cold. The water should also be soft, since plants such as camellias are sensitive to salts in water.

When plants are going to be left unattended for extended periods, such as while a home owner is on vacation, creative methods of watering should be devised to avoid wilting of plants. For example, plants may be enclosed in a plastic bag to retain moisture (figure 10-19). A wick system may also be used (figure 10-20).

10.3.12 FERTILIZING HOUSEPLANTS

All plants need balanced nutrition to grow, develop properly, and produce well. The nutrients in the potting soil are gradually depleted as the plant grows and develops. Unlike field plants, whose fertilizer needs usually consist of only macronutrient elements (nitrogen, phosphorus, and potassium), potted plants need both macro- and micronutrients (3.4) as part of a fertilizer program. Both types of nutrients are required because whereas mineral soil has some native nutrients, greenhouse soilless mixes are often deficient in nutrients, especially micronutrients.

Fertilizers for indoor use are available in several forms—solids, liquids, powders, crystals, or granules. Powders and crystals, which are dissolved in water before being applied to the soil, are called *soluble fertilizers*. Fertilizer sticks and spikes are examples of solid fertilizers packaged in cylindrical shapes similar to pencils that are inserted into the soil for gradual release of their nutrients (figure 10-21). Such fertilizers may also be packaged as pills. Specially coated fertilizer granules are commonly used in feeding houseplants. These solids are called slow-release fertilizers (3.4.2). Finally, fertilizers may be applied in liquid form as foliar sprays (3.4.4) to plant leaves. While all plants grown in soil respond to fertilizers placed in the growing medium, foliar application is especially beneficial to plants such as epiphytes (e.g., bromeliads [chapter 24]), which absorb little nutrition through their roots. The concentration of foliar sprays should be carefully selected to avoid scorching the leaves.

Timing of feeding is essential in houseplant nutrition. It should be remembered that other growth factors (e.g., light, temperature, and water) must be adequately supplied for fertilizers to be effective. Some home growers misdiagnose plant problems, thinking a poorly growing plant is starving when it actually is overwatered or not receiving adequate light or warmth. As previously mentioned, fertilization should take into account the growth phases of the plant. When a plant is growing actively, it needs more nutrition; when dormant, nutrients should be reduced or eliminated altogether.

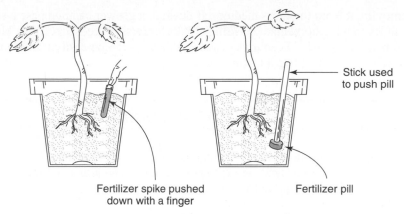

Stick used
to push pill

Fertilizer spike pushed
down with a finger

Fertilizer pill

FIGURE 10-21 Using fertilizer spikes and pills in fertilizing potted plants.

10.4: POTTING MEDIA

Potting media are discussed more fully in chapter 3. Potting mixtures differ widely in constitution. Some commercial mixes are designed for specific purposes (e.g., mixes for use in the terrarium or for seed germination) and others for general purposes. It is critical that a potting medium be sterilized to kill pathogens. The mixes may be soil based (i.e., they include natural soil) or soilless (containing no natural soil ingredients). The grower can sterilize his or her own homemade medium by placing an aluminum-covered tray of the mix in a conventional oven and baking at 82°C (179.6°F) for about an hour. Such an undertaking can be very messy. Soilless mixes are lighter in weight and easier to handle than soil-based mixes, but they lack nutrients. When used for potting top-heavy plants, the plants can be toppled easily. A simple homemade recipe for a mix may consist of the following:

1. One part sterilized soil plus one part medium-grade peat moss plus one part fine perlite
2. One part coarse peat moss plus one part medium-grade vermiculite plus one part medium-grade perlite

These mixes should be supplemented with balanced fertilizer. A grower may also find it convenient to use a compressed peat moss disk for starting certain seeds (figure 10-22). The advantage of this method of raising seedlings is that the seedling can be directly transplanted

> **Peat Pellets**
> *Compressed peat moss disks used to start seeds or root cuttings and planted with the seedlings.*

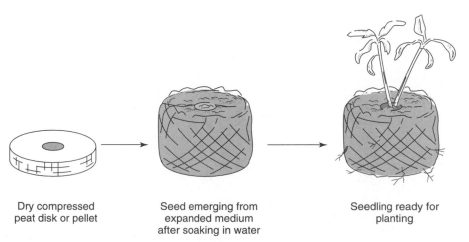

Dry compressed
peat disk or pellet

Seed emerging from
expanded medium
after soaking in water

Seedling ready for
planting

FIGURE 10-22 Seedlings can be raised in compressed peat disks.

without removing it from the moss medium. If desired, a grower can purchase a simple mechanical unit for making homemade seedling blocks for raising seedlings. These blocks work similarly to peat moss disks. Seedlings raised in this way can be easily transplanted with a ball of soil, which increases transplanting success.

10.5: POTTING PLANTS

After the appropriate potting mixture has been determined and purchased or prepared, and the right container chosen, the next step is to pot the plant. The drainage hole or holes in the bottom of the pot should be partially covered with pieces of broken pot (crock) or small stones. Care should be taken not to plug the hole (figure 10-23). Potting mix is then added to the container. A hole is made in the soil to receive the plant roots and then patted firm to keep the plant erect. Additional soil, if needed, should be added, but at least 1/2 to 1 inch (1.3 to 2.54 centimeters) of space should be left at the top of the soil to hold water during watering (sometimes called the *headspace* of the container). It should also be noted that the soil will settle with deep watering.

10.6: REPOTTING

10.6.1 WHEN NEEDED

Repotting is actually transplanting from one pot to another. Sometimes plants are repotted because a grower desires to change containers for cosmetic or aesthetic reasons. However, there are certain times during the growth of a plant when it needs to be repotted for better growth and development. One such occasion and perhaps the chief reason for repotting is when a plant grows too large for its container. When this happens, the roots become *pot-bound* (roots grow over each other and around the bottom of the pot, forming a ball) (figure 10-24). Many roots at this stage are not in contact with soil and thus not aiding in nutrient or water absorption. Pot-bound plants grow slowly even with good nutrition and watering. Roots often grow through the drainage holes in the bottom of the container. To find out whether a plant is pot-bound, it should be removed, along with the soil, by positioning its stem between two fingers

> **Pot-bound**
> *The growing of roots over each other and around the bottom of the pot as a result of limited space.*

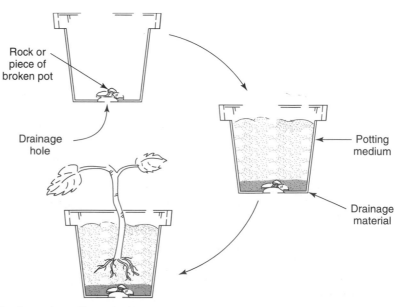

FIGURE 10-23 Steps in potting plants.

FIGURE 10-24 A potted plant showing pot-bound roots.

of one hand and turning the pot over into the palm. Gently tapping on the bottom of the pot or tapping the pot against the edge of a table can help to dislodge the soil. In the case of larger plants, a small block of wood may be used to tap the side of the pot while it is lying on its side. A sharp edge (e.g., the blade of a knife) may be run around the inside wall of the pot to break loose any attachment to the walls. Watering plants before repotting is recommended to aid in their removal from the pot.

10.6.2 CHOOSING A POT OR CONTAINER

The plant height (size) in relation to pot size is important for aesthetic reasons. The grown plant height should be about two times the pot height. An oversized pot not only wastes soil mix but also increases the risk of overwatering to the detriment of the small plant. Pots come in different shapes, sizes, and materials. The strategy of repotting involves changing to a bigger pot size in a stepwise fashion.

Old pots should be cleaned and disinfected (use household bleach or germicidal soap) before reuse. When cleaning glazed clay pots, they should be soaked in water for several minutes to remove all air bubbles in the clay. When dry clay pots are used for repotting, the clay tends to absorb moisture rapidly from the soil.

Containers come in all shapes and sizes. In effect, any receptacle may be used to grow plants, provided adequate provision is made for drainage and the container is convenient to use. Pots are also chosen to complement the design of the room.

Container Materials

There are two basic materials used in making horticultural pots. Each has advantages and disadvantages.

Clay *Clay* (or earthenware) used to be the industry standard for pots but has been replaced with newly developed material. Clay, being a natural material, "breathes," or is porous, allowing water to evaporate from its surface. As such, it reduces the danger of waterlogging from overwatering. Clay pots are heavier, more sturdy, and able to support large plants without toppling over (figure 10-25). However, clay pots are also bulky to handle and breakable,

FIGURE 10-25 Clay pots.

requiring care in handling. Because it is capable of absorbing mineral salts and water from the medium, the surfaces of clay pots often show unsightly whitish marks from salt deposits. During repotting or topdressing, these marks should be scrubbed off by using household bleach and then rinsed in vinegar. Such marks are not associated with glazed clay pots.

Pot-bound root growth occurs more rapidly in plants grown in clay pots than those in plastic pots because the porosity of clay allows air to reach plant roots more readily than those in plastic (where air is obtained from the open top only). Roots thus tend to grow rapidly toward the wall of the pot where, upon meeting the obstacle, they begin to circle around on the surface of the wall.

Plastic   *Plastic pots* are very popular today and are available in a wide range of colors, thickness, durability, shapes, and sizes (figure 10-26). Plastic pots are generally less breakable than clay pots. Even though molded polystyrene pots are used in some situations, hard plastics are most common because they are lightweight and easy to handle. Plant roots grow more evenly in a plastic container. One problem with this synthetic material is that it is not porous and thus plants grown in plastic pots are prone to waterlogging because they lack the ability to absorb moisture and lose it through evaporation. Although overwatering plastic-potted plants may be a problem in winter when drying is slow, plastic pots are advantageous in summer when water stress is most common.

Plastic pots are intolerant of the high temperatures needed for sterilization and thus are best sterilized by using chemicals. Pots may be soaked for about 10 minutes in a commercial disinfectant solution (e.g., Floralife or Green-Shield), rinsed in water, and then air dried. This chemical sterilization is not as effective as steam sterilization. For better results, plastic pots should be washed to remove all dirt before being treated with chemicals. Plastic pots are also readily toppled when plants grow larger.

Other Materials   Apart from clay and plastic, another synthetic material called *styrofoam* is used in the horticultural industry for plant culture. This material is known for its insulation quality and hence is used especially when keeping the soil warm is a priority.

Shapes and Sizes

The shape of a pot is largely a matter of personal preference. Most pots have a circular lateral cross section, but other shapes occur (figure 10-27). Sometimes rectangular pots are used to grow mass displays of plants that hang from places such as the balconies of apartments.

In terms of size, there are four general types of pots:

1.   *Standard pots.* Standard pots vary in size but are characterized by having a height that equals the pot width. This shape is not suitable for growing tall plants since such pots are then prone to tipping over.

Standard Pots
A pot design in which the height equals the width.

FIGURE 10-26   Plastic pots.

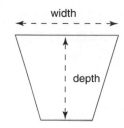

width

depth

Typical pot: depth = width
Range = 1 1/2 – 15 inches (3.81–38.1 cm)

Azalea pot   Standard pot   Square pot

FIGURE 10-27   Pot shapes and sizes.

2.   *Azalea pots.* Many flowering plants are grown in azalea pots. They are more stable than standard pots because they stand three-fourths as high as they are wide.

3.   *Bulb pots.* Bulb pots are also called *half pots* because they stand half as high as they are wide. They are widely used in the propagation of plants. They may be used to grow plants with shallow roots to maturity or plants that may be displayed in mass form (e.g., zebrina, tradescantia, daffodil, hyacinth, and tulip).

4.   *Rose pots.* Rose pots are one and a half times as tall as they are wide and are used for growing deeply rooted plants.

Within each type, pots are referred to by their height (e.g., 6- or 9-inch [15.2- or 22.9-centimeter] pots). Nurseries that produce bedding plants and vegetable seedlings for sale often grow these in small peat pots or plastic containers in packs of 6 or 12 plants (*community packs*) (figure 10-28).

Some plants may be repotted or shifted several times before they are settled in a more permanent container. When plants grow too large for the standard sizes of pots, they may be transplanted to larger containers called *tubs* that may be made out of wood or plastic (figure 10-29). Even though changing pot sizes as plants grow may be an unpleasant chore, it is safer to "pot up," progressively increasing the pot size as the plant increases in size, than to grow

> **Azalea Pot**
> *A pot design in which the height equals three-fourths of the width.*

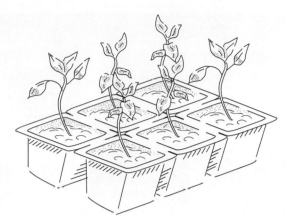

FIGURE 10-28 A community pack of seedlings.

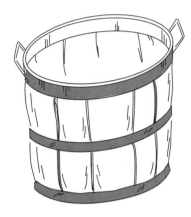

One-bushel wooden container Large plastic container

FIGURE 10-29 Large containers or tubs.

plants in pots that are too small. A small plant in a large pot not only looks awkward but also is prone to overwatering and possibly death. Further, it is a waste of the potting medium.

10.6.3 REPOTTING PROCESS

In repotting, the drainage hole in the container should be covered, as described earlier. As insurance against overwatering, the bottoms of pots with drip pans or those placed in saucers should be lined with gravel or pebbles. The potting mix should be slightly moistened before use. First, some soil is placed at the bottom of the pot. The plant is then removed, along with a ball of soil from the old pot, and all pot-bound roots straightened out before setting on the moist soil in the receiving pot. The remaining space is filled with more of the fresh potting medium (figure 10-30). In some cases, adding more soil to fill up the empty spaces may be cumbersome. In such situations, the old pot may be placed in the new one and the space around it filled to create a mold into which the ball of soil from the transplant will fit. The plant should be placed no deeper than the depth used in the previous pot.

10.6.4 TOPDRESSING

> **Topdressing**
> Applying fertilizer to the surface of the soil while the plants are growing.

The practice of topdressing in potted horticultural plants involves providing plants with fresh growing media without transplanting when the biggest pot size desired or available has been used. It entails scraping off about 1 to 2 inches (2.54 to 5.1 centimeters) of topsoil and replacing with a fresh soil mix. Where it has been determined that roots are pot-bound, they should be pruned to remove the excess roots before topping with fresh soil. Pruning and topdressing should be done with care to avoid damaging the plant.

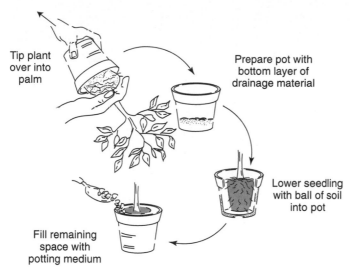

Tip plant over into palm

Prepare pot with bottom layer of drainage material

Lower seedling with ball of soil into pot

Fill remaining space with potting medium

FIGURE 10-30 Steps in repotting large plants.

10.7: PROVIDING SUPPORT

Many flowering houseplants are freestanding and self-supporting. However, plants sometimes need additional support when they produce heavy flower heads that cause the stem to bend (e.g., cineraria and tomato). Other plants (e.g., fatshedera) may have slender stems that need additional support. Another instance where houseplants may need reinforcement in their support system is when they grow large but have brittle stems (e.g., impatiens).

In all of these instances, additional support may be provided by tying the weak stem to a stake with a string. Bamboo provides strong stakes. It may be split into smaller pieces and remain capable of providing good support. Whenever necessary, several stakes may be used to support a single plant in a pot. To keep the aesthetic value, stakes and twine used for tying should be positioned strategically and all loose ends neatly removed. Knots should not be tied too tightly around the stem. The idea is to provide additional support, not to bunch branches together.

Climbing plants (e.g., ivy, depladenia, and hoya), are known to produce vines. The grower should decide whether the vines will be allowed to hang freely or whether the plant will be encouraged to climb on some support (e.g., a wall or pole). The grower may purchase wire frames designed in a wide variety of shapes for *training* vine plants (e.g., philodendron and cissus) or create his or her own support frame (figure 10-31). When plants are trained on supports, they must be tied, as when using stakes.

10.8: DISEASES AND PESTS OF HOUSEPLANTS

When it comes to houseplant diseases and pests, the key is to prevent their occurrence in the first place. Since plants cohabit with humans in small and enclosed spaces, diseases and insect pests should not be allowed to infest the environment. It is most undesirable to be confronted with a situation in which pesticides, which are toxic to humans as well as the pests they control, must be used in the house.

Diseases and pests of houseplants are often traceable to improper growing conditions (e.g., poor lighting, improper temperature, insufficient or excessive moisture, poor nutrition, still air, and high humidity). As such, before a home owner scrambles to implement disease-control measures, it is best to first check whether plants are receiving proper care. Maintenance of strict phytosanitary conditions in the home is important. Where possible, leaves should be cleaned to remove dust, grease, and environmental pollutants that settle on them. Plants may

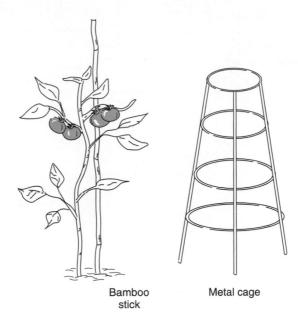

Bamboo
stick

Metal cage

FIGURE 10-31 Wire supports for plants.

be cleaned with plain or soapy water, and care should be taken not to leave drops of water on parts of the plant after washing. Diseases and pests of houseplants are discussed more fully in chapter 7.

10.9: COMMON SYMPTOMS OF ILL HEALTH IN HOUSEPLANTS

Houseplants under stress from inadequate or improper levels of nutrition and other growth factors may exhibit one or more of the following common symptoms associated with ill health in plants.

10.9.1 YELLOWING OF LEAVES (CHLOROSIS)

Yellowing of leaves is usually a sign that plants are in distress. When older leaves yellow, it could be due to nitrogen deficiency. However, any disease that interrupts the flow of nutrients may cause leaves to yellow. When young leaves yellow, the plants may be under stress from poor drainage. When both young and old leaves yellow, the cause could be poor lighting or drought (underwatering). Chlorosis occurring with leaf drop could signal cold temperatures or poor aeration due to waterlogging.

10.9.2 STUNTED GROWTH

Whenever plants show stunted growth with small leaves and poor color, the condition could be due to one or a combination of the following: moisture stress, improper temperature, roots that are pot-bound, and restricted growth.

10.9.3 FOLIAR BURNS

Burns on leaves—when they occur as patches without any definite pattern—could be caused by fungal pathogens or could be the result of burns from foliar application of an insecticide. When plants have experienced severe drought, they may recover but the tissues at the edge may die, especially in young leaves. When the edge burn occurs in older leaves, it could be caused by excessive fertilization, leading to accumulation of salts in the soil. Other symptoms and possible causes are described in table 10-5.

Table 10-5 Common Problems of Houseplants and Their Management

Water-Related Problems

Symptom:	Plant wilting; wilting is intensified with continued irrigation
Possible cause:	Medium is saturated or drainage is poor
Suggested action:	Repot properly for improved drainage; reduce frequency of watering
Symptom:	Plant wilting but recovers with watering
Possible cause:	Medium was left to dry for too long; plant may be pot-bound
Suggested action:	Increase watering frequency; repot into a larger pot
Symptom:	Leaf edges or tips appear scorched
Possible cause:	Insufficient or infrequent watering; medium left dry for a long period
Suggested action:	Water more frequently and more thoroughly each time
Symptom:	Plant is stunted in growth; leaves are of reduced size
Possible cause:	Insufficient watering frequency and quantity of water applied
Suggested action:	Water plant at proper frequency and apply adequate water each time

Light-Related Problems

Symptom:	Etiolated plant growth
Possible cause:	Insufficient light intensity
Suggested action:	Relocate plant into brighter area of room or provide artificial supplemental lighting
Symptom:	Chlorosis (especially in older plants)
Possible cause:	Insufficient lighting
Suggested action:	Relocate to brighter part of room
Symptom:	Bleaching or discoloration of leaves
Possible cause:	Plant is intolerant of direct sunlight
Suggested action:	Relocate to part of room with lower light condition
Symptom:	Flowering is poor or drastically reduced
Possible cause:	Low light intensity
Suggested action:	Relocate to area with proper light condition
Symptom:	Leaf abscission accompanied by chlorosis
Possible cause:	High light intensity
Suggested action:	Relocate plant to lower light area

Nutrition-Related Problems

Symptom:	Uniform chlorosis of leaves on the whole plant
Possible cause:	Nitrogen deficiency
Suggested action:	Apply or increase rate of application of nitrogen fertilizer
Symptom:	Stunted growth of plant
Possible cause:	Inadequate fertilization
Suggested action:	Apply fertilizers
Symptom:	Leaf tips and edges brown or scorched
Possible cause:	Excessive fertilization
Suggested action:	Flush excess salts with large quantities of water; adjust fertilizer rate

Temperature-Related Problems

Symptom:	Slow growth
Possible cause:	Low temperature
Suggested action:	Place in warmer area of room
Symptom:	Leaf abscission
Possible cause:	Drop in temperature
Suggested action:	Protect plant from cold, chilling temperature
Symptom:	Chlorosis with abscission
Possible cause:	High temperature
Suggested action:	Lower temperature; do not locate near heating system

Pathogenic Problems

Symptom:	Gray, fluffy mold
Possible cause:	Gray mold attack due to high humidity
Suggested action:	Reduce residual moisture on leaf; mist spray only lightly; aerate; remove affected parts
Symptom:	Leaf spot
Possible cause:	Fungal or bacterial infection due to high humidity or residual moisture on foliage
Suggested action:	Do not mist spray foliage; keep leaves dry by watering media without wetting foliage; remove affected leaves
Symptom:	Blackening and rotting at the base of the stem
Possible cause:	Black stem rot from gray mold attack; results from overwatering or poor drainage of potting medium
Suggested action:	Remove plant and examine root; if rot is extensive, discard; otherwise treat with fungicide and repot in well-draining medium
Symptom:	Stunted growth; mottling of leaves; distortion of leaves
Possible cause:	Viral infection; due to attack of sucking insects (e.g., aphids)
Suggested action:	Destroy infected plant
Symptom:	Healthy-looking plant starts to wilt in spite of watering and fertilizing; roots are knotted
Possible cause:	Root-knot nematode attack
Suggested action:	Destroy plant and discard potting medium; use sterilized medium for potting plants

Pest Problems

Symptom:	A fine web coating on leaf undersides
Possible cause:	Attack of spider mites; favored under hot, dry air conditions
Suggested action:	Mist plant; cut affected part and discard
Symptom:	Holes in leaves
Possible cause:	Attack of chewing insects (e.g., caterpillar and earwig)
Suggested action:	Remove pest and destroy
Symptom:	Rolling of leaves
Possible cause:	Leaf roller attack
Suggested action:	Locate and destroy moths

10.10: COMMON HOUSEPLANTS

A select number of houseplants are presented in figure 10-32, a-z on the following pages.

(a) Common name: Croton
Scientific name: *Codiaeum variegatum*
Care: Water—keep soil thoroughly moist.
    Light—keep in bright light (partial sun) to
    maintain color.
Comments: It is fairly easy to grow but needs attention
to keep in good condition. Avoiding water stress is
critical. Low light causes poor leaf color.

(b) Common name: Wandering Jew
Scientific name: *Zebrina pendula*
Care: Water—keep soil moist.
    Light—place in indirect sunlight.
Comments: It is easy to grow; makes an excellent
hanging basket plant.

(c) Common name: Jade plant
Scientific name: *Crassula argentea*
Care: Water—drench and keep dry between waterings.
    Light—place in partial light.
Comments: It is easy to grow and slow growing.

(d) Common name: Banana
Scientific name: *Musa acuminata*
Care: Water—keep soil moist.
    Light—place in full sun.
Comments: It is difficult to grow and maintain in good
condition. Plant must be displayed outdoors in summer
to take advantage of the sun. It will produce fruits and
then die after about six to seven years. It can be
replanted from new shoots that develop from the base of
the old stem.

(e) Common name: Snake plant; mother-in-law's tongue
Scientific name: *Sanservieria trifasciata*
Care: Water—drench and allow to dry between
    waterings.
    Light—can tolerate direct or indirect light.
Comments: Easy to grow and care for. It is difficult to
kill through neglect. Can survive very low light
intensities; nematodes possible pests.

(f) Common name: Variegated peperomia
Scientific name: *Peperomia obtusifolia* 'Variegata'
Care: Water—water as needed, but keep soil dry between
    waterings.
    Light—place in bright, partial light.
Comments: It is easy to grow. Cultivars abound. Low
lighting induces etiolated growth that reduces visual appeal.

(g) Common name: Aparagus fern
Scientific name: *Asparagus densiflorus* 'Sprengeri'
Care: Water—drench and allow to dry between waterings.
    Light—display in indirect light.
Comments: It is not difficult to grow, but improper
watering or exposure to poor light causes leaf yellowing
and drop of the needlelike foliage. It is frequently
displayed in hanging baskets.

(h) Common name: Geranium
Scientific name: *Pelargonium x hortorum*
Care: Water—drench and keep fairly dry between waterings.
    Light:—place in direct sunlight.
Comments: It is easy to grow and commonly displayed as
a potted plant.

(i) Common name: Cactus
Scientific name: *Cerus* spp.
Care: Water—keep soil dry.
    Light—keep in full sunlight.
Comments: A wide variety of species occur with wide
variation in size, shape, and other features. Cacti are easy
to grow and slow growing. Desert cacti are adapted to
dry conditions.

(j) Common name: Boston fern
Scientific name: *Nephrolepis exaltata*
Care: Water—keep soil moist; mist frequently.
    Light—keep in indirect (partial) sunlight.
Comments: It is easy to grow and commonly displayed
in hanging baskets. Avoid excessive touching of the
frond, which induces discoloration (browning). Low
lighting causes yellowing of fronds.

(k) Common name: Prayer plant
Scientific name: *Maranta leuconeura*
Care: Water—keep soil moist.
        Light—place in partial sunlight.
Comments: This plant is not recommended for the
novice. It is fairly difficult to grow and maintain in good
condition. It can be displayed in a hanging basket or pot.

(m) Common name: Corn plant
Scientific name: *Dracaena fragrans*
Care: Water—keep soil thoroughly moist.
        Light—adapted to low light levels.
Comments: Several canes are usually planted in each pot.

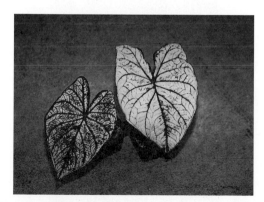

(l) Common name: Caladium
Scientific name: *Caladium* spp.
Care: Water—keep soil fairly moist and well drained.
        Light—prefers moderate sunlight.
Comments: The plant is easy to maintain when growing
but requires a rest period annually. The bulblike tuber
should be removed when the shoot withers, stored in a
cool, dry place, and repotted after about three months of
storage. The plant is available in a variety of leaf color
patterns.

(n) Common name: Pothos
Scientific name: *Epipremnum aureum*
Care: Water—drench and let dry between waterings.
        Light—adapted to low light.
Comments: It is easy to grow and commonly displayed
in hanging baskets. Cultivars vary in leaf variegation and
general color. Overwatering causes lower leaves to
yellow and drop.

(o) Common name: Ponytail palm
Scientific name: *Beaucarnea recurvata*
Care: Water—keep dry between waterings.
   Light—place in full sunlight or partial shade.
Comments: Plant produces a bulblike base from which long, narrow, and cascading leaves arise.

(q) Common name: Canary date palm
Scientific name: *Phoenix canariensis*
Care: Water—water sparingly.
   Light—place in direct sunlight.
Comments: Very hardy plant.

(p) Common name: Spider plant
Scientific name: *Chlorophyatum comosum*
Care: Water—keep soil thoroughly moist.
   Light—place in partial light.
Comments: It is easy to grow. Display on pedestal or in hanging basket to allow hanging room for new plants that develop at the shoot tips. Plant will flower under short photoperiod. It is sensitive to fluoride, which is present in some municipal water supplies.

(r) Common name: Rubber tree
Scientific Name: *Ficus elasticus*
Care: Water—needs a good watering regime; soil should be only slightly dry.
   Light—requires good lighting, medium to full sun.
Comments: Easy to grow and care for. Wipe dust off leaves to keep shinny. Underwatering can cause leaf drop. Disease problems are not significant.

(s) Common name: Swiss cheese plant
Scientific name: *Monstera deliciosa*
Care: Water—water sparingly, and allow to dry between
     waterings.
     Light—place in bright, filtered light.
Comments: Young leaves may be unbroken. The plant
produces aerial roots and can be made to grow on a moss
pole.

(u) Common name: English ivy
Scientific name: *Hedera helix*
Care: Water—drench and keep moderately dry between
     waterings.
     Light—place in partial sunlight.
Comments: It is easy to grow and can be displayed in a
hanging basket or trained on a pole.

(t) Common name: Dumbcane
Scientific name: *Dieffenbachia* spp.
Care: Water—drench and allow soil to dry between
     waterings.
     Light—filtered light preferred; avoid direct sun.
Comments: Easy to grow and care for. Sap contains
calcium oxalate, which causes inflammation of the skin.
If ingested, it can cause temporary loss of speech.
Possible pests include mealybug and mosaic virus.

(v) Common name: Umbrella plant
Scientific name: *Schefflera digitata*
Care: Water—drench and keep moderately dry between
     waterings.
     Light—place in partial sunlight.
Comments: It is easy to grow and can attain a large size.

(w) Common name: Hen and chickens
Scientific name: *Escheveria peacockii*
Care: Water—drench and keep dry between waterings.
    Light—prefers direct sunlight.
Comments: It is slow growing and easy to grow.

(y) Common name: Philodendron
Scientific name: *Philodendron selloum*
Care: Water—water moderately and allow soil surface to
    dry between waterings.
    Light—place in bright, filtered light.
Comments: Avoid direct light. This plant is nonclimbing.

(x) Common name: Pineapple
Scientific name: *Ananas cosmosus*
Care: Water—keep soil moist.
    Light—place in partial or full sunlight.
Comments: This is a tropical plant that bears edible
fruits. The leaf tip is sharp and pointed, and the edges are
prickly.

(z) Common name: Desert fan palms
Scientific name: *Washingtonia filifera*
Care: Water—water thoroughly during active growing
    period.
    Light—keep in sunny window.
Comments: Move outdoors during summer. Do not
damage thicker roots during repotting.

SUMMARY

Many plants can be grown indoors, provided the conditions for plant growth are adequately provided. Plants can be grown indoors to enhance the appearances of rooms. They can be strategically positioned to perform other functions such as dividing a room, directing traffic, and hiding unsightly areas. Indoor plant displays can be greatly enhanced by using ornate containers and placing them under a spotlight, for example. It is important that plants be brought indoors from the nursery in good condition. In winter and under severe weather conditions, plants need protection during transport from the nursery to the home. Various parts of the room in a house receive different amounts of light. As such, plants should be located carefully so that they receive the appropriate amount and quality of light. If supplemental lighting is required, artificial light can be provided by using a variety of light source types, such as incandescent light, florescent light, or mercury vapor light. Florescent light sources are most efficient and recommended for home use. Incandescent light generates too much heat in the process of providing light. A room temperature of between 18 and 24°C (64.4 and 75.2°F) is adequate for most plants.

Many houseplants die as a result of excessive watering. The containers and media used should permit free drainage. Plants should be watered only when they need it. They may be watered from below by placing the pot in water or from above by using a watering can. As much as possible, water should not be left on the leaves, because this condition invites disease organisms. Since nurseries frequently use soilless media, plants need to be fertilized at periodic intervals to ensure good growth; either solid or liquid fertilizer preparations may be used. Potted plants outgrow their containers with time and need to be repotted into larger containers. As plants grow larger, some may require artificial support to stand erect. With good care and management, indoor plants seldom become diseased.

REFERENCES AND SUGGESTED READING

Briggs, G. B., and C. L. Calvin. 1987. Indoor plants. New York: John Wiley & Sons.

Crockett, J. U. 1972. Foliage houseplants. New York: Time-Life.

Perkins, H. O., ed. 1975. Houseplants: A handbook. Brooklyn, N.Y.: Brooklyn Botanical Gardens.

Reader's Digest. 1990. Success with houseplants. New York: Reader's Digest Association.

Wright, M., ed. 1979. The complete indoor gardener. New York: Random House.

PRACTICAL EXPERIENCE

Purpose: To study the effect of varying the levels of plant environmental growth factors on the growth of plants.

Materials and methods: Raise tomato seedlings and transplant individual plants into 6-inch (15.2-centimeter) pots (or smaller). Select 12 pots with plants of equal size and group into sets of 3 pots. Place the pots in an area in the greenhouse where they will be equally exposed to the environmental conditions. Apply four levels of a growth factor (fertilizer, moisture, light, and temperature) to the 12 pots, each set of 3 pots receiving only one of the levels. One of the levels should be "zero level"—that is, it should have only the normal conditions of the environment (e.g., no fertilizer at room temperature). This is called the control. Suggested levels are 0, $1x$, $2x$, and $3x$ (where x is a unit of the factor being applied). Your instructor will guide you to make the right choices.

Measure (nondestructively) plant characteristics of interest (e.g., height) at weekly intervals. Your instructor will assist you in applying appropriate statistical procedures to summarize the data and plot them graphically. Interpret your results.

OUTCOMES ASSESSMENT

PART A

Please answer true (T) or false (F) for the following statements.

1. T F   The average light intensity in a house is about 55 lux.
2. T F   Greenhouses have higher light intensities than homes.
3. T F   North-facing windows do not receive direct sunlight.
4. T F   Container material can affect plant moisture needs.
5. T F   It is safest to water according to a fixed schedule.
6. T F   When watering houseplants, it is best to use cold water.
7. T F   Bulb pots are also called half pots.
8. T F   Pots are described by their height.
9. T F   Deciduous plants grow continuously.
10. T F   Overwatering can cause plants to wilt.

PART B

Please answer the following questions.

1. List three factors that determine what plants are chosen for indoor use.
   _____ _____ _____
2. Name any two specific factors for locating plants in the room.
   _____ _____
3. Plants generally need_____hours of light per day for proper growth and development.
4. _____ lights are the most efficient of all light sources and recommended for use at home.
5. What is a major disadvantage of incandescent light in providing supplemental lighting to houseplants? _____
6. The instrument used to measure humidity is called a _____.
7. A relative humidity of _____ is generally adequate for houseplants.
8. Fertilizer in liquid form may be applied to leaves as_____.
9. Rose pots are _____ times as tall as they are wide.
10. List 10 common houseplants.
   _____ _____ _____ _____ _____
   _____ _____ _____ _____ _____

PART C

---

1. List and discuss any three functional roles of houseplants.
2. Describe two specific ways in which plant display may be enhanced.
3. Describe three specific precautions to observe to prevent adverse temperatures from damaging houseplants.

4. Describe two ways in which relative supplemental humidity may be provided.
5. Discuss the role of the plant growth cycle in plant water needs.
6. Tap water may be injurious to certain plants. Explain.
7. Describe the composition of a good potting medium.
8. Compare and contrast clay and plastic as potting plant container materials.
9. What conditions could cause yellowing in houseplants?

CHAPTER 11

Controlled-Environment Horticulture

PURPOSE

This chapter is designed to show that plants may be grown outside of their natural environments by providing all of the growth requirements needed and to describe the facilities required for such a plant cultivation practice.

EXPECTED OUTCOMES

After studying this chapter, the student should be able to

1. Discuss different designs, construction materials, and locations of greenhouses.
2. Describe the methods of controlling indoor plant growth factors (light, temperature, moisture, nutrients, and air) for the benefit of crops.
3. Describe how greenhouses are used in the production of horticultural plants.
4. Compare and contrast greenhouse and field production of crops.

KEY TERMS

Bag culture	Growth regulators	Proportioner
Conduction	Hydroponics	Quonset
Convection	Incandescent	Radiation
Day neutral	Infiltration	Ramblers
Disbudding	Long day	Reinforced fiberglass
Everblooming roses	Nonrecycling system	Sand culture
Floating system	Nutrient-film technique	Short day
Floribundas	Photoperiod	Soilless media
Gericke's system	Pinching	Substrate culture
Grandifloras	Polyanthas	Thermostat
Greenhouse	Prefinished seedlings	Water culture

SECTION 1 CONTROLLED-ENVIRONMENT FACILITIES AND THEIR OPERATION

OVERVIEW

Plants, like other living organisms, have certain requirements for growth. Climatic conditions are not uniform throughout the world. As such, certain plants are grown or found in nature only in certain regions (i.e., plants are adapted to certain environments). Some plants are more restricted in their range of adaptation. The general principle in choosing plants is that if you desire to grow a plant outside of its region of natural adaptation, you must provide all of the necessary growth requirements (above- and below-ground conditions [chapter 3]) in the new growing environment. Sometimes supplementation of natural conditions such as the provision of additional light is necessary.

In tropical regions of the world where growing seasons are longer, growing flowers indoors is not as popular as growing them outdoors. Outdoor flower gardens can be enjoyed for longer periods of time. On the contrary, the growing season in temperate climates is much shorter. Flowers do not grow outdoors in the cold winter months, which may last more than six months in some areas. In such regions, people desire to grow plants indoors under artificial conditions.

To enjoy flowers or horticultural products out of season, plants must be grown in a *controlled environment,* meaning that humans, not nature, determine how the conditions change. Growers can create stable microclimates that are ideal for specific plants. Otherwise, flowers or crops can be produced elsewhere in due season and imported into an area where the plant is out of season. Although some of this import-export trade in flowers and other horticultural products occurs, the ready availability of controlled-environment structures or facilities called *greenhouses* has spawned an industry that produces off-season ornamentals and vegetables for local and distant markets. This chapter is devoted to the greenhouse industry and discusses its advantages, limitations, design, operation, and maintenance. The culture of plants in liquid media *(hydroponics),* perhaps the ultimate in controlled-environment production, is also discussed.

11.1: WHAT IS A GREENHOUSE?

A greenhouse is a specially constructed building for growing plants under controlled conditions. It is covered with a transparent material and as such permits entry of natural light. The building has no green color but perhaps gets its name from the fact that (green) plants are grown in it. Greenhouses differ in design, size, and the extent of environmental control. Some of the simplest ones are capable of controlling temperature and light. Others are fitted with state-of-the-art computer-based equipment for controlling humidity, light, temperature, nutrients, and soil moisture. In Europe, a greenhouse is called a *glasshouse.*

Small-scale and simply equipped greenhouses are used for the domestic culture of houseplants. Greenhouses are a necessary feature of the horticultural nursery operation, even when plants are grown in season. Commercial greenhouses are widely utilized to produce premium-quality fruits, vegetables, and ornamentals by providing optimal growth conditions for these plants.

> **Greenhouse**
> A structure with transparent covering that is used for growing plants under controllable conditions.

11.2: GREENHOUSE DESIGN AND CONSTRUCTION

The most prominent feature of a greenhouse is how it is designed to take advantage of sunlight. As such, except for the foundation, a short wall (called a *curtain wall*) erected above it, and the metallic frame, a greenhouse consists of a transparent material (e.g., glass, plastic film, or fiberglass-reinforced plastic) that freely admits natural light.

11.2.1 TYPES OF GREENHOUSES

There are three basic types of greenhouses: attached, detached (freestanding), and connected. These types of greenhouses are constructed according to one of several styles:

1. Even-span
2. Uneven span
3. Lean-to
4. Quonset
5. Gothic arch
6. Curvilinear
7. Curved eave
8. Dome

The last four styles are less common than those preceding.

Attached Greenhouses

A greenhouse is attached if part of it is connected to a building. *Attached greenhouse* designs and construction are usually simple. They are less expensive to construct because one side is preexisting, which cuts down on the amount of materials needed. However, because they are connected to existing structures, their sizes and uses are affected by the characteristics of the buildings to which they are attached. The buildings may shade the greenhouse at some time of day. Further, light control and ventilation may be problematic. Attached greenhouses are not used for commercial production but are found in homes, commercial buildings, garden centers, and where plant displaying is needed.

A style of greenhouse that is specifically designed to be attached is the *lean-to green-house* (figure 11-1). The ridge of the roof is attached to the preexisting wall such that the roof slopes away from the wall. Lean-to greenhouses are small in size and best located on the south side of the building to take advantage of sunlight. Where more space exists, the even-span style (see later section) may be adopted. A lean-to greenhouse may also be window mounted.

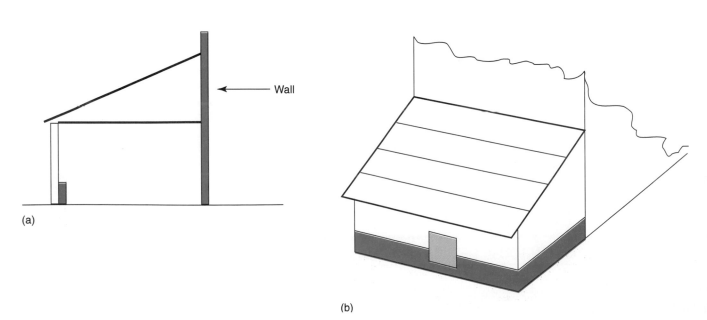

(a)

(b)

FIGURE 11-1 A lean-to greenhouse: (a) side view and (b) front view.

Detached Greenhouses

As the name implies, *detached greenhouses* are designed to be freestanding and thus are sometimes called by that name. None of the walls or the roof are attached to a preexisting structure. A detached greenhouse can thus be located such that it takes maximum advantage of environmental factors such as light and wind. Environmental regulation is easier in detached than in lean-to styles. The most common style of greenhouse is the *freestanding even-span greenhouse.* Also called the *A-frame,* this style consist of a symmetrical roof. The even-span, or A-frame design, is the most common design for a glass greenhouse. An even-span design has a symmetrical roof whose slopes have equal pitch and width. The American-style A-frame design has a larger roof surface area (figure 11-2), and the Dutch design has small gables and a smaller roof surface area (figure 11-3).

Another detached greenhouse style is the *uneven-span* design (figure 11-4). This design has asymmetrical roof slopes of unequal pitch and width. While it is adaptable to hillsides or slopes, it does not readily lend itself to modern greenhouse automation and as such is not commonly used.

Greenhouse roofs may be arched, as in the *Quonset* design (figure 11-5) or the *Gothic arch* design (figure 11-6). The former design is quite commonly used, while the latter is not.

Detached greenhouses have several advantages. Environmental control is easier and can be programmed to meet the needs of a specific production operation. Ventilation is easy to implement, thus limiting carbon dioxide buildup, which can be a limiting factor in plant growth during winter. General operation and maintenance is relatively easier in detached greenhouse designs than attached designs. However, because of their generally vaulted ceilings, they are less energy efficient, increasing operating costs in winter.

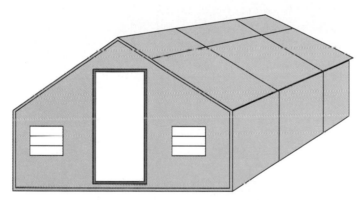

FIGURE 11-2 A freestanding even-span American design greenhouse.

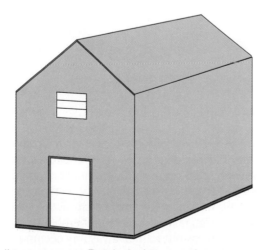

FIGURE 11-3 A freestanding even-span Dutch design greenhouse.

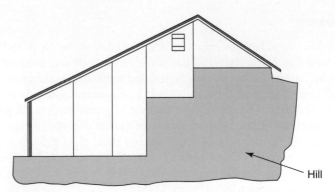

FIGURE 11-4   An uneven-span greenhouse design.

FIGURE 11-5   A Quonset greenhouse design.

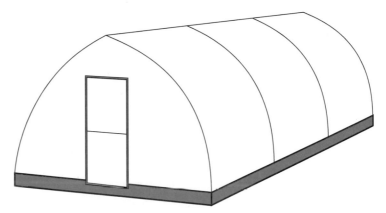

FIGURE 11-6   A Gothic arch–style greenhouse design.

Connected Greenhouses

Several greenhouses of one style may be joined together to form a *connected greenhouse.*
These greenhouse units are joined along the eaves to create a large, undivided space for a
large operation. This arrangement makes the buildings more economical to heat on a per-unit-
area basis. Since the junction between two adjacent eaves creates a gutter, *ridge-and-furrow*
(figure 11-7) designs may be in danger of stress from accumulation of snow where this
weather pattern exists. Their design takes this potential problem into account by the installation of heating pipes in these depressions to melt away any accumulation of snow when it occurs. Ridge-and-furrow designs are suited to greenhouse production enterprises that require

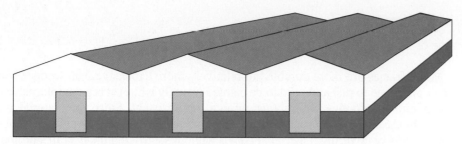

FIGURE 11-7 A ridge-and-furrow greenhouse range.

similar environments. When used for smaller projects requiring unique environmental conditions, this type of greenhouse must be partitioned. When Quonset greenhouse units are connected, they form a *barrel-vault greenhouse* (figure 11-8). Similarly, several lean-to greenhouse units may be connected to form a saw-tooth greenhouse (figure 11-9). *Saw-tooth greenhouses* are used in places such as Texas, Florida, and California, where the climate is mild. Large nurseries, for example, may construct a number of greenhouses on one site to form what is collectively called a *greenhouse range.*

The construction costs of connected greenhouses are higher than freestanding designs. The roofs are longer and thus require more structural strength in the framework to support the building. They are also lower, and hence the volume of air space is less, decreasing the amount of carbon dioxide available for plants in winter.

Greenhouses may also be categorized based on the material used in their construction. These materials may further be grouped according to those used in the framework of the structure and those used for framing or covering the structure. Greenhouses may be arranged in multiple units to form large complexes. All of these styles, arrangements, and types of materials have advantages and disadvantages. These characteristics are discussed further later in this section.

> **Greenhouse Range**
> A collective term for two or more greenhouses at a single location belonging to the same owner.

11.2.2 GREENHOUSE CONSTRUCTION

The material used for greenhouse construction must be strong, light, durable, easy to maintain, and inexpensive. It is important that the frame cast little shadow.

FIGURE 11-8 A barrel-vault greenhouse range.

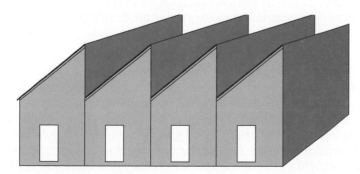

FIGURE 11-9 A saw-tooth greenhouse range.

Material for Framework

The frame of a greenhouse may be made of metal or wood.

Metal   Metal frames are more durable but relatively more expensive than wooden frames. Iron frames are prone to rust and need to be painted (usually white) at regular intervals to prevent rust. Aluminum frames are lightweight and rust resistant. Since the material is very strong, greenhouse designs incorporate fewer and more widely spaced *sash bars* (beams used to support glazing or covering material [11.2.4]) without sacrificing the overall sturdiness of the structure. Further, wider spacing of the sash bars means less shading from obstruction of incoming light.

Wood   Wood was used in early greenhouse designs. It is relatively less expensive but also less durable than metal. Wood decays over time and is susceptible to insect attacks (e.g., termites). Like iron, wood requires painting (greenhouse paint) to protect it from decay and insect attack. Durable species of wood are redwood and cedarwood. For longer life, wood may be treated with preservatives before use. Mercury-based paints are toxic to plants and must not be used. Similarly, pentachlorophenol and creosote wood treatments are toxic to plants. Since wood is not as strong as metal, spacing of sashes in wooden greenhouses is much closer, especially if heavyweight glazing material (e.g., glass) is to be used. Closer sash spacing means an increased shading from obstruction of incoming light.

11.2.3 FRAME DESIGN

There are two basic frame designs, A-frame and arched-frame greenhouses.

A-Frame Greenhouses

In A-frame design, most of the weight of the greenhouse rests on the *side posts,* which are often encased in concrete. These erect posts support the *truss* (consisting of *rafter, strut,* and *chords*). The trusses on either side meet at the peak (ridge) of the roof. Series of trusses are connected by long bars *(purlins)* that run the length of the greenhouse. The end view presents an A-shaped structure called a *gable* (figure 11-10). The bottom 2 to 3 feet (0.6 to 0.93 meters) of the greenhouse above the ground consists of a wall called the curtain wall, which is made out of materials such as cement or concrete blocks. The curtain wall is the structure to which heating pipes are usually attached. Sash bars are attached to purlins as anchors for panes.

Arched-Frame Greenhouses

In arched-frame greenhouse designs, the trusses are pipes that are bent into an arch and connected by purlins (figure 11-11). This design is called a Quonset.

11.2.4 GLAZING (COVERING) MATERIAL

The most important role of glazing or covering material is transmittance of light. No material can transmit 100 percent of all of the light that strikes its surface. Hopefully, most of the light will be transmitted through the material, but some of it will be absorbed or reflected back into the atmosphere. The materials used for glazing are described in the following sections.

Glass

Glass is about 70 percent silica dioxide (SiO_2). However, it contains other oxides, including iron trioxide (Fe_2O_3), which may be present at a level between 0 and 1.15 percent. The amount of light glass transmits depends on the iron content; the higher the iron content, the lower the transmittance. Glass used to glaze greenhouses does not transmit light in the ultraviolet (UV) range of the light spectrum. Glass greenhouses, the first to be commercially constructed and still popular today, are durable, lasting more than 25 years. Glass transmits about 90 percent of the sunlight that strikes its surface. Further, UV light does not adversely affect glass as it does other materials such as plastics. Plastics may warp, but glass does not.

| Glazing |
| A transparent material used to cover a greenhouse frame. |

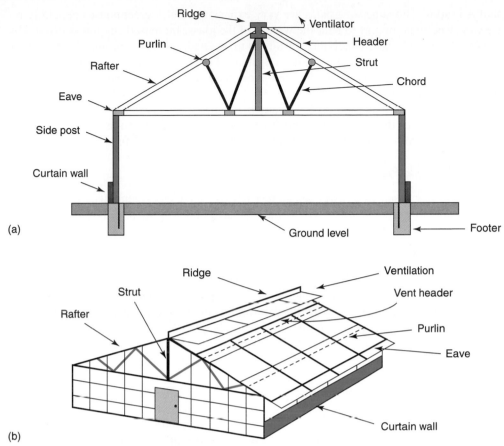

FIGURE 11-10 A-frame greenhouse structure: (a) end view and (b) general view.

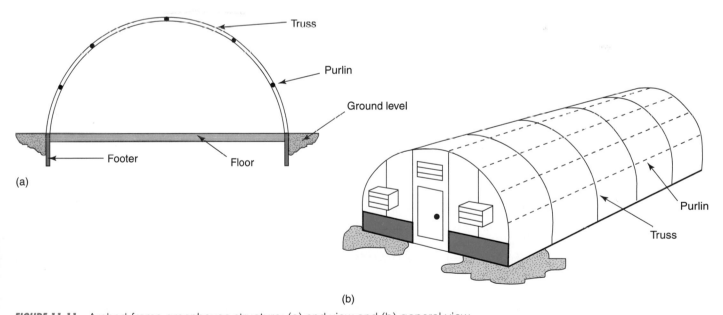

FIGURE 11-11 Arched-frame greenhouse structure: (a) end view and (b) general view.

However, glass greenhouses are expensive to construct, maintain, and operate. Glass breaks easily, is heavy, and must be handled with care. It does not retain heat well.

Glass is available in several grades and weights, including single-strength, double-strength, heavy-sheet, polished-plate, and heavy-plate glass. For greenhouse glazing, the B grade is most commonly used, even though it is less efficient in light transmittance than the

AA and A grades. The superior grades are very expensive. Glass greenhouses generally require extra structural support to bear the weight of the glass. Increased support is offered by closer spacing of sash bars, which then results in more shading. To overcome shading, large and more expensive glass panes may be used (instead of the traditional size of 16×18 inches, some panes are 32×36 inches). Wider glass panes reduce loss of heat through the junctions between adjacent panes.

Large growers of roses and other holiday plants use glass greenhouses. Glass also has limitations in terms of design of the greenhouse. Four basic styles are in used in glass greenhouses—lean-to, even-span, uneven-span, and ridge-and-furrow designs.

Film Plastic (Flexible Plastic Film)

Film-plastic greenhouses have fewer design restrictions than glass greenhouses and are lightweight. They are very inexpensive to construct and are becoming increasingly more popular in use. A double-layer, film-plastic greenhouse can be operated at over 30 percent less than the cost of operating a comparable greenhouse. Another advantage of film-plastic greenhouses is their adaptability to temporary or short-term use. If a controlled environment is required for only a short period for a specific purpose (as is the case in bedding plant production), a film-plastic greenhouse can be quickly constructed and dismantled when not needed. Further, it provides new persons in the horticultural business a less-expensive means of entering the field. Film-plastic greenhouses have disadvantages. They are less durable than glass and may require periodic recovering (e.g., once every three years). The common types of flexible plastic films are described in the following sections.

Polyethylene   The most common but least durable plastic film material is *polyethylene*. This material remains flexible at low temperatures. Polyethylene is permeable to gases such as oxygen and carbon dioxide. Light transmittance through this material is about 5 to 10 percent less than through glass. Ultraviolet light makes plastic material brittle over time, even in the case of the highest-quality, light-resistant, 0.15-millimeter-thick (0.006-inch-thick) plastic. Life expectancy is generally increased by using thicker plastic films or those containing antioxidants and UV inhibitors. These chemical additives reduce the weathering rate of the material.

Woven Polyethylene   *Woven polyethylene* is found in greenhouses equipped with retractable roofs. This type of roof is made of woven polyethylene, a UV-resistant material capable of preventing water condensation on the inner surface of the roof. It transmits light at the rate of 80 percent of light reaching its surface. High-quality crops are produced under this roof, and the amount of light is closely monitored. The material is quite durable, lasting about five to seven years.

Polyvinyl Fluoride   Another film material, made out of *polyvinyl fluoride,* is also available. This material is very durable (lasts over 10 years), has excellent light transmission properties, and is resistant to UV radiation. It transmits light at a level equivalent to that of glass.

Polyvinyl Chloride   *Polyvinyl chloride* films are more durable than polyethylene films. Unfortunately, vinyl films hold an electrical charge that attracts and holds dust particles over time, thereby reducing light transmittance. Further, it becomes soft on warm days and brittle when temperatures are low. The material transmits long-wavelength infrared radiation at a greatly reduced level, thus reducing heat loss at night.

Ethylene–Vinyl Acetate Copolymers   Ethylene–vinyl acetate copolymers are very expensive and thus not widely used as glazing materials. They are durable, of high light diffusion, and less susceptible to weathering than other materials.

Polymethyl Methacrylate (Acrylic)   *Acrylic* is an excellent glazing material. It is weather resistant, lighter than glass, and of comparable light transmission with other materials. Structurally, this material consists of plastic to which acrylic has been chemically bonded. It is more expensive than glass. As a glazing material, it has a tendency to turn yellowish after a period.

Earlier greenhouse designs adapted to plastic films included the A-frame and scissor truss film-plastic types. These designs were temporary structures and built on wooden frames. They had to be protected and preserved through regular painting and wood treatments. Some of the chemicals used (e.g., creosote and pentachlorophenol) were discovered to be toxic to plants. A safer wood treatment is copper naphthenate. More durable and modern designs are made of trusses constructed from metal pipes. A popular one is the Quonset greenhouse. It is very durable, inexpensive, and adaptable to *double-layer covering* with film plastic. A dead-air space is created between the two layers by inflating with a squirrel cage fan. The dead-air space has insulating properties and also prolongs the life of the greenhouse.

Since greenhouses are constructed to be as airtight as possible, condensation of moisture occurs in the internal environment. When plastic coverings are used, droplets of moisture often form on their water-repellent surfaces. Although this condensation poses no immediate danger, when the droplets fall on leaves, a disease-promoting condition is created. The water repellency can be eliminated by spraying detergents on the film surface.

Fiberglass-Reinforced Plastic

Fiberglass-reinforced plastic is a semirigid glazing material and can be bent. It is used in constructing Quonset and even-span greenhouses. The light transmission of this material when newly installed is near the quality of glass. However, with time, the light transmittance reduces as a result of etching and accumulation of dust. Fiberglass-reinforced plastic lasts about 10 to 15 years. The fiberglass content disperses light such that its intensity in the greenhouse is more uniform. Greenhouses constructed from this material are also easier to cool than glass greenhouses. They are available in different colors. However, the clear type permits the greatest light transmission, at a level equivalent to that of glass. They are available in flat or corrugated forms. Greenhouses constructed from fiberglass-reinforced plastic and other plastics are more expensive to insure because they are prone to extreme heat or fire damage. These materials are said to be *thermosetting,* or heat setting. However, fire-retardant fiberglass-reinforced material is available.

Rigid Sheet Plastics

Rigid plastics, when cut into panes (resembling glass panes), may be used as glazing materials for greenhouses. The most popular rigid plastic materials are acrylic and polycarbonates and are available as single- and double-layered rolls. Double-layered types are more durable and increase the energy efficiency of greenhouses by about 50 percent over glass greenhouses. Rigid plastic and glass are the most expensive glazing materials in use. Acrylic may last up to 25 years, and polycarbonate may last between 10 and 15 years. Because they are relatively less heavy than glass, they require less support and hence are adaptable to wider sash bar spacing. Using fewer sash bars reduces the amount of shading.

Saran Plastic Mesh

Saran-glazed greenhouses are used where the intensity of sunlight is very high. This mesh plastic film is used to provide shade, and hence the greenhouses glazed with this material are sometimes described as saran shade houses. They are used in areas such as Florida, Texas, Hawaii, and California. The material is available in a range of colors, thicknesses, and closeness of weave. These factors affect the degree of shading provided by the material.

11.2.5 COLD FRAMES AND HOTBEDS

Cold frames and *hotbeds* are simple climate-controlling structures for growing plants on a limited scale. The difference between a cold frame and a hotbed is that the latter is simply a cold frame fitted with a heating system. Heat may be supplied by electrical cables beneath the soil or steam run through pipes along the wall. Environmental control is limited to opening and closing the structure or rolling away the covering for aeration and temperature modification (figure 11-12). The designs are variable. They may either be attached to a regular greenhouse or erected separately. The glazing may be clear plastic or glass. Cold frames are

> **Cold Frame**
> An enclosed, unheated, covered frame used for growing and protecting young plants in early spring, and for hardening off seedlings.

> **Hotbed**
> A bed of soil enclosed in transparent material and heated to provide a warm medium for germination of seeds or rooting of cuttings.

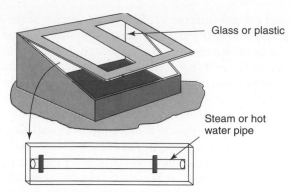

Glass or plastic

Steam or hot
water pipe

FIGURE 11-12 Cold frame and hotbed.

heated by sunlight, which makes their operation very inexpensive. However, their use is limited since environment cannot be controlled. Cold frames may be used for hardening purposes, rooting of hardwood cuttings, raising vegetable seedlings, and limited production of crops such as lettuce, radish, cucumber, and sweet potato. Placing a thermometer inside of a cold frame to monitor the environmental temperature is recommended. For cool-season crops, the structure should be ventilated whenever the temperature reaches 21°C (70°F). Warm-season crops can tolerate a higher temperature, and ventilation is thus needed only when the temperature rises to about 30°C (86°F). Cold frames are not as popular as they once were, largely because greenhouses that provide a broader range of environmental control are used widely.

11.2.6 LOCATING A GREENHOUSE

Greenhouse production is initially capital intensive and hence must be embarked upon only after good planning. The following are important factors for consideration in locating a greenhouse.

Market

The grower must first identify the potential market in terms of its size and distance from the production site.

Accessibility

Greenhouse accessibility is closely related to its potential market. The production site should be readily accessible to the primary customers. If a retail operation is intended, the greenhouse should be located where the general public can readily reach the facility. Certain production operations require that products be delivered promptly to sales outlets. Transportation between greenhouse site and markets should be reliable and convenient. Cut flowers may be able to survive several days of refrigeration with little loss in quality. It is important that supplies for production be delivered on schedule even in inclement weather.

Greenhouses should be readily accessible by a reliable means of transportation because production inputs (including soil mixes, fertilizers, pesticides, and seed) and the harvested produce must be hauled to and from the greenhouse. Locating a greenhouse enterprise near markets significantly reduces operational costs. *Bedding* and *potted plants* (chapter 13, module 1) are expensive to transport; consequently, these enterprises should be located near primary market outlets if possible.

Climatic Conditions

The patterns of weather factors including light, rainfall distribution and other precipitation (e.g., snow, ice, and sleet), and winds affect production costs (e.g., heating, cooling, and lighting). High elevations may provide cleaner air but are colder, requiring more heating in the cold season. Establishing a production enterprise for a crop that requires warm conditions in an area that is mostly cold will increase heating costs (unless the area has a great potential market to offset the additional costs).

Topography

Topography affects the drainage of the area. Greenhouses use large amounts of water and must be located on soils that drain freely. Further, it is easier to mechanize operations if the site is flat. It is more difficult to maneuver on slopes than on flat land. Construction costs may not vary, but it is easier to automate a greenhouse built on level ground than one built on a hill.

Utilities

Another factor to consider in locating a greenhouse is the source of water. Greenhouses use large amounts of water for a variety of activities such as watering plants, washing, and maintaining high humidity inside of the facility. If an urban treated water supply is not accessible, an alternate source of water must be found (e.g., a well). It is critical that the source of water be reliable to provide water year-round. The success of certain production practices depends on the availability of water. The quality of water is also critical, since certain plants are adversely affected by specific pollutants (e.g., fertilizers and pesticides). Pollutants can make water acidic or alkaline. Supplemental light is also required, as is a source of heat for times when the temperature drops below a desirable level. Providing artificial light and heat requires a source of energy. The greenhouse should be located where there is ready access to a reliable and economic energy (fuel) supply.

Labor Supply

The location of a greenhouse should also take into account the kind and availability of labor. Certain chores in the greenhouse are not automated but require some level of skill. Not all greenhouse operations are readily amenable to automation.

Types of Production Enterprises

The type of production enterprise is related to the accessibility factors. Bulky products (e.g., potted plants) are expensive to transport over long distances.

Zoning Laws

Various localities have zoning regulations regarding location of an agricultural enterprise and building codes.

Future Expansion

For a commercial venture, it is advisable to acquire more land than needed immediately. This extra land will allow future expansion to be undertaken as needed.

11.2.7 GREENHOUSE ORIENTATION

The location of a greenhouse is critical to its efficiency. The orientation of the structure has a bearing on the temperature variations experienced within a greenhouse. It is important that tall structures (such as tall trees and buildings) that might cast shade (especially on the south side of the greenhouse) not be in the vicinity. This natural light advantage is diminished when greenhouses are located in areas of intense fog or highly cloudy areas. Shadows are also cast by the frames used for construction. To reduce this occurrence, the ridge of greenhouses in regions above 40 degrees north latitude should be oriented in an east-west direction so that the low angle of winter light has a wider area to enter the house from the side rather than from the end.

11.2.8 GREENHOUSE BENCHES AND BEDS

Greenhouse production occurs either in *ground beds* (in the ground) or on *benches* (raised platforms). The design and layout of these structures should facilitate greenhouse operations and make the most efficient use of space.

Beds and benches in a greenhouse are located so as to allow personnel to freely move around and work and also to move greenhouse equipment such as carts and trolleys. To accommodate equipment, the principal aisles should be about 3 to 4 feet (0.93 to 1.24 meters) wide. Some arrangements are more efficient than others in terms of the efficient use of space. The more usable space, the greater the profits from a greenhouse operation.

In terms of where the planting media and containers are located, three strategies may be adopted in a greenhouse production enterprise.

No Bench

No bench, or floor benching, is the practice in which production takes place directly on the floor of the greenhouse. The floor may be covered with gravel or concrete and should be well draining. It is best to have a gentle slope in the floor so that excess irrigation water drains into a gutter and is carried out of the facility. Gravel covers and porous concrete are prone to weed infestation. Bedding and seasonal plants such as poinsettia are commonly grown on floor benches. Aisle space is created by the way flats or containers are arranged on the floor (figure 11-13). If done properly, up to about 90 percent of the floor space can be utilized for production.

Raised Benches

Raised benches are suited to potted plant production. Most greenhouses have raised benches of a wide variety of designs and constructions. Some of them are makeshift and temporary, consisting of brick legs and movable bench tops. The top may be of wood, concrete, or wire mesh and may or may not have side boards (figure 11-14). Metal benches are most common. Molded plastic is sometimes used to make troughs in which potted plants are grown. This material is also used in the construction of benches for ebb-and-flow irrigation (11.3.3). Notwithstanding the material used, the bench must have a system for draining properly. If wood is used, cedar, redwood, and cypress make good bench materials because they resist decay. Wooden benches may be painted with copper naphthenate preservative to prevent decay. Redwood has natural preservatives that are corrosive to iron and steel; as such, nails and other construction materials that come into contact with this wood should be of different materials such as aluminum or zinc.

The height of the bench above the floor should be such that cultural operations (e.g., pinching, spraying, harvesting, and staking) are facilitated. Width of the bench is also important. It should be narrow (3 to 6 feet or 0.93 to 1.86 meters) enough to permit pots located in the middle rows to be easily reached. Air should be able to move freely around and under the bench, as well as around the pots on the bench.

FIGURE 11-13 A no-bench greenhouse production system showing plants placed directly on the floor.

FIGURE 11-14 Raised bench.

Ground Benches or Beds

Plants to be grown for several years that will grow tall in the process (e.g., cut flower plants such as roses) are planted in ground beds (figure 11-15). Ground beds vary in design and construction. Using ground beds can be problematic from the standpoint of disease control. If the ground bed has no real bottom in terms of depth, diseases such as bacterial wilt are hard to control because of the impracticality of thorough pasteurization of the soil to a reasonable depth. To correct this problem, concrete bins may be constructed in the ground to hold the soil and to facilitate periodic pasteurization to control soilborne diseases. These concrete bins have V-shaped bottoms and drain holes for good drainage. They should be about 6 to 12 inches (15.2 to 30.5 centimeters) deep, depending on the plant to be grown. When drainage is poor, drainage tiles may be installed and overlaid with gravel before topping with the root medium. Walkways should be strategically located between beds to allow gardeners easy access to the beds to prepare them, plant the crop, care for it, and harvest the produce. These spaces should be graded so that water flows away from the beds to reduce contamination from pathogens carried on the wheels of carts and wheelbarrows and the shoes of gardeners.

11.2.9 LAYOUT (ARRANGEMENT)

The five types of bench arrangements commonly used in greenhouses are described in the following sections.

Longitudinal Arrangement

In a longitudinal arrangement, beds or benches are constructed to run the full length of the greenhouse in several rows (figure 11-16). This arrangement is associated with cut flower production (22.1). It is easier to conduct mechanized operations with this type of arrangement, which provides for long, uninterrupted production areas. However, moving across the facility is hampered and requires workers to go all of the way to one end of one row to make a turn to the next row.

Cross-Benching

Cross-benching is like the longitudinal arrangement except that the orientation of the benches are not lengthwise with respect to the greenhouse but are arranged crosswise (figure 11-17). The benches are shorter and aisles numerous. Although the aisle space significantly reduces the usable area of the greenhouse, movement around the greenhouse is easier with this arrangement.

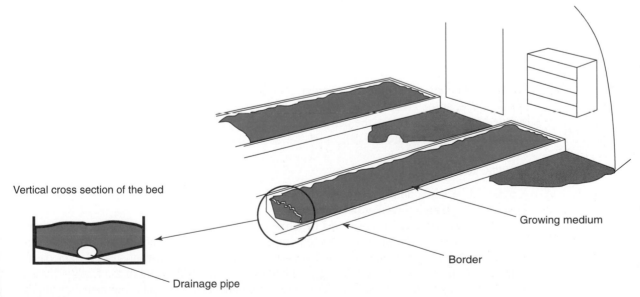

Vertical cross section of the bed

Growing medium

Border

Drainage pipe

FIGURE 11-15 A ground bench design.

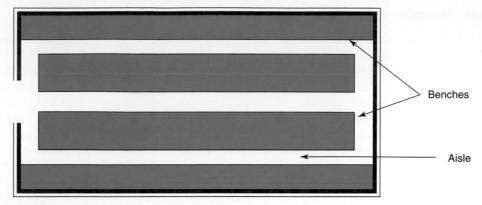

FIGURE 11-16 Longitudinal layout of greenhouse benches.

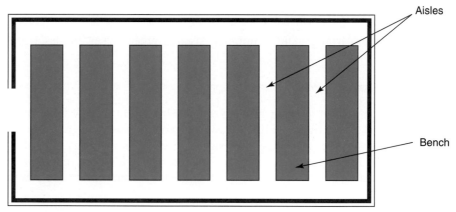

FIGURE 11-17 Cross-benching layout of greenhouse benches.

Peninsula Arrangement

The difference between the peninsula arrangement and cross-benching is the presence of a primary central aisle in the former that runs the entire length of the greenhouse. The primary aisle in cross-benching runs along the wall (figure 11-18).

Movable Benches

Movable benches are especially popular container production enterprises. To maximize the use of space, some greenhouse designs include movable benches and one aisle. When work on one bench is completed, the bench is mechanically moved so that work can be completed on the next bench (figure 11-19).

Pyramid Bench

A pyramid bench is an arrangement in which benches are placed in tiers. It is an ideal configuration for hanging basket production.

Whether concrete bins or raised benches are used, the layout is important. As previously indicated, aisles must be made to allow equipment to be moved around and to provide working room around the bench while maximizing the use of floor space. In greenhouses designed especially for cut flower production, the ground bed may run the full length of the facility. Designs for nutriculture usually have special designs to accommodate the special equipment they require.

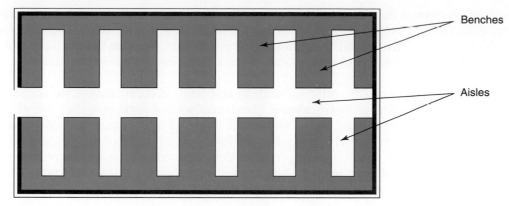

FIGURE 11-18   Peninsula arrangement of greenhouse benches.

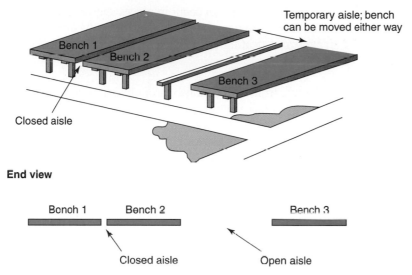

FIGURE 11-19   A movable bench design.

11.3: INTERNAL ENVIRONMENTAL CONTROL

Greenhouses are controlled-environment facilities because the user is able to adjust at least some of the plant environmental growth factors to meet specific needs. The factors and their control are discussed in the following sections.

11.3.1 TEMPERATURE

The daytime temperature in greenhouses is usually higher than the nighttime temperature. Room temperature in most greenhouses is maintained at about 55 to 65°F or 13 to 18.5°C (at night) and 10 to 15°F higher during the day. Maintaining temperature is the next most expensive operational cost after labor. The greenhouse is essentially a giant solar collector. The sunlight energy that enters the greenhouse during the day is trapped (*greenhouse effect* [3.2.1]) and used in heating up the contents of the facility.

How Heat Is Lost

Temperature control is perhaps the major reason for greenhouse use. The goals of heating a greenhouse are to provide heat at the appropriate time and in the appropriate amount, distribute it effectively through the facility, and conserve it. The ideal situation is to maintain a stable air temperature in the greenhouse by adding heat at the same rate at which it is lost. Fuel cost is a big contributor to high overhead in greenhouse enterprises. Therefore, heat loss should be minimized. Heat is lost from greenhouses by three ways: *conduction, infiltration,* and *radiation.*

Conduction   Conduction is the principal mode of heat loss and occurs through the material used in framing and glazing the greenhouse. Metals conduct heat faster than glass, and glass conducts heat faster than plastic. Heat loss is related to surface area. Thus, a corrugated (corrugation gives it more surface area) fiberglass-reinforced plastic greenhouse loses more heat than a flat-plate plastic one. If a glazing material loses heat rapidly, it is more expensive to heat a greenhouse constructed out of it. An estimated 40 percent of savings on heating costs may be realized with a properly constructed double-layered plastic covering because of the insulating property of dead-air space.

Infiltration   Greenhouse heat is lost through cracks and holes that occur in the structure. This mode of heat loss is called infiltration. Cracks occur in places such as the area around closed doors and improperly closed vents. Anytime doors are opened, fresh cool air enters the greenhouse. This influx of air increases heating costs. Older and poorly maintained greenhouses often have air leakage problems.

Radiation   Heat can also be lost from a greenhouse through radiation. Radiation loss is minimal and occurs as heat energy is lost from warmer objects to colder objects. Polyethylene covering can lose large amounts of heat through radiation, whereas glass and fiberglass-reinforced plastic lose virtually no energy to radiation. Water on the plastic covering can reduce this energy loss in polyethylene greenhouses.

Heating a Greenhouse

Types of Fuel   Three types of fuels are commonly used to heat greenhouses. The most popular is *natural gas.* It is relatively inexpensive, burns clean, and is delivered to the facility directly via pipes in most cases, thus eliminating storage and delivery costs. Its heat value is about 1,000 Btu per cubic foot. (One Btu is the amount of heat required to raise the temperature of one pound of water by 1°F.) *Fuel oil* is second in popularity to natural gas. It is often used as a backup fuel where natural gas is used. A grade 2 oil has about 140,000 Btu per gallon. Apart from the storage required when fuel oil is used, its viscosity is affected by temperature. As such, oil does not flow properly at low temperatures, the time when heat is needed the most. Further, it has undesirable ash as a product of combustion. The third and least-preferred heating fuel is *coal.* It produces considerable pollution when burned. The ash produced from burning coal poses a disposal problem. If the coal has a high sulfur content, pollution is an even greater problem, requiring additional equipment to be installed in some cases to reduce air pollution. Because exhaust fumes are toxic to greenhouse plants, it is critical that during greenhouse construction the exhaust stack be located downstream of the wind. Such a location ensures that fumes have little chance of polluting the greenhouse environment.

In choosing a fuel type, one should consider the following:

1. *Availability.* The source should be readily available year-round.
2. *Delivery or transportation.* The cost for delivery of the fuel is important. If gas is used, will it be delivered by truck or piped to the site?
3. *Storage.* If gas or fossil fuel is selected, there must be a storage facility to hold it while it is being used.
4. *Special equipment needs.* Gas must be stored under pressure in a container.

Heating Systems

Greenhouse heat is circulated through several different kinds of media.

Hot Water Heating Systems   The fuels described earlier are used to heat water, which becomes the medium through which heat is circulated throughout the greenhouse. Hot water systems are adapted for use in small greenhouses. The temperature of water may be varied as needed. The disadvantage of this system is that an elaborate network of pipes is usually needed to carry the hot water from a boiler throughout the facility. Further, if a gravity flow return system is installed, gravity causes cold water to flow back into the boiler, thus reducing its efficiency. Modern systems utilize forced-water circulation. Apart from being more expensive to heat and maintain a desired temperature, its heat value is lower than that of steam. Further, hot water is not amenable to use in pasteurization. The system is adapted to small greenhouses because it is difficult to transfer water over long distances without losing heat (temperature drops over long distances).

Steam Heating Systems   Steam can be heated to a higher temperature (212 to 215°F or 100 to 101.7°C) than hot water. Smaller pipes are needed to transport steam over long distances and hence can be efficiently used in large greenhouses. In large greenhouses, the steam pressure at the boiler may be as high as 120 psi (pounds per square inch). Even though steam can be transported over long distances, it condenses in the pipes; thus, provision must be made to drain and recirculate the water for reheating. Steam is very efficient for pasteurization.

Radiant Energy Heating System

INFRARED RADIANT HEATER   Heating a greenhouse by infrared radiation is very economical. Reductions in fuel bills of about 30 to 50 percent have been reported. Heat is not conducted through any medium but transmitted directly to plants (or other objects) without even warming the surrounding air. As such, while plants receive the desired temperature, the general greenhouse atmosphere may be several degrees colder than would be the case if hot water or steam were used. Even though infrared heaters are highly efficient, the equipment or sources of the radiation must be located directly above the plant or object to be warmed. Failure to provide for such placement will result in *cold spots* (pockets of low temperature) in the facility. Further, as plants grow bigger, they tend to block the radiation from reaching the soil, leaving it cold.

SOLAR RADIATION SYSTEM   Like infrared radiation, solar heaters are nonpolluting. The initial cost of solar collectors is high, but they are cost-effective once installed. A major disadvantage with solar heating is its weather dependency. Clouds limit the effectiveness of this system.

Cold Spots
Pockets of low temperature in an enclosed area due to nonuniform distribution of heat.

Coal   The third and least-preferred heating fuel is coal. It produces considerable pollution when burned. The ash produced from burning poses a disposal problem. If coal has a high sulfur content, pollution is even greater, and laws may require that additional equipment be installed to reduce air pollution. Since exhaust fumes are toxic to greenhouse plants, it is critical that the smoke stack be located downstream of the wind, preventing the fumes from polluting the greenhouse environment.

Design of Heat Distribution System

Heat is provided in a greenhouse by using a *heater.* In terms of the nature of the source of heat, a heater may be *localized* or *centralized.*

Localized Heaters   The various designs of localized heating systems are described in the following sections.

UNIT HEATERS   Unit heaters are also called *forced-air heaters.* They burn fuel or circulate hot water or steam in a chamber and then depend on a fan to blow on the hot surface to spread the warm air either vertically or horizontally through the greenhouse (figure 11-20). The more popular models expel heated air horizontally and are called *horizontal units* (as opposed to *vertical units,* which are suspended from above). Because such heaters use oxygen for combustion, there is a potential danger in their use in airtight greenhouses. All of the available

FIGURE 11-20 A forced-air unit heating system.

oxygen could be used up such that if the burner has been on for a long time (e.g., overnight), carbon monoxide could accumulate in the greenhouse to dangerous levels and be life threatening to anyone entering the building. Further, in more airtight greenhouses such as those with plastic covering, oxygen shortage has been blamed for burners going out during the night. To avoid these situations, greenhouses that use unit heaters are designed to have a 1 square inch (6.5 square centimeter) opening in the structure to let fresh air in at all times. Unit heaters are readily amenable to automation. All heaters that use open flames to burn fuel require adequate ventilation through a chimney pipe vented outside of the greenhouse. These units are also called *venting heaters.*

CONVECTION HEATERS   Convection heaters are another class of localized heaters that, like unit heaters, burn a variety of fuels in a fire box. Heat is distributed by convection current, a natural process. The pattern of airflow depends on the location and arrangement of heaters. Generally, in convection current flow, warm air rises and cold air drops (figure 11-21). These heaters are suitable for small-scale use and not readily amenable to automation; however, they are relatively inexpensive. Whenever fuel is burned, it is critical that the exhaust system be effective and efficient in removing all products of combustion from the greenhouse. Such materials include products of incomplete combustion such as ethylene gas, which is harmful to plants. Other exhaust products include sulfur dioxide gas, which when dissolved in water forms corrosive sulfuric acid that scorches plants on contact.

RADIANT HEATERS   Radiant heaters do not burn fuel. They do not warm greenhouse air directly but depend on radiant energy (infrared) to warm objects in their path, which in turn warm the air in their vicinity. Radiant heaters, which are low-energy heaters, are reported to be over 30 percent more efficient than other sources of greenhouse heat. Since air is not heated directly as in conventional systems, the air temperature in a greenhouse heated by this source may be as much as 4°C (7°F) cooler than that in conventionally heated environments. Problems with condensation on surfaces, a common occurrence under warm air conditions, are reduced and subsequently the incidence of diseases associated with condensation is also lowered when using radiant heaters. Further, the greenhouse environment is heated uniformly with little heat stratification. When using radiant heaters, the fans often installed to aid in air circulation under conventional heating systems should not be used. These fans tend to cool the plants and

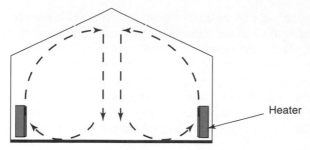

FIGURE 11-21 Convection current in a greenhouse.

FIGURE 11-22 A centralized heating system.

other surfaces and move the heated air around them away. Since the air temperature is lower, the temperature differential between the outside and inside air is lower, resulting in less heat loss. Radiant heaters consume about 75 percent less energy than conventional heaters.

When heaters that combust fuels are used, the potential for injury to people and plants from the pollutants they release into the air is real and must be monitored. In addition to carbon monoxide, products of incomplete combustion such as ethylene can injure sensitive plants such as chrysanthemum. Fuel impurities such as sulfur are burned and converted to sulfur dioxide, which, when dissolved in moisture, produces the corrosive sulfuric acid that burns leaves.

Centralized Heaters Centralized heating systems require a greater investment than localized heaters and are economical for large operations. A boiler produces the heat source, which boils the water that is pumped through pipes and other accessories laid through the greenhouse (e.g., mounted on curtain walls) (figure 11-22). Pipes may also be located under benches or hung over the plants from the roof. Heat is lost through the plumbing and other parts of the boiler. To increase the surface area for heating, square metal fins may be attached to the pipes in series. Centralized heaters are more efficient than localized heaters. The boiler should be located in the service building, where it is dryer, to prolong the life of the unit. Centralized heating may be accomplished by circulating steam instead of boiled water. A steam system is more common in large ranges since it takes a large boiler and large amounts of water to supply adequate amounts of boiled water to heat a large range.

Certain boilers are classified as those that permit the grower to reduce the level of water to generate steam for pasteurization purposes. This process is called *trimming for steam* and is used when the greenhouse operation has no demand for hot water (e.g., in late spring or summer), since the heating system must be shut off during the process.

Solar Heaters   A greenhouse by nature is a giant solar convertor, only less efficient. Sunlight enters the facility through all of the transparent parts and is intercepted by objects in its path (e.g., the floor, soil, plants, and benches). These "collectors" store the energy for a short period and then release it into the atmosphere. Some modern greenhouses depend on solar heat collectors for heating. The initial investment for a solar heating system is high. As such, low-temperature solar systems are often preferred and used in conjunction with a conventional heating system for additional heating, especially in the winter season. A complete system consists of a collector, heat storage unit, heat exchanger, and control panel.

Biotherm Heating   The floor of the greenhouse can be heated by laying down rows of flexible tubes. This strategy of heating a greenhouse, in which the soil is the primary target, favors enterprises in which pots and trays are placed directly on the floor (as often occurs in bedding plant production). The floor of the bench can also be heated by a similar method. This method of heating the greenhouse is called *biotherm heating*. Propagation benches benefit from heating from below.

Improving Heat Distribution

Notwithstanding the system of heating, it is crucial that heat generated be uniformly distributed throughout the greenhouse to prevent cold spots from occurring. Heat distribution is a more significant problem in large greenhouses than in small ones. When central heaters are used, heating pipes should be strategically located throughout the facility to effect uniform heating. Where cold spots occur, nearby plants grow slowly. The heating system should provide adequate heat not only for the above-ground plant parts but also for the root zone. In the case of unit heaters, attachments such as perforated polyethylene tubing located above the plants are used to aid in the distribution of warm air (figure 11-23). The other important factor is that heaters should supply heat at a rate to offset that which is lost by conduction, infiltration, and radiation.

Regulation of Heat

Thermostat
A device used to regulate temperature.

The regulation of heat is as important as its distribution. Heat cannot be continuously generated unless the conditions are such that it is lost as rapidly as it is produced. *Thermostats* are used to regulate heat in the greenhouse (figure 11-24). In commercial greenhouses, aspirated thermostats provide effective control. This unit monitors the temperature of a continuous airflow caused by a tiny electrical fan. The greenhouse air is thus sampled continuously for more uniform control. Thermostats should be located at the height of growing plants so as to monitor the actual conditions of the plants. When the right temperature is attained, the heat should

FIGURE 11-23 Perforated polyethylene tubing used for heat distribution in a greenhouse.

FIGURE 11-24 A thermostat for heat regulation.

be turned off. To be on the safe side, emergency units should be provided as standby heaters in case the primary heat source fails during a crop production cycle. Large greenhouses are divided into sections that are monitored separately. These sections enable the greenhouse to be used for different enterprises simultaneously.

Conserving Heat

Heating a greenhouse is very expensive, and as such every possible measure should be taken to conserve heat. The greenhouse should be inspected regularly to replace broken glasses, seal leaks, and clean boilers and heaters. As previously described, when polyethylene is used to glaze the frame of a greenhouse, a second sheet (i.e., double layer) creates a dead-air space between the layers for additional insulation.

Since most of the heat lost from a greenhouse is dispersed through the roof, the installation of an interior ceiling that can be drawn at night and opened during the day can conserve up to 30 percent of heat. For much greater heat conservation, *thermal screens* or sheets may be installed in the facility. These screens are installed on the ceilings and walls. Double thermal sheets are known to reduce fuel costs by more than 50 percent. Heat can be conserved by using fans installed in the ceiling to distribute the heat that would have been lost through the roof.

Ventilating and Cooling a Greenhouse

Greenhouse Ventilation Systems   Ventilation is required to reduce excessive heat and humidity buildup in a greenhouse and aerate the environment with fresh air and fresh supplies of carbon dioxide for plant use. Since plants release oxygen and take in carbon dioxide, the carbon dioxide concentration in a closed system such as a greenhouse declines with time. The location of vents depends on the design and construction of the greenhouse. In an A-frame greenhouse, vents may be located in the ridge section of a side wall (figure 11-25). Vents may be operated manually or be automated. Quonset greenhouses use forced-air ventilation systems or side vents. Exhaust fans installed at one end of the greenhouse draw in fresh air through louvers installed at the opposite end (figure 11-26). It should be pointed out that ventilation not only refreshes the greenhouse environment but also cools it.

Greenhouse Cooling Systems   Although heat is required in winter and cool periods of the growing season, greenhouses need cooling in summer and warm periods. On the average, the ambient temperature in a greenhouse is about 11°C (20°F) higher than the outside temperature. Plants are damaged by both excessive heat and cold. Excessive heat can cause low crop yield through high flower abortion and drop. Greenhouses have vents that help to circulate

FIGURE 11-25   Greenhouse vents may be located on the side wall or in the roof.

FIGURE 11-26   An exhaust system for greenhouses.

fresh air and provide cooling. However, in certain situations venting is not sufficient to reduce temperatures and cooling systems must be installed.

The most common way of cooling a greenhouse is by using an *evaporative cooling system* (figure 11-27). This system involves drawing air through pads soaked with water, and hence is also called a *fan-and-pad cooling system*. The cooling pads are soaked with water dripping from above. These cross-fluted cellulose materials retain moisture quite well. Excelsior pads may also be used. The excess water is drained into a lower trough (sump) and recirculated by a pump. Warm air from outside is forced through the pads by the drawing action of fans at opposite ends of the greenhouse. In the process, some of the trickling water evaporates. Water evaporation removes heat from the air. The fan-and-pad cooling system uses the principle of evaporative cooling. To be effective, all incoming air should be drawn through the cooling pad. Cross-fluted cellulose pads (which look like corrugated cardboard), last longer than excelsior pads and are more popular.

Controlling Temperature

Controlling temperature should be distinguished from heating the greenhouse. Under conditions in which no heat is required, the temperature of the greenhouse should be maintained at an optimal level. The rule of thumb is to operate a greenhouse at a daytime temperature of 3 to 6°C (5 to 10°F) higher than the nighttime temperature on cloudy days and even higher (8°C or 15°F) on clear days. Since increasing the temperature accelerates growth, it must be done with care so as not to compromise the quality of the produce. Fast growth can cause spindly or thin stems and small flowers. Raising the temperature on a clear, bright day is necessary because light is not a limiting factor for photosynthesis under such conditions. Failure to raise the temperature may cause heat to become a limiting factor.

11.3.2 LIGHT

Light is required for photosynthesis and other growth activities. As previously explained, the light required by plants is described in terms of intensity, quality, and duration (chapter 2).

FIGURE 11-27 An evaporative cooling system for greenhouse cooling.

Intensity

On a clear summer day, plants in a greenhouse may receive as many as 12,000 foot-candles (or 129 kilolux) of light, an excessive amount for most crops, since by 3,000 foot-candles (32.3 kilolux), most plant leaves in the direct path of incoming light are light saturated and cannot increase their photosynthetic rate. Whole plants can utilize about 10,000 foot-candles (108 kilolux). Sunlight intensity is also seasonal, being less in winter and more in summer. Crops have different preferences and tolerances for light intensity. For example, foliage (non-flowering) plants are scorched and fade in light intensities above 2,000 to 3,000 foot-candles (21.5 to 32.3 kilolux). African violets lose green color at an even lower light intensity (1,500 foot-candles). Poinsettia plants are a darker shade of green when light is reduced. Also, plants such as geranium and chrysanthemum require shading in cultivation to prevent petal burn. Shading to reduce light by about 40 percent from midspring to midfall is helpful to prevent chloroplast suppression in most greenhouse plants. Greenhouse light intensity can be manipulated in several ways.

Shading   Shading the greenhouse to reduce excessive light is accomplished in several ways. Glass greenhouses may be sprayed with paint (whitewash) in summer to reflect light and reduce its intensity. Commercial shading paints may be purchased. A screen made from fabric may be more convenient to use if only a section of the greenhouse requires shading. Durable fabrics include polyester and polypropylene; they are available in different densities of weave for providing degrees of shading from 20 to 90 percent, 50 percent being the most common. Some modern greenhouses have retractable shades that are used as needed. Saran (a synthetic woven fabric [11.2.4]) may be used to cover plants in sections of a greenhouse or bench. Also, aluminized strips may be used to create an interior ceiling that can be drawn on sunny days.

Undesirable shading may be inherent in the design of the greenhouse. Materials for construction and design of the frame, cooling systems, and plumbing and light fixtures that are located overhead intercept incoming light. The narrower the frame material, the better. Overall, these obstructions may reduce greenhouse light intensity by 30 percent or more. Further, seasonal and daily light intensity reductions may be caused by the orientation of the greenhouse and the characteristics of the location.

Plant Density   Plants may also be spaced to utilize incoming light. In summer, closer spacing should be used, whereas wider spacing is recommended in winter. Spacing–light intensity interaction can be manipulated to control the size of plants, since the same amount of dry matter will be produced under an available light intensity. Therefore, to obtain larger plants, space widely; to obtain smaller plants, space closely.

Washing   To increase light intensity, the glass covering of the greenhouse should be cleaned. Dust particles on the glass plates reduce intensity by up to about 20 percent. Washing by running water through a hose may not be sufficient. A cleaning solution should be used (e.g., 11 pounds [5.0 kilograms] of oxalic acid dissolved in 33 gallons [150.2 liters] of water may be sprayed onto a damp, dirty greenhouse and hosed down with water after three days). Other greenhouse types also need periodic cleaning.

Supplemental Lighting

In winter or darker periods (middle of fall to early spring season), supplemental lighting may be required by certain plants for good growth and quality. When needed, it may be supplied by using a variety of sources, categorized as follows.

Incandescent Lamp   Incandescent lamps have tungsten filaments (figure 11-28). They are not desirable for greenhouse use because, in addition to light, they generate excessive heat. Incandescent lamps characteristically emit a high proportion of red and far red light, which produces abnormal growth in some plants (e.g., soft growth or induced tallness). They are very inefficient, converting less than 10 percent of the electrical energy consumed into light.

FIGURE 11-28 Incandescent lights.

Fluorescent Lamps   Fluorescent lamps are energy efficient (20 percent electrical energy converted to light) and produce little heat. However, because they are low-power sources, many of them are often required to produce the desired intensity. Consequently, a large number of fixtures must be installed. These installations end up blocking incoming natural light. They are commonly used in growth rooms where seeds are germinated or in small greenhouses (figure 11-29). The cool white model, which provides light of predominantly blue wavelength, is commonly used. However, fluorescent tubes with capacities to emit a superior quality of light for photosynthesis are available. One class, called *plant growth A,* produces light in the red region of the light spectrum; another class of tubes, *plant growth B,* has capabilities for emitting radiation beyond the 700-nanometer wavelength.

High-Intensity-Discharge Lamps   As their name indicates, high-intensity-discharge lamps can generate intense light (figure 11-30). There are several types and designs available: high-pressure mercury discharge, high-pressure metal halide, and high-pressure sodium light. In certain types, power ratings may be as high as 2,000 watts. The high-pressure sodium lamps are less expensive than the others to install and operate and can generate up to 1,000 watts of power for 24,000 hours. The light quality emitted is in the 700- to 800-nanometer spectral range. Low-pressure versions of the high-pressure sodium lamps are available. Considered the most efficient lamps for supplemental greenhouse lighting (27 percent electrical energy converted to light), they are available in power ratings of up to the popular 180-watt size. However, these lamps are deficient in quality of light produced and do not emit enough light in the 700- to 850-nanometer range. Therefore, when used alone, plants often develop abnormally (e.g., pale foliage in lettuce and petunia). When used in combination with daylight or another appropriate source, plant growth problems are eliminated. Supplemental lights in growth rooms can double up as heat sources, except in very cold seasons.

Height of Light

The various types of lamps emit light at different intensities, ranging from about 2,000 to 11,000 lux (185 to 1,000 foot-candles), or even higher in some cases, as in growth rooms. The effect of each source depends not only on the power rating of the unit but also on the arrangement and height above the plants. For example, if a 4-foot-wide (1.2-meter-wide) bed of plants is to be illuminated using one row of 60-watt bulbs spaced 4 feet (1.2 meters) apart, the arrangement should be hung no more than 5 feet (1.5 meters) above the soil. Fluorescent tubes are used widely in growth rooms because they provide uniform light intensity over a wide area. Using cool white lamps (at 125 watts), seven tubes mounted 1.5 to 2 feet (0.45 to 0.62 meters) above the soil provide low-intensity light for a 3.5-foot-wide (1.1-meter-wide) bench.

FIGURE 11-29 Fluorescent lights.

FIGURE 11-30 High-intensity-discharge lamp.

Incandescent light installed at about 2 to 3 feet above the plants to generate 10 foot-candles of light intensity is referred to as *standard mum lighting*. At this intensity, flower formation is prevented so that plant growth is stimulated during the short nights of summer. Poinsettia plants can be kept vegetatively by using this treatment. It is important that light intensity be monitored periodically to ensure that the crop receives at least 10 foot-candles of light. Sometimes reflectors (e.g., aluminum foil) may be installed above the lightbulbs to direct more light to plants. High-density lamps are usually encased in reflectors. They have a higher output and are hence hung higher above the plants.

Duration

Duration of light *(photoperiod)* is a precise time-tracking mechanism (chapter 2). Even changes between duration of light and darkness that deviate minutely could spell disaster in certain situations. Plants are traditionally categorized as *long day, short day,* or *day neutral* (3.2.1). As already indicated, this is a misnomer since the tracking mechanism actually relates to the duration of darkness rather than daylight.

The importance of controlling photoperiod lies in the consequences of not doing so. In poinsettia, the red (or other color) pigmentation in the bracts depends on photoperiod, similar to flowering in chrysanthemum. Short nights are required for asters and lettuce to bolt (form stems and flower). Under short-night conditions, dahlias are prevented from producing tubers and thus produce flowers instead.

On the other hand, plants such as kalanchoe, azalea, and chrysanthemum flower when nights are long. When nights are long, dahlias form tubers instead of flowers; bryophyllum leaves produce plantlets along the margin when nights are long (figure 11-31). For some plants, the duration of night length is not inhibitory to flowering, but the process occurs much more quickly under ideal conditions. These plants are called *facultative short-night* (e.g., carnations) or *facultative long-night* (e.g., Rieger begonias) plants. As indicated previously, photoperiod is influenced by other factors such as temperature, species, and maturity. Plants such as roses are day neutral.

> **Facultative Long- and Short-night Plants**
> Plants that do not require a specific duration of dark period for a response to occur, but will respond faster if the dark period is extended or shortened, respectively.

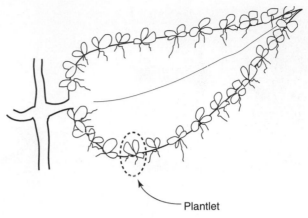

FIGURE 11-31 Bryophyllum leaf producing plantlets in the leaf margin. These plantlets can be used for propagation.

Extending Day Length   To control photoperiod, incandescent lamps are best for extending the day length in short-night treatments because they emit light in the red wavelength of the light spectrum. The Pr *phytochrome* (red phytochrome) responds to this light. To reduce the cost of short-night treatment, intermittent lighting (called *cyclic lighting,* or *flash lighting* [3.2.1]) is sometimes used instead of continuous lighting over the prescribed duration of light interruption. The duration of short-night treatment depends on whether the crop is planted in winter or summer. Nights are naturally short in summer, and therefore no interruption may be necessary then. Short-night treatment involving turning lights on for two to eight weeks is necessary during the winter. The treatment may be administered by turning lights on during the late afternoon to extend the day or interrupting the dark period during the middle of the night (for a shorter duration). The number of hours of supplemental light also increases toward December. The latitude plays a significant role in the duration of light interruption, since northern latitudes experience shorter summer nights and longer winter nights.

Decreasing Day Length   Plants that require decreasing day length treatment flower only when they are exposed to extended periods of darkness. In commercial greenhouses, black cloth is used to block out light at the desired time and removed after the required period (usually from 7 P.M. to 7 A.M.). Light blocking may be done mechanically (very laborious) or automatically. Care must be taken to provide complete darkness since leaks that let in light can cause aberrant developments such as hollow flower buds (crown buds).

11.3.3   WATER

Water is critical to the quality of plant products produced under greenhouse conditions. Plants are adversely affected by both excessive and inadequate moisture supplies. Overwatering can injure plant roots by creating anaerobic conditions in the root zone. Because air pores are occupied by water, plants may wilt and die or become stunted in growth. Too-frequent watering may produce excessive tissue succulence, making plants weak. Moisture stress from infrequent watering may cause plants to wilt or grow at a slow rate. Plant leaves may become small and the general plant stature stunted, with short internodes.

A successful watering regime depends on a well-constituted growing medium. The medium must drain freely while holding adequate moisture for plant growth. The purpose of watering is not to wet the soil surface but to move water into the root zone. The growing medium should be watered thoroughly at each application, requiring that excess water be drained out of the pot. This practice also prevents the buildup of excessive amounts of salts from fertilizer application. Excessive watering is not due to one application but rather is the result of watering repeatedly when it is not needed. When watering frequency is too high, the soil is prone to waterlogging and poor aeration. Roots need to breathe and thus must not be constantly underwater. A good watering regime is one that applies water in a timely fashion, just before plants go into moisture stress. This stage must be ascertained through keen observation and experience. It is recommended that about 10 percent of water applied to a pot drain out of the bottom of the container. This amount ensures that the soil is thoroughly wet, and aids in flushing out excessive salts that may have accumulated as a result of fertilizer applied over a period. A 6-inch (15.2-centimeter) azalea pot should receive 10 to 20 ounces (283.5 to 567 grams) of water, an amount that must be adjusted according to the kind of soil (soilless or real soil mix).

Source of Water

The quality of local water depends on its source because of groundwater pollution problems and water treatment programs. Even where domestic water is used for irrigation, it should be noted that cities treat their water differently. Some add fluoride (called *fluoridation*) to reduce tooth decay in humans. Unfortunately, many plants, including the spider plant *(Chlorophytum),* corn plant *(Drocaena fragrans),* and spineless yucca *(Yucca elephantipes),* are sensitive to and injured by fluoride. Most of these plants belong to the families Liliaceae and Marantaceae. Chlorine is less of a problem but nonetheless injurious to some plants, such as roses, when the concentration reaches about 0.4 parts per million (ppm). When the source of water is a well, the level of soluble salts (sodium and boron) is often high. Many arid coastal regions have high boron toxicity; liming of soil reduces the potential for this problem. Bicarbonate is another water pollutant that affects plant growth.

Water Application Methods

Hand Watering   The age-old method of hand watering by using a watering container or water hose is appropriate for small-scale watering. When a large area is involved, commercial companies often use automated systems. Even then, hand watering may be best and is thus used on certain occasions. The devices used are equipped with nozzles designed for a variety of situations. Some provide fine, misty sprays and others coarser sprays. Watering seedlings and newly sown seeds requires a fine spray. Applying water uniformly is difficult when hand watering. One of the advantages of hand watering is the instant judgment of the operator as to whether a plant needs water. Also, a keen operator can observe any problems (e.g., diseases and insect pests) during hand watering.

Automatic Watering Systems   Automatic watering systems differ in cost, efficiency, and flexibility. Some are semiautomatic and must be switched on and off. In a truly automatic system, the operation is controlled by a programmable timer, which allows watering to occur for a specified period.

TUBE WATERING   Growers use sophisticated methods of water application for premium-quality produce. The tube watering system is the most widely used system of watering for potted plants. When flowers are grown for the fresh flower market, they must be clean. Water is administered only to the soil in the bed without splashing on the plants, which reduces the chance for disease spread. By using microtubes, individual pots can be automatically watered to provide water in the right amounts and at the right frequency for high quality. Two commonly used tube watering systems are the Chapin and Stuppy systems. The difference between the two lies in the weights at the end of each tube. The Chapin system uses a lead weight, and the Stuppy uses plastic (figure 11-32). Tube watering is also amenable to the application of water-soluble chemicals (e.g., growth hormones and fertilizers). The system is expensive to install and not flexible. Pots on the bench have to be uniform in size, and the tubes must be inspected periodically to ensure that no blockage exists.

CAPILLARY MAT   The capillary mat is a subirrigation system in which potted plants are placed on a water-absorbing fiber mat covered with a perforated plastic sheet (figure 11-33). Water is absorbed by capillary action through the drainage holes in the bottoms of the pots. The mat is kept moist by tubes installed on the bench. Potted plants such as African violets are watered by this method. A capillary mat is flexible and easy to install. Different pot sizes can be accommodated simultaneously. Further, it is easy to rearrange plants as required by growth.

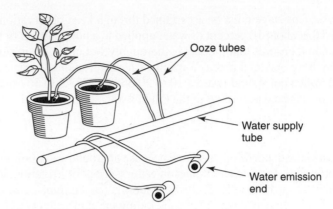

Ooze tubes

Water supply tube

Water emission end

FIGURE 11-32 Microirrigation system. Also called the tube or "spaghetti" watering system.

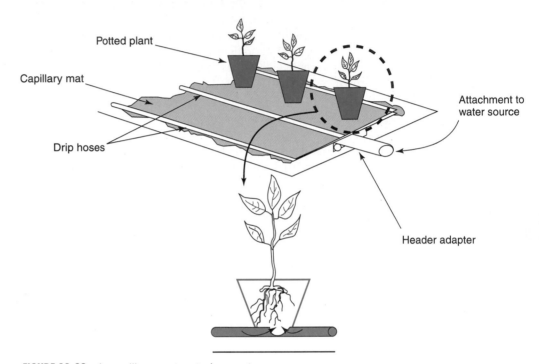

Potted plant

Capillary mat

Drip hoses

Attachment to water source

Header adapter

FIGURE 11-33 A capillary mat watering system.

For the system to work properly, the mat must be placed on a level surface. One problem with using a capillary mat is the growth of algae.

OVERHEAD SPRINKLERS   When used on ground beds, nozzles are mounted on risers whose height depends on the mature height of the crop (figure 11-34). For plants on benches, the sprinkler nozzles may be suspended above the plants. Overhead sprinklers wet the foliage and predispose it to diseases. Watering should be done in the morning to give plant foliage a chance to dry during the day. This system is used widely for watering bedding plants.

PERIMETER WATERING   The cut flower industry cannot afford to have blemishes on foliage and flowers. The perimeter watering system is a method of greenhouse irrigation in which water is provided without wetting the foliage of the plants (figure 11-35). Sprinkler nozzles are installed along the perimeter of the bench. These nozzles deliver a flat spray, and thus the foliage and the part of the plant stem to be included in cut flowers are not wet. The water must be sprinkled at a high enough pressure to reach plants in the center of the bench.

ISRAELI DRIP SYSTEM (FOR HANGING BASKETS)   The Israeli drip system of watering hanging baskets involves hanging plants on a support pipe at intervals that coincide with drip points on a

FIGURE 11-34 An overhead irrigation system for watering plants on greenhouse benches.

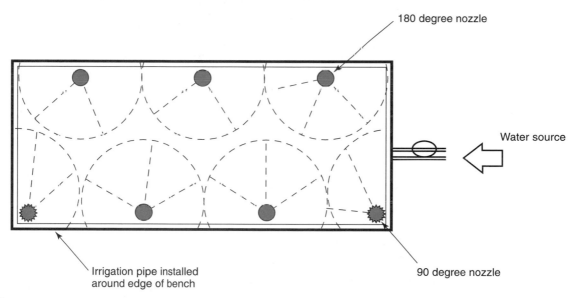

180 degree nozzle

Water source

Irrigation pipe installed
around edge of bench

90 degree nozzle

FIGURE 11-35 Perimeter watering of a greenhouse bench.

plastic pipe located under the support pipe (figure 11-36). The disadvantage of this system is that water drips on foliage, predisposing it to disease. To minimize this problem, plants should be watered early in the morning so that they have a long time to dry before nightfall.

MISTING Misting systems provide very fine sprays of water. They are used on plant propagation benches or beds. Mists are produced from sprinklers fitted with nozzles for very fine sprays.

POLYETHYLENE TUBING Perforated plastic pipes may be used to water cut flowers. Pipes are laid between rows of plants in the bed.

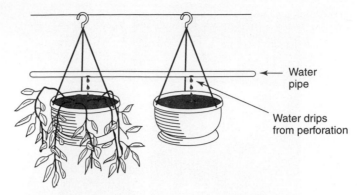

FIGURE 11-36 An Israeli drip irrigation system for hanging baskets.

11.3.4 GREENHOUSE FERTILIZATION

Plants grow in a restrictive environment in a greenhouse. The volume of the growing medium in a pot is only a fraction of that available to plants in the field. Further, the growing medium may be soilless and hence require nutritional supplementation, especially micronutrients. As compared to field fertilization of crops, greenhouse crops sometimes receive a hundredfold of the fertilizer applied in the field. Organic fertilizers are not commonly used in greenhouse plant production because they do not lend themselves to automation (e.g., application through irrigation water). They tend to have odors that make their use under enclosed conditions unpleasant. Further, it is difficult to apply exact concentrations of nutrient elements when using organic fertilizers. Fertilizers are discussed in detail in chapter 3.

Inorganic Fertilizers

Inorganic fertilizers are convenient to apply and amenable to a variety of methods of application, including through irrigation water. Inorganic fertilizers can be applied in exact amounts as needed. Dry and liquid fertilizers are used in the greenhouse, the more common being liquid fertilizers, which are applied through the irrigation water. Another form of fertilizer used in greenhouses is slow-release fertilizer (3.4.2); it is highly desirable because nutrients are released over a period of time and thus better utilized.

Methods of Application of Liquid Fertilizer

Liquid fertilizers may be applied to plants in the greenhouse in several ways.

Constant Feed   Constant-feed application entails administering low concentrations of fertilizer each time the plant is irrigated. It mimics the slow-release fertilizer action and is the most popular method of greenhouse fertilization. This method is desirable because plants receive a fairly constant supply of nutrients in the soil for sustained growth and development.

Intermittent Feed   Greenhouse plants may be fertilized according to a periodic schedule such as weekly, biweekly, or monthly. The disadvantage of this method is that the high level of nutrition available at the time of application gradually decreases over time until the next application. Plant growth is thus not sustained at one level but fluctuates.

Automation

To eliminate the tedium in preparing fertilizer solutions each time an application is to be made, growers usually prepare concentrated stock solutions that are stored in stock tanks. By using devices called fertilizer injectors (also commonly called *proportioners,* since most of them are of this type), the desired rate of fertilizer is applied through the irrigation water (figure 11-37).

> **Proportioner**
> A device used in a fertigation system to control the rate of fertilizer applied through the system.

FIGURE 11-37   A greenhouse fertigation system.

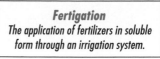

Fertigation
The application of fertilizers in soluble
form through an irrigation system.

Measuring Fertilizer Concentrations

Nutrients may be applied as solids or liquids. When in solid form, they are either mixed in with the medium or applied in the form of slow-release fertilizer. Fertilizer is often applied through the irrigation water; these fertilizers must thus be water soluble. They are formulated as high concentrations of the elements they contain and must be diluted before use. The most common greenhouse fertilizer grade is 20-20-20.

Liquid fertilizer concentrations used in greenhouses are measured in parts per million. To prepare a stock solution to a desired concentration, the grower needs to know the proportioner ratio, stock tank volume, rate of application intended, and fertilizer analysis of the fertilizer source.

The formula for calculating stock concentrations is as follows:

1. ounces/100 gallons of water = ppm desired/% element $\times$ correction factor $\times$ 0.75

    where correction factors are % P = % P_2O_5 $\times$ 0.44

    $$\% \text{ K} = \% \text{ K}_2\text{O} \times 0.83$$

    (No correction is needed for nitrogen.)

    Example: A nursery grower would like to fertilize his plants at the rate of 275 ppm of nitrogen. He purchases a complete fertilizer of analysis 20:10:20 and intends to use a 1:200 proportioner for the application. His stock tank can hold 50 gallons (227 liters) of liquid. How many pounds of the fertilizer should be dissolved in his stock tank?

    Solution:

    ounces/100 gallons = 275/20 $\times$ 1 $\times$ 0.75

    $\qquad\qquad\qquad$ = 275/15

    $\qquad\qquad\qquad$ = 18.33 pounds

2. Preparing the stock solution.

    pounds to add to stock tank:

    = ounces/100 gallons $\times$ second number of injector $\times$ stock tank volume (gallons)/100 $\times$ 16

    = 18.3 $\times$ 200 $\times$ 50/100 $\times$ 16

    = 183,000/1,600

    = 114.4 pounds

Proportioners are calibrated in terms of the ratio of dilution of concentrated fertilizer with irrigation water. Some have fixed ratios (e.g., 1:100 or 1:200), and others are variable. They differ in mechanisms of operation, proportioning ratio, and cost. A relatively inexpensive type is the Hozon proportioner, which operates on the Venturi principle (figure 11-38). Water enters the unit at very high velocity, generating a suction in a feeder line that is dipped into the soluble fertilizer concentrate. The suction draws the fertilizer into the watering hose, where it mixes with tap water according to a predetermined concentration. Proportioners may lose their accuracy over time and require periodic callibration.

Problem of Accumulation of Soluble Salts   Pot media are prone to accumulation of soluble salts, which may result from factors such as poor drainage, excessive fertilization, or insufficient irrigation (resulting in inefficient leaching). Such accumulation may result in toxicity or deficiency of certain minerals. Monitoring the soluble-salt content of growing media, which can be done with an instrument called a *solubridge,* is important.

Fertilizers differ in solubility and *salt index* (a measure of the effect of a fertilizer on the soil solution). Potassium chloride has a salt index of 116, and urea, ammonia, and regular superphosphate have salt indices of 75, 47, and 8, respectively. Fertilizers of low salt indices are preferred to reduce salt accumulation. A greenhouse water supply may contribute to the amount of soluble salts in the soil or growing medium.

11.3.5 CARBON DIOXIDE FERTILIZATION

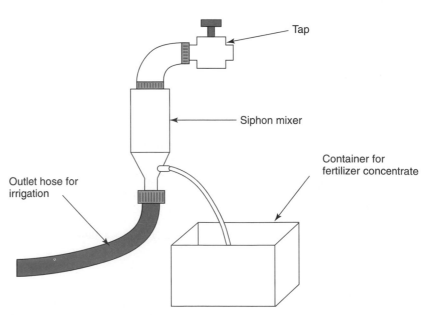

Carbon Dioxide Fertilization
Deliberately increasing the carbon dioxide concentration in the air of the greenhouse to increase the rate of photosynthesis.

Plants utilize carbon dioxide and water in the presence of light to manufacture food by the process of photosynthesis (4.3.1). Carbon dioxide occurs in the atmosphere in a low concentration of about 350 ppm. As long as adequate ventilation and freely circulating air are present, greenhouses receive sufficient amounts of carbon dioxide for photosynthesis. However, in winter, greenhouse vents are closed for more efficient heating. Because plants use carbon dioxide (animals and other combustion processes produce the gas as by-products of respiration [4.2.2] or burning of fuel), the supply of this gas in an airtight greenhouse is limited, which in turn decreases the rate of photosynthesis. Under such conditions, the greenhouse environment may be enriched with supplemental carbon dioxide by burning natural gas in an open flame or using compressed gas or dry ice (solid, frozen carbon dioxide that is used in other laboratory procedures as well). Propane gas burners may also be used to generate carbon dioxide. The goal of carbon dioxide fertilization is to raise the concentration of the gas to about 1,000 to 1,500 ppm. This high level of carbon dioxide must be provided along with bright light (e.g.,

Tap

Siphon mixer

Container for fertilizer concentrate

Outlet hose for irrigation

FIGURE 11-38   A Venturi-type proportioner.

bright daylight) for it to be beneficial to plants. In addition to light, heat is a limiting factor to the effectiveness of carbon dioxide fertilization. Lettuce plant weight increased by more than 30 percent at 1,600 ppm of carbon dioxide. Certain crops have been known to flower earlier under carbon dioxide fertilization. Some of the equipment used to provide additional carbon dioxide may release undesirable toxic gases such as carbon monoxide and ethylene during combustion of the fuel; such gases are toxic to both plants and animals.

11.4: GREENHOUSE PESTS

Greenhouse pests are discussed in chapter 6. The use of pesticides (especially those that are volatile) in enclosed places is a potential health hazard to humans present in those areas. Every effort should be made to minimize the introduction of pathogens and weeds into a greenhouse. Weeds may grow in greenhouses with gravel or porous concrete floors. Apart from being unsightly, weeds may harbor insects and other pests. Weed seeds may be blown into the house or carried in potting media and impure crop seed. If required, only chemicals approved for greenhouse use (e.g., Roundup) may be used. Common greenhouse diseases include root rot, damping-off, botrytis blight, powdery mildew, and root-knot nematodes. Disease incidence can be reduced through strict observance of sanitation. Pots and all containers should be sterilized before reuse. After each crop cycle, the benches should be scrubbed and sterilized. Unsterilized media should not be allowed into the greenhouse. It is recommended that a greenhouse operator maintain a regular schedule of preventive programs. Since many greenhouse plants are grown largely for aesthetic uses, any blemish reduces the price that can be obtained for a product. A wide range of pests are found in the greenhouse, including aphids, fungus gnats, thrips, mealybugs, leaf miners, mites, and whiteflies.

Sometimes the grower has no choice but to use pesticides. On such occasions, the safest products and those recommended for greenhouse use must be selected. After a spray or fog application of a pesticide, the greenhouse must be aerated before people are allowed to work in the facility. Greenhouses may attract rodents (especially mice and rats), depending on the activities taking place in the facility and how it is kept. Baits and traps may be placed at strategic places to catch these rodents.

An effective approach to controlling pests in the greenhouse is the adoption of greenhouse-specific integrated pest management (IPM) programs. Such programs entail a combination of strategies, namely, sanitation, physical control, biological control, and use of pesticides. It should be pointed out that IPM will work properly in a greenhouse if the facility is used for a specific plant or crop or can be partitioned into sections for specific activities. IPM programs may include regular fumigation, washing and disinfecting floors, spraying, and other sanitary precautions. Plant debris should be removed and disposed of without delay, since debris may harbor pests.

One way of keeping plants healthy is to provide adequate growth factors (nutrients, light, temperature, water, and air). Healthy plants are more equipped to resist attacks from pests. To prevent entry of flying insects, ventilators may be covered with fine mesh screens. Other physical control measures include the use of yellow sticky traps. These traps have a dual purpose—monitoring the level of insect infestation and reducing their population (since once captured they die). For ground beds, plastic mulches may be used to suppress weeds.

Improved cultivars of horticultural plants with resistance to various diseases and pests have been developed by plant breeders. Some cultivars are better suited to greenhouse production than others. Tools should be cleaned and properly stored after use. Phytosanitary observance is also critical. All plant remains must be discarded.

SUMMARY

Plants may be grown in or out of season by growing them under controlled-environment conditions in greenhouses. Under indoor conditions, some or all natural growth requirements may either be supplemented or controlled in order to provide the appropriate amounts at the right times. Greenhouses differ in design, efficiency, and cost of construction and operation. Film-plastic models are cheaper to construct but less durable. More expensive and durable materials are glass and fiberglass-reinforced plastics.

An advantage of growing plants in an enclosed environment is that the optimal growth factors can be provided with minimal fluctuations. Heat during the cold season is provided by using heaters that burn fuel (e.g., unit heaters) or those that do not burn fuel (e.g., radiant heaters). Greenhouse heat may be lost through conduction, infiltration, or radiation, conduction being the major avenue. In summer, greenhouses are cooled by using cooling systems such as the evaporative cooling system. Light intensity, which increases during summer, is controlled by shading with fabric or whitewashing the glass outer covering of the greenhouse. During the winter months when light intensity decreases, the paint is washed away. Sunlight is supplemented with artificial lights from a variety of sources, including incandescent lights, fluorescent lamps, and high-intensity-discharge lamps.

Water used in greenhouses often comes from the domestic supply system. Care must be taken not to overwater greenhouse plants, which are generally planted in pots. The soil must be freely draining. The nutrient supply is important since the growing medium may be a soilless mixture, which will be low in nutrients or have none at all. Since the growing plant has a limited soil volume from which to forage for essential nutrients, the soil quickly becomes depleted of minerals and needs replenishing on a regular basis. Slow-release fertilizers are a good choice for greenhouse fertilizer applications.

REFERENCES AND SUGGESTED READING

McMahon, R. W. 1992. An introduction to greenhouse production. Columbus Ohio Agriculture Education Curricular Material Services.

Nelson, P. V. 1985. Greenhouse operation and management, 3rd ed. Retson, Va.: Retson Publishing.

Ortho Book Staff. 1979. How to build and use greenhouses. San Francisco: Ortho Books.

Sunset Book Editors. 1976. Greenhouse gardening. Menlo Park, Calif.: Sunset-Lane.

Walls, I. G. 1991. The complete book on the greenhouse. London: Wardlock.

PRACTICAL EXPERIENCE

Conduct a field trip to commercial or research greenhouses. On the trip, find out about the greenhouse design type, materials used, purpose of the unit, how it is cooled and heated, various types of automation systems, layout of the benches, kinds of plants grown, and so forth.

OUTCOMES ASSESSMENT

PART A

Please answer true (T) or false (F) for the following statements.

1. T F   Glass greenhouses were the first to be constructed commercially.
2. T F   Most of the heat in a greenhouse is lost through conduction.
3. T F   Thermostats are devices used to regulate light intensity.
4. T F   Sunlight intensity is seasonal.
5. T F   To reduce shadows cast by greenhouse frames, the ridges should be oriented in an east-west direction in regions above 40 degrees north latitude.

6. T F Flat plastic greenhouse frames lose more heat than corrugated plastic frames.
7. T F Fluorescent lamps generate more heat than incandescent lamps.
8. T F Overwatering may cause a plant to die.
9. T F A 1:100 proportioner delivers a higher concentration of fertilizer than a 1:200 proportioner.

PART B

Please answer the following questions.

1. List three materials used in the construction of greenhouses.
   _____ _____ _____

2. Give one advantage and disadvantage of each of the materials listed in question 1 for constructing a greenhouse.
   _____ _____
   _____ _____
   _____ _____

3. The most common way of cooling a greenhouse is by using an _____.
4. Give two examples of how seasonal light intensity in a greenhouse may be changed.
   _____ _____

5. Give one example of each of the following:
   a. Day-neutral plant_____
   b. Short-day plant _____
   c. Long-day plant_____
6. List three common greenhouse diseases.
   _____ _____

PART C

1. Describe how heat is lost through conduction from a greenhouse.
2. Distinguish between unit heaters and convection heaters in terms of how they operate.
3. Using two examples, describe how greenhouse horticultural plants are affected by changes in light intensity.
4. Describe, giving examples, the consequences of not controlling photoperiod in horticultural plant production.
5. Describe how day length is shortened in greenhouses for crop production.
6. Describe two specific measures one may take to reduce greenhouse disease incidence.
7. Describe the dangers associated with using heaters that combust fuels.

SECTION 2 GREENHOUSE PRODUCTION

OVERVIEW

Three primary factors should be considered when planning a horticultural production enterprise: *product quality, production cost,* and *transportation cost.* Whereas floricultural plants may be successfully produced in the field, their quality is often not consistent over time because of unpredictable and sometimes hard-to-control environmental factors. Certain plants are more environmentally labile than others. For example, chrysanthemums of acceptable quality may be produced in the field (as in Florida and California). However, roses produced in the field have inferior quality to those produced under greenhouse conditions. Due to bulk, potted plants are produced near major markets (i.e., localized competition). A successful production enterprise should produce high-quality plants at low production and transportation costs.

Floricultural plants adapted for greenhouse production may be grouped as follows:

<table>
<tr><td>

Cut Flowers
Flowers grown specifically to be cut and displayed in vases or for other uses.

</td><td>

1. *Cut flowers.* Some plants are grown so that their fresh flowers may be harvested (cut) and sold.

2. *Potted plants.* There are two classes of potted plants:
   a. *Flowering plants.* Flowering plants are flower-bearing plants grown in pots.
   b. *Foliage or green plants.* Foliage plants are nonflowering plants grown for the beauty of their foliage.

</td></tr>
</table>

Bedding Plants
Largely annual plants produced especially for planting in beds and in containers.

3. *Bedding plants.* Bedding plants are grown to the seedling stage and sold to be transplanted to beds and pots. There are two classes:
   a. *Ornamental plants.* Ornamentals include certain vegetables that are grown for aesthetic purposes.
   b. *Vegetables.* Vegetables are grown to maturity and harvested for sale.

Another basis for grouping greenhouse plants is how they are grown. Plants may be grown in containers (pots), on benches or ground beds, in hanging baskets, or in soilless media (hydroponics).

11.5: Production Costs

The cost of greenhouse production depends on many factors, the major ones including local climate, cost of labor, cost of materials, marketing costs, method of production, and scale of production. The costs associated with these factors may be grouped into direct, indirect, and marketing costs.

11.5.1 DIRECT COSTS

1. *Materials.* Cost for materials includes that for seeds and other planting materials, chemicals (including fertilizers, pesticides, and hormones), growing media, and containers (e.g., pots, trays, and flats).

2. *Labor.* The size of the labor force depends on the level of automation. Some activities are hard to automate and hence must be done by humans. These activities include spraying, preparation of the growing media, setting up containers on benches, harvesting, and cleaning.

11.5.2 INDIRECT COSTS

Indirect costs are generally described as *overhead costs* and include equipment depreciation, utility fees (e.g., water, fuel, and electricity), taxes, and administrative costs. Overhead costs may be about 25 to 40 percent of total production costs.

11.5.3 MARKETING

A major marketing cost is transportation. Others are packaging and advertising. Certain floricultural products require special packaging before marketing. Distribution and marketing costs may vary between 5 and 15 percent of total production costs.

11.6: PRODUCTION REGIONS OF THE UNITED STATES

U.S. Department of Agriculture (USDA) data indicate that bedding plants are the most important greenhouse-produced floricultural plants, accounting for about 35 percent of total production. They are followed by potted flowering plants (24.2 percent), foliage plants (18.5 percent), cut flowers (18.2 percent), and cut cultivated greens (4.0 percent) (table 11-1). The most important production areas in order of volume are California, Florida, New York, Texas, and Ohio, the first three states accounting for about 50 percent of total production. *Plug technology* has greatly facilitated the production of bedding plants and green potted plants. The most important states in this regard are California (16.9 percent), Michigan (8.6 percent), Texas (8.1 percent), New York (7.4 percent), and Ohio (7.3 percent). The most important production states for green plants are Florida, California, and Texas, whereas cut flowers are produced on the largest scale in Colorado, California, and Florida. Colorado and California lead in carnation production; most cut flower roses are produced in California.

> **Plug System**
> A highly automated seedling production system in which seeds are direct-seeded into individual cells in a plug tray (or sheet) by an automatic seeder and covered by a germination mix; little seedlings are transplanted into flats or bedding plant packs and grown to size for sale.

11.7: PRODUCTION AND MARKETING STRATEGIES

Greenhouse producers operate one of three types of floricultural businesses. One group of producers *(growers)* limit their enterprise to producing plants, leaving the selling to others such as wholesalers. They usually concentrate on the production of one or a few types of plants. Some producers *(grower wholesalers)* produce some of their products but also purchase from other growers and sell on a wholesale basis. In addition to plants, they may also sell floricultural production accessories such as containers and wrappers. The third category of growers *(grower retailers)* market their products through a network of outlets. They may sell to other retailers or wholesalers. Brokers acting as sales agents arrange sales between growers and wholesalers. Flowers may also be auctioned. Full-service retailers add value to the fresh products through professional arranging (chapter 22) or wrapping potted plants and tying with a bow, the function of the traditional florist. Florist prices are high because of their overhead costs. Floricultural products may be sold by the mass marketing strategy of selling in high-traffic locations (e.g., malls, airport terminals, street corners, and bus terminals). Prices at these locations are usually low.

Table 11-1 Selected Popular Greenhouse-Produced Foliage Plants

Plant	Scientific Name
Boston fern	*Nephrolepis exaltata*
Dumbcane	*Dieffenbachia* spp.
Snake plant	*Sansevieria trifasciata*
Croton	*Codiaeium variegatum*
Medicine plant	*Aloe vera*
Pothos	*Scindapus aureus*
Philodendron	*Philodendron* spp.
English ivy	*Hedera helix*
Umbrella tree	*Schefflera digitata*
Rubber plant	*Ficus elastica*

11.8: Growing Plants for Festive Occasions

Although floricultural plants are sold year-round, some types enjoy boosts in sales on special occasions. For example, roses sell well around Valentine's Day, Easter lilies during the Easter season, poinsettias near Christmas, and carnations at graduation. Weddings and funerals are also special occasions in which large amounts of flowers are used.

11.9: The Role of Imports

Modern delivery systems comprised of efficient transportation (jet) and storage facilities make it easy to transport fresh flowers across continents in a short period. Further, production costs are relatively lower in certain areas with appropriate climates. In terms of greenhouse production, the Netherlands has the largest number of these climate-controlling units. Cut flowers imported into the United States include carnation (80 percent), mum (57 percent), and rose (42 percent). Other plants involved in the import trade include alstromeria, tulip, daffodil, freesia, potea, and cut palm. Major exporters of floricultural products to the United States are the Netherlands, Spain, Italy, Israel, Kenya, Turkey, Brazil, Colombia, Costa Rica, Ecuador, Mexico, Peru, and Japan. To decrease the competition from imports, U.S. producers will have to reduce production costs (e.g., through improved and more efficient management of the enterprise, marketing, high-quality products, and increased productivity for a steady supply of flowers).

Greenhouses may be used to produce crops on a small or large scale. In fact, for certain crops, commercial production is primarily done in a controlled environment for best quality. Greenhouse production has advantages and disadvantages.

11.10: Importance of Greenhouses in Plant Production

Some of the major advantages and disadvantages of greenhouse production are as follows:

11.10.1 ADVANTAGES

1. Greenhouses may be used to provide a head start on field production. Seedlings of plants that are transplanted may be raised in the greenhouse in the off-season and timed to coincide with the beginning of the growing season so that early crops can be produced for premium prices.

2. Crops and ornamental plants may be produced in the off-season to prolong the availability of fresh produce or products.

3. Produce quality is usually high because growers can manipulate the growing environment for optimal production conditions.

4. Tropical crops can be grown in temperate zones and vice versa.

11.10.2 DISADVANTAGES

1. The initial costs of buildings and equipment are high.

2. Greenhouses operate at very high overhead (e.g., heating, cooling, and lighting), and thus their use is limited to the production of high-premium crops and plants.

3. Greenhouse production is not suitable for all crops and is economically applicable to small-sized plants (e.g., vegetables, ornamentals, and herbaceous plants) but not fruit trees.

11.11: The Concept and Application of DIF

Growing plants in the greenhouse offers opportunities for growers to manipulate the growing environment to control plant growth and development. Traditionally, greenhouse producers grow plants at a lower (cooler) nighttime temperature than daytime temperature, the rationale being that a cooler night temperature reduces respiration (4.3.2), thus conserving carbohydrates. For example, on a clear, sunny day, the nighttime temperature may be 60°F (15.5°C) while the daytime temperature would be about 75°F (24°C), a 15°F difference. On a cloudy day, indoor temperature is usually lower during the daytime (about 70°F or 21°C), the night temperature remaining 60°F (15.5°C). This high-day–low-night temperature regime allows plants to photosynthesize and grow (stem elongation) more during the daytime and less at night.

Scientists have discovered that going against tradition and growing plants at a lower temperature during the day than at night is an effective strategy for controlling plant height (reduced growth or reduced stem elongation). This concept is called *DIF,* the difference between daytime and nighttime temperatures. A positive DIF indicates a lower nighttime than daytime temperature (in the previous example, daytime temperature [70°F (21°C] minus nighttime temperature [60°F (15.5°C)] equals a positive DIF [10°F (5.5°C)]). Similarly, when the nighttime temperature is higher than the daytime temperature, a negative DIF occurs.

Negative DIF treatments are now applied in the commercial production of certain crops to induce an effect similar to the application of growth retardants (4.6). This treatment makes plants short, compact, and fuller in form. Applying the negative DIF concept is problematic in warm climates and under conditions where daytime temperature control is erratic. However, the scientists who formulated this concept also discovered that giving plants a two-hour cool temperature treatment immediately after sunrise (the time when temperatures are usually the lowest in a 24-hour period) has an effect equivalent to growing plants at that cool temperature all day long. This treatment can be applied in spring or summer by ventilating the greenhouse and cooling during that two-hour period. In winter, the thermostat may be set to delay the change from night temperature to day temperature to achieve the desired DIF effect.

Although many species exhibit no adverse side effects from the negative DIF treatment, others, including some Easter lily cultivars, may show curly or wilting leaves. Fortunately, these symptoms are usually temporary, especially if the negative DIF treatment is applied when the plant is young. The desirable DIF range is less than 8°F (4.4°C). Extreme temperature differences may permanently damage plants.

MODULE 1 GROWING POTTED PLANTS: POINSETTIA

Greenhouse production of potted plants is illustrated here by the production of poinsettia. It should be emphasized that the procedure that follows is especially designed to illustrate the various activities involved in the culture of this plant. Greenhouses differ in specific details and may perform additional activities not mentioned in this section or omit some of those described.

Common names: Poinsettia, Christmas star, lobster plant, Mexican flame leaf

Scientific name: *Euphorbia pulcherima* Willd.

Poinsettia belongs to the family Euphorbiaceae, which is characterized by latex that oozes from wounded plants (figure 11-39). Traditionally associated with Christmas in the United States, it is also called Christmas star, lobster plant, or Mexican flame leaf. It was introduced into the United States from Mexico. This winter flowering shrub is characterized by its colored and showy bracts that surround the cluster of greenish-yellow true flowers. The

FIGURE 11-39 Poinsettia plants.

challenge in growing this plant is in inducing the top bracts to change color by manipulating the growing conditions. The most common color is red, but cultivars that produce pink, yellow, white, and variegated bracts are also available. Poinsettias are short-day plants and as such flower only when they receive a period in which each day has an uninterrupted darkness of at least 14 hours (3.2.1). Without flower initiation, the plants remain vegetative and unattractive. Nursery growers have developed poinsettia growing into a science so that plants of the right size and shape can be produced for a target date. The management practice is beyond the capability of the casual home grower.

11.12: POINSETTIA CULTURE

11.12.1 PLANTING MATERIAL

The grower may obtain callused and rooted cuttings from other nurseries or produce them locally from *stock plants*. Using stock plants is a less expensive way of propagation than using rooted cuttings and ensures that materials are available on time. Terminal stem cuttings are obtained from succulent plants. It is important that these cuttings be as uniform as possible. They may be rooted in flats containing rooting media or in foam cubes, strips, or wedges (figure 9.11). After cutting, the pieces are misted to stop the flow of latex from the cut surface of the parent plant. The piece is also dipped in water to stem the latex flow and then allowed to dry for a few minutes before planting it in the rooting mix. Rooting occurs within three to six weeks. Cuttings should be obtained in late spring or early summer so that plants will have adequate growing time before December.

The cuttings root rapidly when kept under low-pressure mist. The mist may contain a low concentration of fertilizer. Alternately, 2 pounds (0.9 kilograms) of 20:20:20 per 100 gallons (378.51 liters) of water may be applied about five days after planting the cuttings and repeated every three days. The room temperature should be about 65°F (18.5°C). A bottom heat of 72 to 75°F (22 to 24°C) helps roots to form rapidly.

Once adequately rooted, these seedlings should be repotted into desirable pots according to the production scheme. Some growers reduce the labor required for transplanting by rooting cuttings directly in the final ("finish") containers.

Pots and Potting Mixes

The size at the time of sale depends on the nutrition and conditions of culture but is also significantly influenced by pot size and spacing of pots in the greenhouse. Poinsettias are sold in azalea pots of sizes ranging from 4 to 10 inches (10.2 to 25.4 centimeters). Some pots are large enough to hold several small plants. The number of cuttings per pot depends on the pot size and production system (single stem or pinched). For example, 5-inch (12.7-centimeter) pots may hold about two to three single-stem plants but only one pinched plant (pinching encourages lateral branching for a fuller plant appearance). Small potted plants may also be arranged together in floral designs for more effective displays. Commercial growing mixes can be used to cultivate the plants, or a grower may create his or her own mixes from soil or nonsoil components. The mix must be well draining and free from diseases and pests. Peat-based soil mixes are commonly used by many producers. The pH of the media should be about 5.5 to 6.0. After transplanting the rooted cuttings into pots, they should be watered thoroughly. The pots may be set on benches or on the floor.

11.12.2 SPECIAL PRODUCTION TECHNIQUES

Photoperiod Control

Poinsettias require adequate light for good growth to obtain the appropriate size and quality. However, to develop the most desired and showy bract color, a period of short days (long nights) is required to initiate flower bud development. This photoperiod occurs naturally in late November to early December. However, when plants are grown in greenhouses in which supplemental light is needed for other purposes, special efforts must be made to provide darkness for the poinsettia plants. Cultivars vary in the time when this treatment is needed and its duration. Greenhouse operators provide the appropriate quality and quantity of light for the desired period and then turn lights off for the long-night treatment. A section of the greenhouse may be curtained off with black cloth during the dark treatment period, which may be from 6 P.M. to 8 A.M. each day.

Pinching

Pinching is a horticultural operation in which the terminal growing end of a plant is removed (figure 4.20). Pinching can be done manually. When plants with apical dominance are pinched, lateral buds are encouraged to grow, resulting in full rather than tall, narrow plants (single stem). Pinched plants produce multiple terminal growths that bear flowers and hence increase the number of showy bracts on the plant. Plants are pinched about two to three weeks after transplanting.

Growth Regulation (Height Control)

Since greenhouse lighting may be inadequate, plants tend to grow spindly, with increased internode lengths. To obtain more compact and fuller plants, some poinsettia growers treat their plants with appropriate growth regulators to reduce internode elongation. The hormone also has the added benefit of intensifying plant color, making the bracts darker and more attractive. Examples of effective growth regulators are Cycocel, Bonzi, A-REST, and B-Nine. Plant growth regulators are applied between pinching and the flower initiation stage. Application is best made in the morning. Cycocel may be applied as a soil drench.

Temperature Control

Proper temperature is required for flower bud initiation and development of bracts. A nighttime temperature of 65°F (18.3°C) and daytime temperature of between 70 and 75°F (21.1

and 23.9°C) should be maintained. After bracts start to develop, a cooler nighttime temperature of about 60°F (15.6°C) is desirable for high-quality results.

Other Production Activities

Watering   Plants should be watered as needed. Overwatering is detrimental to the growth of plants, but severe wilting causes leaves to senesce and drop. When plants are grown on the floor, the pots may be placed on plastic water collectors for additional irrigation.

Fertilizing   A regular schedule for fertilizing should be developed for supplying complete nutrition to the plants throughout the growing period. Poinsettias require large amounts of nitrogen and potassium. The fertilizers used should be high in these elements (e.g., 20:10:0) and should be applied at the rate of about 300 ppm/plant in soilless media culture. This rate should be reduced as the plants mature. Slow-release fertilizers (e.g., 14:14:14 Osmocote) may be used.

Spacing   Spacing should be adequate for air circulation and high penetration of light. Spacing depends on pot size and the number of plants per pot. For example, 4-inch (10.2-centimeter) pots are spaced in 9 × 9-inch (22.9 × 22.9-centimeter) areas, while 8-inch pots are spaced in 19 × 19-inch (48.3 × 48.3-centimeter) areas (when growing a pinched crop).

Diseases and Pests   Important pests of poinsettias include whiteflies, spider mites, mealybugs, and fungus gnats. Insects must be controlled as they occur by appropriate chemicals. Observance of strict phytosanitation in the greenhouse reduces pest problems. Root rot is a common problem that can be controlled by drenching soil with fungicides. In humid conditions, gray mold may occur on plants. Fungicides such as Benlate used as a drench or spray control fungal diseases.

11.12.3   A PRODUCTION SCHEME

A production schedule is developed with the following critical considerations in mind:

1.   Target period for sale: December 10–15. To have plants ready by this time, they should be started at least three months before this date.

2.   Planting material.
   a.   *Cultivars.* Cultivars are categorized according to duration of long-night treatment needed for blooming. For example, a 10-week cultivar requires 10 weeks of long nights to bloom. The different cultivars vary in color (commonly red, but also other shades of red, white, and speckled), leaf shape, plant size, and general appeal.

   b.   *Rooted cuttings.* A grower may purchase rooted cuttings (about 2 to 3 inches or 5.1 to 7.6 centimeters in size) from another nursery. This material is needed by those who desire to raise their own planting materials.

   c.   *Prefinished plants.* Prefinished plants are usually advanced in the stage of production, having received critical treatments of photoperiod and sometimes growth regulators. Some of the plants may already have been pinched. The grower needs to maintain the plants at appropriate temperature and light and provide adequate nutrition.

3.   Size of plants. The size of plants depends on the size of pots, number of plants per pot, and spacing during production. The common pot sizes are 4, 5, and 6 inches (10.2, 12.7, and 15.2 centimeters). Larger sizes are also used. Smaller pots may be grouped to create basket displays. The key is that pinched plants require more room to grow. For example, while a 4-inch (10.2-centimeter) pot with a single plant may

require about 9 × 9-inch (22.9 × 22.9-centimeter) spacing, the same plant, when pinched, may require 15 × 15-inch (38.1 × 38.1-centimeter) spacing.

4. Starting date. The different cultivars on the market require slightly different care in cultivation. It is important to obtain specific instructions from the source nursery (if one does not produce cuttings locally) to know when to start and at what stages to administer treatments such as growth regulators, fertilizers, pinching, and finishing temperatures.

11.12.4 PURCHASE AND HOME CARE

Poinsettias are seasonal plants, and hence timing is of the essence in their production. In an attempt to beat the competition, some growers end up delivering low-quality plants. Retailers may not store the plants under optimal conditions, so plants may deteriorate while in the store. Buyers should select plants with good bract color, good foliage color, and good true flower development. Old plants show browning and yellowing foliage. The plants should be displayed in a cool area that is adequately lit. The soil should not be too dry.

When carrying the plant home, care should be taken not to break the branches or leaves. If the plant is purchased on a cold day, with a temperature below 50°F (10°C), it should be protected from the cold by covering it with a sleeve or placing it in a box. Once at home, the plant should be placed where it will be out of the path of drafts from heaters or cold air from an opened door. Adequate light should be provided, since poor lighting causes leaves to drop. Moisture should be provided only when needed. No fertilization is necessary if the plant is to be used only for the season.

SUMMARY

Poinsettias are associated with the Christmas season. They are short-day plants and hence require short-day treatment in production for them to develop the showy colors in their bracts. The colorful parts of the plants are not petals but bracts. Because of the specific requirements for a good product, poinsettias cannot be successfully grown by a home grower. During nursery production, producers may pinch the plants to make them branch and look fuller. They also apply growth regulators to reduce growth rate to obtain shorter internodes and more compact plants. The sizes and shapes of plants depend on these practices as well as spacing in production. When poinsettias are brought home, they should be provided with adequate light and kept out of the path of drafts from heaters.

REFERENCES AND SUGGESTED READING

Ecke, P. Jr., ed. 1976. The poinsettia manual, 2d ed. Eucinitas, Calif.: Paul Ecke.

Reader's Digest. 1979. Success with houseplants. New York: Reader's Digest Association.

OUTCOMES ASSESSMENT

PART A

Please answer true (T) or false (F) for the following statements.

1. T F   Poinsettias are short-day plants.
2. T F   Poinsettias are associated with the Easter season.
3. T F   Pinching makes plants grow taller.

4. T F Poinsettias are propagated from cuttings.
5. T F Pinched plants require less space in which to grow.
6. T F For high quality, a cooler nighttime temperature of about 60°F (15.5°C) is required for finishing in poinsettia production.

PART B

Please answer the following questions.

1. List two other common names of poinsettias.
   _____ _____

2. Give an example of a growth regulator used in poinsettia production.
   _____

3. List four important pests of poinsettia.
   _____ _____
   _____ _____

PART C

1. Describe how the photoperiod requirements for a short-day plant are met in the greenhouse.
2. Describe a production scheme for producing poinsettias for sale in December.

MODULE 2 GROWING BEDDING PLANTS

Bedding plants are widely used in the landscape at home, parks, and commercial facilities. They are annuals that include ornamentals as well as vegetables. Bedding plants are critical to the gardening industry. They are grown in flats, pots, or hanging baskets, in that order of importance. Certain aspects of production are mechanized for increased efficiency and reduced production costs. Common bedding plants in the United States include those listed in table 11-2.

Table 11-2 Selected Common Bedding Annual Plants

Plant	Scientific Name
Ageratum	*Ageratum* spp.
Snapdragon	*Anthirrhinum majus*
Begonia	*Begonia* spp.
Zinnia	*Zinnia elegans*
Pansy	*Viola tricolor*
Verbena	*Verbena* spp.
Marigold	*Tagetes* spp.
Petunia	*Petunia* spp.
Geranium	*Pelargonium* spp.
Impatiens	*Impatiens* spp.
Coleus	*Coleus blumei*
Cockscomb	*Celosia* spp.
Cosmos	*Cosmos* spp.
Larkspur	*Delphinium* spp.

Greenhouse bedding plants are produced in flats (8.12) or plug trays (figure 11-40).

11.13.1 PRODUCTION IN FLATS

Modern flats are plastic (rather than wooden). They are easy to sterilize and clean and less bulky to carry than wooden flats. They are usually shallow (less than 3 inches [7.6 centimeters] deep). When using flats, seeds are usually sown by hand and at high density, which allows a large number of seedlings to be started in one container. However, because of the high density of planting, seedlings may grow spindly from competition when left too long in the flat. Further, when exposed to disease (e.g., damping-off), many plants will be affected in a short time. During transplanting to pots or other containers, seedlings tend to lose roots and may be in danger of damage.

Planting Seed

After filling the flat with a germination medium, seeds may be sown by broadcasting or in rows, the latter being more common. Sowing seeds in rows facilitates transplanting and slows the spread of disease when it occurs. After sowing, the flat should be watered and placed in a warm, humid environment. When sowing fine seeds, care should be taken not to bury the seeds too deeply in the soil. A soil temperature of about 70 to 80°F (21.1 to 26.7°C) is favorable for seed germination.

Transplanting

Seedlings should be transplanted after the first true leaves develop (8.11.2). Timeliness ensures that seedlings will be sufficiently developed to establish quickly.

FIGURE 11-40 Plug tray production of bedding plants.

11.13.2 PRODUCTION IN PLUG TRAYS

Plug trays differ in terms of their cell sizes, some having several hundred cells. The smaller cells are designed for growing plants for a short period of time since they hold only small amounts of soil. The medium used is usually soilless.

Plug trays are invariably seeded by mechanical seeders instead of by hand. Automated seeders are precise and efficient, drastically reducing planting time. After seeding, the trays should be watered gently. Agricloth spread over the surface reduces the impact of water droplets on the soil, which is especially undesirable for small seeds. A light application of fertilizer produces seedlings with more vigor.

11.13.3 TRANSPLANTING

To prepare for transplanting, seedlings should be hardened for a period of time. Transplanting is a very slow and laborious activity. Seedlings may be transplanted into cell packs or pots (8.13). Plants are transplanted as previously described for flat production. The process constitutes what is called *finishing the crop*. After finishing, the plants are ready for sale to consumers.

In plug production, plants may be transplanted in an assembly-line fashion, whereby a conveyor belt moves finishing containers filled with soil along a row of seated workers who then do the transplanting.

11.13.4 CARE

Newly transplanted seedlings require frequent watering, since drought conditions produce poor finished products. Seedling trays should be watered frequently by overhead sprinkler irrigation. It is best to water in the morning or early afternoon to reduce the incidence of disease. Irrigation water must be monitored for appropriate pH and soluble-salts concentration.

Bedding plants also need a good fertilizer application schedule to grow properly. Fertigation is the method most commonly adopted by growers. Greenhouse production is often done in soilless media, which have little or no nutrients for plant growth. Nutrition must therefore be provided on a regular basis. Sometimes growers include a slow-release fertilizer in the growth medium, which eliminates the need for a constant-feed fertilization schedule. Instead, plain tap water should be used in watering.

In terms of temperature, most greenhouse bedding plants can be successfully produced at daytime temperatures of between 70 and 75°F (21.1 and 23.9°C) and nighttime temperatures of between 60 and 65°F (15.6 and 18.3°C). Most bedding plants require full light, and consequently high-intensity-discharge lamps are usually used. These growth factors should be adjusted according to the practice of hardening. A common problem with greenhouse production is excessive height due to inadequate lighting, overfertilization, or overwatering. Chemical growth retardants such as B-Nine are applied to curtail spindly growth and to make plants look fuller and more attractive.

Bedding plants are often attacked by fungi that cause damping-off, the two most common types being *Pythium* and *Rhizoctonia*. Good sanitation in the greenhouse is helpful, as is the use of resistant cultivars. Common pests of bedding plants are aphids, thrips (western flower thrips, or *Frankliniella occidentalis*), two-spotted spider mites, and whiteflies. If pesticides must be used to control pests, those approved for greenhouse use must be selected.

11.13.5 MARKETING

To prepare for marketing, hardened plants are appropriately tagged with labels that show a picture of the flower, its name, and brief cultural requirements. Bedding plants are produced according to seasonal production schedules. Both commercial growers and home growers purchase these plants, which should be ready by the beginning of the planting season in summer or fall. The stage at which plants are commonly sold is close to blooming. As a result, home growers do not have to wait long to enjoy the blooms. The outlets for these products are diversified and include grocery stores, direct sales from greenhouses, and roadside stalls.

PART A

Please answer true (T) or false (F) for the following statements.

1. T F   Tomato is a bedding plant.
2. T F   Bedding plants are grown solely for planting in beds in the landscape.
3. T F   Most bedding plants require shade.
4. T F   Plug production can be done in an assembly-line fashion.

PART B

Please answer the following questions.

1. List five bedding plants.

    _____   _____   _____

    _____   _____

2. A temperature of between _____ and _____ is favorable for seed germination.
3. List three major diseases of bedding plants during greenhouse production.

    _____   _____   _____

4. At what stage should seedlings be transplanted?

    _____

5. For bedding plant production in the greenhouse, seeds are sown in flats or

    _____ .

PART C

1. Describe the horticultural operation of finishing the crop.
2. Discuss the marketing of bedding plants.

MODULE 3   GROWING CUT FLOWERS: ROSES

Common name: Rose

Scientific name: *Rosa* spp.

11.14: CUT FLOWER INDUSTRY

Cut flower production in the United States is ranked fourth in overall plant production, the top three being bedding plants, potted plants, and foliage plants. Rose sales represent about 40 percent of the total for cut flowers. The other important cut flowers are carnations and chrysanthemums. U.S. producers of cut flowers face stiff competition from external producers, such as Colombia, which are able to produce high-quality products at a fraction of the cost of production in the United States. The existence of efficient transportation makes it feasible to deliver fresh flowers to distant markets within a short period.

Roses are among the most popular flowers worldwide. They are often presented to other people as expressions of love or affection or simply as symbols of beauty. To obtain a good crop, roses need great attention in cultivation. Roses grown for the cut flower market are produced mainly in greenhouses in ground beds.

The florist industry is the principal outlet for cut flowers. These products are arranged for display in vases and used in bouquets, spreads, boutonnieres, wreaths, and the like. The home grower may cut flowers from the garden for use indoors in containers. Major greenhouse-grown cut flower species include rose *(Rosa x hybrida),* carnation *(Dianthus caryophyllus),* chrysanthemum, snapdragon *(Anthirrhimum majus),* freesia *(Freesia refracta),* and alstroemeria *(Alstroemeria* spp.). Cut flowers are commonly grown in ground beds.

11.14.1 CLASSES OF ROSES

According to plant growth habits, roses can be classified into two general classes—bush and climbing roses.

Bush Roses

Bush roses are relatively short in height (2 to 6 feet or 0.61 to 1.82 meters) and self-supporting. The different kinds of bush roses vary in degree of hardiness, flower size, flower number, and plant size.

Hybrid Teas   Hybrid teas, the most widely grown roses (figure 11-41), are everblooming and widely used in the cut flower industry. Hybrid teas vary in height, being as short as 2 feet (0.61 meters) and reaching heights in excess of 6 feet (1.82 meters) in certain varieties. In season, fresh roses can be harvested each month from hybrid tea bushes. Most hybrid tea cultivars require special care in winter.

Floribundas   Floribundas are vigorous bush roses that are prolific bloomers. Their flowers are borne in clusters. They make good bedding plants and require less care in cultivation than hybrid teas. Floribundas are also very hardy plants.

FIGURE 11-41   A rose flower.

Grandifloras Cultivars of grandifloras are hybrids of hybrid teas and floribundas. Grandifloras are large and vigorous plants and produce flowers in clusters. They are used in the cut flower industry and are also good as bedding plants.

Polyanthas Polyanthas are hardy dwarf plants that require relatively little care in production. Their flowers are small and produced in large clusters.

Miniature Roses As the name implies, miniature roses are dwarf species that stand a few inches to about a foot above the ground. They are good species for rock gardens and for use as border or edging plants. Their flowers are also small in size.

Other kinds of bush roses are trees or standard roses, shrub roses, old-fashioned roses, and hybrid perpetual roses.

Climbing Roses

Climbing roses require physical support for their long and slender canes. Such support may be in the form of stakes, trellises, fences, or walls. They can be trained on these structures in cultivation to enhance their display. Without support, they can be good ground covers. Like bush roses, there are different kinds of climbing roses:

1. *Ramblers.* Ramblers are vigorous roses that sometimes produce about 20 feet (6.2 meters) of new growth each season. They bear small flowers in clusters and are usually winter-hardy plants.

2. *Large flowering climbers.* Large flowering climbers are slower growing than ramblers but have large flowers that make them good cut flower plants.

3. *Everblooming climbers (pillar roses).* Pillar roses are winter-hardy plants and prolific bloomers.

4. *Trailing roses.* Trailing roses are characterized by long canes that can trail on walls or the ground. Their flowers are fragrant.

5. *Climbing versions of bushes.* A number of roses are climbing varieties of bush types such as Floribunda, polyantha, and teas.

11.14.2 ROSES FOR CUT FLOWERS

Roses developed for greenhouse production (sometimes called *greenhouse roses [Rosa x hybrida]*) may be grouped into two types—hybrid teas and floribundas.

Hybrid Teas

Hybrid tea roses are characterized by long stems and a single large terminal flower on each stem. Their flower shapes range from singles to high-centered doubles and "very doubles." In terms of color, hybrid teas can be white, pink, red, or brilliant yellow. More than 80 percent of roses grown in the United States are hybrid teas.

Floribundas

Floribundas are descendants of hybrid teas and polyanthas. They produce abundant flowers that are small in size and occur in clusters. They are also called *sweetheart roses*. To suppress the blooming of lateral buds, these buds may be pinched. Some floribundas have a very double-flower shape. They bloom throughout the season and are available in white, yellow, orange, pink, lavender, and red.

11.14.3 ROSE CULTURE

Site and Soil

Greenhouse roses are planted in ground benches (11.2.8), as already indicated. Production is a perennial operation. The soil, which is steam pasteurized before starting a new crop, should be well drained, fertile, and well aerated. Slightly acidic soil (pH of 6.0 to 6.5) is preferred. Loamy soils are best for planting roses. Sphagnum moss, vermiculite, or other materials may be added to improve soil drainage.

Planting

Roses are propagated from cuttings, grafting, or budding (9.1). Plants from budding are most commonly used. The understock widely used in budding in the United States is *Rosa chinensis* var. manetti. Planting materials for roses may be obtained from a nursery as bare roots or dormant roses. Spacing between plants depends on plant vigor and adult size and may vary between 1 and 2 feet (0.31 and 0.62 meters). The size of the hole for planting should be wide and deep enough to comfortably hold the roots. Before setting the plant, the roots should be pruned to remove damaged ones and untangle the remainder. A core of soil is created in the pit from the soil dug from it. The roots are spread over the core such that top roots are about 3 inches (7.6 centimeters) below the soil surface. The hole should be filled and the soil patted firm. Spacing in the bed is 12 × 12 inches (30 × 30 centimeters). After watering, additional soil may be added to the planting hole. Roses should be planted properly, since they remain in production for about 10 years before a new crop is planted.

Support

A grid of wires or plastic mesh is used to support cut flowers in ground bench production. When the plants are young, a grid of smaller size (e.g., 8 × 8 inches or 20.3 × 20.3 centimeters) is laid over the plants in the bed. Tiers of wire grids are added at about 12- to 18-inch (30.5- to 45.7-centimeter) intervals as the plants grow to keep the stems straight (figure 11-42).

11.14.4 CARE

Mulching

Rose plants benefit from mulching, which prevents compaction while conserving moisture.

FIGURE 11-42   Wire grids used in ground bed culture.

Carbon Dioxide Fertilization

Carbon dioxide fertilization (11.3.5) has been reported to reduce the number of blind shoots and increase stem length and weight, as well as the number of petals on the flower. Plants undergoing carbon dioxide fertilization were also found to mature more quickly in winter.

Watering

Roses require a good moisture supply for optimum growth. It is better to water deeply when needed than to provide frequent light waterings. Water should be directed toward the soil rather than the plant. Cut flowers should be watered with systems that do not wet the flowers or foliage, such as perimeter irrigation units (figure 11.35). Moisture on leaves and flowers invites disease organisms, such as *Botrytis*. Where fertigation or hand watering is used, traces of residue may be left on the leaves, which is undesirable in cut flower production.

Fertilizing

Good nutrition is required for high-quality rose production. Vigorous types need more nitrogen fertilization, but excess nitrogen is undesirable. Moderate amounts of complete fertilizer may be used when needed after first conducting a soil test and establishing production goals.

Temperature and Light

Daytime temperatures depend on whether the cut flower is a cool- or warm-season species (1.3.5). Generally, nighttime temperatures are about 5 to 10°F (2.5 to 5.2°C) lower than daytime temperatures. Roses prefer about 65°F (18.3°C) for daytime temperatures. Cut flowers generally should be grown under full light intensity to produce quality flowers. In summer, shading may be necessary to prevent scorching or petal burn.

Disbudding

Disbudding is a practice adopted for improving the size and quality of roses. Hybrid teas can be disbudded to produce large, single-stem flowers.

> **Disbudding**
> The process of removing flower buds, usually lateral ones, in order to improve the size of the remaining bud or buds.

Pruning

Roses are pruned (15.1) for the same reason as other species—to increase flower size, open up the canopy, remove dead and diseased parts, and improve plant shape. Rose plants are cut back each year after the production season to control plant height.

Diseases and Pests

The presence of diseases and pests (7.1) can be reduced significantly by observing basic greenhouse sanitation during cultivation. Dead leaves and weeds should be removed. Flower and foliage should be kept dry. Resistant varieties should be used, where available. Powdery mildew, a serious problem in rose cultivation, develops when foliage is wet. Aphids and red spider mites are also significant problems in cut rose culture.

Harvesting

Harvesting should be done sparingly during the first year of growth. Flowers are best harvested just as the petals start to unfold. This stage is especially desirable when flowers are transported over long distances. Such products are more resistant to damage during shipment. The leaves on the lower third of the stem of the flower should be stripped away. For long vase life, harvested flowers should be placed in warm water immediately after cutting. If they are to be transported over long distances, refrigeration is required. Cut flowers should be stored in a cool environment of about 35 to 40°F (1.67 to 4.44°C) while awaiting sale.

FIGURE 11-43 Cut flowers sleeved and ready for sale.

Postharvest Handling

Before marketing, cut flowers are graded and grouped. The criteria for this activity differ from one species to another and include flower diameter, stem length, and number of flowers per stem. Roses are graded by stem length and then bunched. Bunches are then sleeved for protection before shipment (figure 11-43). Floral preservatives may be added to the holding water of cut flowers to prolong their lives.

SUMMARY

Roses are popular garden flowers. There are two general classes of roses—bush and climbing. The most widely grown are the hybrid teas, which are bush roses. Because they are everblooming, they are very important in the cut flower industry. They are sun-loving plants. Bare root seedlings may be purchased from nurseries for home planting. For best results, soil nutrition levels should be high. Flower size and quality are improved by the practice of disbudding. Roses are best harvested just as petals start to unfurl.

REFERENCES AND SUGGESTED READING

Langhans, R. W., and J. W. Mastalerz, eds. 1969. Roses: A manual on the culture, management, diseases, insects, economics, and breeding of greenhouse roses. Ithaca, N.Y.: Cornell University.

Reader's Digest. 1979. Success with houseplants. New York: Reader's Digest Association.

OUTCOMES ASSESSMENT

PART A

Please answer true (T) or false (F) for the following statements.

1.   T F   Bush roses are the most commonly grown class of roses.
2.   T F   Floribundas are a kind of climbing rose.
3.   T F   Ramblers are a kind of bush rose.

4. T F Roses may be purchased as bare root plants from the nursery for planting.
5. T F Rose plants are not pruned.
6. T F For long vase life, harvested roses should be placed in cold water immediately after harvesting.

PART B

Please answer the following questions.

1. _____ are the most widely grown kind of bush roses.
2. List four kinds of bush roses.

    _____ _____ _____ _____

3. The class of roses that requires physical support in cultivation is called

    _____.

4. What are the classes of roses used most in cut flower production?

    _____

PART C

1. Discuss the use of roses in the landscape.
2. Describe how roses are cared for in greenhouse production.
3. What is disbudding and what is its role in rose production?

MODULE 4 HYDROPONIC PRODUCTION

11.15: BASIC PRINCIPLES

Soil is a medium in which plants may grow. It provides the minerals needed for plant growth and development. When adequate moisture is present, these minerals are dissolved in water and can be absorbed by roots. When plants are grown in another kind of medium, all of the nutrients must be provided in the irrigation water. Over the years, a wide variety of attempts (with varying degrees of success) have been made at growing plants in soilless media (i.e., no real soil is used). The most successful attempt, which perhaps ushered in modern soilless culture, is credited to W. E. Gericke, who in 1936 grew a wide variety of crops in water supplemented with plant growth nutrients. He called this method of growing plants *hydroponics,* or *water culture.* Plants can be cultured in a variety of other inert substances besides water, including rockwool, sand *(sand culture),* or air *(aeroponics).* All of these methods (including hydroponics) are called *nutriculture.* The substances used as media are called inert because they do not add to or alter plant nutrient level in any way (unlike peat moss and natural soil, which are both biologically and chemically active).

> **Hydroponics**
> *The culture of plants in which the root medium is exclusively water fortified with dissolved nutrients.*

11.16: TYPES

There are two classes of soilless plant culture: *water culture* and *substrate culture.*

11.16.1 WATER CULTURE

The three requirements for successful water culture are as follows:

Soilless Culture
The collectivity of methods used to grow plants in media that do not consist of mineral soil.

1. *Root aeration.* The root environment must be aerated to prevent anaerobic respiration (due to waterlogging) from occurring. Preventing anaerobic respiration may include aerating the nutrient solution (as done in domestic aquarium tanks) or circulating the nutrients using a pump. Systems that use a continuous flow of nutrients provide the best cultural environment for plants.

2. *Root darkness.* Algae growth, which occurs around the roots when exposed to light, interferes with root function. Darkness around roots eliminates or reduces this problem.

3. *Physical support.* A system must have a means of holding plants erect in water.

In terms of design, there are two basic types of water culture systems:

1. *Nonrecycling system.* In the nonrecycling system, excess nutrient solution is drained out of the container and lost.

2. *Recycling system.* Excess solution is collected for reuse in the recycling system.

Examples of water culture systems are described in the following sections.

The Gericke's System

The Gericke's system is designed so that plant roots grow through a wire mesh into the nutrient solution, in which they are either partially or fully submerged. This solution may be static or circulated on a continuous basis. Early designs were unsatisfactory because they lacked the air space that allows the solution to be aerated. Apart from poor aeration, another undesirable aspect of the system is the lack of self-support for seedlings. Users need to provide support for plants, which increases production costs. Commercial application of the Gericke's system is currently limited. For better support, some users include a solid substrate (e.g., soil, peat, or straw) in the seedbed.

The Floating Systems

The floating, raceway, or raft water culture system was first developed for lettuce production. Plants are raised in water beds about 6 to 8 inches (15 to 20 centimeters) deep. The nutrient solution in the bed is circulated through a tank. Before returning the nutrient solution to the bed, it is aerated, sterilized (by UV radiation), and chilled. The pH and electrical conductivity of the solution are also regularly monitored. Between crops, the system needs to be sterilized by using bleach, for example.

Lettuce is seeded into a peat plug mix and watered by using a capillary mat. When plants are about 12 to 14 days old, they are transplanted into beds. Planting is best done in the late evening when seedlings are bare rooted. The seedlings are planted in 1-inch-deep (2.5 centimeters) holes in a styrofoam board (raft) after placing them in paper supports. Lettuce is ready for harvesting about 32 days after transplanting.

Floating systems provide some support for plants by using lightweight material that floats (commonly a sheet of polystyrene). Some growers use thick (2.5-centimeter or 1-inch) plastic material for support. Crops that have been grown successfully by this system include chard and strawberry.

Deep Recirculating Water System

Several variations of the Gericke's system are in use. These modern versions are designed to overcome the shortcomings of the previous system. The variations in design are also wide. For example, in the M system, the nutrient solution is circulated by a pump through an air mixer and then returned to the bed via holes in pipes located in the bottom of the container.

Ebb-and-Flow Systems

Ebb-and-flow systems are a type of subirrigation. They consist of a shallow bed, into which nutrient solution is pumped to a depth of about 1 inch (2.5 centimeters). The solution remains in the trough for about 20 minutes and then is drained back into the tank. Flats of plants are set in the shallow bed. Irrigation cycles can be automated; seedlings can be irrigated in plug trays or set in rockwool blocks.

Vegetables such as tomato, pepper, lettuce, broccoli, and cucumber may be propagated using the ebb-and-flow system. Similarly, rose cuttings and cuttings for hanging baskets may be propagated by this method.

Aeroponics

Aeroponics is used on a limited scale for commercial production. This system involves suspending bare plant roots in a mist of nutrient solution.

Nutrient-Film Technique

In the nutrient-film technique (NFT), there is no solid rooting medium. Plants' roots are placed in a shallow stream of recirculating water containing dissolved nutrients (confined to troughs or gullies). Plants produce a thin root mat, part of which is located above the solution. The technique derives its name from the film of nutrient solution on the parts of roots exposed to air. Because their roots are only partly submerged, plants are never in danger of the consequences of waterlogging. By using shallow water, the young seedlings can be set in the water in their propagation blocks or pots without any need for additional support. Furthermore, there is no need for the deep and heavy beds required by other systems.

The basic layout is shown in figure 11-44. The catchment tank is located in the lowest point of the setup. It contains the dilute nutrient solution that is pumped up to the upper ends of the gullies. The gullies must be laid carefully and the flow rate of the pump set properly to prevent irregularities in the flow (i.e., there has to be a uniform gradient of about 1 percent slope). The NFT has other advantages that make it attractive. There is no need for pasteurization, the plastic film used for lining the trough is removed after each production cycle, and it is amenable to automation. Crops being grown in this system include tomato, lettuce, chrysanthemum, snapdragon, and cucumber.

The circulating pump used should be able to tolerate low levels of corrosion from the dilute nutrient solution and be sturdy enough to supply uninterrupted power. Pump failure can be disastrous, so commercial growers often use more than one pump (one serves as a backup). The NFT system may be assembled on the concrete floor or a raised platform. The channels

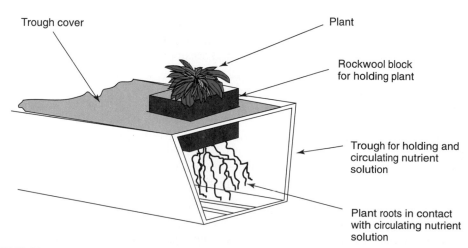

FIGURE 11-44 A nutrient-film technique (NFT) hydroponic system.

require a covering, which helps to reduce evaporation of the nutrient solution, control root temperature, and prevent entry of light (which promotes algae growth). This covering should have reflective properties (not black) to reflect light and thereby reduce the danger of temperature buildup in the troughs, which could damage roots. The nutrient solution is circulated at a rate of about 2 liters (0.5 gallons) per minute. This rate of circulation is required for good aeration. The solution is collected in a sump by gravity.

Nutrient-film fertilizer may be purchased from chemical companies. The solution becomes spent with time and must be replaced periodically (e.g., biweekly). While in use, the pH and electrical conductivity must be monitored on a regular basis (daily). Desirable pH is between 6.0 and 6.5. To adjust pH toward acidity, sulfuric acid may be used; potassium hydroxide is used to increase pH. The electrical conductivity (soluble salts) of an NFT solution may be about 3 millimhos. With recycling and use, the soluble-salt concentration decreases. Replenishing is required when electrical conductivity decreases to about 2 millimhos.

The troughs may be prefabricated or can be formed by the grower from polyethylene sheets. The type of material used is critical. Some have phytotoxic effects that may kill plants or stunt their growth. Flexible polyvinyl chloride pipes and galvanized pipes have potential to cause zinc poisoning, and copper pipes should be avoided. Plants to be grown are raised in media that are lose and will permit transplanting into the NFT channels without debris attached to roots, which can block the nutrient flow in the system. Materials that may be used include peat moss, Oasis foam, and blocks of rockwool. The seedlings are set in rooting blocks.

> **Millimho**
> A millimho is 1/100th of an mho (opposite of ohms, a unit of electrical resistance) or mho/cm $\times 10^{-3}$.

11.16.2 SUBSTRATE CULTURE

Under the substrate system, plant roots are surrounded by either inert or natural organic material. These substrates provide only limited physical support.

Soilless Substrate Culture Systems

Inert Substrate Noncirculating Systems   Materials that may be used as substrates include sand, vermiculite, horticultural rockwool, and perlite. The critical factor in the choice of material is that it be able to hold moisture and yet drain adequately for good aeration. Systems that use these materials are generally referred to as *sand cultures*. They are not widely used commercially since it is not easy to maintain a good balance for drainage, water retention, and aeration.

Commonly used containers include those made of concrete for permanent troughs or beds. Other materials such as timber coated with asphalt or fiberglass may be used. The nutrient solution is applied by using perforated pipes with suitable jets in fixed irrigation models or directly to the soil surface. Sand beds are often sterilized annually to control soilborne diseases and pests.

Shallow-bedded sand cultures are used for shallow-rooted crops. When producing tall-growing row crops such as tomato and cucumber, bags are used as containers (figure 11-45). Drip irrigation systems are used to apply the nutrient solutions. Rockwool (mineral wool) culture is also used for row crops, such as pepper, and ornamentals such as roses that garner premium prices.

Natural Organic Substrates   Natural organic substrates such as peat, sawdust, and wood bark are used as plant growth media. They are less bulky than inert substrates and absorb and retain moisture quite well. Before use, these substrates are fortified with nutrients. The material may be placed in beds or bags *(bag culture).* When using the latter, potted plants, grown in plastic pots with slated bottoms, are set in commercially prepared growing bags. Plants may be watered by using the drip system of irrigation.

Nutrient Solutions Used in Soilless Culture

Just as soil nutrients are needed for plant growth, nutrient solutions used in soilless culture must be complete, having adequate amounts of both macronutrients and micronutrients. The

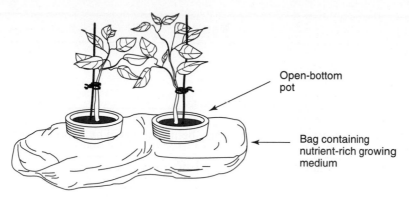

FIGURE 11-45 Bag culture.

formulation of nutrient solutions should take into consideration the plant, the age or stage of growth, the environment (in terms of light intensity and temperature), and the type of soilless culture system.

SUMMARY

Soil has traditionally been the primary source of plant growth nutrients. It also provides physical support for plants. However, technological advances have enabled plants to be grown in media that contain no real or true soils. This soilless culture may take place in water (hydroponics or water culture) or in soilless substrates (bag culture). Water culture has two basic types—recycling and nonrecycling systems (i.e., excess nutrients drain away). Plants may be cultured in inert or organic substrates. Because these substrates provide only physical support, all nutrients must be artificially supplied through irrigation water.

REFERENCES AND SUGGESTED READING

Food and Agriculture Organization (FAO) of the United Nations. 1990. Soilless culture for horticultural crop production: Plant production and protection paper 101. Rome, Italy: Food and Agriculture Organization of the United Nations.

Nicholls, R. 1990. Beginning hydroponics. Philadelphia: Running Press.

Resh, H. M. 1995. Hydroponic food production. Santa Barbara. Woodbridge Press.

OUTCOMES ASSESSMENT

PART A

Please answer true (T) or false (F) for the following statements.

1. T F   No real soil is used in soilless media culture.
2. T F   Bag culture uses organic substrates for plant growth.
3. T F   Irrigation water in bag culture contains soluble major and minor nutrients.

Please answer the following questions.

1. The invention of hydroponics is credited to _____.
2. NFT is the abbreviation for _____.
3. List at least three crops that are currently being cultivated by hydroponic systems.

   _____ _____ _____

4. Give a basic substrate mixture used in substrate culture.

   _____

PART C

1. Discuss the Gericke's system for soilless culture.
2. What are the advantages and disadvantages of hydroponics?

PART 5

GROWING PLANTS OUTDOORS

CHAPTER 12

Principles of Landscaping

PURPOSE

This chapter discusses the principles and steps involved in landscape design.

EXPECTED OUTCOMES

After studying this chapter, the student should be able to

1. Define the term *landscaping* and discuss its categories.
2. List and discuss the basic principles of landscaping.
3. Describe the steps involved in a landscape design.
4. Describe how plants are selected for a landscape.

KEY TERMS

Cold hardiness	Landscaping	Rhythm and line
Focalization	Massing	Simplicity
Interiorscaping	Plantscaping	

OVERVIEW

Using plants outdoors to enhance the general view is not a modern-day invention. Although sometimes plants are used alone, at other times they are used in conjunction with nonplant elements such as sculptures, walkways, and fountains to create a view or environment that is aesthetically pleasing and nurturing to the human spirit (figure 12-1). The key goals of landscaping design may be summed up into two—*function* and *aesthetics*. You may remember from an earlier chapter that horticulture is both a science and an art. Landscaping may be likened to using plants to paint a picture, with the open space as the canvas. One key difference is that unlike painting on canvas, the subjects in the picture are not static but change with time. The changes may be in form, size, and age. It takes time to stabilize plant characteristics, which can be controlled only with good maintenance. However, there is more to landscaping than appearance, even though aesthetics is what is most readily noticeable to most

FIGURE 12-1 A landscape on a college campus combines plants and concrete elements.

people and thus associated with the discipline. The landscape artist, properly called a *landscape designer,* can, with proper choice of plants and design or arrangement, elicit certain responses from people that use the landscaped area.

12.1: WHAT IS LANDSCAPING?

<table>
<tr><td>

Landscaping
The use of plants and inanimate objects outdoors to fulfill aesthetic and functional purposes.

</td></tr>
</table>

Landscaping may be defined as the use of plants outdoors to fulfill aesthetic and functional purposes. The term is identified with the outdoors, even though plants can be used to accomplish similar objectives indoors (sometimes called *interiorscaping* or *plantscaping*). Landscaping is an activity in which beauty, as well as function, may be determined by the customer. To one person, landscaping may mean a couple of fruit trees or just plants on the property. To another customer, plants in the landscape must not only be carefully selected but also strategically arranged.

12.1.1 GOALS OF LANDSCAPING

In fulfilling aesthetic and functional purposes, landscaping may be specifically used to accomplish the following:

1. Enhance the aesthetic appeal of an area. Home and business environments can be beautified to make them more attractive and nurturing to the human spirit. Home owners can enjoy their surroundings and feel relaxed in the presence of an appealing display. People can walk in parks and horticultural gardens to enjoy the beauty of the exhibits.

2. Enhance the neighborhood and increase property value. Homes with curb appeal have higher property values on the real estate market. Landscaping can transform a simple structure into an attractive one.

3. Blend concrete and architectural creations into the natural scenery. Buildings tend to have sharp geometric edges that can be softened with plants. Brick and mortar in a city can be overwhelming and excessively artificial. Plants can be used to introduce life into the area.

4. Provide privacy by shielding the general public from selected areas such as the backyards of homes, utility substations, and patios.

5. Control vehicular and pedestrian traffic. Just as pavements indicate where people should walk, trees, flower beds, and other feature can be used to discourage people

from making undesirable shortcuts across lawns. Trees and other plants on a median in the street prevent drivers from driving over the structure.

6. Hide unsightly conditions in the area. Plants can be used to create a wall around, for example, junkyards and storage areas.

7. Modify environmental factors. Trees can be planted to serve as windbreaks to reduce wind speeds, for example. Similarly, plants can be located to provide shade, block undesirable light, and modify local temperature.

8. Create recreational grounds to provide places for relaxation and community interaction.

9. Provide hobby activities for home owners. People can care for their gardens, water plants, and partake in other activities for exercise and enjoyment.

10. Improve and conserve natural resources by reducing soil erosion, for example.

11. Provide therapeutic relief. Enjoying the landscape can be relaxing. Horticultural therapy is discussed more fully in section 1.6.

12. Reduce noise and environmental pollution. Plants in the landscape can be used to absorb noise.

These and other purposes of landscaping are discussed in this chapter.

12.2: CATEGORIES OF LANDSCAPING

The categories of landscaping do not have fixed boundaries but may overlap. They serve to show what is required in planning a landscape design. The general principles of landscape design are applied in each case to provide the best results by integrating function and aesthetics. In terms of customers or users and their needs or preferences, landscaping may be tailored to four categories of need, as described in the following sections.

12.2.1 RESIDENTIAL LANDSCAPING

Residential or *home landscaping* is geared toward individual home owner and neighborhood needs. Developers may establish a theme for a residential project and landscape the area accordingly. For example, a residential community called Pine Acres may have a large number and variety of pine trees, whereas Cedarville may have a large number of cedar trees. Some developers landscape homes before sale. In such cases, they impose their theme on home owners and the community. However, once purchased, a home owner may add to the landscaping on the property. One problem results when neighbors become concerned by a nearby home owner's exotic design, which may drastically offset the general neighborhood landscape. When home owners are involved in the design of their homes, they are often also involved in the landscape design. Residential landscaping often has a strong personal touch to it, reflecting the taste and needs of the home owner.

12.2.2 PUBLIC LANDSCAPING

Cities are more than brick and mortar. They are designed to look beautiful, and a large part of this goal is accomplished by blending architectural design with an effective and visually pleasing landscape design. Plants are used to enhance the frontage of edifices and to beautify the interior of offices and indoor open spaces.

Parks are designed and located for a variety of purposes. In the middle of sprawling skyscrapers, small parks may be located in strategic places to provide periodic and temporary rest and relaxation to weary pedestrians. Inner-city squares and plazas are designed for this purpose as well. Residential areas often have neighborhood playgrounds, parks, trails, pools, and ponds. On a much larger scale, a city may have community parks that include ball parks, tennis courts, and other recreational facilities.

At the state and national levels, large tracts of land may be developed for public use (e.g., fenced off as a nature reserve). Nature reserves usually include wildlife. Landscapes designed with the public in mind are found near schools, libraries, museums, and colleges.

A large city or community often has a grounds and gardens department responsible for designing, installing, and maintaining public grounds. Public landscaping may be in the form of trees planted along streets and flowers planted in the medians or on street corners. It may also take the form of a recreational park, where a piece of land is developed for residents to use during their free time. The personal or individual preference element associated with residential landscaping is not a debatable goal in the public arena since public areas are designed for a broad spectrum of people with broad backgrounds and preferences. This is not to say that individuals cannot enjoy public parks but rather that they must use what is offered.

12.2.3 COMMERCIAL LANDSCAPING

In a way, *commercial landscaping* has a public element, since businesses are open to the general public. Commercial places often have lots of space for parking. Both high- and low-traffic areas exist on company premises. Some businesses display merchandise in their windows and hence avoid obstructing the view in those areas. Commercial landscaping is found in places such as shopping malls, hotels, banks, and restaurants. When designing with the public in mind, one important though subtle factor to consider is safety. The landscaping installed should not pose a danger to the business's workers or patrons.

12.2.4 SPECIALTY LANDSCAPING

Specialty landscaping is found in places such as zoological gardens and botanical gardens, where formal designs are often used. A botanical garden is designed to exhibit a large variety of plant types. Their design usually has a strong educational component, plant species often being identified by their common and scientific names. However, the exhibits are organized and displayed in a manner that is very attractive to viewers. Golf courses are landscaped to provide the obstacles necessary for the game as well as for scenic beauty. Zoological gardens adopt landscape designs that are functional with respect to the animals on display. A jungle environment may be created in an inner-city zoo to simulate the natural habitat of jungle animals.

A commercial form of specialty landscaping is exemplified by *theme parks*. These amusement centers have landscape designs for the general compound in addition to unique landscaping designs as part of the key attraction.

12.3: LANDSCAPE DESIGNING

Landscape Architect
A professional who designs plans for the installation of plants and inanimate objects outdoors to fulfill aesthetic and functional purposes.

A landscape design should be prepared before installation of the landscape. Planning and design are both an art and a science. The environment must be thoroughly understood, including the area's topography, soil, and climate. The materials to be used should be selected properly and located to achieve the desired purpose. On a large scale, four professionals, the *landscape architect, landscape contractor, landscape maintenance supervisor,* and *nurseryman* work together to execute a landscape project.

12.3.1 LANDSCAPE PROFESSIONALS

The Landscape Architect

The American Society of Landscape Architects defines *landscape architecture* as "the art of design, planning, or management of the land, arrangement of natural and man-made elements thereon, through application of cultural and scientific knowledge, with concern for resource conservation, and stewardship, to the end that the resultant environment serves useful and enjoyable purpose" (Section 2, Article II, Constitution of the American Society of Landscape Architects).

The landscape architect is thus a service provider who advises a client about plans that can enhance the client's environment. He or she is the consultant who provides the *site plan* and *planting plan* for the project. The landscape architect also provides detailed guidelines about how to install the landscape and then oversees the project to completion. Portions of the project are subcontracted to other professionals.

Landscape Contractor

The landscape contractor is a person who, using the architectural plans, installs the structures and plants in the landscape. It is important that the landscape contractor and architect thoroughly understand one another and agree on the practices and their implementation at a particular time. The landscape architect may recommend a specific contractor to a client.

Nurseryman

After the plans have been clearly understood, the landscape architect contracts with suppliers for the materials to be used in a project. Plant materials are obtained from a nurseryman. The landscape architect may participate at this level to ensure that the most suitable materials are selected for use. After installation of the landscape, the landscape architect inspects the work and approves it if it is done according to plan.

Landscape Maintenance Supervisor

The maintenance supervisor is responsible for managing the finished project to ensure that it becomes properly established according to plan and to the satisfaction of the client.

12.3.2 ELEMENTS OF DESIGN

Beauty is influenced by culture, traditions, and personal experiences; the elements that create beauty can be learned. The landscape designer or architect uses plants as the primary objects in creating a design. Plant species have certain characteristics or features that influence how they are used in a landscape. By themselves, each of these features is beautiful and desirable in its own way. However, when more than one feature is found in a place, their interrelatedness and interconnection become significant in how the viewer senses the visual scene. Choice of features can either enhance or diminish the overall appeal of the display. To make appropriate choices, one needs to understand the nature of the features.

Consideration of plant features is important in the design of the landscape. They provide the aesthetic basis of landscape design. They are also called *elements of effect* because they create moods or feelings in the viewer.

Color

People respond differently to color. Certain colors—reds, oranges, and yellows—are described as warm colors and appear to advance toward the viewer. Cool colors—blues and greens—tend to recede in a landscape composition. The attributes of color are hue, value, and intensity or chroma. Some colors are enhanced when viewed against certain background colors, and others clash. Gray is a neutral color and, along with a cool color such as dark blue, can create an effective background. Some colors are overwhelming and loud, and others are soft and mild. Flowering plants have a wide array of colors in their petals. Some flowers have solid colors, while others are variegated.

Choice and arrangement of colors in the landscape are critical considerations to the overall visual appeal. Flower color is affected by light. Shade subdues and sunlight intensifies colors, which should be kept in mind when locating plants in the landscape.

Flowers are not the only sources of color in the landscape. Leaves, although predominantly green, may have variegation and seasonal color changes in autumn. Tree barks also produce variety in color.

> **Site Plan**
> A drawing of the locations of plants and inanimate objects in a landscape.

> **Planting Plan**
> A drawing specifying by means of symbols the types and names of plant species, the quantities, and their locations in a landscape.

> **Elements of Effect**
> Elements in a landscape design that elicit a response by creating moods or feelings in the viewer.

Texture

Texture refers to an object's feel with respect to the sense of touch. Visual surface characteristics may project a soothing image or a harsh, coarse feeling to the viewer. A grass blade has a different texture from a herbaceous plant (figure 12-2). Texture effect is pronounced when contrast exists in the display. For example, placing plants next to a wall enhances the textural differences between two contrasting types.

Texture may change with the season. While a deciduous tree has its leaves, it presents a smooth texture. However, when the leaves fall in autumn, the bare branches present a rough or coarse texture. Tree barks vary in texture, some being smooth and others, such as oak, thick and rough (figure 2-7).

Texture in the landscape is not limited to plants. Other materials, such as those used as mulch, have varying textures. Gravel is fine in relation to rocks, and sawdust is fine when compared with tree bark. Texture is also modified by distance. A coarse-textured plant viewed at a distance may not appear as coarse. One of the goals in designing a landscape is for the elements to flow or blend together without abrupt changes. Gradual transition is always more aesthetically pleasing than sudden changes. To shift from a fine texture to a coarse one, the landscape architect should insert a transitional element between the two extremes. When materials in the landscape change on a seasonal basis, plants should be chosen such that changes

(a)

(c)

(b)

FIGURE 12-2 The occurrence of texture among plants. (a) Contrast among succulents. A broadleaf (b) and grass (c) display a sharp contrast in leaf form.

are sequential. This practice is accomplished by thoughtful selection of plants so that mixtures are planted (e.g., deciduous and evergreen). Flowering plants can be selected to produce sequential flowering so that the landscape can be enjoyed across seasons.

Form

Form is a three-dimensional attribute. The outline of a plant against the sky, it depends on the structure and shape of the plant. For example, trees may be conical, columnar, spherical, and so on (figure 12-3). Plants of different forms may be grouped and arranged in a certain fashion to create a different overall form in the landscape. For example, a landscape architect may group a number of plants with narrow, columnar form to create a horizontal form. This strategy is used in creating a hedge (fig. 15-17) around a house. Plants have form because they have visual mass and thus occupy visual space. Form is therefore directly related to one's ability to see it. The art of topiary (fig. 15-12) (pruning plants with dense foliage into shapes) is a way of creating artificial form in plants.

Form
The three-dimensional shape of the plant canopy.

Line

The effect of *line* is accomplished through the arrangement of objects. Line is a *boundary element* in design. Shape and structure are defined by lines. A design element line, when used effectively, has the capacity for eliciting emotional responses, making one display appear elegant and another disorganized. As indicated previously, form is a three-dimensional attribute, but it can be interpreted as one-dimensional by line. Line is the means by which form guides the eye.

Natural lines occur in nature but often are complex. Line is a design tool that a landscape architect uses to create and control patterns in the landscape. Formal, straight lines are found in pavement design. The monotony can be broken by arranging pavement elements (e.g., bricks) in a design or pattern that softens the effect of straight lines.

Arrangement of landscape objects in a line can be used to direct viewers to the focal point (12.3.3) in the design. Lines do not have to be straight, and, when they are straight, they do not have to be continuous. A designer can deliberately introduce breaks to cause the viewer to pause and change views. Lines can be contoured or curvilinear to slow down viewing and encourage the viewer to spend more time to appreciate a particular display.

Landscape architects combine a knowledge of plant science, art, and creativity to design a functional and aesthetically pleasing product that meets the consumer's approval. It is not enough to know that a plant looks pretty; to serve a functional role, it may be necessary to know whether it is evergreen or deciduous, for example. A deciduous tree planted at the southwest corner of a house will provide shade in summer (when the sunlight is bright and intense), but when it sheds its leaves in winter (when it is overcast and not bright) it will allow light to reach the house, thus serving a dual role.

Line
The one dimensional effect produced by arranging three dimensional objects in a certain fashion.

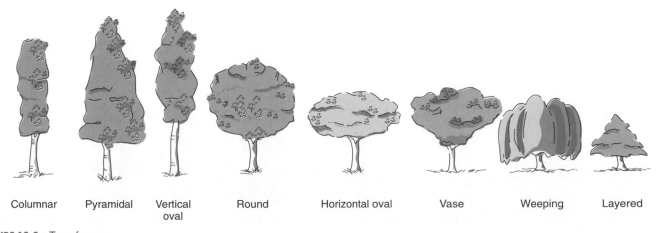

| Columnar | Pyramidal | Vertical oval | Round | Horizontal oval | Vase | Weeping | Layered |

FIGURE 12-3 Tree forms.

12.3.3 PRINCIPLES OF DESIGN

Regardless of the category of landscaping, the observance of five basic principles of design, *simplicity, balance, focalization, rhythm and line,* and *scale* or *proportion,* is necessary. These principles are applied in relation to the elements of design described previously. Landscape design pertains to the arrangement of objects to accomplish a purpose in the landscape. Since these objects may vary in color, form, and texture, the quality of results of the arrangement depends on creativity and successful application of the basic principles of design.

Simplicity

> **Simplicity**
> A landscape design principle that employs a number of strategies to reduce excessive variation and distractions in the landscape.

A landscape designer is a three-dimensional artist working especially with plants (but also with other objects). Like canvas artists, landscape designers must create a theme, subject, or title for a project. Certain themes are more readily understood than others by the viewer. People tend to enjoy a painting much better when they understand it, but some paintings are abstract and difficult to understand even though they may have overall appeal. The elements in the composition should blend together, an aspect of design referred to as its overall *unity.* Also, the viewer should be assisted in understanding and enjoying the creation of the designer.

People who visit parks and botanical gardens expect to see and are awed by the variety of specimens displayed. Variety is thus a very important element in landscape design. The question is, how much variety should be present? Too little is monotonous and can be unattractive, but too much can be confusing and detract from the viewer's enjoyment. The annual tulip festival in Holland, Michigan, for example, offers displays of tremendous variety in the plants. There are vast acreages of tulips in all possible colors, shapes, and sizes. Although this exhibit is successful, another exhibit consisting of only junipers may not be as attractive because of the lack of adequate variety. The landscape architect can enhance the existing variety by the way plants are displayed in the landscape. By using the design element of *repetition* or *massing* (figure 12-4) he or she can give meaning and expression to variety in the landscape by controlling and limiting variety so as to introduce order in the design. Rather than displaying a single plant species in 10 different colors, groups of 10 or more plants of each color will produce a better visual impact on the viewer.

Distractions in the display should be eliminated, which may mean using edging to straighten out a lawn or flower bed. These strategies, when used appropriately, achieve the goal of *simplicity* in a design.

Balance

> **Balance**
> A landscape design principle that presents an equal visual weight of elements to a viewer.

The concept of material *balance* implies stability resulting from equal distribution of weight around a central axis. Balance in horticulture refers to the visual weight a viewer is presented with by the materials in the general design. The distribution of the design materials should not be skewed. The viewer should have a sense that the amounts of things to see on both sides in the viewing frame are fairly equal. The two halves of the design need not be identical, just balanced. When the design is such that the materials on one side of the viewing frame are identical (mirror image) to what are on the opposite side, it is said to be *symmetrical* and formal (figure 12-5); if the balance is achieved by different materials, it is called *asymmetrical* and informal (figure 12-6). Most designs are asymmetrical.

FIGURE 12-4 The concept of repetition or massing enhances the simplicity in landscape design.

FIGURE 12-5   A demonstration of symmetrical balance in landscape design.

FIGURE 12-6   A demonstration of asymmetrical balance in landscape design.

Focalization

The design principle of *focalization* or *emphasis* is employed to satisfy an expectation of the viewers while fulfilling the "vanity" of the designer. It may be described as the center-of-attraction principle. Instead of the viewer's eyes wandering aimlessly to and fro, designs are composed around central pieces on which viewers focus first before looking around. Some natural focal points exist in the landscape. For example, a residential landscape design, at least from a distance, should draw attention to the point of entry into the house, which is usually the front door. If there are fountains, sculptures, or exotic or uniquely attractive plants (*specimen plants*) that one wants to emphasize, the designer may make these objects the focal points (figure 12-7). Focalization is in effect a strategy for guiding the viewer to the must-see exhibits in the landscape. A landscape can be planned such that the focal point changes with the seasons, perhaps by strategically locating a flowering plant and a deciduous plant in the landscape. When the interest in the deciduous plant fades away after a spectacular display of fall colors, the flowering plant will create interest in spring with its attractive blooms.

> **Focalization**
> A landscape design principle that creates interest and accent for a particular arrangement.

> **Specimen Plant**
> A plant with attractive features that is used to fulfill the principle of focalization or emphasis in an arrangement.

FIGURE 12-7 Using a fountain as focal point in a landscape design.

Rhythm and Line

Panoramic view of a landscape design is often possible from one or several strategic viewpoints such as above or overhead (which very few people have the privilege to experience), especially if the design is not a straight-line display of plants. An elaborate and extensive design may incorporate several concepts into one overall picture. To avoid a disjointed display and create a sense of flow or continuity, the different concepts or sections of the landscape should be linked, such as by use of beds of appropriate plants. This linkage creates movement in the design. The principle is visible on avenues and in parks where the same kinds of trees are arranged in long rows (figure 12-8). It is applicable to the home landscape where the display in the front of the house can be linked to the rear display by avoiding an abrupt ending in front. Instead, plant types are repeated or extended from the front bed or display around the corner. This layout suggests to the viewer that the design continues around the house.

Scale or Proportion

Scale, or proportion, is not only desired with respect to the plants in the landscape but also in terms of their relationship with other structures and the general functionality of the design. In choosing plants for the landscape, it is important to take into account their size at maturity. Since inanimate objects such as sculptures and houses maintain their sizes, it is important that trees located in their vicinity do not overshadow and make them inconspicuous when they are fully grown (figure 12-9).

12.3.4 LANDSCAPE APPRECIATION

When people walk through a landscape, they perceive their environment through the engagement of their senses—smell, touch, sight, and sound (not tasting, since it could be dangerous to their health). Landscape architects design the landscape to evoke a certain response. The senses generally interact to communicate specific feelings to viewers.

What viewers perceive is influenced by the time, place, and prevailing circumstances. A walk in the park on a sunny day is different from a walk in the same park on a cloudy day. On a sunny day, colors are brighter and the visual appeal of flowers is enhanced. On a hot afternoon, one is more likely to walk in the shade and view from a distance. On a windy day, one is likely to be influenced by the swaying of branches and leaf movements. In autumn, the ruffle of leaves as one walks over them is likely to attract attention.

Apart from these extraneous factors, the perception of the landscape is influenced by the individual's state of mind and the purpose of taking a walk in the park. A happy person responds to the sights and sounds of a park differently than someone in a sad mood. Further, if one is walking alone, there is no distraction from conversation as would occur if one were

FIGURE 12-8   A demonstration of rhythm and line in landscape design.

Good choice and arrangement of plants

Poor choice and arrangement of plants; house overwhelmed by plants

FIGURE 12-9   Scale and proportion.

in the company of another. A person who is knowledgeable about landscape design or plant botany may walk through the park with an intellectual approach.

The challenge in landscape design is to combine intuition, creativity, training, and experience to create a design that meets the needs of the client.

12.4: PLANNING A RESIDENTIAL LANDSCAPE

Residential designs are usually small-scale designs. Viewers are much closer to plants and hence perceive certain details. Further, plant masses are smaller, though still effective. Larger plants could easily overwhelm the arrangement because of the scale of the landscape.

12.4.1 DESIGN CONSIDERATIONS

Landscaping is an expensive undertaking. Unless one purchases a prelandscaped home, some landscaping will be required, even if it involves installing only a lawn or a few trees. A home owner may design his or her own landscape and may or may not need help in installing it, depending on the background and experience of the home owner. Alternately, a home owner may develop the design and contract its installation to a professional or commit the entire project (design and installation) to a professional service. The expense obviously increases as the involvement of the owner decreases, but the quality of work likewise increases with more professional involvement.

Notwithstanding which route the home owner takes to landscape the property, three considerations are central to all landscape undertakings:

1. *The design should satisfy the home owner's preference.* The home owner is paying for the job and must live with it indefinitely, so the design must consider his or her preferences. In a way, the customer is always right. However, this is not to say that a professional should stand by and allow the home owner to make mistakes without offering professional advice.

2. *The design should be aesthetically pleasing.* The view of the landscape must be aesthetically pleasing to the home owner. Some if not all home owners are pleased to be complimented on the beauty of their landscape designs, much the way they enjoy receiving complements on the interior decor of their homes. A landscape in effect is an extension of the indoor environment, a kind of outdoor room. If one's home is located on a large plot by itself, there is more room to be creative and fancy in the design. However, when it is located on a block in an urban area where the residential area was developed under a specific theme, the plot shapes and sizes may be very similar. Whatever the landscape design installed, it must take into consideration the general neighborhood and especially the immediate neighbors.

3. *The design should be functional or practical.* Landscaping is devoted to developing the outdoor environment of a residence. It must serve the practical purposes that the home owner desires such as creating barriers, offering shade, and providing privacy. The finished work should be easy to maintain; otherwise, the home owner must make a prior commitment to paying for the required care.

12.4.2 DESIGNING FOR RESIDENTIAL PURPOSES

Planning the Design

Assessment of the Home Owner's Needs   To satisfy the customer, the first step in planning a landscape project is to consult with the home owner to learn firsthand about his or her preferences and what is expected from the completed project. At this fact-finding session with the client, the landscape designer should obtain information pertaining to the following:

1. Family size and ages of members. If there are children, does the home owner desire to have a playground?

2. Life-style of the family. Do they prefer the outdoors or indoors for recreation? Do they love a closed (private) or open (outdoor) environment? How frequently do they entertain? Are they avid gardeners? Do they prefer traditional or contemporary designs?

3. Are there any special preferences for certain plants?

4. How much work are they willing to put into (or willing to pay for) the maintenance of the landscape?

5. Do they desire to have an all-season or seasonal landscape?

6. How much are they willing to spend on the project?

7. Do they prefer walking through the landscape or enjoying it from a distance?

8. What time of day might they use the grounds?

9. Are there any service or utility needs?

10. Are there any nonplant installations (e.g., walkways and statues)?

11. Must a permanent irrigation system be installed?

12. Do they want to conform to neighborhood patterns or be unique?

One critical factor that a client must understand during the landscape planning period is that the landscape has a maturity period. Depending on the nature of the project, the landscape may mature and be ready for enjoyment a short time after installation, or it may take a while. Because some clients want to see results almost instantly, a designer may have to compromise and create a design with instant appeal.

Site Analysis   Site analysis may be described as a survey of a landscape project site to ascertain the presence, distribution, and characteristics of its natural and man-made features and the environmental conditions prevailing at the site. It is important to know all of the existing characteristics of the site since they control the success or failure of the project. The designer must recognize and work with these characteristics to achieve a product that is desirable and sustainable.

A site analysis helps the landscape architect to formulate an appropriate strategy for enhancing the site. Designing for a rural setting requires considerations different from those for an urban setting. Most landscape designs are conceptually artificial in the sense that the finished product shows a great deal of order and tidiness. Further, the product is maintained by a regular cosmetic regimen to keep the curves and lines well defined and the site clean. Another approach to design—the naturalistic approach—depends primarily on nature to make the major choices in plant selection and placement. The species used are not only adapted to the area but also compatible with each other. The result is a product that is self-sustaining.

After a needs analysis has been conducted, the designer should visit the site for a thorough study of what is available. The information to be gathered on this visit includes the following:

1. General setting of the property in relation to others in the neighborhood (i.e., the landscape of the neighborhood). Some real estate developers install certain landscape features as a standard part of the design in a neighborhood. Certain neighborhoods have regulations regarding fencing and activities on the public side of the property.

2. Size of property (i.e., lot size, dimensions, or space to be utilized for project). The size of the property determines to some extent the variety, size, and number of plants that can be installed. Playgrounds, pools, fountains, and other similar structures may be practical only on large properties.

3. Architectural design. It is important that the design complement the architecture of the home. The size of the home is an important consideration for the purposes of the proportionality of the design.

4. A record of the land characteristics regarding the terrain (slopes and flat parts), drainage (any wet spots or areas in the flood plain), soil characteristics (texture, structure, pH, and organic matter content), rocks, and any natural features (e.g., creek or stream).

5. Existing landscape. It is important to know how many of the existing materials on the property will be retained in the design. For example, a rocky area may be developed into a rock garden. A mature tree or wooded area may be retained.

6. Existing functionality of the design. It is important to decide which sides offer the best views and what elements can be enhanced, removed, or hidden. If a client has recreational facilities (e.g., swimming pool), installing large trees on the south and west sides of the property is undesirable because of the shade they cast and the litter they produce.

> **Site Analysis**
> An inventory of existing natural and man-made features at a proposed landscape site.

7.   Climatic conditions of the area. One should know the general climate and microclimate of the area. Plant species to be installed depend on their adaptation to cold weather. Local rainfall distribution and winds should also be considered.

8.   Site plan. A site plan showing walkways, fences, sewer lines, and other utility lines should be made. Buried and overhead cables should be included as well.

Preliminary Design   When the customer's needs are known and the site characteristics have been ascertained, the designer may proceed to the drawing board to create a drawing, starting with sketches and moving on to a scale drawing. The designer should pay attention to the three basic functional areas of the house (figure 12-10):

a.   Public area (the entryway area [front door] that often faces the street)
b.   Private living area (patio or deck)
c.   Service or utility area (e.g., vegetable garden, pet house, and storage)

1.   *Public area.* Most home owners are most concerned about the public area—the part of the house most visible to passersby and that first approached by visitors to the property (curb appeal). This side of the house creates the first impression. It often indicates what the visitor will see indoors. An entryway made of concrete or brick usually leads to the door. This area could be straight from the street or designed to curve on approach to the door. Some interest may be introduced into the design by mounding or terracing a flat lot. Flower beds may aid in creating this effect. The design should complement the architecture of the home. This area is truly public and thus some communities restrict what can be done next to the street, especially regarding the erection of walls or the installation of fences and hedges.

2.   *Private living area.* The private living area is where decks and patios are located. If shading is important to the home owner, the patio should be located on the east side of the house, so that it will receive the warmth of the early morning sun and be shaded by the house at noon. If sunlight is desired most of the time, the patio should be located on the south side of the house. The private living area should be screened from the view of neighbors to be fully functional.

3.   *Service and utility area.* The service and utility area may be described as the odds-and-ends section of the landscape. Although important activities occur or structures appear in this area, the home owner does not want them to be in view of the public. In fact, the service area is a second private area. This section is always in the backyard

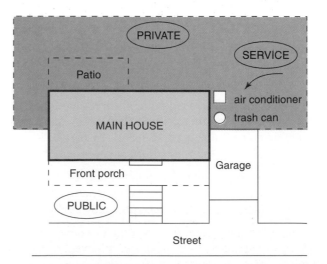

FIGURE 12-10   The three basic functional areas of a home to consider in a home landscape design: private area, public area, and service area.

of the house and may be fenced off. It is not meant to be aesthetically enjoyed but, as its name implies, is a real service area for things such as a doghouse, garbage cans, clotheslines, storage shed, and vegetable garden.

Landscaping of Planned Residential Developments

Some modern developers purchase a large tract of land and designate it for residential development. Streets and sewage and other utility lines are laid down. However, in addition to the infrastructure, developers deliberately set aside portions of the land to be developed into recreational facilities (e.g., golf course, park, tennis court, or swimming pool). They also leave a stretch of land called a *greenbelt* to be an open space where no construction occurs. This part of the land is usually unsuited for residential construction. The development may also include walkways and bicycle paths to be utilized by the residents. Greenbelts are used to control construction (and thus prevent sprawling) and to enhance the neighborhood.

> **Greenbelt**
> *A stretch of land that is left construction-free by real estate developers in a residential area for functional purposes.*

12.5: PLANNING A NONRESIDENTIAL LANDSCAPE

The principles described for residential landscape designing apply to nonresidential landscaping as well. A variety of scenarios with unique design needs occur in residential landscaping.

12.5.1 LANDSCAPING SCHOOL GROUNDS

School grounds require space in which children can roam freely. Trees often dominate the landscape, with some functional plants (such as shrubs planted as foundational plants) and a limited area for annuals or bedding plants. The goal of landscaping school grounds is to achieve low-maintenance and childproof areas. Many schools operate on a small budget and thus cannot afford expensive grounds-maintenance services. Because children are bound to play around trees, species selected for school grounds should be resistant to rough, mechanical treatment.

12.5.2 LANDSCAPING COLLEGE CAMPUSES

Unlike primary, middle, and high schools, college campuses can often afford grounds-maintenance services. The goal in landscaping a college campus is to provide visual continuity by linking the diverse structures on-site by using trees, ground covers, and shrubbery. Since funds are available, campuses often can afford high-maintenance landscapes. On a campus with a horticulture academic program, the campus provides an opportunity for installing plant materials in the general landscape for instructional purposes. In other words, the campus may be turned into a kind of giant arboretum.

12.5.3 LANDSCAPING PARKS AND RECREATIONAL AREAS

A key goal in designing a landscape for public use such as a park is public safety. Appropriate plants should be selected so that the park remains open, without hiding places that may encourage criminal activities. A park may have a significant number of nonplant materials in the landscape, including fountains, benches, playground equipment, and statues. Plants are used to complement these objects. Trees and shrubs grouped and spaced randomly are effective ways of landscaping parks.

12.5.4 LANDSCAPING URBAN CENTERS

Modern urban centers are characterized by steel and concrete structures. Soil for planting is very limited since the areas that are not streets have concrete pavement for use by pedestrians. Container planting (*raised planters*) enables soil to be imported into downtown areas for planting trees. The plant species selected should be adapted to growing in a confined area and tolerant of urban air pollution. Trees may be planted along streets or in the median. Urban species should be slow growing and tolerant of moisture stress. Some irrigating is usually needed to keep plants growing in a healthy way. In the business district, trees should not block window displays. Further, the trees selected should not interfere with power lines or light poles. With planning to provide appropriate structural support, irrigation, and drainage, plants may be planted on roofs (*roof planting*).

12.5.5 LANDSCAPING SHOPPING CENTERS

Shopping centers offer a variety of services and merchandise to the general public and thus attract many shoppers to the complex. The rationale of their design is to sustain the interest of the shopper. A great deal of money is invested in plants, many of which are specimen plants (which provide focal points). Because of the shelter provided by shopping centers from the elements, a wide variety of plants can be used in interiorscaping these complexes. Annuals and perennials are used in the design. Skylight and artificial lighting are provided for the proper growth and development of plants. These features contribute to the overhead costs of operating these complexes, but such costs are offset by increased patronage because of the attractiveness of the surroundings.

12.5.6 LANDSCAPING CEMETERIES

Modern cemeteries do not use the ornate, elaborate, and massive headstones that were popular in the past. The cost of maintaining a cemetery is high, and thus live flowers such as annuals are not commonly used. Normally, a well-kept lawn is the primary landscape element, with a few trees planted in an informal design. Trees should preferably have horizontal form (figure 12-3).

12.6: PLANTS IN THE LANDSCAPE

Up to this point in the designing of a landscape, no plant materials have been specifically included in the design. Space has been allocated for walkways, driveways, private areas, and utility areas. It is now time to allocate plants strategically to complete the design objectives. Plants in the landscape serve more than aesthetic purposes. If selected and located judiciously, they can provide functionality, as described in a later section. On paper, the architect represents plants by graphic symbols. There are symbols for trees, shrubs, flowers, buildings, and all of the elements in the design. A completed landscape design includes the land dimensions and all of the symbols drawn proportionally, or to *scale*. This scale drawing shows what materials are involved in the design, where they are to be located, and how they are to be installed on the property. The landscape architect is now ready to translate the paper drawing into real results on the grounds after obtaining the home owner's approval of the proposed design.

12.6.1 SELECTING PLANTS

A home owner may prefer a certain plant species, but a number of factors determine what can be planted. To make a wise choice, one must know the plants and how they behave at various stages in their life cycles. The following are important factors to consider in deciding which plant species to include.

Climate

The two critical climatic elements to consider are temperature and moisture. The species selected must be adapted to the local area with respect to critical elements. In terms of temperature, *cold-hardiness* is the most important factor. If a plant type cannot tolerate cold, it will not only cease to grow in winter but suffer severe damage, leading to deadwood that must be

---

Cold-hardiness
The ability of a plant to survive the minimum temperature of the growing area.

pruned. Plants need water for good growth. Some areas are climatically drier than others, requiring supplemental irrigation to maintain a decent landscape. To reduce the need for irrigation, drought-resistant varieties or species should be used.

Soil Reaction (pH)

If the property is located on acidic soils, species such as azalea and rhododendron are appropriate, but others such as carnation, dahlia, and buttercup will not grow properly.

Sunlight

The general amount of sunlight the house receives is important in choosing plant species for the landscape. If the general area is open, sun-loving plants should be selected. Some homes are located in heavily wooded environments, in which case shade-loving plants should be considered. Even if the house is located in an open area, the architecture may cause certain parts of the land to receive less light. The areas under large trees are shaded and thus require shade-loving plants.

Features at Maturity

The location of a plant depends on its size, shape, maturity, form, and habit. The plant may be small in initial stages of growth but after about two to four years of growing could be a huge bush or a tall tree. Some plants grow slowly, while others grow rapidly. Some plants have narrow profiles, and others spread their canopies (figure 12-3).

Maintenance Level

Plants differ in the level of care they need to grow and produce the best results. The home owner should declare a certain level of commitment to maintenance of the landscape after installation. Some owners handle the maintenance themselves; others may not want such responsibility but may or may not be prepared to pay for someone to care for the plants. A poorly maintained landscape is unsightly and wastes the funds required for its installation. The plants used must meet the level of care the home owner is willing to provide.

Customer Preferences

The home owner pays for his or her needs to be met by the landscape developer. Therefore, every effort should be made to satisfy the customer. If the customer likes yellow flowers, blue flowering species should not be selected without permission.

12.6.2 PREPARING PLANTING PLANS

Preparing planting plans is an activity that requires drafting and graphics skills. It is a job for professionals. A plan uses symbols to represent different kinds of plants (e.g., shrubs, deciduous plants, and evergreen trees) (figure 12-11). The plan is a medium of practical communication between the landscape architect and the landscape contractor describing the designer's intentions. Such a plan includes specifications as to spacing between plants and where other structures in the landscape are to be installed. Planting plans should, whenever possible, include the name of each plant. The scientific name of the plant is used in all descriptions. A plant list should then be added to indicate the common names, qualities, sizes, and other specifications of all plants listed on the plan (table 12-1).

All clients may not be able to interpret a planting plan. In such a case, a *perspective sketch,* which shows three dimensionally how the finished project will look, may be produced. The planting plans also contain additional sheets that describe in more graphic detail how plants are to be planted, staked, guyed, and so forth. Plans are made on a vellum, linen, or Mylar surface for durability. The client should receive a copy of the original plan on Mylar or sepia paper.

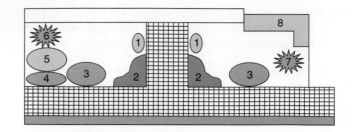

1. Pink cosmos
2. Daisies
3. Blue campanulas
4. Geraniums
5. Blue canterbury bells

6. Pine or other evergreen
7. Short flowering tree
8. Hedge

FIGURE 12-11 A sample planting plan for a home landscape project.

Table 12-1 A Sample Plant List Showing the Variety of Specifications According to the Type of Plant and How It Is Produced in the Greenhouse or Nursery for Sale

Key	Plant	Quantity	Size
AE	*Aspidistra elatior* (cast-iron plant)	10	10-inch pot
EI	*Hedera helix* (English ivy)	60	2-1/4-inch pot
FE	*Ficus elastica* (rubber plant)	4	15 gallons
MY	*Myoporum parvifolium* (Myoporum)	100	1 gallon
WR	*Washingtonia robusta* (Mexican fan palm)	2	12 feet
AS	*Aster* spp. (aster 'Chorister')	10	15 inches
AB	*Abies concolor* (white fir)	2	8 feet
FU	*Fraximus undei* (Shamel ash)	4	36-inch box

Plant Arrangement in the Landscape

Plants are carefully selected and strategically arranged in the landscape for maximum aesthetic and functional effects. They are arranged according to a design concept called the *outdoor room* concept; the landscape is designed as an extension of the indoor environment. The boundaries of the outdoor room are provided through the creation of structures equivalent to the floor, ceiling, and walls of an indoor environment. An outdoor floor is created by establishing a turf (Chapter 14) or some other appropriate ground covering (e.g., natural materials such as gravel, sand, or wood or synthetic materials such as concrete). The outdoor ceiling establishes the height of the landscape display and is provided through the planting of trees or the installation of structures such as a covered patio. The walls of an outdoor room may be provided by constructing a fence, creating a hedge or flower beds, or using shrubs.

The landscape architect selects and allocates plants to the site according to the outdoor room concept by adopting four basic plant arrangements:

1. *Corner planting.* Corner planting starts with creating a bed in the corner of the building or the area to be landscaped. By so doing, the corners of the outdoor room are demarcated. This bed contains a few species of plants, with the tallest in the back. The designer may create interest in the design by including specimen plants located in the corner created by the intersection of the two walls or corner lines.

2. *Foundation planting.* As the name implies, certain plants are located close to the foundation or walls of the building. One should be careful to locate the tallest plants in the corners and the shortest below the window so the view is not obstructed. Generally, low-growing plants are used as foundation plants.

3. *Line planting.* Line planting is also done in beds along the property line. The function of line planting is to create a screen to provide privacy. A variety of shrubs are usually used for this purpose. These plants provide the walls to the outdoor room.

4. *Accent planting.* Accent planting has primarily an aesthetic purpose in the landscape. As the name indicates, it enhances the area by its attractiveness in the color of blooms or leaves, form, or some other unique characteristic. Single specimen plants are often strategically located in the landscape. Sometimes plants and nonplant materials (e.g., statues) may be used in combination. Accent planting may also be created by massing flowering herbs in a bed.

Projected Cost

Cost estimation, made after a landscape plan has been prepared, is done by both landscape architects and contractors. The contractor should attempt to obtain the most accurate prices possible to avoid underbidding or overbidding and losing the project or losing money on the project. For large projects, specific components can be bid on separately. Otherwise, the contractor who wins the bid can subcontract various portions in which his or her expertise is required. For example, the irrigation system may be subcontracted to a company that specializes in that area. Similarly, driveway construction and other concrete structures may be subcontracted to a masonry company.

The plants required should be readily obtained at a reasonable price. A cost analysis made before submitting a bid should address the following:

1. Variable costs (or direct costs), including materials, labor, equipment rentals, and so forth.

2. Overhead costs (fixed costs), including costs that will be incurred whether or not the bid is accepted (e.g., fringe benefits, management expenses, and salaries of employees).

3. Profit, which varies according to the specific business owner.

12.7: OTHER FUNCTIONAL USES OF PLANTS IN THE LANDSCAPE

12.7.1 PRIVACY FROM ADJOINING PROPERTY

No matter how good one's neighbors, home owners usually demand some privacy from adjoining property. A fence may be installed as a divider. High-growing shrubs can be planted to create a dividing wall. Plants behind bathroom and bedroom windows may be selected to be aesthetically pleasing as well as adding privacy from neighbors.

12.7.2 CLIMATE MODERATION

Trees with large canopies provide shade and protection from radiation from the sun on hot days. At night, however, radiation from the soil is reduced, creating a microenvironment under the canopy in which temperatures are more even. By planting evergreen trees or shrubs close to west- or south-facing walls, a dead-air space can be created between the "plant wall" and the wall of the house (figure 12-12). This space has insulation capability similar to the dead space deliberately created by some architects in home design. The effect of this condition is that summer temperatures are moderated and kept even. Ivy-covered concrete walls (or those covered by any other climber) reduce the temperature of the wall relative to the bare wall when exposed to full sunlight.

12.7.3 GLARE REDUCTION

A streetlight in front of a house may provide light in undesirable parts of the house. Blinds may have to be pulled down at night. A tree can be located in the path of the light rays to block

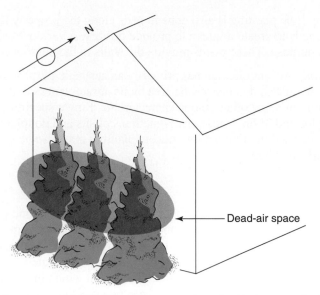

FIGURE 12-12 Planting trees close to the wall can create a dead-air space for temperature modification.

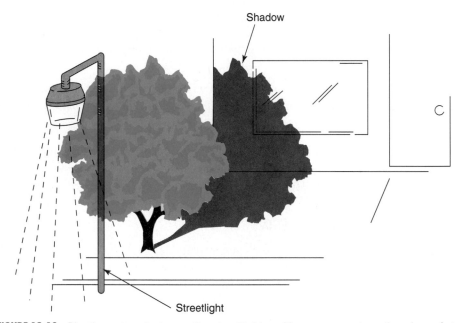

FIGURE 12-13 Planting a tree between the streetlight and home can reduce the glare of streetlights.

or reduce the amount of light reaching the house (figure 12-13). Another light moderation strategy is to plant deciduous trees near the sunniest area of the house to reduce light and provide shade. In winter, these plants shed their leaves, allowing more light to reach the house through the bare canopy. To reduce glare throughout the day as the sun's position changes with the earth's rotation, plants of different heights may be planted in the path of the sun, the shortest plants nearest to the sun.

12.7.4 WINDBREAKS

Plants such as coniferous evergreens are effective windbreaks. They reduce the speed of strong winds considerably, so that a cool breeze reaches the house rather than a storm (figure 12-14). The amount of wind reduction depends particularly on the height and shape of plants and the density and width of planting. Multiple rows of plants may be used for more effec-

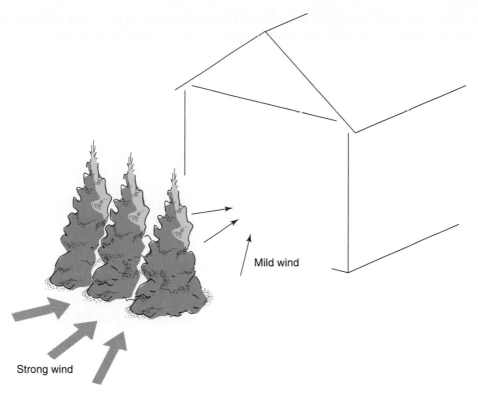

Strong wind

Mild wind

FIGURE 12-14 Using plants as windbreaks to reduce wind speed.

tive reduction. Multiple rows of plant types (e.g., a shrub followed by a tall pine followed by a tall tree) provide a dense wall with minimum penetrability by wind.

12.7.5 OUTDOOR ROOMS

Plants may be planted to create an enclosure or partial enclosure and thereby provide some privacy. A lawn provides the floor of this outdoor room, and the canopy of closely planted trees provides the ceiling. In certain parks, vines are planted and trained on overhead, open wooden structures so that eventually a "roof" is formed.

12.8: XERISCAPING

Xeriscaping is a technique employed in landscape design to create a product that is water efficient in its maintenance. This technique is especially desirable in areas where water should be used sparingly because of scarcity or high cost. Essentially, xeriscaping is a conservation practice.

In using xeriscaping, three general water zones (hydrozones) are recognized, based on the water needs of plants. The levels of moisture in each of these zones are supposed to sustain plants adapted to the particular moisture regime. Irrigation is required only when moisture stress sets in. Zoning the landscape in this way eliminates the traditional, wasteful way of watering the whole landscape each time irrigation is needed. The water zones are classified as follows:

> **Xeriscaping**
> A landscape installation strategy that groups plants according to water needs in order to improve irrigation efficiency.

1. *Very low water zone.* Plants located in the very low water zone are drought resistant and require irrigation primarily during the period of establishment. Once established, such plants are adapted to the local moisture supply and need no supplemental irrigation unless they are under extreme moisture stress. Desert cacti and other species native to the locality succeed under these conditions. A very low water zone will not support annual bedding plants.

2. *Low water zone.* Plants assigned to the low water zone require supplemental moisture to grow and develop properly. This additional moisture is needed infrequently. Shrubs and ground covers can thrive in low water zones.

3. *Moderate water zone.* Plants in the moderate water zone cannot perform well without supplemental irrigation provided at more frequent intervals than for plants in the low water zone. Species in this category are annual plants.

It is undesirable to have plants with high water demands in a landscape in an area where water is scarce and expensive. The goal in xeriscaping is to use plants with low water needs. Consequently, turfgrasses and annual flowering plants are used to a limited extent in a xeriscape because they have a high water demand.

A successful xeriscape requires a good design, careful plant selection, and observance of good conservation practices. Tillage practices that improve water retention should be adopted (3.9). To conserve moisture, mulching is a major activity in the maintenance of a xeriscape. Mulching controls weeds, which deplete the soil moisture, and reduces evaporation from the soil as well as erosion. Since turfgrass is a very popular landscape element, drought-resistant species should be selected for use in a xeriscape. The use of ground covers is encouraged for reducing evaporative losses.

When moisture is needed, high-efficiency irrigation systems should be used. The drip (3.8.11) method of irrigation is the most water efficient, but it is not adapted to watering all plant types in the xeriscape. Highly efficient sprinkler systems (low volume) can be installed to reduce water use. Plants should be watered when most of the water will be retained in the soil (early morning). Using a watering hose or watering can is wasteful under these circumstances.

After installation, a good maintenance schedule is necessary for success of a xeriscape. Weeds and excessive plant growth must be controlled (pruning [chapter 15]).

SUMMARY

Landscaping enhances the surroundings, be they residential, commercial, or public areas. Landscaping means different things to different people. However, a good landscape is the result of careful planning and implementation of a sound design that includes the proper choice and location of plants, the functionality of the design, and the general aesthetic appeal. Whether done by professionals or by the home owner, five basic principles should be followed for success: simplicity, balance, focalization, rhythm and line, and scale or proportion. If contracted to a professional landscape designer, the design process includes an assessment of the owner's needs, site analysis, and preliminary design. Plants are selected according to climatic adaptation, local soil characteristics, features and size at maturity, and customer preference and level of maintenance desired.

REFERENCES AND SUGGESTED READING

Carpenter, P. L., and T. D. Walker. 1990. Plants in the landscape, 2d ed. New York W. H. Freeman.

Crockett, J. V. 1971. Landscape gardening. New York: Time-Life.

Hartmann, H. I., A. M. Kofranek, V. E. Rubatzky, and W. J. Flocker. 1988. Plant science: Growth, development and utilization of cultivated plants, 2d ed. Englewood Cliffs, N.J.: Prentice-Hall.

Rice, L. W., and R. P. Rice Jr. 1993. Practical horticulture, 2d ed. Englewood Cliffs, N.J.: Prentice-Hall.

Robinette, G. O. 1972. Plants, people, and environmental quality. Washington, D.C.: U.S. Government Printing Office.

Schroeder, C. B., E. D. Seagle, L. M. Felton, J. M. Ruter, W. T. Kelly, and G. Krewer. 1997. Introduction to horticulture: Science and technology, 2d ed. Danville, Il: Interstate Publishers, Inc.

Outcomes Assessment

Part A

Please answer true (T) or false (F) for the following statements.

1.   T F   Using plants outdoors to fulfill aesthetic and other functional purposes is called plantscaping.
2.   T F   Balance in landscape design can be symmetrical or asymmetrical.
3.   T F   Landscaping of playgrounds and recreational parks is categorized as commercial landscaping.
4.   T F   The part of the house containing the patio is called the private area.
5.   T F   Cold hardiness is a measure of a plant's ability to survive in winter.

Part B

Please answer the following questions.

1.   List three principles of landscape design.

   _____   _____   _____

2.   A person who designs a landscape is called a _____.
3.   Garbage cans and storage sheds are likely to be found in the functional area of the house called _____.
4.   List four pieces of information a person designing a landscape for a residential area will obtain from a customer during the planning stage.

   _____   _____

   _____   _____

5.   List any three factors that affect the selection of plants for the landscape.

   _____   _____   _____

Part C
---

1.   Describe how the principle of focalization is implemented in a landscape design and installation.
2.   Explain the principle of balance in landscape design.
3.   Explain the role of climate in landscape design.
4.   Discuss what goes into designing the public area of a house.

CHAPTER 13

Nursery Production and Installation of the Landscape

PURPOSE

This chapter is designed to discuss the nursery and its role in landscaping, and how various landscape plants are installed and maintained.

EXPECTED OUTCOMES

After studying this chapter, the student should be able to

1. Describe how a site for locating a nursery is selected.
2. List the categories of plants produced by a nursery business.
3. Discuss container and field nursery production systems.
4. List at least five plants in each of the categories of bedding plants, ground covers, ornamental grasses, trees, shrubs, and plants with underground modified structures.
5. Discuss the principles of designing a flower garden.
6. Describe how to install bedding plants, ground covers, ornamental grasses, trees, shrubs, and plants with underground modified structures in the landscape.
7. Discuss the maintenance of bedding plants, ground covers, ornamental grasses, trees, shrubs, and plants with underground modified structures.

KEY TERMS

Balled and burlapped	Corms	Hardening off
Bamboos	Division	Holder pots
Bare roots	Double potting	Naturalizing
Berm	Flats	Nurseryman
Bulbs	Ground cloth	Overwintering
Container bed	Ground cover	Pinching
Continual bloom garden	Guying	Plastic cell packs

Rhizomes	Shade houses	Tubers
Rush	True grasses	Xeriscaping
Sedge		

OVERVIEW

Plants used in the landscape may be grouped into certain operational categories—trees, shrubs, vines, bedding plants, ground covers, bulbs, corms, tubers, and rhizomes. Plants in each group have common methods of propagation and require similar care. They differ in various characteristics including adult size, shape, color, texture, and growth environment. These categories of plants have certain specific roles in the landscape. When different categories are used simultaneously in the landscape, they have to be properly located to be effective and functional. This chapter is devoted to discussing how these categories of plants are installed and used in creating effective designs in the landscape. The role of the nursery and its operations are also discussed.

13.1: LANDSCAPE CONSTRUCTION

Once a planting plan (figure 12-11) has been prepared, the installation of the landscape can begin. Frequently, however, the site cannot be planted without some modification. A landscape usually consists of more than just plants. Most of the static features are best installed before planting.

13.1.1 HARDSCAPING

Hardscaping is a term used for the installation of hard or static features in the landscape. These features range from simple ones such as fences, walls, patios, and walks to major constructions such as fountains and pools. These hard features can significantly enhance the landscape and increase functionality and property value. Walks may be constructed out of concrete, stone, brick, or some other similar material. Concrete is easy to install and inexpensive; however, it has low aesthetic value and creates little interest in the landscape. Bricks themselves are expensive, as is their installation; they are used for walks, driveways, and patios. A layer of sand is spread on the ground before bricks are laid. With creativity, a work of art can be created with bricks to add tremendous interest to the general landscape design. Stones add a natural touch to the landscape, but they are very expensive to install. In the category of permanent material are decorative patio stones. These stones are made by molding crushed stones and other materials into attractive patterns that can be laid like bricks.

> **Hardscaping**
> The installation of non-plant elements in the landscape.

To add beauty to the landscape at night, some home owners install low-voltage lamps along sidewalks and among plants for ornamental purposes. This use of light is called *night-lighting*. Other hardscaping activities include the construction of a patio or swimming pool on the private side of a house.

> **Night-lighting**
> The installation of special lights to illuminate and beautify the landscape.

Erecting a fence around a property is a common activity undertaken by home owners. The style of fencing chosen depends on the kind of privacy and security desired. The fence may be solid or with breaks. Chain-link fences are inexpensive but provide the least amount of privacy. While some people are interested only in demarcating the property boundaries, others use decorative fences to enhance the landscape. A common type of fence is the stockade type, which can be purchased in easy-to-install 6 × 8-foot (height × width) panels from a local lumber store. For durability, the wood in the panel may be treated with pesticides and thus cost a little more than the untreated type. To create a sturdier fence, steel pipe fence posts may be used.

Constructing the hard features before planting prevents workers from trampling on and damaging established plants. Sometimes large implements (earth-moving equipment) may be involved in the construction, in which case it is best to have no plants obstructing the operation. It is important that the topsoil removed during the construction of the home be replaced.

Terrain

Modifications of the site before planting are necessary to ensure that plants are grown in an environment in which they will perform well. Modifications may be made to introduce additional interest in the landscape by creating variations. Instead of planting on flat areas, mounds may be constructed at certain parts and the area contoured where a slope occurs. On a steep slope, railroad ties may be used to create embankments. As previously mentioned, whenever an earth-moving operation is undertaken in the landscape, it is important to stockpile the topsoil for later use. Mounding may not be necessary in a landscape. Instead, holes or depressions may have to be filled.

Drainage

An important site modification that may be necessary is drainage. Information about *internal soil drainage* at the site can be obtained from the National Resources Conservation Service (NRCS). This property of the soil depends on its type (3.2.1). To avoid the issue of poor internal drainage and other soil-related problems, it is best to check the suitability of the building site before construction. *Surface drainage* is easier to handle than internal soil drainage. Drainage channels can be created to remove from the property excess water resulting from rains. It is critical that the drainage always be away from the foundation of the building, flower beds, and other such structures. Drainage channels can be incorporated during the final grading stage. Where adequate surface drainage occurs, the need for internal drainage is minimal.

Subsurface Irrigation

The landscape can be watered in a variety of ways. However, if a *subsurface irrigation system* (pipes buried in the ground with nozzles that pop up when in use) is to be used, it is best to establish it before installing the turf. An irrigation engineer often provides the best advice as to the most suitable system to install and the site modification necessary to accommodate the system. If the site slopes, pressure will be gained or lost, depending on the part of the slope in question. The type of irrigation system installed depends on the plant type to be irrigated, soil type, wind factors, source of water, terrain, and cost. Irrigation is discussed in detail elsewhere in this text (3.2.1).

Soil Fertility

Even though fertilizers can be applied to plants later on, it is important to start with good soil conditions at the time of planting. A soil test should be performed to determine the native nutrition status. The topsoil should first be replaced where it was removed; otherwise, the turf will be installed on subsoil, which offers less nutrition. If soil amendments are needed, they should be undertaken before the final grading for planting the turf. Whereas liming increases soil pH, sulfur application lowers it (3.2.1).

MODULE 1   NURSERY PRODUCTION

Nurseries produce the plant materials used in the landscape. They also produce seedlings for gardens, fruit seedlings for orchards, and tree seedlings for forests and other uses. A good landscape design is only as good as the quality of materials used to implement it.

13.2: THE ROLE OF THE NURSERY WORKER

The success of the landscape industry depends on the *nurseryman*. The landscape designer depends on the nurseryman to do the following:

1. Supply the correct plants in terms of species, size, and other specified characteristics. The nurseryman should be knowledgeable about the species and be exact in labeling plants. It would be unfortunate to purchase seedlings that are supposed to produce red flowers only to find out midseason that they produce white flowers.

2. Supply healthy plants. The plant material provided should be of high quality and free from diseases and insect pests. Plants grown in the field and potted plants should be free from weeds.

3. Supply materials on a timely basis. The materials should be ready when they are needed to avoid delays in completing the project.

4. Deliver well packaged plant materials. Plants harvested with a ball of soil around the root should be properly packed. Cracked soil balls can seriously jeopardize the survival of plants in the landscape. Plants should be protected from bruising and damage to the bark.

5. Provide at least basic instructions for minimal care of the plants.

13.3: LOCATION OF THE NURSERY

Just as in locating a greenhouse, a nursery site should be selected after careful analysis, considering both economic and ecological factors. The ecological or environmental considerations include soil and climate. Since production is largely under open-air environments (i.e., not controlled like a greenhouse), the factors are more critical now than when considering the location of a greenhouse. These factors are discussed in detail in chapter 11. Thus, discussions in this module are limited to certain aspects of these factors and how they specifically affect nursery production. These factors are described in the following sections.

13.3.1 CLIMATIC CONSIDERATIONS

Important climatic factors to consider in locating a nursery include temperature, rainfall, wind, light, and air pollution.

Temperature

Nurseries produce young plants or seedlings, which are more sensitive than older plants to changes in climatic conditions. Seedlings are generally intolerant of rapid changes in temperature. If a nursery is being considered for the West Coast of the United States (e.g., California) where winters are mild and the growing season longer, the need for winter protection of plants (overwintering) is not critical for container culture. On the other hand, an East Coast production enterprise should consider erecting structures for winter protection of plants in containers.

Rainfall

The rainfall pattern for an area should be well understood. Certain operations cannot be delayed in a nursery enterprise. Rainfall can be supplemented with irrigation if needed, but rains that come during field preparations (e.g., tillage and making beds) or planting time, could be very problematic for an operation. Production schedules are delayed because, for example, the field may be too wet to prepare it for planting. Areas that are prone to unpredictable severe weather should be avoided. If seedlings are damaged by hail or storms, for example, the nursery is likely to take a loss in revenue because the plants may be too old to sell by the time the damage is corrected.

Wind

Nursery plants need to be sheltered from strong winds, which can topple plants in containers and damage young plants. Because plants in the nursery are meant to be around for only a short period of time before being marketed, spacing is closer than it should be in the landscape. Consequently, a delay in selling could make plants compete for space and become top heavy and prone to being blown down by even slight winds. Further, the containers used are generally relatively light to facilitate transportation, which contributes to their susceptibility to wind effects. To overcome this problem, a nursery should be located in an area where natural windbreaks occur; otherwise, artificial windbreaks must be installed.

Light

Unlike a greenhouse, in which supplemental lighting can be provided, field production relies solely on sunlight. Shade houses (11.2.4) can be erected for plants that need such conditions. Unwanted shading sometimes is experienced where trees are used to provide shelter from the winds.

Air Pollution

A nursery should be located where it will not suffer from air pollution. If it has to be established in a heavily industrialized area, it should be located upwind to escape the pollutants emitted from the various industrial facilities.

13.3.2 SOIL FACTORS

Soil factors that affect the location of a nursery include drainage, topography, soil texture and structure, and soil fertility.

Drainage

The proposed site for a nursery should be naturally well drained. If it is not, artificial drainage is required, adding to production costs. Drainage is required for good aeration of the soil and to reduce the incidence of soilborne diseases. Well-drained soils allow rapid soil warming for early production in spring.

Topography

The terrain of the site has implications in drainage, soil erosion, ease of land preparation, and general ease of production operations. If the site is rolling, use of machinery is hampered and irrigation systems are more difficult to install and operate. It may become necessary to spend additional money to level or terrace the site to facilitate production operations. On a rolling site, low parts of the land are likely to experience drainage problems and be susceptible to frost damage.

Soil Texture and Structure

Field production of seedlings for sale requires that plants be dug up at some point. If bare-root (13.7) production is intended, a soil with loose structure (sand or loam) is preferred. For balled and burlapped (13.7.2) production, the soil should be cohesive enough to form a ball around the roots. A properly textured soil also drains well, holds moisture at a desirable level, and is easy to work and well aerated.

Soil Fertility

The soil must be fertilized sooner or later for optimal production. However, the site should have some native fertility and be responsive to fertilization. The higher the soil quality (in terms of organic matter content, pH, and nutrition), the fewer the initial amendments.

13.3.3 SUPPLEMENTAL WATER SUPPLY

A nursery operation cannot be 100 percent rain fed. For timely production, water must be provided when it is needed by the plants. Depending on the location, the nursery relies on groundwater (well), lake, river, or domestic water supplies. These sources vary in cost, quality, accessibility, and reliability of supply. Many nurseries depend on wells for irrigation.

13.4: SITE PREPARATION

Architects and civil engineers help in developing and designing facilities for a nursery site. After surveying the land and developing appropriate plans, the site should be cleared of trees and other undesirable objects. Some leveling may be required. Access roads should be developed to facilitate the movement of nursery equipment and vehicles. The appropriate irrigation systems should be installed to provide a reliable and timely supply of water for plants. A container production operation has higher water demands than ground bed production. If the land survey indicates poor drainage, appropriate drainage systems should also be constructed.

13.5: NURSERY STRUCTURES

The type of structures needed depends on the region in which the nursery is located. As previously indicated, an area's climate may require the provision of overwintering facilities. Some plants may need shelter from intense sunlight, necessitating the construction of shade houses. Because some greenhouse-type production may be necessary to successfully produce certain types of plants, nurseries may construct greenhouses (11.1). For propagation, a cold frame or hotbed may be required at a nursery site.

A variety of storage facilities are needed on-site. Storage is needed for supplies, including seed, chemicals (e.g., fertilizers and pesticides), equipment, and temporarily for planting materials awaiting shipment. These facilities are in addition to basic ones such as preparation rooms (for mixing and potting), propagation houses, and other administrative rooms.

13.6: ECONOMIC CONSIDERATIONS

For profitability, several economic considerations should be taken into account in deciding on the best place to locate a nursery.

13.6.1 MARKETS

Nursery products may be transported to near or far markets. These products are generally bulky, whether container or field grown. Nurseries should be located near highways, if possible, to make them readily accessible to customers. Whereas some nurseries serve local markets, others serve clients out of state and long distances from the production sites. If the operation is large, the company may consider operating its own transportation system. Sometimes nurseries deliver large plants in mechanical augers after digging. This undertaking is economical only if the nursery is located close to the market or clients.

13.6.2 LAND

Virgin land costs more to develop into a usable site than an area that has previously been in cultivation. It is advantageous to acquire a large piece of land and expand the operation as time goes by. Land near metropolitan areas may be expensive. Land in rural areas may be cheaper, but transporting materials and products to and from the nursery and markets would be costly.

13.6.3 LABOR

Nursery production is seasonal in terms of labor needs. Container production is more labor intensive than field production. A limited number of permanent staff should be employed, with a seasonal labor pool readily available.

13.7: TYPES OF NURSERY PRODUCTION

> **Container Nursery**
> A nursery where plants are raised in containers.

There are two types of nursery production—*container* and *field*.

13.7.1 CONTAINER NURSERY

In container nursery operations, plants are grown and marketed in containers that differ in sizes and types according to the species and marketable size desired (figure 13-1). Containers are discussed more fully in chapter 10. Similarly, the media ingredients and mixes discussed for greenhouse production apply here. The conditions described for successful container culture are also the same for the nursery operation.

Seeds and cuttings (chapter 8) may be used for propagation. Nurseries may maintain blocks of parent stock in the field, which are plants used as a source of materials for propagation. Parent stock may be maintained less expensively in containers than in fields. These stock plants must be well maintained to produce healthy planting materials. Container nursery production is done on a large scale in many states, especially Florida, Michigan, Pennsylvania, Virginia, and Ohio. In greenhouses, pots are usually arranged on benches (and sometimes on the floor) during production. In container nurseries, container beds are created such that plants of similar size that require the same cultural conditions are grouped together during production. These beds vary in design and size depending on the sizes of pots, spacing, shade house and overwintering facility installed, and irrigation system used.

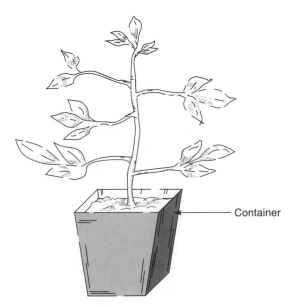

FIGURE 13-1 Container-grown tree seedling.

The site is leveled in one or two basic ways (figure 13-2). In one design, the center of the bed is raised so that excess irrigation water flows away to the edges of the bed. This design may be undesirable because it compels workers to walk through mud to get to the bed. In the other design, the slope is toward the center of the bed. The bed is covered with natural materials, such as gravel or crushed sea shells, or artificial materials, such as black ground cloth or black polyethylene. Black cloth is made of durable natural materials. The problem with gravel and shells is that they allow weeds to grow through, and plant roots may also grow through drainage holes in the pots and through the gravel layer into the ground, making moving plants around problematic.

Container plant roots are less resistant to cold than field roots because they are less protected than roots in the ground. The ground is warmer in winter than the soil in pots. Thus, in container culture, plants need protection from the cold. One way to provide winter protection is to crowd plants together and wrap the pots in black plastic (figure 13-3). In place of plastic, an outer border of containers filled with growing media can be used (figure 13-4). White polyethylene-glazed overwintering houses are widely used for winter protection of plants in nurseries.

One strategy of growing container plants with winter protection is called the *double pot* (pot-in-pot) *production system,* whereby larger pots (*holder pots*) are placed in holes in the ground and buried up to their lips (figure 13-5). These pots then become holes in which containerized plants are grown through the season. The use of holder pots eliminates the need for additional winter protection. However, this year-round insulation may expose plant roots to high temperatures during the summer season.

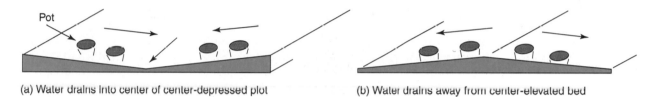

(a) Water drains into center of center-depressed plot   (b) Water drains away from center-elevated bed

FIGURE 13-2 Ways of leveling a field nursery for pot culture.

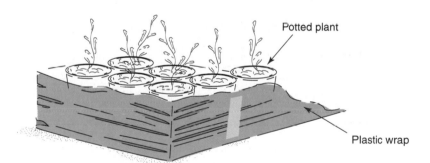

Potted plant

Plastic wrap

FIGURE 13-3 Wrapping pots in plastic for winter protection.

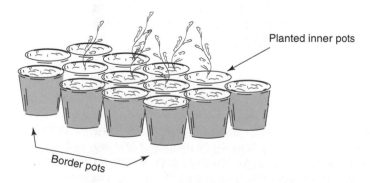

Planted inner pots

Border pots

FIGURE 13-4 Using pots filled with soil for winter protection of nursery potted plants.

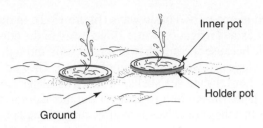

Inner pot

Holder pot

Ground

FIGURE 13-5 Using holder pots in a field nursery.

13.7.2 FIELD NURSERY

A fundamental difference between container and field nurseries is that in field nurseries plants are grown to the desired size in ground beds. This system is often used for producing shade trees (e.g., red maple, pin oak, green ash, honey locust, white ash, and red oak), flowering trees (e.g., crab apple, redbud, flowering plum, and flowering dogwood), and evergreen and deciduous shrubs.

The soil is broken up and prepared for planting by plowing and harrowing. The land is divided into sections, with turf grass–covered access ways between sections. Planting materials may be cuttings or grafted plants or from tissue culture. Trees are planted in rows under two basic systems, which differ according to how they are harvested for the market (13.19).

Bare-Root Trees

Bare-root *A tree or shrub seedling that is offered for sale without soil around its roots.*

Bare-root trees are dug up without soil around the roots (figure 13-6) (13.19.1); they do not store well and are prone to transplanting shock (8.13). This system is suited to small trees and shrubs. Bare-root plants are lighter and easier to transport than balled and burlapped trees.

Balled and Burlapped Trees

Balled and Burlapped *A tree or shrub seedling that is offered for sale with a ball of soil around its roots and wrapped in burlap.*

Balled and burlapped systems require plants to be planted in the ground and dug with a ball of soil around the roots (figure 13-7) (13.19.1). Mechanical harvesters (hydraulic augers or tree spades) are used for this purpose. The ball of soil is wrapped with burlap material and tied. If they are not needed immediately, balled and burlapped plants can be stored for a period of time. Harvesting by tree spade can severely prune roots, jeopardizing their establishment in the field. Balled and burlapped plants are best harvested when the soil is moist.

13.7.3 RETAIL NURSERY

A retail nursery or garden nursery represents the retail outlet of the nursery industry. Garden centers sell nursery products (planting materials) and production materials (pesticides, simple tools, fertilizers, garden or landscape furniture, and various horticultural literature). The personnel at these facilities include a manager for administrative and general oversight purposes and a plant technician who is knowledgeable in a variety of plant problems and capable of advising customers. A garden center may employ a landscape designer who can design small-scale projects for home owners.

SUMMARY

Nurseries are depended on to provide planting materials for the landscape. The plants they specialize in are perennials—shrubs, trees, and fruit trees. The site for the nursery should be carefully selected following the guidelines utilized in selecting a greenhouse site. The site may require some preplanting preparation, including leveling and installation of drainage systems to ensure good drainage. Greenhouse-type structures are often erected for certain production operations and for protecting plants in adverse weather.

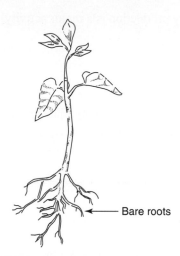

FIGURE 13-6   A bare-root seedling.

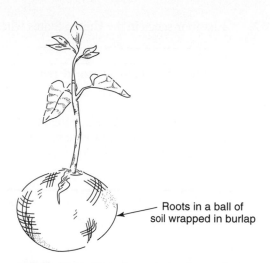

FIGURE 13-7   Balled and burlapped tree seedling.

The two basic nursery production types are container and field. In container production, the prepared ground is covered with gravel or another suitable natural material or artificial materials such as black ground cloths. Plants produced in the field are marketed as either bare-root or balled and burlapped plants.

REFERENCES AND SUGGESTED READING

Schroeder, C. B., E. D. Seagle, L. R. Felton, L. M. Ruter, W. M. Kelly, and G. Krewer. 1997. Introduction to horticulture: Science and technology, 2d ed. Danville, Il: Interstate Publishers, Inc.

OUTCOMES ASSESSMENT

PART A

Please answer true (T) or false (F) for the following statements.

1.   T F   The need for winter protection is greater for field plants than for container plants.
2.   T F   Bedding plants are the major plants produced by field nurseries.
3.   T F   Greenhouses may be installed on nursery sites for special purposes.
4.   T F   Balled and burlapped plants are best harvested when the soil is dry.

PART B

Please answer the following questions.

1.   List the types of plants produced by a nurseryman.

2. List four states in the United States where nursery production is a major activity.

    _____ _____

    _____ _____

3. List four factors to be considered in locating a nursery.

    _____ _____

    _____ _____

4. The production system in which plants receive winter protection by placing potted plants in holes in the ground is called _____.

5. The equipment used to dig large field plants is called_____.

PART C

1. Discuss the importance of temperature as a consideration in locating a nursery.
2. Compare and contrast container and field production systems in a nursery operation.

MODULE 2 BEDDING PLANTS

OVERVIEW

The term *bedding plants* is used to refer to a broad range of largely annual plants (i.e., herbs, vegetables, and flowering ornamentals) that are grown especially in flower beds (but also in containers such as hanging baskets and window boxes). The bedding plant industry has grown tremendously in recent years due in part to convenient packaging of the seedlings by nursery growers. This packaging increases home owners' success in growing bedding plants. One factor limiting success with bedding plants is transplanting shock. Previously, seedlings were raised in seedling trays or flats. The high density of sowing did not allow seedlings to be dug out individually with adequate soil attached to the roots, which predisposed the seedlings to wilting from moisture stress. Today, nursery and greenhouse producers raise seedlings in plastic cell packs, each seedling occupying a cell with adequate soil around its roots. Home growers can purchase packs of seedlings and transplant them conveniently. The continued efforts of breeders make it possible to have a wide variety (in terms of shape, size, color, and maturity) of bedding plants to meet different customer tastes and climatic conditions. Hybrid cultivars of these plants produce superior flowers or edible produce. Ground covers are generally low-growing plants with the ability to spread. They include shrubs, vines, grasses, and perennials.

13.8: ROLE OF ANNUAL BEDDING PLANTS

Vegetable bedding plants, discussed in chapter 17, are usually restricted to gardens, where they are grown for their edible parts. Some ornamental types of certain vegetables exist. The use of bedding plants discussed in this section is limited to ornamental plants.

The use of *ornamental annual bedding plants* is based on their size or height, form (spreading or trailing versus erect), foliage, light requirements (shade, partial shade, or high light), and color of flowers, among others.

Because of their relatively small size, bedding plants can be readily and effectively raised to form a patch of uniform display of color, texture, or other desired characteristics. They can be used to fill open spaces between or around objects. For example, bedding plants may be grown around a sculpture or fountain in the landscape. These plants may be grown in the foreground of permanent shrubbery or along fences and walls. When grown in containers, bedding plants may be displayed in windows or hanging baskets on patios. They may be used to accentuate the border of the walkway. It is easy to be creative with bedding plants. Even the gardener prefers a particular species; plant breeders have developed numerous cultivars of many species to provide for variety in the landscape, such as by using cultivars with different colors in the design. Some bedding plants may serve the dual purpose of providing interest in the landscape and being used for *cut flowers* (11.7) for indoor use. In short, bedding plants can be used in a variety of ways to beautify the landscape.

13.9: DESIGNING A FLOWER GARDEN

To maximize the aesthetic value of flowers, it is important to choose the right plants for the area and also locate or display them attractively in the landscape. In fact, simple or rather unattractive flowers can be greatly enhanced by the way they are displayed. Instead of locating plants haphazardly in the landscape, specific designs can be created and adopted in planting the flower garden. The following is a discussion on the principles involved in designing a successful flower garden.

13.9.1 SELECTING A SITE

A garden site should be freely draining since waterlogged conditions are intolerable to most plants and a nuisance to gardeners as they care for their plants. High soil moisture levels may cause bulbs and roots to rot before they sprout. Where soils drain slowly or poorly, raised beds may be used to improve drainage and provide a warm seedbed. The site of a garden should be strategic. The purpose of planting a garden is to make it readily visible in order to be enjoyed. It is easier to enjoy a garden for a prolonged period if it is located within view of where the gardener usually sits to relax (e.g., a deck or patio). Because it is also gratifying if others can enjoy the garden, it may be located on the public side of a house.

Apart from water and air, light is another factor that is critical to the success of plants. Most flowering plants grow best under conditions of full sunlight. Shading causes some plants to produce more foliage than flowers or to grow spindly, weak, and unattractive.

13.9.2 CHOOSING PLANTS

Some gardeners will grow any plant they find pretty. Others are very particular about their choice of plants and plant only certain species and colors. Gardeners often like to grow a variety of types (e.g., bulbs, herbaceous plants, and annuals). In the case of mixtures, care must be taken in designing the garden so that the different species blend well together. Different species may have different physical features and environmental needs, requiring that the layout be designed such that no component is obscured or diminished. General guidelines for choosing plants are as follows:

1. *Height.* It is important to know the plant height at adult stage so that taller species are planted in the background and shorter ones in front. This way, all plants can be seen very clearly in the display. If the contrast is too great, one species may overwhelm the others.

2. *Habit or form.* Some species stand erect, and others are procumbent (spreading). Many hanging basket species produce long vines that hang down from the container.

3. *Color.* Flowers come in a large variety and intensity of colors. A successful flower garden displays species whose colors blend well. A good display of colors does not necessarily mean contrasting colors. Some stunning and effective displays involve shades of one or a few colors.

4. *Sunlight required.* When plants are located close to the house or trees, shading (partial light) becomes an important consideration. Inadequate sunlight reduces flowering capacity.

5. *Blooming.* Species bloom at different times and for different periods. If the gardener desires continual blooming in the landscape, the species should be selected such that they bloom in succession throughout the season.

6. *Water needs.* The domestic water supply is the source of water for home gardens. In summer when demand for water is at its peak, some areas experience water shortage. Plant breeders develop drought-resistant cultivars of various bedding plants that may be used in such drought-prone regions. Certain species require more water for growth than others. When growing mixtures, water can be used more efficiently by grouping plants with similar needs, the principle of *xeriscaping* (12.8).

7. *General care.* Some plant species require constant vigilance for best results, while others need minimal attention. Generally, annual bedding plants need more hands-on involvement of the grower than do trees and shrubs. The gardener should choose plants he or she is ready to care for adequately.

13.9.3 PREPARING THE BED

The soil should be dug up and turned over to a depth of about 12 to 18 inches (30.5 to 45.7 centimeters) with a spade or rototiller. During the tillage operations, compost or peat should be incorporated to improve drainage and moisture retention. It may be necessary to purchase topsoil to mix into the native soil to improve its physical structure. A complete fertilizer (e.g., 10:10:10) should be spread and mixed in at the rate of about 2 pounds per 100 square feet (1 kilogram per 9 square meters) of land. Other soil amendments such as liming may be necessary where the soil is acidic. A pH of 6.5 is preferred by most annual species.

Where drainage is a problem, raised beds may be constructed for bedding plants. Beds should not be designed to have straight edges. Straight lines are not only hard to maintain but also less attractive than those with sweeping curves (figure 13-8). A variety of materials (e.g., bricks, plastic, or boards) may be used to edge beds to define their perimeter and to avoid encroachment of lawn grass (figure 13-9).

Flower beds may be attached to hedges or fences and thereby serve as borders. Beds may be constructed to be unattached (freestanding or island) in the landscape (figure 13-10), thus providing viewing opportunities from all angles, as opposed to from the front only in the case of border beds.

13.9.4 SEEDING

Bedding plants may be *direct seeded* or *transplanted.*

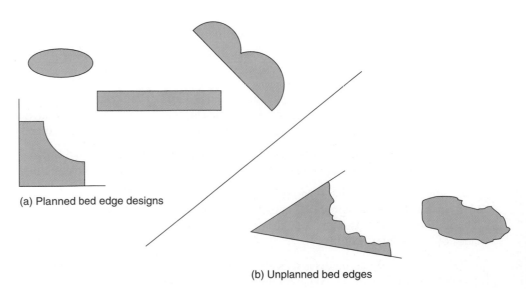

(a) Planned bed edge designs

(b) Unplanned bed edges

FIGURE 13-8 Variations in edging of beds in the landscape.

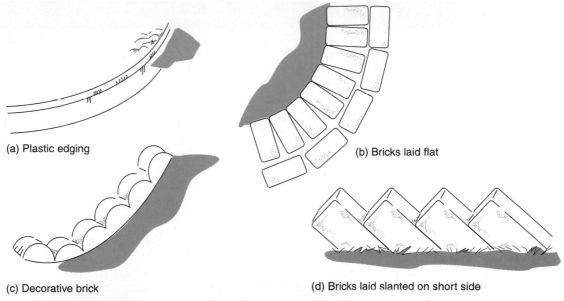

(a) Plastic edging

(b) Bricks laid flat

(c) Decorative brick

(d) Bricks laid slanted on short side

FIGURE 13-9 Edging materials for beds in the landscape.

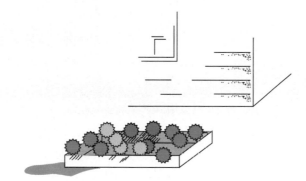

FIGURE 13-10 Freestanding flower bed.

Direct Seeding

Direct seeding is done in early spring when there is no danger of frost. The bed should be prepared to a fine tilth. The proper spacing and planting depth should be observed for good germination. Annual plants such as cleome, poppy, baby's breath, and stock may be direct seeded (table 13-1). It is important that the seeds used be viable in order to obtain a good stand. If possible, fresh seeds should be purchased each season. The bed should be watered after seeding and kept moist but not wet (which promotes rotting).

Transplanting

Raising seedlings for transplanting is an activity that requires good strategy. Seedlings are raised indoors while the weather outside is cold and timed to coincide with early spring. Seeds are sown in flats containing a lightweight soil medium; the medium should be pasteurized to avoid damping-off of seedlings. Such a medium is freely draining but also has good moisture-holding capacity, which makes transplanting easier. Small seeds are sown in rows in flats (figure 18-12). When they attain a good size to be conveniently handled, they are transplanted individually into peat pots or cell packs (figure 13-11). When sowing larger seeds, the cells or peat pots are arranged in the flat before filling with soil. Seeds should be sown such that each pot eventually holds one or two successful seedlings. Examples of transplanted bedding plants are presented in table 13-2.

Table 13-1	Selected Direct-Seeded Bedding Plants
Plant	**Scientific Name**
Spider flower	*Cleome* spp.
Coreopsis	*Coreopsis* spp.
Baby's breath	*Gypsophila* spp.
Sunflower	*Helianthus annuus*
Poppy	*Papaver* spp.
Forget-me-not	*Myosotis* spp.
Zinnia	*Zinnia elegans*
Marigold	*Tagetes* spp.
Nasturtium	*Tropaeolum majus*

FIGURE 13-11 Bedding plants raised in cell packs.

Freshly sown seeds should be thoroughly watered and kept moist. A spray nozzle should be used to avoid burying small seeds too deeply in the soil. Also, the flat may be covered with a sheet of glass to reduce the light and transpiration, and also maintain a high relative humidity. Poor lighting, however, induces etiolation. Overcrowding caused by a high seeding rate also results in plant etiolation (figure 3-9). Alternatively, the seeding flat should be periodically misted. Seedlings should be hardened off for better transplanting success. Transplanting should not be delayed.

The advantage of using peat pots is that the pots can be inserted directly into the soil. Before transplanting, the plants should be thoroughly watered. The rim of the pot should then be torn off and the bottom opened up before inserting it into the soil (8.13). Plants should be set according to recommended spacing. A starter solution may be applied at planting to promote good and quick establishment. The solution is usually high in phosphorus, but soon after the seedling becomes established, nitrogen should be applied.

Commercial greenhouses grow and sell bedding plants. Using greenhouse plants is a convenient way to establish bedding plants if one does not want to go through the process of raising seedlings. For best results, one may invest in a *propagation case,* which may be heated and enclosed to keep the humidity at a desirable level for germination (figure 13-12).

> **Propagation Case**
> *A transparent and enclosed container used for raising seedlings.*

13.9.5 CARE

The flower bed should be kept clean of weeds, which not only compete with bedding plants for nutrients but also are unsightly and reduce the aesthetic value of the display. It is important to maintain a good watering schedule to avoid drying of plants, which can induce premature flower drop. Application of a mulch helps to control weeds and also reduce moisture loss. Fertilizers should be applied judiciously so that the vegetative cycle is not overextended

Table 13-2 Selected Bedding Plants That Are Transplanted

Plant	Scientific Name
Petunia	*Petunia* spp.
Pansy	*Viola tricolor*
Dahlia	*Dahlia* spp.
Lavender	*Lavandula officinalis*
Larkspur	*Delphinium* spp.
Ageratum	*Ageratum* spp.
Snapdragon	*Antirrhinum majus*
Coleus	*Coleus blumei*
Cosmos	*Cosmos* spp.
Impatiens	*Impatiens* spp.

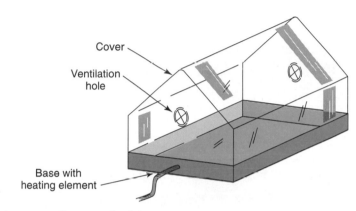

FIGURE 13-12 A propagation case for home use.

to the detriment of flowering. Water-soluble fertilizers may be applied in the irrigation water. If insects or diseases occur, they should be controlled by using appropriate and safe pesticides. In commercial nurseries, some growers apply a hormone (e.g., B-nine or diaminozone) to control growth in bedding plants so that they become stocky, compact, and appealing. These features may be obtained by *pinching* (removing the terminal bud) to induce lateral branching (figure 4-20). Since many annual flowering plants die after their fruits mature (i.e., monocarpic), they can be maintained in bloom for a long time by regularly removing faded flowers and thereby preventing fruiting.

13.10: MAKING A GARDEN PLAN

A garden style may be either *formal* or *informal*. Style is a function of the type of plants, the nonplant materials (e.g., pavements, statues, patios, and walls), the arrangement of plants, and how the overall design harmonizes with the house or building and the general surroundings. The choice of style for a home garden is strongly influenced by the home owner's personality.

Formal gardens have their origins in European culture, the Italians and French being noted for some of the most elegant designs. These gardens tend to emphasize geometry and symmetry (figure 13-13). Plants are set in the landscape in predetermined patterns. Because of the rigidity in design, formal gardens must be well maintained to remain beautiful. Hedges should be well manicured, with their edges trimmed straight. Frequently, the arrangements are made to be symmetrical around a focal point. They are expensive to install and maintain. Informal gardens, on the other hand, are designed to imitate nature by being asymmetrical, with irregularity in the way plants and other objects are located. Straight, rigid lines are not

> **Formal Garden**
> A stylized garden in which arrangement of the materials emphasizes symmetry and geometry.

> **Informal Garden**
> A garden in which the materials are located without emphasis on regularity and symmetry.

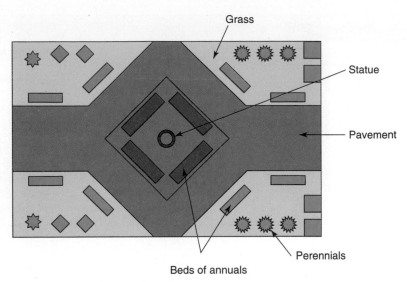

FIGURE 13-13 A formal garden. The emphasis is on symmetry in the location of plants and other landscape design elements.

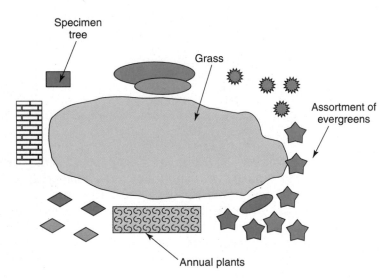

FIGURE 13-14 An informal garden. It lacks symmetry in arrangement of landscape elements.

an objective of the design (figure 13-14). This style originated in the Eastern cultures of Japan and China. The lack of symmetry does not imply lack of balance in the design (figure 12-6).

The most popular ornamental bedding plants include geranium, zinnia, petunia, marigold, impatiens, and begonia. When planning a garden design, the color intensity of flowers is a critical consideration. The impact of this factor on the general appeal of the design is influenced by the prevailing climate. Some colors develop best when formed under intense sunlight (e.g., marigolds). Certain colors overshadow others when they occur side by side. Another strategy used to improve display is to draw attention to plant size. A large plant located next to a small one may produce a situation in which the latter may be diminished by the former. One strategy to reduce this effect is to mass plants (group like plants together). If one desires to include garden ornaments (e.g., sculptures), care should be taken to avoid using an oversized piece that overwhelms the surrounding plants.

For best results, a garden design should be approached methodically. Drawing a field plan before going out to plant helps one to visualize the garden on paper before installing it.

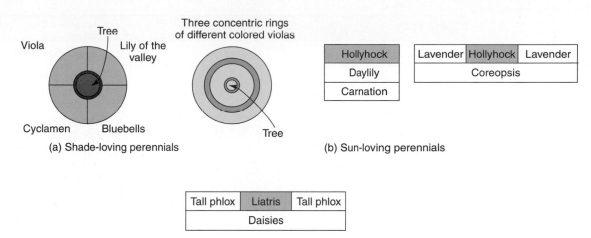

FIGURE 13-15 Samples of simple garden designs.

A plan can be designed and drawn on a computer using commercial software. The purpose of the garden should be in the forefront of the planning activity. For example, annual bedding plants are widely used to fill spaces in the landscape. While waiting for perennials to grow to blooming stage or, as in the case of bulbs, waiting for the next season to arrive, annuals may be planted so that the space is not left vacant. It is important that plants in the landscape complement each other. As such, one should know what already exists before adding new plants. There may already be a hedge or shrub border against which annuals may be planted.

Flower gardens may also be designed with specific themes in mind. One may capitalize on floral characteristics to create a theme. Certain flowers bloom in the evening, such as evening primrose *(Oenothera* spp.), angel's trumpet *(Datura* spp.), pastel evening stock *(Mathiola longipetala),* and four-o'clocks *(Mirabilis jalopa).* A garden that comes alive in the evening could be a viable theme. A grass garden using species such as bromegrass *(Bromus madritenis),* broom corn millet *(Panicum miliaceum),* Job's tears *(Croix lacryma Jobi),* and foxtail millet *(Betana italica)* is also a viable garden theme. Fragrance is another design possibility, whereby flowers such as pastel evening stock *(Mathiola longipetala),* petunia *(Petunia x hybrida),* and flowering tobacco *(Nicotiana sylvestris* and *N. alata)* that produce sweet fragrances are incorporated into a garden design. Some simple garden designs are presented in figure 13-15.

13.11: HANGING BASKETS

Certain annual bedding plants may be grown in containers that are hung from the ceiling in a room, patio, or main entrance to a house. The design and construction of a hanging basket are discussed in chapter 10.

13.12: COMMON PERENNIAL BEDDING PLANTS

Bedding plants are mostly annuals. However, certain garden designs include perennial species. In fact, growing mixtures of species is a strategy for creating a *continual bloom* garden in which plants bloom in succession. This way, the garden can be enjoyed for a long period. Bulbs are included in mixtures for their early bloom in fall; perennials bloom in late spring and throughout the summer. Popular flowering perennials are described in table 13-3.

Continual Bloom Garden
A garden planted with a mixture of plant species that bloom in different seasons so that flowers occur all year round.

Table 13-3 Selected Perennial Bedding Plants

Plant	Scientific Name
Canterbury bells	*Campanula medium*
Carnation	*Dianthus caryophyllus*
Columbine	*Aquilegia* spp.
Daisy, Shasta	*Chrysanthemum maximum*
Delphinium	*Delphinium elatum*
Foxglove	*Digitalis purpurea*
Daylily	*Hemerocallis* spp.
Iris	*Iris* spp.
Peony	*Paeonia* spp.
Sweet pea	*Lathyrus latifolius*
Sweet william	*Dianthus barbatus*
Lupine	*Lupinus polyphyllus*
Daisy, golden	*Anthemis tinctoria*

SUMMARY

Bedding plants are largely annual plants that are grown in beds. Whereas vegetable bedding plants are grown for food, ornamental bedding plants are grown for aesthetic reasons in the landscape. These plants are generally small in size. Flower beds should be well drained and exposed to sunlight. Plants selected for use should be adapted to the environment. Further, one should also consider the adult height, habit or form, color, blooming habit, water and light needs, and general care required. Bedding plants may be direct seeded or transplanted. Garden plants can be designed along certain themes such as flower color, fragrance, or species (e.g., grass garden). Annual gardens, which may include perennial species, generally require a degree of involvement by the grower.

REFERENCES AND SUGGESTED READING

Crockett, J. A. 1981. Crockett's flower garden. New York: Crown.

Hill, L., and N. Hill. 1988. Successful perennial gardening: A practical guide. Pownal, Vt.: Garden Way.

Rice, L. W., and P. R. Rice Jr. 1993. Practical horticulture, 2d ed. Englewood Cliffs, N.J.: Prentice-Hall.

Sunset Magazine and Book Editors. 1981. Annuals and perennials, 2d ed. Menlo Park, Calif.: Sunset-Lane.

OUTCOMES ASSESSMENT

PART A

Please answer true (T) or false (F) for the following statements.

1. T F   All bedding plants are annuals.
2. T F   There are vegetable annual bedding plants.
3. T F   Flower beds may be freestanding or attached.
4. T F   Generally, annual bedding plants need less hands-on involvement than trees and shrubs.

5. T F Annual bedding plants may be planted in hanging baskets.
6. T F Marigolds develop their best color intensity when grown in the shade.

PART B

Please answer the following questions.

1. Seedlings may be raised in trays called _____.
2. List five factors to consider in choosing annual bedding plants.

    _____ _____ _____

    _____ _____

3. Removing the terminal buds to induce branching in plants is called _____.
4. Bedding plants may be direct seeded or _____.
5. List four ornamental vegetable plants.

    _____ _____

    _____ _____

6. List five annual bedding plants.

    _____ _____ _____

    _____ _____

PART C

1. Explain the strategy of xeriscaping.
2. Describe a strategy for creating a continual bloom garden.
3. Design an annual bedding plant garden.

MODULE 3 GROUND COVERS AND ORNAMENTAL GRASSES

OVERVIEW

Even though in the broadest sense any plant that spreads its foliage over the soil can be called a *ground cover,* operationally the term is reserved for low, spreading plants (less than 2 feet or 60 centimeters tall). This group of plants includes shrubs (e.g., dwarf yew, creeping juniper, and dwarf azalea), vines (e.g., English ivy), perennials (e.g., lily of the valley and twining strawberry), and grasses. Some species such as daylilies (*Hemerocallis* spp.) are clump forming, and others such as ajuga *(Ajuga reptaris)* are rhizomatous or stoloniferous.

Ground covers are used in a variety of ways. For example, they may be used to reduce soil erosion on steep slopes; cover rocky and rough areas that are difficult to mow; cover up areas under trees that are poorly lit; hide unattractive parts of the landscape; or enhance the aesthetic value of the landscape by their attractive form, foliage, and flowers. These plants generally have the capacity to grow in a manner as to blanket the ground and frequently are also quick spreading. They do not require the regular maintenance needed by some landscape plants.

> **Ground Cover**
> *A low-growing plant that spreads and forms a mat-like growth over an area.*

13.13: Choosing a Ground Cover

Ground covers should be chosen to meet certain conditions:

1. *The hardiness zone.* Unless the selections are adapted to the local climate, they will not perform to their full capacity. Like other plants, some species are adapted to cool climates and others prefer hot and dry climates. Some plants such as hostas are adapted to a wide variety of climates.

2. *The site characteristics.* A good site analysis is needed to find out the specific problems and how to select a ground cover to solve them. This analysis often includes a soil test (3.2.1) to determine the nutritional status of the soil in the area. If an area in the landscape is bare, it might be due to a number of factors, such as shade, shallow soil, low nutrients, improper pH, and others. The soil may be sandy and drain too quickly or clay and hold too much water. The soil may have an adequate amount of nutrients to support growth but a pH that does not make the nutrients available to plants.

   Table 13-4 provides some suggestions of plants to grow in shady conditions such as under trees. Junipers and cinquefoil *(Potentilla fruticola)* perform well if the area is a dry hillside of poor fertility. Table 13-5 lists some species adapted to steep and rocky slopes. One hard-to-manage area in the landscape is the wet patch that will not support a lawn. For such spots, possible choices include the creeping Jenny *(Lysimachia spp.),* certain ferns, blue flag *(Iris versicolor),* and Japanese primrose *(Primula japonica).*

3. *Special effects.* Ground covers are used for more than protection of bare ground in the landscape. Some species have qualities to make them appropriate for use as a focal point in the landscape. Apart from green foliage, ground covers can be used for specific roles in the landscape. Ground covers with attractive colors include shrubby cinquefoil, lenten rose *(Helleborus orientalis),* Bethlehem sage *(Pulmonia saccha-*

Table 13-4 Selected Ground Covers Adapted to Shade

Plant	Scientific Name	Zone	Comments
Bishop's weed	*Aegopodium podagraria*	3–9	Fast spreading; 12 inches (30 centimeters) high
Ajuga	*Ajuga reptans*	3–8	Short (3 inches or 7.7 centimeters); forms solid carpet
European ginger	*Asarum europaeum*	4–8	5 to 7 inches (12.5 to 17.5 centimeters) high; leathery leaves
Chinese astilbe	*Astilbe chinensis*	5–8	Flowers in summer; 8 to 12 inches (20 to 30 centimeters) high
Winter creeper	*Euonymus fortunei*	5–9	Evergreen; long climbing vines; 1 to 2 feet (30 to 60 centimeters) high
Sweet woodruff	*Galium odoratum*	3–8	Fast spreading; 6 to 12 inches (15 to 30 centimeters) high; sweet fragrance when dried
English ivy	*Hedera helix*	5–9	Excellent spreader; 6 inches (15 centimeters) high
Hosta	*Hosta* hybrids	3–8	Easy to grow; versatile; 6 to 24 inches (15 to 60 centimeters) high
Strawberry geranium	*Saxifraga stolonifera*	6–9	Good houseplant; 6 inches (15 centimeters) high
Common periwinkle	*Vinca minor*	3–9	Quick spreading; 6 to 10 inches (15 to 25 centimeters) high
Japanese pachysandra	*Pachysandra terminalis*	4–8	Popular; good spreader; 8 to 10 inches (20 to 25 centimeters) high

Table 13-5 Selected Ground Covers Adapted to Steep Slopes and Rocky Areas

Plant	Scientific Name	Zone	Comments
Wooly yarrow	*Achillea tomentosa*	3–7	Short (2 inches or 5 centimeters); full sun
Saint-John's-wort	*Hypericum calycinum*	6–8	Easy to grow; vigorous; 12 to 18 inches (30 to 45 centimeters) high
Japanese pachysandra	*Pachysandra terminalis*	4–8	Popular; good spreader; 8 to 10 inches (20 to 25 centimeters) high
Creeping juniper	*Juniperus horizontalis*	3–9	Erosion control; good for foundation planting; 6 to 18 inches (15 to 45 centimeters) high
English ivy	*Hedera helix*	5–9	Excellent spreader; 6 inches (15 centimeters) high
Common periwinkle	*Vinca minor*	3–9	Quick spreading; 6 to 10 inches (15 to 25 centimeters) high
Ajuga	*Ajuga reptans*	3–8	Short (3 inches or 7.7 centimeters); forms solid carpet

rata), Japanese primrose *(Primula japonica),* and moss phlox *(Phlox subulata).* Other ground covers have fragrant flowers, such as lily of the valley *(Convallaria majalis)* and sweet violet *(Viola ordorata).* Thyme, a popular herb, can also be used as an effective ground cover. Ground covers can serve to tie together contrasting areas (e.g., turf and shrubs) in the landscape.

4. *Installation and maintenance.* Ground covers are utilized in areas of the landscape that are difficult to reach or manage. The plants installed should require relatively little maintenance and be easy to install. They should be chosen for their high probability of success in an area.

5. *Size of space available.* Since ground covers are by nature quick spreading, it is important that the species selected not overwhelm the area where they are installed. Additional maintenance is required to keep the ground cover in control. Plant selection is especially important when ground covers are planted in combination. A list of some aggressive spreaders is presented in table 13-6.

13.13.1 SOIL PREPARATION AND PLANTING

Even though ground covers are perceived as plants used in marginal parts of the landscape, they nonetheless require adequate soil preparation for good establishment. Further, if the purpose of using the ground cover is for something other than environmental protection, the site should be appropriately prepared. The ground should be cleared, dug up, and prepared like a bed. For example, sandy soil should receive an application of organic matter; drainage in clay areas can be improved by using raised beds for planting. A popular ground cover such as periwinkle *(Vinca minor)* cannot be grown in poorly draining soils or soils with a low pH (less than 6.5).

Species such as crown vetch *(Coronilla varia),* fringe cups *(Tellima grandiflora),* and mother-of-thyme *(Thymus serpyllum)* are relatively easy to start from seed. Potted plants, though expensive, are easier to handle than seed. For low-budget planting, bare root stock may be used to establish the ground cover. However, such plants need more care to be successful. Small shrubs such as creeping juniper *(Juniperus horizontalis)* may be planted as bare roots. They must be watered thoroughly after planting and mulched to retain moisture and suppress weeds.

Ground covers are seeded on slopes where manual planting is a challenge or a large area must be planted. Steep slopes (more than 1:1) along roads may be conveniently seeded by using a *hydroseeder* or *hydromulcher.* Seed is mixed in a water slurry (which may include some fertilizer) and pumped under high pressure onto the slope. Hydromulching includes wood cellulose fiber mulch in the slurry.

Table 13-6 Selected Ground Covers That Are Aggressive Spreaders

Plant	Scientific Name	Zone	Comments
Mazus	*Mazus reptans*	5–8	Forms quick ground mat; 1 to 2 inches (2.5 to 5 centimeters) high
Creeping jenny	*Lysimachia nummularia*	3–8	Stems root rapidly as they spread; 4 to 8 inches (10 to 20 centimeters) high
Creeping lilyturf	*Liriope spicata*	5–10	Rapid rooting; 4 to 8 inches (10 to 20 centimeters) high
Crown vetch	*Coronilla varia*	3–9	Tough ground cover; 18 inches (45 centimeters) high
Ajuga	*Ajuga reptans*	3–8	Short (3 inches or 7.7 centimeters); forms solid carpet
Bishop's weed	*Aegopodium podagraria*	3–9	Fast spreading; 12 inches (30 centimeters) high
English ivy	*Hedera helix*	5–9	Excellent spreader; 6 inches (15 centimeters) high
Sweet violet	*Viola odorata*	6–8	Fragrant; runners; 5 to 8 inches (12.5 to 20 centimeters) high

Potted ground cover plants are planted like other potted plants. The seedling should be set in the hole no deeper than it was in the pot. Potted cuttings of species such as English ivy *(Hedera helix)* and periwinkle may be used to establish ground covers. Cuttings are spaced closer than potted plants.

13.13.2 CARE

Water is critical during early establishment. Seeds, divisions, and transplants should not be allowed to dry. Mulching helps to prevent drying. Plants should not be overwatered. Fertilizer application of about 10 pounds of 20:10:10 analysis may be applied to promote growth and development. Ground covers spread over the soil surface, and thus irrigation must follow immediately after an application of fertilizer to wash away residue on leaves and prevent fertilizer burn.

Mulching at planting is necessary to control weeds and thereby give the ground cover time to establish. The mulch may have to be renewed annually to keep weeds under control. If weeds appear, they should be removed. Watering is necessary when a prolonged dry spell occurs. Fertilizing is not critical unless the soil is particularly poor in nutrients. Sandy soils generally need fertilizer supplementation more frequently than clay soils.

The pruning of ground covers may be necessary after years of growing under favorable conditions. Under such conditions, growth must be controlled by pruning shrubby species such as junipers and vines such as English ivy *(Hedera helix)* in early spring. Pruning reduces overcrowding, which causes undesirable competition among plants and subsequent loss of vigor. Renewal pruning, which helps to rejuvenate the ground cover, may be accomplished by mowing vines with a trimmer.

Ground covers may be attacked by spider mites, Japanese and other beetles, caterpillars, powdery mildew, aphids, and other pests. However, diseases and insects are usually not a problem for ground covers in the landscape. When conifers are grown by the roadside in cold climates, the salts used on icy roads in winter may damage the needles.

If the ground cover is not uniform (empty spots occur), the spaces can be filled by vegetatively propagating the species using applicable methods. For example, creeping junipers *(Juniperus horizontalis)* can be layered, and moss phlox *(Phlox subulata)* can be propagated by cuttings.

Grass is one of the basic elements in a landscape design. The establishment of a lawn is discussed in detail in chapter 14. The discussion in this module is limited to grasses planted in spots in the landscape for decorative purposes.

Grasses used as ornamentals in the landscape may be grouped into four categories: *true grasses, sedges, rushes,* and *bamboos,* the last three sometimes being described as grass relatives (figure 13-16). They are used in the landscape for the color of their foliage and the variety in their sizes, shapes, and forms. These plants are easy to grow and care for in the landscape. They can be grown in beds or in containers. Like ground covers, certain grasses are clump forming, whereas others form stolons or rhizomes. Some form low mounds (e.g., fountain grass [*Pennisetum alopecuroides*]).

13.14.1 CHOOSING GRASSES

As discussed in chapter 14, turfgrasses may be placed into two general categories: *cool* and *warm season.* Warm-season grasses become dormant in winter and turn brown. Cool-season grasses tend to be evergreen or semievergreen. The space to be allocated to the grass must always be considered. Clump-forming species tend to be contained and less aggressive or invasive. A limited number of species, including blue lyme grass (*Elymus arenarius* 'Glaucus'), are notorious for being very aggressive. When such species are grown, they may be held in check by planting them next to pavements or installing edging strips. Apart from climate, soil pH, salinity, and drainage should be considered in making choices. Table 13-7 presents a list of popular ornamental grasses. A number of annual grasses can be grown in the garden; they can be cut and dried or used fresh for cut flowers.

13.14.2 SOIL PREPARATION AND PLANTING

If many grass clumps are to be planted, a bed may be prepared for the purpose. For a few clumps, the spots for planting should be prepared in the same way as a flower bed. It is critical that the area be free of weeds. Regular lawn grasses should not be allowed to invade the space of ornamental grasses since they are difficult to control after the invasion. Aggressive species such as quack grass (*Agropyron repens*) and popular lawn grasses such as bermudagrass (*Cynodon dactylon*) should be kept away from ornamental species.

Ornamental grasses should be allocated ample space for growth. Clump-forming grasses can be assigned space since their adult size is predictable. When planting grasses for ground cover, spacing should be closer than when planting a few clumps. Grass for planting may be purchased as container-grown or bare-root plants. Thorough watering and mulching are needed after planting. Some grasses may be started from seed. Species that are easy to start from seed include velvet grass (*Holcus lanatus*), northern seas oats (*Chasmanthium latifolium*), Indian grass (*Sorghasrum nutans*), blue fescue (*Festuca cinera*), and fountain grass (*Pennisetum alopecuroides*). Some species may have to be started in a nursery indoors and transplanted into the landscape later.

Grasses may be planted alone or in combination with other bedding plants or ground covers. For shady areas, sedges adapted to shade may be combined with shade-loving ground

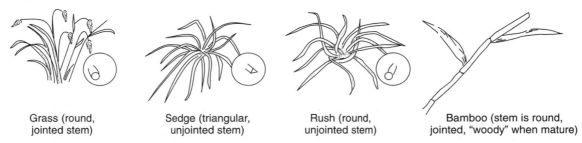

Grass (round, jointed stem) Sedge (triangular, unjointed stem) Rush (round, unjointed stem) Bamboo (stem is round, jointed, "woody" when mature)

FIGURE 13-16   Comparison of forms of grass, sedge, rush, and bamboo.

Table 13-7 Selected Ornamental Grasses

Plant	Scientific Name	Zone	Remarks
Shade loving			
Variegated Hakone grass	*Hakonechloa macra* (Aureola)	6–9	Slow spreading; 18 to 24 inches (45 to 60 centimeters) high
Snowy woodrush	*Luzula nivea*	4–9	One of the best woodrushes; 9 to 12 inches (22 to 30 centimeters) high
Bottlebrush grass	*Hystrix patula*	5–9	Cool season; 1 to 2 feet (30 to 60 centimeters) high
Tufted hairgrass	*Deschampsia caespitosa*	4–9	Blooms early in spring; 2 feet (60 centimeters) high
Dry, sunny sites			
Ravenna grass	*Erianthus ravennae*	5–10	Warm season; perennial; 5 feet (1.5 meter) high
Indian grass	*Sorghastrum nutans*	4–9	Showy fall colors; 3 feet (90 centimeters) high
Little bluestem	*Schizachyrium scoparium*	3–10	Good ground cover, erosion control; 1 foot (30 centimeters) high
Side oats gramma	*Bouteloua curtipendula*	4–9	Good meadow plant; 2 feet (60 centimeters) high
Prairie dropseed	*Sporobulus heterolepis*	3–9	Drought resistant; 1 foot (30 centimeters) high
Moisture loving			
Bulbous oat grass	*Arrhenatherum elatius*	4–9	Clump forming; 1 foot (30 centimeters) high
Prairie cord grass	*Spartina pectinata*	4–9	Deciduous; 2 to 5 feet (60 to 150 centimeters) high
Giant reed	*Arundo donax*	7–10	Warm season; 6 to 20 feet (1.8 to 6 meters) high
Quacking grass	*Briza media*	4–10	1 to 2 feet (30 to 60 centimeters) high; easy to grow
Showy grasses			
White-striped ribbon grass	*Phalaris arundinacea*	4–9	Invasive; 2 to 3 feet (60 to 90 centimeters) high
Fountain grass	*Pennisetum alopecuroides*	5–9	Adaptable; 3 feet (90 centimeters) high
Pampas grass	*Cortaderia sello ana*	8–10	One of the showiest; sharp leaves; 5 to 12 feet (1.5 to 3.6 meters) high

covers such as hosta and lily of the valley. In sunny areas, poppy, peony, and chrysanthemum may be mingled with ornamental grasses. Grasses can be grown in combination with bulbs (e.g., tulip, crocus, or daffodil). Grasses may also be planted in containers. Small- and medium-sized grasses may be planted in decorative pots and other containers. Like all potted plants, moisture management is the key to success.

13.14.3 CARE

When grasses are planted in the garden, watering is critical, especially in the first year of establishment, when it should be done fairly frequently. After establishment, watering is required less frequently. Similarly, only minimal fertilization is required. Organic fertilizer may be used as needed, and regular mulching is recommended. To keep the form and shape, grasses may be trimmed, clipped, or pruned periodically. Such measures may be taken at any time during the year to remove unwanted growth, but in late winter or early spring, grasses may be cut more severely. Cool-season grasses may be pruned to about two-thirds of their height; warm-season grasses may be cut even closer to the ground to leave about 4 inches (10.2 centimeters) of growth.

Grasses usually can remain at the same spot and in good health for a long time. Stoloniferous and rhizomatous species are able to spread, but clump-forming species often exhibit signs of aging, losing quality. When this occurs, the clumps should be dug up and divided into smaller clumps. Overfertilization predisposes plants to disease. Generally, grasses have few pests, provided care is taken to select the proper species and they are planted after adequate site preparation. Beetles, aphids, scale and sucking insects; moles, mice, and other rodents; and other pests and diseases may attack grasses to varying extents. However, grasses stay relatively trouble free in the landscape with minimal care.

13.15: BAMBOOS

Bamboos originate in the tropics and semitropics. They are woody, producing hardwood canes or culms that vary in color and size. They may be as short as 12 inches (30 centimeters) or more than 100 feet (30 meters) tall. Once established, they cannot be eradicated. They spread by rhizomes and can rapidly invade an area. Some species form clumps. Short-growing bamboos may be grown in containers. Hardy bamboos that are adapted to climate zone 6 include fountain bamboo *(Fargesia nitida), Shibatae kumasaca,* and *Sasa palmata.*

13.16: SEDGES

Sedges may be distinguished from true grasses by their angular stems filled with pith. They are generally shade-loving plants and used where grasses may not grow well. They are adapted to climate zones 4 to 9. Sedges may grow vertically or have a weeping plant form. They also vary in foliage color. Sedges may be used in a variety of ways to accent the landscape, and some are good ground covers.

13.17: RUSHES

Rushes differ from grasses and sedges by having cylindrical stems that are stiff and solid. They prefer shade and moist soil but can be grown in drier parts of the landscape. Popular ornamental rushes include big rushes *(Junus* spp.) and wood rushes *(Carex glauca).*

SUMMARY

Ground covers are low-growing, spreading plants that stand less than 2 feet (60 centimeters) tall. They may be shrubs, vines, grasses, or other perennials. They are able to grow in shade and other hard-to-manage areas in the landscape and have attractive foliage of varying color. Ornamental grasses are used for their shapes, sizes, and forms. Some are cool-season and others warm-season plants. They may form clumps or spread aggressively. Bamboos, the largest in the grass group, are woody and can grow to heights in excess of 100 feet (30 meters). Other ornamental grasses are sedges and rushes.

REFERENCES AND SUGGESTED READING

Hill, L., and N. Hill. 1995. Lawns, grasses, and ground covers. Emmaus, Pa, Pennsylvania Rodale Press.

PART A

Please answer true (T) or false (F) for the following statements.

1. T F Bamboo is grass.
2. T F Grasses may be planted along with bedding plants.
3. T F Sedges have cylindrical stems.
4. T F Ground covers are slow spreading.

PART B

Please answer the following questions.

1. Give an example of each of the following:
    a. Grass for shady sites _____
    b. Grass for dry sites _____
2. Distinguish between a sedge and rush.

    _____

3. Describe three specific roles (uses) of ground covers in the landscape.

    _____ _____ _____

PART C

1. Discuss how grasses are selected for the landscape.
2. Describe the care of ground covers.

MODULE 4 TREES AND SHRUBS

Trees in the landscape may be classified as *narrowleaf* or *broadleaf,* and *evergreen* or *deciduous.* Long-lived, trees are large plants in their adult stages. Narrowleaf evergreen plants are particularly popular in landscapes.

13.18: CHOOSING TREES

Trees in the landscape have both functional and aesthetic roles. Because trees are large and long-lived, they must be selected and located with care. In selecting trees for the landscape, numerous factors should be considered. Some factors are critical because they determine the success and survival of the plant at the site; other factors deal with economics and aesthetics.

13.18.1 SITE AND SPACE

The space available for landscaping determines the size of plants one can install. At maturity, trees are the largest plants in the landscape and as such need a relatively large space per plant to grow properly. In certain areas, home owners are not allowed to plant any trees at all or only certain ones in front of their houses if the property is situated next to a street. Further, species such as birch, elm, and sycamore are adversely affected by street lighting. Other species such as little leaf linden *(Tilia cordata)* and pin oak *(Quercus palustris)* are excellent street plants.

13.18.2 ADAPTATION

Regardless of the attractiveness of a particular tree, if it does not grow well in the intended environment, one should not purchase it. Some trees prefer temperate conditions and others tropical. The species should be selected with the hardiness zone (3.2.1) in mind. In terms of adaptation, the most important environmental factor is temperature. Every so often, unseasonable weather conditions occur in an area. The winters may be unusually cold or warm, which may cause damage from winter chilling or prevent blooming. Unless one is willing to invest in irrigation, the trees selected should be tolerant of moisture stress, which is prevalent in summer. For example, the weeping European birch *(Betula pendula)* is adapted to zone 2, red sunset *(Acer rubrum)* to zone 3, redbud *(Cercis canadensis)* to zone 4, Japanese maple *(Acer palmatum)* to zone 5, pecan *(Carya illinoinensis)* to zone 6, and live oak *(Quercus virginiana)* to zone 7.

13.18.3 PLANT SIZE AND CHARACTERISTICS AT MATURITY

Even though tree growth may be controlled by pruning, it is best to select a tree whose mature characteristics will not pose maintenance problems. For example, certain trees branch extensively. If located near the property boundary, its branches may grow over the fence and into the neighbor's yard. This overextension could initiate turf wars between neighbors, especially in the fall season when leaves drop. The flowering dogwood *(Cornus florida)* has a horizontal branching pattern, and shademaster honeylocust *(Gleditisa triacanthos)* has open branching. Even if one is able to control the top of the tree by pruning, the roots may spread and damage pipelines, house foundations, pavements, or drainage systems. Large trees may grow and touch overhead electrical cables or shade the lawn and thereby suppress grass growth.

13.18.4 MAINTENANCE

Some species are low maintenance, requiring no pruning or other landscape maintenance practices. Others must be nurtured with care. Deciduous trees shed their leaves annually, necessitating fall cleanup.

13.18.5 AESTHETICS

The shape of the canopy (Figure 12.3), the foliage (e.g., texture, color, and size), and the general appearance of a tree may be used to create dramatic scenery in the landscape. The inflorescence of flowering trees can be stunning in beauty and design. A weeping tree form is always attractive in the landscape; cut leaf weeping birch *(Betula pendula* 'Laciniata') and weeping European birch *(Betula pendula)* both have weeping forms, the former being oval and the latter pyramidal.

13.18.6 SEASONAL ENJOYMENT

Avid plant lovers look forward to seasonal changes in the plant cycle and their enjoyment from viewing the changes. The fall colors produced by trees such as dogwood, oak, ash, aspen, and maple are very impressive. Flowering plants such as flowering crab apple, magnolia, redbud, and flowering dogwood produce bursts of very attractive flowers that many look forward to each year.

13.18.7 FRAGRANCE

The inflorescence of some plants produce very pleasant fragrances. A good example is the amur maple *(Acer ginnala),* with purplish flowers, which is often used as a specimen plant.

13.18.8 FRUITING

Certain trees have the capacity to produce edible fruits (e.g., apple, cherry, plum, and peach). Other fruiting trees such as mulberry and olive, when used in the landscape, should be discouraged from flowering because they litter the ground with many dropped fruits.

13.18.9 COST

Tree species vary in cost. Specimen trees are usually relatively more expensive than regular plants. Cost also depends on whether the tree is sold as balled and burlapped or bare root (13.7.2).

13.18.10 USE

Trees may be chosen for use as specimen trees (highly ornamental) or for other practical roles such as providing shade or serving as a windbreak. Pecan (*Carya illinoinensis*) is an excellent nut tree, and copper beech (*Fagus sylvatica* 'Atropurpurea') has attractive foliage. The shademaster honeylocust *(Gleditisa triacanthos)* is an excellent shade tree, and sweet bay *(Magnolia virginiana)* has very attractive flowers and is used as a specimen tree.

13.18.11 ROOTING CHARACTERISTICS

Some trees produce roots close to the soil surface, which can crack pavements or driveways (figure 13-17). Other trees have aggressively growing roots that can clog drainage pipes.

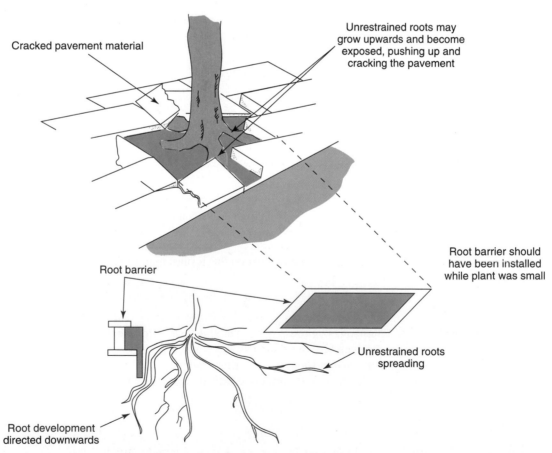

FIGURE 13-17 Roots of an adult tree cracking a pavement due to lack of root deflector.

13.18.12 WILDLIFE ATTRACTION

Trees in the landscape and other ornamentals attract birds into the area. Fruiting plants attract birds more than nonfruiting types. Examples of good wildlife-attracting trees are *Crataegus phaenopyrum,* which produces bright-red fruits in fall, and American beech *(Fagus grandifolia).*

13.18.13 DISEASES

Certain species have devastating disease problems, making their use problematic. For example, the American elm *(Ulmus americana)* is susceptible to Dutch elm disease. A tree that matures only to be attacked by diseases must be removed from the landscape at high cost.

13.18.14 ALLERGIES

Flowering trees that produce copious amounts of pollen may pose allergy problems in affected home owners.

13.19: PURCHASING TREES FOR PLANTING

Care must be taken to select very healthy plants for installing in the landscape. The following are factors to consider in purchasing trees for use in the landscape.

13.19.1 TYPE OF SEEDLING

Nurseries raise tree seedlings for sale to home owners and landscape contractors. In terms of how they are grown and packaged for sale, there are three classes of tree seedlings—bare-root trees, balled and burlapped trees, and container-grown trees (13.2).

Bare-Root Trees

As the name indicates, a bare-root tree seedling is dug up without a ball of soil around the roots (13.7.2). Seedlings are prepared in this way when the plants are dormant (i.e., winter). Deciduous plants are usually sold as bare-root seedlings. Conifer seedlings that are not more than three years old are also frequently sold this way. Bare-root plants are inexpensive to raise and purchase. Seedlings tend to lose some of their roots when uprooted, which places them in jeopardy of dying during transplantation. If it cannot be planted immediately, the bare-root seedling should be *heeled in* (roots covered up in a shallow trench) (13.20).

Balled and Burlapped Trees

A tree seedling may be harvested for transplanting by digging a certain distance around the trunk and lifting it with a ball of soil around the roots (13.7.2). The soil is then tightly wrapped in a burlap sack, thus the name balled and burlapped (B&B) seedling. Balling and burlapping is an expensive operation, which increases the price of such seedlings. Evergreen trees (narrowleaf and broadleaf) are marketed in this way. Balling and burlapping can be done any time of year. Deciduous trees may sometimes be handled by this method. Transporting balled and burlapped trees is difficult because of their bulk, but the seedlings establish quickly. The operation is mechanized, especially when seedlings are large. Large trees, as previously indicated, are not burlapped but carried in mechanical spades to the site at which they are to be transplanted.

Container-Grown Trees

Tree seedlings may be raised in containers (e.g., plastic, concrete, or wooden) for sale (13.7.1). Seedlings may be raised in this restricted soil environment for 12 months or more depending on the species. An overgrown plant may experience pot-bounding (i.e., roots grow over each other in circles). Another problem with container-grown plants is rapid soil drying.

Potted plants should be irrigated on a frequent schedule. An advantage of container plants is that they can be readily relocated in the nursery in adverse weather, such as winter, into a place where they can be protected from the cold. This transportability allows a wide variety of species to be raised and also makes seedlings available year-round. Container plants are easy to transport over long distances.

13.19.2 CHARACTERISTICS OF A GOOD TREE SEEDLING

The method of raising notwithstanding, the buyer should always purchase seedlings from a reputable nursery and select healthy plants. For container plants, one should not select those with exposed roots on the soil surface, which indicates pot-bounding and overgrown seedlings. Such plants establish very poorly in the field. Plants with large tops may not be advantageous, since they may be prone to being blown down by the wind. Such seedlings also may be too old and may not establish properly. Like all seedlings, they should be disease free and without physical injury, have a well-developed stem and good branching, and be generally healthy and vigorous.

13.20: PREPLANTING STORAGE

It is best to prepare the site before bringing in purchased seedlings from the nursery. This preparation eliminates additional care of plants and handling in the preplanting stage. Sometimes a short period of storage may be required while waiting for the proper soil and weather conditions for planting. Preplanting care basically involves prevention of drying. In bare-root seedlings, the seedlings are uprooted while the plant is dormant. Nonetheless, many roots are lost in the process, and freshly bruised surfaces may provide avenues for moisture loss. Burlapped seedlings have some soil around the root to hold moisture for a short period but eventually will begin to dry out. Bare-root and burlapped seedlings may be heeled in while in storage. Heeling in may be accomplished in several ways in the soil or by using a variety of materials. The roots may be placed in a shallow ditch or covered in mulching material (e.g., straw, wood bark, peat moss, and well-rotted sawdust) that can hold moisture. The roots are kept moist by frequent watering. To reduce evaporation, the storage environment should be kept very humid. The seedlings should be protected from direct sunlight to reduce the rate of transpiration.

13.21: PLANTING TREES

The success of a transplanted tree seedling depends on the timing of planting, seedling preplanting preparation, soil preparation, and planting technique, among other factors.

13.21.1 TIMING

Spring offers the best conditions for planting many tree species. The conditions at this time provide adequate moisture and warm soils for plant root establishment and growth. The relatively cool temperatures of spring minimize moisture loss through transpiration and thus reduce the incidence of transplanting shock. The cool temperatures of fall (in areas where winters are mild) also provide good conditions for planting trees. The limitation to planting under this condition is frozen soil. Many deciduous and evergreen broadleaf trees as well as conifers may be planted during this period. Planting in summer is most challenging because of the high temperatures and intense sunlight that induce rapid drying of plant tissue. Some container plants may be planted during this period but will require great care for success.

13.21.2 DIGGING THE HOLE

Some general guidelines should be observed in digging a hole for planting trees. First, the hole should be large enough to contain the plant roots without the need to squeeze or pack them tightly. There should be ample room around the roots for soil to be added. The hole should be at least 12 inches wider than the ball of roots. The depth of the hole should be such that when the plant is set, the original soil level on the plant is maintained after filling the hole with soil. It should be about 6 inches deeper than the soil ball around the roots. The topsoil should be carefully piled up near the hole in a separate heap from the subsoil, which is called the *backfill soil*. This practice may not be possible when a tractor-mounted power take-off is used to dig the hole.

Backfill Soil
Topsoil dug out of a hole and used to fill it during seedling transplanting.

13.21.3 PLANT PREPARATION

Bare-root plants should be sent to the field in a container of water to avoid drying roots. The roots are then carefully pruned before setting in the hole. In container plants, the pot-bound roots should be straightened and spread out before placing them in the hole.

13.21.4 PLANTING

Bare-root plants need to be held while setting them in the hole. First some backfill soil is placed in the hole. The plant is then held such that the crown is slightly above the soil level, and the hole is refilled with topsoil or, as some growers prefer, amended soil (soil mixed with organic matter and sometimes fertilizer). After filling, the soil should be packed firmly by tamping with a foot or stick, making sure the trunk remains erect (figure 13-18). In the case of burlapped seedlings, the ball is set in the hole before untying the ropes. A broken soil ball may result in the seedling's death. The crown should also be set above the soil line. The burlap material is usually biodegradable and need not be removed before planting. Instead, only the overhanging top is trimmed off (figure 13-19). The remainder of the space is filled with topsoil and patted firm. If the burlap material is not biodegradable, it must be removed by making a slit in the bottom of the wrapping and pulling it up after properly setting the seedling in the hole. Under no circumstances should a plant be planted at a depth lower than it was before transplanting. Plants are easily killed when seedlings are planted too deeply.

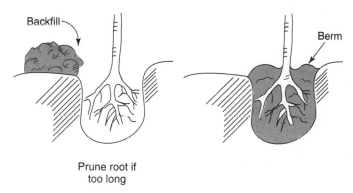

FIGURE 13-18 Planting a bare-root tree seedling.

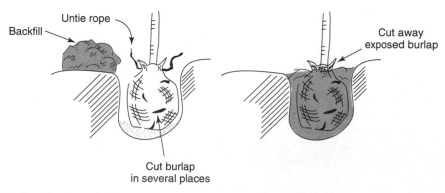

FIGURE 13-19 Planting a balled and burlapped tree seedling.

13.22.1 INSTALLING A BERM

After firming the soil, a water-retaining wall or ridged structure called a *berm* may be installed by using the excess soil to form a ring around the trunk (figure 13-20). This structure forms a basin to hold water around the base of the tree.

Berm
A circular ridge of soil constructed around the base of a newly transplanted tree to hold water.

13.22.2 MULCHING

A mulch should be placed around the base of the stem to control weeds and retain moisture in the bare soil. Mulching also prevents the soil from cracking and aids in soil infiltration by water and rapid root development.

13.22.3 STAKING AND ANCHORING (OR GUYING)

Balled and burlapped seedlings are often self-supporting. Newly planted tree seedlings, especially bare-root seedlings, are prone to toppling by the wind. Without additional support, the tree may be tilted by winds. A stake, which is often a metallic rod, is positioned close to where the stem will be and driven into the soil before planting the tree. The stake should be positioned on the west or northwest side of the trunk. The stem height of the stake should be such that after tying the stem it remains upright. Any bare-root seedling taller than 8 feet (2.4 meters) should be staked. Tying should not completely immobilize the tree but permit some degree of movement with the wind, which helps in the development of a strong trunk. To prevent injury to the tree, the bare wire used for tying should be covered in a piece of rubber hose. Sometimes, double stakes may be used (figure 13-21). In this case, the stakes are fixed outside of the planting hole on opposite sides of the stem. In place of staking, a newly planted tree may be anchored in place by using three well-positioned *guy wires* (figure 13-22). These anchors may be removed after about a year or two (sooner for small trees).

13.22.4 WRAPPING

Trees (especially those with sensitive bark such as *Acer rubrum*) often need to be protected against trunk damage from sunlight and cold in cold climates. They are protected by wrapping the trunk with strips of burlap or tree wrapping paper after treating with an insecticide (figure 13-23). This material is left in place for about 12 months. The wrapping also reduces moisture loss. Without the stabilizing effect of the wrapper, rapid changes in temperature in winter will cause the bark of the tree to crack or become sun scalded. The paper used must not be black or dark colored.

13.22.5 TRUNK PAINTING

Good tree seedlings should have small shading branches on the lower part of the trunk. When these are not present, trees may be protected from sun damage by painting about 6 to 12

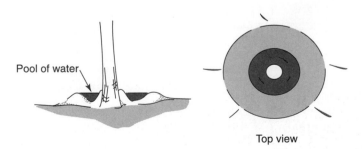

Pool of water

Top view

FIGURE 13-20 A berm.

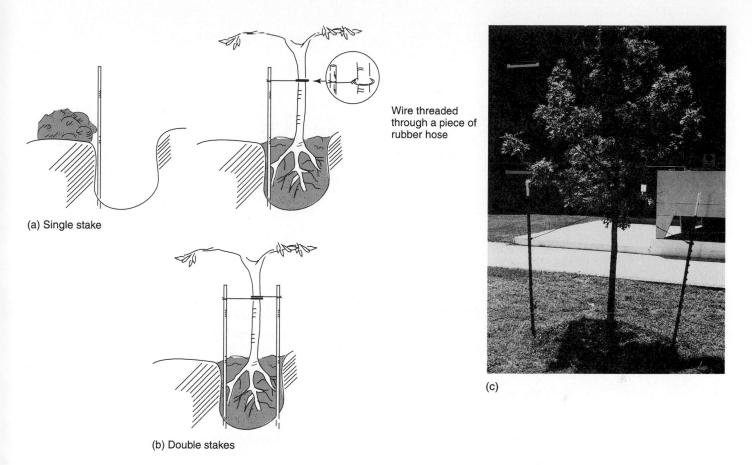

(a) Single stake

Wire threaded
through a piece of
rubber hose

(b) Double stakes

(c)

FIGURE 13-21 Staking newly planted trees in the landscape: (a) single stake and (b) and (c) double stakes.

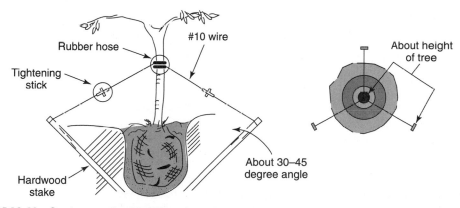

Rubber hose

#10 wire

Tightening
stick

About height
of tree

Hardwood
stake

About 30–45
degree angle

FIGURE 13-22 Guying newly planted trees in the landscape.

inches (15.2 to 30.5 centimeters) of the base of the trunk with diluted white latex paint or lime (whitewash). It should be mentioned that shading branches should not be allowed to grow into main branches but should be removed in due course.

13.22.6 PRUNING

In bare-root seedlings, about 30 percent of the top should be removed at transplanting for good and quick establishment. Because of root mass loss, bare-root seedlings need severe pruning to minimize water loss. Thinning of the top also reduces the dangers of toppling from wind.

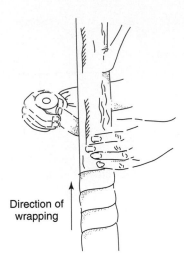

Direction of wrapping

FIGURE 13-23 Tree wrapping.

13.22.7 ANTITRANSPIRANTS

To reduce transpiration and subsequently transplanting shock, antitranspirants may be sprayed with a foliar application of antitranspirants before transplanting. This treatment is used when large plants are being transplanted. It has only a temporary effect.

13.22.8 WATERING

New transplants should be watered deeply for quick establishment. However, care should be exercised to avoid waterlogging of the soil. Although watering next to the trunk is proper during tree establishment, after the tree has grown older it should be watered only in the area outside of the *drip line* where the roots occur (16.1.2).

13.22.9 FERTILIZATION

Nitrate fertilizers (3 to 6 pounds per 100 square feet or 1.53 to 2.7 kilograms per 9 square meters) may be applied in moderate amounts after planting. Adding fertilizer directly to the backfill soil may scourge the roots.

13.22.10 INSTALLING A WIRE MESH

To protect against pests such as rodents, transplants may be encircled with a wire mesh, plastic, or metal guard installed around the tree trunk. These tree guards also protect against accidental damage from mowers.

13.22.11 INSTALLING A ROOT DEFLECTOR

Trees that produce roots close to the soil surface should not be planted near pavement. If they have to be planted near such materials, damage to the pavement may be prevented by installing a root deflector (a sheet of impervious material inserted between the pavement and the roots to force the roots near the surface to grow downward and away from the pavement material) (13.18.11).

13.23: SELECTED TREES FOR THE LANDSCAPE

13.23.1 SELECTED NARROWLEAF EVERGREENS

Examples of common narrow-leaf evergreens are listed in table 13-8. They differ in plant height and leaf characteristics, some being clusters of needles and others scalelike. Plants in this category make good *foundation plants* (located near the building) and screen or hedge plants.

Table 13-8 Selected Narrowleaf Evergreens

Plant	Hardiness Zone
Low growing	
Sargent juniper *(Juniperus chinensis Sargentii)*	4
Spring heath *(Erica carnea)*	5
Spreading English yew *(Taxus baccata repandens)*	4
Blue rug juniper *(Juniperus horizontalis wiltoni)*	2
High growing	
Mugho pine *(Pinus mugo Mughus)*	2
Hicks yew *(Taxus media hicksii)*	4
Leyland cypress *(Cypressocyparis leylandi)*	4
English yew *(Taxus baccata)*	6

Table 13-9 Selected Broadleaf Evergreens

Plant	Hardiness Zone
Azalea *(Rhododendron spp.)*	4–8
Holly, English *(Ilex aquifolium)*	6–9
Holly, American *(Ilex opaca)*	6–9
Holly, Japanese *(Ilex crenata)*	6–9
Holly, Chinese *(Ilex cornuta)*	6–9
Magnolia, Southern *(Magnolia grandiflora)*	7–10
Nandina *(Nandina domestica)*	7–8
Privet *(Ligustrum japonicum)*	6–7
Barberry *(Barberis darvinii)*	4–9
Boxwood, English *(Buxus sempevirens)*	6–8
Oak, Southern live *(Quercus virginiana)*	9–10
Orange *(Citrus sinensis)*	9–10
Weeping fig *(Ficus benjamina)*	10

13.23.2 SELECTED BROADLEAF EVERGREENS

Broadleaf evergreens are used in the landscape in much the same way as narrowleaf evergreens. Some are good foundation plants (e.g., azalea and Japanese holly) and others good specimen plants (e.g., magnolia and English holly). Japanese holly also makes a good hedge. Rhododendrons are frequently located in the corner of the area. Examples of broadleaf evergreens are given in table 13-9.

13.23.3 DECIDUOUS TREES

Deciduous trees vary widely in adult height, some being less than 30 feet (9.1 meters) and others more than 70 feet (21.2 meters) tall. They are used to provide shade in the landscape and for ornamental purposes through flowering or fall color changes. The characteristics of a selected deciduous tree used in the landscape are described in table 13-10.

13.24: USING TREES IN THE LANDSCAPE

Apart from adaptation, other factors affecting use of deciduous species in the landscape include plant height, fall color, tree form (e.g., conical or rounded), leaf texture, and growth rate (slow or fast). Some species, such as Japanese maple *(Acer palmatum)* and flowering dogwood *(Cornus florida),* make good specimen plants; copper beech *(Fagus sylvatica)* is a good accent plant.

Table 13-10 Selected Deciduous Trees for the Landscape

Plant	Hardiness Zone
Alder, gray *(Alnus incana)*	3–8
Ash, Arizona *(Fraxinus velutina)*	8–10
Aspen, quaking *(Populus tremuloides)*	3–8
Beech, European *(Fragus sylvatica)*	5–10
Birch, European white *(Betula pendula)*	3–9
Chestnut, Chinese *(Castanea mollisima)*	5–9
Crab apple, Siberian *(Malus baccata)*	3–8
Dogwood, flowering *(Cornus florida)*	7–9
Elm, Chinese *(Ulmus parvifolia)*	7–10
Fig, common *(Ficus carica)*	7–10
Honey locust *(Gladitsia triacanthos)*	5–9
Larch, European *(Larix decidua)*	3–8
Sycamore *(Platanus x acerifolia)*	5–10
Poplar, Lombardy *(Populus nigra 'Italica')*	2–10
Redbud, western *(Cercis occidentalis)*	6–9
Walnut, English *(Juglans regia)*	7–9
Willow, weeping *(Salix alba)*	3–10

13.25: INSTALLATION OF SHRUBS

13.25.1 SHRUBS VERSUS TREES

Shrubs play a major role in the landscape. In terms of nomenclature, it is common for the scientific name of the shrub to be part of the common name of the plant; for example, common camellia is *Camellia japonica* L., glossy abelia is *Abelia x grandiflora,* and forsythia is *Forsythia x intermedia.*

Shrubs are distinguishable from trees in several ways (13.18):

1. *Height.* Shrubs are usually low growing (less than 10 feet or 3 meters).

2. *Central axis.* Trees usually have one stem or trunk, and shrubs produce multiple stems from a low crown.

3. *Size.* Overall, shrubs are smaller and (as individuals) require less space than trees.

4. *Branches.* Shrubs branch more profusely than trees, starting low on the stem. Shrubs may be deciduous or evergreen plants. Broadleaf or narrowleaf species, such as oleander and Russian olive, may be grown to be shrubs or trees depending on how they are trained when they are young. Shrubs may also be flowering or nonflowering.

13.25.2 CHOOSING SHRUBS

The factors for consideration described for choosing trees (i.e., climatic adaptation, plant size, characteristics at maturity, maintenance, aesthetics, seasonal appeal, use, and cost) also apply to shrubs.

13.25.3 PURCHASING SHRUBS FOR PLANTING

Just like trees, shrubs may be purchased as bare-root, balled and burlapped, or container-grown seedlings (13.9).

13.25.4 PLANTING AND IMMEDIATE CARE

The methods of planting trees are applicable to shrubs (13.2; 13.22). Since shrubs are small in size, staking or anchoring is not necessary; besides, the multiple stems do not have to grow vertically.

Table 13-11 Selected Deciduous Shrubs

Plant	Hardiness Zone
Low to medium growing	
Cotoneaster (*Cotoneaster horizontalis*)	5
Daphne (*Daphne genkwa*)	5
Cinquefoil (*Potentilla fruiticosa*)	2
Abelia (*Abelia grandiflora*)	6
Flowering almond (*Prunus glandulosa*)	4
Butterfly bush (*Buddleia davidii*)	6
Barberry (*Barberis thunbergi*)	6
High growing	
Arrowwood (*Viburnum dentatum*)	3
Cranberry bush (*Viburnum opulus*)	3
Lilac, Chinese (*Syringa chinensis*)	5
Bottlebrush buckeye (*Aesculus parviflora*)	5
Mock orange (*Philadelphus* spp.)	6
Winterberry (*Ilex verticillata*)	4

13.26: USING SHRUBS IN THE LANDSCAPE

Unlike trees, shrubs are not grown for shade. Structurally, they provide bulk and mass in the landscape. Shrubs may be grouped into three size classes (table 13-11):

1. Small: less than 3 feet (1 meter)
2. Medium: 3 to 6 feet (1 to 2 meters)
3. Large: 6 to 12 feet (3 to 6 meters)

As with trees, when shrubs are being included in the landscape, the appropriate size should be selected so that at maturity the plant will not overwhelm the building. Small- and medium-sized shrubs are suited to traditional uses; large species fit best in commercial landscapes.

Shrubs and trees have some common uses. Shrubs and trees are major flowering species that add color to the landscape. Their leaves vary in shape, size, and color, and they also influence the texture of the landscape design. Many shrubs produce attractive fall colors, and fruiting types attract wildlife.

In addition, many shrubs produce fragrant flowers. Other functional uses include as hedges, screens, erosion control on slopes, foundation plantings, and background material in the landscape. They are readily massed and pruned to form a contiguous plant wall or hedge. The lower branches and some of the multiple stems may be removed so that one trunk is trained to give the shrub the appearance of a tree.

SUMMARY

Trees and shrubs are perennial plants. The largest plants in the landscape, trees may be deciduous or evergreen and broadleaf or narrowleaf. Principles for selecting annual bedding plants apply to selecting trees and shrubs. The critical caution is that once planted, trees remain for a long time in their spots (i.e., until they are cut down or die) and as such should be selected and located with care. Some trees and shrubs flower; they are purchased from the nursery as either bare-root, balled and burlapped, or container-raised plants. After planting, trees may be staked or guyed to hold them upright. Certain trees change color during the fall season and can be strategically located to enhance the landscape during the season. Trees provide shade, and shrubs are common hedge plants.

REFERENCES AND SUGGESTED READING

American Horticultural Society. 1982. Shrubs and hedges. Alexandria, Va.: American Horticultural Society.

Rice, L. W., and P. R. Rice Jr. 1993. Practical horticulture, 2d ed. Englewood Cliffs, N.J.: Prentice-Hall.

Time-Life Editors. 1989. Evergreen shrubs. Alexandria, Va.: Time-Life.

Zucker, I., and F. Derek. 1990. Flowering shrubs and small trees. New York: Freedman.

OUTCOMES ASSESSMENT

PART A

Please answer true (T) or false (F) for the following statements.

1. T F   A deciduous tree loses its leaves at some time during the year.
2. T F   When planting a balled and burlapped seedling, the wrapping should be removed completely.
3. T F   Tree trunks may be caged in wire mesh to ward off rodents.
4. T F   When planting a tree, the hole should be filled with soil to a level above the original soil level on the seedling trunk.
5. T F   Deciduous trees are often sold as bare-root seedlings.

PART B

Please answer the following questions.

1. Give four factors to consider in choosing trees for the landscape.
    _____ _____
    _____ _____

2. List five deciduous trees and five evergreen trees.
    _____ _____ _____ _____ _____
    _____ _____ _____ _____ _____

3. List five evergreen shrubs.
    _____ _____ _____ _____ _____

4. A ridged structure installed around a tree to hold in water is called a_____.
5. Tree trunks may be wrapped in paper to _____.
6. If a bare-root seedling cannot be planted immediately, it should be _____.
7. List five trees that are noted for their fall colors.
    _____ _____ _____ _____ _____

PART C
---

1. Briefly describe how a balled and burlapped tree is planted.
2. What is the purpose of painting the trunks of trees?
3. What is the role of a tree root deflector in the landscape?
4. Give three specific reasons why certain trees should not be located too close to a building or pavement.

MODULE 5 BULBS, CORMS, TUBERS, AND RHIZOMES

Plants with underground structures (e.g., bulbs, corms, tubers, and rhizomes) have a wide variety of uses in landscape design. These plants are modified roots, stems, and leaves that have the capacity to store large amounts of water and plant food (9.10). Sometimes the term *bulb* is used loosely to include all of the categories of plants described.

13.27: ROLE IN THE LANDSCAPE

Bulbs are used to add color to the landscape. They are very effective when placed in a bed with perennial plants. They are most effective when massed. Bulbs such as tulips have a spectacular array of colors. Flowers of bulbs, as well as plant form and height, vary widely. For example, lilies are tall and should be planted behind short bulbs such as grape hyacinth. Fall-planted bulbs provide landscape color in early spring. Some bulbs are planted in spring (table 13-12). After the fall crops have withered, summer annuals may be planted in the same area as the bulb. Certain species of bulbs such as *Crocus* and *Galanthus* may be planted to *naturalize* in the lawn. In naturalizing, these bulbs are dormant in fall and summer so that the lawn can be enjoyed during that period. In winter and early spring, when the lawn is dormant, these naturalized bulbs flower to keep the space attractive year-round. Species that lend themselves to this treatment include crocus, daffodil, scilla, early narcissus, spring start flowers, and glory-of-the-snow. When a lawn is planted in this way, it should not be mowed until the bulb begins to wither, which may take 8 to 12 weeks after peak flowering.

> **Naturalized Plant**
> A plant introduced from one environment into another in which the plant has become established and adapted to a given region by growing there for many generations.

13.28: ESTABLISHMENT

13.28.1 SITE SELECTION AND PREPARATION

Bulbs prefer neutral soils that are well drained and loamy. Acidic soils (less than pH 5.9) should be limed. If necessary, the organic matter content of the soil may be increased by adding compost or sphagnum moss. The bed should be well prepared and loose since some bulbs are planted as deep as 10 inches (25.4 centimeters) in the soil. Some form of complete fertilizer (10:10:10) or slow-release bulb fertilizer should be incorporated into the soil during preparation for planting. Because bulbs generally prefer sunny and warm locations, the planting site should experience full sun for at least four hours daily and be protected from strong winds. Beds that face south or west receive adequate exposure to light.

13.28.2 PLANTING BULBS

As indicated previously, some bulbs are planted in spring and others in fall. Bulbs differ widely in depth of planting (figure 13-24). Begonias are planted shallowly (about 1 inch or

Table 13-12 Spring-Planted Bulbs	
Plant	*Scientific Name*
Giant allium	*Allium giganteum*
Amaryllis	*Hippeastrum x hybridum*
Peruvian daffodil	*Hymenocallis narcissiflora*
Dutch iris	*Iris hollandica*
Tiger flower	*Tigridia pavonia*

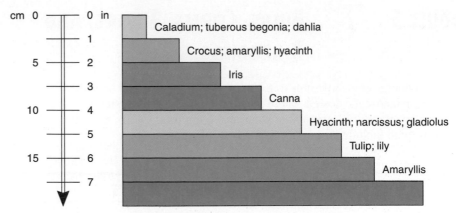

FIGURE 13-24 Planting depth for various plants with underground swollen structures.

2.45 centimeters deep) and daffodils deeply (about 10 inches or 25.4 centimeters deep). It is important that planting depth be uniform in the bed to promote even flowering. Spacing is also variable. Since bulbs have more aesthetic appeal when massed, close planting is desirable. For large bulbs such as daffodils, five bulbs per square foot (0.18 square meter) is adequate; 10 to 15 bulbs per square foot (0.18 square meter) is recommended for small bulbs such as crocus. However, the deeper they are planted, the wider they are spaced. Spacing is also influenced by the garden plan. Bulbs such as daffodils are easy to orient in the soil since they show roots at the bottom. Others, such as dahlia and anemone, do not have clear upward-facing parts. If roots are hard to find, the bulb can be planted sideways. Bulbs can be planted with the aid of a trowel or spade. A *bulb planter* (hand and foot models are available) facilitates the planting activity. It is used to remove plugs of soil from the ground, into which the bulbs are placed and covered up. The recommended depth of planting and spacing are indicated on the packaging. The plot should be watered well after planting.

13.28.3 PLANTING BULBLETS AND BULBILS

Species such as fritillary, lily, and onion, produce underground offspring called *bulblets*. In lily, bulblet formation may be encouraged by removing the flowers and flower buds. Also, certain structures called *bulbils* may form in the leaf axils and be harvested for use in planting. Since these structures are small, they are planted at a depth of about 2 to 3 inches (5.1 to 7.6 centimeters).

13.28.4 PLANTING CORMELS

Cormels are produced by species including freesia and gladiolus. After digging up the corms (figure 2-17) and cormels at the end of the fall growing season, the cormels are picked off and the top growth of the corm cut off. These parts are treated with a fungicide and stored in a fresh paper bag containing dry vermiculite in a cool place (50°F or 10°C) through the winter. In spring, the corms are planted in the same way as bulbs. The cormels are planted in a nursery (similar to bulblets) and transplanted after about two years.

13.28.5 PLANTING SCALY BULBS

Scaly bulbs (figure 2-19) (such as lilies) have plump leaves or scales that can be picked apart. The outermost ring of overlapping leaves is used for propagation. It is important that each scale have at least some basal tissue for rooting to occur. These loose scales are rinsed in tap water and air dried. After shaking in a plastic bag filled with fungicide, the scales are planted in a flat containing moist sand or vermiculite. The flat is placed in a plastic bag with holes for ventilation, and the setup located in a bright and warm place (70°F or 21.1°C). Rooting occurs in about six weeks. These rooted scales are planted in the same way as bulblets.

13.28.6 PLANTING SLICED TUNICATE BULBS

Tunicate bulbs (figure 2-18) are dug up when their top foliage withers in late summer or early fall (for the spring bloomers). The bulbs are split in half and planted in a flat in the same manner as the scales of the scaly bulbs. Rooting occurs within six to eight weeks.

13.28.7 SCORING AND SCOOPING

Scoring and *scooping* are applicable to tunicate bulbs. The object of these techniques is to induce the formation of bulblets. Scoring entails making three half-inch cuts across the basal plate with a sharp knife; in scooping, the basal plate is sliced away. The cut surface is treated with a fungicide, placed on dry vermiculite or sand, and subject to warm conditions (65 to 70°F or 18.3 to 21.1°C) to promote callus formation. Callus is formed in about two weeks, after which the bulbs are incubated in a dark, warm (85°F or 29.4°C), humid place for six to eight weeks. The bulblets formed need to be nursed for about three years before transplanting.

13.29: GROWING BULBS HYDROPONICALLY

Bulbs such as hyacinth and some tulips may be propagated by placing the bulb in a vase full of water such that the basal plate touches the water. *Hydroponics* on a large scale is discussed in chapter 11 (11.8).

13.30: TREATMENT AFTER DORMANCY BEGINS

Bulbs wither and die after flowering. Some species are annual in the sense that they have to be replanted each season. These species are dug up after they wither since they are not hardy and are damaged by cold soil temperature (table 13-13). Digging and preparation for seasonal storage are initiated when the rest period starts. Tender bulbs include caladium, canna, tulip, and hyacinth. Perennial bulbs include crocus, daffodil, and lily. Perennials are planted for naturalizing and remain in the same area year after year.

Bulbs are dug up using a spading fork or nursery spade. The bulbs should be not be injured in the process. Loose soil should be removed immediately after digging; the bulbs are then allowed to dry for a few days so that the remaining soil may be removed easily. It is a good practice to dust the bulbs with appropriate fungicides to reduce spoilage from mold attack.

Bulbs should be separated (offsets broken off from the parent bulb) and the broken surface dried before storage. Bulbs such as *Dahlia, Canna*, and *Caladium* require about two to three days of curing at about 60 to 70°F (15.6 to 21.1°C); others, including *Gladiolus, Freesia*, and *Watsonia*, require two to three weeks of curing to heal the bruised surfaces. If bulbs require long storage, they may be stored, after burying in dry sphagnum moss, at a temperature of about 35°F (1.67°C). Not all bulb species are tolerant of this treatment.

Table 13-13 Selected Cold-Sensitive Bulbs That Must Be Dug Up after the Season

Plant	Scientific Name
Lily of the Nile	*Agapanthus africanus*
Garden canna, common	*Canna x generalis*
Elephant ear	*Caladium bicolor*
Dahlia, garden	*Dahlia* spp.
Freesia	*Freesia x hybrida*
Gladiolus, garden	*Gladiolus* spp.

13.31: Diseases and Insect Pests

Bulbs are attacked by a variety of insect pests and diseases. Observance of good phytosanitation and the use of healthy bulbs for planting reduces the incidence of diseases and pests. Bulb diseases include soft rot, basal rot, blight, nematode, and mosaic. Thrips, mites, and aphids are among the common insect pests of bulbs.

13.32: Growing Bulbs Indoors

Forcing
A cultural manipulation used to hasten flowering or plant growth outside their natural season.

Bulbs can be planted in pots. It is possible to treat certain species such that they flower early or outside of their natural conditions. The cold treatment for this purpose is called *forcing*. Spring-blossoming bulbs, that respond well to forcing include daffodil, tulip, hyacinth, and crocus. Bulbs to be forced are planted in a plot with an appropriate potting medium (e.g., 3:1:1 of peat to vermiculite to perlite plus a teaspoonful of 10:10:10 fertilizer or slow-release bulb fertilizer). The bulb are not completely covered with the medium, but the top parts are slightly exposed. They are watered from below and placed in a cold environment (e.g., cold frame or refrigerator) at about 50°F (10°C) for about 9 to 12 weeks, after which the pots are transferred to a cool environment maintained at 60°F (15.6°C). This treatment induces flowering in the bulbs. Different species are treated differently during the forcing process. Various cultivars should not be forced in the same container, since they may bloom at different times.

Forcing can be done in a trench in the ground. The container should have drainage holes. The forcing mix is placed in the container at an appropriate depth after which about 1 inch (2.54 centimeters) of mix is spread on top. A trench is dug about 6 to 12 inches (15.2 to 30.5 centimeters) deep and should be freely draining. After placing the container in the trench, soil is used to fill in the space. The site is then mulched and watered periodically. The bulbs are ready to be dug up when shoots begin to appear. The pits or containers are then placed in a warm place to await flowering.

13.33: Tubers

Tuberous ornamental species include begonia, cyclamen, caladium, and dioescorea. Most tubers do not produce offsets and have to be propagated by other means. A tuber may be divided such that each section contains an eye, or bud (9.10.2). The cut surface is dusted with a fungicide such as captan before planting. Some tubers produce tiny tubers that are used for propagating without dividing the tuber. Cyclamen is propagated from seed.

Summary

Certain horticultural plants produce underground structures (e.g., bulbs, tubers, rhizomes, and corms) by which they may be propagated. Bulbs add color to the landscape and are most effective when massed in the plot. Bulbs may be planted indoors in containers. They may be planted as naturalizing plants in lawns to make use of the plot when the grass becomes dormant in winter. Tubers generally do not produce offsets and are propagated by division.

References and Suggested Reading

Hartman, H. T., and D. E. Kester. 1983. Plant propagation: Principles and practices, 4th ed. Englewood Cliffs, N.J.: Prentice-Hall.

Hartman, H. T., A. M. Kofranek, V. E. Rubatzky, and W. L. Flocker. 1988. Plant science: Growth, development, and utilization of cultivated plants, 2d ed. Englewood Cliffs, N.J.: Prentice-Hall.

Sunset Magazine and Book Editors. 1985. Bulbs for all seasons, 4th ed. Menlo Park, Calif.: Sunset-Lane.

OUTCOMES ASSESSMENT

PART A

Please answer true (T) or false (F) for the following statements.

1. T F   A tuber is a swollen stem.
2. T F   Gladiolus is propagated by bulb.
3. T F   Cyclamen is propagated by seed.
4. T F   Tulips have rhizomes.
5. T F   Bulbs are outdoor plants.

PART B

Please answer the following questions.

1. The process of cold temperature treatment of bulbs to make them flower is called

    _____ .

2. List five bulbs.

    _____ _____ _____

3. List five tuberous ornamentals and garden crops.

    _____ _____ _____

4. List three diseases and pests of bulbs.

    _____ _____ _____

PART C
---

1. Describe the process of forcing bulbs.
2. Describe how annual bulbs are handled after dormancy begins.

CHAPTER **14**

Lawn Establishment and Maintenance

PURPOSE

This chapter discusses the principles of establishing and maintaining a lawn. The basic equipment used for planting and caring for a lawn is also presented.

EXPECTED OUTCOMES

After studying this chapter, the student should be able to

1. Distinguish between the terms *lawn* and *turf*.
2. List three examples each of popular cool- and warm-season turfgrasses.
3. Discuss the guidelines for selecting turfgrass species.
4. Describe how a lawn may be established by seeding, sodding, sprigging, or plugging.
5. List advantages and disadvantages of each method of lawn establishment.
6. Discuss how a lawn mowing schedule is developed and implemented.
7. Describe how pests and diseases are controlled in a lawn.

KEY TERMS

Apomixis	Plugging	Thatch
Blend	Sod	Turf
Colonizing species	Stolonizing	Turfgrass
Lawn		

OVERVIEW

Before starting a discussion on lawns and turfgrasses, it is important to distinguish between the terms *lawn* and *turf*. Ordinarily, grass used in the landscape is referred to by most people as a lawn; grass used on a football field or golf course, however, is referred to as turf. Technically, the definition of *lawn* is a piece of land on which grass grows, including residential, commercial, and recreational areas. The term *turf* is used by horticulturists to refer to grass that is mowed and maintained, which again includes the uses in residential, commer-

FIGURE 14-1 A lawn.

cial, and recreational areas (figure 14-1). A lawn installed as part of a landscape is seeded to *turfgrass* so that it can be mowed and cared for.

Lawns are very important in the landscape for several reasons. For many, the aesthetic value is most readily appreciated, but there are also secondary benefits. A lawn provides a relatively inexpensive ground cover that protects the soil against erosion. While conserving soil, this "natural carpet" also reduces dust on a dry day, mud on a wet day, and heat and glare from the ground on a sunny day. Lawns have recreational use in both residential settings and public places where people gather to relax or play.

A healthy and beautiful lawn is obtained through good installation and effective management practices. It starts with the proper turfgrass species and is followed by good management practices including mowing, irrigating, fertilizing, and pest control. The culture and management of turfgrasses often impose stress on the growth of individual shoots. However, it is the totality (population) rather than the individuality that counts in turf culture. To begin, turfgrass is cultivated to provide a dense ground cover (a carpet), and thus seeding rate is excessive, resulting in crowding. Turfgrasses are mowed regularly, reducing the photosynthetic surface of the plants. Under limited moisture conditions, there are more shoots per unit area to compete for the available moisture. Careful stress management is, thus critical to good turfgrass culture.

14.1: PURPOSE OF LAWNS

Turf has three main functions in the landscape—ornamental, sports and recreation, and utility.

14.1.1 ORNAMENTAL

A lawn is often the least one can provided by way of landscaping a home environment. In this capacity, turf has a decorative value and enhances the home or site where it is installed. Turf is found not only in the home landscape but also in public parks and commercial building sites.

14.1.2 SPORTS AND RECREATION

Many sports surfaces, including tennis courts; golf courses; and soccer, football, athletic, and baseball fields, are covered with turf. Horse racing tracks and polo grounds may also be turf covered. Community and home playgrounds use turf to some extent as ground cover.

14.1.3 UTILITY

Bare areas on slopes along roads, ponds, and other construction work sites are prone to soil erosion. To prevent this potential soil loss, turf is used in an ecological role to stabilize the soil.

The ornamental role of turf depends on the visual quality it projects. Visual quality is a function of several factors including texture (fine or coarse), color (pale or rich green), uniformity or evenness in appearance, growth habit, and other characteristics. To be used in sports fields, a turf's functionality depends on its resistance to wear (capacity to rebound after traffic, or its elasticity and resilience). To be used to cover bare grounds, the utility quality of a turf depends on its aggressivity, or colonizing habit. These factors are further discussed in this chapter.

14.2: ESTABLISHING A LAWN

Once established, a lawn can remain indefinitely, provided a sufficient management schedule is adopted. It is therefore important that a good deal of planning go into its establishment.

14.2.1 REGIONAL ADAPTATION OF SPECIES

Grass species should be selected based on regional adaptation, use of the lawn, maintenance level required, and of course aesthetic value. Like other plants, turfgrass species differ in their climatic growth requirements. Some species prefer cooler temperatures and others warmer conditions. Turfgrass species may be divided into two general groups on the basis of climatic adaptations (figure 14-2).

Cool-Season (Temperate) Grasses

In the United States, cool-season grasses are grown in the northern states where the temperature ranges between 15 and 24°C (60 and 75°F) during the growing season. These grasses perform well during the mild temperatures of spring and fall. They may become dormant during the hot temperatures of summer and be killed altogether if extreme heat persists. Active growth of these grasses ceases when the temperature rises above 80 to 90°F (27 to 29°C) in summer. Examples of cool-season grasses are bentgrass, annual ryegrass, bluegrass, red fescue, tall fescue, and perennial ryegrass. Cool-season grasses are intolerant of heavy shade.

Warm-Season (Subtropical) Grasses

Warm-season grasses are grown in the warmer southern areas where the temperature ranges between 27 and 30°C (80 and 95°F). Soil temperature should average about 60°F (15.5°C) for best growth. Warm-season grasses become dormant when cold temperatures arrive and are susceptible to permanent cold injury during seedling establishment. Examples are bermudagrass, bahiagrass, zoysia, carpetgrass, buffalograss, Saint Augustinegrass, grama, and kikuyugrass. To keep a warm-season lawn looking green throughout the winter season, some home owners overseed their lawns with a cool-season species such as fescue or ryegrass.

Transitional Zone

States located in the region in which these two zones of adaptation meet straddle the two climates and are described as belonging to the *transitional zone*. In this region, cool-season

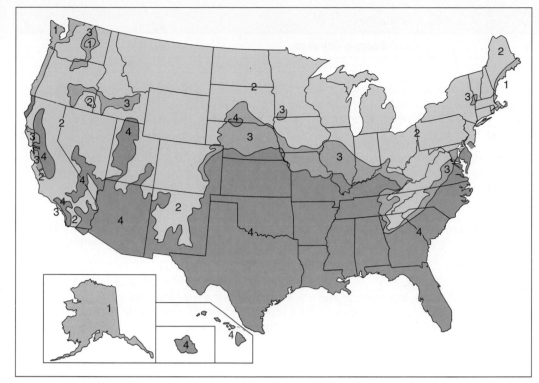

FIGURE 14-2 Areas of turfgrass adaptation: 1 and 2 = cool-season grasses, 3 = transitional zone, 4 = warm-season grasses.

species perform well during the cooler months of the year but cease to grow or are even damaged during the high-temperature periods of summer. Likewise, warm-season species perform poorly under the cold conditions of winter, often turning brown.

14.2.2 CHOOSING TURFGRASS SPECIES

Apart from adaptation, turfgrasses differ according to growth habit and other anatomical features that dictate how they are used and maintained (table 14-1).

Growth Habit

Turfgrasses are the most durable ground covers. The base of the grass is critical to its survival and should be well protected from damage. Each grass blade grows upward from the base (not the tip). In terms of growth habits, grass may be described as *bunching* or *creeping* (figure 14-3). Bunching grasses grow upright in clumps. They spread by means of new growth called tillers, which arise from the crown. Grasses adapted to cool climates such as fescue and ryegrass have bunching habits. Creeping grasses spread far by means of modified stems, which could be stolons or rhizomes. Examples of creeping grasses are those adapted to warm climates such as bermudagrass and kikuyugrass.

Texture

Grasses differ in texture, some being fine, with narrow blades, as in bluegrass, bermudagrass, or zoysia, or coarse, as in Saint Augustinegrass, bahiagrass, and tall fescue (figure 14-4). Fine-textured grasses such as bentgrass are used on golf courses in the cool-season zones of the northern United States. Likewise, fine-textured bluegrass is used on golf coarse fairways in these areas. These grasses also make excellent lawns. Bluegrasses are the most commonly used cool-season grasses. In the South, fine-textured bermudagrass is the most commonly used grass. It is found on golf fairways and greens as well as lawns. Coarse-textured bahiagrass is used on playgrounds and along roads.

Table 14-1 Selected Turfgrasses: Their Adaptation, Characteristics, and Uses

Cool-Season Grasses

1. Colonial bentgrass (*Agrostis tenuis*)   Zone of adaptation: 3–8

This grass is shallow rooted, aggressive, and pale colored. It is tolerant of light shade. It is drought prone and intolerant of heavy traffic. If mowed more than 1 inch (2.5 centimeters) high, the lawn looks puffy and coarse textured. It is suitable for use as utility turf and decorative lawns. Examples of cultivars are Astoria and Exeter.

2. Tall fescue (*Festuca arundinaceae*)   Zone of adaptation: 2–7

This multipurpose grass is drought resistant, deep rooted, and robust. It is coarse textured, with a bunch growth habit. It is best grown in pure stands and provides erosion control. It is usable as playground or sports turf and best mowed higher than 1.5 inches (3.8 centimeters). Examples of cultivars are Colchise, Arid, Bonanza, Mustang, and Olympic.

3. Perennial ryegrass (*Lolium perenne*)   Zone of adaptation: 3–7

Ryegrass is an excellent all-purpose grass. It has good texture and color similar to bluegrass. It is tolerant of high traffic and is used as sports turf or in utility or decorative lawns. It is best mowed higher than 1.5 inches (3.8 centimeters). Examples of cultivars are Derby, Sunrise, Tara, Citation II, and Prelude.

4. Kentucky bluegrass (*Poa pratensis*)   Zone of adaptation: 4–7

This grass makes one of the most beautiful lawns. It is fine textured and has perhaps the best lawn color but is susceptible to diseases. The grass is used for sports turf and decorative lawns. It is best to mow above 1.5 inches (3.8 centimeters) [e.g., 2.5 inches (6.4 centimeters)]. Examples of cultivars are Touchdown, Adelphi, Baron, and Pennstar.

5. Annual ryegrass (*Lolium multiflorum*)   Zone of adaptation 3–7

This grass is short lived (dies after one season) and is used commonly to overseed a lawn during the winter in the South. It is winter hardy and quite disease tolerant. Mowing height may be about 2 inches (5.1 centimeters). Examples of cultivars are Gulf and Tifton 1.

Warm-Season Grasses

1. Bermudagrass (*Cynodon dactylon*)   Zone of adaptation 7–9

Bermudagrass is a dense-growing, vigorous grass. It propagates by rhizomes and stolons. It is drought and salt tolerant but turns brown in winter. If desired, it may be overseeded in winter for color in the landscape. Bermudagrass builds up thatch rapidly and requires frequent dethatching. It is best when mowed low (1/4 to 1 inch [0.6 to 2.5 centimeters]). It is a multipurpose grass that is used as a decorative lawn or on sports fields. It makes good turf grass for putting on the golf course. Examples of cultivars are Texturf 10, Cheyenne, Tifway, and Common.

2. Centipedegrass (*Eremochloa ophiouroides*)   Zone of adaptation 7–9

Like bermudagrass, centipedegrass is dense growing and vigorous. It is a relatively low-maintenance grass that can perform well on marginal soils. It is also a heavy thatch builder. It can be used for a decorative or utility lawn and responds to low mowing (1 inch [2.5 centimeters]). Examples of cultivars are Centennial, Centiseed, and Oaklawn.

3. Bahiagrass (*Pasalum notatum*)   Zone of adaptation 3–9

This grass is very difficult to mow and produces a coarse, low-quality turf. Because of its coarse, open growth pattern, it is readily invaded by weeds. Bahiagrass may be used for decorative and utility lawns, and is good for erosion control. It is best mowed high (2 to 2.5 inches [5.1 to 6.4 centimeters]). Examples of cultivars are Paraguay, Saurae, Argentine, and Pensacola.

4. Saint Augustinegrass (*Stenotaphrum secondatum*)   Zone of adaptation 6–9

This grass has a dense, vigorous, and coarse growth habit and propagates by stolons. It is adapted to shade but is disease prone. When used as a decorative or utility lawn, it should be mowed high (2 to 2.5 inches [5.1 to 6.4 centimeters]). Examples of cultivars are Servile, Sunclipse, Bitter blue, and Floratam.

5. Buffalograss (*Buchloe dactyloides*)   Zone of adaptation 3–9

This a slow-growing and fine-textured grass adapted to dry and hot areas. It is clump-forming and needs little fertilization and moisture. It can be used for decorative lawns and for erosion control. Mowing height is between 1.5 and 2 inches (3.8 and 5.1 centimeters). Examples of cultivars are Bison and Prairie.

6. Zoysia (*Zoysia japonica*)   Zone of adaptation 6–9

This tough, very-dense-growing grass has good color and fine texture. It is vegetatively propagated by stolons or rhizomes. It is good for erosion control but can be used for a decorative or utility lawn. Examples of cultivars are Emerald, Meyer, and Midwest.

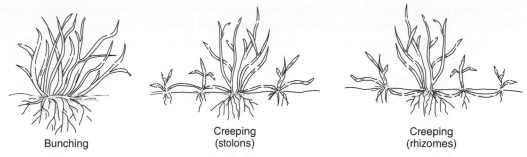

FIGURE 14-3 Types of growth habit of grasses: bunching or creeping.

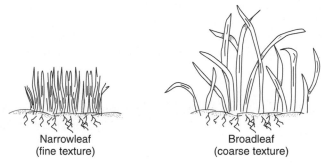

FIGURE 14-4 Texture of grasses.

Competitiveness

Turfgrass species differ in aggressiveness. Some species are described as colonizing species because they are aggressive and quickly take over the area where they are introduced. For example, in the cool-season zone, bentgrass in a bluegrass lawn poses a major problem because the former is very competitive. The conditions under which grasses are grown affect their competitive abilities. When bluegrass occurs in a mixture that is planted in full sun, the bluegrass will soon become the dominant species. Similarly, when fine fescues are mixed with bluegrasses under full sun conditions, the fescues tend to be disadvantaged. However, under less favorable growing conditions (like shade), the fescues are superior to bluegrasses. Bermudagrass is a good competitor provided it is grown in full sun. The stolons of zoysia have been known to be so aggressive that they grow under edging material (border-control materials) to invade adjacent plots.

Resistance to Wear

Grasses differ in degree of resistance to wear. Saint Augustinegrass is intolerant of heavy use. On the other hand, tall fescue is rugged and very tolerant of traffic, making it a popular choice for playgrounds, athletic fields, and areas where pedestrian use is heavy. In the warm-season zone, bahiagrass is used on playgrounds and along roads. Similarly, bermudagrass is a high-wear grass.

Maintenance

Once established, turfgrasses differ in their requirements for care to remain attractive. In this regard, certain species are low maintenance and others high maintenance. Bentgrasses (colonial and creeping) produce a thick, matlike growth and must be dethatched (removal of thatch or dead grass accumulation) on a regular basis. When warm temperatures occur, bentgrasses are prone to disease attack. A lawn of bentgrasses should be mowed low (3/4 to 1 inch [1.9 to 2.5 centimeters]) and frequently to maintain a good appearance. Because of their high maintenance requirements, bentgrasses are best planted as a pure stand rather than in mixtures. Bluegrasses are also high-maintenance species that require frequent mowing. They vary in susceptibility to diseases and adaptation and thus are best established as mixtures (of different cultivars) or blends (of different species) (14.3.2). Tall fescue is resistant to drought and most turfgrass diseases and insect pests. It is low maintenance and prefers to be mowed high (2 to 3 inches).

High-maintenance, warm-season grasses include bahiagrass and bermudagrass. They are susceptible to diseases and insect pests and require frequent mowing to maintain a good appearance. Bermudagrass should be fertilized frequently for good growth. It responds well to low mowing (3/4 to 1 inch high). Zoysia is also high maintenance and susceptible to diseases and insect pests. Centipedegrass is a relatively low-maintenance turfgrass that responds well to good fertilization.

14.2.3 CHOOSING GRASS FOR CHALLENGING SITES

Most turfgrasses prefer bright and sunny sites and are generally intolerant of shade. Lawn sites should be well drained and of good fertility. When lawns are established in less-than-ideal locations, they grow poorly and less uniformly. However, with planning, correct choices, and good care, it is possible to establish a fairly good lawn in these challenging areas. Some of these challenging locations are discussed in the following sections.

Drought Prone

Water is a critical factor in lawn maintenance. In dry areas, home owners may have to water their lawns twice daily on some days to create a healthy and lush lawn. Bermudagrass or buffalograss may be selected for drought-prone, hot-summer areas. For similar conditions in temperate zones, wheatgrass or tall fescue may be chosen.

Shade

Maintaining a lawn under trees is a challenge. Lighting under such a condition is filtered and diffuse. Grasses (and, for that matter, other plants) grow spindly and weak under reduced light. Shade-tolerant grasses include Pennlawn red fescue and chewing fescue in the cool-season zone. For warm-season zones in the South, Saint Augustinegrass is a good choice.

Extreme Soil Reaction

Proper soil acidity or alkalinity is required to make soil nutrients available at safe levels. Where extreme conditions occur, species that are tolerant of those conditions should be selected. In the cool-season zone, chewing fescue, Canada bluegrass, and hard fescue are acid tolerant. Perennial ryegrass, wheatgrass, and bermudagrass are tolerant of high-alkaline soils and adapted to warm regions.

Salinity

Salinity is a problem in areas where salt is used as a road treatment during winter. At the end of the season, melting snow carries toxic levels of salts, which can be injurious to lawns. Fescues and Saint augustinegrass have some salt tolerance.

Heavy Use

Lawns on playgrounds and sports fields come under intense traffic and rough use and should be able to withstand considerable wear and tear. In cool-season areas, tall fescue and perennial ryegrass may be used; zoysia, bermudagrass, and bahiagrass may be used in warm regions of the South.

14.3: PLANTING TURFGRASS

Several methods are used for establishing new lawns, each with advantages and disadvantages. The most common methods are the use of *seed* and *sod*; others are *plugging* and *sprigging* (or *stolonizing*). These methods differ in cost and how quickly a lawn can be started.

14.3.1 SOIL PREPARATION

Preparing the soil at the site where the lawn will be installed is the first major undertaking; it involves several activities including clearing of debris, grading, providing good drainage, amending soil fertility, providing a good seedbed, and controlling weeds. As previously indicated, the primary objective of a builder is first to ensure the structural integrity of the building. Therefore, loose topsoil is moved for the foundation of the building to be established. Incidentally, the topsoil is the medium in which plants grow. If efforts are not made to return it, successful plant cultivation will be difficult without soil amendments. The topsoil moved is usually replaced such that water will drain away from the foundation. The area left for landscaping should be cleared of all rocks and large plant debris. Topsoil should be redistributed and graded to no more than a 15 percent slope. This degree of slope provides good drainage without severe consequences of erosion. It also ensures the safety of operators during mechanical mowing. Some soils may require additional provision for effective drainage, such as the installation of drainage tiles. It may be necessary in some cases to haul in additional topsoil so that the depth of topsoil is at least 6 inches (15.2 centimeters).

A soil test is always recommended before using a piece of land for cultivation. It is easier to establish a lawn on good soil than to amend the soil when the lawn is well established. Soil pH, which should be about 6.5, can be corrected for acidity by applying lime to raise it or sulfur to lower it. Sulfur should be added to a lawn with caution. Approximately 10 pounds (4.5 kilograms) of garden lime per 100 square feet (9 square meters) is required to raise soil pH by one unit. Soil fertility may be boosted by applying a starter amount of fertilizer. To improve its organic matter content, organic amendments such as peat or other wastes may be incorporated into the topsoil. *Starter fertilizers* for lawns with high phosphorus content may be used. Fertilizers should be spread as evenly as possible.

For seeding, it is important that the bed be of a very fine tilth for the tiny grass seeds to germinate properly. A rake is useful not only for removing clods and large debris for a fine tilth but also for incorporating starter fertilizer. If subsurface drainage tiles and underground irrigation systems are to be installed, they should be in place before the final seedbed preparation.

14.3.2 SEED SELECTION

As previously discussed (14.2.2), choosing turfgrass species depends on several factors— adaptation, aesthetics, maintenance level, cost, competitiveness, and resistance to wear. In addition, the planting material can be obtained in several forms, based on genetic and physical constitution. These qualities depend on the source (supplier) of the seed.

Seed Constitution

When seeding a bare soil, it is important that the cultivar be aggressive (a colonizing species) so that it will be quickly established before weed species emerge. Another consideration in choosing a seed for planting is the genetic purity of the seed. Seeds are sold as *straight* (or *pure*), consisting of only one species or cultivar, or as a *blend* (consisting of a mixture of two or more cultivars of the same grass species).

> **Blend**
> Seed consisting of a mixture of two or more cultivars of the same grass species.

Seeds may also be sold as a *mixture* of two or more species. When a mixture is chosen as planting material, it should contain at least 60 percent of the *permanent species* desired and other species of similar characteristics. That is, if bermudagrass is the permanent species, 60 percent of the mixture does not have to be only bermudagrass but could also include species such as fescue that have similar characteristics. Sometimes, to obtain a quick cover of the ground, a seed mixture containing a species such as ryegrass that establishes rapidly may be used. The ryegrass is the *temporary species* and thus the area must be *overseeded* with the desired permanent mixture at a later date. Some cultivars are clones and hence genetically pure. Clonal cultivars are common for species such as Kentucky bluegrass because of the phenomenon of *apomixis* (the production of seed without fertilization). Such genetically uniform cultivars have a disadvantage in that the entire lawn responds similarly to any adverse environmental factors (e.g., disease) and can be completely destroyed by a single attack. This uniform susceptibility is why bluegrass seed is commonly sold as a blend of several cultivars (e.g., 'America,' 'Manhattan,' and 'Princeton'). Blends are formulated for a variety of growing conditions (e.g., shade or marginal soil) and available for species besides bluegrass. The rationale for constituting a good permanent mixture is to include a species for beauty (e.g., 50 percent consisting of bluegrass), a species for toughness and disease and pest resistance (e.g., 25 percent consisting of fescue), and a species for quick establishment and durability (e.g., 25 percent consisting of perennial ryegrass). Examples of good mixtures are as follows:

1. Kentucky bluegrass and red fescue (for dry, shady, and marginal soils)
2. Kentucky bluegrass and ryegrass
3. Kentucky bluegrass, fescue, and ryegrass

Pure seed may be used for establishing Bermudagrass or carpetgrass. If a mixture contains bentgrass, it must be cared for properly to prevent this grass from becoming a weed in the lawn.

Since several species are adapted to a single region, the choice of one species or cultivar over another is influenced by other properties of the species, such as tolerance to drought, shade, cold, and diseases prevalent in the region; tolerable mowing height; and appearance. Some species are low maintenance, and others require more than the regular care needed for a good lawn.

Source

Seed should always be purchased from a reputable source, and certified seed is preferred. It is worth the investment to pay a little more for quality seed than to use bargain seed. The seed industry has an obligation to declare certain facts to the customer. This information may differ from place to place but generally includes the following:

1. Company name and address
2. Cultivar name
3. Percent germination
4. Date of testing.
5. Purity (proportion of usable grass seed)
6. Nongrass seed
7. Inert material (rocks and other debris)
8. The seed lot number may be included

The seed should be free from weeds (especially noxious ones) as much as possible and have a high germination percentage (at least 80 percent). If the germination percentage is low, the seeding rate should be increased. Whenever possible, freshly harvested seed should be purchased.

Time to Sow

The best time to sow grass seed differs from one region to another and is chosen strategically to benefit from the weed cycle and adverse weather. Timely sowing ensures grass establishment before weed seeds germinate or cold weather sets in. In the cool-season zone, the best time to sow temperate grasses is late summer or early fall, which allows the lawn to be established before freezing temperatures set in. Furthermore, when grasses are sown in fall, the need to irrigate is reduced because of cooler temperatures and less evaporation. In the warm-season zone of the South, a good strategy is to sow in early summer after clearing away the weeds. Sometimes delays in construction projects may not allow the timely sowing of seeds. The result is poor seed germination and thus poor lawn establishment.

Seeding Rate

Grass seeds are very tiny. Species such as bentgrass may have more than 10 million seeds per kilogram (4.8 million seeds per pound). The objective of seeding is to obtain a quick establishment and produce a dense lawn resembling a carpet. However, overseeding leads to overcrowding and additional stress to plants, which consequently grow poorly and delay the attainment of a quality lawn. Similarly, low plant populations delay the covering of the ground, allowing weeds to infect the lawn. Sowing at the rate of 1 pound per 1,000 square feet (0.45 kilograms per 90 square meters) will suffice for most grasses that are sown in due season. Tall fescue and perennial ryegrass require much higher rates of between 3 and 6 pounds per 1,000 square feet (1.35 and 2.7 kilograms per 90 square meters). Seed companies indicate the recommended seeding rates on package labels.

Seeding

Small plots may be effectively seeded by hand, even though this method has the disadvantage of uneven spreading. Mechanical seeders, which should be used for large areas, distribute seed by one of three ways—drill, gravity feed, or broadcast. If a narrow strip is to be seeded, using a drill may be most appropriate. However, drilling places seeds in narrow strips, leaving wide bare spaces between them. Because ground covering is slow, weed control is required during the establishment of a lawn. Gravity feeders may be manually pushed or tractor operated. Similarly, broadcast spreaders may be portable or tractor mounted. For very large areas, helicopters may be used to broadcast the seed. To reach steep slopes, hydroseeding or hydromulching may be used (8.11.1). When one of these methods is used, frequent watering is needed during lawn establishment. To increase the seeding efficiency for uniform coverage, the recommended seeding rate may be divided into two so that one-half is distributed in one direction on the first pass and the other spread crosswise on the next pass (figure 14-5). Mixing grass seed with sand in a mechanical spreader has been found to aid in the even spreading of seed. Since grass seeds are so fine and lightweight, sowing on windy days should be avoided. Furthermore, because they are so tiny, grass seeds should not be seeded too deeply in the soil (less than 1 centimeter or 0.4 inch). Proper depth of planting is accomplished by lightly raking the area after sowing.

Mulching and Firming

Mulching a freshly sown lawn has several purposes:

1. To protect grass seeds from predators (birds)
2. To prevent seeds from washing away during irrigation
3. To conserve moisture for germination
4. To control erosion of soil

The mulch used should be spread thinly so that emerging seedlings are not impeded. Straw mulches are commonly used, although other coarse plant materials are also appropriate. To aid in germination, the mulched area is firmed by rolling over it with a lightweight roller. Rolling firms the soil for seeds to effectively imbibe moisture while also anchoring them in the soil.

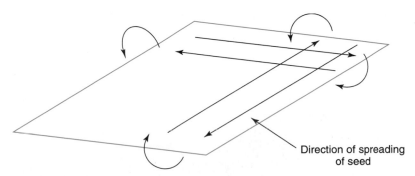

Direction of spreading of seed

FIGURE 14-5 Seeding of a lawn by the crosswise method.

Watering and Fertilizing

Keeping the newly seeded lawn moist during seed establishment is vital. After the first watering, frequent irrigation (about two times daily) should be provided until germination and seedling establishment is attained. A brief drought spell during the germination period can wipe out an entire lawn. Excessive moisture predisposes the plants to disease, especially damping-off caused by a number of fungi. For most species, germination starts within about a week of planting and continues for another week. Slow-germinating species such as bluegrass may have an extended germination period of several weeks; the frequent watering schedule thus lasts a little longer in such species. The frequency of watering is reduced gradually after germination is completed, but supplemental nitrogen may be required soon after that. An application of nitrogen at the rate of about 0.5 pounds per 1,000 square feet (0.23 kilograms per 90 square meters) will maintain seedling vigor.

Advantages and Disadvantages of Seeding

Seeds are less expensive and less bulky than sod. Plots can be seeded easily by the owner. However, seeds establish rather slowly.

14.4: PLANTING GRASS BY VEGETATIVE METHODS

Grass may be established in the landscape by vegetative methods, the most common including sodding, plugging, sprigging, and stolonizing.

14.4.1 SODDING

> **Sod**
> A shallow (1-3 inches) of topsoil that is bound by grass roots and usually harvested and sold in narrow strips.

Sodding is the establishment of a lawn by using sod. Sod is grass that is specially cultivated, mowed, and cut into strips (1 to 2 × 4 to 6 inches) like pieces of carpet, including about 1 to 2 inches (2.5 to 5.1 centimeters) of roots. Long strips are rolled, and short ones are sold flat. These strips are stacked on pallets and transported to the site where the lawn is to be established. Establishing a lawn in this way provides an instant ground cover. Sodding may be likened to *transplanting* of seedlings, except that it involves a large mass of seedlings. Sodding involves activities similar to those of seeding. Although it is the most expensive means of establishing grass, sodding produces the most rapid results and is the least problematic.

Source of Sod

As with seed, sod suppliers differ in the quality of product they sell. Certified sod producers should be used as suppliers of the planting material, since poor-quality sod has a high weed infestation. The best cultivar adapted to the region should be selected. Whenever possible, freshly cut sod should be used in lawn establishment.

FIGURE 14-6 Laying sod.

Soil Preparation

The site is prepared for sodding in the same fashion as for direct seeding. Erosion is not a problem, which allows steep slopes to be planted, as long as the lawn is not intended to be mowed, especially by a tractor. The soil surface should provide a loose medium in which roots from the sod may be established. The soil bordering walkways should be piled up to a lower depth so as to accommodate the thickness of the sod. The seedbed should be moistened before laying rolls of sod. To achieve a good level of moisture, the plot may be irrigated several days before the planting date.

Installation

Soil fertility may be boosted by a starter application of nitrogen and phosphorus. Sod laid in fall or spring establishes more quickly than that laid in summer or winter. However, sod of subtropical grasses establish well in summer. It is important to lay sod as soon as possible after harvesting, which means that the sod should be delivered after the ground has been prepared. If sod is left to sit in a stack, it will rapidly deteriorate because heat builds up in the pile. In case of an unexpected delay, the edges of the stack should be sprinkled lightly with water to prevent drying. High-quality sod should be free of weeds. Similarly, the ground should be well prepared and weed free. To suppress weeds, sod should be tightly laid, edge to edge, without gaps between adjacent strips or between rolls and structures such as concrete walkways against which they are laid (figure 14-6). It is best to lay the rolls in a staggered, checkerboard pattern. A sharp knife should be kept handy to cut smaller strips to fit challenging areas. Staggering of sod strips prevents the formation of noticeable lines in the lawn, especially in the early stages after establishment. If narrow strips of bare soil remain after laying a full piece, it is better to cut a small strip to complete the job than to stretch a larger piece. If sod is stretched, the edges will shrink later to expose gaps between the pieces for weeds to grow through. As in direct seeding, newly laid sod should be rolled over with a roller to ensure good contact between roots and soil (figure 14-7).

Immediate Care

The first thing to do after laying sod is to water it thoroughly to wet the topsoil for root establishment. Thinly cut sod tends to dry more quickly than thick sod but also establishes more quickly. Even though freshly laid sod gives the appearance of an established lawn, all heavy traffic should be kept off of the lawn until it has rooted. Because of the thick soil cover sod produces, the soil does not dry up quickly. However, the area should be watered frequently during the first several weeks to ensure proper establishment.

Fertilizing

Applying small amounts of lawn fertilizer after the sod has rooted is helpful for good establishment.

FIGURE 14-7   Rolling newly laid sod.

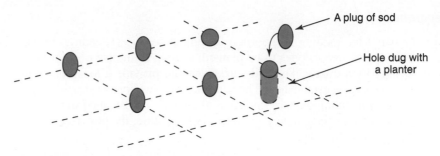

A plug of sod

Hole dug with a planter

FIGURE 14-8 Plugging method of lawn establishment.

Advantages and Disadvantages

The major advantages to using sod are that it provides instant cover (instant lawn), promotes a high-quality lawn, is good for a quick fix of high-traffic areas of already-established lawns, and is easy to lay.

The disadvantages of using sod include that it is bulky to handle, expensive, and once purchased must be laid without delay.

14.4.2 PLUGGING

Plugging is a method of lawn establishment that involves the transplanting of small pieces of sod plugs into holes in the seedbed (figure 14-8). The holes are spaced about 6 to 12 inches (15.2 to 30.5 centimeters) apart, depending on how the species spreads. Zoysia spreads more slowly than bermudagrass and Saint Augustinegrass and should be spaced much closer (6 inches). Plugging is a labor-intensive operation and takes time to cover the plot with grass; fortunately, it be done mechanically. After the operation, the plot may be rolled or firmed to provide good soil contact with plant roots. This method of planting sod is sometimes called *spot sodding.* Timing is critical to the success of plugging. Warm-season turfgrasses must be planted about two or more months before the first frost of fall for good establishment.

14.4.3 SPRIGGING

Pieces of short stems or runners (called *sprigs*) may also be used to establish a lawn. This method of lawn establishment, called *sprigging,* is accomplished by placing the sprigs in shallow (1 to 2 inches or 2.5 to 5.1 centimeters) furrows at about 4- to 6-inch (10.2- to 15.2-centimeter) spacing. The sprigs are covered such that at least one-fourth of the material is above ground. Again, it is important to allow about two months before the onset of adverse weather (e.g., frost) for establishment.

14.4.4 STOLONIZING

Stolonizing is a form of sprigging in which the recommended number of sprigs are spread uniformly over the seedbed. Stolonizing is sometimes called *broadcast sprigging.* After spreading, the sprigs are partially covered with soil by disking or rolling. This method is usually adopted for planting large areas.

14.5: TURF MANAGEMENT

The grass should be allowed to be well established and grow to a height of about 1 or 2 inches (2.5 or 5.1 centimeters) above the mowing height for the period or season and species before the first mowing. It is important that watering be suspended for the topsoil to dry slightly before mowing.

Once established, a lawn should be maintained according to a schedule that promotes healthy growth of the grass while serving the purpose or purposes for which it was established. In putting together a management program, factors to be considered include the type

---

> **Plugging**
> *A method of lawn establishment involving the use of small pieces of sod.*

of grass, climatic zone, use of turf, equipment, and owner's level of commitment to maintenance. For example, more frequent watering may be required in areas that are prone to drought; the cultivar or composition of the lawn material affects the height of mowing and response to fertilizer and watering. High-traffic areas such as playgrounds are maintained differently from lawns established primarily for aesthetic purposes. Some home owners are willing to put in time and effort (or pay for such) to have the "best lawn on the block"; others just want to have grass growing around the house.

Whatever the management practice, it should include certain basic maintenance activities: mowing, watering, fertilizing, weed control, and disease and pest control. If diligently pursued, these practices will result in a healthy and attractive lawn. It should be stressed that a good lawn does not come naturally. The home owner has to invest in some kind of regular maintenance schedule. Such maintenance should start after installation and be sustained thereafter. A variety of aggressive grasses and plant species can invade even the very best lawn if the maintenance schedule is irregular. Just as neglect can lead to disastrous consequences, overmanagement is equally undesirable and largely wasteful. As stated earlier in this chapter, with good management, a lawn can remain indefinitely once installed.

Turf culture essentially entails the management of stress in grasses for healthy growth and development. First, an unusually high number of plants are maintained per unit area, creating a situation in which competition occurs for plant growth requirements. Second, the photosynthetic area required for food manufacture is drastically reduced periodically. Good growth and development are therefore critical in turf management and are attained through three primary practices: *mowing, watering,* and *fertilizing.*

14.5.1 MOWING

A home owner may not remember or care to provide supplemental irrigation and fertility to the lawn, but an unkept lawn may invite problems with neighbors. Mowing a lawn is the least that can be done, and even this is often a chore for many. To be effective, mowing should be done with the proper equipment, at the right frequency, and to the proper height.

Domestic Mowing Equipment

Domestic mowers are of two types: *reel* and *rotary* (figure 14-9). They come in different sizes, power levels, and designs. Some require the user to push, and others are motorized so the user need only steer the equipment. Some domestic mowers are mounted on small tractors. Even though the type of mower is the home owner's choice, rotary mowers are more effective than reel mowers in mowing tall grasses. The rotary mower has a horizontal blade fixed on a vertical axis that rotates at high speeds. The reel, on the other hand, has one stationary blade (called a bed knife) and a set of blades arranged in a helical fashion that gather the grass and shear it off against the bed knife. In terms of results, reel mowers produce the best mowing finish. They are excellent for use around homes, resorts, and other areas where noise is a problem. They are also used to mow golf course fairways.

Some mowers are designed to not only cut the grass but also to reduce it to mulch. These implements are called *mulching mowers.* Dual-type mowers allow owners to change the mowing type from regular to mulching. A good mower should have sharp blades because dull blades produce a rough cut and bruise the leaves, which may discolor at the tips and not recover quickly from mowing. Sharp blades are more critical in the case of rotary mowers, which cut by impact and even under the best conditions are likely to produce ragged cuts. A maintenance schedule for mowers should include changing or sharpening the blades as needed. Operator care is also critical in the use of rotary mowers because of the high energy required by the rotating axle to move the blades.

Commercial Mowers

When the lawn to be mowed is expansive and a fine finish is not the goal (such as mowing roadsides), a *flail mower* is used (figure 14-10). It has the capacity to mulch and mow tall grasses and weeds. Such commercial mowers are designed to have broad coverage with each pass of

FIGURE 14-9 A rotary mower. The bag may be removed to convert it into a mulching mower.

FIGURE 14-10 A flail mower.

the equipment. The *sickle bar mower* is used where the clippings are going to be collected, because it does not mulch but arranges the cut leaves such that they can be readily collected.

Mowing Frequency

Since mowing has physiological implications, it must be done judiciously. The rule of thumb, if one exists, is that at any mowing time, no more than 30 percent of the leaf should be removed. If the lawn is left to grow too tall, it will be necessary to cut more than 50 percent of the leaf in some cases, thus drastically offsetting the photosynthetic machinery and limiting translocation of food to the roots for proper growth. Based on the grass variety and season, it is desirable to establish a height and adopt a mowing schedule to maintain it. In this way, plants are able to adapt to the routine and not experience shock. When plants do not have an adequate leaf surface to support the photosynthetic requirements, they grow pale or yellowish until new growth occurs. To maintain an attractive lawn, the kind that is the envy of the neighborhood, a mowing frequency of once a week should be maintained. When the lawn overgrows, it encourages the growth of tall weed species.

Mowing Height

Mowing height depends largely on the species and the culture and varies from 1/2 to 4 inches (1.3 to 10.2 centimeters). Bentgrass should be mowed low. A low cut gives a lawn a carpet-like appearance, but, unfortunately, not all turf species are amenable to cuts of less than 1 inch (2.5 centimeters) in height. The opposite of a carpet appearance is puffiness, which occurs when grasses such as fescue and bluegrass are mowed higher than 3 inches (7.6 centimeters). Under such conditions, the grass becomes more open and exposes more of its stems. It is not uncommon to see two adjacent and well-kept lawns mowed at different heights, which happens when two neighbors have different preferences, one for a low cut and the other for a high cut. Low mowing should be accompanied by other maintenance activities such as effective weed control to suppress weeds since the lawn is unable to suppress weeds through shading.

Clippings and Thatch

In a mowing activity, the operator, depending on the type and design of the equipment, may choose between allowing the fresh clippings to fall on the ground or collecting (bagging) and discarding them. Mulching mowers are designed to cut the clippings into small pieces so they do not lie on the lawn but fall to the ground. Clippings on the ground eventually decompose to improve soil fertility. The undecomposed plant organic material (thatch) can build up over a period. (14.5.7). This buildup of thatch is attributed more to other lawn maintenance activities such as watering and fertilizing than to clippings per se. If a good mowing schedule is adopted, the clippings from each mowing will be small and fall to the ground without problem. However, in overgrown lawns, mowing leaves behind visible clumps of clippings that are not only unsightly but also detrimental to the growth of the grass by reducing the photosynthetic surface and producing disease-causing conditions. On such occasions, bagging of clippings is not only desirable but necessary. Some mowers are fitted with containers for bagging the clippings. Bagging clippings is additional work that many home owners would rather not add to the chore of mowing. When thatch buildup becomes excessive, it must be reduced, in part because insects and other pests thrive in thatch.

> **Thatch**
> Accumulated undecomposed organic material between the turf and soil surface.

Edging

Edging is a cosmetic activity in which the edge of a lawn, especially next to a walkway or driveway, is trimmed in a straight line or smooth fashion. Edging machines may be purchased separately for this purpose (figure 14-11). However, a combination trimmer-edger may also be purchased for tidying up a lawn near walls and in places that are too hard to reach with a mower. Gas-powered and electric trimmers and mechanical edgers are available.

Mowing Tips

For best mowing results, an operator should observe certain guidelines including the following.

1. Grass should not be mowed when it is wet. The mower performs more efficiently on dry grass. Wet lawn mowing leads to soil compaction, and the mower may clog up frequently and require interruptions in the mowing operation to clean the system. These problems extend the time required for mowing. Further, wet grass does not fall to the ground but forms clumps.

2. A mowing pattern should be developed. Mowing the area in one direction one time and at right angles the next time reduces compaction of the soil.

3. A mower with sharp blades should be used at each mowing.

4. An overgrown lawn should be mowed at least two times over. The first time, the mower should be set to mow to a higher height to avoid clogging the machine. It may be wise to bag at least the first round of clippings.

5. Each pass should slightly overlap the previous one to ensure 100 percent coverage of the lawn.

6. When mowing a rough lawn with debris and rocks, it is best to remove these objects before beginning and also to wear goggles.

FIGURE 14-11 An edger.

7. A large lawn may be mowed in a spiral pattern. This method eliminates backtracking and stops, as well as direction changes. The recommended practice is to mow clockwise in the first instance and counterclockwise the second time.

14.5.2 WATERING

Watering is another seasonal chore that many home owners in areas experiencing seasonal drought feel obligated to perform. Where the drought spell is protracted and occurs regularly, some home owners invest in permanent automatic irrigation systems, which are discussed in detail in chapter 3. Pipes are buried during the ground preparation for the installation of the lawn. Pop-up sprinklers, which are located at strategic spots on the lawn below the mowing height, pop up when irrigation is needed. Lawns can also be watered with lawn sprinklers that oscillate. Such units are fed by water hoses connected to the outside taps of homes.

Irrigation, unlike mowing, is not done according to a set schedule but is provided only when needed. The lawn needs watering just before wilting sets in. Watering should be thorough so as to wet at least the top 4 inches (10.2 centimeters) of soil where most grass roots are found. It is important that watering be thorough since light sprinkling or partial wetting encourages roots to grow up (not down) the soil surface in search of water, thereby making plants more susceptible to drought.

At each watering time, at least 1 inch (2.5 centimeters) of water should be supplied. A homemade rain gauge in the form of a can placed on the lawn within the area of coverage of the sprinkler may be used. When the water in the can rises to the level of 1 inch (2.5 centimeters), it is time to move the sprinkler to another location or shut it off. The amount and rate of application are affected by soil characteristics. Sandy soils drain fast and clays drain slowly. The infiltration rate in sandy soils is faster than in clay soils.

The best periods to apply water are early in the morning and late in the afternoon. Evaporation of water, which is a major source of water waste in irrigation, is minimal during these times. Overwatering a lawn is wasteful, but inadequate moisture may encourage the growth of aggressive and hardy weeds. Healthy grass growth shades out weeds in the lawn and impedes their growth. A combination of the alertness of the home owner and the nature (genetics and botany) of the species, as well as familiarity with the regional climate and weather patterns, is helpful in the judicious application of water. Information on turf culture is available from local extension agents and horticulturalists. The home owner should know the characteristics of the turf cultivar in the lawn. Some cultivars are shallow rooted and benefit from shallow soaking, while others are deep rooted and benefit from deep soaking. By listening to weather forecasts, one can take advantage of the rain or know when to increase or decrease the frequency of watering.

Turfgrasses differ in their response to and recovery from adverse weather. Extreme conditions of temperature or drought cause browning or yellowing of the grass. Lawns under stress become dormant but recover with varying degrees of success when the stressor is removed. A soaking rain after a drought rejuvenates lawns quite successfully, provided the proper species is used.

14.5.3 FERTILIZING

A good green color is often associated with a healthy lawn. In trying to maintain this color, some home owners may go overboard and be in danger of overfertilizing their lawns. Since the main purpose of fertilizing is to promote vegetative growth, the principal component of a fertilizer analysis is nitrogen; however, excessive nitrogen may predispose plants to diseases. Fertilizers specially formulated for lawns may be purchased from nursery shops. Some of these preparations contain pesticides (e.g., herbicides, fungicides, and insecticides) and must be used cautiously.

14.5.4 WEED CONTROL

The secret to a weed-free lawn is keeping the turfgrass healthy so that it develops into a thick carpet. If the sod is not laid properly, the gaps between strips provide room for weeds to grow. If the lawn is not mowed at the appropriate frequency and to the appropriate height, weed growth is encouraged. Poor nutrition predisposes grasses to disease and weakens them, making them less competitive against weeds.

When weeds appear, they should be removed before they set seed. Weed species such as crabgrass are seasonal in occurrence, and thus when they infest a lawn they die after the season is over, leaving bare spots in the lawn. Preemergence or postemergence chemicals may be used. A popular chemical, 2,4-D, is used to selectively control broadleaf weeds in lawns. Common lawn weeds include dandelion, plantain, burclover, and puncture vine. Table 14-2 presents a sample of effective lawn herbicides against common weeds. The final choice of a herbicide and for that matter any pesticide is according to regional or state recommendations.

A few days before spraying a herbicide, the lawn should be mowed and watered well. Lawns should not be watered after application of a herbicide, since the chemical will be washed away. For a small lawn, a backpack tank sprayer or compression tank sprayer may be used. This type of sprayer may hold up to 5 gallons (19 liters) of fluid.

14.5.5 DISEASES AND PESTS

Lawn diseases are frequently fungal in origin. They are readily spread by people and pests as they walk on the lawn and shake off the fungal spores on their bodies. Overfertilization and excessive moisture predispose turfgrasses to diseases. Table 14-3 summarizes the characteristics of lawn insect pests and diseases. The Japanese beetle is the most devastating lawn insect pest. Its destructive stage in the life cycle is the larva (or grub), the same stage biological control of this pest is most effective. The dormant form of a bacterium called *Bacillus popilliae,* in the form of an insecticidal powder (sometimes called milky spore powder), is applied to spots under the grass. Patches of the lawn are lifted up using a spade. After application, the lawn should be watered thoroughly.

14.5.6 AERATION

After a lawn has been walked over for a long period, the soil underneath becomes compacted, depriving the lawn grass of air and impeding water infiltration. To increase aeration and

> **Aeration**
> A method of improving soil infiltration and air penetration of compacted soil by, for example, removing small cores of soil.

Table 14-2 Some Common Weed Situations in the Lawn and Turf and Suggestions for Their Control

	Preemergence	
Common weed	Crabgrass	
	Foxtail	
	Annual bluegrass	
	Barnyard grass	
Control	Use Bensulide, Dacthal, Siduron, or Benefin	

	Young Seedlings of Turfgrass	
Common weeds	A wide variety of broadleafs are common at this stage	
Control	Bromoxynill	

	Postemergence (Established Lawn or Turf)	
Common weeds	Grasses:	Nutsedge
		Crabgrass
		Quackgrass
Control		Use Bentazone, Fenoxaprop-ethyl, or Glyphosate (spot treatment)
	Broadleafs:	Black medic
		Chickweed
		Dandelion
		Plantain
		Clover
Control		Use 2,4-D amine

	Nonselective Control of Grass Weeds (Many Species)	
To remove existing grass to establish new turf		
Control	Use Glyphosate	

Note: Some of these chemicals are more effective against certain weeds than others; some may damage desired grasses. Labels must be read and directions followed very carefully.

Table 14-3 Some Common Lawn and Turf Diseases and Suggestions for Their Control

1. General problems (all grasses affected)	
Nematodes	No chemical control
Fairy rings	No effective chemical control; remove infested patch and reseed
Algae	Use Mancozeb as needed
Toadstools and mushrooms	Drench spots with Dinocarp or Thiram
Slime mold	Use Mancozeb as needed
2. Fusarium blight	Avoid excessive thatch buildup; use Benomyl
3. Brown patch (*Rhizoctonia*)	Use Benomyl, Mancozeb, or Anilazine; avoid excessive nitrogen application
4. Pythium blight (*Pythium*)	Use Zineb or Thiram; treat promptly
5. Dollar spot (*Sclerotinia*)	Use Anilazine or Chlorothalonil
6. Crown rot (*Helminthosporium*)	Use Anilazine or Chlorothalonil

drainage, cores of soil may be removed using an aerifying machine called a *plunger* or *core aerator*. The holes poked in the soil may be left to fill up naturally or may be plugged by adding a topdressing material such as peat moss. It is best to aerate a lawn when it is actively growing.

14.5.7 DETHATCHING

Thatch is an accumulation of old, dead grass; bits of unraked leaf; and other plant material above the soil (figure 14-12). Beyond just being unsightly, excessive thatch reduces oxygen

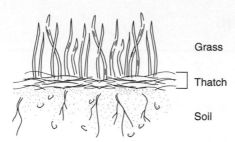

FIGURE 14-12 Thatch in a lawn.

and moisture entry into the soil and harbors pests and diseases. Warm-season grass species produce more thatch than cool-season species. When thatch accumulates to more than 1/2 inch (1.3 centimeters), it should be removed. Lawns are dethatched when the grass is actively growing, which allows for quick recovery. A variety of dethatching tools exists, ranging from a simple *thatch rake* to *motorized dethatchers*. A dethatching rake is first run in one direction and then repeated in a perpendicular direction.

> **Dethatching**
> *The removal of thatch.*

SUMMARY

A lawn, basic to most landscape designs, is established by using turfgrasses (grasses grown to be mowed and maintained). Certain turfgrass species, such as bluegrass and ryegrass, are adapted to cool climates. Other grasses (e.g., Bermudagrass and buffalograss) prefer warm climates. Between these two climatic zones is a transitional one in which either class of grass grows well. A new lawn may be established by seed, sod, or sprigs. The selection of species should take into account the use for which the lawn is intended and the maintenance level the home owner is willing to provide. Seeds are less expensive and less bulky than sod.

Once established, a lawn needs regular maintenance to keep it healthy and attractive. The three primary elements of turf management are mowing, watering, and fertilizing. Lawn mowers vary in design and efficiency. Mowing should be done in a timely fashion, under the proper conditions, to the correct height, and at the right frequency.

REFERENCES AND SELECTED READING

Beard J. B. 1973. Turfgrass science and culture. Englewood Cliffs, N.J.: Prentice-Hall.

Carpenter, P. L., and T. D. Walker. 1990. Plants in the landscape. New York: W. H. Freeman.

Crockert, J. V. 1971. Landscaping and ground covers. New York: Time-Life.

Hartmann, H. T., A. M. Kofranek, V. E. Rubatzky, and W. J. Flocker. 1988. Plant science: Growth, development, and utilization of cultivated plants, 2d ed. Englewood Cliffs, NJ: Prentice-Hall.

MacCaskey, M. 1987. All about lawns. San Francisco: Ortho Books.

Turgeon, A. J. 1985. Turfgrass management. Reston, Va.: Reston Publishing.

PRACTICAL EXPERIENCE

1. Visit a local golf course to see the types of grass used and how the course is maintained (watering, fertilizing, mowing, and disease and pest control).
2. Tour local residential and commercial areas. Observe how the lawn is featured in the landscape. Look for differences in style of mowing, quality of maintenance, species, color, mowing height, watering methods, and other characteristics.
3. Visit a department store to see the various kinds of lawn maintenance tools and machinery available.

PART A

Please answer true (T) or false (F) for the following statements.

1.　T　F　Bluegrass is a cool-season crop.
2.　T　F　Most turfgrasses prefer a soil pH of 6 to 6.5.
3.　T　F　Pieces of stems or runners used to establish a lawn are called plugs.
4.　T　F　Rotary mowers are more effective than reel mowers in mowing tall grasses.
5.　T　F　The rule of thumb in mowing is not to remove more than 30 percent of the leaf at any mowing.
6.　T　F　The best time to water a lawn is about noon.

PART B

Please answer the following questions.

1.　Give two examples for each of the following:
　　a.　Cool-season grasses_____　　_____
　　b.　Warm-season grasses _____　　_____
2.　Give two reasons why mulching is required after freshly sowing seeds in a lawn.
　　_____　　_____
3.　Grass that is specifically grown, mowed, and cut into strips is called _____.
4.　What are the three primary lawn maintenance practices?
　　_____　　_____　　_____

PART C

1.　Describe the role of a lawn in the landscape.
2.　Describe any two specific mowing strategies that make mowing operation results more desirable.
3.　Discuss the principles behind choosing a mowing height.

CHAPTER 15

Pruning

PURPOSE

This chapter discusses the importance and methods of pruning horticultural plants.

EXPECTED OUTCOMES

After studying this chapter, the student should be able to

1. List and discuss the general purposes of pruning.
2. List the basic pruning tools and their uses.
3. List and discuss the basic strategies of pruning.
4. Describe how roots, fruit trees, ornamental trees, and shrubs are pruned.
5. Describe specialty pruning strategies such as espalier, topiary, and pollarding.

KEY TERMS

Central leader	Espalier	Open center
Coppice	Limbing up	Pergola
Desuckering	Modified central leader	Pollarding

OVERVIEW

Plants have different growth habits and produce different adult forms. Uncontrolled, plants produce vegetative growth in response to the environmental provisions for growth. Branches form profusely and grow upward in search of light. Strong winds may twist limbs of plants and sometimes even break them off of the stem. The general appearance of plants under such conditions is not always appealing to humans. Under cultivation, humans employ a variety of procedures to manage plant growth and development for a number of reasons including the improvement of aesthetics and productivity.

Management of plant growth entails removing excessive and undesirable growth and structures by cutting, a procedure called *pruning*. Pruning is an art and a science. Manipulating the growth of plants in this way requires an understanding of plant botany and physiology. It is important to know plant structure and growth habits, as well as how plants

> **Pruning**
> The technique of cutting selected plant parts to accomplish a desired purpose.

respond to their environment and to removal of vegetative growth. The attractiveness of plants after pruning depends on the gardener's creativity and understanding of plant form and texture. Pruning is a standard cultural practice in orchards and vineyards, as well as in landscape management. The principles of pruning are generally the same in all situations. However, the specific methods or techniques are varied, depending on the species and the goal of pruning; that is, apples, citrus, grapes, and roses are pruned in different ways.

Pruning is sometimes done in conjunction with another horticultural procedure called *training*. Training involves cutting, repositioning, and guiding the course of development of branches and limbs of plants according to a specific objective. During training, limbs may be bent and tied to support structures or even removed altogether, resulting in creative and attractive shapes. In the landscape, aesthetics appears to be more important than productivity, while the reverse is true in orchard management.

15.1: GENERAL PRINCIPLES OF PRUNING AND TRAINING

The success of pruning and training plants depends on the understanding and observance of certain principles:

1. *Evaluate the whole plant.* Pruning affects the entire plant, whether physiologically or physically. Removing a limb may change plant form or shape and may also affect the plant's capacity for performing certain physiological functions such as photosynthesis and transpiration. By assessing the whole plant, one can make a decision as to which part of the plant to cut to give the overall best results.

2. *Think before you cut.* Cutting is an irreversible operation. Thus, a limb should be cut only when there is a good reason to do so. It is best to cut in stages, especially when one is relatively inexperienced in pruning. More of the branch or plant part can be removed if the first cut is not adequate.

3. *Apical dominance is broken when a stem is cut.* Apical dominance (15.3) resides in the terminal bud and gives direction to plant growth. While the apical bud is present and in control, lateral buds are suppressed. Breaking apical dominance stimulates new growth. New growth tends to arise below a wound on the stem because of interference in apical dominance. Apical dominance is strongly associated with vertical growth. Thus, any attempt to alter vertical growth induces a response in plants to correct the change. For example, when a branch that is growing vertically is bent and forced to remain horizontal, the buds in the leaf axis are stimulated to grow vertically. This happens because apical dominance is reduced by changing the vertical growth to horizontal growth. The new side shoots are likely to develop into reproductive shoots, flowers, and fruits.

4. *Pruning invigorates regrowth.* When pruning is used to reduce the size of a plant, it encourages vigorous new growth. The more severe the pruning, the more vigorous the regrowth. Thus, if a plant is growing in an unbalanced fashion, for example, the weaker side should be pruned severely to encourage vigorous new growth. For this reason, it may be best to select plants that will fit the available space when they are mature, thereby eliminating pruning that would produce vigorous new growth.

5. *Pruning can be used to direct growth.* By removing apical dominance, the direction of growth is transported to the topmost lateral bud. Thus, by selecting which bud will become the topmost bud, the regrowth is given direction because buds are positioned to face certain directions (15.5).

6. *Timing is critical.* Plants have different flowering habits that must be considered in pruning. Shrubs that flower in late summer and autumn produce flowers on the

current season's growth. These plants are pruned in spring so that they will produce vigorous shoots that will flower later the same year. On the other hand, shrubs that flower in spring or early summer produce flowers on the previous season's growth and thus are pruned after flowering. This way, the new growth has time to develop and be ready for flowering in the following year. Certain plants lose much sap when cut. Such plants should not be pruned in spring when sap production is at its peak.

7. *Pruning can be used to create special effects.* With pruning and training, plants can be manipulated to produce unique shapes and forms in the landscape. Such techniques include pollarding and coppicing (15.11).

15.2: OBJECTIVES OF PRUNING

Although the manner of pruning varies, the general objectives remain similar. All of the objectives may not be required or accomplished in any one particular instance, since pruning may be used for a specific purpose at a particular stage in the growth of a plant.

The four general purposes of pruning are: plant sanitation, aesthetics, reproduction, and physiology.

15.2.1 PLANT SANITATION

Pruning may be undertaken to remove plant parts that create an unsanitary condition. Specific actions geared toward improved plant sanitation include the following.

1. Broken branches and dead tissue on plants provide surfaces on which disease-causing organisms grow and thus jeopardize the health of plants. Broken branches pose grave danger to people.

2. Diseased plant parts may be removed to prevent the spread of infection.

3. The plant may be cleaned by removing unsightly dried or dead parts.

4. The canopy can be opened up so that air can circulate freely and thereby reduce humid conditions that predispose plants to disease.

5. An open canopy enables effective spraying by allowing pesticides to penetrate the canopy.

6. Removal of excessive undergrowth not only keeps the landscape clean but also reduces hazards from brush fires.

15.2.2 AESTHETIC OBJECTIVES

In ornamental horticulture, the visual appeal of plants is of paramount importance to gardeners, especially if plants are in the landscape. Pruning is employed for shaping the form of the plant. After determining the desired shape, branches are strategically removed or their growth controlled to maintain the shape.

In *formal gardens* (15.16) or on certain public grounds such as those found in zoological or botanical gardens, certain plant species are grown, trained, and trimmed into geometric figures or readily recognizable or abstract shapes. This art form is called *topiary* (15.11).

Pruned plants can by themselves be attractive elements in the general landscape. However, more pleasing components can be produced if the style of pruning takes into account other elements in the landscape such as the architecture of the house, the terrain, and artificial structures such as statues and fountains.

15.2.3 REPRODUCTIVE OBJECTIVES

Pruning may be undertaken to enhance the reproductive capacity of the plant in several ways:

1. The canopy may be opened up by cutting out the branches in the center. This allows light to penetrate the canopy for fruiting to occur on inner branches.

2. Fruiting may be regulated by encouraging the growth and development of fruiting shoots while reducing nonreproductive shoots.

3. Pruning may be done to balance reproductive and vegetative growth for optimal yield.

4. Pruning may be done to reduce the number of fruiting branches per plant (*thinning out*) in order to increase fruit size and quality.

5. Generally, proper pruning enables a fruit-bearing tree to produce higher-quality fruits over a longer period. Flowering plants also produce bigger flowers over a longer period of the plant's life when pruned.

6. Properly pruned trees have good fruit distribution throughout the plant canopy (not only at the edges) and bear fruit of good size, color, texture, and sugar content.

7. Pruning allows the gardener easier access to fruits during harvesting.

15.2.4 PHYSIOLOGICAL OBJECTIVES

Pruning, if not done judiciously, may have adverse physiological consequences on plant growth and development. However, pruning may be employed to manipulate plant physiology in a variety of ways to enhance the performance of the plant:

1. Pruning roots before transplanting reduces the chance of transplanting shock.

2. Pruning shoot tips in species with apical dominance (e.g., apple, pear, and cherry) induces lateral branching and thereby prevents the tree from growing straight without sufficient branching.

3. Pruning deciduous species during the dormant period in winter conserves the plant's stored food for use in spring for vigorous new growth.

4. Severe pruning may have a dwarfing effect on a plant by reducing total vegetative growth.

5. Older plants may be rejuvenated by pruning to stimulate new growth.

Apart from these four general purposes of pruning, the procedure may be employed on specific occasions for practical reasons. For example, when plants grow bigger and exhibit destructive tendencies such as roots cracking pavements or foundations of buildings, roots clogging sewage pipes, and branches destroying the roof or touching utility cables, the affected plant parts need to be pruned to contain the plant in the available space.

15.3: PLANT RESPONSE TO PRUNING

Removing vegetative parts of plants affects certain plant physiological processes and subsequently growth and development. The two basic plant responses to pruning are described in the following sections.

15.3.1 INTERFERENCE WITH APICAL DOMINANCE

As previously stated, apical dominance of terminal buds suppresses the growth of lateral buds. The removal of terminal buds removes this inhibitory effect, allowing lateral buds to

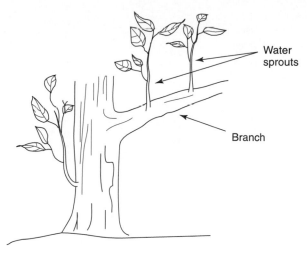

FIGURE 15-1 Water sprouts growing on a branch.

grow. In certain plants, gardeners deliberately remove terminal buds (pinching or *pinch pruning*) to encourage lateral bud growth so that plants look fuller and more appealing. The vertical shoots that arise on the upper side of branches (called *water sprouts*) are an example of a plant response to interference with the process of apical dominance (figure 15-1).

15.3.2 GROWTH STIMULATION

A physiological balance exists between the top and bottom growths of plants. Removing parts of the top growth upsets this balance. Plants respond with a burst of new growth, especially just below the cut. In spite of the new growth, pruned plants do not exceed the size of the plant before pruning. This dwarfing effect on plant size occurs because the amount of new growth does not match what was removed in addition to what would have been produced by the plant in its prepruned form. Similar to the practical application of interference with apical dominance, gardeners employ this dwarfing effect of pruning in creating art forms. The epitome of such an art form is the Japanese art of dwarfing plants called *bonsai* (chapter 26).

15.4: PRUNING TOOLS

Most pruning tools are handheld and operated manually. A motorized chain saw may be used for cutting large branches. Cuts are made by either sawing or scissor action of cutting tools. Pruning tools are hence a variety of saws and shears designed to cut different sizes of limbs at different heights and different locations on the plant. When the plant is tall and the limbs are hard to reach, pruning aids may be used to lift the operator to the desired height. Some of the common pruning tools are described in the following sections.

15.4.1 SHEARS

Figure 15-2 shows a variety of shears.

Hand Shears (Hand Pruners)

Hand shears are the most commonly used pruning implements. They can be held in one hand and are available in two basic designs. One design has a true scissor-cutting action produced by two cutting edges or blades (*bypass action* type). The other design uses one sharp blade that cuts against a metal piece (*anvil* type). The limitation of this tool is that it can only cut limbs that are less than 1/2 inch (1.3 centimeters) thick.

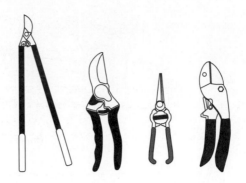

FIGURE 15-2 A variety of shears used in pruning. The type used depends on the size of the branch.

Lopping Shears (Loppers)

Lopping shears are designed to cut larger branches (up to 2 inches or 5.2 centimeters in diameter) by a two-handed action. Their long handles provide the needed leverage for cutting thicker limbs. The long handles also extend the reach of the operator so that limbs located high on the stem may be pruned without using a pruning aid such as a ladder or lift. Short-handled loppers are available. The common loppers have hinge action, but more expensive designs with different types of action are available.

Hedge Shears

Two-handed tools, hedge shears are designed for trimming and shaping hedges and ground covers. Manual models are common, inexpensive, and easy to use. Electric models are also easy to use and have a fast-cutting action. However, they may be frequently jammed by twigs during operation and also be limited in operation by the length of the extension cord.

15.4.2 SAWS

A saw may be designed to cut only on the forward stroke and thus make it easier to maneuver when cutting limbs located high on a tree. A saw may also be fine toothed for smooth, close cuts and is especially appropriate for cutting deadwood. Coarse-toothed saws are easier to use on greenwood. Common horticultural saws can be placed into four general categories: manual saws, power saws, pole saws, and pole pruners.

Manual Saws

Folding Saw   The folding saw can be folded to make it even smaller and more convenient to carry around (figure 15-3). It is fine toothed and used for small branches.

Rigid-Handle Curved Saw   The rigid-handle curved saw may be designed to have all lance teeth for cutting deadwood or raker teeth (a deep slot after every five even-sized teeth to carry away sawdust) for greenwood.

Bow Saw   The blade on the bow saw is thin and replaceable. It can be used to make quick cuts of even large branches but is restricted to use in unobstructed areas because of the pronounced arching of the bow.

Tree Surgery Saw   The tree surgery saw comes closest to the common carpenter's saw but differs in that it is designed to cut only on the forward stroke. It is more difficult to use than the other types and requires more effort because of its fine teeth.

Two-Edge Saw   Because of its design, the two-edge saw can cut on both edges and therefore must be used with great care.

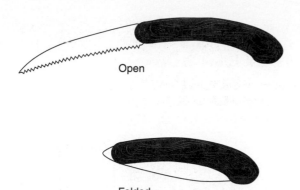

Open

Folded

FIGURE 15-3 A folding saw for manual cutting of small branches.

Power Saws

Motorized saws, or *chain saws,* are easy to use and very efficient. They can be used to cut all sizes of plant limbs. However, they pose a great danger and can inflict serious injuries to operators if not handled properly. According to the source of power, there are two basic models:

1. *Electric-powered chain saws.* Electric power makes this model of chain saw quiet during operation. It is easy to use. Models with cords are limited by the length of the extension cord.

2. *Gasoline-powered chain saws.* Gasoline models are completely portable, larger in size, and generally more expensive than electric-powered chain saws. They also require more maintenance and are noisy during operation.

Pole Saws

A pole saw has a small curved saw blade mounted on the end of an extendable pole. This type allows branches on tall trees to be easily pruned while the operator is standing on the ground.

Pole Pruners

The pole pruner has a J-shaped hook mounted on a pole along with a saw blade, which is operated by a rope or pull rod. The hook is used to grab and hold the branch while the saw cuts it (figure 15-4). A combination pole saw and pruner may be purchased.

15.4.3 PRUNING KNIFE AND RASP

The pruning knife is very sharp and is used when minor pruning involving a few small branches is needed. The knife may also be used to clean and smooth large cut surfaces. A rasp is like a file and is used for shaping or smoothing tree wounds.

15.4.4 LADDERS

Ladders, tools for extending the reach of an operator during pruning, should be used with care. They must be set up properly to ensure good stability. Leaning while standing on a ladder should be avoided. Of the different types of ladders available, those most recommended include the following:

1. *Orchard ladder.* This three-legged tool has added stability from its wide stance (figure 15-5).

2. *Extension ladder.* This ladder can be adjusted in length and is useful for reaching high places on a tree.

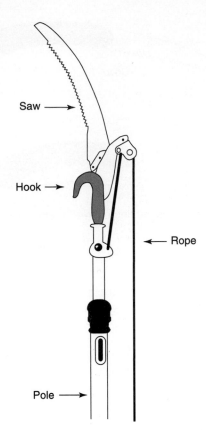

FIGURE 15-4 A pole pruner has a long reach and allows branches high in the tree to be pruned while the caretaker is standing on the ground.

FIGURE 15-5 An orchard ladder has a tripod for added stability.

15.5.1 GENERAL GUIDELINES

The exact way a particular pruning method is implemented is determined by a number of factors including species, goal to be accomplished, whether the plant is grown strictly for ornamental purposes or for producing fruits, and the environment in which the plant is growing. Fruit trees are pruned differently from ornamental landscape trees and bushes; flowering plants are pruned differently from foliage plants.

In selecting and implementing a method of pruning, one should consider not only the desired outcome but also how the species responds to pruning, especially in terms of the extent of pruning. Further, one method may be suited to one species but not to another.

Notwithstanding the method, certain general guidelines may be followed for successful pruning:

1. Have clearly defined goals.

2. Prune at the appropriate time. Some plants may be pruned any time of the year, while others are best pruned when dormant (late fall to early winter).

3. Proceed cautiously. Take time to look at the plant to determine which limb needs to be pruned. It is better to cut less and revisit the plant later for further pruning than to cut too much in one instance.

4. Use sharp tools and make clean cuts. Avoid tearing off the bark of the plant. Clean cuts heal much faster and reduce the chance of disease infection.

5. Prune the parts that must be pruned first. Phytosanitation is important in any pruning operation. All dead or dying parts and broken limbs should be removed.

6. Branches that grow inward toward the center of the canopy are prime candidates for pruning. Outward-growing branches should be encouraged. Branches that are acutely angled with respect to the central axis are also candidates for pruning.

7. Look for abnormal growths. Species have certain natural forms, and as such pruning is more successful when the natural tendencies of plants are taken into account. For example, it is easier to prune a species with a cone-shaped canopy to remain cone-shaped than to force it to assume a spherical shape.

8. For fruiting plants, it is critical to know the fruiting habits (i.e., lateral or terminal) and identify and distinguish between vegetative and fruit buds or branches.

9. Seek the help of a professional arborist if a large limb or high branches require the use of a ladder.

15.5.2 CUTTING

Cutting is the primary activity in pruning. After determining which branch or limb to cut, the next critical step is deciding how to cut it. A wrong cut may ruin a bud or defeat the purpose of pruning. The way a plant part is cut depends on the size of the part and the location of the cut to be made. The following are some general guidelines for cutting:

1. A cut should be clean (figure 15-6), which is made possible by using a sharp tool. If a saw is used, any rough edge should be smoothed with a pruning knife. A clean cut accelerates callus formation for quick healing.

2. When removing a branch, the cut should be close to the main branch or stem so as to leave very little stub (figure 15-7). Without an active bud, a stub will gradually wilt and die back. Deadwood can become infected with disease.

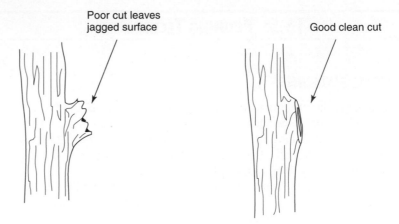

Poor cut leaves
jagged surface

Good clean cut

FIGURE 15-6   Pruning cuts must be clean for proper healing.

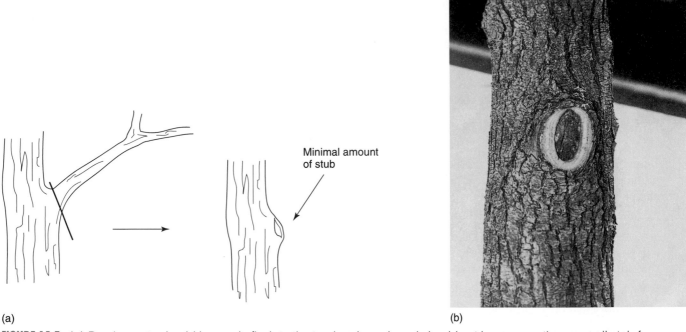

Minimal amount
of stub

(a)

(b)

FIGURE 15-7   (a) Pruning cuts should be made flush to the trunk or branch and should not leave more than a small stub for proper healing. (b) A well-healed wound.

3.   Avoid splitting and tearing (which are caused by the weight of the branch). When cutting a large branch that cannot be cut shears, it is best to cut in stages (figure 15-8). First, make a cut on the underside of the branch, about 6 inches (15.2 centimeters) away from the trunk or main branch. This cut should go only about halfway deep. Next, make a cut on the top side at about 1 to 2 inches (2.54 to 5.1 centimeters) away from the first cut and away from the main stem or trunk. When this second cut has reached about halfway through, the branch may be easily snapped off. The remaining stub is then removed closer to the trunk or main branch without danger of tearing.

4.   The cutting angle is important for two main reasons. Too sharp of a cutting angle produces a larger surface area, thus increasing the time for healing and predisposing the wound to infection (figure 15-9). All cuts die back from the surface, but the

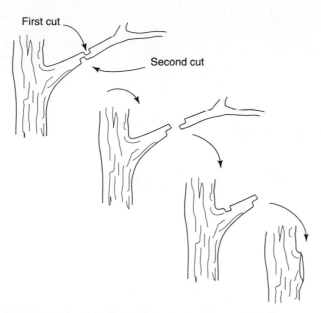

FIGURE 15-8 Large branches should be cut in stages to prevent tearing.

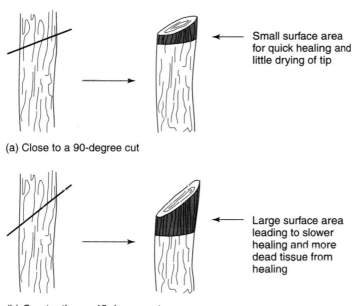

Small surface area for quick healing and little drying of tip

(a) Close to a 90-degree cut

Large surface area leading to slower healing and more dead tissue from healing

(b) Greater than a 45-degree cut

FIGURE 15-9 (a) Pruning cuts should be made as close to a 90-degree angle as possible. (b) Cut angles greater than 45 degrees create large surface areas, slow healing, and lead to a large amount of dead tissue during the healing process.

dieback is more extensive for a sharp-angled cut. A 45-degree or less slanted cut is recommended.

5. The distance of the cut from a bud is important. Since all cuts die back a little, cutting too close to a bud (or cutting at a sharp angle) might cause the dieback to kill the nearest bud in the process (figure 15-10).

6. The direction of cut determines the direction of the new growth from the immediate bud. Slanted cuts are approximately parallel to the direction of the tip of the immediate bud below. An outward slant (upper tip pointing out of the canopy) indicates that the

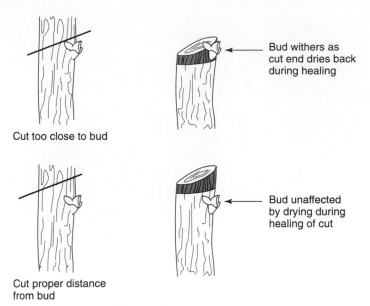

Cut too close to bud

Bud withers as
cut end dries back
during healing

Cut proper distance
from bud

Bud unaffected
by drying during
healing of cut

FIGURE 15-10 Distance of cut from the nearest bud is critical to the survival of the bud. If a cut is too close to the bud, it might wither with the advance of drying tissue as part of the healing process.

bud immediately below will produce an outward growth (figure 15-11). This direction of growth is desirable and is one of the objectives of pruning—opening up the canopy for ventilation and light penetration. An inward growth achieves the opposite effect. Therefore, when pruning, the operator should select the appropriate bud above which a cut is to be made. It should be mentioned that there may be situations in which pruning to obtain a tight center may be preferable. If the plant has opposite buds, one of them should be removed to leave the outward-pointing bud.

7. The ideal cutting position is close to a bud at a fork or branch or just above the collar of a branch.

8. When cutting with shears or pruners, the thin blade should be nearest the bud or the trunk, which permits the desired cut.

15.5.3 CARING FOR WOUNDS

Plants have natural defense mechanisms that are triggered upon injury. First, a layer of waterproof material called *suberin* is produced to protect the exposed tissues of the wound from drying. Next, *callus* growth occurs to produce new tissue to close up or heal the wound. It is important that wound closure occur without delay to avoid invasion by decay organisms. The rate of healing depends on a number of factors:

> **Suberin**
> *A hydrophobic material that occurs on the inner surface of a cell wall.*

1. *Size of wound.* A large wound heals slowly.

2. *Location of wound.* Generally, wounds occurring at lower levels on the plant heal more quickly than those at higher levels.

3. *Number of wounds.* A plant with few wounds heals more quickly than one with numerous wounds.

4. *Species.* Relatively longer-lived trees such as oaks tend to heal more quickly than short-lived trees such as willows.

5. *Age of plant and plant vigor.* Younger plants have more vitality and thus heal more quickly than older plants. Further, when buds and leaves on a plant are expanding, wounds tend to heal slowly because plant energies at that stage are channelled into growth activities.

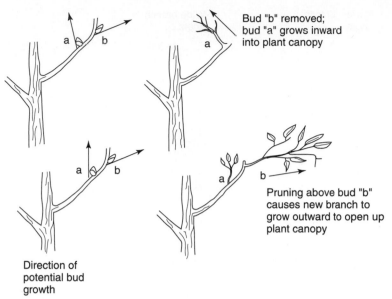

Bud "b" removed;
bud "a" grows inward
into plant canopy

Pruning above bud "b"
causes new branch to
grow outward to open up
plant canopy

Direction of
potential bud
growth

FIGURE 15-11 The direction of new growth after pruning depends on the potential direction of growth of the bud closest to the cut. If an outward-pointing bud is the last bud, the shoot will grow outward. An inward-pointing bud should be targeted with care since it defeats the purpose of pruning in certain cases by encouraging new growth to crowd the inner part of the tree canopy.

6. *Environmental conditions.* Temperature and moisture conditions play a role in the healing process of wounds. Early spring provides the best conditions for wound closure, and wounds heal slowly in late fall and early winter when physiological processes slow down.

7. *Use of wound paint.* Wounds may be treated with pesticides and other chemicals to prevent or slow the rate of rotting. Wood paints may include asphalt-based dressings, antibacterial preparations, and fungicides. When rotting occurs to the extent that cavities are formed in the tree trunk, such holes may be plugged with materials such as concrete, asphalt, and polyurethane foam.

15.5.4 WHEN TO PRUNE

Pruning may be done during or after the dormant season when flowering occurs. On certain occasions, plants are pruned during both the dormant and active seasons. Species differ regarding the best time to prune them. For the same species, pruning may be done at different times during phases of plant growth and according to the production schedule. Species that flower in spring (e.g., lilac, forsythia, and magnolia) develop flower buds during the previous growing season. Such species are pruned after flowering. On the other hand, species that flower in summer or fall (e.g., rose, croton, blueberry, and dogwood) develop flower buds in the current season's growth (i.e., on the new growth). As such, it is best to prune them before the new growth begins in spring.

15.5.5 WHEN CUTTING IS NOT DESIRABLE

Pruning involves cutting plant parts. Some plants (such as pears) tend to develop a narrow profile such that the branches do not spread but grow upright, nearly parallel to the stem. In such situations, removing a branch might do more harm than good in terms of productivity. A solution to this problem is to open up the canopy by inserting boards (*spreader boards* or *branch spreaders*) between adjacent upright limbs to push them apart (figure 15-12).

> **Spreader Board**
> A flat piece of board with v-cuts at both ends that is forced between the stem and a branch with a narrow crutch to widen it.

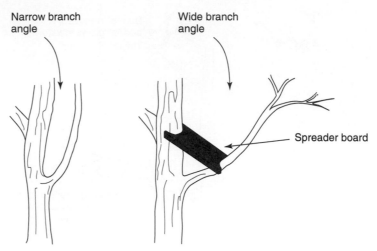

FIGURE 15-12 A spreader board may be used to open up the canopy of a tree for the development of strong and wide crotches.

15.6: STRATEGIES FOR PRUNING ABOVE-GROUND PLANT PARTS

The three general strategies for pruning above-ground plant parts are described in the following sections.

15.6.1 THINNING OUT

> **Thinning**
> Removal of excess vegetative growth to open the plant canopy and reduce the number of fruiting branches for larger fruits.

The principal objective of *thinning* is to open up the plant canopy for light to penetrate to lower branches for better fruit set and increased productivity. Without adequate light, shaded branches become unproductive but nonetheless use up nutritional resources available to the plant. In thinning, the operator strategically removes certain branches, such as those that are inward growing. The operation involves limited or no trimming at all so that the general shape of the plant is preserved. The limbs are evenly spaced on the stem, and the low and unproductive ones are removed (figure 15-13). Thinning is a common orchard management practice for keeping fruit trees in the best shape for high productivity.

15.6.2 HEADING BACK

The method of heading back involves the removal of the terminal parts of branches (figure 15-14). Even though it appears as though the plant is being trimmed, it is not done haphazardly. Its primary effect is to promote secondary branching. A plant pruned by this method becomes fuller because of the additional growth. It is an easy method to employ but care must be taken to trim the plant such that the topmost bud (at the end of the branch below the cut) points in the right direction (to produce outward growth). After this procedure, the plant is reduced in height and size and may have a new shape.

15.6.3 RENEWAL PRUNING

The first two types, thinning and heading back, are the basic methods of pruning. Renewal pruning is employed to rejuvenate old plants by removing old, unproductive branches (figure 15-15), which allows for fresh and vigorous replacement growth. Flowering shrubs may be rejuvenated in this way. Fruit trees in the landscape neglected by previous home owners may require renewal pruning by new owners.

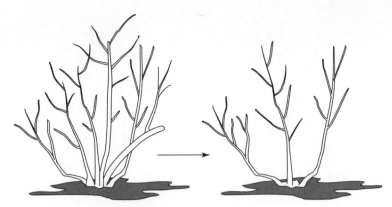

FIGURE 15-13 Thinning out is one of the two basic cuts in pruning. It is used to remove excessive growth and old plant limbs, thereby opening up the plant canopy.

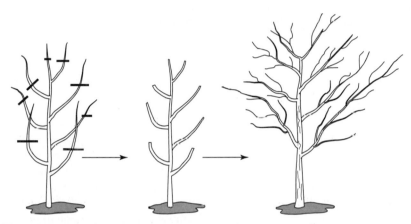

FIGURE 15-14 Heading back is one of the two basic cuts in pruning. It is used to reduce the length of plant limbs. Heading back encourages a burst of new growth and branching.

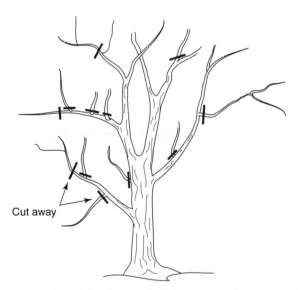

Cut away

FIGURE 15-15 Renewal pruning of an old tree removes old and unproductive or damaged branches and heads back limbs to encourage new growth.

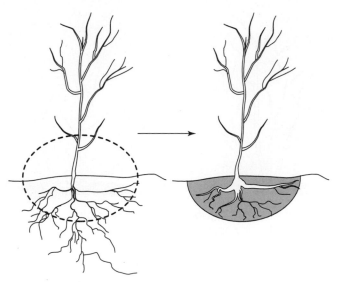

FIGURE 15-16 Root pruning is accomplished by using a spade for a small plant or using commercial tree augers for large plants. Bare-root plants may be pruned before planting to remove damaged roots.

15.7: PRUNING ROOTS

Unlike the pruning of above-ground plant parts done as part of the plant management operation, roots are pruned once during the plant establishment process. Roots of large plants are pruned immediately or shortly before transplanting (figure 15-16). If pruned long before transplanting, the roots have the opportunity to develop a new mass of secondary roots for better establishment when transplanted. Sometimes pruning is necessary to reduce the root size in order to fit the hole for transplanting. Tractor-mounted mechanical root pruners are used by large commercial nurseries that produce and sell large plants.

15.8: TRAINING PLANTS

Training **Training**
A system of plant management that involves pruning and trying to create and maintain desirable plant size and shape.

Training, a horticultural activity performed while the plant is young, entails laying the basic architectural framework for the shape the plant will assume at maturity. Further, it is easy to manipulate the limbs of young plants with little danger of breaking them. Plants are trained according to a predetermined strategy to accomplish specific purposes. A successful program combines a good understanding and application of the practices of pruning and plant nutrition and the use of physical supports in some cases. Plant training requirements vary according to the determined objectives. Hedge plants are rarely trained. Small fruits such as grapes require elaborate training for good production (chapter 19). Sometimes it is best to allow plants to grow and develop to assume their natural shapes and forms without any interference. Certain species cannot tolerate limb manipulation without adverse consequences. After developing the foundation for the adult plant shape and form, the gardener must prune or reposition limbs periodically to maintain these characteristics.

15.9: TRAINING AND PRUNING ORNAMENTAL TREES

The primary goals of training ornamental trees are to develop a strong tree trunk, develop an attractive plant form, and establish a heading height.

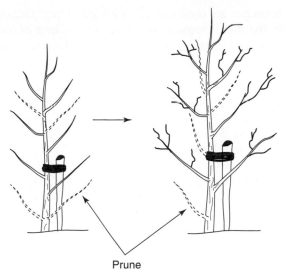

Prune

FIGURE 15-17 Feathered tree form development.

15.9.1 DEVELOPING A STRONG TRUNK

The training strategy for developing a strong trunk is called the *central leader* (15.12). One strong upward-growing branch is identified and encouraged to grow to become the central axis of the tree. One way of developing a strong trunk is as follows:

> **Central Leader**
> *The main upright shoot of a tree.*

1. Cutting back (heading back) the plant in the first year will allow new growth to occur just below the cut. Unless a young tree is weak, it should not be staked. If required, staking should be loose to allow the trunk free movement in the wind in order to develop strength.

2. A strong upward-growing branch is identified in the second year, and other branches (competing leaders) below the selected leader are removed or pruned back. Water sprouts will develop as a result.

3. In the third and subsequent years, the central leader is encouraged to maintain its dominant position and the development of secondary branches is controlled.

4. Spreaders may be necessary to widen narrow crotch angles between secondary branches and the trunk.

15.9.2 DEVELOPING TREE FORM

Ornamental trees take time to develop their form, and therefore growers should not be overly concerned about unattractive forms in the early years of their growth. It is important that about 30 percent of the plant's foliage be located in the lower half of the tree during the first three to four years of growth. The lower branches are progressively removed as the plant grows to establish a desirable clearance beneath the limbs. By adopting this strategy, the tree develops a straight, strong, tapered, and attractive trunk. Ornamental trees can be trained to have one of several forms, described in the following sections.

Feathered Form

The goal of training a feathered tree form is to maintain a simple main stem with well-spaced laterals for good balance. Once a central leader has been identified, crossing laterals should be pruned by cutting them to the main stem (figure 15-17). A desirable clearance is established beneath the limbs by removing the lowest branches. If staking is needed, a low stake allows the main stem to flex for strength development. In the second and subsequent years, regrowth at the stem base is pruned along with any vertical shoots that compete with the central leader. Crossing branches are removed as they occur. Once adequate strength has been developed in the trunk, the stake should be removed. This system of training is the easiest to perform. Evergreen trees are commonly trained to have a feathered form.

Standard Form

A standard tree form can develop naturally. There are two variations of this form. The central leader standard is a modification of the feathered form but with more clearance beneath the limbs (figure 15-18). Laterals on the lowest one-third of the tree are pruned in stages. In the first year, laterals in the top one-third are left untouched, with the exception of dead branches. Those in the midsection are shortened by half, and those in the lower one-third are removed completely. In the next two years, the pattern in year one is repeated, except that the laterals that were shortened in the previous year are removed. In addition, any existing cross branches are removed in the top section of the tree. This process continues until a desired length of clear stem is attained.

Branched-Head Standard

In the branched-head standard, the tree is trained as a central leader in the first two to three years before the leader is cut back such that three to four strong laterals remain (figure 15-19). The growth of these branches is controlled by pruning to outward-facing buds to open the crown of the tree.

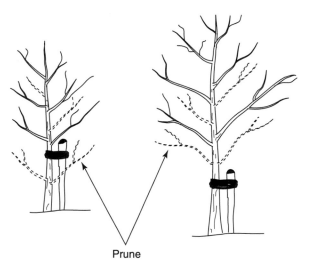

Prune

FIGURE 15-18 Development of a standard tree form.

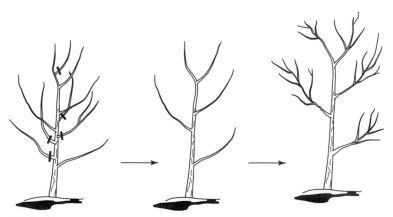

FIGURE 15-19 Development of a branched-head standard.

Weeping Standard

Weeping tree forms may be natural or grafted. The tree in natural weeping standard form is trained as a single stem to a suitable height and then the branches are allowed to arch down (figure 15-20). Commonly, a high stake is needed to support the downward-arching branches until a strong trunk has developed. In the top-grafted weeping standard, a short stem is grafted onto the stock.

Multistemed Tree Form

Certain trees naturally develop multiple stems and branches that are very low on the base of the tree. They also tend to produce suckers. To create this form, the stem of a young tree (about two years old) is cut as close to the ground as desired (figure 15-21). This practice induces new shoot growth. Two to three healthy shoots, well spaced around the base, are selected and retained while all others are pruned. Any regrowths are removed in subsequent years.

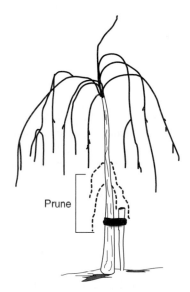

FIGURE 15-20 Development of a weeping standard. This tree form has a high heading height to allow the branches room to hang.

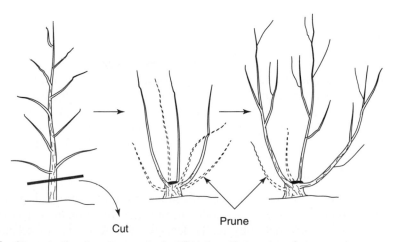

FIGURE 15-21 Steps in the development of artificial multistem tree form.

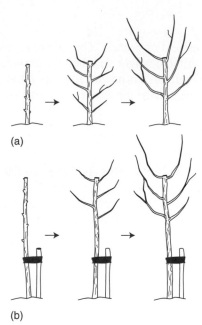

(a)

(b)

FIGURE 15-22 Developing heading height of trees: (a) low heading height and (b) high heading height.

15.9.3 DEVELOPING HEAD HEIGHT

Scaffold Branch
The main branch growing from the trunk of a tree.

The head height determines the height of scaffold branches above the ground in the adult stage. Whether a low or high head is chosen depends on the natural form of the tree and the desired amount of clearance beneath the tree. Generally, trees with narrow forms (such as conifers) have a low head height, whereas spreading trees have a higher head height (figure 15-22). Trees with weeping forms (such as weeping willow) need very high head height for proper display of their low-hanging branches and good clearance. For a low adult-stage head height, the initial heading height should be about 2 to 3 feet (0.61 to 0.91 meter) above the ground. If a high head height is desired, the initial heading may be 3 to 5 feet (0.91 to 1.5 meters) or even higher.

15.10: MAINTENANCE OF ESTABLISHED TREES

Once established, ornamental trees need occasional pruning to remove unwanted growth, correct growth, or repair damage. Specific pruning activities include those described in the following sections.

15.10.1 REMOVING SHOOTS

Epicormic Shoots
Shoots that develop from previously dormant buds under tree bark after being stimulated by external factors such as wounding.

A tree has dormant buds under its bark that may be stimulated to produce shoots on the stem. These shoots are called *epicormic shoots* (or water sprouts or water shoots) (15.3.1). They are undesirable because they deplete plants of food and also distract from the tree's form and general attractiveness. Sometimes shoots arise from the roots of adult trees such as aspen *(Populus tremula)* as part of their natural development. Suckers may arise when certain plants such as black poplar *(Poplar nigra)* are wounded. These shoots should be removed.

15.10.2 REPAIRING DAMAGE

Trees in the landscape may suffer damage from a variety of sources. Lightning, strong winds, hail, ice, and other weather-related damage may occur. Humans may vandalize and landscape maintenance equipment physically damage trees. Animals in the landscape may also cause injury. Broken branches and torn barks need to be repaired by cutting them away.

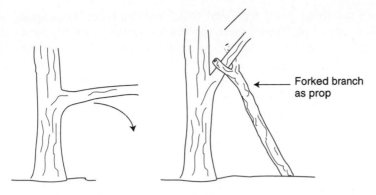

FIGURE 15-23 A low-hanging branch may be propped up by using a forked branch.

15.10.3 REMOVING DISEASED BRANCHES

Deadwood provides a hiding place for insects and tissue for disease organisms including fungi. Any diseased branches should be removed.

15.10.4 PROVIDING PHYSICAL SUPPORT

Sometimes tree branches become excessively inclined and need to be physically propped up to avoid snapping (figure 15-23). Propping with a forked branch may help to enhance the appearance of a tree by improving its form.

15.11: RENOVATING ESTABLISHED TREES

The purpose of renovation is to rejuvenate a tree that has been neglected for a long time in order to restore tree form. Sometimes a tree may have to be felled. Pruning for renovation may involve either minor or major cuts. Trees are usually renovated when they are dormant. To avoid disease infection, wounds should be properly treated for quick healing.

A tree may grow unevenly in terms of form or shape because it encounters an obstruction that suppresses normal growth and development on one side of the tree. When the obstruction is removed, the tree may be renovated to stimulate growth on the suppressed part. As previously indicated, vigorous growth is restored to the suppressed side by pruning that area.

An old tree may require severe pruning to reduce its size, a procedure called *crown reduction.* Neglected trees may develop excessive branches, congesting the canopy. Such trees may undergo *crown thinning* to open up the canopy. Neglect may cause lateral branches to grow on the lower part of the trunk. Through the process of *crown lifting,* the lower branches may be removed to create a higher clearance below a tree.

15.12: SPECIAL TRAINING AND PRUNING TECHNIQUES

Sometimes trees may be cut back drastically to allow for new growth. The severe pruning may be used to create unique and fascinating structures in the landscape. Examples of such artistic creations are described in the following sections. It should be noted that few species will tolerate this degree of pruning.

15.12.1 POLLARDING

Pollarding entails a severe pruning of the plant after it has attained its maximum desirable height. Trees may be headed when they attain a height of 8 to 12 feet (2.4 to 3.6 meters). The tree top is cut back drastically during its dormant period (winter). Scaffold branches are headed when they are about 2 to 5 feet (0.61 to 1.5 meters) long. When spring starts, new buds sprout below each cut. These water sprouts are removed each year as they recur. Repeated pruning leaves the tops of branches in clumps of growth resembling stubs (figure 15-24). Infection by decay organisms is minimal because the branches that are pruned are small in diameter and hence incur only small wounds. This look is desirable in spring, but in the fall and winter seasons, the plants look unattractive without the foliage, though the stumps create some interest in the landscape. Ornamental trees that may be pollarded successfully include elm, sycamore, poplar, and willow. The colorful stems of dogwood (*Cornus* spp.) and willow (*Salix* spp.) add attraction to the stumps in winter.

15.12.2 COPPICING

In coppicing, pruning is even more extreme than in pollarding. The tree is cut back to leave a short stump (forming the stool) from which new growth emerges (figure 15-25). Trees such as *Eucalyptus* and *Paulownia* are amenable to such treatment. Coppicing and pollarding both have dwarfing effects on plants.

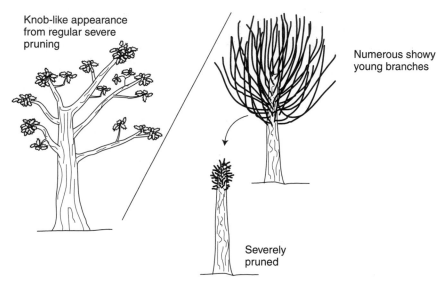

FIGURE 15-24 Pollarding. This severe pruning technique creates knoblike and stubby branches. During the spring, numerous new shoots develop from the stubs that produce a spectacular display of branching in winter when the leaves have fallen.

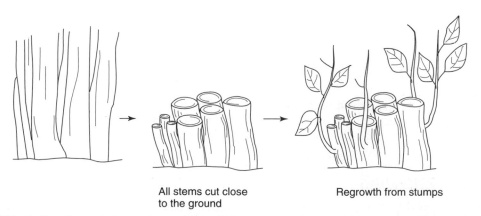

FIGURE 15-25 The technique of coppicing. Once the coppice stool has produced new shoots, thinning may be done to remove excessive growth.

15.12.3 PLEACHING

Pleaching is a technique used to weave together the branches of a row of trees that, with appropriate pruning, can develop into a hedge. Species adapted to this kind of treatment include linden, hornbeam, and holly.

15.12.4 TOPIARY

Topiary may be described as the art of plant sculpture in which plants are trained and pruned into formal shapes. These shapes may be abstract or geometric and sometimes may be readily recognizable objects such as animals. Topiary may be created in container plants, in the garden, or in the general landscape. Sometimes the top of a hedge (15.16) is capped with a topiary.

The simplest designs are those closest to the natural shape of the plant. To create complex forms, a metal framework is first designed and placed over the plant so that it grows into and over it (figure 15-26). The plant is then carefully clipped to shape, following the outline of the framework. Horticultural techniques such as pinching, training, and tying are used to encourage dense growth to cover and hide the framework. When pruning, a straightedge and other guides often are required for accuracy. Without such guides one may cut too much on one side, which may require that the whole piece be reworked to obtain the geometric symmetry desired.

Geometric shapes are difficult to create and maintain. To keep them attractive, the sides and surfaces must be properly cut to the symmetrical, sharp, and well-angled shapes. Round shapes are relatively easier to create and maintain. It is very important to use sharp tools in pruning and for the operator to exercise patience, proceeding cautiously. Common and more or less standard shapes are the poodle or cake stand and the spiral (figure 15-27). To create complex and irregular shapes, one must be very creative and patient. The species most frequently used for topiary include those that are easy to train such as yew *(Taxus),* boxwood *(Boxus),* and bay *(Laurus nobilis).* These plants are evergreen and long-lived.

> **Topiary**
> Training and pruning of plants into formal shapes, sometimes geometric or abstract but highly stylized.

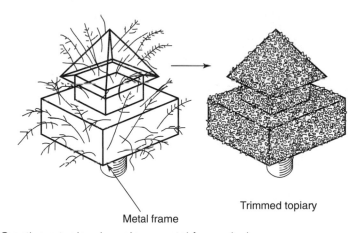

Metal frame

Trimmed topiary

FIGURE 15-26 Creating a topiary by using a metal frame design.

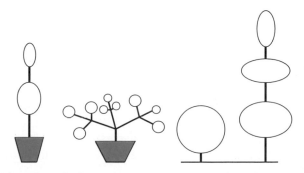

FIGURE 15-27 Pompon topiary designs.

Topiaries need to be maintained to keep them in form and attractive. Like all plants, they require fertilizing and watering to enable regrowth after cutting. They also need frequent routines of maintenance clipping during the growing season. The frequency of clipping depends on the species, plant vigor, form, design, and demands of the owner. If owned by a perfectionist who must have the plant in great shape all of the time, very frequent pruning will be required. Species such as the yew can be presentable with about two clippings per year. However, if the design is geometric, a monthly schedule may be required to remove any new growth. Over time, certain portions of the plant may become damaged and require repair. The damaged part should be removed and a nearby shoot trained to fill the gap. Depending on the damage, it may take up to several years to restore the topiary to its original form.

15.13: TRAINING AND PRUNING FRUIT TREES

Fruit-bearing tree training goals are similar to those of ornamental trees. These goals include the following:

1. To develop strong branches to bear the weight of fruits.

2. To properly space and retain an appropriate number of branches for enhanced productivity.

3. To control the time of first fruiting. While fruiting should not be delayed, the quality of fruits and the duration of fruiting period are increased when trees are not allowed to fruit in the early years (three to five years). The heavy weight of fruits may injure young plant limbs and jeopardize productivity in later years.

4. To facilitate production operations such as harvesting, pruning, and spraying.

5. To produce attractive tree shapes, which is particularly important if fruit trees occur in the landscape.

6. To provide physical support for weak stems. In dwarf and semidwarf fruit trees, some kind of support system is needed to aid the tree in bearing the weight of heavy fruits. The limbs may be propped with a notched-end piece of wood.

7. To confine the tree to the space available.

Training of fruit trees may involve developing wider crotches, perhaps by using spreaders. A crotch alteration should be done when the trees are young. The degree of spreading or bending affects shoot growth and development. If too wide (e.g., 90 degrees), water sprout growth is stimulated. These shoots should be pruned. Narrower crotches (about 30 degrees) slightly suppress terminal growth while increasing the number and length of side branches. Moderate angles (45 to 60 degrees) have an intermediate effect between the narrow and wide crotches. Whenever physical support is needed, tying is often involved. Tree trunks should not be too tightly restrained to a support but allowed some room to move in the wind. Wind movement allows the trunk to develop strength in its wood for additional support. Wide-angled crotches encourage the development of strength in the wood of the limb and thereby improve its resistance to damage from the weight of fruits or snow accumulation. Narrow-angled crotches are prone to damage because of weakness resulting from the formation of *bark inclusions* (figure 15-28).

15.13.1 TRAINING SYSTEMS FOR FRUIT TREES

Pruning and training of fruit-bearing trees are done together to increase productivity and ease harvesting and other operations. These practices start early in the growth of plants, a strategy that prevents young trees from overbearing too early, an event that adversely affects plant structure and productivity later in life.

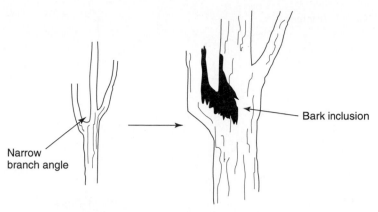

Narrow branch angle

Bark inclusion

FIGURE 15-28 Development of bark inclusion occurs when a branch angle is too narrow. It results in the development of weak crotches.

Not all fruit trees require regular pruning. Some simply cannot be pruned because of their very tall adult height. Deciduous fruit trees shed their leaves and are dormant during the winter season. Pruning objectives include establishing a strong, straight trunk and well-positioned primary branches. The manner of pruning also takes into account the age of the plants, fruiting habits, and how fruits are harvested. In species such as citrus, almond, and peach that are harvested mechanically by trunk shakers, the trunk should be well developed and the first *scaffold branch* located high enough to allow the machines to be used without interference. Another important strategy is to prevent one branch or limb from developing directly over another. This practice guards against shading of lower limbs. Since optimum light distribution is key to high productivity, keeping higher limbs upright permits light to reach the lower limbs (which are more stretched out). Fruit trees should be pruned lightly. If young trees are heavily pruned, their growth is retarded, making them dwarfish or delaying bearing by one to several years. Three training and pruning strategies are employed in tree production—central leader, modified central leader, and vase (multiple leader or open center).

Central Leader

Dwarf and semidwarf fruit trees are most suited for the central leader system, which produces narrow and conical-shaped trees with tiers of scaffold branches. Apples and sweet cherries are naturally adapted to this training system. The acute angle increases productivity while easing harvesting from the inner parts of the plant. Training starts with identifying the first tier of scaffold branches in the first dormant season and, where necessary, spreading them out to create moderate (45- to 60-degree) crotches. The branches are then headed to stimulate branching (figure 15-29). During the next dormant season, the activities are repeated, making sure to head at a height that leaves ample space (2 to 3 feet or 0.61 to 0.91 meter) according to tree size) between the lower tiers of scaffolds and the ones to be developed above. About a third of the terminal shoot of the branches are pruned each year. Three tiers of scaffolds are adequate for 12-foot-tall (3.6-meter-tall) trees. To prevent the central leader from bending and ceasing terminal growth, fruiting on the upper third should be discourage. Water sprouts should be pruned when they develop. Nut trees are also trained and pruned by this method. In later years, pruning should be used to keep the central leader at the desired height. Suckers, broken branches, deadwood, and low branches should be removed. The pyramidal shape of the tree should be maintained.

Modified Central Leader

The modified central leader training system proceeds like the central leader system in the early stages to allow for the formation of strong scaffold branches. After the final tier of scaffolds is formed, the central leader is removed to create an open center (figure 15-30).

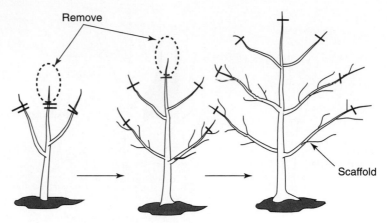

FIGURE 15-29 Training and pruning a central leader.

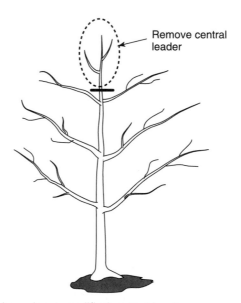

FIGURE 15-30 Training and pruning a modified central leader.

Vase (Multiple Leader or Open Center)

Training plants according to the vase or open center system starts with heading when the plant is about 18 to 30 inches (0.5 to 0.8 meters) above the ground (figure 15-31). Heading stimulates new growth in the formative training period. In the first dormant season, three to four branches are selected to form the permanent scaffold branches and the remainder pruned. These branches should be adequately spaced (6 to 8 feet or 1.8 to 2.4 meters) to allow room for growth. If too many branches are retained, the productivity of the tree will be reduced significantly since fruits are not borne on scaffold branches but rather on smaller secondary and tertiary branches. Further, too many primary branches will crowd the center of the plant and cause fruiting to occur near the tips of branches. A problem with the vase training system is the tendency of naturally upright trees to close up the open center with the passage of time. Such plants bend outward to an undesirable extent under the load of fruits. In the second season, the primary scaffolds are usually pruned to varying lengths. Scaffolds are developed on the primary branches in a similar manner in the second dormant season. The vase system is very popular in orchards.

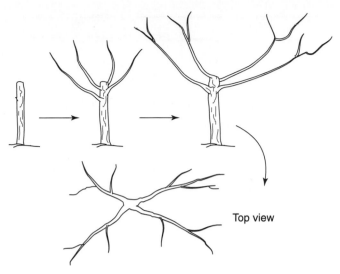

FIGURE 15-31   Vase or open center training and pruning.

15.13.2   TRELLIS TRAINING SYSTEM

Trellis training involves tying limbs of plants to wires strung between posts or constructed against a wall or fence. Materials for tying range from wires to masking tape and should be of a material that will not damage the limbs. Common trellis training systems are described in the following sections.

Espalier

The goal of espalier training is to create a work of art confined to a specific space; as such, it requires heavy pruning to control shoots on top of horizontal limbs. The tree forms produced are usually two-dimensional since frequently a wall or fence provides the background of the trellis. All designs require some support system, at least in the formative training. When forming espaliers, one should take into account the species and the vigor of the plant. The space between horizontal limbs, for example, is wider for peach trees (20 to 24 inches or 50.8 to 61 centimeters) than for apple trees (12 inches or 30.5 centimeters). About three tiers of limbs are adequate for small plants, but vigorous plants can have more. All bending should be done when the limbs are still flexible. A balance should be maintained between limbs on either side of the central axis by keeping opposite limbs at each tier equal in length. The limbs should be tied snugly to the support or trellis, as already mentioned. Espaliers may be formal or informal in design.

> **Espalier**
> A training system in which the tree or shrub is made to grow flat against a wall or on a trellis, often in formal branch patterns.

Horizontal Espalier   The horizontal espalier training system consists of a set of horizontal wire trellises attached to walls, fences, or posts. When starting from a *whip* (seedling with a single slender stem), it is cut just below the first wire. This practice causes new shoots to develop from lateral buds. Two lateral shoots of equal vigor are selected and trained on stakes tied diagonally to the wires (figure 15-32). These stakes are later (in summer) lowered to the horizontal wires and tied. Stakes may not be necessary. Horizontal limbs are developed by first heading the plant just below each wire. Growth toward the wall should be pruned. Horizontal espaliers produce uneven vigor in the plant. The whole process is repeated to form the next tier of lateral branches. The ends of the lower scaffold branches usually lose vigor.

Once established, side shoots will be produced, first on the lower limbs. These shoots are pruned in summer to form fruiting spurs in the next season. As fruiting spurs increase in number over the years, they should be thinned out to avoid overcrowding and reduced plant vigor.

Palmette   The palmette training system consists of several designs—*baldessari, v-form, verrier, oblique,* and *candelabra.* This system is a variation of the espalier system whereby plants are trained at about a 40-degree angle instead of having horizontal branches. The can-

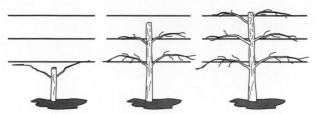

FIGURE 15-32 Training and pruning of a horizontal espalier.

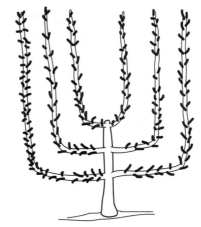

FIGURE 15-33 Training and pruning of a candelabra palmette.

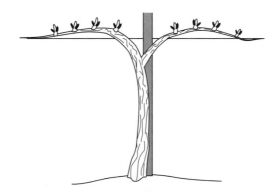

FIGURE 15-34 Training and pruning of a cordon with a spur.

delabra palmette training system uses a lattice framework consisting of horizontal and vertical arms to create balanced and attractive trees (figure 15-33).

Cordon

The cordon training system requires the use of wire, similar to a horizontal espalier system. A single main stem is tied at an oblique angle to the wires. Cordons perform best if the angle is oblique. Laterals, which grow from the main system, are pruned to form fruit-bearing spurs (figure 15-34). Double cordons (U) and four-armed Kniffen are also used in training plants (figure 15-35).

FIGURE 15-35 A four-armed Kniffen training system.

15.13.3 PRUNING EVERGREEN BROADLEAF FRUIT TREES

Broadleaf evergreen fruit trees are pruned only lightly, especially once established. Species such as citrus and avocado are rarely pruned. Light pruning may be done to control plant height, remove deadwood, and induce fresh growth, as in species such as lemon, coffee, olive, and mango. Citrus and lemon fruits are heavy, and hence fruiting should be encouraged on strong branches to prevent breakage. The productivity of citrus trees declines with age, requiring that trees be rejuvenated by topping and hedging to remove old and weak limbs.

15.14: COMMON TREE PROBLEMS

As trees and shrubs grow and age, they develop characteristics that reduce their visual appeal or pose problems in the general environment. Many of these problems can be corrected by pruning. The major ones include the following:

1. *Excessive height.* Ornamental trees in open space are normally allowed to grow freely to attain maximum height. Excessive tree height becomes a problem if it interferes with utility lines or overwhelms structures such as buildings. When this happens, the height of the tree may be reduced by *drop crotching* (cutting a main branch on the leader back to a lower crotch) (figure 15-36).

2. *Excessive spread.* When large branches located high on the tree spread excessively, they become prone to damage by wind. The condition may also cause the tree to be deformed and lose its visual appeal. When branches spread excessively, limbs from the highest and outermost parts should be pruned.

3. *Low-hanging limbs.* Low-hanging limbs pose clearance problems for humans and vehicles. It is desirable to establish clearance early in the growth of the plant.

4. *Deadwood.* When deadwood occurs, it should be removed to prevent disease infestation and spread. Deadwood also detracts from the beauty of a tree.

5. *Overcrowding of trees.* As trees grow and mature, they take up more space in the landscape. That is why it is critical to know the mature characteristics of trees before installing them in the landscape. Overcrowding can reduce the aesthetic value of plants considerably. It also causes trees to compete for light and thereby grow excessively tall. Their natural shapes and forms are often ruined. The remedy to overcrowding is complete removal of trees or pruning of limbs.

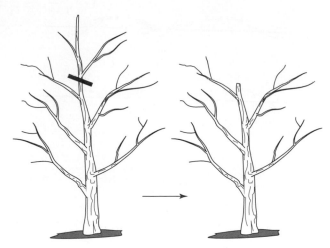

FIGURE 15-36 Drop crotching performed to lower the height of a tree.

6. *Forked trunk.* A forked trunk stands the danger of splitting in strong winds to produce unsightly results. Unless there is an overwhelming advantage to maintaining two leaders, landscape trees should be trained early to have one trunk and well-spaced scaffold branches. Only a few (four to six) scaffold branches should be maintained. All other branches arising directly from the trunk should be removed.

15.15: TRAINING AND PRUNING SMALL FRUIT TREES

Small fruit trees may be trained and pruned to be standard sizes, thereby bearing fruits at a higher level for easy picking. Small fruit trees require some pruning to bear quality, large fruits and have high yield. Two groups of small fruit trees may be identified in terms of pruning and training needs—cane and bush fruits.

15.15.1 CANE FRUITS

Cane fruits are temperate fruits and include raspberry, blackberry, and hybrid berries (e.g., boysenberry, sunberry, and tayberry). They bear their fruits on long canes. These cane fruits (sometimes called *bramble fruits*) are pruned in summer or winter to remove all of the canes that fruited in the previous season and any unwanted canes and suckers.

> **Brambles**
> The collective name for the fruits in the genus Rubus.

An important consideration as in all pruning operations is knowing where fruits will be borne on a plant. Some species, including many brambles, bear their fruits on canes produced in the previous season. Pruning removes all of the canes that fruited in the previous season. Pruning is also done in summer or winter when the plant is dormant. Also, weak canes and broken branches are cut off. Under such circumstances, plants must be pruned judiciously such that enough buds are left from the current season's growth for production in the next season.

Cane fruits can be trained on trellises or on a fence or wired wall. Fruiting and new canes occur simultaneously. Once harvested, the fruited canes are removed so that the new canes can be positioned for fruiting in the next season (figure 15-37). The goal of pruning is to ensure that year-old canes are in position each season. Pruning of individual species differs slightly.

15.15.2 BUSH FRUITS

Popular bush fruits include currant (black, red, and white), gooseberry, and blueberry. The goal of pruning these crops is to remove older and less productive wood to allow new and more vigorous shoots to grow (figure 15-38). Horizontally growing wood is removed and

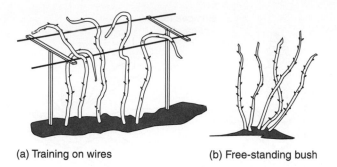

(a) Training on wires   (b) Free-standing bush

FIGURE 15-37 Training and pruning small fruits: cane may be (a) trained on wires or (b) cut back to be freestanding in certain species.

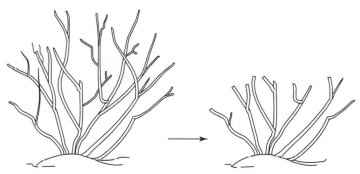

FIGURE 15-38 Pruning bushes involves both thinning cuts and heading back.

the bush thinned to avoid overcrowding and improve circulation of air. This aeration decreases disease incidence. Black currant *(Ribes nigrum)* is produced as a "stooled" bush whereby the old wood is cut back close to the ground to allow a set of new shoots to develop. The highbush blueberry *(Vaccinium corymbossum)* can remain productive for a long time, and thus removing fruited wood is not necessary for several years.

15.16: PRUNING ORNAMENTAL PLANTS

15.16.1 PRUNING CONIFERS

Conifers are cone-bearing plants. They may be low growing, as in prostrate junipers, and useful as ground covers. Some conifers are used as hedge plants.

Ground Covers and Hedges

Conifer hedges require pruning to maintain the desired shape, while ground covers seldom need it. Conifers have different branching habits. Some plants branch only once a year (as in pine, spruce, and fir), starting with the beginning of the season's growth. This growth pattern results in circular (whorl) growth of branches at the growing tip. These plants lack latent buds on old wood and as such will not regrow when cut back severely. The *candles* (new growth) on the whorls may be shortened or pinched back while the needles are small to control growth (figure 15-39). Junipers, on the other hand, branch as growth proceeds and produce fresh growth at the point of pruning. Junipers may be thinned or sheared.

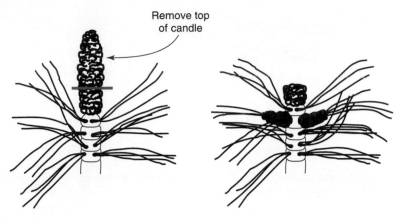

FIGURE 15-39 Pruning conifers.

Pruning Coniferous Trees

Coniferous trees commonly develop a single dominant central axis with an overall narrow pyramidal shape. This shape makes it impossible to shorten a mature plant without destroying its natural look. Some conifers such as junipers and digger pines have multiple central leaders and are more tolerant of height reduction. Conifers are amenable to pruning to fit symmetrical shapes. When pruning conifers, one should always cut back to visible buds and leave no stocks. Conifers tend to grow low branches. Pruning of some of these branches may be necessary to provide for clearance under the tree. Conifers are best pruned in spring or early summer, when most (e.g., pine, cypress, cedar, and spruce) experience rapid growth. Some new growths (candles) may be removed or cut in half.

Pruning Christmas Trees

Christmas trees are characterized by a conical shape. To attain this desirable shape, growers maintain a schedule of shearing to produce plants that are symmetrical and of uniform canopy and density and with uniform spacing between branches. Shearing of Christmas trees may begin in the third or fourth growing season, depending on the species. It is important to shear pines in their active growing periods or no new buds will form near cuts and dieback may occur. In the case of species such as fir and spruce, shearing causes shoot growth to cease while buds form. These species may be sheared after new growth begins. Pines tend to lose their shape readily (they grow less tight) and hence require more frequent shearing than fir and spruce.

The density of the canopy depends on how the central leader is headed. The first heading is done when the pine is about 12 inches (30.5 centimeters) tall. This lead shoot is headed again in the next and subsequent seasons (up to four years) in the same manner. If the lead shoot is cut short, the tree will be denser and more attractive but attain harvest height after additional growing seasons. Whorls of branches develop in tiers as the tree grows. Side shoots should be headed to about half the length of the lead shoot. The sides are sheared to approximately 50 percent taper. To facilitate harvesting, handling, and installation of the tree for display, up to about 12 inches (30.5 centimeters) of the trunk from the ground is pruned to leave a clean base.

Pines characteristically have a conical or pyramidal shape. A significant difference between pines and broadleaf trees is that pines mostly have one axis that does not branch if cut back. Once a limb is removed, no replacement limb will be produced. When pruning, no stubs should be left. The limb should be cut back to a visible live bud. Christmas trees are *sheared* to obtain an attractive symmetrical and conical shape with uniformly dense foliage (figure 15-40). Pine, spruce, and fir produce circular growth of branches (whorls) and branch once a year. To obtain a compact plant, the tips of the branches may be pinched. Once formed, the internode between whorls is fixed and cannot be shortened.

FIGURE 15-40 Shearing of Christmas trees. Shearing is designed to create a symmetrical shape with uniform distribution of branches and foliage throughout the plant. Plants are sheared to about 50 percent taper. The maximum width should be about 40 to 60 percent of the height of the plant.

15.16.2 PRUNING NONCONIFEROUS SHRUBS

Shrub problems in the landscape include the following:

1. *Plant overgrowth.* Shrubs may grow such that the canopy is too large. When this occurs, the size of the plant may be reduced by carefully removing the long side branches (thinning) or removing all large branches in the top of the shrub *(dehorning).* Thinning should be done such that the shape of the plant is not destroyed. Plants that sprout very easily from old wood are amenable to dehorning, which is a severe pruning technique.

2. *Canopy too dense.* Dense shrubs cannot receive light in the center of the canopy. This situation may be corrected by thinning spindly growths and the top of the shrub.

3. *Canopy too loose.* A loose canopy results from spindly, weak branches, which tend to sprawl. These branches should be removed or shortened.

Shrubs, unlike trees, are low growing in terms of height and have small stems. Several stems often arise from the ground and are bunched together. This close arrangement encourages upward growth as the response to competition for light. The foliage at the bottom part of the plant is shaded, leading to poor growth in those parts and bare stems. Shrubs are pruned by removing old stems from the ground level to open up the clump by thinning out.

Some smaller branches and twigs are also removed. Shrubs may be headed back slightly to encourage new growth in the right direction by making appropriate cuts. Overgrown shrubs may be pruned back drastically to rejuvenate them through production of fresh growth.

15.17: PRUNING HEDGES

Hedges are planted for several purposes, including to demarcate property boundaries, to hide utility areas, or to provide decorative borders. Hedge plants are shrubs that are planted very close together. The plants may be allowed to grow tall or dwarfed through frequent trimming. For rapid results, a fast-growing species may be selected, but slow-growing hedge plants live longer and produce dense hedges that require less frequent clipping.

There are two basic styles of hedges:

1. *Formal hedges.* Formal hedges are created by planting and shearing plants according to geometric shapes (figure 15-41). For best effect, formal hedge species should have

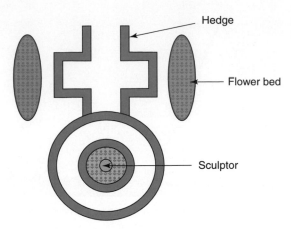

FIGURE 15-41 A formal hedge. The emphasis is on symmetry and details in shearing to keep the symmetry and cosmetic appearance.

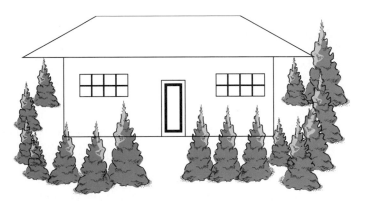

FIGURE 15-42 An informal hedge. The emphasis is more on privacy and less on symmetry.

a dense growth habit and tolerate close clipping. It is important that hedges be uniformly dense from top to bottom. Formal styles of hedges are more difficult to maintain. Hedges should be sheared on a regular basis. Because the shapes and sizes are monitored regularly, formal hedges require less space in the landscape.

2. *Informal hedges.* Plants may be allowed to grow freely according to their natural shapes, and pruned only to restrict size and maintain an attractive visual appearance (figure 15-42).

15.17.1 CLIPPING OR SHEARING OF HEDGES

Plants vary in their tolerance of shearing, density of growth, leaf size, growth rate, and adaptation. The best hedge plants should be upright, small leaved, and slow growing. For formal hedges, the species should be tolerant of frequent shearing (e.g., boxwood, holly, and privet). Flowering shrubs (e.g., camellia, oleander, and azalea) may be used as informal hedge plants. Conifers that sprout from old wood are good hedge plants. To have a good hedge, the plants should be shaped early. Frequent trimming of hedge plants in early seasons promotes the development of a dense hedge.

The shapes of columns of trimmed hedges are varied. Individual plants are not trimmed but rather the whole bunch is trimmed to blend together as a solid column. A very common yet improper hedge form is one that is wide and flat at the top, resulting eventually in thinning in the foliage at the bottom (figure 15-43). A columnar form that is as wide at the top as it is at the bottom is more desirable. The best shape is a pyramid-like form that is wider at the

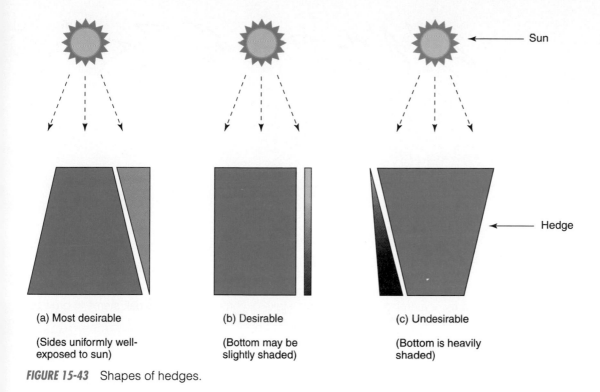

Sun

Hedge

(a) Most desirable

(Sides uniformly well-exposed to sun)

(b) Desirable

(Bottom may be slightly shaded)

(c) Undesirable

(Bottom is heavily shaded)

FIGURE 15-43 Shapes of hedges.

base than at the top. If straight hedges are desired, one may guide the trimming operation by stretching a string between stakes along the hedge.

15.17.2 RENOVATION OF A HEDGE

Hedges may last for many years provided the plants are fertilized and properly maintained. Renovation may involve severely cutting back an overgrown hedge. Species such as yew *(Taxus),* holly *(Ilex),* and hornbeam *(Carpinus)* tolerate such *hard pruning.* Whenever severe pruning is involved, it is best to spread out the operation over several seasons; that is, one should not reduce plant height and width at the same time. Even when reducing only the width of the hedge, it is best to start with one side of the hedge in one season and the other in the next season. Some kinds of damage may be repaired by removing the older material and replacing it with younger plants.

15.18: ORNAMENTAL STEMS

Apart from creating a work of art with the foliage of plants, the stems of certain shrubs may be twisted or woven together (figure 15-44). This training must be done while the plants are young and flexible. To create the single stem shape, the young leader of the shrub is trained around a strong and sturdy pole. All growths on the stem are then removed. When the plant develops adequate strength, the pole is removed. Similarly, the barley sugar stem involves training two stems simultaneously around a pole. A more fancy and eye-catching stem effect is produced by plaiting three young and flexible stems to form a braid. To obtain multiple stems, the central axis is cut back like a coppice (15.11) to produce new shoots. As the plants grow, the braided stems become bound together in a natural graft by fusing at all of the sites where contact is made between stems, as in approach grafting (9.8.1). A pole may be needed to provide temporary support until the stem is strong enough to be self-supporting. The foliage may then be trained and pruned into attractive shapes. Shrubs that can be manipulated in this way include boxwood *(Buxus),* bay *(Laurus nobilis),* and weeping fig *(Ficus benjamina).*

FIGURE 15-44 Woven and twisted stem designs. Shrubs may be trained to stand up as single stemmed trees by reinforcing the stem with a stake or combing several weak stems to grow as one strong stem through braiding.

15.19: TRAINING AND PRUNING CLIMBING PLANTS

Climbing plants are a versatile group of landscape plants that can be utilized in a variety of creative ways. They require physical support and training to control and manage their fast-growing shoots (figure 15-45). Climbers can be trained along walls and fences or to climb poles and other nearby plants. They are utilized in decorative topiary (15.12.4) and as free-standing specimen plants growing over a frame.

Climbing plants differ in how they attach themselves to physical supports, which affects the way in which they are pruned. Ivy has aerial roots (figure 2-38) by which it clings to supports. *Parthenocissus tricuspidata* uses adhesive pads that are touch activated to hold onto objects. Unfortunately, these climbing aids remain part of the physical support even after the climbing plant is dead, leaving unsightly marks. Plants such as climbing roses and bougainvillea utilize thorns to hook onto the support, and *Wisteria* simply twines around the support upon contact. Some plants, such as *Passiflora racemosa,* climb by coiling their tendrils around objects. Some species neither twine nor cling but simply scramble over objects by producing long shoots; once over the object, they cascade down.

Plants that climb by twining or tendrils can be grown near posts, pillars, fences, or walls provided these supports have objects or materials (such as netting or wires) to which these plants can anchor themselves. On the other hand, plants that cling to their support need flat surfaces such as fences, tree trunks, and walls or masonry. When trellises or other wires and fixings are provided for twiners, it is important that they not be attached directly to the wall or surface. Some room is needed for the plant to move to twine around the surface. Further, the space provides a means of aeration to reduce disease incidence.

After selecting the appropriate species and the area to plant it, the plant should be set at least 6 to 9 inches (15.2 to 22.9 centimeters) away from the base of the support. The growing shoot is led to the support by providing it with an angled stake on which to climb. Once on the support, it is allowed to grow without much interference in the first year. Thereafter, the shoot should be trained in the desired direction. Where a shoot is growing at a rapid pace without producing side shoots, it should be pruned to stimulate branching.

During the dormant season, deadwood should be removed. Any shoot that has overgrown the space allotted should be cut back. Excessive growth should be removed to avoid congestion.

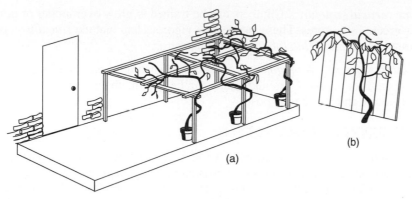

FIGURE 15-45 Climbers can be trained on physical structures such as (a) the frame of a patio or (b) a fence.

Training is important to guide the growth of the plant and to contain it within desirable boundaries. Vines that bloom on the previous season's wood should be pruned after they bloom in spring so that new growth will occur during the summer period to produce flower buds for the next spring. Those that bloom on the current season's growth should be pruned before growth starts in spring to stimulate growth later in spring for more flowers. It is important to contain vines, especially aggressive ones such as trumpet vines, so that they do not damage a home's roof and gutters. When vines are old and show deadwood, renewal pruning may be necessary.

15.19.1 USING CLIMBERS IN THE LANDSCAPE

Vines may be used in a variety of ways to enhance the landscape and for other functional purposes:

1. *Ground cover.* Vines may be used as a ground cover by allowing the plants to trail on the ground. Species that can be used in this way include ivy *(Hedera), Vitis, Akebia,* and *Ampelopsis.*

2. *Hedge plants.* Climbers with thick, woody stems, including climbing roses, bougainvillea, and *Hedera,* can be trained over appropriate structures such as tightly stretched wire to form a fence.

3. *Container grown.* Climbers can be grown in containers and trained over wire supports to form a variety of decorative shapes (figure 15-46). Climbers such as philodendrons and epipremnums that produce aerial roots can be used in creating moss poles.

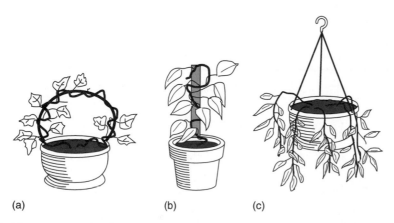

FIGURE 15-46 Vines may be trained (a) on metal frames, (b) on moss poles, or (c) in hanging baskets.

4. *Cover vertical structures.* Climbers can be trained to grow over arches or other vertically erected structures. These species are vigorous and include *Humulus lupulus* 'Aureus,' *Adlumia, Mina lobata,* and *Codonopsis.*

5. *Climbing a pergola.* A *pergola* is a structure consisting of pillars or posts linked by cross beams or arches. Climbing plants can be grown, one at each post, to climb over the structure and around the posts.

15.19.2 CLIMBERS AS STANDARDS

Climbers generally do not have enough wood to be self-supporting even when mature. However, some plants such as *Wisteria sinensis* and *Jasminum polyanthrum* can be trained to stand up as single-trunk, self-supporting plants. Most species trained as standards never achieve self-support and have to rely indefinitely on a support of some sort, usually metallic. Standard climbers are also used as border plants around patios.

SUMMARY

Perennial plants, especially, require periodic management in which parts of the plant are selectively clipped to control growth and remove unproductive and diseased parts. This activity, called pruning, is designed to improve and sustain quality plant products and aesthetic value. The four general purposes of pruning are phytosanitary, aesthetic, reproductive, and physiological. Trees and shrubs are pruned differently. Fruit trees are pruned to improve light penetration into the canopy and to remove unproductive and diseased limbs. Pruning strategies include central leader, modified central leader, and open center types. Plants may be thinned out, headed back, or renewed by pruning methods. Hedges and vines are pruned differently from trees. Sometimes the technique of training is combined with pruning to force plants to assume certain forms, called the espalier method of pruning. This plant art form reaches a high level in topiary, the art of training and pruning trees to assume geometric or other recognizable shapes.

REFERENCES AND SUGGESTED READING

American Horticultural Society. 1980. Pruning. Alexandria, Va.: American Horticultural Society.

Brickell, C., and D. Joyce. 1996. Pruning and training. Alexandria, Va.: American Horticultural Society.

Harris, R. W. 1992. Arboriculture: Integrated management of landscape trees, shrubs, and vines, 2d ed. Englwood Cliffs, N.J.: Prentice-Hall.

OUTCOMES ASSESSMENT

PART A

Please answer true (T) or false (F) for the following statements.

1. T F Only above-ground plant parts can be pruned.
2. T F When removing a tree limb, it is best to leave a short stump.
3. T F An outward slanting cut above a bud will cause the emerging branch to grow outward.
4. T F A slanted cut of 45 degrees is recommended when pruning.
5. T F Training of plants involves tying.

6. T F Species that flower in spring are usually pruned after flowering.
7. T F Lopping shears are used to prune small limbs.

PART B

Please answer the following questions.

1. List the four general purposes of pruning.

   _____ _____

   _____ _____

2. Training and pruning trees to assume geometric and readily recognizable shapes is called _____.

3. _____ is the pruning method of removing water sprouts.

4. Rejuvenation of citrus trees involves_____.

PART C

1. Describe how cuts should be made during pruning.
2. Describe the pruning method of heading back.
3. Describe the pruning method of central leader.
4. What pruning activities are done to accomplish the purpose of phytosanitation?
5. Describe the best shape for pruning a hedge.
6. List five pruning tools and describe their uses.

CHAPTER 16

Maintenance of the Landscape and Garden

PURPOSE

This chapter discusses how a landscape and home garden are maintained during growing and the off-season, the tools used, and how they are maintained.

EXPECTED OUTCOMES

After studying this chapter, the student should be able to

1. Discuss the regular care of plants in the landscape and garden.
2. Discuss seasonal care of landscape plants.
3. List landscape and garden maintenance tools and how they are used.
4. Describe how to maintain landscape and garden maintenance tools.

KEY TERMS

Bagging mower	Line and blade trimmer	Tillage tools
Edger	Mulching mower	Trickle irrigation
Irrigation tools	Sprinkler irrigation	

OVERVIEW

A landscape is installed to serve functional and aesthetic purposes. To be aesthetically pleasing, it should be regularly maintained. The home garden is on the property and must be kept clean. Further, a garden must be cared for properly to obtain quality produce and good harvest. As previously mentioned (chapter 12), one crucial piece of information a landscape designer needs to obtain from a customer is the level of maintenance desired. It makes no sense to install a landscape that cannot be maintained. Certain landscape designs are high maintenance and require many hours of work and or a substantial monetary investment to maintain the initial level of quality and elegance. However, a decent and relatively inexpensive landscape can be maintained at a low cost. To start with, modern technology makes it possible to install an automatic sprinkler system in the landscape to facilitate watering, which is one of the major chores in landscape maintenance. Other choices in landscape design require assistance from experts who understand the plants and their habits and other characteristics of importance to the design. For a low-maintenance landscape, the home owner should choose plants that require less of the customary chores associated with maintaining a landscape such as mowing, pest-control measures, and pruning.

In addition to automation of irrigation, the low-maintenance landscape may have a reduced lawn area by including more ground covers. To reduce the need for weeding, mulching could be used whenever possible to cover bare soil. A careful choice of cultivars enables those best adapted and resistant to local diseases and pests to be planted to eliminate the need for pest control. Lawns and annual flowering plants are high maintenance. Suitable shrubs may be selected and used instead of annual and perennial flowering species. Deciduous plants shed their leaves in fall, creating a seasonal chore of raking and collecting leaves. Evergreens may be substituted for deciduous plants. Whether low or high maintenance, every landscape needs care at some point. Some care is required routinely during the growing period. Other kinds of care are required on a seasonal basis. Landscape maintenance is imperative when the home is located in a rural or urban neighborhood. An unkept landscape not only stands out like a sore thumb, but the offending home owner is likely to hear about it from neighbors. Many of the tools described here are also used in the garden for similar operations.

16.1: REGULAR MAINTENANCE

The principal regular and general maintenance activities required by all landscapes and gardens are watering, weed control, fertilizing, pruning, and pest control, in that approximate order of regularity and importance. The order is approximate since landscape designs differ, some emphasizing annual flowers that need frequent watering and others emphasizing trees and shrubs. Further, watering is crucial in summer, while weeds are more of a problem in spring. Also, home owners may not wish to grow crops that require pruning, for example.

16.1.1 PRUNING

Pruning is a common landscape maintenance activity, especially where perennials are involved. Pruning as a maintenance activity is done according to a schedule or as needed. In chapter 15, the use of training and pruning to establish tree form, heading height, and other features is discussed. Once a form has been established, periodic pruning is required to maintain this form because the plant continues to grow and increase in height and size.

Scheduled pruning is especially critical in some fruit trees and other fruit-bearing plants where it is done after each fruiting cycle. However, in other cases, plant height and size must be controlled to contain a plant in the allotted space. Depending on the location of the plant, space may not be a problem, and thus the plant may be allowed to grow to its fullest potential size with little interference from the grower. Trees that grow along streets and especially those under power lines require pruning to keep them from making contact with the lines. Such pruning may be done on an annual basis after trees have reached a certain height.

Unscheduled maintenance is needed in the landscape when weather-related damage (e.g., from strong winds, lightning, or snow) or equipment damage (e.g., from lawn maintenance equipment such as mowers) occurs. Damage may also occur from vandalism. In such situations, branches may be broken and whole plants even uprooted in some instances. Pruning is thus required to remove broken branches and clean wounds for proper healing.

A topiary (15.11) requires regular maintenance to maintain its shape. An established topiary may require nothing more than shearing or trimming new growth. Other specialty pruning such as pollarding (15.11) requires scheduled maintenance. Hedges also require periodic trimming to keep them in shape. The frequency of trimming depends on the plant species and the type of hedge (formal or informal [15.16]).

Pruning for sanitation is needed to remove dead branches and blight from the landscape. Occasionally, plants that have been allowed to grow naturally may develop excessive branches or shoots. Such plants must be thinned out (15.5) to open up the plant canopy for aeration. Other plants may need heading back (15.5) to control growth or rejuvenation after a long period of neglect.

When vines are involved, wayward growth must be redirected and the plant restricted to the space allotted. Thus vines require occasional light pruning.

16.1.2 WATERING

Watering is a landscape maintenance activity that is frequently performed improperly. One of the causes is that for many plant growers watering simply means wetting the soil surface. Effective watering, however, requires that water be delivered in adequate amounts to the root zone of plants ("recharging" the root zone). It is important to note that the depth of rooting is influenced by moisture supply. Roots grow toward water, so if a plant is watered lightly, the roots stay near the soil surface as opposed to growing down when an area is soaked deeply. In fact, a little watering can be dangerous to most plants. In adult trees, watering should be limited to around the *drip line* where most of the roots occur (figure 16-1).

Factors Affecting Watering Frequency

The frequency of watering depends on a combination of several factors:

1. *Species.* Certain species such as cacti are drought tolerant, as are deep-rooted plants that are adapted to tropical and dry climates. Drought-tolerant cultivars of various horticultural plants have been bred by scientists. Annual bedding plants, including garden vegetables such as tomato, pepper, and cole crops (17.24), need to be watered frequently.

2. *Soil properties.* Sandy soils are well aerated, drain freely, and have lower moisture-retention capacities than loams or clays (3.3.1). When sandy soils are used for any type of plant production, watering is a major production input. Clay soils are more prone to waterlogging than sandy soils. The moisture-retention capacity of sandy soils may be improved by incorporating organic matter and using mulch (21.1.5).

3. *Stage in plant growth.* Frequent watering is needed during the period of plant establishment. In some cases, plants may require watering twice a day. Once established, watering frequency may be reduced according to the capacity of the plants to forage for water. Water is needed at critical times such as flowering and fruiting. Drought at these times can cause flower and fruit drop.

Drip Line
An imaginary circle that indicates the outermost boundary of the canopy of a tree; fertilizer is applied to holes dug around the drip line.

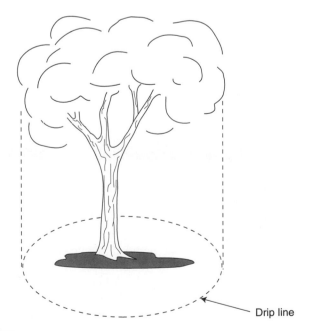

FIGURE 16-1 Drip line.

4. *Season.* Certain trees and shrubs can be solely rain fed during their cultivation. However, annual species and smaller plants need supplemental moisture to survive the summer. In summer, especially in areas with less than 40 inches of annual rainfall, lawns need to be irrigated more frequently, sometimes every other day or even daily if a luxuriant lawn is desired.

5. *Conservation practice.* Installation of a mulch reduces evaporation and increases soil water retention. Similarly, the presence of a ground cover (13.12) improves water retention. The concept and practice of xeriscaping (12.8) minimizes the use of water in the landscape.

Methods of Watering

Hand Watering   Plants can be hand watered by using a *watering can* when a few plants or a small area is involved and a small amount of water is needed (figure 16-2). Watering cans differ not only in capacity but also in length and size of spout. Some cans have no nozzle for sprinkling water, while others have different kinds of nozzles for delivering a variety of spray patterns and degrees of fineness. When large amounts of water are needed (e.g., for watering trees) a *garden hose* may be used. It is moved from plant to plant after leaving it to soak the ground for a period at each spot (figure 16-3).

Sprinklers   *Sprinkler irrigation* methods are popular for maintaining lawns and flower beds (16.2). The characteristic of this method that makes it conducive to watering densely planted areas is uniform wetting of the entire area of coverage. Sprinkling can be wasteful since it cannot usually be confined completely within the boundaries of the area to be watered. Further, a significant amount of water is lost through evaporation.

Drip   *Drip* or *trickle* irrigation (3.8.11) provides for spot application of water using a simple feeder line. Water is delivered at low pressure and in drops or trickles to plants by connecting microtubules from the main line to individual plants. Sometimes perforations may be made in a plastic conduit at spacing that matches plant spacing. This method, which eliminates the need for microtubules, is used for watering trees and shrubs. By applying water only to spots where it is needed, less water is lost through evaporation. Further, only the target plants receive moisture, consequently starving the weeds. It should be noted that microtubules can become clogged if the water supply contains salts and debris.

FIGURE 16-2 A watering can.

FIGURE 16-3   A garden hose.

Irrigation Kits

For the do-it-yourself grower, misting, watering, and drip irrigation kits can be purchased for use. These units are designed for application in greenhouses, nurseries, and general landscapes. A unit consists of the pipes and outlets (risers, sprinklers, or tubules) and all necessary connectors. Installation of irrigation systems in the landscape and garden is discussed later in this chapter.

Conserving Soil Moisture

Water loss in the landscape during irrigation can be minimized by observing several simple practices:

1. Water in the early morning when the temperature is lowest to reduce evaporative loss.
2. To avoid runoff, apply water at the rate of soil infiltration.
3. Improve the water-retention capacity of soil by incorporating organic matter as needed.
4. Do not overwater. Water that cannot be used by plants is wasted.
5. Adjust timers on automatic sprinklers on a seasonal basis to avoid use when water is not needed.
6. Wetting of soil should be confined to the diameter of the plant canopy.
7. Use a mulch, whenever appropriate, to reduce evaporation.
8. Apply the concept of hydrozoning in designing the landscape.

16.1.3 FERTILIZING

Adequate nutrition is required for good growth and resistance to diseases and pests. Starter applications assist in the quick establishment of new plants. However, once established, many plants respond to some kind of fertilization. Fertilizers are needed as active growth begins (in spring) and also to sustain vegetative and reproductive growth. Water-soluble fertilizers may be applied through the irrigation water, and dry materials may be surface applied (as top-dressing or sidedressing).

Trees are less frequently fertilized in the landscape. If needed, trees may be given moderate amounts of nitrogen fertilizer (e.g., 1/2 to 1 cup of ammonium sulfate per tree, depending on size). Shrubs are more frequently fertilized with complete fertilizers containing more nitrogen (e.g., 15:5:5) at a similar rate as for trees. In the case of ground covers, 2 to 4 pounds per 1,000 square feet (0.9 to 1.8 kilograms per 90 square meters) per year will suffice.

16.1.4 WEED CONTROL

As much as possible, herbicides should be avoided in the landscape, since people play on lawns and handle plants around the home. Cultural methods of weed control are preferred around the home. In the first place, weeds should not be allowed to overgrow and set seed before they are controlled. A hoe may be used to remove certain weeds, while others may be pulled. Cultivation with a hoe not only controls weeds but also improves soil structure. Ground cover is effective for suppressing weeds and may be used to cover bare soil. Mulching (plastic, organic, or rock) may be used for decorative purposes, while also serving as an effective ground cover to control weeds and conserve soil moisture. Mulching may be used judiciously to create a low-maintenance landscape. In terms of mulching, synthetic mulches such as plastics and landscape fabrics are the most effective for weed control in the landscape. The major drawback of these materials is that they are unattractive; however, they can be covered with a layer of decorative rock or organic mulch. Table 16.1 lists a few examples of weed situations and their chemical control.

16.1.5 DISEASE AND INSECT PEST CONTROL

Landscape plants are attacked by a variety of insects (including mites and sucking insects) and diseases (including rusts and fungal types). In most cases, these pest attacks are not sig-

Table 16.1 Selected Weed Problems in the Landscape and Their Management

Problem	Example of Herbicide and Its Action
Localized weeds	Amitrol: This nonselective herbicide is effective on young, growing weeds
Annual grass weeds in flower bed	Eptam (EPTC): Not effective against established grass
Weeds around woody ornamentals	Glyphosate (Roundup): Effective against actively growing weeds
Grass weeds	Fusilade: Effective against most actively growing grasses

nificant, except for the fact that blemishes distract from the aesthetic value of ornamental plants. Table 7.5 (chapter 7) describes the characteristics and control measures against some of these pests.

16.2: SELECTING AN IRRIGATION SYSTEM FOR THE LANDSCAPE

To make the proper choice of an irrigation system for the landscape or garden, a variety of factors should be considered including the size of the area, the type of plants growing there (e.g., trees, vegetable crops, or lawn), the rainfall pattern in the region, the cost of the system, the source of water, the topography, and the view of the area (public or private side of the house).

16.2.1 SOIL MOISTURE DETERMINATION

To determine how much moisture the site receives during the season, one may install a *rain gauge* in an open area in the landscape if there is enough space. One may also obtain such information from the local or national weather service publications. Soil moisture can be measured by using various instruments. A soil *moisture meter,* or *tensiometer,* may be used to measure the amount of water in a given soil at a particular time. The tensiometer is an instrument that measures the force with which moisture is held to soil particles. The higher the tension, the less moisture available to plants. Generally, if a site has not received any rainfall after about 10 days, the soil moisture level is low. Sandy soils may reach this status much sooner than a soil with high clay content or one with high moisture-holding capacity.

> *Rain Gauge*
> *An instrument installed in the ground in an open space for measuring the amount of rainfall in an area.*

16.2.2 BEST TIME OF DAY TO IRRIGATE

The critical considerations in watering plants in the landscape or garden are to deliver adequate moisture to the root zone, reduce waste, and avoid the persistence of a microclimate conducive to disease organisms. Water should be delivered at a rate at which it can infiltrate the soil and for a long enough time to achieve deep wetting of the root zone. Such watering can happen any time of day. In terms of waste, certain times of day, especially around noon, provide the greatest opportunity for water loss from the soil and other surfaces due to *evapotranspiration* (see next section). Contrary to popular belief, plant leaves will not be scalded when they are watered at high noon. Rather, moisture loss is very high at this time of day. The other important consideration is avoiding the persistence of high humidity and high moisture, which predisposes plants to diseases. Watering plants such that there will be time for leaves to dry before nightfall reduces disease incidence. Most landscapes are watered in the early part of day for the reasons given.

16.2.3 EVAPOTRANSPIRATION AND ITS EFFECT ON IRRIGATION

Moisture in the root zone of the soil is lost through plants by the process of transpiration (4.3.3) through plant surfaces. Soil moisture is also lost from the soil when water vapor is lost by evaporation from the soil surface. These two processes produce a combined effect, called evapotranspiration, that is responsible for most of the water removal from the soil. The efficiency of irrigation is dependent on evapotranspiration. To minimize this factor, water must be applied at the right time of day by the most water-efficient method and protected from rapid loss. Water applied to bare soil is rapidly lost by evaporation from the soil surface. Mulching, where practical, may be applied to reduce this moisture loss. Irrigation in the early morning allows water to seep into the soil and thereby reduces the loss that occurs at high noon. Methods of irrigation vary in water use efficiency; drip irrigation provides water in the most efficient manner (3.8.11).

16.2.4 SIZE OF AREA TO IRRIGATE

A small area can be readily irrigated using a watering can, since hand watering under such conditions is not tedious. It may require one or two trips to the tap to refill the can to complete the job. However, hand watering of a large area can be such an unpleasant chore that as the watering proceeds the quality of the work diminishes. Plants watered last may not receive sufficient water or be watered frequently enough. For a large area, an automatic system is desired.

16.2.5 SOURCE OF WATER

For home gardening, municipal water is the most frequent source of water for irrigation. The advantage of this source is that it is ready for use, provided the user has a means of connecting to it, eliminating the initial cost of providing an irrigation system. The source is reliable, since water is needed each day for a variety of uses in the home; the city ensures a continued supply. The disadvantages of this source of water are its high cost, the possibility of rationing in certain areas at times when plants need water the most, and the sensitivity of certain plants to municipal water treatments (chlorine and fluoride).

For a large landscape, a well may be an alternative source of water. Wells are practical where the water table is at a readily accessible level. Sinking a well is expensive, and the quality of water is affected by soil and rock materials and may yield hard water (containing rock mineral deposits). Underground water may be contaminated by soil surface pollutants. Three types of wells may be used for irrigation. *Driven wells* are relatively inexpensive and easy to install. The unit is small and thus unobtrusive in the landscape. However, it yields a limited amount of water and is prone to blockage from debris over time. Driven wells are practical in areas where the water table is high. A well point must be found, which is perhaps the most difficult aspect of the whole process. Once found, shafts are driven into the soil using a sledgehammer. Shafts may each be about 5 feet long (1.5 meters) and are driven one after another, connecting the sections as one proceeds (figure 16-4). *Dug wells* take up more space but also yield more water. If the site is properly selected, a dug well produces good-quality water year-round. A pump is needed to lift the water for use; dug wells may be lined with rock or concrete. The third type of well, a *drilled well,* is rather expensive and not a common option for home owners. It is installed by using large drilling equipment similar to that used for prospecting oil.

> **Driven Well**
> A method of drawing ground water by driving shafts into the water table.

16.2.6 SYSTEMS OF IRRIGATION

The four systems of irrigation for the landscape and garden may be operated on the surface or underground.

Surface Irrigation Systems

There are three general surface irrigation systems used in the landscape. They may be stationary or mobile and may involve the use of simple or sophisticated equipment. The types include overhead, mobile, and drip irrigation.

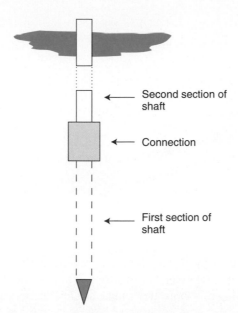

FIGURE 16-4   A driven well. Shafts are driven into the ground with a sledgehammer.

Second section of shaft

Connection

First section of shaft

Overhead Irrigation   Overhead irrigation systems are easy to install and do not require trenches to be dug in the landscape. Water is usually moved under high pressure through a simple or more elaborate network of pipes. For home gardens and landscapes, the pressure used is from the municipal water source. Municipal water is delivered at very high pressure (more than 80 psi), which can damage equipment unless it is reduced by installing pressure-reducing valves. A network of pipes is installed on-site and fitted with sprinkler heads to deliver a spray of water to the area. The pipe installation is often obtrusive, especially if used on the public side of the house. This system can be used in the garden and the landscape to water flower beds but is less suited for use on lawns. The network of pipes is more suited for use in the garden on the private side of the house where it can be left in place as long as needed.

Mobile Irrigation   A mobile irrigation system is designed to be used temporarily in a location and then moved to another area or removed for storage. It may be as simple as a garden hose sprinkler (figure 16-5). There are different designs of this kind of sprinkler, which is easy to use and adapted to irrigating lawns, vegetable gardens, and flower beds. Their use is limited by the length of the hose to which the sprinkler is attached. For irrigating a large area, a *drum-mounted hose* (figure 16-6) may be used. A disadvantage of this system is that it must be set up for use and then removed afterward or moved periodically until the area is completely irrigated. Certain models allow the sprinkler to move *(moving sprinklers)* automatically across the landscape, thus eliminating the need for human intervention. A *perforated hose (porous hose)* may also be used to distribute water to plants in the landscape.

Drip Irrigation   Drip irrigation or microirrigation is the most efficient irrigation system in terms of water used (3.8.11). Water is applied in trickles to the spots where it is needed. This system is especially suitable for watering trees but is not appropriate for watering lawns. It is relatively easy to install and may be as simple as a hose with holes at specific points with or without adapters for dispensing water.

Subsurface or Underground Irrigation Systems

Underground irrigation systems are suitable for watering lawns and also trees in the landscape (figure 16-7). The pipes are buried in the ground permanently, making this system unobtrusive and usable on the public side of a house. This system is readily amenable to automation. Initial

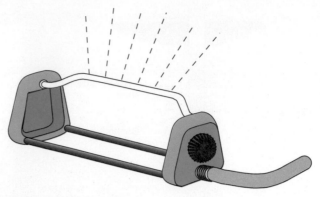

FIGURE 16-5 An oscillating garden sprinkler.

FIGURE 16-6 A drum-mounted water hose system.

installation costs may be high, depending on the particular unit selected. It may not be convenient or practical to install a subsurface irrigation system in a garden that is tilled frequently. Installing it under such circumstances requires that pipes be buried under the alleys between sections of plots in the garden. In areas where the ground is prone to freezing, the use of the system may be problematic in winter. This system also depends on high water pressure to operate.

It is important that an automatic lawn irrigation system be carefully designed for efficient and effective high water use. The topography of the area, soil water-holding capacity, and plant needs must be considered in the design. One factor of paramount importance is the *precipitation rate,* the measure of the amount of rainfall received by the area within one hour. A high precipitation rate is considered to occur when more than 1 inch (2.5 centimeters) of rain falls within an hour; a low rate is less than 1 inch per hour. An area of predominantly sandy soil has a high soil infiltration rate and thus requires a high precipitation rate to satisfy plants' water needs. Using this information, one can select a system that will not be overworked (because of use for an excessively long period at one time).

Sprinkler heads used in subirrigation systems may be divided into three groups according to precipitation rates. Low-precipitation-rate sprinklers are recommended for areas that are prone to erosion such as slopes; high-precipitation-rate sprinklers are used where the

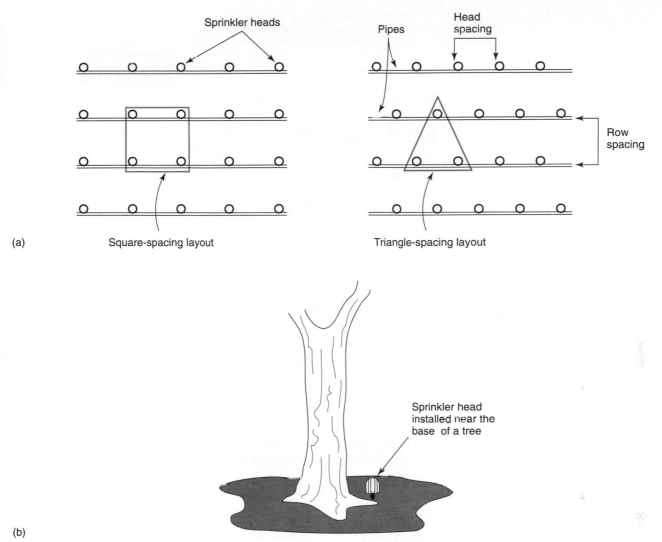

(a)

Sprinkler heads

Pipes

Head spacing

Square-spacing layout

Triangle-spacing layout

Row spacing

(b)

Sprinkler head installed near the base of a tree

FIGURE 16-7 Layout of the sprinklers in either the square or rectangular design.

soil can tolerate a large amount of water in a short period. Moderate-precipitation-rate sprinklers are run for a longer time and can have variable output. They are located in areas that are average in infiltration rate. It should be mentioned that although a home owner can self-install an underground irrigation system, certain decisions are best made by a professional, unless one has sufficient knowledge about the system. For example, one should know the *running time* for the system, a calculation based on the precipitation rates, which in turn is dependent on the infiltration rate of the soil in the area. Further, one should determine where to locate the sprinklers in the landscape. A sprinkler system does not utilize identical sprinkler heads throughout the landscape, which are varied according to site characteristics. The general rule is that the precipitation rate of a particular system is inversely related to the head spacing (spacing between sprinkler heads on a lateral) or row spacing (between lateral pipes). For best results, a design should ensure even or balanced water coverage. Uniform coverage requires that sprinkler heads be properly matched; failure to do so will result in dry spots in the lawn (figure 16-8).

There are a variety of sprinkler heads to select from to suit the needs of the landscape. The heads vary in versatility, cost, design, and other features. For areas where high precipitation is needed, one may choose a *fixed-spray sprinkler,* a *bubbler,* or a *microsprayer.* Bubblers are able to soak the soil deeply in a small area. In areas where low to medium precipitation is needed, the choice can be made from *rotating-stream heads, stream-spray heads, impact-rotor heads, gear-rotor heads,* and *pop-up heads.* Pop-up sprinklers are very unobtrusive in the

> **Nozzle**
> A device through which a spray is delivered; it is designed to deliver a spray in a certain pattern, rate, and uniformity.

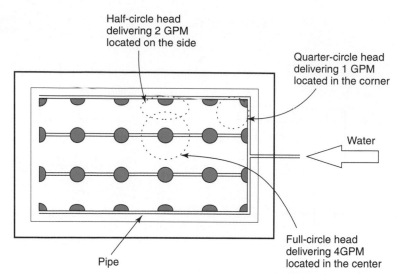

Half-circle head delivering 2 GPM located on the side

Quarter-circle head delivering 1 GPM located in the corner

Water

Full-circle head delivering 4GPM located in the center

Pipe

FIGURE 16-8 Sprinklers in an irrigation network should be arranged for balanced coverage of the area to be watered. GPM = gallons per minute.

landscape. They pop up under pressure when the system is turned on and sink back into their shells out of view after use. Some of these heads can be installed at angles to enable effective and efficient watering of slopes. Some are designed to water a variety of plant arrangements, including low plants in the foreground and tall plants in the background.

An irrigation system is not complete without a control panel and valves. This aspect of the system varies widely, and the choice is best left to professionals. The purpose of this section, then, is to make the reader knowledgeable about what an underground system consists of in order to be able to participate in its design and supervise its installation to a reasonable degree.

16.2.7 PUMPS FOR IRRIGATION SYSTEMS

Landscape and garden irrigation systems depend on pressure to move and distribute water. If one is depending on a municipal water supply for irrigation, there is no need to obtain and install a pump for irrigation. A pump is invariably required to lift water from a low level (e.g., a well) and distribute it throughout the landscape in pipes. There are several types of pumps from which to choose, a decision that must be made with great care for best results. *Effluent pumps* (*utility* or *sump pumps*) are relatively inexpensive. They are designed for use in pumping water that contains no solids (e.g., pumping water from a basement after flooding or emptying a pond or pool). For irrigation purposes, this pump is adequate when the source of irrigation water is a shallow pond or well. A disadvantage with this pump is the need for an outside power supply. Also, the pump works at a constant, nonadjustable water pressure.

Single-pipe jet pumps (or *shallow-well jet pumps*) have adjustable water pressure capability. Designed for use in shallow wells of less than 25 feet (7.5 meters) deep, they are not submersible and are more expensive than effluent pumps. However, they are automatic and not restricted by power cables and the need for an outside power supply. To utilize water from a well that is deeper than 25 feet (up to about 125 feet), a *dual-pipe jet pump (deep-well jet pump)* is needed. This pump is not submersible and is less efficient than the more expensive *submersible deep-well pump,* which can pump water from a depth of about 500 feet (150 meters).

16.3: SEASONAL MAINTENANCE

As cold temperatures arrive, plant growth processes decline. Deciduous plants shed their leaves and enter a dormant phase in their life cycles. However, evergreens continue to be active. Lawns turn brown and unattractive. Whether dormant or active, perennial landscape plants need some care to sustain them through the harsh winter conditions. This care is not only nutritional but also protective against physical and physiological damage.

16.3.1 NUTRITIONAL CARE

Watering

All plants, especially evergreens, need water, which should be provided below the frozen soil or frost line. Plants in containers need special attention year-round because they are usually planted in soilless mixes that do not hold water well.

Fertilizing

Trees, shrubs, and vines need fertilizing in the fall season to prepare them for winter. These plants should be fertilized before freezing weather arrives.

16.3.2 PROTECTION

A variety of plant protective measures are employed by horticulturists and home owners during the winter. For example, rose bushes are pruned and then immediately covered and wrapped with straw, plastic sheeting, or other suitable material throughout the winter. Perennial ground covers planted near the roadside need shielding from the corrosive action of salts used for deicing roads in winter. This protection is provided in the form of tarpaulin shields placed between the plants and the roads to protect the plants from the splashes caused by traffic (figure 16-9). The barks of certain tree species such as maple and elm crack under extremely cold conditions. To prevent cracking and the attack of rodents, tree trunks may be girdled with aluminum sheets or paper (figure 16-10).

16.3.3 WINTER DAMAGE

The harsh conditions of winter take their toll on plants. Strong winds cause branches of plants to break. Although some plants recover well, others do not. In evergreens such as jasmine and albelia, the damaged plant tops should be removed after new growth has appeared at the base. In shrubs, all deadwood should be removed. With fertilization, these plants will grow again after pruning.

16.3.4 SPECIAL CARE FOR ANNUAL BEDDING PLANTS

In addition to the care already mentioned, annual flowering plants can be improved by pinching terminal buds of herbaceous plants to make them develop more attractive and fuller forms and shapes. The flowering period of plants may be prolonged by removing faded blossoms.

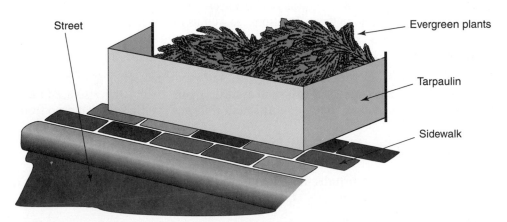

FIGURE 16-9 A tarpaulin shield may be installed to protect evergreens planted near the street from the splashing of salty water from melting snow.

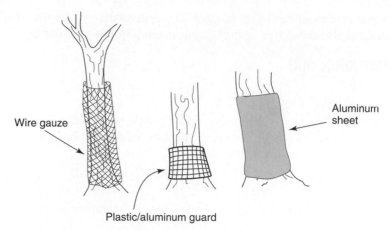

Wire gauze

Plastic/aluminum guard

Aluminum sheet

FIGURE 16-10 Tree trunks may be protected by installing a wire mesh or aluminum sheet guard.

16.4: TOOLS FOR THE LANDSCAPE

Landscape maintenance tools may be hand or power operated. They may be small enough to be handheld or easily carried around. Some tools are multipurpose, and others are designed for specific uses. Most landscape tools are within the range of affordability of most home owners. The most common tools and their uses are described in the following sections. They have been placed in certain categories according to primary use but can also be used for other appropriate functions.

16.4.1 LAWN MAINTENANCE TOOLS

Lawn Mower

Lawn mowers are power tools (a few are mechanical) and vary in size, power rating, technical design, and features. Mowers and mowing are further discussed in chapter 14 (14.6.2).

Lawn Edger

Lawns look more attractive when the edges have been trimmed after mowing. A tool called the edger is used to trim the grass along the edge of the sidewalk (14.6.2).

Line and Blade Trimmer

Regular lawn mowers have some limitations in maneuverability and as such are unable to completely clear areas around tree trunks and under low-hanging branches. The line and blade trimmer is a tool used in these situations, as well as for other trimming jobs (14.6.2).

Backpack Blower

A backpack blower is used to blow clippings that fall on the pavement after mowing back onto the lawn or to be gathered and collected as trash (figure 16-11). It is also used for other general cleaning jobs such as removing trash from walkways and leaves that drop from deciduous trees in fall.

16.4.2 TILLAGE TOOLS

Tillage is any activity that disturbs the soil. Soil-working tools vary in size and design (figure 16-12):

1. *Hand trowel.* The hand trowel is used for transplanting seedlings and for digging small holes for other planting operations.

FIGURE 16-11 A portable motorized backpack blower.

(a) Hand rake; weeder; hand trowel

(b) Garden hoe; spade; rake; spading fork

FIGURE 16-12 Selected garden tools.

2. *Nursery spade.* The spade is a digging tool. It may be used in preparing planting beds and digging plants for transplanting.
3. *Spade fork.* The spading fork has sturdier tines and can be used for digging and turning the soil in the bed and the compost heap. It is also used for moving refuse.
4. *Garden hoe.* A hoe is used for shallow cultivation and for weeding. It can also be used for making rows for sowing seeds.
5. *Cultivator.* The cultivator is used to stir soil in beds.
6. *Weeder.* A weeder is a tool used to individually remove unwanted plants.
7. *Tillers.* Tillers are used in preparing the land or plot for planting. When a large bed or plot is to be tilled, motorized tillers may be used (figure 16-13).

FIGURE 16-13 A garden tiller.

16.4.3 IRRIGATION TOOLS

Watering Can

Watering cans can be made of metal or plastic. The spout may be short and stout or long and slender and may be capped with a nozzle for either fine and gentle spray or coarse and harsh spray. This method of watering is adequate when only a few plants are to be watered and only small amounts of water are needed.

Garden Hose

Hoses are used for a continuous supply of water to plants. They vary in length and quality of material. Low-quality material makes the hose prone to kinking. Hoses are capped with a variety of nozzles (e.g., pistol nozzle for a focused jet stream of water and fan nozzle for a gentle spray). Other devices may be attached to the ends of hoses for certain applications such as *fertigation* (administering fertilizers through irrigation water).

Garden Sprinklers

Garden sprinklers vary in design. The sprinkler unit is connected to a hose, which in turn is connected to a tap. The oscillating sprinkler design of an overhead sprinkler is convenient for watering a wide area such as a lawn. Where plants in the landscape are tall, sprinkler risers may be used.

16.4.4 RAKES

A variety of rakes are shown in figure 16-14. They include the following:

1. *Garden rake.* A garden rake may be used for a variety of purposes in the landscape or garden. In tillage operations, it is used as a secondary tillage tool to produce a fine tilth for a seedbed. It may also be used for removing a thin layer of trash such as thatch in a lawn.

FIGURE 16-14   Assortment of rakes.

2.   *Leaf rake.* Structurally, a leaf rake is weaker and lighter than the garden rake. It is used for light jobs such as gathering dry leaves from a lawn.

16.4.5 PRUNING TOOLS

The basic pruning tools, which are described in more detail in chapter 15, include the following:

1.   *Hand pruners or shears.* Hand pruners are designed for cutting small branches of less than 3/4 inch (1.9 centimeters) in diameter.

2.   *Lopping shears.* These shears, also called loppers, are used for cutting larger branches than those cut by hand shears. They have longer handles.

3.   *Pruning saw.* Saws for pruning are designed to cut either dry or wet wood. Dry wood saws have a finer set of teeth. The pruning saw has several designs; one type can be folded. Pruning saws usually have curved and narrower blades than a regular carpenter's saw. Pole-mounted pruning saws may be used to reach branches on the top parts of plants.

Tools vary in quality in terms of materials used for their construction and durability of design. For example, one model of a hand trowel is designed such that the unit is one piece, while another has a wooden handle into which the metallic part is inserted. In the latter design, the blade has a tendency to become loose with use and age. A variety of safety features are incorporated into certain designs, which a customer should look for in purchasing a tool.

16.5: MAINTENANCE OF GARDEN TOOLS

The life of a tool depends on using it properly and only for appropriate tasks, as well as maintaining it correctly. Many of the tools described in this chapter require only simple maintenance. Motorized tools usually come with manufacturers' instructions for a maintenance schedule.

16.5.1 GENERAL MAINTENANCE

All tools gather dust after some time and need to be cleaned. After use, all tools should be cleaned, especially those that come in contact with soil or moisture. Water may be used to wash dirt away from tools, but they should be dried before storage to prevent rusting. A lawn mower should be cleaned after use by scraping and washing away trapped grass and dirt. Tools with wooden handles may be given a polish of boiled linseed oil to extend their life. A light spray of WD-40 oil may be made part of a general maintenance schedule to keep rust away from moving parts of garden equipment and to keep them in good condition. In addition to dusting, certain tools require additional specific maintenance operations, as described in the following sections.

16.5.2 POWER TOOL MAINTENANCE

The manufacturers' instructions for use and maintenance should be followed when dealing with power tools. Instructions may include changing the oil, filter, and spark plugs according to a schedule and greasing certain parts. These activities can be undertaken by the owner.

16.5.3 CUTTING TOOL MAINTENANCE

Cutting tools are designed for clean cuts. A dull blade on a lawn mower bruises the grass as it cuts, causing browning of the leaf blade edges. A dull cutting tool is also more difficult to use and may require the operator to apply more pressure to make a cut. After use, the blade should be cleaned and dried. Periodically, the blade should be sharpened with the appropriate tool and to the recommended extent. Most tools need not be razor sharp.

16.5.4 WATERING TOOL MAINTENANCE

Metallic watering cans should be drained before storage to prevent rusting. Garden hoses should be drained and coiled after use. Because prolonged exposure to sunlight is damaging to the hose, it should be stored in a shady place.

16.5.5 SEASONAL STORAGE OF TOOLS

Some tools require special long-term storage care. For example, lawn mowers should be stored without gas during the winter months when they are not used. Hoses should be disconnected from the taps in winter and stored indoors.

SUMMARY

There are two categories of landscape maintenance: regular and seasonal. Regular maintenance includes watering, weed control, fertilizing, pruning, and pest control. Watering can be accomplished by one of three basic methods: hand watering, sprinkler irrigation, or drip irrigation. To conserve moisture in the landscape, plants should be watered in the morning or evening when the rate of evaporation is lowest. The bare soil may be mulched, which reduces evaporation and controls weeds. Whereas annuals require fertilization for healthy growth, trees are fertilized only sparingly. Seasonal care involves providing fertilizer and water below the frost line, especially for evergreen plants. Certain plants such as roses require special protection from the cold by being bundled in mulch. To prevent cracking, tree trunks of sensitive trees such as elm and maple may be wrapped in paper or girdled with an aluminum sheet. A variety of tools, including the lawn mower, tiller, rake, hoe, garden hose, spade, and others are used in landscape maintenance. These tools should be maintained on a regular basis to make them function properly and for long life.

REFERENCES AND SUGGESTED READING

Harris, R. W. 1992. Arboriculture: Integrated management of landscape trees, shrubs, and vines, 2d ed. Englewood Cliffs, N.J.: Prentice-Hall.

Pirone, P. P. 1988. Tree maintenance, 6th ed. New York: Oxford University Press.

Sinnes, A. C. 1982. Easy maintenance gardening. San Francisco: Ortho Books.

PART A

Please answer true (T) or false (F) for the following statements.

1. T F Mulching can be done to suppress weeds.
2. T F Evergreen trees require watering in winter.
3. T F Annual flowering plants and lawns are low-maintenance landscape items.
4. T F Fertilizers may be applied through irrigation water.
5. T F A nursery spade is used for weeding.

PART B

Please answer the following questions.

1. List any four important regular landscape maintenance activities.
    _____ _____
    _____ _____

2. List any three lawn maintenance tools.
    _____ _____ _____

3. List any four tillage tools.
    _____ _____
    _____ _____

4. List any three pruning tools.
    _____ _____ _____

PART C

1. Discuss two factors that affect the frequency of watering landscape plants.
2. Discuss water-conservation practices that can be implemented in the landscape.
3. Distinguish between sprinkler and drip irrigation methods.
4. Describe how water hoses should be stored.

PART 6

GROWING GARDEN CROPS

CHAPTER **17**

Growing Vegetables

PURPOSE

This chapter discusses the general principles of designing, planting, and caring for a vegetable garden, as well as the cultivation of selected vegetables.

EXPECTED OUTCOMES

After studying this chapter, the student should be able to

1. Describe the characteristics of a home garden.
2. Discuss the benefits of a home garden.
3. Design a vegetable garden.
4. List and describe the basic tools required by a gardener.
5. Discuss the choice of garden site and vegetable varieties to grow.
6. Describe how selected vegetables are cultivated.

KEY TERMS

Blanching	Growth cycle	Slips
Blossom end rot	Hardy	Soil test
Bush cultivars	Head	Summer squash
Cage	Monoecious	Tender
Capsicum	Parthenocarpic fruits	Trellis
Cole crops	Photoperiod	Vernalization
Curd	Physiological maturity	Vine ripened
Dry sets	Pole cultivars	Winter squash

OVERVIEW

Small-scale vegetable production is an activity that may be undertaken year-round (except in extreme weather) by people in a wide variety of situations. It is applicable to rural as well as urban dwellers. The size of the project depends on what is available in terms of land and how much time and effort one is willing to devote to the activity. Urban dwellers may be limited to only a fraction of an acre, whereas rural dwellers may have more than an acre with which to work. The space available may have to be devoted to both ornamentals and vegetables, since people who enjoy gardening usually like to grow some flowers as well.

Gardeners have a wide variety of crops from which to choose, including roots, cole crops, cucurbits, and solanaceous crops. There are crops for the cool season and others for the warm season. Small-scale vegetable farming can be undertaken as a hobby, where the primary purpose of the project is recreational while also providing fresh produce for the table. Larger-scale vegetable farming may be used to produce more than one needs so that the surplus can be marketed for supplemental income. This kind of small-scale production, whether as a hobby or for commercial purposes, is commonly called *home gardening* or *backyard farming*.

When growing to sell, postharvest handling and marketing become very important considerations in planning a garden. In herb gardening (chapter 18), one has to decide whether to grow mainly for culinary use or also for ornamental purposes. Some herbs may be grown indoors in pots for ready access for culinary use.

MODULE 1 GENERAL PRINCIPLES OF VEGETABLE GARDENING

17.1: CHARACTERISTICS OF A HOME GARDEN

No two home gardens are exactly alike. Home gardens reflect the creativity and idiosyncrasies of the gardener. Generally, they tend to be more cosmetic than commercial or large-scale farming enterprises. One reason may be that they are often located on the premises of the dwelling place and in a residential area so that they have to be kept clean. Also, since it is near the house, the gardener is able to visit it more frequently than if it were located in a distant place. The general characteristics of a home garden include the following:

1. It is usually located in close vicinity of a home (often in the backyard).
2. The garden is usually less than 1 acre (0.4 hectares) in size.
3. More than one crop is usually grown.
4. Most of the crops grown are annuals.
5. Labor for the garden is supplied solely by the gardener and his or her family.
6. It involves the use of hand implements. When the plot is large, simple powered machines such as a garden cultivator may be used.
7. It depends on the home water supply for additional moisture.

17.2: BENEFITS OF A HOME GARDEN

Avid gardeners usually make it a point to show their projects to visitors or to brag about their accomplishments in conversations (sometimes with a bit of exaggeration, for example, about the size of the tomato fruits or the yield of some crops). As a hobby or an income earner, a garden has a number of benefits to the gardener, including the following:

1. It brings pleasure, satisfaction, and a sense of achievement.
2. It reduces the grocery budget.
3. It is a source of fresh produce for the table.
4. It provides a means of exercising the body.
5. When engaged in by the family, it provides valuable time of family interaction.
6. Gardeners may participate in clubs in the community for social interaction.

17.3: Choosing a Site

People in urban areas who live in apartments may not have access to land in the immediate vicinity of their residences. Some property owners are gracious enough to provide tenants with a vacant lot that may be devoted to gardening projects by those who wish to do so. Some avid gardeners go to varying extents to grow plants, even in high-rise apartments, by using pots and wooden boxes in their balconies, for example.

In choosing a site for a garden, certain general guidelines may be followed. Even vegetables that require partial shade should not be denied sunlight, especially at the midday period. It should be kept in mind that longer shadows tend to be cast in northern than in southern areas. Further, spring and autumn shadows are longer everywhere than the shadows of June and July. To make maximum use of light, the orientation of the beds in which crops are planted is critical. Following are some general guidelines for site selection:

1. *Adequate sunlight.* The garden should be located in a place where it will receive full, direct sunlight or at least partial sunlight. Gardens should not be planted in the close vicinity of trees, not only because they cast shade but also because they compete with the garden crops for sunlight and with their extensive root systems deplete the soil of nutrients and moisture.

2. *Well-drained soil.* Plant roots need to breathe. Because gardeners tend to visit their gardens on a frequent basis (daily in most cases), gardens receive heavy traffic from humans and are prone to soil compaction. The soil structure, especially in the paths, can be ruined and thereby cause poor drainage and the formation of puddles after rainfall. Loamy soils are best for gardening (3.3.1) because they drain well and are also easy to work.

3. *Source of water supply.* Many garden crops require a good supply of moisture for growth and quality produce. The primary water source for garden crops is the home water supply from the tap. This supply may be accessed by means of a long water hose. If the garden is located too far away, a watering can may have to be used for watering, making the project more tedious.

4. *Flat or gentle slopes.* Whenever possible, steep slopes should be avoided for a garden project. Steep slopes are prone to soil erosion once the natural vegetative cover has been removed for farming.

5. *Windbreaks.* Garden crops are tender and should be sheltered from strong winds. The garden receives some protection from winds if it is located on the sheltered side of a hedge or a row of trees acting as a windbreak (12.7).

Because a garden is only a miniature farming operation, the preceding restrictions may be altered. Characteristics that normally would be a major problem for large-scale farming are much easier to correct or manage in home gardens. For example, if the land is not well drained, garden crops may be planted on raised beds, an activity that is not difficult on a small scale. Where the land slopes, beds should be oriented across the slope. Shade-loving plants should be allocated to sections of the land where there is partial lighting. If the land is far from the house and not within reach of a watering hose of reasonable length, water may be carried to the garden.

17.4: Designing a Vegetable Garden

After choosing a site, a decision must be made about which crops and how much to grow. The following are some factors that influence decision making in this respect.

17.4.1 CHOOSING WHAT TO GROW

The choice of crops to grow depends on several factors including those described in the following sections.

Gardener's Preference

Gardeners grow what they want to grow. Since gardening has a strong hobby element, people devote time to what they enjoy doing. They do not grow crops just to be growing something.

Crop Adaptation

Even though a grower's preference comes first, this factor may be overruled by the reality of what *can* be grown. If one likes strawberries but the environment cannot support this crop, it ceases to be an option (unless artificial growing conditions, such as greenhouse facilities, will be provided). Even when a crop can be grown in a general region, specific varieties or cultivars are adapted to particular areas (chapter 1). An additional operational classification is important in choosing the season and planting dates for crops. Based on the temperature plants can endure during the growing season, vegetables may be classified into four groups:

> **Hardy Plants**
> Plants adapted to cold temperatures (or other adverse climatic conditions).

> **Tender Plants**
> Plants that are sensitive to cold temperatures (or other adverse climatic conditions).

1. *Hardy.* Hardy plants are frost resistant and may be planted before the last killing frost in the region.
2. *Semihardy.* Plants that are tolerant of light frost but will succumb to severe frost are said to be semihardy.
3. *Tender.* Tender plants will grow in cool weather just like semihardy plants but will be severely damaged or killed by even a light frost.
4. *Very tender.* Plants in this category range from very sensitive to intolerant of cold temperatures; even cool weather could kill them.

Market

If an objective of the garden project is to provide supplemental income, it is important that the crops chosen be readily marketable. At the very least, friends and neighbors should be interested in purchasing small quantities of the produce.

Culture

Some crops are easier to grow than others in terms of land preparation, plant care, harvesting, and postharvest handling. Once planted, some crops require only minimal care and attention. Others may require special attention such as staking, pruning, and other activities and expenses that some growers may not be willing to provide. Before deciding on a crop, it is important for the gardener to find out as much as possible about it.

Growth Cycle

Some plants are annuals, while others are biennials or perennials. Certain cultivars of crops are early and others late maturing.

Plant Form and Size

Some plants grow erect, while others climb or are runners. Each type has ideal space requirements. Some plants are small, and others grow to be large bushes. Knowing the crops in this way helps in locating plants in the garden. You do not want plants shading or climbing onto others. Further, tall plants may cast shadows over short ones (unless strategically located in the garden).

17.4.2 GARDEN PLAN

Crop rotation is key to a good garden design. It entails shifting crops from one location in the garden to another according to a strategic and predetermined sequence (figure 17-1). This strategy has several advantages:

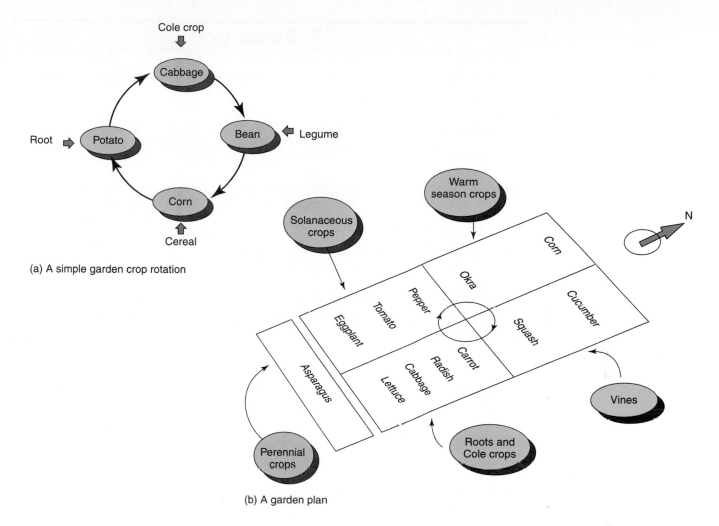

FIGURE 17-1 A crop rotation and garden plan. In locating plants, attention must be paid to the heights to avoid shading problems. Gardens frequently have more than the four crops that are often used to illustrate the concept of crop rotation.

1. *It helps to control insect pests and diseases.* Diseases and pests tend to be associated with a group or family of plants. Their population soars when the host plants are planted year after year in the same location. This is especially true of pests that inhabit the soil and overwinter. For example, whiteflies, Colorado potato beetle, and hornworms are associated with the family Solanaceae, which includes potato, eggplant, tomato, and pepper. If such crops are shifted to another location in the garden and replaced with an unrelated species such as corn, the pests will be starved to death and their numbers drastically reduced.

2. *Plants feed at different levels and at different intensities.* Some crops are generally heavy feeders (e.g., corn). Others consume large amounts of specific nutrients (e.g., leaf crops such as lettuce consume large amounts of nitrogen). By rotating crops, the soil fertility will be properly utilized.

3. *Variety of crops may be produced.* Since gardeners frequently grow a fairly large variety of crops, grouping crops according to some sound and strategic basis is an important consideration in the design of a garden plan. For example, cucumber and summer squash belong together; eggplant, tomato, and pepper are related; and carrot and beet (which attract the same pests) can be grouped together. Because a gardener may want to grow perennial vegetables, such species (e.g., rhubarb and asparagus) should be separated from the annuals and not included in the crop rotation cycle.

17.5: GARDEN TOOLS

The tools used for gardening are simple. Many garden tools are also used in the landscape (chapter 16).

17.6: GROWING POPULAR GARDEN CROPS

Thanks to plant breeders, certain garden crop species can be grown almost anywhere because they have developed a wide range of cultivars for various adaptation zones. In this section, the cultivation of selected commonly grown vegetables is briefly described. It is important to note that due to regional differences, additional practices may be adopted in certain cropping zones for best results. In fact, a discussion of crop production can only be made in general terms since recommendations are specific to the site. For example, some chemicals are banned in certain states, and some diseases or pests are a problem only in particular areas. Fertilizer recommendations depend not only on the crop but also on the desired yield, soil pH, and current soil fertility characteristics (from a *soil test*). Planting dates and duration of growth periods differ from one region to another. For best results, the gardener should consult local extension services for additional information regarding the best cultural practices suited to the region of interest.

17.7: GARDEN INSECT AND DISEASE CONTROL

Garden plants are plagued by a host of insect pests and diseases. A variety of strategies may be adopted to minimize pest infestation. The control of pests is discussed in detail in chapter 7. In addition to these general strategies, the gardener may make use of the repellent action of scents exuded by certain plants to reduce pest attacks. For example, aphids are known to be repelled by garlic, and radishes repel beetles that attack plants such as tomato, squash, and eggplant as well as mites. Other repellant actions of plants are presented in table 17-1.

17.8: USE OF HERBICIDES IN THE GARDEN

Weeds can be a problem in gardens that are neglected. Because of the small size of backyard gardens and the frequent visitation of the gardener, weeds can be kept under control by uprooting them with bare hands or weeding with a hoe. The use of chemicals may become necessary when the size of the garden is increased. Herbicides for use in the garden are described in chapter 16.

Table 17-1 Selected Plants Known to Repel Insects and Other Pests

Plant	Insect Repelled
Basil (*Ocimum basilicum*)	Flies and mosquitoes
Marigold (*Tagetes* spp.)	Many insects
Garlic (*Allium sativum*)	Many pests
Mint (*Mentha piperita*)	Cabbage moths
Onion (*Allium cepa*)	Ants
Radish (*Raphanus sativus*)	Many insects
Rosemary (*Rosemarinus officinalis*)	Cabbage moths, bean beetles, and carrot flies
Tansy (*Tanacetum vulgare*)	Beetles and flying insects

17.9: GARDEN PLANNING GUIDE

A garden planning guide is a regional affair and depends on the climate and weather patterns. Gardening may be done in fall, spring, or summer. Local extension services can provide appropriate guidance for particular areas.

17.10: COMMON GARDEN PROBLEMS

A garden may suffer from neglect and general improper management. Table 17-2 presents some of the common problems and how they can be corrected or avoided.

Table 17-2 Common Garden Problems and Their Causes and Control

Stage	Possible Cause	Suggested Control
Pregermination		
Seed rot	Seed rot	Treat seed before planting; use captan or thiram
	Seed corn maggot attack	Treat before planting
	Soil too wet	Plant on raised beds for improved drainage
	Seed not viable	Use fresh seed for planting
Seed eaten	Rodent attack	Use baits, repellents, ultrasound devices, or traps
	Bird attack	Use baits, traps, or ultrasound devices
Germination		
Seedlings die	Damping-off	Treat seed before planting; use captan, thiram, or Zineb or apply fungicidal soil drench
Seedlings cut off	Chewing insects (e.g., ant and caterpillar such as cabbage looper, slug, and snail)	Remove by hand; spray pyrethrins
Seedlings wither	Aphid and thrip attack	Spray malathion or rotenone
Vegetative growth		
Foliage chewed	Caterpillar, beetle, or cricket	Use rotenone, carbaryl, or Diazinon
Leaves tunneled	Leaf miner	Apply malathion
Leaves discolored or withered	Aphid or spider mite	Apply ethion, rotenone, or malathion
Roots chewed	Grub or beetle	Apply Diazinon as preplant
Flowering and fruiting		
Flower drop or damage	Aphid, thrip, caterpillar, or mite	Apply disulfoton, Docifol, or malathion
Fruit tunneled	Corn earworm	Apply carbaryl or Diazinon
Fruit chewed	Fruitworm or beetle	Apply carbaryl or malathion
Weeds		Chemical control not recommended; remove mechanically (e.g., with hoe)

17.11: INCREASING EARLINESS OF VEGETABLE CROPS

The early bird catches the worm, according to the adage. Vegetable growers may want to take advantage of the preseason or early season limited supply of fresh produce to obtain premium prices by taking certain calculated risks to produce an early crop. Particular strategies may increase the production cost of the crop. Some activities and provisions help in this regard, the key ones including the following:

1. *Site selection.* Where the growing season is preceded by cold weather, the site chosen for vegetable production should be such that it encourages rapid warming of the soil. A piece of land that slopes (about 20 degrees) to the south absorbs more of the early sunlight and warms up to a greater depth. Soils that are sandy (light) and drain freely also warm up much faster than cold (clay) soils.

2. *Cultivar selection.* Early maturing cultivars should be chosen.

3. *Use transplants.* Instead of direct seeding, plants can be started indoors in a nursery during the adverse weather conditions preceding the growing season. This practice gives plants a head start on growth and reduces the field growing time. Although crops such as pepper and tomato are routinely transplanted, vines such as cucumber and muskmelon are not.

4. *Protection from weather elements.* Planting on slopes helps to protect against frost damage. The soil may be irrigated the day before a predicted frost to increase the soil's capacity to retain heat. Windbreaks reduce wind speeds and soil loss while increasing soil temperature. Some growers use plant covers to shield plants from inclement weather.

5. *Mulching.* Mulching controls weeds, keeps soil warm, and increases soil water retention. Crops such as tomato, watermelon, and pepper tend to mature earlier when mulched.

6. *Adequate nutrition.* While overfertilization is detrimental to crops and often delays maturity, providing adequate nutrition encourages rapid plant development. The quality of produce is high with sufficient nutrition. Plants need to be watered adequately to ensure a proper rate of growth.

17.12: EXTENDING THE HARVEST

A garden's harvest may be extended such that the grower is able to harvest fresh produce over a prolonged period. This strategy maximizes the use of garden space and increases productivity per unit area. It entails overlapping the planting of vegetable crops by planting crops in combination. It makes use of the fact that certain crops are early maturing while others are late maturing. Early season and early maturing crops are planted first, but before they mature, they are interplanted with midseason crops. There is minimal competition between the two sets of plants. The same effect can be realized with one or several crops by staggering the planting dates. The grower may be able to purchase different cultivars of a crop with varying maturity dates (early, medium, and late). Some crops that can be interplanted are cucumber and pepper, corn and squash, broccoli and pepper, lettuce and onion, and carrot and beet.

SUMMARY

Some home owners devote a portion of the backyard of the home to growing vegetables and other bedding plants for fresh produce for the table or use in the home. Gardening may be undertaken as a hobby or may be on a scale large enough to produce extra for sale. A garden should be located where it will receive adequate sunlight and where the soil drains well. The grower may consult planting dates and crop adaptation maps and charts produced by the U.S. Department of Agriculture (USDA) and local extension agencies in selecting appropriate crops, varieties, and planting dates for the locality. Some plants are hardy and others tender. Gardening requires simple handheld tools.

REFERENCES AND SUGGESTED READING

Abraham, G. 1977. The green thumb garden handbook. Englewood Cliffs, N.J.: Prentice-Hall.

Knoot, J. E. 1988. Handbook for vegetable growers, 3d ed. New York: John Wiley & Sons.

Newcomb, D., and K. Newcomb. 1989. The complete vegetable gardener's source book. Englewood Cliffs, N.J.: Prentice-Hall.

Ortho Books. 1981. All about vegetables. San Francisco: Ortho Books.

Rice, L. W., and R. P. Rice. 1993. Practical horticulture, 2d ed. Englewood Cliffs, N.J.: Prentice-Hall.

Yamaguchi, M. 1983. World vegetables. Westport, Conn.: AVI Publishing.

OUTCOMES ASSESSMENT

PART A

Please answer true (T) or false (F) for the following statements.

1. T F No two home gardens are exactly alike.
2. T F A hardy plant is frost tolerant.
3. T F Most garden plants are perennial in duration of growth.
4. T F A garden spade is used for weeding.
5. T F Fertilizer recommendations for a home garden depend on the soil pH.

PART B

Please answer the following questions.

1. List three characteristics that are typical of home gardens.
   _____ _____ _____

2. List three specific benefits a home garden brings to a home owner.
   _____ _____ _____

3. List five popular garden crops.
   _____ _____ _____
   _____ _____

PART C

---

1. Discuss the importance of adequate sunlight in choosing a garden site.
2. Distinguish between a hardy and tender vegetable plant.

3. How do plant form and size affect the design of a home garden?
4. List four common garden tools and describe their uses.
5. Classify garden plants according to their adaptation to temperature.

MODULE 2 GROWING CROPS WITH UNDERGROUND EDIBLE PARTS

When growing plants for their underground parts it is difficult to follow the development of these economic parts since they are buried in the ground. Above-ground parts are often monitored and used as indices of what is happening underground. The underground parts may be modified roots, stems, or leaves. Soils that are loose, free from obstructions, and well drained are desirable for good development of underground plant parts. The discussion in this module is general. Specific local or regional cultural practices may differ to some extent.

17.13: CARROT

Common name: Carrot

Scientific name: *Daucus carota* L.

Family name: Apiaceae

Carrots are a cool-season crop but are adapted to all climatic zones. They perform best at temperatures between 40 and 80°F (4.44 and 26.7°C). Popular members of the carrot family include dill, parsley, celery, and parsnip. This underground swollen root is characteristically bright orange in color, long, slender, and slightly tapered (figure 17-2). Other shapes

FIGURE 17-2 Carrot *(Daucus carota).*

and sizes are available. Carrot seeds are tiny, and the crop is intolerant of extreme temperatures. Carrots are used fresh in salads or cooked.

17.13.1 SOIL AND LAND PREPARATION

The soil should be deep, loose, well drained, and free from obstructions such as stones. Clay soils are undesirable since they crust upon drying and impede germination. If this soil type must be used, cultivars with shorter roots should be planted. Obstacles in the soil cause deformed and unattractive roots. The addition of organic matter aids in good rooting. Since the seeds are tiny, they should be sown in a bed of fine tilth.

17.13.2 PLANTING

Carrots may be direct seeded or transplanted, even though the latter practice can cause roots to deform. Seeds should be planted shallowly (about 0.5 inch or 1.3 centimeters) for good germination. When direct seeded, thinning is often necessary to obtain proper spacing for good growth and yield. The final stand should be about 1.5 to 2 inches (3.8 to 5.1 centimeters) apart in the rows. Carrot seeds germinate slowly and require a long time to become established. Cultivars are chosen depending on the intended use of the produce.

17.13.3 CARE

Good soil moisture is required for germination, as well as high crop yield. In drier regions, supplemental moisture is often required to produce a good crop. It is important to moisten the soil evenly to prevent splitting of the carrot roots. Weeds should be controlled promptly. The root should be completely buried under soil at all times; otherwise, the exposed tops of the roots will turn green (in an attempt to photosynthesize). To avoid this discoloring, soil should be heaped over the crown of the plants. Carrots respond to nitrogen fertilization; when needed, nitrogen should be applied in moderate amounts as a sidedress. Diseases and pests of the crop include carrot rust fly, nematodes, aphids, bacterial blight, and cercospora leaf blight.

17.13.4 HARVESTING

Carrots are hand harvested, but the process is made easier by first using a tool (e.g., garden fork or spade) to undercut the roots. This activity loosens the roots from the soil and makes it easier to pull them out by the leaves. The roots are washed and graded. Some markets prefer that the carrots have leaves attached. Carrots may be stored, if needed, for a short period at high humidity (90 percent or more) and cold temperature (0°C or 32°F).

17.14: SWEET POTATO

Common name: Sweet potato

Scientific name: *Ipomea batatas* Lam.

Family name: Convolvulaceae

Sweet potato is a warm-season root crop that is native to tropical America. It can be grown as an annual (in temperate and subtropical climates) or perennial (in tropical and subtropical climates). The crop is heat tolerant and drought resistant. Sometimes sweet potatoes are incorrectly called *yams,* which are an entirely different species of plants. Yams belong to the genus *Dioscorea,* and some belong to the species *alata.* To grow well, one warm growing season of at least three-month duration is required. The sweet potato swollen root may have moist flesh (soft) or dry flesh (firm). The moist-flesh types are sweeter and have average to deep orange flesh color. The dry-flesh types have white or yellow flesh color and yellow skin and are less sweet.

17.14.1 SOIL AND LAND PREPARATION

The site should be well drained, preferably sand or sandy loam, for good root development; potatoes prefer slight soil acidity. Where drainage is poor, the crop should be planted on raised beds. The soil should be deeply plowed and harrowed to remove clods and stones that could obstruct root formation.

17.14.2 PLANTING

The sweet potato crop is grown by using sprouts *(slips)* that are produced from tuberous roots (9.12). Professional growers raise these slips for sale to home growers. Small-scale home gardeners hand plant the slips, but a mechanical planter may be used for large-scale planting. Spacing of 1 to 1.5 inches (2.54 to 3.81 centimeters) within rows and 3 to 3.5 feet (0.91 to 1.5 meters) between rows is recommended.

17.14.3 CARE

Weeds should be controlled by hand weeding. Care should be taken not to damage the roots of the crop. Since sweet potato responds to fertilization, a moderate application of a complete fertilizer (with more phosphorus and potassium, such as 5:10:15) is recommended. Overfertilization causes excessive vegetative growth and reduced root development. A split application may be done by applying about 0.5 pounds per 25 feet (0.23 kilograms per 7.6 meters) at planting and the same amount as a sidedress. The crop requires good moisture through the growing season.

Important diseases of sweet potato are fusarium wilt, black rot, cercospora leaf spot, and stem rot. The root-knot nematode is often a problem and may be controlled by fumigation. Important insect pests of potato include sweet potato weevils, whiteflies, flea beetles, and wireworms.

17.14.4 HARVESTING

It takes about 130 to 150 days for the sweet potato crop to attain full maturity and its highest quality. For easier harvesting, the vines are removed by cutting before digging. Some cultural systems allow frost to kill the vines. However, low temperatures (10°C or 50°F) may damage roots (chilling injury) and induce rotting. Further, if sweet potatoes are not harvested before the vines start to decay, the decay may spread to the tuberous roots. Tubers are cleaner and harvesting easier if the soil is dry. The skin of the tubers is easily bruised, and thus care must be exercised in harvesting to avoid damage that causes reduced produce appeal. Potatoes may be stored at high relative humidity (80 to 85 percent) and at temperatures of about 13 to 15°C (55.4 to 59°F) for a short period.

17.15: RADISH

Common name: Radish

Scientific name: *Raphanus sativus* L.

Family name: Cruciferae

Radish is a popular early spring vegetable. It is very easy to grow, requiring only three to six weeks of growth after the seedling stage. This is a cool-season crop and does not tolerate heat (requires less than 25°C or 77°F). High temperatures may cause plants to bolt or the tissue to be pithy and poor in quality. Cultivars differ in time to maturity, skin color (usually red, but mixed colors occur), pungency, and size. Winter cultivars are larger in size and more pungent. Radish may be used fresh in salads, cooked, or pickled.

17.15.1 SOIL AND LAND PREPARATION

Radish may be planted on raised or flat beds. The soil should be well drained and fertile. Sandy loams are preferred, but the plant grows well in a wide variety of soils, provided it is deep enough to contain the root size and allow good tuber formation.

17.15.2 PLANTING

Plants are spaced closely in the row (three to four seeds per inch or 2.52 centimeters) and about 10 to 12 inches (25.4 to 30.5 centimeters) between rows. Seeds are planted into a seedbed of fine tilth.

17.15.3 CARE

Weeds should be controlled by hand hoeing. Produce quality may be improved through fertilization. About 1 pound (0.45 kilogram) of 10:10:10 per 50-foot (15.2-meter) row of plants applied at planting is helpful. Good fertility is needed to produce a good crop. Important diseases include *Phytophthora, Rhizoctonia, Pythium,* anthracnose, powdery mildew, and rust. Insect pests of radish include aphids, flea beetles, and cabbage root maggots.

17.15.4 HARVESTING

Radish is harvested by pulling mature plants. The tops may be cut off or left in place, depending on the market. Roots are washed, graded, and bagged in plastic. Early cultivars may be ready within three to four weeks if the temperature remains normal. Maturity is delayed by two to four weeks by cold temperatures. Winter cultivars are ready in 50 to 90 days. If harvesting is delayed, the roots may become pithy and bitter. Winter cultivars store better than summer cultivars. Harvested radish is held in storage at cool temperatures.

17.16: ONION

Common name: Onion

Scientific name: *Allium cepa* L.

Family name: Amaryllidaceae

Onions are biennial cool-season plants. Cool temperatures (55 to 77°F or 12.8 to 25°C) are important in the early stages of plant growth, but later—especially at harvesting and postharvest curing—warm temperatures and low relative humidity are required. Unlike other plants with underground parts, the formation of an onion bulb is initiated mainly by day length *(photoperiod)* and not the age of the plant. Photoperiod needs depend on the cultivar and usually range between 12 and 15 hours (i.e., there are short-day and long-day cultivars). If onions are grown for seed, *vernalization* (cold treatment) is required for bolting and flowering. Cultivars differ in bulb size, shape, pungency, and exterior color (figure 1-18). The color may be green, red, yellow, white, or purple. Onions may be bulbing (producing single large or small bulbs) or bunching (producing tiny bulbs or no bulbs at all). They may be grown for their edible and tubular green stems that grow in a bunch from the base of the plant.

17.16.1 SOIL AND LAND PREPARATION

Onions do best in soils that are deep and friable, with good moisture retention. Sandy loams and soils with good organic matter are desirable. The soil is first deeply plowed and second tilled to create a good seedbed. If soil is acidic, it should be limed so that the pH is above 6.5.

17.16.2 PLANTING

The crop may be direct seeded at 20 to 30 plants per meter (3.3 feet) of row. Rows are spaced 30 to 60 centimeters (24 to 48 inches) apart. Home growers often establish their crop by *dry sets* (small bulbs). Using bulbs has been found to produce higher yields and an earlier maturing crop. Sometimes a field may be planted by transplanting small seedlings.

17.16.3 CARE

Weeds may devastate an onion field in the early part of growth since onions are slow starters. Weed control during this period, which often takes the form of hoeing, is critical to success. Fertilizers may be applied in moderate amounts as needed. Sometimes sidedressing with nitrogen is helpful. Since onions are shallowly rooted, irrigation should be provided at regular intervals for good crop yield. Onions are attacked by root-knot nematodes and insect pests including onion aphids, cutworms, and onion maggots. Important diseases include downy mildew, white rot, smut, yellow dwarf virus, and fusarium wilt.

17.16.4 HARVESTING

> **Scallions**
> Harvested immature onion plants (without bulb development) for use as fresh vegetables.

Onions may be harvested as green immature plants (called *scallions*) for the fresh market for use in salads; however, most onions are harvested as mature dry bulbs. Onions are harvested when the top foliage starts to dry and fall over. A tool such as a garden fork (or onion plow in commercial production) is used to loosen the bulbs by passing under the rows. The plants are pulled and laid in windrows for drying and curing in the field or gathered into crates for drying and curing off of the field. Improper handling at harvesting can bruise the bulbs.

To keep well, onions should be properly stored in ventilated rooms. They are usually sold in nets or open mesh sacks for this reason. Onions store better if they are harvested as mature bulbs. To slow sprouting, onions should be stored at close to 0°C (32°F) and a relative humidity of about 60 percent.

SUMMARY

Plants whose economic and edible parts are underground structures require loose and deep soil for good development. They may require a cool season for growth or may be warm-season plants. They may be bulbs or roots and may be used fresh or cooked. Some types require special conditions for development (such as long days for bulb initiation and vernalization for bolting and flowering in onions). Although roots such as radish mature within eight weeks, others, including sweet potato, may require about 130 to 150 days for maturity.

OUTCOMES ASSESSMENT

PART A

Please answer true (T) or false (F) for the following statements.

1. T F *Daucus carrota* L. is the scientific name for onion.
2. T F Carrot is a warm-season crop.
3. T F Clay soils are best for crops with underground edible parts.
4. T F Moist-flesh sweet potato is sweeter than dry-flesh types.
5. T F Radish cultivars with large roots tend to be less pungent than those with small roots.
6. T F Dry sets are used by home growers to establish onion plots in the garden.
7. T F Sweet potato is intolerant of frost.
8. T F Onions are annual plants.

Please answer the following questions.

1. List five crops with underground edible parts.
_____ _____ _____
_____ _____

2. List two important diseases of carrots.
_____ _____

3. Sweet potatoes may be planted by using sprouts called _____.
4. Name any two important insect pests of sweet potato.
_____ _____

5. List four ways in which onion cultivars may differ.
_____ _____
_____ _____

6. The formation of onion bulbs is initiated by _____ and not the age of the plant.

1. Describe how high temperatures affect radish in the field.
2. Describe how tubers and other underground economic plant parts are stored after harvesting.
3. Describe the role of light in the production of onions.

MODULE 3 GROWING CUCURBITS

Cucurbits are a class of plants that produce fruits with a hard outer cover (1.3.4). This module discusses the culture of selected crops in general terms only. Local and regional cultural differences may occur.

> **Cucurbits**
> *Plants in the gourd family with fruits that have a hard outer cover.*

17.17: CUCUMBER

Common name: Cucumber

Scientific name: *Cucumis sativus* L.

Family name: Cucurbitaceae

Cucumber is a warm-season annual vegetable (figure 17-3). It has a short growing season and as such is able to grow in most places without fear of frost. Cucumber produces long and cylindrical fruits that are commonly used in salads but also are pickled. It is grown commercially in both the field and greenhouse.

17.17.1 SOIL AND LAND PREPARATION

Cucumbers need to grow in deep soil since they are deep rooted. The soil should be well drained and slightly acidic to neutral.

FIGURE 17-3 Cucumber *(Cucumis sativus)*.

17.17.2 PLANTING

The choice of cultivars depends on the use intended for the crop. Those grown for the fresh market (called slicing cucumber) are long (up to about 10 inches or 25.4 centimeters) and straight, with dark-green and smooth skin. The cultivars grown for processing (e.g., pickling) have lighter-colored and thinner skins that are rough. The fruits are about 5 inches (12.7 centimeters) long. Some are smaller and more expensive. The cucumber plant is monoecious in flowering habit. Some flowers are staminate (male) and others pistillate (female); the first flowers tend to be male. High-yielding gynoecious hybrids are available. Seedless *(partheno-carpic)* (5.15) cultivars are grown in greenhouses and preferred for pickling. The crop is propagated from seed and spaced 2 to 3 feet (0.61 to 0.91 meter) apart in rows to accommodate the vine-producing nature of the plant. Early maturing (about 60 days) bush cultivars are available. Vines may be allowed to trail on the ground or on a support such as a trellis, pole, or fence.

17.17.3 CARE

Nitrogen fertilization increases yield. About 3 to 5 pounds (1.35 to 2.25 kilograms) of 10-10-10 fertilizer per 25 to 50 feet (7.6 to 15.2 meters) may be applied on rows. Important insect pests of cucumber include the cucumber beetle, which spreads bacterial wilt, a devastating pest; aphids; leaf hoppers; and vectors for viral diseases such as cucumber mosaic. The crop is also attacked by diseases such as powdery and downy mildews, fusarium wilt, bacterial wilt, anthracnose, and angular leaf spot.

17.17.4 HARVESTING

It is important to harvest the cucumber crop before it is overgrown. The best quality for fresh market is crispness. Overmature fruits have large seeds and low quality. Smaller fruits fetch high market prices but lower yield. The crop may be hand picked or harvested by machines. Usually, more than one round of harvesting is done; machine harvesting requires special, high-yielding, single-harvest cultivars. Freshly harvested fruits may be stored for a short period at temperatures of 7 to 10°C (45 to 50°F) and high relative humidity.

17.18: WATERMELON

Common name:	Watermelon
Scientific name:	*Citrulis lunatus* Thunb.
Family name:	Cucurbitaceae

Watermelons are warm-season annuals (figure 1-3). They are less tolerant of cold than other cucurbits such as cucumber and cantaloupe. They have long prostrate vine growth and thus require a lot of garden space. For good yield, the plant prefers warm temperatures of between 25 and 30°C (77 and 86°F) or warmer and requires a growing season of 12 to 16 weeks for production. It is not a particularly high-economic-value crop. Watermelons differ in size, shape, exterior coloration (striped, black, and green), rind thickness, and flesh color (usually red). Varieties with seeds also vary in seed size and number.

17.18.1 SOIL AND LAND PREPARATION

Like cucumber, watermelon prefers well-drained soils of good fertility and light texture. Soil reaction should be neutral or slightly acidic.

17.18.2 PLANTING

Watermelons are spaced at between 6 and 10 feet (1.8 and 3 meters) between rows and about 2 to 3.5 feet (0.61 to 1.1 meters) within rows. Bush types that demand less space are gradually being introduced into cultivation. The crop is direct seeded, planting four to six seeds per hill and thinning to two to three after germination.

17.18.3 CARE

It is important to cultivate or use herbicides to control weeds, especially in the early stages of establishment, since wide spacing places the young plants at a competitive disadvantage against weeds. Once established, the plants are able to suppress weeds. The watermelon is desired largely as a refreshing source of tasty water. To produce this juicy flesh, it utilizes large amounts of moisture. Water is most critical during flowering and fruit development. Root-knot nematodes are a serious soilborne pest problem. In addition, aphids, mites, cutworms, and leaf hoppers are significant insect pests of watermelon. Fertilizers increase yield and quality; when needed, moderate amounts of nitrogen and phosphorus may be applied.

17.18.4 HARVESTING

Watermelons are ready to harvest when the tendrils die and the parts of the fruit in direct contact with soil change color from green to creamy white (spot coloration); the sugar content of the fruit should be at its highest concentration. The best way to determine readiness for harvesting is to sample the field. Watermelons are harvested by hand by cutting the fruit from the vine. If storage is needed, the fruits may be placed in a cold environment at about 5 to 10°C (41 to 50°F) for several weeks.

17.19: MUSKMELON

Common names:	Muskmelon, cantaloupe
Scientific name:	*Cucumis melo* L. Reticulatus group
Family name:	Cucurbitaceae

Muskmelons are warm-season plants (figure 17-4). They grow best where the temperature in the growing season is above 30°C (86°F) and the season is frost free. Fruit size and shape are variable among cultivars. Cultivars also differ in flesh color (e.g., orange, green, and yellow), size of seed cavity, flesh thickness, rind thickness, and ribbing of the outer surface (smooth to deeply ribbed).

17.19.1 SOIL AND LAND PREPARATION

Soil requirements are similar to those for watermelon. The crop is usually planted on wide raised beds.

FIGURE 17-4 Muskmelon *(Cucumis melo)*.

17.19.2 PLANTING

Muskmelons are direct seeded at a wide spacing of 5 to 6 feet (1.5 to 1.8 meters) between rows and 2 to 3 feet (0.63 to 0.91 meter) within rows.

17.19.3 CARE

Production of muskmelons requires an abundant moisture supply. However, soil should be relatively dry during ripening. Nitrogen, phosphorus, and potassium fertilizers supplied in moderate amounts (50 to 100 kilograms or 110 to 220 pounds per acre) are beneficial. For early maturity, some growers mulch or use hot caps (to cover plants to provide heat) for accelerated growth. Important diseases of muskmelons include fusarium and verticillium wilts and powdery mildews. Root-knot nematodes are common, as are insect pests such as cucumber beetles, aphids, leaf miners, and wireworms. A number of viral attacks, including watermelon viruses 1 and 2 and squash mosaic virus, may reduce yield.

17.19.4 HARVESTING

Fruits are ready for harvesting when they readily detach from the vine. Similar to watermelon, harvesting at peak maturity is essential for high-quality produce. It should be noted that when harvested at the fully ripe stage, good handling becomes critical for retaining quality. The fruits at this stage begin to soften and can be damaged easily during transportation. Cool storage may be used as needed.

17.20: SQUASH

Common name: Squash

Scientific name: *Curcubita* spp. L.

Family name: Cucurbitaceae

Squash is easy to grow and varies widely in fruit characteristics (figure 17-5). Certain cultivars called *summer squash* are harvested as immature fruits with soft rinds and look like cucumber. Winter cultivars are more varied in shape and size. Sometimes pumpkins are classified as *winter squash* (hard rinds). Squash and pumpkins are closely associated and have a lot in common (especially winter squash and pumpkins).

FIGURE 17-5 Squash (*Cucurbita* spp.).

17.20.1 SOIL AND LAND PREPARATION

The soil should be well drained. Summer squash does well on light-textured soil, and winter squash is most successful on fine-textured soil. Recommended soil pH range is between 5.5 and 7.5. The soil should be deeply plowed to accommodate the deep roots of many cultivars. The seedbed should be fine for direct planting.

17.20.2 PLANTING

Squash is direct seeded. Summer squash cultivars are usually bushy in growth habit and also early maturing. The zucchini type, which is very popular, is small. Squash has a wide variety of fruit shapes—flattish, curved, spherical, long, short, blunt, and tapered. In terms of color, squash comes in all shades of green; solid colors of black, yellow, or white; and mixtures. Winter squash is usually larger than summer squash and has a hard rind and well-developed seeds. Summer squash is harvested when the rind is soft and seeds immature. Some winter squashes are durable and sometimes used as ornamentals because of their fruit shapes and characteristics. Winter cultivars require between 75 and 100 days to attain full maturity. Spacing is 3 to 4 feet (0.91 to 1.2 meters) between rows and 12 to 15 feet (3.6 to 4.6 meters) within rows for summer cultivars and wider (5 to 6 × 2 feet) (1.5 to 1.8 × 0.61 meters) for winter cultivars.

17.20.3 CARE

Weed control is necessary in early stages of establishment, but, like other creepers, squash is able to suppress weeds through shading when it has developed adequate foliage. Squash responds to fertilization; an application of manure or moderate amounts of complete fertilizers that are high in potassium produces good results. Squash is affected by root-knot nematodes and a variety of insect pests including squash vine borers, aphids, and the 12-spotted and the striped cucumber beetles. Because adequate soil moisture at flowering stage and fruiting is critical for a good crop, irrigation may be necessary, especially in summer cultivation.

17.20.4 HARVESTING

It is important to harvest summer squash on time to avoid poor fruit quality due to hard rinds and seeds. Squash are harvested by hand, accomplished by cutting the fruit from the stem. Some summer squash is cooked or used fresh; winter squash is usually boiled or baked. Squash may be preserved through canning. Winter squash is especially durable and can be stored at about 50 to 55°F (10 to 13°C) and 50 to 75 percent relative humidity for a long time. Summer squash is less durable and stores for about two weeks at 45 to 50°F (7 to 10 °C) and about 80 to 95 percent relative humidity.

SUMMARY

Cucurbits have a characteristic hard rind. They belong to the gourd family. Some modern cultivars have no seeds. The rind differs in thickness, color, smoothness, and texture. The flesh also differs in color and thickness. Cucurbits are creepers, and consequently are widely spaced in cultivation. Bush types of some species are now available. Cucurbits are attacked by a wide variety of insects, viruses, and fungi.

OUTCOMES ASSESSMENT

PART A

Please answer true (T) or false (F) for the following statements.

1. T F Cucumber is a warm-season plant.
2. T F Slicing cucumber cultivars have thin, light-colored, and rough skin.
3. T F Watermelon is also called *Citrulis lunatus* Thunb.
4. T F Muskmelon is a cool-season plant.
5. T F Winter squash and pumpkin are similar in characteristics.
6. T F Summer squash is more durable (stores better) than winter squash.
7. T F If a fruit has no seed, it is described as parthenocarpic.
8. T F The cucumber plant is dioecious.

PART B

Please answer the following questions.

1. Describe how watermelon cultivars differ.
   _____
2. Muskmelon is also commonly called _____.
3. The best way to determine watermelon ripeness is _____.

PART C

1. Why are cucumber and watermelon usually widely spaced in the field?
2. Discuss the strategies of harvesting muskmelon.

MODULE 4 GROWING LEGUMES

Legumes are noted for the relatively high protein content of their seed and high nitrogen concentration of vegetative parts. Apart from their distinguishing flower characteristics, legumes are also noted for their capacity to live symbiotically with bacteria in their roots, resulting in the fixation of atmospheric nitrogen by the *Rhizobia* for plant use. They are beneficial to the garden soil since they improve soil fertility. Legumes planted as vegetables may be used as pulses (dry grain) or premature edible pods.

17.21: Snap or Green Bean

Common name:	Snap or green bean
Scientific name:	*Phaseolus vulgaris* L.
Family name:	Fabaceae

Snap beans, or green beans, are grown for their immature fresh pods and seeds. It is a warm-season, self-pollinated crop. They are called green beans because most are green. However, other colors (yellow and purple) exist but are not common.

17.21.1 SOIL AND LAND PREPARATION

Soils should be well drained and slightly acidic to neutral in reaction. Heavy soils impede germination and are not desirable. Snap beans are sensitive to boron and salt.

17.21.2 PLANTING

Snap beans are direct seeded. Home growers may use indeterminate cultivars (pole beans) to produce hand-harvested beans over a prolonged period. Determinate types (bush beans) are preferred for mechanically harvested commercial production. The pods vary in color (green or yellow) and shape (e.g., flat, oval, or round in cross section). The seeds may have light- or dark-colored testa. Home gardeners may also stagger the planting to obtain fresh produce over a longer period. Snap beans should be planted only when the soil is warm (about 20°C or 68°F) to avoid decay of seed as a result of slow germination in cold soil. Plants grown for hand harvesting are spaced 30 inches (76.2 centimeters) apart between rows and about 2 inches (0.61 meter) within rows.

17.21.3 CARE

Weeds should be controlled. Fertilizers, when needed, should have phosphorus and little nitrogen to avoid undesired vegetative growth that may reduce pod yield and delay maturity. Bacterial diseases such as halo blight, common blight, and bacterial wilt affect snap beans. The major viral diseases include cucumber mosaic and common bean mosaic. A variety of insect pests are found on snap beans, including aphids, leaf hoppers, spider mites, cutworms, and Mexican bean beetles. Fungal diseases such as rusts, white mold, anthracnose, powdery mildew, and fusarium wilt attack snap beans in the field.

17.21.4 HARVESTING

Snap beans may be harvested 50 to 65 days after planting. They are ready for harvesting when they have attained physiological maturity, about two weeks after flowering. If harvesting is delayed, the pods overmature and become fibrous and have larger than desirable seeds. Snap beans store fresh for a short period. When stored at a high relative humidity (95 percent) and cool temperature (45 to 50°F or 7.22 to 10°C), they may stay fresh for about two weeks.

17.22: Lima Bean

Common name:	Lima bean
Scientific name:	*Phaseolus limensis* Macf.
Family name:	Fabaceae

Lima beans are warm-season legumes that may be grown for the fresh market (not very common) or harvested as dry beans. Yields are drastically reduced through young fruit drop when temperatures fluctuate between low and high temperatures.

17.22.1 SOIL AND LAND PREPARATION

Soil and land preparation are the same as for snap bean.

17.22.2 PLANTING

When a heavy soil is used for planting, the seeds should be sown shallowly or covered lightly so as not to impede germination. Bush cultivars are spaced 3 feet (0.91 meter) between rows and 3 to 4 inches (7.6 to 10.2 centimeters) between plants in rows. Pole cultivars need physical support, which is provided in the form of trellises or poles on which the vines climb. They are spaced more widely than bush cultivars at 3 feet (0.91 meter) between hills of three to four plants or, when using trellises, 6 to 12 inches (15.2 to 30.5 centimeters) in rows.

17.22.3 CARE

Weeds need to be controlled, but fertilizers, when applied, should be low in nitrogen to avoid vegetative growth at the expense of pod production (e.g., 5:10:10). The disease and insect pests listed for snap bean also apply to lima bean.

17.22.4 HARVESTING

Lima beans are harvested when mature (about 80 to 100 days). At this stage the color of the seed is getting ready to change to white. To keep the plant productive, pods should be picked as they mature. They can be cooked fresh, processed (cured), frozen, or dried. Dry seeds store well, and fresh seeds require cold storage at about 0°C (32°F) and 90 percent relative humidity.

17.23: PEA

Common name: Pea, garden pea

Scientific name: *Pisum sativum* L.

Family name: Fabaceae

Apart from being a popular vegetable, the pea plant is most noted for its role in the discovery of the laws of heredity. Mendel, a noted geneticist, made his landmark discoveries while studying the pea plant. This cool-season plant is grown for the fresh market, in which the immature pods are eaten, or for seed.

17.23.1 SOIL AND LAND PREPARATION

The preferred soil is one that is fertile, well drained, and light textured. Soils rich in organic matter are desirable. Because the pea plant is intolerant of high soil acidity, soil pH should be above 5.5 and up to 7.0. The soil should be well plowed and of good tilth.

17.23.2 PLANTING

The grower may choose either bush (more popular) or vine cultivars. The seed may be wrinkled or smooth, the former having higher sugar-to-starch content. Pea quality is considerably lowered when temperature increases during cultivation. High temperatures cause starch to accumulate and the fiber content of seeds and pods to increase. Plants are spaced about 2 inches (5.1 centimeters) in rows and 2 to 3 feet (0.61 to 0.91 meter) between rows. Plants may also be placed in double rows spaced about 6 inches (15.2 centimeters) apart.

17.23.3 CARE

Like all legumes, when fertilizers are needed, they should be applied at a low rate of nitrogen to avoid unwanted vegetative growth. Pea plants are attacked by insect pests including mites, pea weevils, pea aphids, and seed corn maggots. Nematodes are also a problem where they occur in significant numbers. Common pea plant diseases include powdery mildew, downy mildew, fusarium and phytium rots, rust, bacterial blight, septorial blight, anthracnose, and bean yellow mosaic.

17.23.4 HARVESTING

When harvested at an immature stage, the seeds have high moisture content and are also tender. Fresh peas are hand harvested. Dry peas and peas for processing are harvested mechanically on a large scale. The seed color of mature peas may be dark or pale green. When storage is needed, peas store well under similar conditions of temperature and relative humidity as fresh lima beans (about 0°C or 32°F and 90 percent relative humidity).

SUMMARY

Legumes produce pods that contain one to several seeds. They are noted for their capacity to form a symbiotic association with bacteria in their roots, the result being the fixation of atmospheric nitrogen for host plant use. Legumes are also noted for the high protein content of their seeds and leaves. Popular garden species are harvested fresh or dry. Determinate and indeterminate cultivars exist.

OUTCOMES ASSESSMENT

PART A

Please answer true (T) or false (F) for the following statements.

1. T F *Rhizobia* inhabits the roots of legumes.
2. T F Legumes are a good source of protein.
3. T F Bush cultivars are spaced more widely than pole cultivars.
4. T F Snap beans are transplanted rather than direct seeded.
5. T F Nitrogen is the most needed soil nutrient for legumes.

PART B

Please answer the following questions.

1. The association between bacteria and legumes for their mutual benefit is called

   _____.

2. List five common garden legumes.

   _____ _____ _____

   _____ _____

3. Cultivars that need physical support are sometimes called _____.
4. *Phaseolus limensis* Macf. is the scientific name for _____.

PART C
---

1. Describe how symbiotic relationships are formed between bacteria and legumes.
2. Discuss some strategies a home gardener may adopt to obtain fresh produce from the garden over a long period.

MODULE 5 GROWING COLE CROPS

Cole crops belong to the taxonomic order of Cruciferae. Most cole crops are vegetables, including broccoli, Brussels sprout, cauliflower, and cabbage. Other brassicas are rape and canola, two crops that are noted for their oil (low in polyunsaturated fats [4.2.2]).

17.24: CABBAGE

Common name: Cabbage

Scientific name: *Brassica oleraceae*

Family name: Brassicaceae

Cabbage is a widely grown cool-season crop that is very popular with gardeners (figure 1-20). The crop is grown for its leaves. At the terminal bud the leaves are arranged into a ball of tightly wrapped and overlapping leaves called the *head*. The fresh crop is used in salads but may also be cooked.

17.24.1 SOIL AND LAND PREPARATION

Soils for cabbage should be well drained and fertile. For direct seeding, the soil should have a fine tilth. Low ridges or beds may be prepared for planting.

17.24.2 PLANTING

Cabbage may be direct seeded or transplanted by using 4- to 6-week-old seedlings. Cultivars vary in time to maturity, head size at maturity, form or shape of head, and color of head (e.g., pale, dark, blue-green, or red). Some cultivars produce a more complete head than others. The leaves vary in texture (smooth or coarse). Hybrid cultivars are superior in yield to nonhybrid cultivars. Recommended spacing is 3 × 1 feet (0.91 × 0.3 meter) (between rows by within rows), which may be varied according to the size of the mature head.

17.24.3 CARE

Weed control is necessary, but deep tillage should be avoided. Fertilizers (especially nitrogen) promote rapid growth, high yield, and high produce quality. A sidedressing of nitrogen is desirable after the heads have formed to about half the size at maturity. Cabbage insect pests include cabbage worms, aphids, seed corn maggots, and cutworms. Cabbage is susceptible to diseases such as downy mildew, black rot, alternaria leaf spots, and club root. Root-knot nematodes and cysts may be a problem in certain soils.

17.24.4 HARVESTING

Delayed harvesting may cause the head to split. Cabbage is ready to harvest when the head has attained its expected full size and is firm. The crop is hand harvested by cutting the stem below the head but including a few of the loose outer leaves. In large-scale production, mechanical harvesters may be used where the produce is intended for processing. To preserve quality, cabbage may be stored for several months at high relative humidity (95 percent) and low temperature (0°C or 32°F).

17.25: BROCCOLI

Common name: Broccoli

Scientific name: *Brassica oleracea* L.

Family name: Brassicaceae

Broccoli is a cool-season crop. The edible part of broccoli is the inflorescence, which is borne at the terminal bud, much like a head in the case of cabbage (figure 1-20).

17.25.1 SOIL AND LAND PREPARATION

The soil should be well drained and of medium texture, good fertility, and high organic matter concentration.

17.25.2 PLANTING

Broccoli may be direct seeded or transplanted after nursing seedlings for three to four weeks. Plant spacing in the field may be 12 to 18 inches (30.5 to 45.7 centimeters) within rows and 2 to 3 feet (0.61 to 0.91 meter) between rows. Cultivars differ in inflorescence characteristics (e.g., shape, size, color, compactness, degree of branching, and size of flower buds). Early and late cultivars, as well as hybrids, are available.

17.25.3 CARE

Weeds, diseases, and pests should be controlled. Diseases of broccoli include club root, downy mildew, and seedling damping-off caused by *Pythium* and *Rhizoctonia* spp. The crop is attacked by insects including cutworms, aphids, cabbage worms, corn earworms, and flea beetles. An application of moderate amounts of a complete fertilizer increases yield.

17.25.4 HARVESTING

Broccoli is harvested before the flower buds open. The edible part (the inflorescence or head) should be compact. The crop is harvested over a period of time—the terminal head harvested first, giving room for secondary heads to develop and grow. Harvesting is accomplished by cutting the head by hand. After harvesting, the heads should be stored properly by cooling immediately since the quality deteriorates rapidly. If longer storage is required, broccoli may be stored like cabbage at high relative humidity (95 percent) and low temperature (0°C or 32°F).

17.26: CAULIFLOWER

Common name: Cauliflower

Scientific name: *Brassica oleracea* var. *botrytis*

Family name: Brassicaceae

Cauliflower is adapted to moderately cool climates (figure 17-6). It is much more difficult to grow than other cole crops such as cabbage, since production requirements must be met to a higher degree for good harvest. High temperatures, which cause poor head formation, and frost are undesirable.

17.26.1 SOIL AND LAND PREPARATION

Soils that are freely draining, fertile, and of medium texture should be selected.

17.26.2 PLANTING

Like the other cole crops, cauliflower cultivars differ in head size, compactness, shape, and amount of foliage cover. The head of the cauliflower is sometimes called the *curd,* which is surrounded by protective leaves that prevent the former from discoloration (or *blanching*) due to direct exposure to sunlight. Hybrid cultivars are available. Plants are spaced 18 to 24 × 36 inches (45.7 to 61 × 91.4 centimeters) (within by between rows).

FIGURE 17-6 Cauliflower (*Brassica oleracea* var. *botrytis*).

17.26.3 CARE

To prevent blanching, some producers tie the basal leaves over the curd to protect it from direct sunlight. Blanched curds are yellowish instead of whitish. High temperatures also cause the curd to be less compact and "ricy" (like grains of rice packed together). Adequate moisture should be provided in cultivation. Cauliflower responds to nitrogen fertilization. Trace elements, especially boron, are utilized in large amounts by the crop. Diseases and pests are similar to those described for broccoli and cabbage. Cauliflower is prone to a variety of other disorders that are nonpathogenic, including hollow stem, curd "riciness," and buttoning of the head (premature development). *Phytophthora* stem and root rot are important diseases of cauliflower.

17.26.4 HARVESTING

Delayed harvesting may also cause the curds to be ricy. Curds are hand harvested by cutting off the mother plant along with some protective lower leaves. The produce may be stored as previously described for other cole crops.

SUMMARY

Garden cole crops produce leaves that are tightly wrapped into a ball called a head or a bunch of inflorescence that is sometimes called a curd, as in cauliflower. Garden cole crops are cool-season plants. Cultivars differ in head characteristics. In the garden, they are usually cultivated in beds containing well-drained soil. If harvesting is delayed, the produce may blanch (discolor) or become coarse and grainy.

OUTCOMES ASSESSMENT

PART A

Please answer true (T) or false (F) for the following statements.

1. T F Most cole crops are vegetables.
2. T F Cabbage heads may be red in color.

3. T F   The edible part of cauliflower is its inflorescence.
4. T F   Garden cole crops are cool-season plants.

PART B

Please answer the following questions.

1. Cabbage has a head; the head in cauliflower is also called a _____.
2. _____ is the discoloration of a head of cauliflower.
3. *Brassica oleracea* is the scientific name for _____.
4. List four cole crops.

   _____   _____
   _____   _____

PART C

1. Describe strategies that may be employed by the grower to prevent deterioration of cole crop produce in the field.
2. Discuss disease and pest problems and their control in cole crop production.

MODULE 6 GROWING SOLANACEOUS CROPS

Solanaceous crops are plants that belong to the family Solanaceae. Some of the most popular and widely adapted vegetable crops belong to this family.

17.27: EGGPLANT

Common names: Eggplant, aubergine

Scientific name: *Solanum melongena* L.

Family name: Solanaceae

Eggplants are warm-season crop plants that are produced worldwide but to a lesser extent than other solanaceous vegetable crops such as pepper and tomato (figure 17-7). Although some eggplant cultivars are still small and egg shaped, many modern cultivars are large and may be round or elongated. The crop is intolerant of frost.

17.27.1 SOIL AND LAND PREPARATION

The eggplant performs well on sandy loams where drainage is good and the soils are light. It tolerates slightly acidic (pH 5.5) to neutral soil reaction. Soil preparation activities for planting include plowing and preparation of seedling beds.

17.27.2 PLANTING

Eggplants are usually transplanted since direct seeding produces slow establishment in the field. Cultivars vary in size, shape, skin color, and other characteristics. Some have a deep purple color while others are creamy; some are globe shaped and large while others are slender and small. Plants may be spaced 2 to 3 feet (0.61 to 0.91 meter) within rows and wider (3 to 4 feet or 0.91 to 1.2 meters) between rows to accommodate their branching habit.

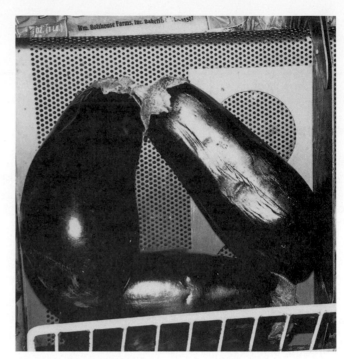

FIGURE 17-7 Eggplant *(Solanum melongena)*.

17.27.3 CARE

The eggplant crop is subject to devastation from a number of insect pests and diseases. Major diseases include verticillium wilt, fusarium wilt, *Phytophthora* fruit rot, anthracnose, phomopsis blight, and cucumber mosaic. Eggplants are also attacked by aphids, leaf hoppers, red spider mites, and flea beetles, among others. Blossom end rot, a physiological disorder, develops under adverse environmental conditions. Weeds are controlled by weeding or using herbicides. Complete fertilizers may be applied in moderate amounts to increase yield.

17.27.4 HARVESTING

Eggplants are best harvested when the fruit size desired has been achieved. If allowed to fully mature, the seeds will be hard. Eggplants are harvested by hand by cutting the fruit stem with a sharp knife. The fruit is attached tightly to the stem and very difficult to pull without damage to either the easily bruised fruit or the stem. For premium price, the fruit should be shiny, firm, and well colored. The fruit is injured by chilling; as such, when short-term storage is needed, eggplants may be stored at 45 to 50°F (7 to 10°C) and 85 percent relative humidity.

17.28: PEPPER

Common name:	Pepper
Scientific name:	*Capsicum annuum* L.
Family name:	Solanaceae

Pepper is a very widely used spice and is often associated with a hot sensation, even though mild and less pungent types occur (figure 17-8). The crop is a perennial shrub that is grown as an herbaceous annual because it is frost sensitive. Solanaceous peppers have several cultivated species, the *C. annuum* being the most commonly grown. Others are *C. frutescens, C. chinense, C. pubescence,* and *C. baccatum.* These species should be distinguished from the common black

FIGURE 17-8   Pepper *(Capsicum annuum).*

pepper *(Piper nigrum),* which is not a solanaceous pepper. Peppers are sensitive to frost, but extreme heat (about 100°F or 38°C) causes reduced yield. Peppers have a wide variety of culinary uses and may be eaten fresh, dried, cooked, or stuffed.

17.28.1 SOIL AND LAND PREPARATION

As with other solanaceous vegetables, a soil that is fertile, light textured, freely draining, and slightly acidic (pH 5.5) to neutral is preferred. The soil should be plowed and the seedbed well prepared.

17.28.2 PLANTING

Peppers differ in shape, size, color, and pungency. In terms of shape, the common ones include bell, pimento, cayenne, jalapeno, cherry, wax, and tabasco peppers. Fruit colors include red, green, yellow, and shades of purple. The fruits may hang down (pendant) or be upright. The fruit walls may be thin or thick. Perhaps the most important consideration in the choice of the type of pepper to plant is the pungency, a characteristic caused by certain chemicals, particularly *capsaicin.* Some types are mild or even sweet, while others are very pungent or hot. Peppers are commonly transplanted for rapid establishment. Spacing of plants in rows may be 1 to 1.5 feet (0.30 to 0.45 meter), while between-row spacing may be 3 feet (0.91 meter), depending on the size of the cultivars at maturity. The soil should be warm at planting to prevent damping-off attack on seedlings.

17.28.3 CARE

Weeds should be controlled in cultivation since the plant has no smothering effect on weeds. Since peppers are naturally perennial in growth habit, a good fertilizer regime should be adopted to keep the plants growing. A complete fertilizer is recommended and should be applied in moderate amounts. Peppers are attacked by a number of viruses such as tobacco mosaic, potato x virus, cucumber mosaic virus, and tomato spotted wilt virus. Nonviral diseases include verticillium wilt, fusarium wilt, bacterial soft rot, bacterial spot, *Cercospora,* anthracnose, and root-knot nematodes. Important insect pests include aphids, pepper weevils, leaf miners, pepper maggots, and flee beetles. In addition to these pathogenic diseases, peppers may succumb to physiological disorders caused by abnormal growth conditions such as injury from cold temperature, sun scald, and blossom end rot.

17.28.4 HARVESTING

Peppers may be harvested fresh or allowed to dry on the plants and harvested for processing. Most harvesting is done by hand. Fresh peppers store well for several weeks at high relative humidity (90 percent) and cool temperature (5 to 7°C or 41 to 45°F). As long as dry pepper is kept dry, it will retain its quality.

Common name:	Tomato
Scientific name:	*Lycopersicum esculentum* Mill.
Family name:	Solanaceae

Tomato is often one of the first crops a home gardener considers in selecting crops for planting. Botanically, the crop is a perennial that is grown as an annual because it is frost sensitive. A warm growing period of 12 to 16 weeks is required for good growth (figure 1-11).

17.29.1 SOIL AND LAND PREPARATION

A well-drained, medium-textured soil, rich in organic matter and slightly acidic to neutral (pH 5.5 to 7.0), is desirable. Land preparation should include incorporating organic matter into the soil.

17.29.2 PLANTING

The many tomato cultivars differ in shape, size, and plant form. The soluble solids content and viscosity differ among cultivars. Those grown for salads are either small in size or large, round, and firm for slicing. Tomato is commonly established from seedlings, even though direct seeding is possible. The plant in cultivation may be freestanding or require support (in the form of a cage, pole, or trellis). Bush cultivars (no support required) may be spaced 2 to 3 × 2 to 3 feet (0.61 to 0.91 × 0.61 to 0.91 meter) (between rows by within rows), while pole types are spaced closer (1 to 2 × 1 to 2 feet) or (0.3 to 0.61 × 0.3 to 0.61 meter).

17.29.3 CARE

Where physical support is used, the branches of the plant may have to either be tied or adjusted periodically to fit into the cage. Suckers, when they appear, are usually removed. Weeds should be controlled by shallow cultivation or herbicides. For good yield, adequate moisture should be supplied throughout the growth period. Fertilizing tomatoes increases crop yield; complete fertilizers rich in phosphorus and potassium may be applied in moderate amounts. Tomatoes are devastated by a significant number of pests and diseases. Insect pests of tomatoes include cutworms, flea beetles, hornworms, and mites. The major diseases include canker, powdery mildew, anthracnose, and botrytis gray mold. Tobacco mosaic virus, cucumber mosaic virus, and root-knot nematodes are also serious problems in production. Nonpathogenic diseases of tomato include blossom end rot (due to calcium deficiency) and sun scald.

17.29.4 HARVESTING

Tomato may be harvested as vine ripened or mature green. The decision to harvest at a particular stage of maturity depends, among other factors, on how soon the fruit will be used. Vine-ripened fruits have well-developed color, are soft, and have good flavor and chemistry (pH, sugar content, and soluble solids). They are meant for immediate use or processing or short-term storage. Mature green fruits have attained physiological maturity and are firm. Fruits harvested at this stage are for use at a later date and may have to be transported. If storage is needed for a short period, the temperature should be above 55°F (13°C) to avoid chilling injury to the fruit. Tomatoes may be used fresh or canned.

SUMMARY

Some of the most popular garden crops, including tomato and pepper, belong to the family Solanaceae. These crops are rich in minerals and vitamins. Some of the solanaceous plant species that are culturally considered annuals are botanically perennials that are adapted for cultivation as annuals. Solanaceous plants are widely adapted but prefer slightly acidic to neutral soils. Their produce may be eaten fresh or processed.

OUTCOMES ASSESSMENT

PART A

Please answer true (T) or false (F) for the following statements.

1. T F   Pepper is a perennial plant.
2. T F   Some perennial plants can be cultivated as annual plants.
3. T F   Tomato is associated with a hot sensation when eaten.
4. T F   Blossom end rot of tomato is caused by bacteria.
5. T F   Jalapeno is a kind of pepper.

PART B

Please answer the following questions.

1. Give two examples of perennial crops that can be grown as annuals.
    _____   _____
2. Give an example of a solanaceous spice.
    _____
3. _____ is the most common species of pepper.
4. The pungency of peppers is due primarily to a chemical called _____.

PART C
---

1. Describe how pepper varieties differ.
2. What are the desirable qualities of vine-ripened tomatoes?

MODULE 7 GROWING MISCELLANEOUS GARDEN CROPS

17.30: SWEET CORN

Common names: Sweet corn, maize

Scientific name: *Zea mays* L. var. *rugosa,* Bonaf.

Family name: Poaceae

Sweet corn is a tropical plant and one of the most popular garden crops. It differs from other corn varieties by its sweetness. Sweet corn is harvested and eaten fresh but may also be canned. It is a monoecious plant (figure 17-9).

FIGURE 17-9 Corn *(Zea mays)*.

17.30.1 SOIL AND LAND PREPARATION

Corn prefers mildly acidic soil of pH 6 to 6.5 that is freely draining. It is not necessary to make a bed for planting sweet corn, since it may be planted on a flat area.

17.30.2 PLANTING

Sweet corn is direct seeded. Plants may be spaced at 3 × 1 feet (0.91 × 0.31 meter) between and within rows. It is better to use several short rows than a few long ones to achieve more complete pollination. Cultivars differ in plant size, ranging from under 3 feet (0.91 meter) to 6 feet (1.8 meters) tall. Hybrid cultivars and highly improved cultivars with increased sugar content (super sweet) are available; they keep their sweetness for up to about two weeks. Most cultivars produce two large cobs per plant. The kernel color may be white or yellow and sometimes bicolored. The bicolored cultivars are more suitable for canning and freezing. Because corn is cross-pollinated, planting super sweet and regular cultivars close together will cause cross-pollination and drastically reduce the sweetness of the former.

17.30.3 CARE

Weeds should be controlled for good yield. In the garden, this may be accomplished by using a hoe or approved herbicides. Corn responds to fertilization, especially nitrogen. After an initial application of a complete fertilizer, additional nitrogen should be provided, preferably in a split application in two sidedressings. Sweet corn uses large amounts of water in production for well-filled and juicy kernels. The corn earworm, which attacks and destroys the kernels directly, is perhaps the most serious pest of sweet corn. Other diseases include blights, rusts, common smut, corn borer, head smut, and maize dwarf virus.

17.30.4 HARVESTING

Sweet corn should be harvested on time. Ripeness is indicated by the dark-brown color of the silk. At this stage, the kernels are well filled with milky sap at its peak sweetness. Because warm temperatures convert the sugar in the kernels to starch, sweet corn should be stored in a cool environment immediately after harvesting or eaten fresh. It may be processed by canning.

CHAPTER 18

Herb Gardening

PURPOSE

This chapter discusses the use of herbs and their cultivation.

EXPECTED OUTCOMES

After studying this chapter, the student should be able to

1. List 10 herbs.
2. Discuss the use of herbs in the landscape.
3. Discuss the medicinal and culinary uses of herbs.
4. Describe the general cultivation and care of herbs.

OVERVIEW

Herbs are plants that include a variety of types—annuals, biennials, perennials, woody, herbaceous, roots, leaves, bulbs, flowering, and nonflowering. They may be grown in the landscape as ornamentals, in gardens (flower and vegetable), or in containers for medicinal and other uses. They are also adapted to a wide variety of growing conditions.

18.1: USES

Characteristics of common herbs are described in table 18-1.

18.1.1 MEDICINAL

Herbal medicine was the primary form of therapy for treatment of diseases in ancient times and still plays a significant role in the treatment of diseases in many cultures today. Medicinal herbs are used in a myriad of ways. They may be brewed to extract the essential chemicals and drunk like a tea. Some are burned as incense or vaporized and inhaled, and others have essential oils that can be extracted and used as ointments or balms. Sometimes herbs are ground into a thick slurry and applied to parts of the body for a variety of purposes. Using herbs for medicinal purposes should only be on the advice of qualified practitioners and is discouraged for pregnant women.

Table 18-1 Popular Herbs and Their Parts Used

Plant	Scientific Name	Parts Used
Bay	*Laurus nobilis*	Leaves
Basil (sweet)	*Ocimum basilicum*	Leaves
Chives	*Allium schoenaprasum*	Leaves
Cumin	*Cuminum cyminum*	Seeds
Coriander	*Coriandrum sativum*	Leaves or seeds
Dill	*Anethum graveolens*	Leaves
Ginger	*Zingiber officinale*	Root
Mint	*Mentha* spp.	Leaves (seasoning or tea)
Marjoram (sweet)	*Origanum marjorana*	Leaves
Oregano	*Origanum vulgare*	Leaves
Parsley	*Petroselinum crispum*	Leaves (seasoning or garnish)
Rosemary	*Rosemarinus officinalis*	Leaves
Sage	*Salvia officinalis*	Leaves
Thyme	*Thymus vulgaris*	Leaves

18.1.2 ORNAMENTAL

Herbs have great decorative value in the landscape. They may be used in edging, for either their attractive foliage (e.g., sage, rosemary, lavender, and thyme) or color (e.g., parsley, chives, and pot marigold). Plants such as sage, hyssop, bushy thyme, and lavender can be trimmed to amenable sizes and attractive shapes.

18.1.3 FRAGRANCE

Part of the use of herbs in the landscape is to provide pleasing scents in the environment. Herbs are known for their scents, which emanate from flowers and foliage. Some of the most notable herbs with sweet scents include thyme, lavender, rosemary, and rosebuds. Certain herbs are cut and dried (leaves and flowers) and included in the mixture used to make potpourri, which is widely used in perfuming indoor environments.

18.1.4 CULINARY

A number of herbs are used in cooking to add their characteristic flavors to foods. Fresh leaves, where available, may be picked and added to foods for this purpose; dried forms may be purchased in stores for culinary use. Flowers of chives and borage, among others, may be used in salads, and seeds of dill, coriander, and cumin are good flavoring aids. Certain herbs are dried and used in brewing teas (herb teas), which have refreshing and medicinal values. Apart from these common uses, certain herbs such as hyssop, clover, thyme, and sage have flowers that attract insects such as bees and butterflies. Others, such as tansy, pyrethrum, and santolina, on the other hand, repel insects with their scent.

18.2: CULTIVATION

18.2.1 SITE SELECTION

Herbs grow vigorously and attain the most flavor when exposed to full sun. The soil must be well drained with a neutral to alkaline soil reaction. Clay and sandy soils should be avoided since they present water-related problems in cultivation, clays being prone to waterlogging and sandy soils being prone to drying out. Many herbs are adapted to marginal soils.

18.2.2 SOIL PREPARATION

Raised beds should be used where drainage is poor. If required, soils should be limed to reduce acidity. Herbs may be grown in containers in the kitchen using regular potting mixes; such herbs are readily accessible for culinary use (figure 18-1).

18.2.3 DESIGN OF AN HERB GARDEN

Herbs, as mentioned previously, have both ornamental and culinary uses. For ornamental purposes, and like other ornamental species, herbs may be set in a formal design in the landscape by judiciously selecting species according to the principles of landscaping (chapter 12). Herbs may be arranged in a fashion similar to bedding plants. In herb gardening, themes are important in creating attractive designs. These themes should take into account the adaptive characteristics, along with growth habits and other botanical features such as flower color. A popular design is the checkerboard, in which the site is divided into squares with walkways between squares and each square planted with one species.

18.2.4 PROPAGATION

The most common method of propagation of herbs is by seed. Herbs have not received the attention that vegetables and many important field crops have from the standpoint of breeding and improvement. As such, wild characteristics may be present in some species, making their germination unpredictable and sometimes very poor. Species with some degree of germination problems include rosemary, lavender, catnip, mint, and winter savoy. Instructions on seed packs should be followed with care. Vegetative propagation (by cuttings or division) may be used in propagating certain seed-bearing plants; it is the only method for seedless species such as French tarragon. Mints are widely used in the confectionery industry and have received some attention in terms of improvement. They are often propagated from division of rootstocks and runners to make them come true to type. Tansy and bergamot are also propagated from root division.

18.2.5 CARE

Herbs, being relatively wild, are quite hardy and can be produced on marginal soils, eliminating or reducing the need for additional fertility in cultivation in many cases. Herbs such as basil, chives, and parsley whose foliage is frequently cut benefit from moderate fertilization. Mulching improves moisture retention and suppresses weeds.

Herbs grown in restricted areas in containers generally need more attention than those grown in beds. They are more likely to experience moisture stress and also more prone to danger from excessive moisture. Herbs that do best indoors are those with preference for low light, unless artificial light supplementation is provided. They may be grown in window boxes or hanging baskets. Few diseases and pests are a problem in herb cultivation. Chives and mints are known to be susceptible to rust attacks. Under damp conditions, fungal diseases such as molds, mildews, and rots may occur on some herbs, as is generally the case in many plants in cultivation.

18.2.6 HARVESTING AND DRYING

Herbs may be used fresh or cut and dried. For culinary uses as seasonings, young leaves should be picked. Plants attain peak flavor when flowers open, which is the best time for harvesting for drying of stems and leaves. The plants to be dried are cut when leaves exhibit no moisture. Slow drying at low temperatures produces the best results. Drying is done not in the sun but in a well-ventilated dry room or in an oven at low temperature. This drying condition helps to retain the green color, essential oils, and aroma of the leaves and stems. Seeds are harvested when brown or black by cutting the stems. The stems are then hung upside down over a receptacle so that when completely dry the seeds will drop into a container (figure 18-2). After drying, the products (seeds and leaves) should be stored in an opaque container or in the dark to retain color and flavor.

FIGURE 18-1 Selected herbs: (a) rosemary, (b) garden sage, (c) catnip, (d) common thyme, (e) English lavender, and (f) basil.

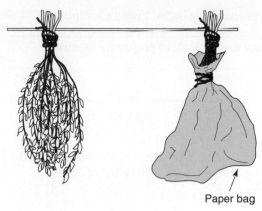

Paper bag

FIGURE 18-2 Methods of drying herbs.

SUMMARY

Herbs are relatively hardy plants that are cultivated for their medicinal value and used as flavorings in food. Some have pleasant scents and attractive blooms and can be grown as landscape plants. They generally have wild tendencies and can tolerate marginal soils. Herbs may be grown in pots or beds. Commonly propagated from seed, they may be used fresh or dried.

REFERENCES AND SUGGESTED READING

Rice, W. R., and R. P. Rice. 1993. Practical horticulture, 2d. ed. Englewoods Cliffs, N.J.: Prentice-Hall.

Stern, K. R. 1997. Introductory plant biology. Dubuque, Iowa: Wm. C. Brown Publishers.

Sunset Magazine and Book Editors. 1972. Herbs: How to grow. Menlo Park, Calif.: Sunset Lane.

OUTCOMES ASSESSMENT

PART A

Please answer true (T) or false (F) for the following statements.

1. T F   Older leaves of herbs are best for use as seasoning in foods.
2. T F   Herbs are relatively wild plants.
3. T F   For best quality, herbs should be dried rapidly.
4. T F   Pyrethrum has a scent attractive to insects.
5. T F   The most common method of propagation of herbs is by seed.

PART B

1. List five herbs of
    a. Medicinal value_____ _____ _____ _____ _____
    b. Ornamental value_____ _____ _____ _____ _____
    c. Good fragrance_____ _____ _____ _____ _____
    d. Culinary value_____ _____ _____ _____ _____

2. List two ways of propagating herbs, giving examples.

_____   _____

3. List three herbs that are difficult to propagate by seed.

_____   _____   _____

PART C

1. Discuss three factors to consider in selecting a site for growing herbs.
2. Why is slow drying desired in the processing of herbs?

PART 7

MISCELLANEOUS TOPICS

CHAPTER 19

Management of Selected Fruit Trees and Small Fruits

SECTION 1 PRODUCTION OF FRUIT AND NUT TREES

PURPOSE

The purpose of this section is to discuss the general principles and cultural practices employed in the establishment of a small-scale fruit orchard.

EXPECTED OUTCOMES

After studying this section, the student should be able to

1. Describe the environmental conditions suited to fruit and nut tree culture.
2. Discuss the general culture of pecan (nut).
3. Discuss the general culture of apple (fleshy fruit).

KEY TERMS

Crotch	Shuck	Topworking
Dichogamous	Spurs	

OVERVIEW

The science and art of producing and marketing fruits and nuts is called pomology (1.3). The classification of fruits was presented in chapter 1 (1.3.4). Vegetable fruits are discussed in chapter 17; these fruits are usually annual plants or cultivated as such (e.g., tomato and pepper). Further, the term *fruits* technically includes nuts, which are dry fruits (1.3.4). However, the terminology *fruits and nuts* is conventionally used to make the distinction between the

> **Fruit Trees**
> *Trees that bear fleshy fruits.*

629

two kinds of fruits. The discussion in this section is limited to fruits borne on trees (fruit trees). Fruits borne on shrubs and herbaceous plants (small fruits) are discussed in section 2.

Fruit-bearing trees, such as orange, pear, and plum, are conventionally called fruit trees. The apple plant is used to represent fleshy fruits, and the pecan is used to represent nuts. Their discussion is preceded by a general overview on locating and establishing an orchard. Coverage of the two plants is intended to show the general activities involved in the production of fruit trees and nuts; the specific production practices likely differ from one production area to another. Each area has its best planting dates, schedule of management operations, and cultivars. However, the principles of production are similar.

19.1: IMPORTANCE OF FRUIT AND NUT TREES

Nut Trees
Trees that bear dry, indehiscent, single-seeded fruits with a hard pericarp or shell.

Fruit and nut trees are utilized for food and are also found in the landscape (1.5). They are used as shade trees and as ornamental plants. As sources of food, fruits are rich in vitamins (A, C, B_6, and folacin) and minerals (potassium, magnesium, copper, and iron). Fruits are important sources of fiber in the diet, and the pectins they contain are known to be effective in controlling blood cholesterol levels. Important temperate fruits include apple, pear, peach, plum, cherry, and apricot. Nuts are dry fruits (1.3.4). Generally, they are also rich in vitamins (riboflavin, thiamin, and niacin) and minerals (calcium, phosphorus, iron, and potassium).

19.2: LOCATING A FRUIT ORCHARD

Orchard
A parcel of land devoted to the cultivation of fruit trees.

The success of an orchard depends on the soil, site, and management of the enterprise. Areas in the United States where commercial orchards occur in large numbers include the Great Lakes region, the Central Valley of California, the Washington Valley, and the Fort Valley Plateau of Georgia.

These areas are characterized by climatic and soil conditions that are ideal for fruit tree production. However, the home grower living outside of these ideal regions can successfully grow adapted fruit trees by observing certain basic factors described in the following sections.

19.2.1 TEMPERATURE

Temperature-related injury to fruit trees can occur in any of the four seasons—winter, spring, summer, or fall. The U.S. Department of Agriculture (USDA) hardiness zone map (3.2.1) should first be consulted in deciding the kind of fruit plants to grow. The effect of temperature is moderated by the presence of features such as hills and large bodies of water (3.2.1). Low winter temperatures can damage not only flower buds but also the whole plant. To prevent frost injuries, low-lying areas should be avoided (3.2.1). The upper parts of hills are preferred for fruit culture.

All fruit crops are prone to damage as a result of a late spring freeze, when blooming occurs. Because the blooms are sensitive to low temperature, they are easily damaged permanently when temperatures drop to about $-2.2°C$ (28°F) or lower. The freeze may occur by one of two mechanisms, which are wind dependent. On a windless, calm night, freezing can occur in spring when cold, heavy air settles in low-lying areas. This type of freeze is called *radiation freeze* (3.2.1). By locating the orchard on higher grounds, damage from this type of freeze is minimized. On the other hand, crops can be exposed to cold temperatures under windy conditions, called *advective freeze*. This wind-aided freeze is problematic because it is difficult to protect crops from it.

When late-maturing cultivars are planted, they are prone to freezing in early fall. Certain fruit trees such as apple are intolerant of the high heat prevalent in summer. The orchard should thus be located on the northern or eastern slopes in regions in the northern hemisphere.

19.2.2 LIGHT

Light is critical to fruit tree productivity, which is highest in full sunlight. It is important for light to penetrate the plant canopy to reach the inner fruiting branches for fruiting to occur; fruit trees must thus be pruned (15.13.1) to open up the canopy to increase reproductive growth. Certain fruit trees exhibit a photoperiodic response by slowing growth in one season or producing flowers only under certain light conditions.

19.2.3 WATER

Fruit trees need adequate soil moisture for proper growth and good yield of high-quality fruits. Fresh fruits may consist of about 90 percent or more water. However, they are intolerant of waterlogged conditions. In fact, a significant amount of fruit production occurs in areas where rainfall is inadequate to sustain production. These dry areas depend on irrigation for successful production. Fruit trees, especially in dry regions, are irrigated by *microirrigation methods* (drip) (3.2.1). It is critical that the trees have adequate moisture during the last 30 days of fruit development. An erratic or insufficient moisture supply leads to reduced productivity, and the fruits grown under such conditions are prone to physiological disorders (4.7).

19.2.4 SOIL

The soil should have a good texture that poses minimal resistance to root penetration and is easy to till (3.3). Sandy loams are ideal for fruit tree production. Fruits such as peach, nectarine, and apricot prefer well-drained, coarse-textured soils in the region of sand or silt loams. Cherry prefers silt loams, while apple grows well on silt loams to clay loams. Pear and quince, on the other hand, prefer finer-textured soils in the region of silt loam to clay. In considering soil, one should pay most attention to the texture of the subsoil. It might be necessary to use a subsoiler (a plow capable of being operated to depths in excess of 15 inches [38.1 centimeters] to break any pans that may occur at the site) as part of the land preparation activity.

The soil used for fruit tree production should be deep (4 to 6 feet deep or 1.21 to 1.82 meters). Peach and nectarine are deeply rooted and require deep soils, while pear and quince are not as deeply rooted. Tree roots prefer freely draining soil. Excessive moisture in the soil in spring (during bud and shoot development) is undesirable; the water table should not be high. Further, accumulated irrigation water or rain water should drain within a few days. Certain fruits such as peach and nectarine perform best under very well drained soil conditions.

19.2.5 SLOPE

The slope of the land should encourage both air and water drainage. Land that is very flat is not desirable unless it occurs at a higher elevation than the surrounding land. Thus, valley floors and river bottoms are not good fruit tree lands, because they are prone to flooding and frost damage. Land with excessive slope is difficult to cultivate and manage and soil erosion is likely. Generally, a 1 to 10 percent slope is acceptable. Further, west slopes should not be selected in cold regions or areas of high elevations since they may not have sufficient heat units (3.2.1) to properly mature the crop. North slopes minimize the potential for plants to be damaged by frost or sunburn. However, these sites provide conditions that delay the blooming of plants and thus also harvesting.

Most fruit trees perform well on slightly acidic soils (pH 5.5 to 6.5). Some plants, including plum, prefer a soil pH of 6 to 7. In terms of choosing a site for fruit tree production, one should pay more attention to the soil's physical characteristics, since soil fertility can be readily amended.

19.2.6 LABOR

For a small garden, labor is usually not a problem. However, for a large operation, labor is needed for fruit harvest and also for various pruning operations. A commercial orchard should be located where seasonal labor is readily available and affordable.

19.2.7 MARKET

Home gardens are designed primarily for home consumption. However, contingency plans should be made to handle surplus produce. The surplus can be preserved by processing it in a variety of ways (20.6). If a large operation wishes to serve the general public, markets and marketing strategies should be carefully considered. Dry fruits can be stored for long periods. However, fresh produce is highly perishable and thus markets must be known before production. Strategies for marketing are further discussed in chapter 20.

19.3: PROPAGATION

Fruit tree production depends on the selection of the right cultivar. The right cultivar is adapted to the production area, resistant to major diseases in the area, high yielding, of desirable quality, and ripens to coincide with market demands. Planting materials are of two basic types—asexual and seed.

19.3.1 ASEXUAL PROPAGATION OF FRUIT TREES

Asexually propagated planting materials are widely used in establishing orchards. These seedlings are produced by either grafting or budding (9.4 and 9.8). In selecting cultivars, one should pay attention not only to the fruit cultivars (the scion or bud cultivar) but also to the rootstock. Rootstocks are used for several purposes, as already described in chapter 9. They are resistant to soilborne diseases and insect pests and tolerant of the local soil conditions (to which the fruiting cultivar is not). Such conditions include pH, salinity, and soil moisture. Sometimes dwarfing rootstocks (9.2.1) or mauling rootstocks are used to dwarf the plant and control its size. Suitable dwarfing rootstocks are available only for certain fruits. Asexual propagation ensures that the fruits produced are true to variety (true to type). Certain fruit species are asexually propagated but without grafting or budding. The planting materials are raised from cuttings, suckers, layers, and other methods (9.1). Species amenable to such procedures include olive, fig, quince, and pomegranate.

19.3.2 PROPAGATION OF TREE FRUITS BY SEED

Propagation of fruit trees by seed results in fruits that are not true to type (due to the consequence of meiosis [5.2.3]). The degree of deviation from type depends on the species in question. Seeds are needed sometimes to produce the rootstock (or understock) used in asexual propagation. Most fruit seeds need special treatment to germinate. The causes of delay in germination for fruit seed are several, the most common being dormancy of the embryo (8.6). When embryo dormancy occurs, the seed must undergo a period of post-harvest physiological modification (after ripening [4.3.4]) at the appropriate temperature and in the presence of air and moisture. In practice, after ripening of seed is accomplished by stratification (8.7.1). It entails mixing seed with a moisture-holding material such as peat, sawdust, or even sand. These materials are also porous enough for aeration. The mixture is then held at a cool temperature in storage for the appropriate duration according to the plant species. Plum and apple seeds require stratification at 34 to 40°F (1 to 5°C) for about 60 to 90 days, while peach is stratified for 75 to 100 days at 32 to 45°F (0 to 7°C). Black walnut and hickory may be stratified at 33 to 50°F (1 to 10°C) for about 60 to 90 days.

Certain seeds experience a delay in germination caused by a hard seed coat. In this instance, the seed may be soaked in hot water (or dropped in boiling water momentarily), mechanically scratched (scarification [8.7.1]), or soaked in sulfuric acid.

19.4: The Annual Cycle of a Fruit Tree

A temperate fruit tree undergoes a certain developmental cycle during which a number of physiological and developmental changes occur. These changes are influenced by the environment. In winter, fruit trees enter a dormant period that affects seeds and buds. This type of dormancy, *endodormancy*, is caused by certain internal, physiological mechanisms. Upon exposure to cold temperature, the dormancy is effectively broken. This *winter chilling requirement* is essential for the plant to be prepared for proper development when spring arrives. For most fruit trees, the winter chilling requirement temperature is 45°F (7°C) or lower. The duration of chilling varies both within and among species.

Springtime brings the warm temperatures and heat units needed for the dormant buds that have been successfully winter chilled to develop into either flowers or shoots. At some point during the growth cycle of the tree, the flowers become pollinated, fertilized, and then produce fruits. Many trees have an inherent capacity to self-regulate the load of fruits borne during the season by the process of self-thinning. Excessive blossoms as well as fruit drop occur at certain times. However, species such as apple and peach are ineffective in self-thinning. Generally, fruits require seed development for fruits to set properly. As previously indicated, fruits differ in growth pattern. Apple development follows the classic sigmoid curve, while peach development follows the double sigmoid curve (4.1.1).

Flower buds for the next year's crop are formed in mid and late summer. Deciduous fruits generally follow this pattern. The environmental conditions must be appropriate for the desired number of fruiting branches to be formed. An unfavorable condition may cause most buds to develop into vegetative buds. Fruit tree flowers are generally perfect (2.8). However, fruits such as walnut and pecan are monoecious (2.9). Endodormancy starts in the late fall, and plants remain in this state until winter chilling occurs to break the dormancy.

19.5: Spacing Fruit Trees

The proper spacing among trees in an orchard is determined by the following:

1. *The adult size of the plant.* Trees grow slowly but eventually occupy a significant amount of space. It is important to know what size the plant will ultimately attain before deciding on plant spacing for the orchard.

2. *Rootstock.* Certain rootstocks, as previously discussed (9.7.2), have the capacity to affect the size of the fruiting cultivar. While some rootstocks have a dwarfing effect (e.g., M9 in apple), others, such as MM1110, enhance the growth of the flowering cultivar.

3. *Growing environment.* The growing environment determines how much of the plant's potential will be achieved in cultivation. Under conditions of proper temperature, high soil fertility, and adequate moisture, plants generally grow large.

4. *Predetermined planting density.* In terms of tree density, three strategies of planting are adopted in orchards. Stone fruits and nuts are highly productive under a *low-density* planting strategy (with a plant population of about 250 trees per hectare [100 trees per acre]). Using dwarfing rootstocks enables the grower to increase plant density because of the size-reducing effect of the rootstock. This practice allows a *high-density* strategy to be adopted whereby a plant population of 500 to 1,235 plants per hectare (200 to 500 plants per acre) can be achieved. This close spacing is employed under intensive plant culture; effective management and high fertility are required for success. Operations using close spacing make use of various plant training systems (15.1), coupled with regular pruning to control growth and plant size. Fruit trees may

also be spaced moderately in the orchard. This *medium-density* spacing is possible if plants are small in adult size. It allows a density of about 250 to 500 plants per hectare (100 to 200 plants per acre).

19.6: Fruit Tree Planting Styles

Trees in an orchard may be arranged in one of several ways, the most common being the *square system*. In this system, all plants are equally spaced between adjacent plants. The *quincunx* arrangement is a variation of the square system whereby the permanent crop is interplanted with a temporary crop that is grown, harvested, and completely removed from the field after several years. Consequently, the open space between trees is utilized until the trees have attained adult size. Other plant arrangements are also in use.

MODULE 1 GROWING PECAN

19.7: Soil

Soil suited to pecan *(Carya illinoensis)* production is similar to that found on river and creek bottoms where pecan is known to occur natively. The soil should be well drained, deep, and have good moisture retention, and the site should be open to full sunlight. Other general conditions for tree fruits previously discussed apply.

19.8: Planting Material

Grafted tree seedlings should be selected when a quick establishment is desired. As always, seedlings should be obtained from a reputable nursery. Bare-root seedlings are commonly used, but they need extensive care before planting to keep them in good condition. Improper preplanting handling may lead to poor establishment or even seedling death. As such, bare roots should be protected from drying and freezing. If dry, the roots should be soaked in water for several hours before planting. Seedlings that are about 5 to 6 feet (1.5 to 1.8 meters) tall should be selected. Container-grown seedlings, which are much easier to handle and establish rapidly, may also be used.

As already indicated, pecan and many other fruit trees do not produce true to type when seeds are used. However, one may start with seeds and then later topwork (15.12) with a desirable cultivar. Ungrafted seedlings may be used in the landscape with minimal training and pruning. An advantage of grafted seedlings is that the fruiting cultivar is usually an improved cultivar that bears early and has superior nut quality.

Pollination is a critical consideration in selecting cultivars for planting. Pecan trees are monoecious (2.9) and *dichogamous* (pollen maturity and shedding are not synchronized with stigma receptivity). Some cultivars produce female flowers that become receptive after the pollen is mature (early-pollen-shedding cultivars), while others produce flowers whose stigmas become receptive before pollen maturity (late-pollen-shedding cultivars). Self-pollination thus occurs only to a limited extent in pecans. To overcome the problems of dichogamy and cross-pollination, pecans should be no more than 300 feet (90 meters) from each other in the landscape. In the orchard, a mixture of four to six cultivars may be planted to ensure that pollen is readily available for the effective pollination of all flowers.

Pecan cultivars should be selected based on adaptability to the region, precocity, time to maturity, and disease resistance, as well as on tree and nut characteristics. A good tree is hardy and a heavy and early bearer. It has a good tree architecture, with strong limb angles (crotches). The nuts are large and have thin, brittle shells for easy cracking; the kernels should be of good color, flavor, and taste, and high in oil content. Precocious cultivars are early bearing (five to seven years); less-precocious types bear in 10 to 12 years.

Some good cultivars include Pawnee, Cheyenne, Sioux, Kiowa, and Mohawk. Sioux is a good cultivar for use as a yard tree, since it has a high nuts per pound value of about 62 and a percent kernel value of about 59 percent. Certain cultivars produce heavily when they are young and decline dramatically with age. However, other cultivars are able to produce consistently at an appreciable level when old.

19.9: PLANTING

The best time for planting pecan varies from one place to another, but it is usually between December and February. Fall and winter planting is best in most cases. As indicated previously, bare-root seedlings must not be allowed to dry before planting. If planting must be delayed, such seedlings should be heeled in (13.20) or covered with moist soil or damp mulching material. The planting hole should be dug large and deep enough to contain the plant roots. As in other tree planting, the stem should be set no lower than the original soil line (13.21). After planting and firming the soil, a berm should be installed around the stem to hold water (13.22) for deep soaking into the soil. Cutting back bare-root seedling foliage by half is recommended to improve chances for survival and good establishment. Further, the tree trunk should be wrapped in paper to protect it from cracking.

Spacing in the orchard should not be less than 35 feet (10.5 meters) since the adult pecan tree has an extensive root system and expansive canopy. When planted in the landscape, pecan should be located far from buildings and walkways to prevent root damage to the structures. Close spacing adversely affects canopy development and form.

19.10: TRAINING AND PRUNING

Ungrafted seedlings usually have natural desirable forms, with well-defined central leaders (15.13.1). Grafted planting material, however, must be trained to develop desirable crotch angles, good branching, and a strong trunk. Pecan should be trained to have one central leader standard. It should not be pruned in spring since the plant bleeds sap profusely; pruning should be done between autumn and midwinter. A few well-spaced scaffold branches are retained, but the shoots on the trunk below these branches are removed. Mature pecan trees do not need pruning and are adversely affected by hard pruning.

19.11: FERTILIZING

Young pecan trees should be fertilized carefully to prevent damage. While nitrogen appears to be the most important nutritional element required by pecan plants, the amount applied depends on the size and age of the plant. When plants are fertilized after June, they tend to be prone to damage by winter freeze. Frequent (biweekly during the growing season) foliar application of zinc in the early growth and development of the plants is recommended. When plants come into bearing, they should be fertilized to support a heavy nut set. Mature trees also benefit from zinc application, especially when applied before bud break. Zinc may be applied by using zinc nitrate (2 to 4 teaspoons per gallon or 10 to 20 milliliters per 3.8 liters of water) or zinc sulfate (2 teaspoons per gallon or 10 milliliters per 3.8 liters of water).

19.12: Irrigation

Young and bearing trees need adequate amounts of water provided on a consistent basis. Water is more critical in summer, when at least 2 inches (5 centimeters) should be applied each week. The root system of pecan is rather extensive, and as such microirrigation systems may be ineffective in summer. Moisture stress may cause premature nut drop or nuts to sprout before harvesting. Drought and heat stress may cause the kernel to shrivel or be poorly filled and the shuck to fail to open. In the landscape where pecans are planted in a lawn, the competitive use of water by the turf grass makes it necessary to increase the amount of water supplied to the trees.

19.13: Pest Control

Pecans cannot compete well with aggressive weeds such as bermudagrass and johnsongrass. Herbicides may be used to control weeds around trees; in the orchard, a ring of 2 to 3 feet (0.6 to 0.9 meters) of clean space should be maintained around trees.

Rodents, birds, and deer prey on pecan. Timeliness of harvest is important to reduce the damage caused by these pests.

19.14: Harvesting and Storage

Pecans are ready for harvesting when the shucks start to open. Even though harvesting and processing is easier when the shucks are wide open, delaying harvest to wait for this stage may allow pests a greater opportunity to damage the nuts. Pecans should not be stored until well dried. The nuts should not be solar dried but rather in a room with adequate ventilation. Drying is especially important when nuts are harvested in the early part of the growing season. A well-dried nut snaps when bent. Shelled nuts should be stored in cold storage separated from other odor-producing materials.

Module 2 Growing Apple

Apples are fruit trees that can be grown in a home orchard, provided the grower is willing to invest time in its proper maintenance and management.

19.15: Site and Soil Preparation

Apple trees prefer full sunlight and should be located on an area of the property where they receive sunlight in the morning and most of the day. Because the apple plant is intolerant of waterlogging, the area must be well drained. It is also important that the soil be free from soilborne diseases, especially cotton rot, which is devastating to the crop. Alkaline soils are prone to the cotton rot disease and should be avoided. Slightly acidic soils (pH of 6.0 to 6.6) are best for apple production. Deep soil cultivation stirs the soil for improved drainage and root development.

19.16: Selecting Planting Materials

Adapted cultivars must always be selected. There are numerous desirable cultivars with a wide variety of tastes, shapes, sizes, colors, and other quality characteristics. Old standards such as Golden Delicious and Red Delicious are still popular, but numerous strains of these standards that are superior in taste and other fruit qualities are available. Once the cultivar has been chosen, the planting material should be obtained from a reputable nursery. For best results, one-year-old *whips* (about 1/2 inch [1.3 centimeters] diameter and 3 feet [0.9 meters] tall) should be selected. At this stage of development, whips have a large number of buds on the lower portion of the trunk from which branches arise and can be properly trained. When older seedlings are used, additional buds will usually have to be induced by cutting the whip back. Seedlings are most successful if they have well-developed rooting systems. *Feathered* trees (15.8) that are about two years old may be used if good trees can be selected; they give a year's advantage in the training process.

The tree's characteristics (e.g., size) and productivity depend on the rootstock and the fruiting cultivar used. Like pecan, apple tree size can be controlled by grafting the fruiting cultivar on a *dwarf, semidwarf,* or *standard* rootstock. Standard rootstock grafts provide protection from the soil pathogen without affecting the tree size. The trees produced may develop to be as tall as 30 feet (9.1 meters). Semidwarf rootstocks reduce plant size by up to about 25 percent of the standard size in similar cultural conditions. Popular semidwarf rootstocks include M27, M7, MM106, and MM111, the latter two producing larger plants than the first two. Further, M7 grafted plants are early bearing. Dwarf rootstocks such as M9 reduce plant size by up to about 70 percent of the standard. These small plants are easier to manage, occupy less space, and come into bearing at the earliest stage. Apple trees raised on dwarf and semidwarf rootstocks are best suited to the home orchard. However, they are more expensive seedlings to purchase. An even more expensive planting material is called the *interstem* tree, in which the planting material consists of a short dwarf rootstock (M9 is the most common dwarf rootstock) intersected between a semidwarf understock and the fruiting cultivar. This more complicated vegetative propagation material has the effect of reducing plant size to between dwarf and semidwarf size. Dwarfed or semidwarfed cultivars are usually trained on trellises or other physical supports (15.13.2).

> **Dwarfing Rootstocks**
> Specially developed rootstocks with the ability to hormonally control plant size and rate of growth.

Other characteristics to consider in choosing planting materials are pollination and spur production. Apple cultivars vary in pollen producing capacity. An apple tree is cross-pollinated, and thus an orchard should include two or more cultivars that have good combining ability (ability to "nick") for a good fruit set. It is important that the cultivars bloom at the same time. Whereas some apple cultivars produce spurs (2.3.2), others do not. Most apples are *spur bearing* and produce fruits on short shoots called spurs. Other cultivars are tip bearing, producing their fruits exclusively or partially on the tips of the branches.

19.17: PLANTING

Bare-root seedlings should be soaked in water for about an hour before planting. Broken roots should be pruned. These seedlings should still be dormant and thus are best planted in early spring. The planting hole should be large enough to contain the roots. Care must be taken to ensure that the graft or bud union is about an inch (2.5 cm) above the soil line when the hole is filled with the backfill soil. The seedling should be watered deeply immediately after planting.

19.18: PRUNING AND TRAINING

Apples are pruned and trained initially in the first year of establishment. Regular pruning is required for good productivity, shape, and size. The adopted pruning and training systems should consider the fruiting habit. Since apple trees vary in growth and fruitfulness, it is important to prune on an individual basis.

Commonly, apple trees are trained and pruned according to the central leader system (15.12.1). Scaffold branches (15.13.1) are developed by first heading the whip back at planting or in the first year, thereby inducing new growth that can be trained. Training should emphasize the development of good crotches. A clothespin may be used in the training process (15.5) to obtain a wide crotch. Apple trees should ideally have about four to five well-spaced scaffold branches. The scaffold branches are developed in layers in the subsequent years. These layers are separated by about 24 to 36 inches (0.61 to 0.91 meters). If needed, spreaders (15.5) may be used to open up the canopy. Generally, terminal growth should be headed back 25 percent each year, making sure that the central leader is always the highest point on the tree and the ends of all branches are maintained below the height of the tree.

Spur-bearing cultivars are generally pruned by shortening the laterals to stimulate the growth of short, fruiting side shoots. Future growth from these side shoots is similarly shortened, resulting in the characteristic knobby spur system. After a period, the plant canopy becomes congested, requiring periodic thinning (15.11). In tip bearers, the fruited shoots should be reduced to renew growth.

19.19: FERTILIZING AND WATERING

Fertilizers should not be applied immediately after planting but when the tree begins to grow. The first application may consist of 10:10:10 fertilizer at the rate of 1 cup (0.24 liters) per tree, for example. This application may be repeated after about six months. Generally, fertilization in subsequent years consists primarily of nitrogen, phosphorus, and potassium, which are especially necessary when a tree matures and comes into bearing. Once the tree is mature, a fertilizer regimen should be developed and implemented for sustained productivity. This regimen must take into account the soil and the plant size (dwarf or standard).

Apple trees require adequate moisture, especially during seedling establishment. Adult plants in the home orchard should be watered weekly. Trees in the landscape may not require separate watering since they can benefit from the moisture supplied to lawns. In drier seasons, additional watering may be necessary.

19.20: FRUIT THINNING

Apple trees are known to produce profusely under good fertility and environmental conditions. If all fruits set are allowed to mature, the quality (shape, color, and size) of the fruits is reduced significantly. Further, the production in the next season is adversely affected since the number of flower buds formed in the next season is reduced. Excess fruits should be removed by thinning early in the season, before the flower buds for the next season have been initiated, which occurs about four to six weeks before full bloom. The process is accomplished by hand and must be done carefully to avoid damaging the spur or the fruits to be left for development.

19.21: PEST CONTROL

Apple trees are attacked by diseases such as apple rust, scab, bitter rot, and fire blight. Insect pests that are of economic importance include spider mites, aphids, coddling moths, and plum curculios. During the stages of early establishment, the area around the seedling should be free from weeds. Diseases and pests should be controlled to ensure a good harvest.

19.22: HARVESTING AND STORAGE

To obtain the best fruit flavor, as well as good storage quality, the fruits should be allowed to mature and ripen properly on the tree before harvesting. The stage of best harvesting depends on the region and the cultivar. Fruits may be harvested with or without an attached stalk, the latter having better storage quality. Harvesting should be done very carefully to avoid damaging the spurs or the fruits. Fresh produce should be stored in a cool place to maintain the quality of the fruit.

SUMMARY

Fruit trees (dry, or nuts, and fleshy) may be grown in the home garden, provided the proper cultivar is selected. The site for fruit trees should be well drained, be exposed to sunlight for most of the day, have a deep soil, and be protected from winter chills. Fruit trees are rich sources of minerals and vitamins. Planting material may be from seed or asexual propagation. Fruit trees do not come true from seed, and thus asexual propagation is frequently used to establish fruit tree orchards. The planting material is grafted or budded. The understock may be standard material without effect on the plant size, or it may have a mauling effect as is the case in dwarf and semidwarf rootstocks. They require much space in the garden or landscape. Many fruit trees are cross-pollinated, and thus an orchard should consist of several cultivars of the fruit tree, some of which are good pollinators. Fruit trees need regular training and pruning to maintain their size and shape and for good productivity; a sufficient moisture supply is necessary for them to produce optimally. Fertilization and pest control should be undertaken for good growth and development. Fruit trees come into fruiting after several years of establishment.

REFERENCE AND SUGGESTED READING

Lipe, J. A., L. A. Stein, G. R. McEachern, J. Begnaud, and S. Helmers. 1995. Home nut production: Pecans. College Station, Tx: Texas Agricultural Extension Service, Texas A&M University System.

OUTCOMES ASSESSMENT

PART A

Please answer true (T) or false (F) for the following statements.

1. T F   Pecan is monoecious.
2. T F   Pecan is native to river and creek bottoms.
3. T F   Pecan is wind pollinated.
4. T F   Apple trees are commonly trained by the vase system.
5. T F   A graft involving an M9 rootstock produces a large tree.
6. T F   Apple trees prefer alkaline soil conditions.
7. T F   A V-crotch angle is best for fruit trees.
8. T F   Standard trees are spaced more widely than apple trees with an M7 rootstock.

PART B

Please answer the following questions.

1. List the three kinds of understocks used in apple propagation.

    _____   _____   _____

2. List two specific symptoms that occur when pecan plants do not receive adequate zinc.

   _____ _____

3. List two diseases each of apple and pecan.

   _____ _____

4. List three specific factors to consider in locating a fruit tree orchard.

   _____ _____ _____

PART C

1. Discuss the importance of a site's slope in the location of a fruit tree orchard.
2. Discuss the pruning of fruit trees.
3. Discuss the role of tree fruits in human nutrition.

SECTION 2 PRODUCTION OF SMALL FRUITS

PURPOSE

The purpose of this section is to describe the general conditions required for the cultivation of small fruits and to present detailed cultural practices for a selected number of small fruits.

EXPECTED OUTCOMES

After studying this section, the student should be able to

1. Define the term *small fruits*.
2. List and discuss the major factors that affect the cultivation of small fruits.
3. Describe the production and management of strawberry, blueberry, and grape.

KEY TERMS

Everbearing	June bearers	Polyploid series
Hill system	Matted rows	Runners

OVERVIEW

Small Fruits
Collective term for plants that produce edible fruits of a relatively small size.

The term *small fruits* is a general one and includes a number of domesticated species that are cultivated for their berrylike fruits. Compared to fruit trees, these species are small in plant size and in terms of growth habit may be herbs, vines, bushes, or shrubs. Their fruit sizes also range from small to medium. The importance of small fruits in horticulture lies in the dual role they play as species in the landscape and sources of food. In the landscape, both the vegetative and reproductive phases of their life cycles have aesthetic appeal. Their relatively small size permits them to be grown in pots. The flowers and fruits are colorful. Small fruits are highly nutritious and are used as snack and dessert foods and in beverages. The small size of the plants permits their inclusion in small vegetable gardens by property owners with limited space.

Small fruit plants are perennial in growth cycle and often woody. The species, cultivated in temperate regions, are dicots. They are frequently propagated vegetatively, a method that is desirable considering that small fruits are highly heterozygous genetically. Unlike fruit trees, most small fruits produce harvestable fruits within one to two years of planting. They are generally easier to cultivate than fruit trees.

Perhaps the oldest small fruit in terms of domestication is the grape. This climbing vine is one of the most important crops economically in the world. It is eaten fresh or processed into beverages. Other small fruits include strawberry, raspberry, blackberry, blueberry, and cranberry.

19.23: SMALL-SCALE PRODUCTION

An estimated 20 percent of the world fruit market is devoted to small fruits. Consequently, the production of this group of crops, especially grapes, is a lucrative commercial venture. The discussions in this chapter are focused on growing a small fruit garden and using small fruit plants in the landscape. Nonetheless, the principles are applicable to both small- and large-scale production.

19.24: GENERAL PRODUCTION PRINCIPLES

Small fruits are relatively easier to establish and have quicker returns on investment than fruit trees. However, adequate planning is required for success. Some general considerations must be taken into account in planning a small-scale enterprise. The following discussion assumes that the grower intends to sell some of the produce.

19.24.1 NATURE OF THE ENTERPRISE AND USE OF PLANTS

Some home owners may have large acreages that they can devote to a small enterprise. It is important to decide the purpose of the project in order to determine the scale to which it will be organized. Certain fruit trees may blend in with the general landscape; however, some home owners may prefer to set aside a parcel of the land solely for the purpose of growing the fruit plants. They may grow fruit for the table or to sell.

19.24.2 CROP SPECIES TO GROW

The critical first step in a small fruit production enterprise is deciding what to grow, since how to grow, where to grow, and other production concerns are directly related to the kind of crop. Although a preference or liking for a particular fruit is an important consideration, growers should spend some time obtaining adequate information about the suitability and profitability of such an enterprise. Extension specialists and county agricultural agents are able to advise on the matter.

19.24.3 CAPITAL INVESTMENT AND RETURNS ON INVESTMENT

Some projects may require large initial investment of capital with slow returns on investments. Prospective growers may thus be indebted for a period of time while establishing the enterprise. If a loan is required, it is important to know how financial institutions respond to such requests. Some institutions may not be willing to finance agricultural projects for borrowers without a proven track record of success.

19.24.4 MARKETS AND MARKETING STRATEGIES

Horticultural produce is perishable and must not be produced in large quantities unless there is a reliable market or storage for it. Contingency plans must be made for times when the regular market may fail or inclement weather may disrupt the operation. If growing for the fresh market, certain cultivars are preferred over others. Fruits for processing may be harvested at a different stage than those for the table. If a u-pick operation (20.7.1) is contemplated, the location of the farm should be readily accessible to customers. If the produce will be transported over long distances to markets, packing and refrigeration or temporary storage may be necessary.

19.24.5 CULTIVARS

The cultivars selected should be adapted to the growing area and preferred by customers at targeted markets. As previously stated, they should be adapted to the marketing strategy and use.

19.24.6 SITE

After determining that the crop and cultivar or cultivars are adapted to the locality, the grower should then consider the site characteristics, including the soil characteristics (e.g., drainage, fertility, reaction, organic matter content, and depth), topography, altitude or elevation, proximity of water bodies, exposure to solar radiation, air movement over the area, precipitation, and temperature. One or more of these factors may limit small fruit production, but temperature is perhaps the most critical climatic factor that limits the adaptation of temperate fruit crops. Cold temperatures are damaging to crops. Fruit crops that have long growing seasons (such as grape and strawberry) must have an adequate frost-free period to produce economically.

19.24.7 LEVEL OF MANAGEMENT

Certain crops are labor intensive while others are not. Some may require training to produce on a large scale. If a crop is labor intensive, the cost of production may be high.

MODULE 1 GROWING STRAWBERRY

19.25: BOTANY

The strawberry (*Fragaria* spp.) is a low-growing herbaceous perennial (figure 19-1). Some woody species of the plant occur in nature. The most widely cultivated is the modern garden strawberry (or cultivated strawberry) *(Fragaria x ananassa),* which is derived from an interspecies hybridization between *F. ananassa* (pineapple strawberry) and *F. virginiana* (scarlet strawberry). Strawberry belongs to the family Rosaceae (rose family) and is a dicot and an angiosperm. Other fruit plants in this family include apple, raspberry, peach, and blackberry. The basic chromosome number of the genus *Fragaria* is $x = 7$ and occurs as a polyploid series from $2n = 14$ (diploid) to $2n = 56$ (octoploid). The modern garden strawberry is an octoploid.

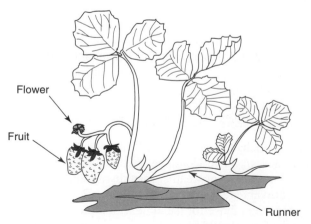

FIGURE 19-1 Strawberry plant.

Morphologically, the strawberry plant has a compressed stem called a *crown* from which leaves emerge. The leaves are compound pinnate and trifoliate, with leaflets that vary in shape from wedge to ovate. During the growing season, the axillary buds of the crown produce prostrate growths called *runners* or *stolons* (2.4.3). Numerous *daughter plants* arise from the nodes of these stolons and continue for as long as favorable cultural conditions prevail. When the bases of these new leaves come into contact with moist soil, *adventitious roots* emerge.

The strawberry has an inflorescence arranged as a *cyme.* The flowers of cultivated types are *perfect,* although *dioecious* species occur (2.8). Further, environmental conditions affect the degree of fertilization and fruit set. Wind and insects aid in the pollination of the flowers. The fruit of the strawberry is complex in structure. The hard indehiscent fruit, called an *achene,* is the true fruit and is referred to as the seed; it is not the edible portion of the fruit (2.10). The edible fleshy berry consists of an enlarged and ripened receptacle around the seed. It takes between 20 and 50 days, depending on the cultivar and growing conditions, for a berry to attain maturity and ripen after pollination. Fruits ripen slowly in the fall season but are larger in size and of higher quality, with regularity in the shape. Spring berries have irregular shapes when the temperature is low but ripen much sooner (about 30 days). The berries are borne in clusters.

19.26: Uses

Nutritionally, strawberries are rich in vitamin C and also provide fiber. They are mostly used as fresh fruits and in numerous food preparations—jams, preserves, pies, cakes, and so forth. They are often added to breakfast cereals.

19.27: Production

19.27.1 ADAPTATION

Strawberry is adapted to the various climatic zones of the United States. However, commercial production is concentrated in two areas—the Pacific Coast and the eastern Midwest.

19.27.2 SITE SELECTION

Site selection is critical to successful strawberry production. The soil should be deep, porous, well drained, and well aerated. These conditions are ideal soil for most crops, but some plants, including strawberry, are less tolerant of deviations from this ideal. The strawberry has two types of roots that develop at different times. The true roots precede the adventitious roots, which are more sensitive to the environment. Loamy soil, especially sandy loam, is desirable because it drains well and has good aeration. Good drainage is critical because fungal diseases (e.g., red steele) that occur in wet soils are devastating to strawberry production. Sandy loams are easier to fumigate, a practice employed to control soilborne diseases.

For good development and high fruit quality, the soil moisture level and fertility are also critical. The soil should have good water-holding capacity and native fertility. Gentle slopes provide for good drainage and aeration and should be selected over flat or steep slopes whenever possible. Frost is an enemy of strawberry production, and low lying areas should thus be avoided. When frost occurs during blossoming, yield is drastically affected because of flower abortion.

For commercial production, it is important and economically sound to search for land with these characteristics; however, a home owner often has little choice in site selection. The good news is that the available soil can be improved for use through the adoption of appropriate strategies. Soils with known disease problems should be avoided. Crop rotation is important since the cropping of solanaceous species such as tomato attracts soilborne diseases such as verticillium wilt.

19.27.3 SOIL PREPARATION

A soil test (3.3.3) should precede any soil preparation activity. Optimal soil pH for strawberry is between 3 and 7, although a more alkaline range of about 4 to 8 is tolerable. Where the soil is heavy (clay), drainage could be a problem. Under such conditions, soil preparation should include creating raised beds or incorporating organic matter to improve drainage and provide a warm seedbed. When sandy soils are used, drainage is not a problem, but fertility could be. Light, sandy soils benefit from the incorporation of organic matter to improve water-holding capacity and fertility and reduce the need for frequent watering.

Since the true roots of strawberry are deeply penetrating, soil preparation should loosen at least the top 12 inches (30.5 centimeters) of soil. The use of ridged beds aids in root penetration of soil. Noxious weeds (e.g., grasses such as bermudagrass) should be removed or controlled, but soilborne diseases and pests (e.g., *Verticillium, Sclerotium,* and grubs) that build up after the cultivation of certain vegetables, sod, and soybean are more of a concern. Before using such land for strawberry cultivation, it should be fumigated or the soil preparation extended for at least two years.

19.27.4 CULTIVAR SELECTION

Growers have a choice of one of three classes of strawberry cultivars, which are distinguished by the frequency of flowering and fruiting within a year. Some cultivars flower and fruit just once, others twice, and still others more than twice a year. Cultivars that flower and fruit once over a year are called *single croppers, June bearers, noneverbearing,* or *short-day* cultivars. The most commonly cultivated cultivars are June bearers. Growing noneverbearing cultivars under cool temperatures can cause them to change their flowering habit to become everbearing.

Everbearing cultivars (also called *rebloomers, perpetual,* or *day-neutral* cultivars) characteristically bear more than once a year. Some modern cultivars with everbearing characteristics also have runnerless traits and are planted much closer together to make up for the fewer runners they produce. The frequency of bearing is an important general consideration in cultivar selection, but additional specific considerations should be made for successful production.

1. *Growing for home use only.* Home growers delight in fresh produce from their gardens. Produce is harvested over a period in small quantities, and excess produce may be sold or processed. In selecting cultivars for the home, a range of maturity classes—early and late—should be grown so that the supply of fresh produce is extended.

2. *Growing for sale.* A production operation intended to produce excess for sale must first have an identified market and understand the nature of the marketing process. If wholesale markets are targeted, each crop harvest must be substantial. Further, some transportation (short and long distance) will be necessary, in which case cultivars that have firm berries should be selected. For roadside marketing, smaller-scale and more periodic harvesting is desirable. Since sales are not as predictable as for wholesale markets, cultivars with good shelf lives should be selected for a roadside enterprise.

3. *Personal or consumer preferences.* The texture, flavor, color, and size of the berries chosen depend on the grower and the intended market.

4. *Climate.* Climate affects the berry's shape, size, flavor, and a variety of physical and chemical characteristics. Temperature is especially critical among environmental factors that affect fruit quality. Because spring frost can devastate an entire production operation, the cultivars chosen should bloom after the last killing frost in a region.

5. *Disease and pest resistance.* Cultivars resistant to major diseases (e.g., fungal and nematodes) and pests are available and should be selected for planting.

19.27.5 PLANTING MATERIAL

Seed and Seeding

Strawberry may be established from seeds, which are the achenes that occur on the surfaces of fruits. These seeds are extracted, cleaned, and often scarified (to induce quick and uniform germination). Diploid species are propagated by this method. However, planting of the crop is not by direct seeding. Seeds are sowed in germination trays and the seedlings transplanted into the field.

Runner Plants

Runner plants (or daughter plants) are produced in cycles. The earlier cycle runners are more developed, having thicker crowns and well-developed root systems, leaves, and flower buds, and survive transplanting better than less-developed plants. Runner production occurs in an environment of long days and high temperatures.

> **Runner Plant**
> A plant with a slender, prostrate, above-ground stem.

Source

The highest-quality planting material should always be obtained and used for planting a crop. Disease-free and healthy plants should be purchased from a reputable nursery. The field must be ready before a purchase is made so that plants do not have to sit around long after purchase. In case of an unanticipated delay in planting, seedlings may be held in short-term cold storage. They may also be heeled in by burying their roots in shallow trenches that are kept moist.

19.27.6 PRODUCTION SYSTEMS

The types of production systems vary in cost in terms of structural provisions and labor and maintenance expenses. Annual, or single cropping, systems must be reestablished each season, making them more expensive. On the other hand, perennial, or perpetual, systems that produce for several years need to be established only once and at low density and hence low cost. However, they must be maintained throughout each year at additional cost and are prone to disease and pest buildup that continues season after season and may affect produce quality.

Strawberry may be produced under an *open system,* where plants receive minimum protection from the environment (e.g., winter mulch). The crop may be produced under a *protected system* involving the use of protective facilities that mimic greenhouse conditions and provide more consistent cultural conditions for plants.

Strawberry may be established in spring, summer, or winter. Spring planting means lower-priced planting materials. Planting in summer or winter involves planting stocks that have been further handled by the supplier by way of providing cold storage. In terms of land preparation, each season has its own challenges, the summer being drier for ground work.

19.27.7 PLANTING SYSTEMS

Strawberry may be planted in rows on flat areas or raised beds. The spacing between plants depends on the cultivar's ability to run and its vigor. Runnerless cultivars are planted closer together on the bed than those with runners. Where multiple rows are planted on a single bed, adequate spacing between rows must be provided to avoid shading and maximize light interception. Training systems in use include the following (figure 19-2):

1. *Hill system.* The distinguishing characteristic of the hill system is mother plants that are solitary because runners are discouraged and cut off when they form. The hilled mother plants develop multiple crowns and bear well the next growing season. In the hill system, plants may be planted in a single row *(single hill),* two rows *(double hill),* or four rows *(quadruple hills).* Since edge rows intercept more light and yield higher than inner rows, two-row systems are the most economical. Hills are spaced 12 to 18

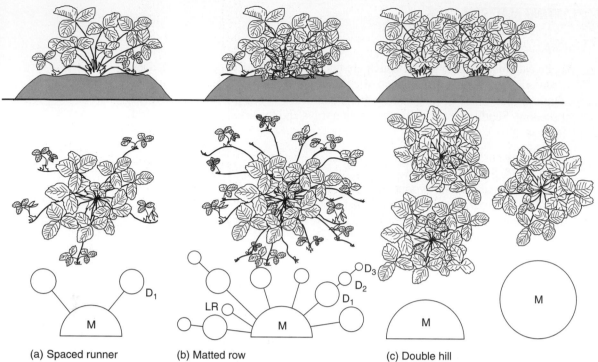

FIGURE 19-2 Common cultural systems for growing strawberry. M = mother plant; LR = lead runner; D₁, D₂, D₃, = successive daugher plants. Source: Galletta, G. J., and D. G. Himetrick (eds). 1990. *Small Fruit Crop Management.* Upper Saddle River, N.J.: Prentice-Hall. Reprinted with permission.

inches (30.5 to 45.7 centimeters) apart between and within rows. The hill system of training is suited for growing spring berries and everbearers.

2.   *Spaced runner.* The spacing between mother plants is greater (18 to 24 inches or 45.7 to 61 centimeters) in this system because runners are encouraged to some extent. When runners radiate to about 2 feet (0.61 meters) away from the mother plant, further growth is discouraged by cutting. The yield obtained from this system is good.

3.   *Matted row.* The matted-row system has the highest plant population per acre. The development of runners is allowed to proceed at random. The initial plants are set 24 inches (61 centimeters) apart in rows with 36 to 48 inches (91.4 to 122 centimeters) between rows. The relatively uninhibited spreading of runners results in a vegetative mat. This method of production is the simplest and most economic.

19.27.8 PLANTING

The plot or field should be weed free at the time of planting. Although commercial farms may mechanize their planting operations, the home garden is planted by hand. The roots of planting material should be kept moist at all times. Roots may be pruned before setting. Plants should be set such that the crown is exposed but the roots are all buried. The soil around the plant is then packed and patted firm. A spade is adequate for the planting operation. Freshly set plants need watering for quick establishment.

19.27.9 POSTPLANTING CARE

Mulching

Weeds decrease produce quality in strawberry production, but mulching helps to suppress weeds. Fumigating before planting not only controls native pathogens such as nematodes but also kills weed seeds. All of the mentioned cultural operations control weeds to varying degrees, but the use of herbicides together with mechanical cultivation is most successful and cost-effective.

Fertilizing

Strawberry is usually fertilized in late August or early September (i.e., late summer or early fall). Low rates of nitrogen (20 to 40 pounds or 9 to 18 kilograms per acre) are applied within a month of planting, depending on the cropping environment. Nitrogen fertilizer affects the fruit flavor and firmness and should be applied judiciously, especially in spring planting, to prevent soft fruits.

Winter Protection

Low-lying areas are prone to frost damage. If left unprotected, blossoms that appear in spring can be severely damaged. Strategies for averting disaster include the use of floating *row covers,* which consist of a transluscent material spread over the row and in direct contact with the growing plants. The cover is held in place by piling soil along the edges. This cover is permeable to water and light and can be left in place throughout the production period to accelerate maturity. When frost is expected, strawberry mulch may be spread over the plants for the same effect. Sprinklers are also used in frost control because the moisture prevents the blossoms from freezing. Repeated freezing and thawing damages strawberry roots, but winter kill can be minimized by deep mulching, as in the use of floating row covers.

Pollination

Commercial producers introduce bees (e.g., one to two hives per acre) to enhance pollination of blossoms for higher yield.

Irrigation

Adequate soil moisture is required for good fruit texture and yield. A variety of methods are used to supply additional moisture, including trickle and overhead sprinklers for production in drier areas.

Harvesting

Most strawberries (about 70 percent) are used as fresh fruit. Whether for fresh market or processing, the only method available for harvesting is by hand. The way fruits are picked depends on the intended use. The berries are easily bruised during harvesting and transportation. Such damage can be reduced by observing simple picking techniques. Optimum flavor is obtained if fruits are allowed to ripen properly. However, the more they ripen, the more delicate and easily damaged they become. When fruits are pulled during picking, they are easily damaged, the damage being of more severe consequence in fruits for fresh market than those for processing. For processing, fruits in some production systems are picked without stem and cap. For fresh fruits, at least the cap and in many cultures a short stem are harvested by pinching off the stem. To accommodate individual preferences and also to reduce the labor required in production, small growers may encourage customers to pick their own fruits.

Renovation

Renovation or renewal of the crop is necessary in cultures in which plants bear over several years. After each harvest, the plants must be rejuvenated. This operation involves weed control, between-row cultivation, thinning of the stand, or even mowing of the plant tops.

SUMMARY

Strawberry is a low-growing herbaceous perennial and is a popular fruit with gardeners. It is adapted to a wide variety of climatic zones. The production site should be well drained and not prone to frost. Some cultivars flower and fruit over a year and are called single croppers; others are everbearing. Strawberry is planted either from seed or runners. Annual, or single cropping, systems must be reestablished each season, and perennial cultivars require rejuvenation on an annual basis. Strawberry may be planted in the field by using one of three training systems—hill, spaced runner, or matted row. When produced on a commercial basis, bees are often introduced into the field to aid in pollination for a good yield.

OUTCOMES ASSESSMENT

PART A

Please answer true (T) or false (F) for the following statements.

1. T F   Modern strawberry is an octoploid.
2. T F   Strawberry is frost insensitive.
3. T F   Everbearers are day neutral.
4. T F   The matted-row system of training strawberries produces the lowest plant density.
5. T F   Strawberry is insect pollinated.
6. T F   Strawberry is rich in vitamin C.

PART B

Please answer the following questions.

1. Strawberry cultivars that flower and fruit over a year are called _____.
2. Strawberry may be established from seed or_____.
3. _____ is the most economical and simplest production method for strawberry.
4. Strawberry has a compressed stem called a_____.

PART C
---

1. Distinguish between the hill and matted-row systems of training strawberry.
2. Discuss the uses of strawberry.
3. Discuss the site characteristics suitable for growing strawberry.

MODULE 2 GROWING GRAPES

19.28: BOTANY

Grape belongs to the genus *Vitis* of the family Vitaceae or vine family. The United States is the center of origin to many species of grape. The *Vitis* genus consists of woody deciduous vines that are perennials. The vines have tendrils that help the grapevine to attach to structures in the vicinity for support (figure 19-3).

The stem of a grape plant is also referred to as a *trunk*. Species differ in leaf shape, color, margin, and surface. The grape flower is an inflorescence located opposite to a leaf. The length of the inflorescence differs among species. The fertilized flowers develop into berries that may contain up to four seeds each. A berry consists of a pulp in which the seeds are embedded; the pulp is wrapped in a thin film of skin that contains the chemicals responsible for the varietal pigmentation *(anthocynins)* and aromatic flavoring compounds.

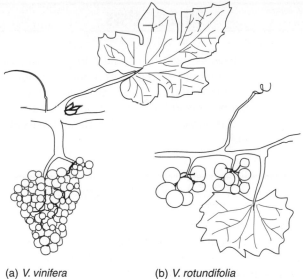

(a) *V. vinifera*   (b) *V. rotundifolia*

FIGURE 19-3   Grapes may be bunching (a) or nonbunching (b).

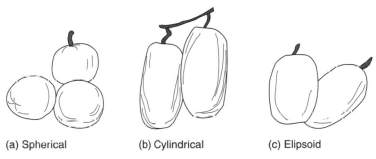

(a) Spherical   (b) Cylindrical   (c) Elipsoid

FIGURE 19-4   Shape and size variation in grapes.

The genus *Vitis* has two subgenera, the *Euvitis* and *Muscadinia*. The muscadine grapes bear berries that ripen over a period and as such are harvested as individual berries. The other group of grapes, called *bunch grapes,* bears berries that ripen uniformly in a bunch and are thus harvested in clusters.

Berries also differ in size, from small currant grapes to medium-sized wine grapes to large-sized table grapes (figure 19-4). Species grown in the United States may belong to one of the following general classes:

1. *Native American grapes.* Native American grapes are resistant to disease and valuable as rootstocks for other classes; they are also used in hybrid programs to breed new cultivars.

2. *Vinifera grapes.* Most of the grape cultivars in use are derived from the *V. vinifera* species.

3. *Hybrid grapes.* Many good cultivars have been produced from a cross between native American species and French vinifera species to produce high-yielding and disease-resistant cultivars of *Muscadinia grapes.*

4. *Muscadine grapes.* Muscadine grapes have smaller clusters than vinifera grapes. An undesirable characteristic is that mature berries do not stay attached to the vine for long but instead drop. Also, grape skins can tear during harvesting, and the grapes ripen over an extended period. Crop loss due to fruit drop may be significant.

19.29: Uses

Grapes have a wide variety of uses as fresh fruits and also processed products. The native species are used especially for improvement purposes in the development of new cultivars and rootstocks. In terms of food, the vinifera cultivars are used in wine production and eaten fresh (table grapes), canned, and processed into juice or raisins. Some cultivars are seedless. vinifera cultivars also make good rootstocks. Muscadine grapes are used extensively in making wine, champagne, and brandy.

19.30: Production

19.30.1 ADAPTATION

Native American grapes are widely adapted to U.S. growing areas and are noted for their inherent disease and pest resistance. vinifera grapes, the most widely cultivated, are relatively long-season cultivars that require high summer temperatures and mild winters for good yield. Low humidity is important, and rain is not necessary during the ripening period. The muscadine grapes are found in the southern United States, especially the cotton belt. They are sensitive to cold and should not be cultivated in areas where temperatures fall below 10°F (−12°C). Currently, grape production is not found in regions that experience less than 45°F (7°C) during the growing season because cool temperatures not only cause poor vegetative growth but also poor fruit quality.

19.30.2 SITE SELECTION

Temperature and length of growing season are the two most critical considerations in choosing a vineyard site. At least a 165-day frost-free period is required for good vine and fruit development. Locating the vineyard on higher grounds and on slopes helps to reduce the danger of frost damage that occurs in low-lying areas. Apart from elevation, large bodies of water moderate the climate, lessening the occurrence of frost in fall and spring. Production in the Great Lakes region takes advantage of the lake effect. Vineyards that are located south or east of large bodies of water experience a moderated local climate.

Good drainage is also critical to the success of a vineyard, making slopes desirable for grape production. The soil should be deep for good rooting and have adequate organic matter. Deep, loamy soils that are well drained and moderately acidic (pH 5.5) to neutral in reaction make good soils for vineyards.

19.30.3 LAND PREPARATION

The soil should be plowed deep enough to break up any hardpan that might impede drainage and root penetration. Other soil preparation operations depend on factors such as methods of irrigation, soil reaction and fertility, drainage, presence of noxious weeds, and serious soil pathogens. Where appropriate, the land should be leveled, drained, limed, or fumigated. Cover crops may be used to boost the organic matter content of the soil.

19.30.4 CULTIVAR SELECTION

Cultivars are selected on the basis of regional adaptation and use. The vineyard should be managed such that consideration is given not only to fruit production but also to vegetative quality for prolonged production. The length of the growing season and minimum winter temperature are paramount considerations in cultivar selection. In regions that have fewer than 150 frost-free days, early maturing cultivars from the American and French hybrid pool should be considered. European cultivars may be grown in regions with 180 or more frost-free days. The vines are easily damaged by cold winter temperatures, necessitating winter hardiness for cultivars grown in cool regions. Most commercial cultivars in the United States will tolerate occasional winter temperatures of −5°C (23°F).

19.30.5 PLANTING MATERIAL

Grapes may be propagated by seed or vegetative means. The most common method of propagation is the use of dormant hardwood cuttings, which should be free of viruses. In some cases, other horticultural propagation methods such as budding, layering, and grafting are employed. Grafting especially is routinely used to propagate cultivars that are susceptible to soilborne diseases such as nematodes (*Meloidogyne* spp.) and phylloxera *(Daktulospharaira vitifoliae)*. Native American species make good rootstocks. *V. vinifera* cultivars routinely require resistant rootstocks. Grafting not only protects against disease but also improves the yield, vigor, and longevity of the vines. The grower should obtain year-old planting stock consisting of rooted cuttings from a reputable nursery. If planting stock will not be planted immediately upon arrival at the site, the material should be placed in cold storage at 36°F (2.22°C) and high humidity. In lieu of storage, the plants may be heeled in as described for strawberry (19.27.5). Plant roots should never be allowed to dry before transplanting.

19.30.6 PLANTING

To reduce transplanting shock and for rapid establishment, both the roots and plants of grape are pruned. Small plots are usually hand planted. Large vineyards may be planted with mechanical planters. Instead of digging individual holes, a quick and efficient way of planting grapes involves setting the roots in about 12-inch-deep (30.5-centimeter-deep) furrows. Once set, the furrow is closed up by disking or plowing in the soil. Whether by hand or mechanical planting, care must be taken to keep the graft junction above soil level. Spacing between rows is about 10 feet (3 meters) and between 8 and 10 feet (2.4 and 3 meters) within the rows. Row spacing is determined to a large extent by the width of the machinery used in various production operations (such as harvesting). Wider spacing makes fruit harvesting easier. Row length is affected by field characteristics such as relief, the size of the plot, and the method of irrigation. The rows should be set in a direction such that plants are able to intercept maximum sunlight. In areas with strong winds, the orientation should not allow the vines to become windbreaks but instead should be parallel to the wind direction.

19.31: MANAGEMENT

Good production depends on how the grape plants are managed after transplanting. The first three years set the pace and management style for the vineyard. Each year has specific objectives. Two major management operations are *trellising* and *pruning* (15.13).

19.31.1 TRELLISING

A trellis is a support system of posts with two to three rows of wire strung tightly between them like a fence (figure 19-5). The vines are tied to the trellis and trained along the wires. Although some growers construct a trellis in the first growing season, others wait much longer. A variety of trellis designs are available, including the four-armed Kniffin, umbrella Kniffin, single curtain cordon, and Geneva double curtain. The vines are tied to the trellis and trained along the wires.

19.31.2 PRUNING AND TRAINING OF VINES

Pruning in the first years is designed to remove excess growth and encourage the development of one cane that will be the primary trunk from which vines will radiate later in growth. The training starts by tying the selected cane to the bottom wire. The pruning schedule adopted depends on the training system to be followed. For bunch growth production, the four-armed Kniffin may be used. During the first year, several rounds of pruning are required to remove flower buds and unwanted vegetative growth, and the vine must be tied to the trellis according to the requirements of the training system adopted. Other activities in the vineyard include weed and pest control, which are accomplished more easily if trellis construction is done early in the first growing season.

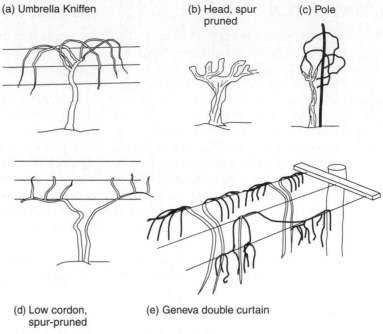

(a) Umbrella Kniffen

(b) Head, spur pruned

(c) Pole

(d) Low cordon, spur-pruned

(e) Geneva double curtain

FIGURE 19-5 Selected methods of training grapes in cultivation.

In the second year, the next round of pruning is done during the dormant period. The selected lead or primary cane (the prospective trunk) is guided to grow up to the top of the trellis. Several lateral secondary canes or vines are permitted to grow to form arms. Flower buds are removed. The third-year training prepares the plants for fruiting; pruning at this time prevents overbearing by thinning out the vines.

Trellis designs vary from one production region to another to accommodate the cultural conditions. Where plants are prone to trunk damage, two trunks may be encouraged to develop. Some growing regions have designs for table grapes that are different from those for wine grapes. No. 9 gauge steel wire is commonly used, but other gauges (nos. 10, 11, and 12) are also used. The general rule is to provide a wire of high tensile strength for the vines to be able to bear the weight of their fruit. For example, nos. 9 and 12 gauge wires may be used in combination with the no. 9 wire on top.

Good production depends on the grower's ability to prune judiciously to improve light interception, shape the vines, maintain good vigor, and control cropping for a high-quality yield of good-sized grapes. A number of other special production practices are employed by growers to improve the quality of produce, including leaf removal, berry and cluster thinning, and girdling.

19.31.3 FERTILIZER

Early spring is a good time to fertilize grapes. Grapes are fertilized by broadcast application because grape roots cover a wide area. The rate of fertilization depends on soil test results.

19.31.4 WEED CONTROL

Mechanical cultivation using a hoe is adequate to control weeds. Chemicals and mulch may also be used.

19.31.5 HARVESTING

As grapes mature and ripen, a number of chemical changes occur that affect fruit quality and flavor. Sugar content increases, as does the pH; warm weather promotes these conditions. If not picked at the right time, berries may split or shrivel in size. However, it is not always easy to tell the correct time for harvesting since fruit color may change prematurely. For table grapes, tasting may be an effective way to tell readiness for harvest. For processing, instruments such as refractometers are used to determine optimal harvesting time.

Picking grapes at the right time is necessary not only because premature or delayed harvesting causes significant yield loss but also because grapes, unlike many other fruits, do not continue to ripen once picked. Grapes for the table are usually hand picked. The entire crop of a vine is not harvested at once, but individual clusters are picked according to the extent of maturity. On the contrary, harvesting for processing is usually done with a mechanical harvester and the entire crop harvested at one time.

19.31.6 DISEASES AND PESTS

Biological enemies of grapes include a wide variety of fungi and numerous insects. For example, the grape shoot is attacked by black rot *(Guignardia bidwellii)* and the root is attacked by Armilaria root rot *(Armilaria mellea)*. The botrytis bunch rot *(Botrytis cinerea)* attacks the berries, and mildews (powder [*Uncinula nector*] and downy [*Plasmopora viticola*]) attack shoot and reproductive structures.

Similarly, a wide variety of insects attack all parts of the plant: rootworms, cane borers, flea beetles, galls, berry moths, leaf hoppers, Japanese beetles, and thrips.

SUMMARY

Many grape species originated in the United States. Grapes cultivated in the United States belong to one of four classes—native American, vinifera, hybrid, or muscadine grapes. The vinifera are the most widely cultivated and are eaten fresh or used in producing raisins or juice. Muscadine types are grown for wine production. Grapes are cold sensitive and not grown in areas where temperatures fall below 10°F (-12°C). The cultivars selected should be locally adapted and capable of prolonged production. Grapes are propagated by seed or cuttings. Good production depends on proper management that includes trellising and pruning as two major activities. For best quality, grapes must be harvested at the right time.

OUTCOMES ASSESSMENT

PART A

Please answer true (T) or false (F) for the following statements.

1. T F   Vinifera grapes have larger clusters than muscadine grapes.
2. T F   Vinifera grapes are the most widely cultivated grape types in the United States.
3. T F   Vinifera grapes are used extensively in making wine and champagne.
4. T F   Raisins are processed grapes.
5. T F   The grape fruit is a berry.

PART B

Please answer the following questions.

1. Grapes belong to the genus _____ .
2. The field where grapes are grown is called the _____ .
3. Grapes are most commonly propagated using_____ .
4. What are the two major management operations required in the vineyard?

   _____   _____
5. Grapes eaten fresh are also called _____ .

1. Describe the four-armed Kniffin pruning and training method in grape production.
2. Describe the role of warm weather in grape production.

MODULE 3 GROWING BLUEBERRY

19.32: BOTANY

Blueberry belongs to the family Evicaceae and genus *Viccinium*. Cultivated species are classified on the basis of height as either *highbush* (5 to 23 feet or 1.5 to 7 meters) or *lowbush* (less than 3 feet or 0.91 meter) (figure 19-6). The cultivated highbush varieties are derived from *V. corymbosum* L. and *V. australe* Small. The most widely grown lowbush species in the United States is *V. augastifolium*.

The blueberry flower is epigynous and borne on a raceme. The florets are urn shaped and usually have inverted pedicels. The fruit is a berry whose size depends on plant vigor; further, the larger the fruit, the more seeds it contains. The mature berry has a blue surface color that is attained by a temperature-sensitive ripening process. Tree-ripened berries have a high soluble sugar content and taste better than room-ripened fruits.

19.33: USES

Blueberry plants are cultivated primarily for their berries, which are eaten fresh or processed, similar to strawberries. In addition to food, blueberries have tremendous ornamental value as landscape plants because of their attractive characteristics across seasons. In fall, blueberry foliage turns an attractive red color, while in spring its blossoms enhance the landscape and release a pleasant fragrance into the air.

19.34: PRODUCTION

19.34.1 ADAPTATION

Commercial production is confined to areas where the crop grows wild. The regions include the Northeastern coastal stretch, north Florida, the Midwest, the Southeast, and the Northwest. There are two primary cultivated species—the *highbush* and *rabbiteye*. Native growing areas

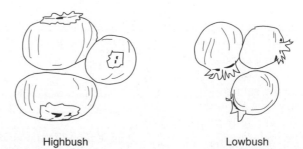

Highbush   Lowbush

FIGURE 19-6 Fruit shape and size of highbush and lowbush blueberries.

for highbush types include Michigan, New Jersey, North Carolina, Arkansas, Washington, and Oregon. Michigan is the leading producer of blueberries in the United States. Rabbiteye types are adapted to southern Georgia, northern Alabama, and northern Florida, but new varieties of highbush can also be grown successfully in the South.

Blueberries prefer acidic soils between 4.5 and 5.2 pH. Sandy loams with low pH are desirable. The soil should drain well, but adequate moisture should be provided during the growing period. A growing season of about 160 days is required for good highbush production. Prolonged exposure to temperatures of below $-20°F$ ($-28.9°C$) damages roots. Once fully opened, flower buds can be killed by even $30°F$ ($-1.11°C$) temperatures. Temperature plays a significant role in the attainment of berry sweetness during the ripening stage. Some heat energy is needed for good fruit quality, which is attained under long days and cool nights. An accumulated chilling period ranging between 250 and 850 hours, depending on the cultivar, is required to break winter dormancy and induce blossoming. Full sunlight is most desirable for high-quality produce. Cloudy and cool summers reduce yield and berry quality.

19.34.2 SITE SELECTION

Drainage is critical to good blueberry cultivation. A well-drained, acidic soil should be selected for highbush blueberry. The area should be open for the crop to receive full sunlight.

19.34.3 SOIL PREPARATION

A soil survey should be conducted and the pH amended (if necessary) to obtain moderately acidic conditions (pH 4.5 to 5.2) by using lime or sulfur, as appropriate. Organic matter should be generously incorporated into the soil to improve drainage and soil structure. With this approach, soil preparation should be done well in advance to allow the organic matter time to properly decompose. A fallow period of one year is recommended on newly cleared land. Since drainage is critical, the land may be graded to 1 foot per 300 feet (0.3 meter per 91 meters). The rows for planting may also be ridged to provide additional drainage.

19.34.4 PLANTING MATERIAL

Blueberry is propagated by cuttings (hardwood from dormant shoots and softwood from the shoots of the first seasonal growth). Hardwood is most commonly used, and cuttings from hardened one-year-old shoots are best. Cuttings are about 1 inch (2.54 centimeters) thick and 5 inches (12.7 centimeters) long and are rooted without the use of rooting hormones in propagation beds consisting of a medium of 1:1 mixture of coarse sand and acidic sphagnum peat moss. The cuttings are not slanted but set vertically in the bed. They are watered and fertilized well (e.g., 1 ounce or 28 grams per gallon or 3.8 liters of weekly application of a complete soluble fertilizer). Commercial nurseries raise rooted cuttings to sell to growers at the age of between two and three years.

19.34.5 PLANTING

The best time to transplant into the field is early spring. Fall-planted fields are prone to frost damage. A desirable plant population is about 1,000 plants per acre. This stand is attained by using a planting spacing of 4.5 × 9 feet (1.4 × 2.7 meters) or 4 × 10 feet (1.2 × 3 meters). Holes large enough for the plant roots are prepared. The plants are not planted deeper than the depth at which they were raised in the nursery. Sidedressing with a complete fertilizer (10:10:10) at the rate of 90 pounds or 40.5 kilograms per acre may be applied after transplanting. To minimize transplanting shock and loss, plant roots should not be exposed to drying; foliage may be reduced by about one-third by pruning.

19.34.6 MULCHING

Since blueberries are grown on sandy loams that drain freely, moisture retention in the soil is important to prevent the need for frequent irrigation. Mulches are very effective for controlling weeds in blueberry fields. If organic mulches are used, adding some nitrogen fertilizer to offset what is used up by microbes that decompose the material is recommended. Straw mulch should be piled at least 4 inches (10.2 centimeters) deep.

19.34.7 FERTILIZING

Fertilizers, when needed, should be applied in acidic forms so as to maintain soil acidity. Nitrogen fertilizers applied should be of the sulfate variety (not nitrate) since the crop is intolerant of chloride and sulfate fertilizers leave acidic residues in the soil. Applications of about 600 pounds per acre of 10:10:10 are commonly used. On sandy soils where the crop is best grown, split application is desirable to minimize loss through leaching. Applications are broadcast around the plants.

19.34.8 IRRIGATION

A blueberry crop should not be exposed to prolonged drought, especially during the picking season. During a dry spell, the crop should be irrigated at 1 to 2 inches (2.54 to 5.1 centimeters) at biweekly intervals.

19.34.9 POLLINATION

Because blueberries are not self-pollinating, they need additional aid in pollination for economic yield. Much of the mature pollen is shed into the air since the flowers hang upside down like bells. Sufficient pollination is effected through the aid of cross-pollinating agents such as bees, which may be introduced into the field at the rate of one to two hives per acre. Also, blueberry fields are planted as alternating rows of two or more cultivars that bloom at the same time for cross-pollination.

19.34.10 PRUNING

A management practice critical for large-scale fruit production is pruning. Pruning is usually started three years after planting. This operation is needed to prevent overbearing, which is a characteristic of most cultivars. Excessive fruiting results in numerous small fruits and reduced plant vigor for the next year's crop since fruits are produced on the previous year's wood. Effective pruning removes weak wood, opens up the canopy for good light penetration, and leads to the production of vigorous growth in the next season. If plants are pruned on a regular schedule, only light pruning is required on each occasion to be effective. Severe pruning reduces the number of bearing sites and thus lessens crop yield, but it also leads to early ripening for premium prices during the early part of the season. Pruning is done when the plants are dormant.

19.34.11 INSECTS AND DISEASES

Some insects that are significant in blueberry cultivation are blueberry blossom weevil (*Anthonomus musculus* Say), blueberry maggot (*Rhagoletis pomonella* Walsh), and blueberry bud mite (*Aceria vaccinii* Keifer). Insect infestation can be minimized through observance of phytosanitary practices. Approved insecticides may be used to chemically control these pests.

In terms of diseases, the stem canker (*Botryosphaeria corticis* Demaree and Wilcox) is devastating to production, especially in North Carolina; resistant cultivars are used in such areas. The mummy berry disease (*Monilinia vaccinii-corymbosi* Read) causes berries to dry out. This fungal disease is a serious problem to growers in New Jersey and the Pacific Northwest. Under cool temperatures and high humidity, blueberry *Botrytis* or gray mold (*B. cinerea*) can be a serious problem. Blueberry production is also impacted by viral diseases such as mosaic, shoestring, blueberry stunt, and necrotic ring spot.

19.34.12 WEED CONTROL

Weeds can be suppressed by a deep mulch application. However, clean cultivation is the most common control measure, although it is labor intensive. Herbicides may also be used for weed control.

19.34.13 BIRD CONTROL

Fruit-eating birds such as starlings and blackbirds may be a serious problem if blueberry production is located in a general farming area. When blueberries are used in the landscape, an advantage to home owners is the variety of birds that are attracted to the property when the plants are fruiting. It may be necessary to use netted cages in areas where bird attack is serious.

19.34.14 HARVESTING

Fruits desired for the fresh market are hand picked. In growing areas such as Michigan, mechanical harvesters are used to harvest a majority of the crop for processing. To retain produce freshness, fruits that will not be used immediately should be hand picked and stored in cold storage at 55°F (12.8°C).

19.34.15 RABBITEYE PRODUCTION

Rabbiteye blueberry varieties are grown in the South, starting at about midwinter. These plants are small in size and are spaced closer than highbush varieties in cultivation. Also, they are generally not pruned during production. Rabbiteye blueberry is cultivated by suckers or offshoots and sometimes by cuttings. Mulching is an effective weed-control strategy in cultivation.

SUMMARY

Blueberry belongs to the family Evicaceae and genus *Viccinium*. Fairly easy to grow in the garden, it is propagated from cuttings. Commercial production in the United States is concentrated in areas where the crop grows wild, including states in the Midwest and on the East and West Coasts. These areas grow highbush varieties of the blueberry. The lowbush types (less than 3 feet [0.9 meters] tall), also called rabbiteye, are found in the southern states. Whereas highbush varieties are planted in early spring, rabbiteye varieties are planted in midwinter. The ripened fruits may be hand picked or machine harvested (mainly for processing). Blueberries make excellent landscape plants because of their attractive red fall colors and fragrant blossoms in spring.

OUTCOMES ASSESSMENT

PART A

Please answer true (T) or false (F) for the following statements.

1. T F Blueberries are acid-loving plants.
2. T F Rabbiteye types of blueberries are grown in the Midwest and on the East Coast.
3. T F Blueberry is propagated by seed.
4. T F Blueberry is a self-pollinating plant.
5. T F The fruits of certain blueberry cultivars drop when ripened.
6. T F Tree-ripened blueberries have lower sugar content than room-ripened fruits.

PART B

Please answer the following questions.

1. List the two classes of blueberries.

    _____   _____

2.   Name five states in the United States that produce blueberry on a commercial scale.

_____   _____   _____

_____   _____

3.   Cultivated highbush varieties are derived from *V.*_____.

PART C

1.   Describe the application of fertilizers in blueberry production.
2.   Describe how pruning is done in blueberry fields for high-quality yields.

CHAPTER 20

Postharvest Handling and Marketing of Horticultural Products

PURPOSE

This chapter discusses the methods used to preserve the quality of horticultural products after harvesting and strategies for marketing products.

EXPECTED OUTCOMES

After studying this chapter, the student should be able to

1. Describe the various methods of cold storage of horticultural products.
2. Describe how products are handled while being transported.
3. Describe methods of processing products.
4. Discuss the advantages and disadvantages of product marketing strategies: u-pick, truck marketing, farmers' market, and wholesale.
5. List and describe the basic elements in a product marketing operation.
6. Discuss the role of middlemen in the marketing process.

KEY TERMS

Brine	Mechanical refrigerators	Sublimation
Forced-air cooler	Pickles	Warehousing
Freeze drying	Solar dehydration	

OVERVIEW

Horticultural crops and ornamental plants are grown with certain markets (consumers) in mind. Whereas some operations are small and limited to home consumption, others are large scale and designed to sell produce. Some horticultural products are harvested as mature and dry items, while others are harvested fresh. Some products are designated for processing and others for table use. Whether planted for the table or processing, the producer may target a multi-tier of consumers—those with discriminating taste and willing to pay premium prices and those looking for bargains. All of these and other factors affect how crops are managed in production and handled after harvesting. If ornamental plants and crops are grown for sale, a paramount consideration in postharvest activities is preservation of product quality so that it reaches the consumer in the best desirable state. The discussion in this chapter focuses on edible products. Handling of nonedible products such as cut flowers is discussed in chapter 22.

20.1: HARVESTING

Since postharvest activities have a significant bearing on how products are handled, it is useful to precede discussions on postharvest activities with how horticultural products are harvested. It is fair to say that, regarding storage performance, if poor-quality products go into storage, poor-quality products come out. Harvesting is a very timely and expensive operation. The difficulty (or ease) of the harvesting operation and how it is done depends on factors such as the crop, the part of economic importance, the growth habit (annual or perennial), the market needs or uses, the maturity pattern, and others. Peppers are harvested differently from apples. Potatoes are dug up (they have underground tubers of economic importance), while oranges are picked (they have fruits borne above ground on branches). In harvesting annuals, the whole plant may be cut or destroyed in the process without any consequence to the enterprise as a whole. On the other hand, when harvesting perennial crops such as apples and oranges, one must be careful to leave the trees healthy for production in subsequent years. The same crop may be harvested in two different ways for two different target markets. High-premium produce for the table is harvested with great care (e.g., hand picked), whereas produce for canning can have bruises and cracks without any adverse consequences. Some cultivars are determinate in growth habit and therefore exhibit even maturity and ripening. In other cultivars and certain species, the product matures at different times and hence requires multiple rounds of harvesting. Growers devise harvesting strategies to suit various production operations in order to cut costs.

20.1.1 WHEN TO HARVEST

From a physiological standpoint, a crop is ready for harvesting when it has attained *physiological maturity* (4.4). At this stage, no additional dry matter accumulates. This stage may have readily recognizable signs in some crops. *Ripening* follows physiological maturity and consists of processes that result in changes such as the fruit's texture, sugar content, flavor, and color. These qualities differ depending on whether the fruits are ripened on the vine or in storage. Generally, while vine-ripened fruits are tastier, they are also more prone to rotting. Growers often have ways of determining readiness for harvesting their crops from experience. On certain occasions, special instruments may have to be used for precision to obtain the highest quality of product. For example, the percentage of soluble solids (primarily sugars) is important in fruits such as pear, while sugar-acid balance is important in blueberry. In certain crops, color development is critical for premium price; growers may thus delay harvesting until the desired color has developed. In certain fruits, the degree of softness (upon applying a gentle squeeze or thumb pressure) is used to indicate readiness for harvest. In pod crops, the way the pod snaps may indicate maturity.

Sometimes crops are harvested prematurely for certain markets and uses. For example, green beans or peas may be harvested when the pods are only partially filled so that they can be eaten fresh. When the pod matures it becomes too stringy (fibrous) to be desirable for use in this fashion. Fiber development is also not desired in celery. In crops such as banana and others that have to be transported over long distances, harvesting is usually done at physiological maturity but before ripening. The fruits are forced to ripen artificially under a controlled environment (enriched with ethylene gas) to obtain the characteristic bright-yellow color for the market. Most horticultural products are highly perishable and should be harvested and used promptly.

> **Vine-ripened Fruits**
> Fruits picked after they are ripe and ready for immediate use.

20.1.2 METHODS OF HARVESTING

Harvesting may be done by hand or as a mechanized operation (figure 20-1). Several factors are considered in deciding on the appropriate method of harvesting a crop. Some crops offer no choice since machines have not yet been developed for harvesting them. In other cases, the product is so delicate that mechanical harvesting becomes a great challenge and is not cost-effective. Where human labor is plentiful and inexpensive, hand picking may be economical. Sometimes the acreage of the crop is too large for hand harvesting to be carried out in a timely fashion. Crops that are commonly machine harvested include corn, tomato, carrot, beet, and potato. In vegetable production, machine-aided hand harvesting is commonly used. Crops such as cauliflower, cabbage, pepper, and broccoli are directly packaged in the field and loaded onto trucks. They are hand harvested and sometimes wrapped and packed in boxes that are then placed on conveyor belts and transported into trucks. Certain crops, such as onion, require some drying or curing in the field after being dug and cut. The harvesting operation is therefore multistage; in the first pass the vegetables are cut or dug, and in the second they are picked up and transported. In the production of nuts such as almonds, fruit shakers are used for shaking down the mature fruits, which are then swept into windrows. The rows are picked up by vacuuming into bins.

Hand harvesting is labor intensive, expensive, and slow, but this produce will usually not have cracks or other injuries associated with harvesting by machines. Hand harvesting also allows fruits and other produce to be picked selectively. This approach saves the extra time and cost of cleaning and sorting after harvesting, which is required for certain crops. Mechanized harvesting, on the other hand, is generally indiscriminate, picking up good and bad fruits along with plant debris. Mechanization is also capital intensive. However, a large acreage can be harvested in a shorter time with mechanization. Further, it reduces the tedium of harvesting crops such as roots and tubers that require digging. Products are more prone to bruising and other physical injuries when mechanically harvested. Plant breeders often breed special cultivars of crops for mechanized harvesting. Mechanized harvesting is also adapted to crops that mature uniformly so that they can be harvested all at once.

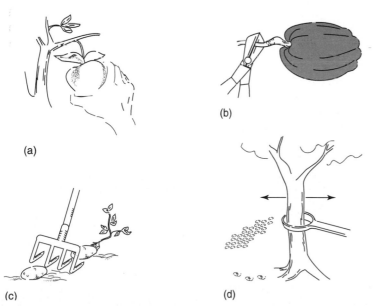

(a)

(b)

(c)

(d)

FIGURE 20-1 Some methods of harvesting horticultural produce: (a) hand picking of fruits, (b) harvesting vining fruits by cutting the vine, (c) lifting underground modified structures such as tubers of potato by the potato rake, and (d) harvesting nuts by mechanical tree shaker.

20.2: POSTHARVEST CHANGES IN PRODUCTS

As mentioned previously, many horticultural products are highly perishable and hence should be handled appropriately to maintain good quality (appearance and condition) for the market. The shelf life of produce is reduced drastically if harvesting occurs under unfavorable conditions. While some crops do not continue ripening after harvest, others do. Products tend to lose moisture from bruises or the point of attachment to the plant, which results in weight loss, wrinkling, and in extreme cases shriveling. Fruits that continue to ripen after harvesting (e.g., pear and banana) experience changes in flavor and sugar content (carbohydrates change to sugars), skin color, and softness. Dormancy may be broken in certain crops to initiate sprouting, as is the case in potatoes. If not properly protected, eventually the perishable product begins to rot.

20.3: HANDLING

Before a horticultural product reaches the table or the consumer, it undergoes a number of postharvest handling processes, as described in the following sections.

20.3.1 WASHING

Vegetables are generally not washed before packaging or marketing. They are usually managed such that they are clean while growing in the field (e.g., by mulching to prevent soil from being splashed onto the harvestable parts during rainfall or irrigation). Where chemicals (i.e., pesticides) are used in production, they should be timed such that no active residue will be on the produce at harvesting. The proper waiting period is indicated on the pesticide label. Fruits such as apples may be washed before packaging or marketing. Root vegetables and crops (e.g., carrot, turnip, and beet), whose commercial parts are developed underground, are generally washed to remove the soil before marketing. Onions and other bulbs are not washed. Because horticultural products from above-ground plant parts are not washed before packaging, the consumer should routinely wash the products before use.

20.3.2 SORTING AND GRADING

Grading
The process of assigning products to categories according to predetermined standards.

Warehousing
Holding of products in storage prior to sale.

Horticultural products destined for the fresh market are delicate and should be handled with great care. They are often hand harvested and hence are more expensive. *Sorting* and *grading* are two postharvest operations designed to group products into quality classes for pricing and use. As indicated earlier in this section, hand harvesting allows only a certain quality of product to be picked, and hence sorting and grading may be done in one operation in the field for certain crops. In fact, some products are packaged as they are picked and transported directly from the field to the intended market or consumers, without *warehousing*.

Harvested products may be hand sorted, especially those destined for the fresh market. Mechanized grading units are also used for certain crops. During the process of sorting, defective and immature products are eliminated, as are diseased products. However, cracked or broken products and those with blemishes are removed but not always discarded. They are placed in a lower-quality category or grade and sold at a lower price. Premium grade, for example, may comprise products with no blemishes, uniform in maturity and size, of good uniform color and firm flesh, and with unbruised, clean skin. A lower grade, on the other hand, may have a mixture of sizes, different maturity days, possibly some cracks in the fruit, and poor color. While lower grades may still be sold on the fresh market, they may also be sold for processing. Each crop has its own quality standards that are used for sorting and grading. Fruits are frequently graded on the basis of size.

Dry products such as nuts and grains store for long periods in dry environments. Fruits and vegetables have a much shorter shelf life, unless preservation measures are taken to prolong it. The storage conditions (especially temperature, humidity, and light) and the kind of crop (regarding quality characteristics and condition at time of storage) affect the duration of storage the crop can withstand before deteriorating (table 20-1). Even under the best conditions, poor-quality produce deteriorates rapidly. The general goals of storage are to slow the rate of respiration occurring in living tissues (which also slows the rate of microbial activity) and to conserve moisture in the tissues (prevent excessive dehydration). These goals are accomplished by providing the appropriate temperature (usually cool to cold), maintaining good levels of oxygen and carbon dioxide, and controlling humidity. Bruised products respire at a higher rate than intact ones, and the areas around the wounds become discolored. Certain crops have an inherent genetic capacity for prolonged storage. Those with dormancy mechanisms have a reduced respiration rate in storage.

As a general rule, cool-season crops are stored at low temperatures (0 to 10°C or 32 to 50°F) while warm-season crops are stored at warmer temperatures (10 to 12°C or 50 to 54°F). However, sweet corn, a warm-season crop, should not be stored at warm temperatures, which causes sugar to convert to starch, an event that reduces the sweetness of the corn. Instead, sweet corn that is not going to be used soon after harvesting should be placed in cool storage. Fresh fruits and vegetables should be stored at high relative humidity to retain their succulence and general quality. Crops such as lettuce and spinach require higher relative humidity (90 to 95 percent) than crops such as garlic and dry onion (70 to 75 percent). Darkness or subdued light should be provided in storage areas. Light may cause products such as potato tubers to green (from the development of chlorophyll). The two general methods of storage for unprocessed products are described in the following sections.

Table 20-1 Selected Vegetables and Fruits and Their Susceptibility to Injury from Cold Temperature When Stored below Freezing

Crop	Susceptibility
Fruits	
Apricot *(Prunus armeniaca)*	VS
Apple *(Pyrus malus)*	MS
Blueberry (*Vaccinium* spp.)	VS
Blackberry *(Rubus argustus)*	VS
Grape (*Vitis* spp.)	MS
Peach *(Prunus persica)*	VS
Plum *(Prunus domestica)*	VS
Squash *(Cucurbita mixta)*	VS
Pear *(Pyrus communis)*	MS
Vegetables	
Asparagus *(Asparagus officinalis)*	VS
Brussels sprouts (*Brassica oleraceae* var. *gemmifera*)	RR
Carrot (with top) *(Daucus carota)*	MS
Cucumber *(Cucumis sativus)*	VS
Celery *(Apium graveolens)*	MS
Cauliflower (*Brassica oleraceae* var. *botrytis*)	MS
Onion (dry) *(Allium cepa)*	MS
Turnip *(Brassica rapa)*	RR
Rutabaga (*Brassica campestris* var. *napobrassica*)	RR
Beet (with top) *(Beta vulgaris)*	RR
Spinach *(Spinacia oleracea)*	MS

VS, very susceptible; MS, moderately susceptible; RR, relatively resistant

20.4.1 LOW-TEMPERATURE METHODS

Fresh products retain the capacity for certain physiological activities such as respiration (4.3.2). Because respiration produces heat, ventilation in storage is critical for fresh produce to prevent excessive heat buildup, which causes rotting. The respiration rate of crops such as spinach is very high. At a given temperature, strawberries can respire about six times as much as lemons. Temperature is known to affect the rate of respiration, lower temperatures slowing all biochemical and enzymatic reactions. Temperate or cool-season crops generally tolerate lower temperatures than tropical crops, which are readily injured by cold.

Whether at home or in a commercial setting, the *mechanical refrigerator* is the mechanism for cooling (figure 20-2). Refrigerated trucks and containers are used to transport fresh horticultural products over long distances. For products such as cut flowers, strawberry, and lettuce that should be stored dry (no contact with moisture), a *forced-air cooler* system is used to pass cooled air through a stack of the product in a cold room. Some commercial growers use *vacuum cooling* for the direct field packaging of leafy crops. *Package icing* involves the use of slush ice for certain crops. In fall, vegetable produce may be stored outside in earthen mounds or trenches.

The rate of respiration is affected by the concentration of carbon dioxide and oxygen in the environment (4.3.2). Where carbon dioxide levels are very high (low oxygen), respiration slows down. The normal levels of oxygen and carbon dioxide in the air are 21 and 0.03 percent, respectively, nitrogen being 78 percent. Incidentally, in an airtight room full of fresh fruits such as apples, the oxygen soon gets used up in respiration and is replaced by the by-product of respiration (carbon dioxide). As such, after a period in storage, the carbon dioxide level reaches about 21 percent (the previous level of oxygen). At this stage, the fruits respire anaerobically (fermentation), a process that produces alcohol. Fermentation is undesirable, and hence growers should ventilate such a storage room before anaerobic respiration sets in. It has been determined that when the carbon dioxide level is raised, fruits can be stored at high temperatures of 37 to 45°F (3 to 7°C) instead of low temperatures of 30 to 32°F (−1 to 0°C).

The gaseous environment during storage can also be enriched with a variety of volatile organic compounds to influence ripening. One of the most common is ethylene, which is used to commercially ripen banana. Bananas are harvested when mature but still green in skin color. When left to ripen on the tree, they split and lose quality in terms of taste and texture. The bunch is cut into "hands" consisting of about 4 to 15 fruits. These are then shipped in polyethylene wraps. At the destination, bananas are kept at temperatures of between 13.5 and 16.5°C (56 and 62°F) and humidity of about 85 to 95 percent. While in this airtight room, 1 percent (1,000 ppm) ethylene gas is introduced for about 24 hours. To initiate ripening, additional ethylene is introduced at the rate of 0.09 cubic meters (3 cubic feet) per 90 cubic meters (3,000 cubic feet) of space. Fruits are ready for sale when they begin to yellow.

FIGURE 20-2 A typical walk-in mechanical refrigerator used in grocery stores.

Certain fruits produce ethylene naturally as they ripen. Since the gas is harmful to cut flowers, fruits (e.g., apples) should not be stored in the same room with cut flowers. The storage environment should be maintained at an appropriate humidity level to prevent excessive moisture loss from fresh products. High humidity predisposes stored products to decay. A relative humidity of 90 to 95 percent is appropriate for most fruits including apple, banana, pear, and pineapple. In the case of leafy vegetables that are prone to wilting, a high relative humidity of 95 to 100 percent is recommended. Examples of such crops are lettuce, broccoli, celery, and root crops such as carrot and turnip. On the other hand, vegetables such as garlic, dry onion, and pumpkin store better at 75 to 85 percent relative humidity. It is important to maintain good ventilation when manipulating relative humidity to prevent condensation and the accumulation of undesirable gases.

20.4.2 LOW-MOISTURE METHODS

Many crops including grape, plum, date, fig, and apple may be preserved for long periods by drying. *Solar dehydration* is a relatively inexpensive method for drying in areas where a long, dry, and reliable sunny period occurs (figure 20-3). The product is spread in appropriate containers and exposed to dry and warm air. For more rapid dehydration of large quantities of product, the *forced hot air* method, which involves air heated to 60 to 70°C (140 to 158°F), is used. This method removes water by dehydration. However, water may also be removed by *sublimation* to ice at temperatures below the freezing point by using the *freeze-drying* method. This method is expensive, but the product quality is restored by rehydration to the level of quality of products stored by freezing. As occurs in cold storage, the oxygen concentration of the storehouse may be reduced by increasing the carbon dioxide concentration. Fresh products stored in conditions of reduced oxygen (as indicated previously) respire slowly and thus deteriorate slowly.

> **Solar Dehydration**
> Drying of produce by direct energy from the sun.

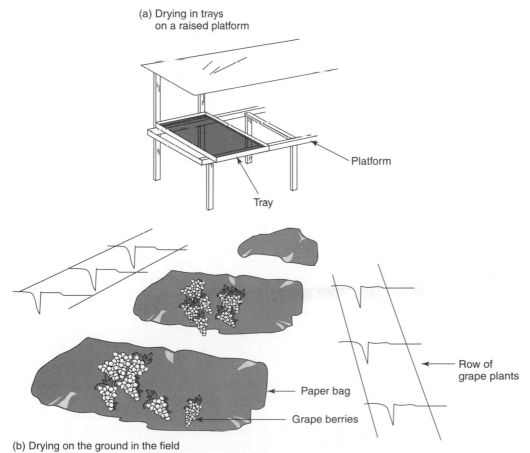

(a) Drying in trays on a raised platform

Platform

Tray

Row of grape plants

Paper bag

Grape berries

(b) Drying on the ground in the field

FIGURE 20-3   Solar drying of produce. Drying can be done in shallow trays arranged on raised platforms. Grapes are commonly dried on brown paper spread between rows.

The succulence of fresh products depends on how well they retain their moisture content. A high relative humidity reduces the rate of water loss from plant tissue, but in the presence of high temperature, the combination might encourage the growth and spread of disease-causing organisms. Storage room ventilation ensures that condensation of moisture does not occur, and that harmful gasses do not build up.

20.5: FUMIGATION

Storage of dry grain and fruits such as grapes and citrus benefits from fumigation to rid the environment of rodents, insect pests, and decay-causing organisms. One of the widely used fumigants is methyl bromide, which is effective against insects in storage houses. Sulfur dioxide is used to protect grapes from decay.

20.6: PROCESSING OF HORTICULTURAL PRODUCTS

20.6.1 FREEZING

One of the quickest and most commonly used methods of crop product preservation is *quick freezing,* whereby a fresh product is kept in a freezer. The main disadvantage of this method is the damage it causes to the physical or structural integrity of some products. For example, frozen tomato does not remain firm after thawing but assumes a soft texture; consequently, use of the product may be limited by freezing. For best results, freezing should be done rapidly. Slow freezing causes larger ice crystals to form in the cells of the tissue and ruptures them. These large reservoirs of water in fractured cells give frozen products a soft texture upon thawing. Rapid freezing results in tiny crystals that do not rupture cells. Quick freezing temperatures are around -29 to $14°C$ (-20 to $40°F$). Stored produce may lose some color, flavor, and nutrients. To protect against dehydration, products to be frozen must be packaged (e.g., in plastic wrap). Failure to do so will lead to *freezer burn,* resulting from sublimation of water to ice, with adverse consequences such as deterioration of flavor, color, and texture.

20.6.2 CANNING

A common household processing method is the preservation of products in sterilized water. *Canning* is another method used in processing (figure 20-4). After placing the products in airtight or hermetically sealed containers, they are sterilized in a pressure cooker. Instead of using water, *brine* (a salt solution) may be used to process vegetables such as onion, beet, and pepper. The intense heat used in sterilization changes some quality traits such as color, texture, and flavor, as well as nutritional value. Low-acid products (pH 4.5 to 7.0) such as vegetables require very high temperatures for sterilization to kill the bacteria that cause food poisoning *(Clostridium botulinum).* Canned products can stay in good condition for several years. However, because heat treatment does not kill all bacteria, spoilage sets in after some time in storage. The salt in canning corrodes the can and reduces the shelf life. Likewise, humidity and high temperatures accelerate spoilage.

20.6.3 FERMENTATION

Fermentation involves bacteria that decompose carbohydrates anaerobically. Some of the products of fermentation prevent the growth of bacteria. The products differ according to the organism, conditions, and duration of the process. Fermentation may produce alcohol and lactic acid, products that affect the flavor of fermented foods. Alcohol may further ferment to produce vinegar. Certain fruit juices are deliberately fermented to produce alcoholic beverages (e.g., grape juice becomes wine). A special fermentation process involving the use of salt is called *pickling*. Vegetables that are pickled include cucumber, onion, cauliflower, and

Pickling
Fermentation of produce in high concentration salt.

FIGURE 20-4 Preservation of fruits. Fruits are processed to varying degrees in canning. They may be ground into sauce, diced, sliced, or canned whole.

tomato; pickled cabbage is called *sauerkraut*. Instead of using bacteria in pickling, pickles may be produced by placing products directly in citric acid or vinegar.

20.6.4 PROCESSING WITH SUGAR

High concentrations of sugar may be used to process certain fruit products. The sugar increases the osmotic pressure to a degree that prohibits microbial activity and thereby reduces spoilage opportunities. Different fruit products may be preserved in this way. When fruit juice is used, the product is called *jelly*. *Jam* involves concentrated fruit, while *marmalade* is sugar-processed citrus fruit and rind. When whole fruits are used, the product is called a *preserve*.

20.7: MARKETING

Marketing in its simplest form entails the supply of satisfactory products by a producer to a consumer at a price acceptable to both. In more advanced market economies (and even in less advanced ones), where division of labor occurs, a host of service providers (called *middlemen*) operate between the producer and the consumer. The services provided include packaging, storage, transportation, financing, and distribution. Sometimes the fresh product changes in nature between the farm gate and the consumer's door, as is the case when middlemen *add value* to the product by processing it into other secondary products. In spite of the activities of middlemen, some growers are able to deal directly with consumers. Notwithstanding the mechanics of marketing, certain general principles and characteristics of a horticultural enterprise should be well understood:

> **Middlemen**
> Persons who provide services that link the producer to the consumer.

> **Add Value**
> The processing of raw products into secondary products, the latter being more expensive than the former.

1. Horticultural products are highly perishable; they lose quality rapidly.
2. Many horticultural products are bulky to transport.
3. Prices for horticultural products are not stable.
4. Some storage may be required in a production enterprise.
5. It is important to identify a market before producing horticultural products.

Ultimately, the market is the most critical consideration in a crop production enterprise. Growers should not grow what they cannot sell, since horticultural products are highly perishable. The various marketing strategies are described in the following sections.

20.7.1 U-PICK

U-pick or *pick-your-own* (PYO) operations are self-harvest enterprises in which the customer walks through the farm to pick the quantity and kinds of items he or she needs. The farmer then prices the harvested items and sells them to the customer. Such operations provide an opportunity to minimize the cost of labor involved in harvesting. They must be readily accessible to customers to be successful. The buyer is provided fresh produce, which he or she handles as desired. The grower should have contingency plans to sell or use what is not sold at the farm.

20.7.2 TRUCK OR ROADSIDE MARKETING

Truck or *roadside marketing* (figure 20-5) involves selling products from the back of a truck or from a roadside stand. The grower transports the products to a strategic spot and displays them in the back of a truck and on tables and in containers arranged nearby. The advantage of a truck operation is that the grower can move the point of sale from place to place. Alternatively, a permanent or movable stall may be erected at a strategic location for use in marketing the products. This method of marketing is usually located at road intersections or other strategic locations along busy roads.

20.7.3 FARMERS' MARKET

Producers may transport their products to a central place where buyers may negotiate the prices of items. Farmers' markets usually convene periodically, such as on a specific day of the week. By setting a time, consumers and growers are able to plan ahead to participate. The variety of items offered for sale is large, and quantities are also large. Because of the competitive nature of marketing, consumers can often purchase items at good prices.

20.7.4 WHOLESALE

Wholesale
The sale of products in large quantities or in bulk.

Growers may have contracts with buyers to supply certain quantities of produce at regular intervals. The contract is for a certain quality and quantity at a predetermined price. This large quantity or bulk sale *(wholesale)* is made to customers (grocery stores) who then retail the items at higher prices. Wholesaling is often undertaken by middlemen, who may package the bulked items into smaller units. Wholesalers must have some kind of storage to hold the products in good condition until they are sold.

20.7.5 RETAILERS

Retailers are people in the marketing process who sell units of products to customers. Retailing may be undertaken by grocery stores (figure 20-6), which require temporary storage and display systems (e.g., shelves, cabinets, and containers).

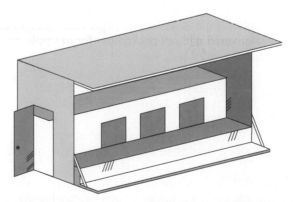

FIGURE 20-5 Fresh horticultural produce is commonly sold in roadside stalls or from trucks.

FIGURE 20-6 Fresh and dried fruits are sold in small quantities or individually by retailers.

20.8: THE ROLE OF MIDDLEMEN IN MARKETING

U-pick and truck sales provide customers direct access to the grower. The product is usually fresh, and the cost is generally lower to the buyer because the grower has less overhead to pass on to consumers. A grower's operation may not be equipped to service individual customers, a role filled by middlemen. Middlemen, acting as intermediaries, help in distributing products to customers over a large area. They provide warehousing and storage facilities and transportation that the grower may not have the resources to own or manage. Some middlemen are processors who are able to preserve the fresh product, thereby extending its shelf life. Because of the added value to fresh farm products, the resulting products usually sell for much higher prices.

20.9: ELEMENTS OF MARKETING

Marketing is a very complex and involved operation. To be successful in marketing agricultural products, the following should be considered:

1. *Packaging.* Many horticultural products are delicate and bruise easily and thus should be packaged or assembled in appropriate containers whenever they are moved or transported. Because some items do not tolerate being piled one upon the other or touching and rubbing against each other, they may be wrapped individually. A variety of packaging materials are available for use (figure 20-7). Agricultural products for processing may be delivered without packaging; for example, tomatoes may be delivered by the truckload directly to processing plants.

2. *Storage.* Proper storage is required at various stages in the marketing process. Anytime the product will not be sold immediately, storage is necessary to preserve quality at least until it is sold. When fresh products have to be moved over long distances, special provisions should be made for storage in transit. In terms of storage conditions, some products simply need adequate ventilation so that heat does not build up in the package. Processed products (e.g., canned foods) often do not require refrigeration but should be stored in a cool, dry place.

3. *Transportation.* Mechanically refrigerated vehicles, containers, and the like are used to transport fresh products over long distances by road, rail, water, and air.

(a)   (b)

FIGURE 20-7   Horticultural produce is packaged in a variety of ways, fresh or processed.

4. *Distribution.* An effective and efficient distribution network is critical to marketing success. Marketing thrives on supply and demand principles. Products should be moved in a timely fashion to where they are needed when they are needed. Retail chains aid in the distribution of products. To keep products fresh, suppliers are first sought in an area close to the customers.

5. *Financing.* Financing is required not only for marketing but also for the entire production operation. Horticultural crop production is a high-risk operation since products are perishable and cannot be stored for long periods of time. Prices are also unstable in this industry.

SUMMARY

Many horticultural products such as vegetables are perishable and should be used soon after harvesting or as appropriate. Products that are destined for the market may be washed, especially underground parts or tree fruits that were sprayed with pesticides before harvesting. Since crop products do not usually have identical features, the grower often has to sort and grade them into quality classes. During transportation or holding (warehousing) before sale, horticultural products often are stored under cold temperature conditions using mechanical refrigerators or air coolers, for example. Some products may be processed by solar dehydration or freeze-drying, pickling or canning, or other methods of preservation. Those who grow crops with the intention of selling have several options in terms of marketing strategies, including u-pick or self-harvest enterprises, truck or roadside marketing, farmers' markets, or wholesale operations. Marketing is complex, involving packaging, storage, transportation, and distribution.

REFERENCES AND SUGGESTED READING

Davis, D. C. 1980. Moisture control and storage systems for vegetable crops. In Drying and storage of agricultural crops, edited by C. W. Hall. Westport, Conn.: AVI Publishing.

Finney, E. E., Jr., ed. 1981. CRC handbook of transportation and marketing in agriculture, Vol. I. Food commodities. Boca Raton, Fla.: CRC Press.

O'Brien, M., B. F. Cargill, and R. B. Friedley. 1983. Principles and practices for harvesting and handling of fruits and nuts. Westport, Conn.: AVI Publishing.

Peleg, K. 1985. Produce handling, packaging, and distribution. Westport, Conn.: AVI Publishing.

Penson, J., R. Pope, and M. Cook. 1986. Introduction to agricultural economics. Englewood Cliffs, N.J.: Prentice-Hall.

OUTCOMES ASSESSMENT

PART A

Please answer true (T) or false (F) for the following statements.

1. T F   All underground edible products are washed before use.
2. T F   Cool-season crops are generally stored at low temperatures.
3. T F   Storage of products is best in strong light.
4. T F   Brine-processed cucumbers are called pickles.
5. T F   Processing adds value to fresh products.

PART B

Please answer the following questions.

1. Crops such as spinach and lettuce require relatively higher or lower relative humidity than dry onion and garlic while in storage? _____
2. The workhorse for cooling in horticultural product storage is _____.
3. Dehydration by using cold temperatures below freezing is called_____ .
4. List three specific marketing strategies for horticultural products.

    _____ _____ _____
5. List four critical elements of developing a marketing strategy.

    _____ _____

    _____ _____

PART C
---

1. Describe the operation of sorting and grading in horticultural production.
2. Describe the effect of warm temperatures on the quality of sweet corn.
3. Describe how deterioration of fresh produce in storage from respiration can be reduced.
4. Why is quick freezing not recommended for produce such as fresh tomatoes?

Organic Farming

PURPOSE

This chapter is designed to show the importance of organic matter in plant nutrition and how it is utilized as the principal source of additional fertility, apart from rock minerals. Methods of composting are also described.

EXPECTED OUTCOMES

After studying this chapter, the student should be able to

1. List the major sources of organic matter for gardening.
2. Describe the importance of organic matter in soil fertility.
3. Discuss the sources and importance of mulches.
4. Discuss the methods of disease and pest control in organic farming.
5. Describe how composting can be done on a small scale.

KEY TERMS

Activators	Catch crop	Mesophiles
Biological control	Compost (composting)	Mineralization
Carbon-to-nitrogen (C:N) ratio	Decomposers	Mulch
	Green manuring	Thermophiles

OVERVIEW

Is *organic farming* a new method of growing crops or an ancient practice? Like many bandwagon terms (such as *biotechnology*), there is always the question of novelty. The term has gained prominence in society in the face of a strong move toward healthy eating and environmental protection. Organic farming is certainly not a new method of growing crops. In fact, it is as old as agriculture itself. To appreciate the resurgence into the limelight of organic farming, it is important to understand how it differs from modern methods of cultivation. Modern methods of plant cultivation are greatly influenced by technological advancements that enable scientists to duplicate or mimic natural conditions. To qualify as an organic producer, one must be duly certified by the state after complying with certain standards.

The fundamental differences between the two systems of plant production lie in the ways in which plants receive nutrition and protection from diseases and pests in cultivation. However, what launched organic farming into the limelight were the side effects of modern crop production practices on the general environment. Natural resources in the soil environment for plant growth are exhaustible. Unless there is a way to replenish the nutrients extracted from the soil by plants, the land soon becomes depleted of plant food and consequently incapable of supporting plant growth. Natural nutrient recycling systems exist to replenish the soil nutrients. These cycles are largely dependent on the activities of soil microbes and occur at a slow pace. Nutrient recycling is the organic way of fertilizing crops. Scientists have identified these plant growth nutrients and produced synthetic versions (chemical fertilizers) that are easier to regulate and apply.

Regarding disease and pest control, organic farming depends on nature for control, while modern farming depends on chemicals. Nature has a way of regulating itself, maintaining balance so that one organism does not dominate the community. Further, prey-predator relationships occur in nature; some potentially harmful organisms fortunately have natural enemies. Pesticides often indiscriminately kill organisms, the good and the bad.

Organic production is not without its problems or critics. It is generally a more expensive cultural practice. Crop yields and quality are significantly lower than under traditional production. People are willing to pay more for organically produced crops because of the fear of chemical residues from the application of pesticides under traditional systems of vegetable production. However, the absence of chemicals in the production system may encourage the presence of natural toxins, which could also be harmful. Disease and insect pest buildup is increased under this system of production. In fact, the benefit to humans of organic production is largely experienced through a safer environment because of the reduced amounts of chemicals entering the soil and polluting the water supply.

This chapter is devoted to the discussion of the principles and practices of organic farming. In practice, organic farming or gardening, like minimum or zero tillage, is a method of cultivation of plants in which inorganic production inputs are eliminated or used at the bare minimum. Nature is depended on to maintain a healthy balance for sustained production. *Sustainable crop production* incorporates the strategies of integrated pest management (IPM) (7.5) and the principles underlying the various *best management practices* to produce crops under an ecologically sound cultural environment. Under this system, the productivity is high even though inputs, synthetically produced and other tillage practices, are minimized.

21.1: PRINCIPLES OF ORGANIC FARMING

21.1.1 ENVIRONMENTAL FACTORS

The environmental factors for production are the same in organic and "inorganic" gardening. Climatic and weather factors (e.g., wind [air], rain [water], light, temperature, and nutrition) are critical to crop production under any system. In addition to these abiotic factors, there are biotic factors, some of which are detrimental to crop production because they are pests. For successful production, the good factors should be provided in adequate amounts and the detrimental factors controlled or eliminated altogether. The following are essential factors to be considered for success in organic farming.

21.1.2 SOIL

The soil comes first in plant production. A major challenge in organic farming is obtaining and sustaining a system that enables the soil to replenish its nutrients after a cycle of crop production. A typical soil consists of 45 percent mineral, 25 percent water, 25 percent air, and 5 percent organic matter (chapter 3). Organic matter, in spite of its small relative proportion, is critical in the nutrient recycling processes that restore fertility to the soil. Plants absorb mineral elements from the soil and fix them as organic matter in the form of plant tissues. These immobilized elements are returned to the soil when plants die and decompose. Since decomposition is required for mineralization of fixed elements (conversion of organic matter into its component inorganic minerals), soil microbes are critical to the success of the nutrient recycling system and must be present in adequate numbers (3.5).

The soil should be properly managed to maintain a good amount of organic matter and other beneficial organisms (e.g., earthworms) and fungi (especially *mycorrhizae* [3.5]). The soil physical conditions should permit good drainage and aeration for the microbes to function properly. Apart from physical characteristics, the soil pH (soil reaction [3.3.3]) is also critical to the survival of soil organisms. A pH of less than 4.5 makes the soil too acidic for soil organisms to thrive. The gardener who wishes to use an acidic soil should consider liming to raise the pH to a desirable level of about 6.5.

The organic matter level in the soil can be boosted by adding *compost* (21.2) to garden soil. Composting is a principal activity undertaken by organic gardeners. Plowing under of certain plant remains after harvesting and *green manuring* help increase the soil organic matter. Organic matter and clay particles are the two *colloids* in the soil that have the capacity to hold minerals in ionic form.

For enhanced microbial activity and an adequate population of good soil organisms, the organic gardener should pay attention to drainage. A well-drained soil has good aeration for aerobic respiration and consequently decomposition of organic matter. Clay soil (heavy soil) is prone to waterlogging. If such soils are to be utilized for crop production, they should be drained (e.g., by tiling). Organic matter may be added to improve drainage. Instead of planting on the flat, raised beds may be used to improve drainage. Raised beds improve soil drainage, and well-drained soil is warm and hence encourages microbial activity.

Although drainage is important, excessive drainage is detrimental to crop production. Soils under such conditions become dry quickly and also lose nutrients through leaching. As much as possible, the farmer should avoid leaving the soil bare and unprotected, especially in the warm season when nutrients are most soluble.

Green Manure

Plants purposely grown to be plowed or mixed into the soil to rot and add organic matter to the soil.

21.1.3 IMPORTANCE OF ORGANIC MATTER

A typical mineral soil consists of 5 percent organic matter (3.3.1). Organic matter has a variety of important roles in agricultural production:

1. *Nutrient supply.* When decomposed, the immobilized nutrients become mineralized for plant use. Organic matter is also a source of food for soil organisms.

2. *Soil moisture retention.* Plant tissue absorbs moisture and makes it available at a later time for plant use.

3. *Aeration.* Plant fibers improve soil structure and open it up for aeration.

4. *Warming of soil.* Organic matter gives soil a darker color that makes it absorb heat. By improving drainage and aeration, the soil becomes warmer.

5. *As a mulch.* Organic matter applied as a mulch protects the soil from destruction. It suppresses weeds, improves water retention, and keeps the soil warm.

21.1.4 SOURCES OF ORGANIC MATTER

Organisms, both plants and animals, provide materials that can be incorporated into the soil to supply organic matter. The major sources of organic matter for gardeners are described in the following sections.

Green Manures

Green manure plants are those planted for the sole purpose of being plowed into the soil at a later date. These plants are usually quick growing and able to establish a good ground cover. They are often legumes (such as alfalfa, clover, and trefoil) and hence have the added advantage of fixing nitrogen. Annual species such as lupin stay in the ground for about two to three months, while perennials such as alfalfa may stay in the ground for over a year. The advantage of these plants is that they act as *catch crops* that utilize the nutrients that might otherwise be leached out of the soil to manufacture food for plant growth and tissue development. All of these nutrients are returned to the soil for plant use on decomposition.

Animal Manures

Animal manures are good sources of organic matter. They are relatively rich in nitrogen, phosphorus, and potassium. The main disadvantage in their use is their bulkiness and salt content. For practical purposes, what is used depends on what is obtainable within close vicinity of the farm; that is, if a crop farm is located near a cattle ranch, cow manure may be the most practical choice. Rather than using manures directly in the field, they may be included in a compost heap. Cow manure is relatively low in nitrogen, phosphorus, and potassium, while poultry manure is very high in nitrogen and phosphorus. Further, while the nitrogen in cow manure is stable, that from poultry manure is highly volatile and is responsible for the ammonia smell associated with poultry manure. Treated sewage sludge is a useful source of plant growth nutrients.

Compost

Compost and composting are discussed in detail in section 21.2.

Mulch

Mulching is the practice of spreading a material over the soil surface. The purposes for mulching are varied and include aesthetic reasons and benefits for the growth and production of plants. Often, a combination of benefits are derived from using a mulch. The materials that can be used are varied.

21.1.5   TYPES OF MULCHES

Mulches can be organic or inorganic in nature (figure 21-1).

Organic Mulches

Common organic mulches include the following:

1.   *Straw.* Straw is the most common type of mulch. It is slow to decay and used widely in vegetable gardening. Straw is obtained from grass plants and has no aesthetic value.

FIGURE 21-1   An organic mulch.

2. *Bark.* Tree bark is obtained from either softwood or hardwood and processed into different sizes. Bark is very resistant to decay and thus remains for a long time in the landscape. It is used for decorative purposes.

3. *Wood chips.* After pruning, the branches may be chipped and used as mulch. Wood chips may be used like bark for aesthetic purposes. They decay slowly and thus last long in the landscape.

4. *Shredded leaves.* In fall, leaves may be gathered and shredded. A disadvantage of this type of mulch is its light weight, making it prone to being blown away by strong winds. Shredded leaves are used in compost heaps.

When using organic mulches, care should be taken in the choice of the source. Certain species such as Douglas fir, walnut, and white pine carry phytotoxins that are injurious to other species. To be safe, it is best to use products from these species in a treated form (included in compost) or limit their application to within the species. Bark mulches may also increase soil acidity.

Inorganic Mulches

Common inorganic mulches include the following:

1. *Plastic.* Polyethylene sheet mulches are usually impermeable to water and durable. Widely used in vegetable production, they have no aesthetic value.

2. *Gravel and stones.* Gravel and crushed stones are used in the landscape as mulches (figure 21-2). They are not used in the vegetable garden. Sometimes they are used indoors in pots, but mostly they are used outdoors.

21.1.6 PURPOSES OF MULCHING

Weed Suppression

A layer of mulch suppresses weeds by preventing them from receiving light. Impermeable plastic sheets provide an effective weed-suppressing material. Cheaper materials such as straw should be applied about 6 to 12 inches (15.2 to 30.5 centimeters) deep for greatest effectiveness. It is best to apply the mulch before the weeds grow. Organic mulches tend to settle with time. Some are less resistant and decompose after a short time. As such, it may be necessary to add fresh material to the original layer to make it effective in controlling weeds.

FIGURE 21-2 Gravel mulch is primarily for decorative purposes.

Moisture Retention

Mulches are effective in reducing soil moisture loss from evaporation. Impermeable plastic mulches are the most effective for retaining moisture. Organic mulches are also effective, but when moisture is applied by irrigation or through rainfall, the amount should be adequate to reach the soil. A sprinkle or light shower over a 12-inch-deep (30.5-centimeter-deep) straw mulch is of no benefit to plants, since the moisture will hardly reach the soil.

Improving Soil Fertility

Permanent organic mulches that are readily decomposed eventually become part of the soil. While in place, they provide an environment that is conducive to the growth and activity of soil organisms. They keep the soil moist and warm and prevent or reduce leaching of soil nutrients.

Aesthetics

In landscaping, mulches are used to enhance the appearance of displays. Inorganic mulches such as gravel and rocks play a role in aesthetics. Plastic mulches may be used, but usually they are overlaid by organic mulches for aesthetic purposes. Barks and wood chips are of great ornamental value as mulches.

Temperature Moderation

A mulch keeps the soil warm for the germination of seeds and increased activity of soil organisms. It also tends to limit soil temperature fluctuations by protecting the soil from excessive heat in the daytime and retaining more heat at night. Mulched soils warm up slowly in spring, an event that may be detrimental to crops. On the other hand, mulched soils cool slowly in fall, allowing plant root activity to continue for a longer time. Also, because the soil surface is covered, heat loss from the soil is eliminated, making the environment above mulches colder than it would be if it had the benefit of radiated heat from the soil. Therefore, while the roots are kept warm, the top parts of the plant may be in danger of damage from cold.

Improved Soil Physical Structure

Mulching prevents the direct impact of water drops by splashing rain or irrigation water from compacting the soil. Soil surface runoff is also impeded. The increased activity of soil organisms keeps a mulched soil in good physical condition.

Produce Quality Enhancement

Fruits and vegetables grown for fresh produce markets are kept clean from the splashing of dirt from irrigation water by spreading a layer of mulch over the bed.

Disease and Pest Control

A layer of impermeable mulch prevents soil-dwelling pathogens from reaching the top parts of the plants. Some may be killed outright if they become trapped by the layer of mulch.

21.1.7 WEED CONTROL IN ORGANIC FARMING

Weeds should be controlled because they compete with cultivated plants for resources (e.g., plant growth nutrients, air, water, and light). Weeds may also harbor diseases and pests (6.1). For the organic farmer, chemical control is not an option. A variety of state-specific laws apply regarding the occasional use of pesticides and for certification as an organic producer. A variety of cultural practices may be adopted to keep weeds under control. Weeds, since they are volunteer plants, grow in environments to which they are best adapted. Certain weeds are adapted to dry areas, while others prefer moist soil. Some are shade loving, while others prefer full sunlight. A gardener can have an idea about the soil or species that may infest the garden based on the conditions under which the garden is operated.

Activities That Encourage Weeds

Certain activities of organic farmers can introduce new weed species or cause dormant mature ones to be revived:

1. Whenever soils are dug up, weed seeds that were buried and dormant become activated.

2. Farmyard manure and sewage sludge, when used raw, contain undigested seeds that spread weeds. It is best to compost such material before use, especially if the source was harvested after blooming.

3. Straw mulches introduce weeds from the seeds of the grass used.

4. Seeds purchased for planting often contain impurities such as stones, pieces of wood or leaves, and weed seeds.

5. If seedlings or other plants are purchased from nurseries of low repute, plant quality may be questionable and the soil may become weed infested.

6. Seeds also have adaptation for dispersion, such as thorns found on sandbur; such seeds can be carried on clothes, shoes, and vehicle wheels, for example. The seeds stick to clothes and animal fur, the dirt on truck tires, and the like.

7. A compost heap goes through a series of temperatures. If the proper peak temperature is not attained, the weed seeds may not die. Such compost is a source of weeds when used as fertilizer.

Cultural Methods of Control

Weed control in organic farming is achieved by using methods including minimal tillage, mulching, crop rotation, cover crop, biological control, and manual weeding.

1. *Minimal tillage.* Avoid digging the soil too deeply during land preparation. Minimal tillage discourages the growth of a burst of weeds.

2. *Mulching.* Mulching can prevent seeds from germinating, but loose mulching tends to encourage weeds. Therefore, for effective weed control, mulch application should be deep enough to prevent light and air from reaching the seeds.

3. *Crop rotation.* Growing crops in a cyclic fashion has the effect of preventing the buildup of diseases and pests as well as weeds that are associated with certain cultural conditions. As previously mentioned, weeds are adapted to certain soil and environmental conditions. Tuber crops are cultivated differently from legumes. Different weed species thrive under each condition, thereby preventing the buildup of pests associated with any one species.

4. *Ground cover or cover crop.* Bare soil can be protected from damage while improving fertility by using beneficial leguminous cover crops such as lespedeza and alfalfa.

5. *Biological control.* Commercial preparations of fungi as weed killers are available to control certain plant species.

6. *Weeding (hoeing).* Weeds are more effectively controlled when they are cut just below the soil surface. It is recommended that weeding be done before weed plants set seed.

21.1.8: ORGANIC PEST CONTROL

The key to controlling pests in organic farming is adopting the proper *pest management* plan. Nature has a built-in mechanism that maintains a balance in populations (7.5). When the balance is offset, the population of one species may explode because the factor (prey) that keeps it in check is lacking. Chemical sprays (pesticides) are often strong enough to collaterally damage nontargeted organisms (7.10). Some pesticides can kill both pests and predators, and frequently more pests than predators survive. Further, some pests develop resistance to the pesticide, requiring growers to change or use higher concentrations of the chemicals.

That prevention is better than cure cannot be overemphasized in organic farming. Some preventive measures should be built into the planning and location of the garden. The following are general guidelines:

1. Select resistant plant cultivars. While a resistant plant may not be 100 percent immune to a pest, it certainly resists devastation from attack.

2. Grow a mix of plant types. Increased diversity from simultaneously cultivating different species in the same area invites a diversity of pests and thus more predators. Each species requires a different microclimatic condition for good growth and attracts different pests and organisms.

3. Arrange species such that closely related ones are not next to each other. Mixing the species physically makes it difficult for pests to spread quickly.

4. Keep a keen eye on the garden to spot the onset of imbalance or explosion of a pest population.

5. Purchase seed from reputable nurseries and use seeds of high purity.

6. Plant at the appropriate time of the season.

7. Use adapted cultivars.

8. Provide good nutrition. High levels of nitrogen make certain plants develop weak stems and soft and succulent tissues that are prone to damage by pests.

9. Observe strict phytosanitation.

10. Adopt crop rotation practices.

21.1.9 BIOLOGICAL CONTROL

The introduction of biological enemies of a pest as a means of controlling them is discussed in detail in chapter 7. For example, several species of ladybirds, larvae of hover flies, and larvae of lacewings are known predators of aphids. Certain plants exude scents that repel pests.

21.2: COMPOSTING

Plants obtain nutrients (inorganic) from the soil and air and convert (fix) them in organic form (e.g., protein and carbohydrates) as plant material through complex biochemical processes (4.2). When they die, the organic matter decomposes, or breaks down, and thus inorganic nutrients that were locked up become available to plants once again. There are well-known natural recycling processes, including the nitrogen, phosphorus, and carbon cycles, in which specific elements undergo alteration of form between inorganic and organic (3.3.2).

Composting is a deliberate activity by gardeners aimed at accelerating what occurs naturally—*rotting,* or the decomposition of organic matter. As previously mentioned, a typical mineral soil consists of about 5 percent organic matter. Organic matter affects both the physical and chemical properties of soil. It improves the aeration and moisture retention of the soil and, through gradual decomposition, releases both major and minor nutrient elements into the soil for plant use. Composting in effect is organic matter recycling. In the soil, compost acts like a source of slow-release fertilizer, in addition to its desirable influence on the physical characteristics of soil.

> **Compost**
> An organic soil amendment consisting of highly decomposed plant organic matter.

21.2.1 PRINCIPLES OF COMPOSTING

To be successful at composting, one needs to understand the underlying science of the biological processes involved. Composting involves both biotic and abiotic factors, the essential ones being decomposers (microorganisms), organic material (plant or animal), environmental factors for growth (of the decomposers), and time (over which decomposition occurs).

21.2.2 DECOMPOSERS

Decomposers are the agents that convert organic matter into compost through the process of decomposition.

Types

There are two groups of decomposers that inhabit the soil—microorganisms and macroorganisms.

Microorganisms   The major microorganisms (or microbes) involved in decomposition are bacteria, fungi, and actinomycetes.

Macroorganisms   The major macroorganisms include earthworms, grubs, and insects. These groups of organisms should be provided the appropriate environmental factors for their growth and development in order to have large enough populations to work effectively in the compost pile. Microbes have four basic requirements for growth:

1.   *Source of energy.* Microbes obtain energy from the carbon inorganic materials. Plant materials differ in their carbon content, and thus proper materials must be selected for the compost pile.

2.   *Source of protein.* The protein source for microbes is the nitrogen from materials such as blood meal, manure, and green vegetation. Protein is required in only small amounts.

3.   *Oxygen.* Aerobic microbes (which use oxygen for respiration) are more effective and efficient decomposers than anaerobic microbes (which do not use oxygen for respiration). As such, a compost pile should be well aerated. When a pile is poorly aerated and thus dominated by anaerobic bacteria, the decomposition process produces foul odors.

4.   *Moisture.* Moisture is required by organisms for metabolism, but excess moisture in the compost pile fills up the air spaces and creates anaerobic conditions.

An active compost heap is in effect an environment teeming with a wide variety of microorganisms that operate in succession, depending on the temperature in the heap. The temperature in the heap changes because the by-product of a metabolic reaction is heat. It is important to note that bacteria operate over a broad spectrum of temperature conditions. Some prefer cool conditions while others prefer warm conditions. A group of microbes called *psychrophiles* can operate at temperatures even below freezing (28°F or -2.22°C) but work best around 55°F (12.8°C). They dominate a compost heap when the temperature is cool (at the initiation of the pile), but soon the by-products of their metabolic activities cause the temperature to increase. A temperature range of between 70 and 90°F (21.1 and 32.2°C) ushers in another group of microbes called *mesophiles.* Mesophiles are the workhorses of composting. However, at 100°F (37.8°C), they are replaced by *thermophiles,* heat-loving bacteria that work to raise the temperature of the compost heap to a much higher level, reaching a peak of about 160°F (71.1°C).

After bacteria have operated on the organic matter, cellulose, lignin, and other hard-to-metabolize substances are left behind in the pile. Fungi and actinomycetes are able to decompose these substances. Their presence in the heap is indicated by the occurrence of whitish strands or cobweblike structures.

Larger microorganisms such as earthworms are important in the compost heap. They feed on organic matter and excrete materials rich in nutrients for plant growth. Earthworms abound in soils that have high microbial activity.

21.2.3 COMPOSTABLE MATERIAL

The quality of compost depends in part on the materials included in the compost heap. The secret to quality is variety. Avoid including too much of one type of material. Good compostable materials include the following:

1. *Grass clippings from mowing a lawn.* A mulching mower may be used to spread fine clippings on the soil. When clippings are bagged, they may be used as a good source of plant material for composting, provided a few cautions are observed. Do not use clippings from a lawn that has been recently sprayed with pesticides. A fresh pile of grass clippings has a tendency to form a slimy and soggy product with a foul odor. It is best to spread grass clippings in thin layers or to mix them in with dry leaves.

2. *Household garbage that is void of fats and oil.* Greasy materials are hard for microbes to metabolize.

3. *Leaves.* During the fall season, leaves drop from trees. Dry leaves may be gathered for use in composting. Leaves decompose slowly and need some help to accelerate their breakdown. Instead of using full-size leaves, they should be chopped before adding them to the compost pile. Leaves should be added in thin layers.

4. *Sawdust.* Sawdust from softwoods, pine, and cedar decompose more quickly than those from hardwood (e.g., birch and oak). When including sawdust, it should be sprinkled lightly and in layers like the other materials.

5. *Straw or hay.* Old (highly weathered), not fresh, hay makes a good compost material. Straw or hay should be chopped before adding it to the pile.

6. *Ash.* Wood ash from the fireplace contains potash and is a good material to include in a compost heap.

21.2.4 MATERIALS TO AVOID

Some materials are undesirable in a compost heap because they either are not biodegradable or produce toxic factors that are harmful to microbes. These materials include

1. *Diseased plants.* All pathogens may not be killed by the heat generated, even at the peak temperature of about 160°F (71.1°C).

2. *All nonbiodegradable material* (e.g., plastics, synthetic cloths, and styrofoam).

3. *Pesticides.* Pesticides should not be used under any circumstance because they destroy the organisms that are the agents of decomposition.

4. *Pet litter.*

Other materials should be used with caution. For example, the plant remains from corn harvest, including cobs and husks, are hard to decompose. If they must be added, they should first be chopped into small pieces.

21.2.5 THE CARBON-TO-NITROGEN RATIO FACTOR

The carbon-to-nitrogen (C:N) factor is a measure of the material's relative proportion of carbon to nitrogen. The higher the value, the lower the nitrogen content and the longer it takes to decompose. A C:N ratio of 30 is best for composting. Straw has a C:N ratio of 80, while sawdust has a ratio of 400. Materials of leguminous origin have a low C:N ratio (e.g., 15). While a high C:N ratio material decomposes slowly, using materials of low C:N ratio produces excess nitrogen that is expelled from the heap in the form of ammonia gas.

21.2.6 TIME FACTOR

Plant breeders in a way cause artificial evolution to occur in their breeding programs (5.7). The difference between their activities and natural evolution is the fact that natural evolution requires an estimated millions of years while a plant breeding program may be completed within 10 years. Natural evolution depends on *natural selection* to guide it, while plant breeding relies on *artificial selection* imposed by the breeder. Similarly, a biodegradable material eventually decomposes, if left in the soil long enough. However, a gardener may not be able to wait for nature to take its course. To accelerate the process of decomposition, a compost pile is inoculated with materials called *activators*. They contain good populations of decomposers and a nitrogen-protein source to sustain the decomposers in the initial stages of composting.

21.2.7 COMPOST ACTIVATORS

There are natural and artificial compost activators.

Natural Activators

Natural activators include the following:

1. *Loamy soil.* The decaying organic component of a loamy soil contains soil microbes.
2. *Compost.* For the person who composts frequently, finished compost from a previous pile may be used to inoculate a fresh compost heap.
3. *Protein meal.* Protein meal may be derived from high-protein plant material, such as alfalfa and cotton seed. Animal sources include fish meal, bone meal, and blood meal.
4. *Manure.* Manure from a variety of farm animals including poultry, cattle, and sheep is a good activator. However, it should not be used fresh, since it is safest when well decomposed. Manure may be decomposed by allowing the fresh substance to sit exposed to the weather for several weeks. Unfortunately, manures contain weed seeds, which could be a problem if the composting process does not reach the peak temperature required to kill weed seeds.

Artificial Activators

Artificial activators include the following:

1. *Fertilizers.* Fertilizers are less efficient than natural activators because they lack protein. Compound fertilizers consisting of nitrogen, phosphorus, and potassium (10:10:10) may be used.
2. *Inoculant.* Commercially prepared dormant bacteria and fungi that are packaged as tablets or granules may be used.

21.2.8 COMPOSTING SYSTEMS

Methods of composting are varied and adaptable to one's needs and situation, once certain general guidelines are observed. There are designs for small-scale and indoor use, and others for large-scale and outdoor production of compost. They also vary in terms of ease of aeration, cost of setup and maintenance, time to completion, quality of product, and odor. The two types of compost systems are categorized on the basis of the receptacle.

No-Container Method (Sheet Composting)

The ultimate goal of composting is to decompose plant material for incorporation into the soil to improve its nutrient level and physical characteristics. The most direct way of accomplishing this is to use in situ composting, in which the plant material is composted in the soil where it will be used. The material used may include leaves, plant residue after harvesting, grass clippings, and manure hauled onto the field. These materials are incorporated by an appropriate implement such as a spade or a mechanical tiller. Another version of this method, called green manuring, involves the growing of leguminous species, such as clover, alfalfa, peas, and soybean, and plowing them under while still fresh.

Sheet composting allows organic materials to be applied to large fields. A negative side of this method is that it takes several months for the incorporated material to decompose. Also, when materials of high C:N ratio are used, the soil experiences a short period of nitrogen deficiency *(nitrogen starvation)* because the existing nitrogen is used in decomposition. Crops therefore should not be planted until several months after composting, when the nitrogen deficiency has been corrected. Another negative factor is that heat does not build up to a level at which weed seeds are killed.

Container Methods

Gardeners also compost in specially constructed containers or pits. These ex situ methods of composting follow certain recipes, depending on the design. The compost is prepared in one place and transported to another for use.

21.2.9 GENERAL PRINCIPLES OF COMPOSTING

The success of composting depends on the observance of certain principles including layering, moisture supply, size of pile, and aeration.

Layering

Since one is encouraged to use a wide variety of materials in a compost pile, the way they are arranged in the pile is important. Materials are not haphazardly mixed up in the pile but rather are placed strategically in layers. For example, layers of dry materials such as straw should be alternated with fresh materials such as clippings and vegetative matter. Materials high in nitrogen (low C:N) should be alternated with high-carbon (high C:N ratio) material. After a number of layers, the activator should be spread evenly before another set of material is added. This pattern is repeated until the container is filled.

Moisture Supply

Water is required by microorganisms for decomposition. Too much of it may cause anaerobic conditions to develop in the pile, while too little slows decomposition. Efforts should be made to provide moisture uniformly throughout the pile, accomplished by moistening the material layer by layer as it is being piled and as needed. A well-moistened compost heap material feels as moist as a wet sponge that has been wrung. Overwatering a compost pile is a waste of water and also causes leaching of nutrients. Rain water is ideal for watering since it contains useful microorganisms, minerals, and oxygen.

Size of Pile

An effective pile size is one that is manageable and self-insulating without causing compaction in the layers. A large pile may cause overheating and anaerobic conditions to prevail in the inner part, a situation that is detrimental to bacteria. A small pile, on the other hand, may be over-ventilated and thus not be able to reach peak temperatures; it may also require artificial insulation. Since a pile should be turned over regularly, a huge pile may be unmanageable.

Aeration

A compost heap should be well ventilated for good growth of aerobic bacteria. A pile may be built around ventilating pipes or a tube of wire mesh. Such a practice may be necessary in methods where the pile is left unturned or turned infrequently.

21.2.10 CONSTRUCTING OUTDOOR COMPOSTING SYSTEMS

Two critical factors to consider in constructing an outdoor composting unit are location and container type and design.

Location

The compost site should be near the garden or a place where it is easy to manage. The pile should not be allowed to dry out, so locating it in the shade in tropical areas is desirable. In the temperate zone, however, sunlight is required to keep the heap warm. It is wise to locate the heap so that the house is not downwind from it; in the case of poor decomposition, the foul odors will not blow into the house with the wind. The site should be well draining.

Container Design

Containers or retaining walls may be constructed out of a wide variety of materials including blocks, plastic, wood, and even straw (baled) (figure 21-3). On a well-drained location, the ground may be used as the bottom wall of the container (figure 21-4). The advantage of using bare earth for the bottom of the container is that natural decomposers in the soil have the opportunity to act on the material in the pile. Wooden and wire bins are also used (figure 21-5).

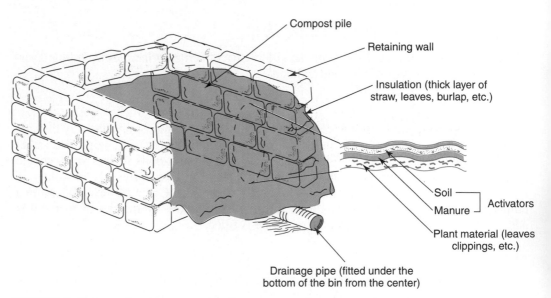

FIGURE 21-3 Composting in a concrete pit.

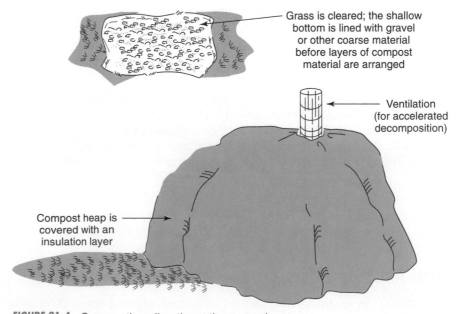

FIGURE 21-4 Composting directly on the ground.

Since wood is biodegradable, a wood preserver or latex paint coating reduces the danger of its decomposition. Wire bins are well ventilated and as such lose heat rapidly. Barrels, drums, tumblers, and plastic bins may be used as containers for composting (figure 21-6).

Pit composting is simply burying the composting materials in a pit dug in the soil until they decompose (figure 21-7). When ready (which may be up to a year), the compost may be dug up and used elsewhere or left in place and crops planted in it. The container in this case is the earth. A variation of this method is *trench composting*, in which the material is buried in long trenches (figure 21-8). Once decomposed, the trench is used as a bed for growing

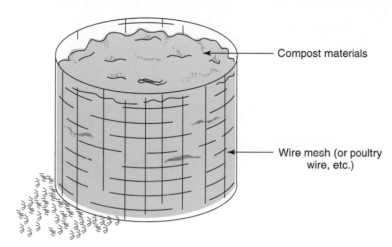

FIGURE 21-5 Composting in a wire basket.

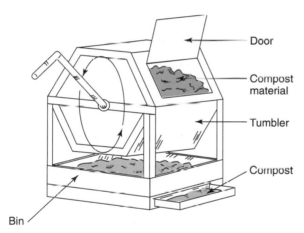

FIGURE 21-6 A compost tumbler.

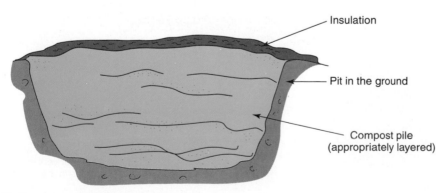

FIGURE 21-7 Pit composting.

crops, while the adjacent row separated by a path is then dug up as a new trench for composting. The compost trench becomes a walkway before it is used for growing crops, thus giving the material two years to decompose properly.

21.2.11 INDOOR COMPOST SYSTEMS

Indoor compost systems are portable and can be readily located and relocated in small areas such as a garage (figure 21-9). Small-scale composters with worms as the principal decomposers are used to decompose domestic waste (kitchen garbage). Their success depends to a large extent on protecting the worms from exposure to extreme temperatures.

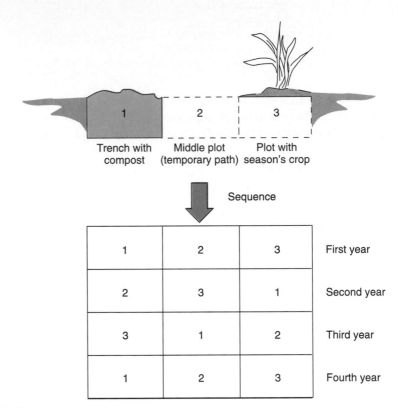

FIGURE 21-8 Trench composting.

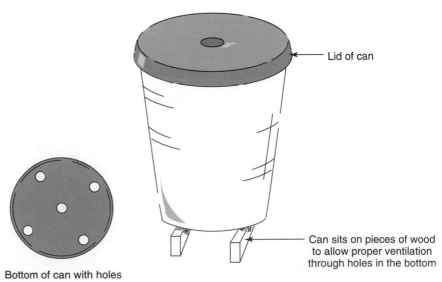

FIGURE 21-9 An indoor composting setup.

21.2.12 MAINTAINING COMPOST PILES

Turn Regularly

Turning the compost heap frequently is a tedious but worthwhile chore since it quickens the rate of decomposition. Home composters may turn their heaps less frequently (every 6 to 12 weeks), unless a foul odor develops earlier. In turning frequently, one should be careful to do it each time after peak temperature has been attained in the pile (to kill weed seeds). Turning too frequently, however, is detrimental to the activity of the decomposers.

Keep Aerated

Aeration is accomplished by turning the heap or poking it. Poor aeration encourages the development of a foul odor.

Keep Moist

It is best to moisten the pile after each layer is added. If watering becomes necessary later on, it should be done very carefully to avoid creating anaerobic conditions in the pile. If the pile is generally damp and warm only on the inside, the pile size may be too small. In this case, it must be rebuilt on a larger scale.

Control Weeds

Weeds should not be allowed to grow on a compost heap since they deprive the pile of moisture and the nutrients being produced through mineralization of the organic matter.

Maintain Heat

A "young" compost heap should heat up quite rapidly. If it remains cool, the nitrogen level in the pile may be low. This factor may be remedied by poking holes in the pile and adding a nitrogen source (e.g., fresh manure or blood meal). Lack of heat may also be due to low moisture content; water may thus be added through holes poked in a dry pile. Sometimes a cool pile may simply indicate that it is time to turn it or that the composting process is complete.

SUMMARY

Organic farming is the practice of growing plants by depending primarily on natural sources of fertility for plant nutrition and nonchemical strategies for controlling diseases and pests. The management of organic matter is a critical activity in organic farming. Organic matter may be obtained from plant or animal sources. To obviate the need to use chemicals in controlling weeds, mulching, crop rotation, use of cover crops, and mechanical weeding are alternative methods employed. The use of disease- and pest-resistant cultivars, observation of strict phytosanitation, and good crop production practices reduce disease incidence in cultivated crops. Composting, an activity in which raw organic matter is processed for use, is a primary activity for organic gardeners.

A successful composting operation depends on the use of compostable material, good activators, proper layering of materials in the pile, and good management of the compost pile in terms of aeration, moisture supply, and temperature. Composting systems are designed for either indoor or outdoor use.

REFERENCES AND SUGGESTED READING

Campbell, S. 1990. Let it rot: The gardener's guide to composting. Pownal, Vt.: Storey Communications.

Minnich, J. 1979. The Rodale guide to composting. Emmaus, Penn.: Rodale Press.

PRACTICAL EXPERIENCE

Construct a small-scale compost pit following the steps described in the text.

PART A

Please answer true (T) or false (F) for the following statements.

1. T F   A typical mineral soil consists of 10 percent organic matter.
2. T F   Mineralization is the process by which nutrient elements in organic matter are released into the soil.
3. T F   Organic matter has colloidal properties.
4. T F   As a mulch, tree bark decays very rapidly.
5. T F   Mulched soils cool slowly in the fall season.
6. T F   Materials with a high C:N ratio have low nitrogen content.
7. T F   Fertilizer may be used as an activator in a compost heap.
8. T F   Mesophiles are bacteria that dominate the microbial population of a compost heap when a pile is first established.
9. T F   Household garbage containing grease is good for the compost pile.
10. T F   Low C:N ratio materials decompose at a faster rate than those with a high C:N ratio.

PART B

Please answer the following questions.

1. _____ is the practice of growing plants for the purpose of plowing them under the soil to increase organic matter and general nutritional levels.
2. List three green manure plants.

_____   _____   _____

3. List three roles of organic matter in gardening.

_____   _____   _____

4. _____ is the most common type of mulch.
5. Name five specific ways in which diseases and pests can be controlled in organic gardening.

_____   _____   _____

6. List any five major decomposers.

_____   _____   _____

7. List any three natural activators for a compost system.

_____   _____   _____

PART C

1. Discuss the use of mulching in weed control.
2. How can the use of farmyard manure introduce weeds into a garden?
3. Discuss the role of temperature in composting.
4. Why is rain water preferred as a moisture source for a compost heap?
5. What could be the cause of a foul odor from a compost heap, and how can it be corrected?

CHAPTER 22

Cut and Dried Flowers: Production and Arranging

PURPOSE

This chapter is designed to discuss the field culture and handling of cut flowers and how they are used in creating floral displays.

EXPECTED OUTCOMES

After studying this chapter, the student should be able to

1. List five plant species that make good cut flowers.
2. List five plant species that make good dried flowers.
3. Describe how cut flowers are managed for longevity.
4. Describe how flowers are dried for preservation.
5. Describe the principles of flower arranging.
6. List five tools or materials used by the florist.
7. Describe how plants are chosen in creating a floral design.
8. List four different floral designs.
9. Describe the steps in creating a floral design.

KEY TERMS

Asymmetrical design   Line flowers   Symmetrical design
Filler flowers   Mass flowers   Vase life
Form flowers

OVERVIEW

Flowering plants are frequently grown outside (in beds) or inside (in containers) to be enjoyed and admired as whole living plants in the landscape or interior plantscape. Flowers are used in a variety of ways to convey sentiments. For most of these uses, flowers are detached from the parent plant and used individually or in groups. As such, *cut flowers* (i.e., detached flowers) are an important part of the horticultural industry. *Florists* are the specialists who use cut flowers in their trade. To add value to cut flowers, florists often use them to create *floral designs* for a variety of occasions. Using cut flowers in this art form is called *flower arranging*. Whereas fresh cut flowers last for only a short period, flowers may be dried and preserved for a long time. Dried flowers may also be arranged into durable arrangements.

Cut flower displays are found in restaurants and homes, as accessories to dressing (corsages and boutonnieres), at weddings (e.g., as a bridal bouquet), at funerals (e.g., as casket spreads and wreaths), on special occasions to express love and appreciation (e.g., Valentine's Day and Mother's Day), and to offer congratulations (e.g., on graduation days). The use of flowers in this fashion is enormous. Some flowers are more suited than others for use as cut flowers. In fact, scientists have developed special qualities in some plants that are used this way to extend their postharvest lives and qualities.

22.1: CUT FLOWER SPECIES

Many species used as cut flowers are herbaceous (table 22-1). However, a number of woody species can be cultivated for cutting (table 22-2). The most popular cut flower species are rose, carnation and chrysanthemum. The popular species (and many others) are predominantly grown in greenhouses for best quality. Greenhouse production of cut flowers has been discussed previously (11.7). Herbaceous plants used for cut flowers are mostly annuals and require planting each season and proper maintenance to ensure high-quality products. Once established, woody plants have the advantage of being perennial and require relatively less maintenance in production as sources of cut flowers. A significant disadvantage in using woody plant species for cut flowers is that they require several years to grow to the stage where they are usable. Species and even cultivars within species differ in *vase life* (the duration of time within which a cut flower retains its desirable qualities before deteriorating), the equivalent of shelf-life for perishable garden products.

Table 22-1 Selected Species Commonly Grown for Use as Cut Flowers

Plant	Scientific Name
African daisy	*Dimorphotheca sinuata*
Astroemeria	*Astroemeria* spp.
Baby's breath	*Gypsophila elegans*
Celosia	*Celosia plumosa*
Cleome	*Cleome hasslerana*
Carnation	*Dianthus caryophyllus*
Cornflower	*Centaurea cyanus*
Cosmos	*Cosmos bipinnatus*
Chrysanthemum	*Chrysanthemum* spp.
Cyclamen	*Cyclamen persicum*
Freesia	*Freesia refracta*
Kalanchoe	*Kalanchoe* spp.
Fountain grass	*Pennisetum setaceum*
Gerbera daisy	*Gerbera jamesonii*
Snapdragon	*Antirrhinum majus*
Zinnia	*Zinnia elegans*
Nasturtium	*Tropaeolum majus*
Globe amaranth	*Gomphrena globosa*

Table 22-2 Selected Woody Species Used for Cut Flowers

Plant	Scientific Name
Rose	*Rosa* spp.
Buttersweet	*Celastrus orbiculatus*
Redbud, eastern	*Cercis canadensis*
Red osier dogwood	*Cornus stolonifera*
Hydrangea	*Hydrangea arborescens*
Nadina	*Nadina domestica*
Pussy willow	*Salix matsudana*
Forsythia	*Forsythia x intermedia*
Hollies	*Ilex* spp.
Virburnum	*Virburnum* spp.
Beauty berry	*Callicarpa americana*
Weigela	*Weigela florida*

22.2: CULTURE

22.2.1 GROWING SCHEDULE

Cultural procedures for cut flowers are the same as described for the production of garden plants in the landscape (chapter 13). The profitability of the enterprise depends on increasing production efficiency to reduce production costs. In field production, effective scheduling of production activities coupled with sound irrigation and fertilizer management are needed for efficient production.

Plants grown for cut flowers require good management in cultivation to produce healthy, disease-free, vigorous plants and durable cut flowers. Since flowers are expected to live for a period after being detached from the parent plants and deprived of new nutrients, it is critical that plants for cut flowers be cultivated with care so that at the time of harvesting the detached portion has appreciable nutritional reserves to depend on for some time. Malnourished plants produce poor-quality cut flowers with short vase lives. For high carbohydrate levels, cut flowers need a good fertilizer program in cultivation, proper temperature and light (quality and intensity), and adequate moisture. Diseases and pests should be effectively controlled since they not only reduce the value of products but also accelerate deterioration and reduce the longevity of cut flowers.

Because species bloom at different times and have different degrees of temperature adaptation, the gardener can maintain a nearly year-round active cutting garden through judicious selection of plants. One such growing schedule is described in figure 22-1. The strategy is to select plants that will provide continuous blooms in the garden.

22.2.2 HARVESTING

Flowers should be harvested at the optimal stage of development and under the best environmental conditions. Whereas some species, such as *Helianthus annus* and *Chrysanthemum* spp., are harvested when fully open, others, including *Hemerocallis* spp. and *Trollus* spp., are best harvested when half open (table 22-3). Because warm temperatures tend to accelerate plant development, summer flowers may have to be harvested at a less mature stage. In terms of time of day, advantage should be taken of morning's low temperatures to conserve tissue moisture content; fresh flowers should be cut in the early morning hours. Cut flower longevity can be extended by placing the flower stems in water immediately after cutting. The water should be turbid, not cold. Water uptake is impeded if the stem is cut with a dull blade, so it is important that all cutting tools be very sharp and all cuts be as clean as possible. Cuts should be angled to create a wider surface area and increased water absorption. Cut flowers may be handled dry (stems

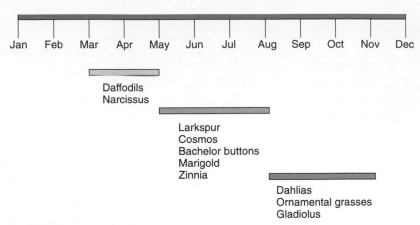

FIGURE 22-1 A growing schedule for cut flowers.

Table 22-3 Best Stage of Harvesting Selected Cut Flowers for Immediate Sale

Fully Open Flowers

Chrysanthemum	*Chrysanthemum* spp.
Stonecrop	*Sedum* spp.
Sunflower	*Helianthus annuus*
Camellia	*Camellia japonica*
Cyclamen	*Cyclamen persicum*
Marigold	*Tagetes erecta*

Partially Open Flowers

Rose	*Rosa* spp. (first two petals starting to unfold)
Daylily	*Hemerocallis* spp.
Fritillaria	*Fritillaria* spp.

50 percent Florets Open

Acacia	*Acacia* spp.
Cockscomb	*Celosia argentea*
Foxglove	*Digitalis purpurea*
Primrose	*Primula* spp.
Amaranthus	*Amaranthus*
Summer phlox	*Phlox paniculata*
Kalanchoe	*Kalanchoe* spp.

Colored Buds

Peony	*Paeonia* spp.
German iris	*Iris germanica*
Poppy	*Papaver* spp.

Source: Extracted and modified from Oklahoma State University Extension Publication no. 6527

not placed in water) during transportation under low temperatures. At the destination, flowers are then rehydrated, but the vase life of such flowers is generally reduced. Flowers should be harvested with care to avoid mechanical damage that not only reduces aesthetic value but also accelerates aging and reduces vase life. Bruised plants also become prone to diseases.

22.2.3 POSTHARVEST MANAGEMENT

To increase cut flower longevity before sale and after purchase, growers and consumers have to manage the environment of the products with respect to temperature, humidity, moisture and noxious gases.

Temperature and Humidity

Wilting reduces the vase life of cut flowers. To reduce respiration and water loss, flowers should be held at temperatures between 32 and 35°F (0 and 7°C) immediately after harvesting until they are ready to be used. A relative humidity of about 90 percent should be maintained. Beading and droplets of water on flowers indicate excessive humidity, a condition that predisposes flowers to fungal attacks such as the gray mold caused by *Botrytis*.

Water

After harvesting, flower life depends on the availability of good-quality water and the capacity to absorb it. Hard water is harmful to cut flowers, as are fluoride, sodium, and sulfate salts when present in high concentrations. Whenever possible, cut flowers should be placed in warm (110°F [43°C]) deionized and distilled water of acidic reaction (pH 3 to 4). The pH of water can be lowered by adding citric acid to it. Acidic water not only improves water uptake but also has antiseptic properties, reducing infection of the cut surface, which is known to clog the xylem vessels and prevent water absorption. Passages are also blocked by trapped air bubbles. To correct these problems, flower stems should be recut under warm water using a sharp cutting instrument before being set in a vase or arranged for display. Some species, such as poppy and dahlia, require a flame treatment in which the cut surface is passed over a flame for a very brief period. This treatment hardens the sap. The milky sap that oozes from plants such as poppy clogs the stems of other species held in the same container unless flame treated. The leaves on the part of the stem that will be submerged in water in the vase should be removed. Depending on the moisture status, rehydration may take between 30 and 60 minutes to restore plants to full turgidity. Flowers that are purchased from a store or transported over a long distance often need rehydration. Flowers obtained from the home garden for use indoors can be maintained fresh by placing them in water immediately after cutting.

Nutrition

The nutrition of plants during growth is critical to the success of flowers after cutting. Flowers must be harvested at the right time, which is when they have accumulated enough sugars and starches to open and sustain them after cutting. During rehydration, sucrose may be added to the holding water to extend vase life and improve flower qualities. Commercially formulated floral preservatives that contain pH adjuster, biocide, wetting agent, food source, and water can be purchased from a local nursery store. Homemade concoctions such as 1 tablespoon of corn syrup plus 10 drops of household bleach in a quart of warm water may be used. Also, one part of soda (lemon-soda) plus two parts of water was developed at the University of California to extend vase life. These preservatives work because the sugar is nutritious and the lime reduces the pH of the water to suppress bacterial growth.

Ethylene Reduction

Ethylene is a by-product of gasoline or propane combustion and is emitted through vehicles' exhaust systems. More importantly, this chemical is a natural plant growth hormone involved in several physiological processes such as fruit ripening, seed maturation, aging, and wound healing. Certain flowers are ethylene sensitive (table 22.4). In the presence of excessive amounts of the gas, flowers of these plants deteriorate rapidly. Ripening fruits and vegetables produce large amounts of ethylene and as such should not be stored or transported with cut flowers (4.6). Treatment with a commercial preparation of *silver thiosulfate* reduces the effects of ethylene on cut flowers.

Diseases

In an attempt to reduce moisture loss, cut flowers should be kept under humid conditions. However, high humidity provides an environment that is conducive to the growth of gray mold (*Botrytis*).

Plant	Scientific Name
Astroemeria	*Astroemeria* spp.
Carnation	*Dianthus caryophylus*
Celosia	*Celosia* spp.
Delphinium	*Delphinium elatum*
Larkspur	*Delphinium* spp.
Freesia	*Freesia* spp.
Gladiolus	*Gladiolus* spp.
Baby's breath	*Gypsophila paniculata*
Lily	*Lilium longiflorum*
Phlox	*Phlox* spp.
Snapdragon	*Antirrhinum majus*
Stock	*Matthiola* spp.
Speedwell	*Veronica spicata*
Moneywort	*Lysimachia nummularia*

Table 22-4 Selected Ethylene-Sensitive Species Used for Cut Flowers

22.3: FLOWER ARRANGING

Flower design and arranging is an art whose success depends on the observation of certain basic principles, similar to those for landscape designing. Throwing a couple of flowers together in a bunch produces a bouquet, but it takes creativity to use flowers to make a statement, influence mood, or enhance the ambience and general decor of a place. Floral designers can create flower arrangements for specific occasions. Arrangements are designed for happy occasions such as weddings and sad ones such as funerals. Flowers such as roses may be displayed individually in tiny vases and thus do not require arranging.

22.3.1 PRINCIPLES OF DESIGN

Even though flower arranging is an art, the following principles underlie basic floral designs. These basic principles are described in detail in chapter 12 (12.3) as pertaining to landscape design.

Balance

Balance in design relates to the display's visual weight projected to the viewer. In a *symmetrical design*, viewers on opposite sides of a display see similar things; *asymmetrical designs* do not offer the same view from two different angles. The centerpiece for a formal dinner may be arranged in a symmetrical design.

Proportion and Scale

Several elements go into producing a proportional design—the characteristics of the flowers, container, table (where applicable), and room. Although tall containers can be used for tall displays, low containers should not hold tall flowers. The general recommendation is that the arrangement be about one and a half to two times the container height or width. The display should not overwhelm the table or the room. An oversized display is out of place in a small room, and a tiny display is ineffective in a large room or on a large table.

Focal Point and Accent

A focal point can be created by including an exotic or very attractive flower in the design. This flower could be described as a conversation piece that draws the immediate attention of viewers. In lieu of such a specimen, a designer can use other techniques such as *repetition* and *massing* to draw attention to a design.

> **Accent Plant**
> *A plant strategically located in a landscape to draw attention to a particular feature in the area.*

FIGURE 22-2 Containers for cut flowers. There is variety and room for creativity in selecting containers.

Contrast

Without contrast, a design can be monotonous and boring. Flowers differ in color, size, texture, and shape; these characteristics should be used to enhance the arrangement.

Unity

If a designer observes the principles of good balance, proportion, scale, and contrast and includes an effective focal point, the resulting creation blends together to produce an effective display that is aesthetically pleasing and functional. Unity is achieved in a design when the viewer gets the sense that all elements are working together. The design elements are not seen individually when there is unity.

The Role of Containers

Containers serve important roles in flower arranging other than just holding flowers. In some displays, they can hardly be seen. In many others, however, they provide a background for the arrangement. The size of the container determines the size of the finished product (remember that the arrangement should be one and a half to two times the size of the container). Containers can be ceramic, plastic, crystal, or some other material (figure 22-2). Bright colors should be avoided and preference given to shades of white, green, gray, or beige. Solid colors should be chosen, although simple patterns may add to the display; elaborate patterns distract from the floral display and should be avoided.

22.3.2 TOOLS AND MATERIALS

The tools used in the cut flower industry are simple:

1. *Cutting tools.* The basic cutting tools are the florist models of hand pruners, wire cutters, knives, and scissors. These instruments should be sharp and capable of providing clean cuts.

2. *Straight wire.* Wires of 16 to 20 gauge are used to provide additional support to the stems of cut flowers and also for creating certain shapes.

3. *Chicken wire.* Chicken wire may be placed in a tall vase to hold stems so that they do not slump and bunch together.

4. *Floral foam.* Instead of using water alone, flowers in certain displays may be arranged in a block of absorbent material called *floral foam.* The foam is soaked to saturation in water before being placed at the bottom of the container. The block then acts as the soil medium in which the cut flowers are "transplanted."

5. *Floral tape.* Floral tape can be used in a variety of ways, such as for creating cells (similar to chicken wire) on top of a vase to separate the flowers and prevent them from bunching together.

6. *Floral clay.* Floral clay is a putty used to hold materials such as chicken wire in place.

> **Floral Foam**
> A block of material with a high capacity for water absorption and retention that provides mechanical support for cut flowers in an arrangement.

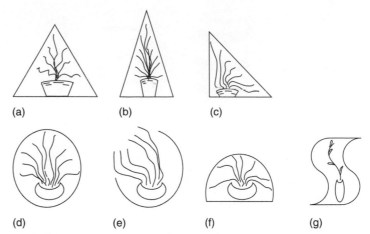

FIGURE 22-3 Selected basic floral shapes: (a) symmetrical triangle, (b) vertical, (c) asymmetrical triangle, (d) round, (e) crescent, (f) half-circle, and (g) Hogarth or S curve. Arrangements that are sometimes described as horizontal are flatter in the center and more drawn out than the half-circle design.

22.3.3 BASIC FLORAL DESIGN

A floral designer should decide on the shape of the overall display to be created before beginning to arrange the flowers. Basic shapes are shown in figure 22-3. The shape of the design is determined by where the arrangement will be placed and the occasion. A round shape is ideal on a table that can be viewed from all directions. For a dinner occasion, the design should be low (e.g., half-circle or horizontal) so that it does not obstruct the view across the table. If the arrangement is intended to be displayed where it can be viewed from only the front, a half triangle design is suitable.

Choosing Flowers

Flowers should be chosen with the occasion and the shape of the design in mind. In certain cultures, specific flowers are associated with funerals; consequently, it would be an unfortunate mistake to present these flowers at a wedding. Based on anatomy, individual flowers are suited for certain roles in a display. Some are used to create the skeletal structure or framework of the basic design and are called *line flowers* (e.g., gladiolus, delphinium, and snapdragon). They establish the height and width of the arrangement. As previously described, the focal point in a display is provided by an exotic, unique, or very attractive flower (e.g., orchid and potea); these flowers are called *form flowers*. The bulk of the remainder of an arrangement is occupied by the base flower, which is most abundant in the overall arrangement. Flowers used in this fashion are called *mass flowers* (e.g., carnation, aster, and rose). After these three categories of flowers have been incorporated, spaces may occur in the design, causing the overall appearance at this stage to be coarse and somewhat fragmented. Sometimes nonplant materials (e.g., wires and foam) used in the design remain exposed in a manner that reduces the aesthetic value. These gaps and rough edges can be covered by using *filler flowers* such as palms, heather, fern, and baby's breath.

Creating a Design

The following steps provide a guide for creating a *low design:*

1. Place a block of floral foam saturated with water at the bottom of the container. It maybe cut (if necessary) to fit the container. The foam should rise about 2 inches (5.1 centimeters) above the rim of the container (figure 22-4).
2. Establish the height and spread of the arrangement.
3. Establish the focal point.
4. Add massing flowers.
5. Add filler flowers.

Line Flower
A flower with a splice appearance.

Form Flower
A flower with a unique shape that is used to provide a focal point in an arrangement.

Mass Flower
A flower that is usually round-shaped and is used most abundantly in an arrangement.

Filler Flowers
A flower used to complete an arrangement by filling in gaps to tie in the various aspects of the design.

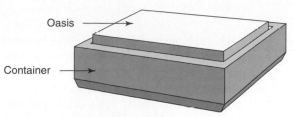

FIGURE 22-4   Fresh flower stems are inserted into a foam soaked in water.

(a)

(b)

FIGURE 22-5   Examples of simple arrangements.

In creating the design, it may be necessary to work by trial and error, cutting and re-cutting, twisting and turning, and doing all that is necessary to produce the most desirable arrangement. If a flower is too long, it should be cut back. Some leaves may have to be removed. Flowers may have to be repositioned in the floral block. It is critical that all cuts be made under water (22.2.3). It is also important to the display that all of the flowers and foliage appear to begin from a common spot. Examples of designs are shown in figure 22-5.

22.4: DRIED FLOWERS

Garden plants brighten the landscape with their wide variety of colors and textures, and many may be cut and used fresh indoors in containers. Some species may be cut and dried and used just like fresh flowers in a variety of arrangements. In this way, flowers can be enjoyed year-round. Cut and dried flowers usually lose their scent but frequently maintain their color. Preparing flowers in this way is a means of preserving them.

22.4.1 DRYING

Flowers may be dried by natural or artificial means. One method may suit particular species more than others. Flowers should be cut in dry weather and when they are not covered with dew.

Natural Drying

To dry naturally, flowers are cut and hung upside down in bunches or singly in a room that is dry and well ventilated until they feel dry and crisp. Species that may be dried naturally include lupins (dry singly), *Gypsophila*, *Delphinium* (dry singly), holly *(Eryngium)*, strawflowers, and statice *(Limonium)*. Some species require special treatment during drying. For example, *Hydrangea* flowers dry to the best quality when dried with the stems in water.

Grasses can be dried naturally, a good species being pampas grass *(Cortaderia selloana)*. Cereal plants such as wheat and barley may be dried and used successfully in arrangements. If the flower stock is too short to be hung, the flowers may be dried in an upright position using long pins (figure 22-6). Plants with fruits or pods may be dried and used in very attractive displays. Such plants should be harvested for drying after the fruits or pods have ripened and begun to turn brown. Plants with pods such as poppy, foxglove, and hollyhock make excellent dry specimens. The Chinese lantern is a perennial favorite for dried flower displays.

Artificial (Chemical) Drying

Instead of air drying, special drying agents may be used to quickly dry cut flowers. Examples of drying agents are silica gel crystals (or powder) and a mixture of silver, sand, and borax. Chemical drying is done in a box. First, about an inch-deep (2.45-centimeters-deep) layer of drying agent is placed in the bottom of the box before placing the plant or flower inside. Additional drying agent is then spread carefully over the top (figure 22-7). The box is set in a warm place for about three days to complete the drying process. Flowers that can be dried chemically include *Delphinium*, *Pelargonium*, rose, *Fuchsia*, and gentian.

Dry flower arrangements often include preserved leaves from trees. Preservation of leaves is accomplished by standing the ends of the petioles in glycerine mixed with water in a ratio of 1:2. To increase the uptake of glycerine, the ends of the petioles are hammered before standing in a container of the mixture (figure 22-8). Leaves of horse chestnut, laurel, oak, and beech can be successfully preserved by glycerine treatment. Flowers treated with glycerine tend to stay open and retain their natural color.

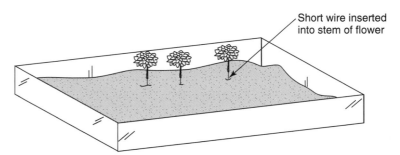

Short wire inserted into stem of flower

FIGURE 22-6 Drying cut flowers naturally.

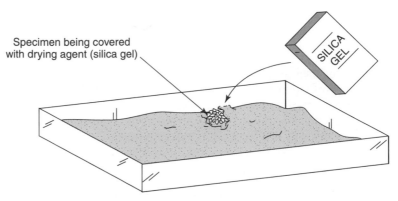

Specimen being covered with drying agent (silica gel)

SILICA GEL

FIGURE 22-7 Drying cut flowers by using chemicals.

22.4.2 PRESSED FLOWERS

Drying flowers preserves them in their natural shapes. Leaves and flowers may also be dried by sandwiching the material between two layers of dry paper (e.g., blotting or newsprint) and applying pressure to the arrangement. The pressure may be applied by piling weights (e.g., heavy books) on the arrangement. Specially designed plant presses may be purchased for use (figure 22-9). The press may be left in place for about four weeks, after which the plant material is dry and flat.

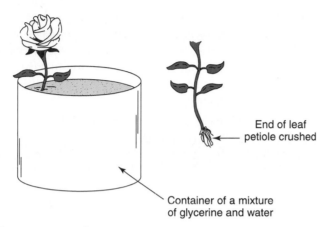

End of leaf
petiole crushed

Container of a mixture
of glycerine and water

FIGURE 22-8 Cut flower preservation.

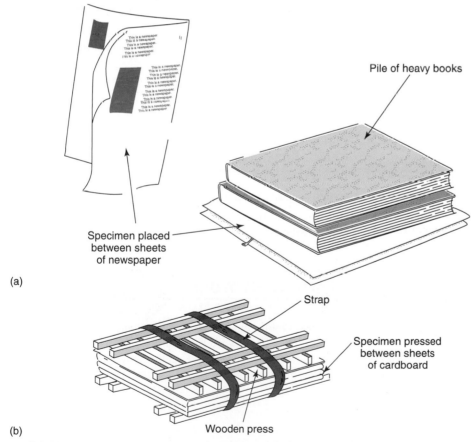

Pile of heavy books

Specimen placed
between sheets
of newspaper

(a)

Strap

Specimen pressed
between sheets
of cardboard

Wooden press

(b)

FIGURE 22-9 Pressing cut flowers and plant specimens: (a) homemade method and (b) commercial press.

22.4.3 DRIED FLOWER ARRANGEMENTS

Dried flowers may be arranged in dry vases just as fresh cut flowers are arranged. They may be arranged in bouquets and wall displays after fastening them to decorative bands. Pressed leaves and flowers may be laminated (which extends the life of the product) and arranged in an album. Dried flowers should be handled with care since they are more delicate than fresh flowers.

SUMMARY

Flowers can be enjoyed in the garden, in the landscape, and as potted plants. They can also be enjoyed as cut flowers when they are detached from the parent plant and displayed in containers. Florists employ basic principles similar to those that govern landscape design (balance, proportion, scale, focal point, accent, contrast, and unity) to create attractive floral designs for a variety of occasions. There are five basic designs in floral arranging. To increase the longevity of cut flowers, they first need to be properly raised in cultivation and then given proper postharvest care that includes provision of water, high humidity, cool temperature, and absence of ethylene gas. In a basic floral design, four categories of flowers are used: Line flowers are used to set the height and width of the arrangement, form flowers are used for creating the focal point, mass flowers provide the bulk of the flowers in the arrangement, and filler flowers are used to dress up the general arrangement to smooth roughness and fill in gaps. Cut flowers may be used fresh or dried. Dried flowers are arranged in the same way as fresh flowers.

REFERENCES AND SUGGESTED READING

Armitage, A. M. 1993. Specialty cut flowers: The production of annuals, perennials, bulbs, and woody plants for fresh and dried cut flowers. Portland, Oreg.: Varsity Press/Timber Press.

Belcher, B. 1993. Creative flower arranging: Floral design for home and flower show. Portland, Oreg.: Varsity Press/Timber Press.

PRACTICAL EXPERIENCE

FIELD TRIP

Visit a florist's shop to observe how a variety of floral arrangements are created. During these trips, note the kinds of equipment used for creating the arrangements, the storage facilities for prolonging the lives of cut flowers, the training required to do the work, and the like.

LABORATORY WORK

---

1. Select any five cut flower species. Obtain cut flowers for these species and conduct an experiment to study the differences in the vase lives of the various species.
2. Collect, press, and dry cut flowers from the landscape. Create an album of dried flowers.
3. Collect and dry flowers from the landscape.

PART A

Please answer true (T) or false (F) for the following statements.

1. T F Only herbaceous species are sources of cut flowers.
2. T F Cut flowers should be displayed in alkaline water.
3. T F Form flowers are used to define the size (height and width) of a floral arrangement.
4. T F A symmetrical floral design presents viewers on opposite sides with similar displays of flowers.
5. T F Sucrose may be added to the holding water to increase vase life.
6. T F Ripening fruits and cut flowers should not be stored together.

PART B

Please answer the following questions.

1. The duration that cut flowers can retain their quality in a container is called _____.
2. Cut flowers may be handled dry under conditions of _____.
3. The specialist who uses cut flowers in his or her trade is called a _____.
4. The art of using cut flowers in creating floral designs is called _____.
5. The flowers that constitute the bulk in a floral display are called _____.
6. A floral foam is used for _____.
7. List two factors that can impede water uptake by cut flowers.

   _____   _____

8. List four principles of floral design.

   _____   _____

   _____   _____

9. List four tools used by the florist.

   _____   _____

   _____   _____

10. List the steps in creating a floral design.

    _____   _____   _____

11. List five good cut flower species.

    _____   _____   _____

    _____   _____

PART C

1. Discuss the issue of ethylene effects in the cut flower industry.
2. Why is it important to cut flowers under water?
3. Explain the rationale behind the formulation of commercial floral preservatives.
4. How is the principle of focal point accomplished in a floral design?

CHAPTER 23

Computers in Horticulture

PURPOSE

This chapter is designed to show how computers are used in modern horticulture to perform a variety of tasks, both mundane and complex, and to discuss future applications.

EXPECTED OUTCOMES

After studying this chapter, the student should be able to

1. Appreciate the significance of computer technology in modern horticulture.
2. Describe specific situations in which computers are used in horticulture, including uses in educational instruction, equipment control, record keeping, modeling, and expert systems.
3. Discuss future application of computer technology in horticulture including robotics and sensor technology.

KEY TERMS

Database	Multimedia	Software
Expert systems	Program	Supercomputers
Internet	Robotics	Word processing
Modeling	Sensor technology	

OVERVIEW

Computers are electronic devices that can be programmed to perform specific functions (figure 23-1). They have the capacity to store large amounts and varieties of information. They are able to retrieve the stored information quickly and also manipulate the data in complex ways based on complex mathematical instructions, doing so at speeds that humans are unable to match. The "engine" that runs a computer is the *chip,* whose design determines what the computer can do and how it performs the task. Computers are incapable of independent thinking and must be provided with specific, accurate, and unambiguous instructions as to how to

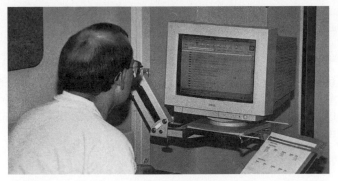

FIGURE 23-1 Computers are available in a wide variety of makes and models.

perform a given task. These instructions are contained in a package called *software* (as opposed to *hardware,* which refers to the machine and its components) or a *program* that is written in a specific *computer language* (e.g., BASIC, COBOL, and C).

Software packages exist for *word processing* (making it possible for the user to operate the computer as an advanced typewriter) and *graphics* (to make drawings and charts). Other software is available for *database* management, whereby the user is able to keep extensive records and manipulate data in numerous ways. In addition to the general-purpose software, numerous other types are designed and developed for specific tasks.

Multimedia systems are a new generation of computer systems that combine traditional computing with sound and other visual effects for enhanced results. Coupled with telephone facilities, computer users are able to link up and share software and data over a *network* that may be local, national, or even worldwide. Anyone with the necessary equipment and know-how can cruise the information superhighway, the current worldwide network for communication via computers.

Technological advancement has made it possible for computers to be portable and much smaller than the first mainframes. For complex tasks, *supercomputers* (such as Cray computers), which are able to handle huge amounts of data and perform mind-boggling calculations at lightning speed, are used. The essential point to note in the use of computers is what is referred to in the industry as "GIGO" (for garbage in garbage out). In other words, do not blame the computer for the outcome; it only does what you tell it to do, except faster and more easily.

> **Software**
> A package of instructions that enables users to work interactively to perform specific tasks on the computer.

> **Hardware**
> The machine and its components.

> **Program**
> A set of instructions designed to guide the computer in performing specific tasks.

23.1: AREAS OF SPECIFIC APPLICATION

Apart from the type of chip they contain, computers are adapted for performing certain functions using appropriate software. Some software is designed for performing specific functions. Special areas of computer application in horticulture are discussed in the following sections.

23.1.1 HORTICULTURAL INSTRUCTION

Computers are widely used in the classroom to facilitate instruction and student experiential learning. A variety of software packages that may be used for instruction in the basic principles of horticulture are available. Some software is designed for users to work interactively with computers. Using computers interactively provides a platform for humans and machines to communicate back and forth, which facilitates the learning process of students who are able to use computers for autotutorial sessions. Computer-based teaching and learning have become commonplace. A number of horticultural topics available in computer format may be used to supplement classroom instruction. Instructors with computer experience often develop course-specific software for their classes.

Current technology enables researchers to communicate with colleagues virtually anywhere in the world. Information that previously could be obtained only through a third party may now be directly accessed from the convenience of an office equipped with a computer and appropriate accessories.

23.1.2 DESIGNING PLANS

Landscape architectural instruction is facilitated by sophisticated computer software (e.g., CAD) for designing landscape plans, instead of the traditional method of drawing by hand. Professional designers may also use computers in their work. Computers may be used by individuals to design landscapes and vegetable garden plans for their homes.

23.1.3 CROP MODELING

Crop modeling is the science of predicting a crop's response to environmental conditions or its productivity in response to levels of production inputs. With the aid of computers and complex mathematical procedures, scientists are able to develop *models* that enable them to estimate a crop's needs to produce a certain response. For example, how will 80 kilograms per hectare (176 pounds per hectare) of nitrogen fertilizer applied as a sidedress four weeks after planting affect the yield of a crop? To make the calculation, a wide variety of data on the different crop characteristics and responses are gathered. The information includes the following:

1. Plant botany
2. Effect of environmental factors on crop growth and development (i.e., effect of light, temperature, moisture, and nutrients)
3. Yield response to environmental factors
4. Performance of the crop under various cultural practices

These data are obtained from research conducted over many years by many scientists. Also, various laws and principles that pertain to the problem are considered in developing the model.

Models are developed on an individual crop basis. Crops with a large body of research findings are the ones for which models are currently available. Modeling is possible because of computer technology, which allows huge amounts of data to be manipulated simultaneously using complicated formulas.

23.1.4 SOIL MODELING

A *soil test* (3.3.) (for nutritional level and pH) is recommended before applying fertilizers to a crop. Based on the results of the test, the soil's physical characteristics (e.g., texture), the temperature, the crop, the desired level of production, and other local production factors, a recommendation is made to the grower. To effectively and efficiently integrate these factors, a model (similar to the one for crops) is developed and utilized to make predictions and appropriate recommendations.

23.1.5 EQUIPMENT AUTOMATION

Controlled-environment production practices depend on the use of automation to monitor and control plant growth factors such as light, moisture, and temperature indoors. Sensors strategically positioned in the room monitor the environment and initiate changes automatically (figure 23.2). Nurseryworkers may not be available around-the-clock to attend to equipment. The crop may require a specific nighttime temperature or light to be turned on for a certain period of time. Equipment automation is not limited to greenhouse equipment but also applies to a variety of field implements. Automated home landscape irrigation systems may be installed to ensure that plants are watered when needed (16.2).

23.1.6 PUBLIC INFORMATION

A large variety of horticultural data in computer format are available for the general public to use free of charge or at minimum cost. For example, the National Agricultural Library (NAL) at Beltsville, a unit of the U.S. Department of Agriculture (USDA), has compiled a compre-

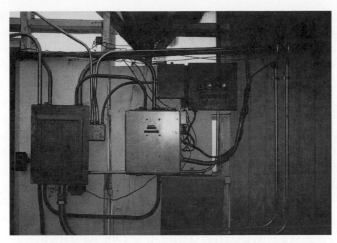

FIGURE 23-2 A variety of implements in the greenhouse are electronically controlled for automation.

hensive set of data on gardening and landscaping called Plant It. This documentation provides information that includes color images of plants, trees, and shrubs and identifies those best suited to the different climatic zones of the United States. Users of the *Internet* are able to access a database that allows assessment of the water pollution potential of more than 200 of the most widely used pesticides (herbicides, fungicides, and insecticides).

23.1.7 RECORD KEEPING AND DATABASES

Good record keeping is critical to the success of any horticultural enterprise. Computers are used to maintain an inventory of an operation and to manage the financial aspects of the business. Growers may use enterprise-specific applications that are designed to facilitate the operation's specific tasks. Users with some familiarity with computers may use general-purpose commercial software packages such as Quicken, D-base, Lotus 1-2-3, and MS Excel to design and customize applications for their individual operations. The home grower ordinarily does not need a computer to manage a garden operation. Record keeping and data management applications are useful when production is on a large scale. A variety of enterprise-specific applications are available through extension offices and commercial outlets for managing greenhouse operations and field production (e.g., fertilizer mixes, irrigation, and pesticide applications).

23.1.8 EXPERT SYSTEMS

Expert systems are knowledge-based decision aids that make use of *artificial intelligence (AI)* technologies. These programs are designed by using information from research and known laws and principles pertaining to a specific area or topic of interest. Using complex mathematics and computer programming methodologies, software developers create a product that can be used to help in making specific decisions. For example, expert systems designed for disease diagnosis take a user-supplied list of symptoms and provide suggestions as to the possible disease that may be present. An example of such a software package is Plant Doc.

> **Expert System**
> A computer-based decision making tool that uses artificial intelligence technology.

23.1.9 COMMUNICATION

In this information age where the world is connected by an elaborate computer network, a wide variety of horticultural information can be accessed via the Internet. Computer bulletin boards post information that growers can use in their enterprises. Public domain software is available to users on a 24-hour basis.

23.1.10 ROBOTICS

Robots are used to perform tasks that may be dangerous or repetitive for humans. A number of universities are engaged in developing and using robots in horticultural research. For example, robots for transplanting are being perfected by Purdue (for pepper) and Rutgers (for marigold and tomato). Oklahoma State University is working on automated tissue propagation systems. Michigan State University and Texas A&M are working on a variety of automated guided vehicles. A number of fruit harvesting prototypes are being developed for tomato, citrus, apple, and cucumber.

23.1.11 SENSOR TECHNOLOGY

A new generation of equipment is being developed based on sensor technology including machine vision, electrical sensors, nuclear magnetic resonance (NMR), and near-infrared (NIR) spectroscopy. The geographic information system (GIS) is a database of land information that has an increasingly wide application.

23.2: HORTICULTURE ON THE INTERNET

The Internet is increasingly becoming the method of communication of choice.

23.2.1 BRIEF BACKGROUND

> **Internet**
> An interconnection of numerous computers that can be accessed over a network.

Basically, the Internet is an interconnection of a multitude of computers that one can access over a network. The concept is not all that new, having been in operation since 1969. However, it has only recently become a household word, largely because literally anyone with a computer and the appropriate electronic communication device and access to a phone may participate in this network. Originally designed to facilitate the exchange of information among scientists and researchers, today any kind of information may be exchanged digitally between any two persons with the appropriate equipment. Information is available on the Internet to match everyone's interests. You can buy and sell over the Internet, conduct a telephone conversation, watch movie clips, advertise, consult electronic bulletin boards, read newspapers, play games, review literature, or, at the very least, just wander about in *cyberspace,* an activity casually called "surfing the 'Net."

Patronage of the Internet has soared because software packages have been developed to make it readily accessible and usable by people with limited familiarity with computers. In 1991 the University of Minnesota launched its contribution to the Internet by introducing *Gopher,* a graphic user interface that allows the user to point and click with a computer mouse instead of typing commands. The Gopher also eliminated the need for users to access each computer independently, allowing them to browse the electronic network as a whole. A user can visit one *site* on the network and skip to another site on a different computer without having to log on anew each time a change is made. This computer hopping can be local or even transcontinental and occurs instantly.

As one travels along in cyberspace, some of the information encountered may be copied for use by printing or *downloading* (transmitting from a remote computer to a local one) to the client's computer. It must be mentioned that a variety of copyright provisions govern the legal use of material displayed on the Internet. In 1992 a Swiss group at CERN, the European Particle Physics Laboratory, introduced the *World Wide Web (WWW),* currently the most popular graphic user interface for navigating the Internet. The WWW uses *hypertext* (standards and protocols consisting of words and images used to link different texts, sounds, images, and movies on computers throughout the world) in its operation. As convenient as the WWW is, the creation of other Internet tools such as Mosaic in 1993 (by the National Center for Superconducting Applications), Netscape in 1994, Microsoft Explorer, and others has made access to the Internet even easier. These software packages (called *browsers*) account for the astonishing growth in use of the Internet. As a matter of clarification, the Internet is the electronic data transportation network, and the Web is the battery of navigational tools used for traveling the information superhighway.

The huge and ever-growing patronage of the Internet is making it more economical for certain activities to be conducted *on-line.* Through *electronic publishing* capabilities, busi-

> **Browser**
> Computer software that enables users to navigate the Internet.

nesses are realizing that it is less expensive to disseminate information on-line than to print and mail it. An advantage of electronic publishing, apart from lower costs, is the ease with which information can be updated (instantly). The interactive feature of electronic communication enables people with questions to pose them to a worldwide audience and receive answers from all over the world. Each subscriber to this worldwide network may contribute information to the pool as described. Information is shared, for example, through the creation of a *home page* or a *Web site* at which the individual subscriber provides whatever information he or she wishes to share. The information may be text, graphics, sounds, or movie clips and is posted in the form of a *page* similar to a book page.

> **Home Page**
> A specific site or address on the Internet that is owned by a subscriber and is the avenue by which information is shared with other users.

23.2.2 INTERNET SERVICES

In subscribing to the Internet, one has four general services from which to choose:

1. *Electronic mail (E-mail).* The E-mail function enables customers to send and receive messages electronically. Electronic conferences and group discussions can be conducted using this feature.

2. *Telnet.* Telnet enables two computers to be connected for computer-to-computer interactive use.

3. *File transfer protocol (FTP).* FTP is used to move files and data from one computer to another (e.g., download books, software, music, graphics, and other documents).

4. *World Wide Web (WWW).* The WWW, or simply the Web, consists of protocols and standards (tools) used to access any information (video, text, or audio) on the Internet.

23.2.3 BEFORE YOU START

Before accessing the Internet, one needs certain basic equipment and to subscribe through commercial Internet providers.

Basic Equipment

A computer equipped with a modem and adequate memory are all that one needs by way of hardware to access the Internet. The primary communication facility needed is a telephone line. Individual preferences determine the kind of computer and accessories. For the best-quality graphics, the computer should be equipped with a super virtual graphics adapter (VGA) card and accompanied by a color monitor of comparable quality. A powerful and fast computer (e.g., 486 or Pentium, Microsoft Windows environment, and 28,000 bps modem or better with V.34 and V.42bis) will make working on the Internet fun. When a document to be accessed includes graphics, slow computers take a long time to retrieve the images, which means high access fees (unless one has a flat-rate subscription). The speed of the modem is a critical determinant of how fast information is sent and received. The capacity of the phone lines, however, limits how much information can be transmitted or received. Further, if one is using a long-distance line to access the Internet, a faster modem is advantageous. Unfortunately, if an Internet provider uses a slow connection speed, a customer's fast modem will not be advantageous in any situation.

> **Modem**
> A telecommunication device that enables a user, via the telephone, to connect a computer to a network of other computers.

Internet Providers

After acquiring the equipment, you will need to subscribe to the Internet through one of several providers, categorized as follows: *on-line services, national Internet access providers,* and *local Internet access providers.* All of these providers have advantages and disadvantages. Local Internet access providers offer local access numbers, thus eliminating the costly long-distance calls. However, some of these providers may not offer certain services, such as those requiring graphic interface, limiting subscribers to text-based services. National on-line services provide easier access to the Internet; these providers, which include America Online

(AOL), CompuServe, and Prodigy, are considered more comprehensive in the services they provide. In addition, they also have their own programming added to the regular services (e.g., *Time Magazine* on-line). Subscribers are required to establish an *Internet account* and choose a *password* to log onto the Internet. The provider supplies a set of instructions for logging on and off the system. The fee charged differs from one provider to another. As time goes by and competition increases, fees are bound to be revised.

Uniform Resource Locators

After signing with an Internet provider, the subscriber is ready to access the Web. In regular computing, software and user files stored in specific locations in the computer's memory are accessed by following specific *paths* to these *addresses*. Similarly, one needs to know addresses and paths to specific sources of information on the Internet. The uniform resource locator (URL) provides a standard format for finding specific locations on the Internet (e.g., "protocol//server-name/path"). The protocol is the method of communication used to access the information from the server. For example, the *hypertext transfer protocol (http)* is used for information transfer within the WWW. A Web site URL may be http://home.netscape.com/welcome.html. *Hypertext markup language (html)* is the programming language used to create a Web page.

23.2.4 SEARCH DIRECTORIES AND ENGINES

<aside>
Search Directory
A database of information organized under specific groups with specific addresses by which they can be accessed.
</aside>

<aside>
Search Engine
Software designed to help a user search for and locate desired information in a directory.
</aside>

After successfully logging onto the Internet, *search directories* and *search engines* are tools used to speed the search for information and resources. Search directories are like directories established on personal computers that describe various registries of Web sites. To utilize a search directory, the user may type in a keyword that best describes the topic of interest. The network then displays the results of the search, indicating the number of *hits* or sites that have information related to the keyword. A search engine searches the Web and other Internet resources such as Usernet newsgroups to find matches for the keyword. Popular search directories are Yahoo, Magellan, and Galaxy. Search engines in use include Alta Vista and Open Text. Research tools such as Excite and Infoseek are both search engines and search directories. These tools vary in the comprehensiveness of their searches. Whereas some allow only one-word (keyword) searches, tools such as Excite allow the use of descriptive phrases to refine the search to include only pertinent and close matches. This search strategy is called *concept-based searching*.

23.2.5 APPLICATIONS IN HORTICULTURAL EDUCATION

The Internet is used in a variety of ways for horticultural instruction. Professors may refer students to specific sites for additional information. Students may search the network for information in writing reports, conducting research, and communicating with colleagues. Selected sites of horticultural interest are presented in table 23-1.

College/University Web Sites

Some of the most informative and elaborate Web sites for educational and research purposes are sponsored by university academic departments and experiment or extension stations. These sites usually reflect the program strengths of the represented institutions and mandates with respect to crops (and animals) that are the focus of research. For example, vegetables are important to programs in Florida and California, and you are unlikely to find any valuable information on citrus from a university in the Midwest.

Government Institutes and Centers

Government agencies and research centers, institutes, or other such bodies tend to provide excellent information on the Internet. The USDA, for example, maintains a variety of Web sites that provide valuable information.

Table 23-1   Selected Internet Sites of Horticultural Interest

Vegetable Crops, Horticulture, and Forestry Information

Address:	gopher://cesgopher.ag.uiuc.edu:70/11/Crops-Horticulture-Forestry
Comments:	A good source of general information on crop production

Fruit Crops

Address:	http://hammock.ifas.ufl.edu/txt/fairs/19943
Comments:	A variety of information on fruit culture: agronomic, management, and postharvest handling

Vegetable Crops from the University of Florida

Address:	http://hammock.ifas.ufl.edu/text/aa/39584.html
Comments:	Good information on vegetable crop production with accompanying publications

Vegetable Gardening Handbook

Address:	http://hammock.ifas.ufl.edu/text/vh/19996.html
Comments:	Information on garden production

Missouri Botanical Gardens Home Page

Address:	http://www.mobot.org/welcome.html
Comments:	Important links to Missouri Historical Society, Center for Plant Conservation, and American Association of Botanical Gardens and Arboreta

The Master Gardener Information

Address:	http://leviathan.tamu.edu:70/1s/mg
Comments:	Provides information on annual and perennial flowering plants, ornamental trees and shrubs, turfgrasses, and vegetables

Newsgroups

Newsgroups are special-interest groups formed to provide a forum for communal interaction. One may locate a newsgroup from the browser's *newsreader* feature, which provides a list of newsgroups organized on a main subject basis. The user may locate a group of interest from the list provided or use a search process to locate one.

Bulletin Boards

Bulletin boards are small-scale information sources often operated by individuals from home computers. They have specialties and report on items such as community events, local weather, and farmers' market news. Although some bulletin boards can be accessed free of charge, others charge a small fee for their services. Some of these bulletin boards are worth the time it takes to search them, but others are not.

SUMMARY

Computer technology has significant applications in horticulture. In education, computers are now routinely used in the classroom for instruction. They are also used in database maintenance and manipulation. Record keeping and financial aspects of a horticultural enterprise are facilitated through the use of computers. Greenhouse automation and equipment calibration for automatic application of chemicals, for example, are routine applications of computers in horticulture.

More sophisticated applications of computers occur in the use of expert systems as knowledge-based decision aids, applications that depend on artificial intelligence technology.

Other applications that are being perfected involve the use of robotics to perform repetitive tasks and sensor technology to develop a new generation of versatile equipment. Currently, it is possible to access an international pool of information via the Internet.

REFERENCES AND SUGGESTED READING

Anonymous. 1994. Computers in agriculture. Proceedings of the 5th International Conference. February 6–9. Orlando, Fla.: American Society of Agricultural Engineering.

Varner, M. A. 1994. Computer applications in agriculture: Encyclopedia of agricultural sciences, Vol. IA. New York: Academic Press.

PRACTICAL EXPERIENCE

Objective: To demonstrate how an expert system works.

1. *Materials:* Software—Plant Doc.
    Method: Attempt to diagnose a number of diseases by entering into the computer different disease symptoms.
2. *Materials:* Software—Plant It.
    Method: Select plants for the landscape.

OUTCOMES ASSESSMENT

PART A

Please answer true (T) or false (F) for the following statements.

1. T F   The chip is the engine that runs a computer.
2. T F   Instructions that direct computer operations are contained in a software package, or program.
3. T F   Computers can be equipped to "think" on a limited basis through artificial intelligence technology.
4. T F   The Internet is an interconnection of computers.

PART B

Please answer the following questions.

1. The Internet tool, Gopher, was developed by _____.
2. The Internet service which enables two users to exchange electronic messages is called _____.
3. List two requirements for using the Internet.

    _____   _____
4. List two Internet access providers.

    _____   _____
5. List two search engines on the Internet.

    _____   _____

1. Discuss specific applications in horticulture that involve computers in automation.
2. Describe how computers are used in modeling applications.
3. Discuss how the Internet impacts the horticultural industry.
4. Describe some of the future applications of computers in horticulture.

CHAPTER 24

Growing Succulents

Purpose

This chapter is designed to discuss how to grow and care for cacti and other succulent plants.

Expected Outcomes

After studying this chapter, the student should be able to

1. Describe the basic characteristics of succulents.
2. Describe how succulents are propagated.
3. Distinguish between desert and jungle cacti.
4. Describe the methods of cacti propagation.
5. Discuss how to care for cacti for healthy growth.
6. Describe how to propagate and care for bromeliads.

Key Terms

Bromeliads	Grafted cacti	Offsets
Desert cacti	Jungle cacti	Succulents
Epiphytes		

MODULE 1 SUCCULENTS IN GENERAL AND BROMELIADS

Overview

> **Succulent**
> *A plant with thickened leaves and other parts adapted for water retention.*

Succulents are plants that are capable of storing large amounts of water in their tissues. The water is stored in either the stems or leaves. Succulents differ widely in the degree of succulence, cacti being very succulent and species such as the agaves having thin leaves. The plant families that contain succulents include Agavaceae, Aizoaceae, Asclepiadaceae, Compositae,

Table 24-1 Selected Popular Succulents (Excluding Cacti and Bromeliads)

Plant	Scientific Name
Houseleeks	*Sempervivum* spp.
Century plant	*Agave americana*
Wart plant	*Haworthia* spp.
Milk bush	*Euphorbia tirucali*
Crown of thorns	*Euphorbia splendens*
Candle plant	*Senecio articulatus*
Mother-in-law's tongue or snake plant	*Sansevieria trifasciata*
Air plant or Panda plant	*Kalanchoe* spp.
Hen and chickens	*Echeveria* spp.
Agave	*Agave angustifolia; A. botterii*
Aloe	*Aloe variegata* and other species
Ox tongue	*Gasteria* spp.
Jade plant	*Crassula argentea*
Burro tail	*Sedum* spp.

FIGURE 24-1 Succulents come in a wide variety of shapes, sizes, and growth forms. They offer variety in choices of display in hanging baskets or standing pots.

Crassulaceae, Euphorbiaceae, Liliaceae, and Cactaceae. Because of their immense capacity to store water, succulents can survive as indoor plants with minimum care. A list of various succulents is provided in table 24-1. Figure 24-1 shows a sample of plant forms represented by succulents.

24.1: PROPAGATION

Succulents may be propagated from seed, cuttings, or offsets.

24.1.1 SEED

Seeds may be nursed in trays or pots containing a rooting medium that drains very well. Many succulent seeds are small and should not be buried during propagation. The soil surface should be of fine tilth. The seeds are sprinkled on the soil surface and then watered. The soil should be kept moist by covering with either a glass plate or clear plastic. Germination occurs within two to three weeks.

24.1.2 CUTTINGS

Leaf cuttings of succulents can be propagated by first rooting them in water or soil. Succulents of the family Crassulaceae are particularly suited for propagation from leaf cuttings. After removing a leaf, the cut surface should be allowed to dry for a few days before planting in the medium.

24.1.3 OFFSETS

Nonbranching succulents often produce offsets that readily propagate the species. Offsets are identical plants that develop as side growths from the mother plant originating from either the main stem or the base of the mother plant, as occurs in bulbs. These young full-fledged plants may be removed after they have grown to a size that will enable them to survive independently of the parent plant. The daughter plants are detached by cutting with a sharp razor blade or knife and inserted into a rooting mixture. They are nursed under warm, medium light conditions with sufficient moisture until they have formed adequate roots. Offsets are commonly produced by cacti, bromeliads, and bulbs.

24.2: GROWTH REQUIREMENTS

24.2.1 LIGHT

An inadequate light supply is usually the cause of poor growth and development of succulents grown indoors. Since an indoor light supply is irregular, succulents grown indoors should be located on the windowsill or near a window where they can receive sunlight. Window-grown plants are prone to phototropic response and should be turned regularly to prevent the foliage from curving toward bright light. It is a good practice to place potted succulents outside for some time during the bright days of spring and summer.

24.2.2 TEMPERATURE

Succulents generally grow well under warm conditions. They are cold sensitive and as such should not be left outside during the cold season.

24.2.3 MOISTURE

By nature, succulents are able to store large amounts of water due to adaptive mechanisms that enable them to survive dry conditions. However, this should not be misconstrued to mean that they do not need watering. On the contrary, a good watering regime increases the success of succulents. Physiologically, they need watering during the period of active growth. The amount provided should be reduced during the rest period of the plant. Since the foliage of most succulents have great aesthetic value, it is best to water potted indoor succulents from below to prevent water from splashing on the plants, which often leaves unsightly markings from the salts in the water (10.3.11).

24.2.4 FERTILIZING

Unless the succulents are fast growing, they are not likely to benefit a great deal from any fertilizer application regimen. However, some nutrition is needed to support basic growth and development. Nitrogenous fertilizers should be used with caution since they may cause weakness and flabbiness in top growth.

24.2.5 DISEASES AND PESTS

Indoor plant pests such as root mealybugs, spider mites, scales, and mealybugs affect succulents because the anatomical arrangement of the leaves is a conducive environment for such pests to hide and thrive. When root mealybugs become a problem, succulents should be repotted after removing the old medium. Observance of phytosanitation reduces the incidence of diseases and pests.

24.2.6 REPOTTING

Like other potted plants, succulents periodically need to be repotted into larger containers. In spite of their succulence, some of these plants are brittle and should be handled with care. The foliage may be cleaned with a mild detergent and water.

24.3: Bromeliads

Bromeliads belong to the family Bromeliaceae. These tropical plants have certain distinguishing features. They have a rosette plant form and absorb food and moisture primarily through their leaves. The name *bromeliad* derives from this aerial feeding characteristic. As such, bromeliads are sometimes called *air plants.* Most bromeliads grown indoors are epiphytes; however, nonepiphytic terrestrial species exist. The most popular terrestrial species is the edible pineapple *(Ananas comosus).* Bromeliads produce brilliantly colored flowers that may appear any time of the year once the plant has attained maturity. Bromeliads also need to have water in the center of the rosette, which is the growing point. The two most common plant shapes are presented in figure 24-2. Popular species are listed in table 24-2.

> **Epiphyte**
> A plant that grows upon another plant but is not a parasite.

24.3.1 CREATING AN EPIPHYTE BRANCH

Epiphytic bromeliads are not planted in soil in pots but are attached to pieces of wood. To create a display, one needs sphagnum moss, a log, cork bark, wire, plastic wrap, and shears. The moss is moistened and wrapped around the base of a bromeliad rosette to cover all of the roots. The moss-covered plant is attached to the log and temporarily secured by tying with a piece of wire. The cork bark is wrapped around the moss and secured with wire permanently. Several rosettes may be prepared and attached to various parts of the log to produce a creative display (figure 24-3). Different species may be used in one display.

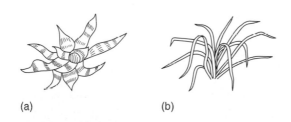

(a)   (b)

(c)

FIGURE 24-2 Typical bromeliad shapes: (a) tightly packed rosette of leaves with a central cup and (b) loosely packed rosette. (c) Photo shows a tightly packed rosette.

Table 24-2 Selected Popular Bromeliads

Plant	Scientific Name
Pineapple	*Ananas comosus*
Urn plant	*Aechmea fasciata*
Patriotic plant	*Aechmea mertensis*
No common name	*Ananas bracteatus* or *A. comosus variegatus*
Angel's tears or queen's tears	*Billbergia nutans*
Orange star	*Guzmania lingulata*
No common name	*Guzmania musaica*
Blushing bromeliad	*Neoregelia carolinae;* other species are *N. spectabilis* and *N. ampullacea*
Earth stars	*Cryptanthus bromelioides, C. bivittatus, C. acaulis,* and others.
Flaming sword	*Vriesia splendens;* other species are *V. hieroglyphica* and *V. fenestralis*
Blushing bride	*Tillandsia ionantha*
Pink quill	*Tillandsia cyanea*
Black Amazon bird nest	*Nidularium innocentii;* another species is *N. fulgens*

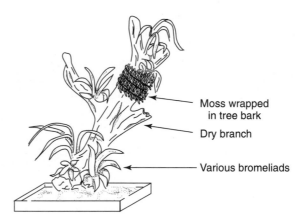

Moss wrapped in tree bark

Dry branch

Various bromeliads

FIGURE 24-3 An epiphytic branch may be used to display a variety of bromeliads.

24.3.2 CARE

Care of succulents includes the following:

1. *Water.* The moss part of the display should be watered when it feels dry, and the middle part of the rosette that holds water should be kept full at all times. The receptacle should be emptied and refilled periodically (monthly).

2. *Light.* The display should be mounted in a part of the house that receives bright filtered light.

3. *Temperature.* Warm temperatures (above 60°F or 15.6°C) and relatively high humidity are required for best growth. In a drier season, a periodic mist application may be needed to provide humid microclimates for bromeliads.

4. *Fertilizing.* Foliar application or addition of liquid fertilizer to the water in the rosette may be helpful during the period of active growth.

24.3.3 POTTED BROMELIADS

Terrestrial species may be potted in a medium that is porous but retains moisture adequately. Lime is not desirable in the medium. Bromeliads do not produce roots profusely and hence do not need to be grown in deep pots.

Propagation

Bromeliads may be raised from seed. The seeds are sowed and cared for as described for succulents. Commercial nurseries raise plants in this way for sale. However, it is easier to propagate bromeliads by using offsets.

Repotting

Plastic pots retain moisture longer than clay pots. Being air plants, repotting to renew the soil is not important in bromeliad culture. Repotting into larger pots is necessary on an infrequent basis.

MODULE 2 CACTI

OVERVIEW

Cacti (plural for *cactus*) are mostly *succulent plants* that belong to the family Cactaceae. They differ from other succulents by having *areoles* that frequently carry spines and also are sites of flower production. There are two kinds of cacti—*desert cacti* and *jungle cacti*—each with its own growth requirements. Desert cacti are adapted to drought conditions *(xerophytic)*, largely because they lack leaves or have greatly reduced leaf structures, thereby eliminating or minimizing stomatal transpiration. The stems of desert cacti are columnar or spherical in shape and also are ribbed. Most of them have spines that are often prickly (figure 24-4).

(a)

(b)

FIGURE 24-4 Desert cacti occur in a wide array of shapes, sizes, and forms.

Jungle cacti, on the other hand, are generally *epiphytic* and adapted to humid and shady conditions. Their stems are often flattened and cylindrical structures. Jungle cacti lack ribs on their stems, which are jointed into long strings (figure 24-5). Their stems and spines are not prickly.

Desert cacti have more aesthetic appeal than jungle species in their vegetative states. However, cacti, under proper conditions of light, produce attractive flowers, the finest and largest ones being produced by jungle species. Because the jungle species are adapted to shady conditions, they are better suited to indoor use. However, due to their stringy stems, they often require staking or physical support in cultivation. They also do well as hanging basket plants. Common cacti suited to indoor cultivation are listed in table 24-3.

24.4: PROPAGATION

Cacti can be raised from seeds, offsets, or cuttings.

24.4.1 SEED

To raise cacti from seed, fresh, viable seed should be obtained from a reputable nursery. A lot of moisture is needed for germination. First, a pot is filled with the soil mixture, leaving about

Cylindrical   Nonsegmented   Segmented

FIGURE 24-5 Basic stem types occurring in jungle cacti.

Table 24-3 Selected Cacti Suitable for Growing Indoors

Plant	Scientific Name
Bunny ears	*Opuntia microdasys;* others are *O. leucotricha,* *O. monacantha,* and *O. tuna*
Sea urchin cactus	*Astrophyton asterias*
Peanut cactus	*Cleistocactus strausii*
No common name	*Cereus peruvianus*
Queen of the night	*Selenicereus grandiflorus*
Rat's tail	*Aporocactus flagelliformis*
Crown cactus	*Rebutia* spp.
Golden barrel cactus or mother-in-law's arm chair	*Echinocactus grusonii*
Bishop's cap	*Astrophytum myriostigma*
Rose-plaid cactus	*Gymnocalycium mihanovichii*
No common name	*Hematocactus stispinus*
Pincushion cactus	*Mammillaria* spp. such as *M. bocasana,* *M. bombycina,* and *M. parkinsonii*
Rainbow cactus	*Echinocereus pectinatus*
Easter cactus	*Rhipsalidopsis* hybrids
Christmas cactus	*Schlumbergera* hybrids

1 inch (2.54 centimeters) of space on top. The medium should be drenched with water. After the excess water has drained away, the seeds are scattered on the soil surface without burying them; large seeds (e.g., those from the opuntia group) may be partially buried. To maintain a moist soil, the pot should be covered (e.g., by tying a piece of plastic sheet over the top); the pot is then placed in a warm area (figure 24-6). After about a year, the seedlings are ready to be transplanted into individual containers.

24.4.2 OFFSETS

As cacti grow, older plants develop plantlets, or *offsets,* around the base of the stem. The offsets can be harvested by either cutting or pulling them from the parent plant. The separation leaves a wound at the point of attachment that needs to be dried for a few days before planting to prevent rotting. Offsets are planted by pushing them slightly into the seed medium (figure 24-7). After planting, they should be watered moderately and kept in medium light until they root, after several weeks. Since cacti have sharp needles, they cannot be handled with bare hands. Gloves, folded paper, thongs, and such items are used while cutting or planting cacti.

24.4.3 CUTTING

Apart from whole plantlets, vegetative pieces can be obtained from the stem for propagation. For species such as opuntia that have segmented stems, individual segments can be removed for planting (figure 24-8). If done carefully, the overall appearance of the plant is not ruined. Nonbranching species such as trichocereus are propagated by cutting off about 2 inches (5.1 centimeters) of the tip of a stem. The piece is planted after the cut end is dried.

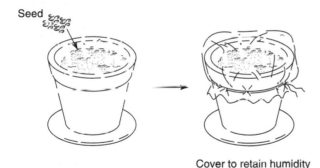

Cover to retain humidity

FIGURE 24-6   Desert cacti may be propagated from seed.

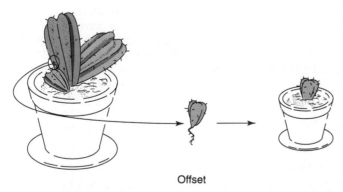

Offset

FIGURE 24-7   Desert cacti may be propagated by using offsets in certain species. This is simply accomplished by using a pair of tongs to carefully pull offsets and planting them in potting media.

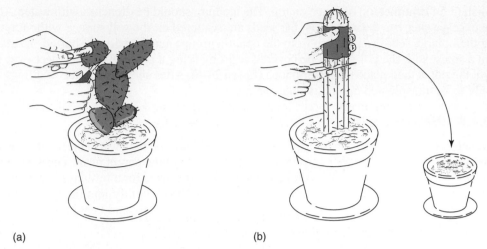

(a)   (b)

FIGURE 24-8 Both desert and jungle cacti may be propagated by cuttings. (a) In segmented or branching species, a branch is simply cut off and planted. (b) In nonbranching species, tip cutting is made. Care must be exercised to avoid injury from prickly spines.

24.5: CARE

24.5.1 WATERING

Many cacti are killed because of improper watering. Failure to reduce watering frequency during the dormant period (fall to spring) leads to overwatering and subsequent injury to the plants. To avoid overwatering, it is best to water cacti from the bottom of the container by placing the pot in a dish of water until the soil is saturated. This method of watering also eliminates the marks often left on plant foliage from contact with hard water. Epiphytic cacti require more water than desert cacti; however, overwatering is generally detrimental to the growth of cacti.

24.5.2 FERTILIZING

Fertilizing is required during the active growing period, especially when the plant is growing in a soilless medium. Nitrogen-rich fertilizers should be avoided since they tend to discourage flowering and instead cause the stem tissue to become soft and weak.

24.5.3 LIGHT

Desert cacti prefer intense light and should be positioned in well-lit places or in windows that receive the most sunlight. Plants placed in windows need to be rotated regularly to prevent their curving toward light. Jungle cacti prefer shade and can be killed if exposed to intense summer sunlight.

24.5.4 TEMPERATURE

Desert cacti are adapted to warm conditions. However, extra indoor heat during the winter period coupled with reduced light may cause cacti to grow spindly and weak. A room temperature of between 45 and 65°F (7.22 and 18.3°C) is desirable for cacti. The low indoor humidity during winter is especially intolerable to jungle cacti, which need to be misted with water frequently during this season.

24.5.5 CLEANING

To keep plants attractive, cacti may be dusted, cleaned with a damp cloth, or washed. These cleaning activities should be done very carefully when handling species with thorns.

24.5.6 REPOTTING

Cacti require repotting every two to four years. To determine whether repotting is due, the plant should be removed from its pot and its roots checked for pot bounding.

24.6: MINIATURE ROCK GARDEN

Cacti are excellent plant species for creating rock gardens. They offer a wide variety of plant shapes and forms and can be used to develop desert gardens. The garden is created in a shallow container. The bottom of the container is lined with gravel or a suitable drainage material and then covered with a potting mixture (figure 24-9). Stones of a variety of shapes and sizes are carefully chosen and placed to create a rocky terrain. A variety of cacti (both desert and jungle) are selected and planted around the stones.

24.7: GRAFTED CACTI

Cacti occur in a spectacular array of forms, as indicated in this chapter. The beauty of these slow-growing plants can be enhanced by grafting different species with contrasting forms and features onto one another (figure 24-10). For example, a columnar species may be topped with a globelike species. Two species are commonly grafted, but some creative growers may

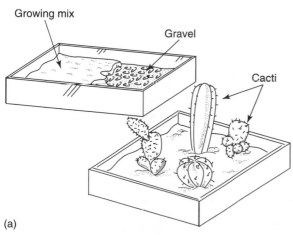

(a)   (b)

FIGURE 24-9 (a) A variety of desert cacti may be planted in a miniature garden. (b) Plant selection should include contrasting types for the best effect.

FIGURE 24-10 Grafted cacti add interest to displays. A graft may be simple, involving just two species, or multiple, with several species growing on one stock.

graft more than two different species on one plant. In some cases, the scions develop into very colorful and attractive ornamental displays.

SUMMARY

Succulents are plants that are able to hold large amounts of water in their leaves or stems. Plant families that have succulents include Cactaceae, Liliaceae, Crassulaceae, and Agavaceae. Cacti are a special group of succulents that frequently carry spines. There are two kinds of cacti: desert (which are xerophytic) and jungle (which are epiphytic). Desert cacti have stems that are either columnar or spherical and may or may not be ribbed. Jungle cacti, on the other hand, have flattened and cylindrical stems that are jointed. Under proper light conditions, cacti may be induced to flower. They can be raised from seed, offsets, or cuttings; improper watering in cultivation is detrimental to cacti plants.

REFERENCES AND SUGGESTED READING

Rice, L. W., and R. P. Rice Jr. 1993. Practical horticulture, 2d ed. Englewood Cliffs, N.J.: Prentice-Hall.

Reader's Digest. 1979. Success with houseplants. New York: Reader's Digest Association.

PRACTICAL EXPERIENCE

Follow the steps in the text to construct an epiphyte branch.

OUTCOMES ASSESSMENT

PART A

Please answer true (T) or false (F) for the following statements.

1.   T F   Desert cacti are xerophytic.
2.   T F   Desert cacti have jointed stems.
3.   T F   Jungle cacti are shade tolerant.

4. T F Cacti do not flower.
5. T F Jungle cacti prefer low humidity in the environment.
6. T F If a cactus is ribbed, it is likely to be a desert species.
7. T F Jungle cacti have prickly spines.
8. T F Pineapple is a succulent.
9. T F Bromeliads are called air plants.

PART B

Please answer the following questions.

1. List three ways in which jungle and desert cacti differ.
   _____ _____ _____
2. Cacti belong to the family _____.
3. List any four plant families that contain succulents.
   _____ _____
   _____ _____
4. Succulents may be raised from seed, _____, or_____.
5. Give one example of a nonbranching cactus species.
   _____
6. Give four examples of bromeliads.
   _____ _____
   _____ _____

PART C

1. Discuss the mineral nutrition of succulents.
2. Which parts of a room are suited to succulents and why?
3. Describe how cacti may be planted from cuttings.
4. Describe how desert cacti adapt to drought conditions.

CHAPTER 25

Terrarium Culture

Purpose

This chapter is designed to discuss the principles and methods of growing plants in restricted environments.

Expected Outcomes

After studying this chapter, the student should be able to

1. Describe how a terrarium works.
2. Discuss the selection of containers and plant species for a terrarium.
3. Discuss the maintenance and care of a terrarium.

Key Terms

Activated charcoal   Phenotypic response   Sphagnum moss
Moisture cycle   Sand art   Terrarium

Overview

In chapter 11, greenhouses were described as controlled-environment structures for growing horticultural plants, sometimes on a commercial scale. Miniature versions of greenhouses are exemplified by indoor plant growth chambers that may be purchased for domestic use (figure 25-1). Like greenhouses, growth chambers are equipped with devices for automatic control of certain growth factors such as light, water, and temperature. Very simple homemade controlled-environment plant growth chambers can be constructed, primarily for aesthetic purposes. Nathaniel B. Ward is credited with inventing the method of growing plants in closed containers. His original invention was based on the principle that when plants grow in an enclosed and restricted environment, moisture from transpiration and evaporation from the soil is not lost but retained through condensation and returned to the soil. This *water cycle,* coupled with adequate provision of other growth factors (3.2, 3.3), enables plants to be grown in a variety of closed containers. This system for plant culture is called a *terrarium* (or *plantarium*) or *plant case.*

FIGURE 25-1 A small plant growth chamber.

25.1: TYPES OF TERRARIUMS

In terms of design, there are two basic types, with some variation. In one design, the container for retaining moisture and maintaining high humidity does not come into direct contact with the growing medium. Instead, a potted plant is placed inside of a larger container with a lid (figure 25-2). A very large container can hold several pots. In the second design, the potting medium is placed directly in the container (figure 25-3). *Bottle gardens* are a very attractive variation of the traditional terrarium and *Wardian case* (named after Dr. Ward). They work on the same principle and are designed like the traditional terrarium. A terrarium may be sealed or openable.

> **Terrarium**
> *A unit with a high capacity to retain moisture that is used for displaying plants.*

25.2: DESIGNING A TERRARIUM OR BOTTLE GARDEN

25.2.1 CONTAINER

The container should be of clear transparent (preferred) or translucent material (glass or plastic). It should be noted that water runs off glass better than plastic. The plant in the container should be clearly visible. Scratches, cloudiness, and water marks distract from the beauty of the display. Glass reduces the amount of light that enters the container, more so if the material is tinted. However, lightly colored material can enhance the display. Shape is limited only by the grower's creativity, since a terrarium is purely for decorative use. Wine glasses, bell jars, glass bowls, glass bottles, and aquarium tanks are a sample of the containers that can be used (figure 25-4). The shape and size of the container and the size of the neck opening determine the sizes and types of plants that can be grown. Further, if the container is tinted, only plants that are adapted to subdued light (e.g., peperomia and pteris fern) can be grown. It is critical that the container be watertight and without cracks. A bottle may be used in either a vertical or horizontal position for a terrarium (figure 25-5).

25.2.2 SELECTING PLANTS

In selecting plants for a terrarium, it should be kept in mind that the primary advantage of a terrarium is maintenance of humidity for plant growth. Cacti and succulents are therefore not recommended, unless the container will be left open (which in a way defeats the primary principle of holding moisture inside). Since the walls of the container reduce the amount of light entering the system, plants that are adapted to low or medium light intensities are desirable. It is de-

FIGURE 25-2 A terrarium display may involve a container-in-container design in which the potted plants are enclosed in an outer container such as a bell jar.

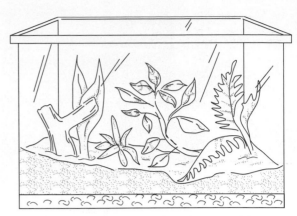

FIGURE 25-3 Commonly a terrarium is created in one container such as an aquarium tank.

FIGURE 25-4 Terrarium containers are varied in shape and size. A key requirement in container type is clarity of material to allow adequate light to reach plants and also for plants to be visible.

(a) Horizontal bottle

(b) Vertical bottle

FIGURE 25-5 Bottle terrarium designs: (a) horizontal bottle and (b) vertical bottle display.

sirable to select plants that are slow growing, such as bonsai (chapter 26). If several plants are to be planted together, it is best to choose plants that vary in leaf shape, size, and color. Small leaves prevent overshadowing of one plant by another and allow all plants in the display to be visible. Terrarium plants are commonly foliage plants, but good plant selection can introduce some color and variety to break the monotony of green by including plants with different leaf arrangements, forms, shapes, and variegation patterns.

Apart from the plant characteristics described, another factor to consider in plant selection is the fact that a terrarium provides a highly restricted growth environment. Therefore, if more than one plant is to be grown, they must all be adapted to the same conditions. Terrariums may be designed according to environmental themes (e.g., tropical plants or ferns). Good terrarium plants include those that originate in the forest region and thus are adapted to humid and shaded conditions. Examples are lichens, ferns, mosses, violets, and ground ivy, which are suited to closed-container terrariums. Cacti and other succulents make good open-container terrarium plants. Table 25-1 provides a list of plants suitable for terrarium culture.

Table 25-1 Selected Plants Suitable for Growing in a Terrarium

Plant	Scientific Name
Maidenhair fern	*Adiantum capillus-veneris* and other species
Spleenwort	*Aspleniun fontanum, A. viride, A. trichomanes*, and *A. ruta-muraria*
Scaly spleenwort	*Ceterach officinarum*
Begonia	*Begonia* spp.
Killarney fern	*Trichomanes speciosum*
Irish moss	*Selaginella* spp.
Button fern	*Pellaea rotundifolia*
Parlor palm	*Chamaedorea elegans*
Lace flower vine	*Episcia dianthiflora*
Table fern	*Pteris* spp.
Hard fern	*Blechnum* spp.
Creeping fig	*Ficus pumila*
Christmas cactus	*Schlumbergera* spp.
Prayer plant	*Maranta leuconeura*
Siper plant	*Sinningia pusilla*

25.3: TOOLS

When planting a large terrarium such as one using a fish tank, ordinary hand tools used in the garden are adequate. However, in bottle gardens, creativity may be necessary in choosing instruments to plant and maintain the unit. The operation becomes more challenging as the neck of the bottle becomes longer and narrower. A number of tools may be improvised for placing growing medium into the container, scooping soil, inserting and planting, firming soil after planting, cutting, cleaning the inside of the container, and retrieving material (figure 25-6). To deliver the planting medium into a bottle, for example, a funnel may be used to direct the soil placement in order to avoid dirtying the inner walls and having to clean them later, sometimes with great difficulty. A disposable funnel can be made out of paper. A spoon with a long handle, or tied to a long stick or wire for increased reach, may be used as a minispade to make planting holes and fill them after transplanting. For firming the soil, the spool from sewing machine thread mounted on a stick or dowel may be used. Long forceps, a forked stick, or a pair of long sticks may be used to deliver plants into planting holes. Dirt on the inner wall of the container may be wiped with a damp piece of cloth or foam attached to the end of a flexible wire. When a leaf or branch must be cut off after planting, a scalpel blade mounted on the end of a stick may be used.

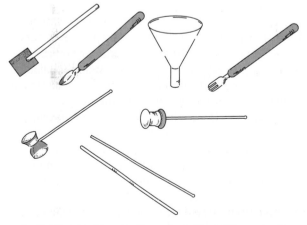

FIGURE 25-6 Assortment of tools used in creating a terrarium. These tools are largely homemade and include cutting, cleaning, scooping, picking, grabbing, and compacting tools.

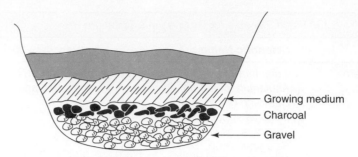

Growing medium
Charcoal
Gravel

FIGURE 25-7 Typical layering of a terrarium. A drainage material lines the bottom, followed by a sprinkle of charcoal for deodorizing, and topped with the growing medium.

25.4: THE PLANTING MEDIUM

Some general principles should be observed in selecting and preparing a growing medium for a terrarium. Good drainage is critical since most plants do not tolerate waterlogging. Good drainage also ensures that plant roots have adequate air for proper growth and development. To provide good drainage, the bottom of the container should first be lined with a layer of coarse material such as washed gravel, sand, or pieces of broken pottery (figure 25-7). To prevent this reservoir for excess water from being destroyed, a layer of sphagnum moss may be spread over the drainage layer as a retaining layer before adding the growing medium. In a closed system, odors tend to build up over time because of the decomposition of organic matter by bacteria. To deodorize the closed environment, a small amount of activated charcoal may be sprinkled over the sphagnum moss. Alternately, lumps of charcoal may be used in place of gravel to line the bottom of the container to about 1 to 2 inches (2.54 to 5.1 centimeters) deep. The planting medium may be purchased from a nursery store or made at home. Fertility management is important in a terrarium, since in a closed environment certain nitrogenous fertilizers tend to release ammonia gases over time; these gases can be toxic to plants and kill them. Using chemical fertilizers in a terrarium, especially a sealed one, is therefore not recommended.

25.5: PLANTING

The tools described earlier (25.3) may be used for planting: the spoon to dig a hole, the forked stick or pair of sticks for lowering the plant into the planting hole, and the dowel for firming the soil around the plant. The hole should be large enough to hold the root mass. After removing the plant from the original container, some of the potting medium should be removed before transplanting. Plants to be located near the side of the bottle should be planted first. For a bottle garden intended for all-around viewing, the tallest specimens should be located in the center and the shortest near the container walls. If the unit is intended to be displayed against a wall, a slope design may be a better choice. In this design, the soil is banked up against the wall rather than leveled off. Plants with a lot of foliage may have to be gently bent to maneuver them through the narrow neck of a bottle. Plants should be arranged such that no plant is obscured by another. The arrangement should pay attention to plant size, and form, and leaf size, shape, and texture so that the overall display is attractive. After transplanting, the plants should be watered by spraying with a mister. If necessary, the inner walls of the container should be wiped with a moist cloth or foam to remove the dirt. The container is then sealed to hold in moisture.

25.6: Care and Maintenance

25.6.1 WATERING

A terrarium should be self-sustaining and able to maintain a good moisture cycle. The evidence of a successful system is the beading of water on the inside of the wall of the container at certain times. A terrarium should not be overwatered. If overwatering is suspected, the terrarium should be unsealed for a period to allow excess moisture to evaporate. Open-top terrariums require periodic watering, the frequency of watering increasing with the size of the container opening. Sealed terrariums hardly need watering; those that require infrequent watering may receive it every two to six months.

25.6.2 LIGHT

Plants vary in light requirements and should be grouped accordingly. Where additional light is needed, the terrarium may be placed under fluorescent light. When placed on the windowsill or close to it, the unit should be regularly rotated to negate the phototropic response of plants, which makes them curve toward light.

25.6.3 TEMPERATURE

A terrarium with a tropical theme requires a warmer temperature than one with a subtropical theme. Ideal temperatures are similar to those required by other indoor plants (about 65 to 75°F or 18.3 to 23.9°C). As previously indicated, potted plants placed on windowsills are subject to greater temperature fluctuations than those placed away from windows.

25.6.4 FERTILIZING

When needed, organic fertilizers should be used. However, if the initial growing medium is well composed, additional fertilizing may not be necessary. Since slow growth is desired, fertilizer application should be minimal. Chemical fertilizers may be used in open terrariums but not in closed types.

25.6.5 PRUNING

When vigorously growing species are used or when growing conditions are very favorable, plants may grow too big too soon and crowd the restricted space in the container. In this situation, excess foliage should be removed. Pruning plants in narrow-necked containers is a challenge but may be accomplished by using a razor blade or scalpel mounted on the end of a long stick. All clippings should be retrieved from the container, lest they decay and cause a foul odor.

25.6.6 DISEASE AND INSECT PEST CONTROL

To reduce the incidence of diseases and insect pests, the growing medium should be sterilized to kill pathogens. The plants used should be disease free, and the foliage should be examined for insects and other pathogens before transplanting into the container. Insect pests found in terrariums include mites, whiteflies, aphids, and mealybugs, which may be controlled by spraying appropriate chemicals.

25.6.7 ENHANCING THE DISPLAY

Certain terrarium designers enhance the display by topping the soil with pebbles or pieces of driftwood. As already mentioned, the containers used should be clear, but light tinting adds beauty to the display, even though it further reduces light penetration.

25.7: Troubleshooting

The following are some common problems associated with terrarium culture, and how they may be remedied.

Symptom	Possible cause	Action
1. Yellowing of foliage	Poor drainage or inadequate fertility	Renew soil and ventilate for evaporation of excess water
2. Poor growth or stunting	Inadequate fertility or poor drainage	Replace soil and add organic fertilizer
3. Browning of foliage	Lack of moisture or burning from direct sunlight	Water plants and move away from direct sunlight
4. Odors	Decomposition of material or presence of nitrogenous fertilizer	Ventilate and add activated charcoal

Summary

Terrarium culture entails growing plants in an enclosed container such that a water cycle is established in the chamber to make further watering unnecessary for an indefinite period of time. A variety of container types and shapes may be used, but clear glass is most preferred for sufficient light provision for the plants. Plants that are adapted to low or medium light intensities are most desirable. The planting medium should drain freely. Chemical fertilizers should be avoided since certain nitrogenous fertilizers release toxic ammonia gases that are injurious to plants. In terms of moisture, a terrarium should be self-sustaining and able to maintain a good moisture cycle. The growing medium and the plants included should be free of diseases.

References and Suggested Reading

Wright, M., ed. 1979. The complete indoor gardener. New York: Random House.

Reader's Digest. 1979. Success with houseplants. New York: Reader's Digest Association.

Outcomes Assessment

Part A

Please answer true (T) or false (F) for the following statements.

1. T F Plants adapted to low to medium light intensities are most desirable for use in terrariums.
2. T F Terrarium plant species should be fast growing.
3. T F Browning of foliage may be due to lack of moisture.
4. T F Terrariums may be sealed or unsealed.
5. T F Terrarium plants may require pruning at some point.
6. T F Flowering plants do not make good terrarium materials.

PART B

Please answer the following questions.

1. The original concept of a terrarium was developed by _____.
2. Using colored sand to decorate the medium base of a terrarium is called _____.
3. List four plants that are good terrarium materials.

   _____ _____

   _____ _____

PART C

1. Discuss the principles for selecting plants for a terrarium.
2. What could be the cause of plants yellowing in a terrarium?
3. Why should one not use chemical fertilizers in a terrarium?
4. What is the role of activated charcoal in a terrarium?
5. Describe the role and nature of a moisture cycle within a terrarium.
6. Why are plant species with small leaves the most desirable terrarium plants?

CHAPTER 26

Bonsai: The Art of Miniature Plant Culture

PURPOSE

This chapter is designed to describe the principles and show the techniques of creating and maintaining miniature plants.

EXPECTED OUTCOMES

After studying this chapter, the student should be able to

1. Describe the art of bonsai.
2. Discuss the principles of bonsai design.
3. Describe the steps involved in creating a bonsai.
4. List plant species that are amenable to this art form.
5. Discuss how to use and care for a bonsai.

KEY TERMS

Bonsai   Pruning   Training
Miniature plants

OVERVIEW

In chapter 5 we learned that the expression of genes in an environment produces a phenotype. For the genetic potential to be fully expressed, optimal environmental conditions must be provided for the development of the genotype. In other words, it is possible to manipulate phenotype to some extent by altering the environment in which an organism is developing. Plants in the landscape are usually nurtured to quickly establish a vigorous vegetative growth and to develop to full size. However, when growing plants indoors, it is desirable to nurture plants such that they grow and develop at a much slower pace. An extreme case in the application of this concept of slowing plant growth and development is found in the Japanese art form called *bonsai* (*bon* meaning tray and *sai* meaning tree). Bonsai is the art of growing and training plants (trees, shrubs, and vines) to be miniatures of their natural forms (figure 26-1). In effect, bonsai is a cultural technique for dwarfing plants. Dwarfing is accomplished through creative design, pruning, and culturing plants in shallow containers.

(a)

(b)

FIGURE 26-1 Bonsai can be created using a variety of species for specific effects. (a) A juniper bonsai shows the traditional bonsai shape. (b) A fruiting bonsai created from an orange plant.

26.1: PRINCIPLES

To successfully produce a bonsai plant, four general principles—plant selection, design, pruning and training, and management—should be understood and observed.

26.1.1 PLANT SELECTION

All plants are not suitable for bonsai. Those with small leaves (deciduous or evergreen) are most desirable. It is also desirable that the species be able to grow in restricted space. Bonsai are essentially outdoor plants, although recently they have been adapted for indoor display. In China and Japan, where the art originated, bonsai plants are obtained from temperate and humid forests. The three groups of plants used in creating bonsai are discussed in the following sections.

Conifers

Conifers are hardy plants and tolerant of the pruning and other manipulations customary with bonsai culture. The beginner may fare best with conifers as a starter material. Conifers require a short period (usually a few years) to be ready for displaying, and therefore keep the enthusiasm of the novice alive. Another desirable characteristic of conifers is that they are mostly evergreen and as such can be enjoyed year-round. A selection of conifers suited to miniaturization is presented in table 26-1.

Deciduous Trees

Deciduous plants shed their leaves in the fall season, after displaying dazzling fall foliage colors. During the dormant period, when all leaves drop, the tree offers a good opportunity for pruning and reshaping. Vigorous growth resumes in spring. As such, deciduous bonsai require regular pruning to keep their leaves small. Species from the Orient are desirable because their leaves are naturally small in size. Table 26-2 presents a selected number of deciduous trees adapted to bonsai culture.

Table 26-1	Selected Conifers Used for Bonsai
Plant	**Scientific Name**
Cedar	*Cedrus* spp.
Silver fir	*Albies alba*
Japanese cedar	*Cryptomeria japonica*
Chinese juniper	*Juniperus chinensis*
Japanese white pine	*Pinus parviflora*
Japanese black pine	*Pinus thubergii*
Spruce	*Picea* spp.
Yew	*Taxus baccata*
Larch	*Larix* spp.
False cypress	*Chamaecyparis* spp.

These are evergreen species (except for the larch, which has deciduous needles). Silver fir has upright cones, and spruce has pendant cones. Cedar has dark green needles, and Japanese white pine has a bluish-green color.

Table 26-2	Selected Deciduous Trees Used for Bonsai
Plant	**Scientific Name**
Trident maple	*Acer trifidum*
Japanese maple	*Acer palmatum*
Chinese elm	*Ulmus parvifolia*
Hornbeam	*Carpinus laxiflora*
Crab apple	*Mallus floribunda*
Black birch	*Betula nigra*
Beech	*Fagus crenata*
Gray-bark elm	*Zelkova serrata*

Ornamental Shrubs

Conifers are most commonly used in creating bonsai. However, ornamental shrubs that bear small fruits and flowers also make good bonsai. Their spectacular display of flowers and fruits is a sight to behold. Ornamental shrubs that are suitable for use in creating bonsai are listed in table 26-3.

26.1.2 DESIGN

Because bonsai is an art form, creativity is critical to the overall appeal of a finished product. Longevity is an important aspect of bonsai; designs are meant to produce plants that appear old, rugged, and weathered. It should be acknowledged that, to a large extent, the art of bonsai imitates nature. Therefore, some designs portray plants responding to the impact of natural forces such as the wind, as is the case in the *Fukinagashi* design and the *Nejikan* twisted trunk design (figure 26-2). Designs may involve a single tree with a single trunk, a single tree with multiple trunks, or a group of trees. Group designs offer an opportunity to create a miniature landscape. Some designs are very creative, with cascading branches that hang over the edge of the container.

26.1.3 PRUNING AND TRAINING

Pruning to control plant size is critical to bonsai culture (26.3). Roots and shoots are judiciously pruned to obtain the planned shape and size and to control the rate of development.

Table 26-3 A Selection of Popular Ornamental Shrubs and Trees Used for Bonsai

Plant	Scientific Name
Azalea	*Azalea* spp.
Rhododendron	*Rhododendron* spp.
Rock cotoneaster	*Cotoneaster horizontalis*
Crab apple	*Malus* spp.
Japanese apricot	*Prunus mume*
Almond	*Prunus amygdalus*
Japanese cherry	*Prunus serrulata*
Wisteria	*Wisteria* spp.
Winter jasmine	*Jasminum nudiflorum*
Japanese camellia	*Camellia japonica*

Plants such as cotoneaster and crab apple produce small flowers and fruits

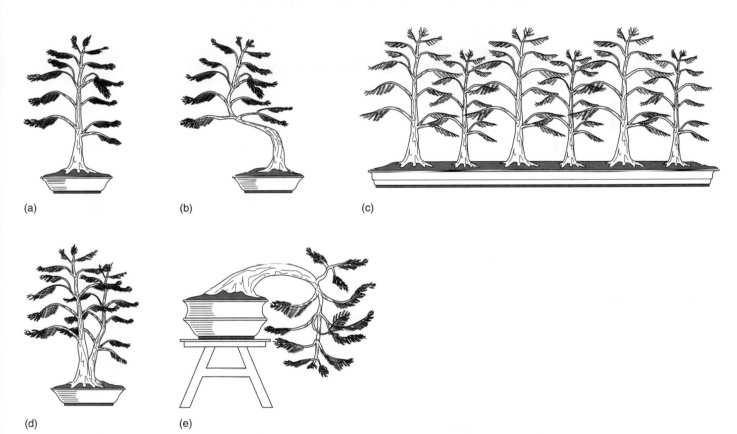

(a)   (b)   (c)

(d)   (e)

FIGURE 26-2 Selected bonsai designs: (a) single straight stem, (b) single curved stem showing more of the desired aged and weathered effect, (c) a bonsai forest involving a number of plants in one container, (d) a branched-stem bonsai, and (e) a cascading design for displaying on pedestals and other higher places.

It is often necessary to use wires during the training process to force the plant to assume a desired shape.

26.1.4 MANAGEMENT

After establishment, pruning is periodically required to contain the roots and maintain the shape of the bonsai. The plant needs to be watered, fertilized, and placed under appropriate environmental conditions at all times.

26.2.1 TOOLS

A variety of pruning and cutting tools are used in bonsai culture. The cutting ends of these tools should be small enough to cut parts of the miniature plants (figure 26-3). A pair of pinchers is used for cutting the taproot, and pliers are used for stripping away the bark of limbs. The clippers are used for shaping the top of the tree. Apart from cutting tools, copper wires of various gauges (10 to 20) are used in creating bonsai plants. The wires are required for forcing plant limbs to assume shapes and to grow in the direction desired by the designer.

26.2.2 CONTAINER

The plant selected can be left in its original container for the initial stages of the culture (the training and primary or initial pruning). After the topwork is completed, it should be transplanted to a bonsai container, which is characteristically very shallow (figure 26-4). The bonsai container is usually round but may also be oblong or rectangular in shape for certain planting designs involving more than one plant. The color of the container should not be bright (browns and greens are preferred). Earthen pots that are unglazed on the inside are best. Drainage holes should be provided in the bottom of the container.

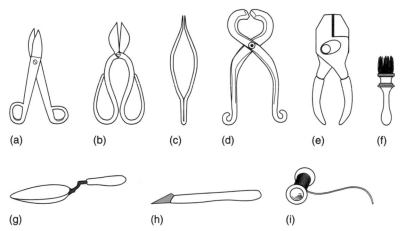

(a)   (b)   (c)   (d)   (e)   (f)

(g)   (h)   (i)

FIGURE 26-3 Tools for creating a bonsai, including a variety of cutting implements (a, b, e), a pair of forceps (c), a pair of pliers (d), a brush (f), a hand trowel (g), a scalpel (h), and a wire (i).

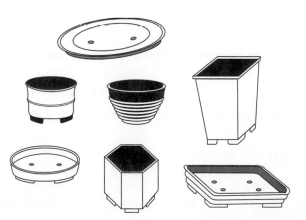

FIGURE 26-4 Typical bonsai containers have an Oriental design. They are usually shallow.

26.2.3 PLANT SELECTION

As previously mentioned, both deciduous and evergreen plants, purchased at a local nursery, can be used for bonsai (26.1.1). Species such as pine, juniper, and pyrachantha are excellent bonsai material. In selecting a plant, it is best to look for one that already has a rugged and irregular appearance in order to have a head start on the design. As indicated, deciduous species shed their leaves at some point. If green leaves are desired year-round, conifers should be selected. Further, conifers require relatively little maintenance.

26.3: TRAINING BONSAI

Plants for bonsai may be collected from nature, raised from seed, or vegetatively propagated.

26.3.1 COLLECTING BONSAI PLANTS FROM NATURE

The Japanese name for collecting ready-made bonsai from natural surroundings is called *Yamadori*. The natural bonsai is called *Yamadori shitate*. These specimens have the advantage of having been molded by the elements (weather-beaten) and usually exhibit some of the features (such as twisted branches, stunting, and aging) that are artificially induced when bonsai are created from scratch. However, should the plant not be perfect, it is difficult to remold it to conform to an acceptable style. Old specimens are not flexible and do not readily submit to twisting and bending without damage.

The best environment under which bonsai material can be collected is one that is harsh (poor to marginal nutrition, shallow soil, unseasonable temperature, and windy conditions). Plants under such conditions do not grow normally and may be stunted and possibly deformed. When such a plant is found, it should be dug up and transplanted in the dormant period. It is important to dig up (rather than pull) all of the roots including a ball of soil. Some root or shoot pruning may be required before transplanting.

26.3.2 GROWING PLANTS FOR BONSAI

Starting from Seed

Starting from seed is a slow process, requiring several years to obtain a plant of decent size that can be trained. Seeds may be collected from the wild or purchased from a nursery. Seeds differ in preplanting preparation, some requiring soaking and others stratification.

Using the Vegetative Propagation Method

Any of the methods for vegetative propagation described in chapter 9 may be used to raise plants for training as bonsai.

26.3.3 PLANT SHAPE AND PRIMARY PRUNING

The selected plant should be examined for desirable natural features such as natural curvature and twists in the stem. The designer should determine the best viewing angle (or front) of the finished product and work to enhance it. The primary branches are also identified, starting with the lowest ones, which determine how low the display will be. All unwanted branches below the lowest ones should be removed flush from the trunk. Sometimes dead branches below the selected lowest one may be retained and incorporated into the overall design. The maximum height of the bonsai must be determined next. The plant should then be pruned to that height. The branches between the top and bottom ones are judiciously pruned such that they alternate along the stem and are shorter at the top than at the bottom. Some downward-pointing branches may be removed. During the initial pruning stage, one or two of the lowest branches may be cut back to leave 2 to 4 inches (5.1 to 10.2 centimeters) of limb that is stripped of its bark to kill it. This action creates deadwood (aging), which enhances the design. The taproot should be removed to encourage the growth of lateral or secondary roots. When starting from seed, pruning is initiated when the seedling is young. The taproot and main stem are pruned to induce lateral branching (figure 26-5).

26.3.4 SECONDARY (MAINTENANCE) PRUNING

Secondary pruning is done very frequently to preserve the plant shape, remove unwanted growth, and control growth (figure 26-6). Leaves are clipped and sometimes completely removed in deciduous species. Further, deciduous species are pinched on a regular basis to remove buds, which has the effect of producing smaller leaves.

26.3.5 WIRING

The central stem may or may not be vertical. Vertical stems are desirable when a miniforest design is being created, whereby several to many bonsai plants are grown in one container (26.1.2). The stem or branches may be forced to assume unnatural shapes by twisting a copper wire around a limb and bending it in the new direction or into a new shape (figure 26-7).

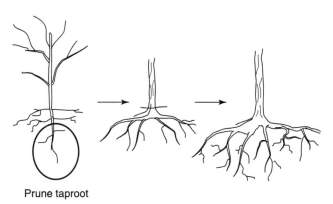

Prune taproot

FIGURE 26-5 Root pruning. One training goal in creating bonsai is to discourage the growth of the taproot. Bonsai plants have a taproot system. Root pruning also slows plant growth.

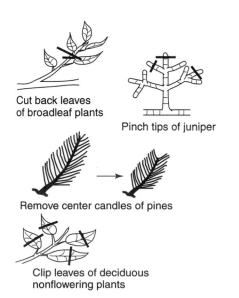

Cut back leaves of broadleaf plants

Pinch tips of juniper

Remove center candles of pines

Clip leaves of deciduous nonflowering plants

FIGURE 26-6 Shoot pruning. Leaves of bonsai need to be pruned at certain times to control growth.

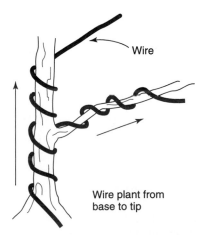

Wire

Wire plant from base to tip

FIGURE 26-7 Wiring, a primary activity in creating bonsai, is done to force branches to grow in a desired manner to create the desired bonsai shape. Wiring must always start from the base of the stem or branch.

Wiring starts from the lower part of the stem or branch and progresses to the tips. Thinner wires are used for thinner limbs. Branches that are close together can be pushed apart by first wiring and then bending. These operations should be done very carefully to avoid bruising the bark or breaking off the limb due to excessive tension from bending or wrapping the wire too tightly around the limb. The ends of the wires should be tucked away from view. Conifers are wired at the end of the dormant period in late winter, while deciduous plants are wired in spring. Wires should be removed after the shape is set (about 6 to 12 months). If left for a long time, the wire may become embedded in the bark of the plant. The base of the trunk may be deliberately wired to induce thickening (figure 26-8). After wiring, bending, and determining the final shape, additional secondary pruning may be required to finish the product. Typical bonsai top shapes are flattened and layered foliage.

26.3.6 AGING

The technique of aging is called *jin.* Although a special knife may be purchased for this procedure, a simple grafting knife may also be used successfully. After stripping off the bark, the bare surface should be polished with fine-grade sandpaper, after which dilute citric acid or a scouring solution may be rubbed on the bare surface to bleach the area. The result gives the appearance of aged plants (figure 26-9).

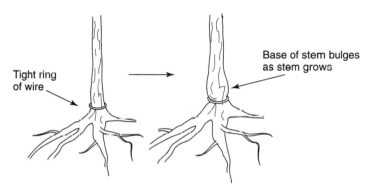

FIGURE 26-8 A technique of creating a bulging stem base is used to create additional interest in bonsai.

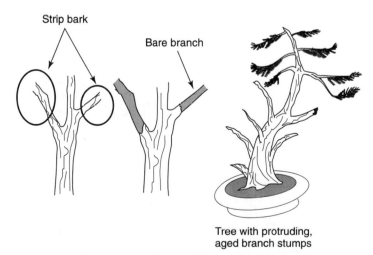

FIGURE 26-9 Aging is a significant aspect of bonsai design. Artificial aging can be induced through a combination of mechanical and chemical treatments.

26.4.1 PREPARING THE CONTAINER

When repotting bonsai, a piece of wire is passed through the holes in the bottom of the container to anchor the plant (figure 26-10). A fine wire mesh is stretched over the drainage hole, and a layer of gravel or potting medium (sterilized mix of equal amounts of coarse sand, soil, and sphagnum moss) is placed to a depth of about 1 inch (2.5 centimeters) in the container.

26.4.2 ROOT PRUNING

The plant is removed from its original container and the roots examined after removing much of the soil. The taproot is cut back to leave a short stump. Other lateral roots are also pruned so that the remaining root mass fits the bonsai container.

26.4.3 SECURING THE PLANT

With a reduced root mass, the plant is often top-heavy and needs to be secured in the shallow container by tying with the anchor wire passed through the drainage holes (figure 26-11). After that, more potting medium is added to cover the roots, slightly mounding it around the stem.

26.4.4 WATERING AND FERTILIZING

Watering and fertilizing are two operations that are very critical to the success of a bonsai. As previously indicated, the miniature plant depends on one's ability to control and slow the

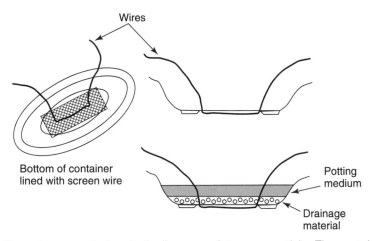

FIGURE 26-10 Bonsai are repotted periodically as a maintenance activity. The container should be lined with a drainage material and then topped with the potting medium.

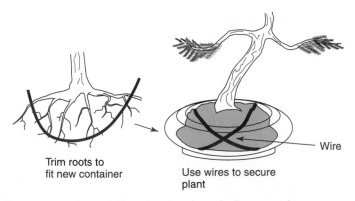

FIGURE 26-11 Wires are used to stabilize the plant in a shallow container.

growth and development of the plant. Growth control relies to a great extent on the management of watering and fertilizing. Since the bonsai container is shallow, the plant should be watered thoroughly, especially in summer. Hard and polluted water should be avoided; rain water is the most recommended but may not always be available. Chlorine in an urban water supply is injurious to plants. If tap water must be used, the container of water should sit open outside to evaporate the chlorine. The spout of the watering can should be capped with a head that delivers a fine spray.

For fertilizing, slow-acting fertilizers are most desirable. Specially constituted bonsai fertilizer may be purchased from a nursery. Organic fertilizer pellets contain nitrogen, phosphorus, and potassium in proportions of 50:30:20 compounded from materials such as bone meal and fish meal. Bonsai are fertilized during the active growing period. Fertilizing of flowering and fruiting bonsai plants should be delayed until after flowering.

26.5: POSTESTABLISHMENT CARE

26.5.1 PRUNING AND REPOTTING

A bonsai plant needs periodic pruning to maintain its shape and control growth. Depending on the species, roots require pruning every three to five years, since they quickly become pot-bound in the restricted area of the bonsai container. The plant to be pruned is removed from the container and the roots trimmed below the soil level. Pruning should also be geared toward providing room for new growth over the next three to five years. The plant should then be resecured (if necessary) by tying with a wire before fresh potting medium is added and the plant watered.

26.5.2 SANITATION

It is critical to maintain clean surroundings and thereby reduce the chances of disease and pest attacks. Because of its small size, any disease or pest attack can be devastating to a bonsai plant. Routine cleaning includes scraping and brushing off moss growth on the trunk and removing liverworts and other weeds from the soil.

26.5.3 PEST CONTROL

Pests of bonsai include aphids (or greenflies), caterpillars, scale insects, red spider mites, and ants. Aphids and mites may be controlled by spraying with insecticides. Ants may attack and destroy roots and should be removed. Caterpillars are readily removed by hand and destroyed.

26.5.4 DISEASE CONTROL

Powdery mildew is a problem where humid conditions prevail. Rust and black spot may also occur. Black rot may be controlled by applying sulfur. Root rot may occur if waterlogged conditions prevail, or if poorly decomposed organic fertilizer is applied. The bonsai plant should be lifted and all contaminated soil brushed away. Canker is effectively controlled by cutting away affected plant parts.

26.5.5 SEASONAL MAINTENANCE

Winter

When winter temperatures become severe (below freezing), protection may be required for outdoor bonsai. Plants may be brought indoors and kept at temperatures below 5°C (41°F). Fertilization should cease during the dormant period. Water should be provided as needed but sparingly. Deciduous plants are best pruned in winter, when they shed all of their leaves. The plants can also be cleaned during winter.

Spring

Outdoor bonsai should be returned to the outside environment only after it is safe to do so (frost free). Plants begin to bud as active growth resumes. Watering should be increased, but very carefully. As spring advances, the growth rate increases, and some pruning may be necessary to prevent excessive growth. When blooms wither, plants should be pruned before new buds appear. Blooms occur on the previous year's wood, and any delay in pruning will prevent blooming in the next year. Coniferous bonsai may be repotted in midspring.

Summer

Foliage growth is intense in early summer. Pinching and pruning are required to control growth. Water is critical in late summer when it is hot; however, hot and humid conditions predispose plants to fungal diseases.

Fall

A key activity in fall is to prepare the plant for winter by boosting its nutrition with fertilizers. Deciduous species display their most intense colors in midfall. Late fall ushers in the time of leaf drop and the approach of winter. As leaves drop, the plants may be pruned. Preparations should be made to protect plants in case of an unexpected frost when bonsai are displayed outdoors.

SUMMARY

Young trees, shrubs, and vines can be manipulated and managed such that they grow to become miniature versions of the adult plants. The technique of miniaturizing plants is called bonsai and originated in China and Japan. Bonsai can live for more than a hundred years. One of the key elements in design is to create the appearance of old age and the effects of weathering. Materials can be selected or manipulated to appear old and weather-beaten. Pruning and nutrition are two major management activities in the successful culture of a bonsai. Both the roots and shoots of a plant are regularly pruned to create the desired shape and to control growth. Bonsai are generally designed for the outdoors, and thus when used indoors, the appropriate adjustments should be made to provide the necessary environmental conditions for proper growth and development.

REFERENCES AND SUGGESTED READING

Koreshoff, D. 1984. Bonsai: Its art, science, history, and philosophy. Beaverton, Oreg.: ISBS/Timber Press.

Pessey, C., and R. Samson. 1989. Bonsai basics: A step-by-step guide to growing, training, and general care. New York: Sterling Publications.

Wright, R., ed. 1975. The complete indoor gardener, revised edition. New York: Random House.

Young, D. 1990. Bonsai: The art and technique. Englewood Cliffs, N.J.: Prentice-Hall.

OUTCOMES ASSESSMENT

PART A

Please answer true (T) or false (F) for the following statements.

1. T F   A bonsai plant can live for more than 100 years.
2. T F   Large-leaved plant species are most suitable for bonsai.

3. T F   Only evergreen plant species can be used in creating bonsai.
4. T F   Clay pots that are glazed on the inside are best for bonsai culture.
5. T F   Bonsai roots need pruning every three to five years.
6. T F   Bonsai is an outdoor plant.

PART B

Please answer the following questions.

1. List the four general principles of bonsai culture.

    _____   _____

    _____   _____

2. Typical bonsai top shapes are _____ and _____.
3. List four plants that make good bonsai material.

    _____   _____

    _____   _____

4. The technique of aging in the creation of bonsai is called _____.

PART C

1. What is the role of copper wire in creating a bonsai?
2. List three bonsai tools and describe how they are used.
3. Why do bonsai designers sometimes strip away the bark of a branch?
4. Describe how a bonsai is repotted.
5. Explain how a plant can be miniaturized.

Appendix A

°C	Known Temperature (°C or °F)	°F	°C	Known Temperature (°C or °F)	°F
−73.33	**−100**	**−148.0**	5.00	41	105.8
−70.56	−95	−139.0	5.56	42	107.6
−67.78	−90	−130.0	6.11	43	109.4
−65.00	−85	−121.0	6.67	44	111.2
−62.22	−80	−112.0	7.22	45	113.0
−59.45	−75	−103.0	7.78	46	114.8
−56.67	−70	−94.0	8.33	47	116.6
−53.89	−65	−85.0	8.89	48	118.4
−51.11	−60	−76.0	9.44	49	120.2
−48.34	−55	−67.0	**10.0**	**50**	**122.0**
−45.56	**−50**	**−58.0**	10.6	51	123.8
−42.78	−45	−49.0	11.1	52	125.6
−40.0	−40	−40.0	11.7	53	127.4
−37.23	−35	−31.0	12.2	54	129.2
−34.44	−30	−22.0	12.8	55	131.0
−31.67	−25	−13.0	13.3	56	132.8
−28.89	−20	−4.0	13.9	57	134.6
−26.12	−15	5.0	14.4	58	136.4
−23.33	−10	14.0	15.0	59	138.2
−20.56	−5	23.0	**15.6**	**60**	**140.0**
−17.8	**0**	**32.0**	16.1	61	141.8
−17.2	**1**	**33.8**	16.7	62	143.6
−16.7	2	35.6	17.2	63	145.4
−16.1	3	37.4	17.8	64	147.2
−15.6	4	39.2	18.3	65	149.0
−15.0	5	41.0	18.9	66	150.8
−14.4	6	42.8	19.4	67	152.6

°C	Known Temperature (°C or °F)	°F	°C	Known Temperature (°C or °F)	°F
−13.9	7	44.6	20.0	68	154.4
−13.3	8	46.4	20.6	69	156.2
−12.8	9	48.2	**21.1**	**70**	**158.0**
−12.2	**10**	**50.0**	21.7	71	159.8
−11.7	11	51.8	22.2	72	161.6
−11.1	12	53.6	22.8	73	163.4
−10.6	13	55.4	23.3	74	165.2
−10.0	14	57.2	23.9	75	167.0
−9.44	15	59.0	24.4	76	168.8
−8.89	16	60.8	25.0	77	170.6
−8.33	17	62.6	25.6	78	172.4
−7.78	18	64.4	26.1	79	174.2
−7.22	19	66.2	**26.7**	**80**	**176.0**
−6.67	**20**	**68.0**	27.2	81	177.8
−6.11	21	69.8	27.8	82	179.6
−5.56	22	71.6	28.3	83	181.4
−5.00	23	73.4	28.9	84	183.2
−4.44	24	75.2	29.4	85	185.0
−3.89	25	77.0	30.0	86	186.8
−3.33	26	78.8	30.6	87	188.6
−2.78	27	80.6	31.1	88	190.4
−2.22	28	82.4	31.7	89	192.2
−1.67	29	84.2	**32.2**	**90**	**194.0**
−1.11	**30**	**86.0**	32.8	91	195.8
−0.56	31	87.8	33.3	92	197.6
0	32	89.6	33.9	93	199.4
0.56	33	91.4	34.4	94	201.2
1.11	34	93.2	35.0	95	203.0
1.67	35	95.0	35.6	96	204.8
2.22	36	96.8	36.1	97	206.6
2.78	37	98.6	36.7	98	208.4
3.33	38	100.4	37.2	99	210.2
3.89	39	102.2	**37.8**	**100**	**212.0**
4.44	**40**	**105.2**			

To make conversions not included in the table use the following formulas:

1. $°F = 9/5(°C + 32)$
2. $°C = 5/9(°F − 32)$

Appendix B

Metric Conversion Chart		
Known	**Multiplier**	**Desired**
Length		
inches	2.54	centimeters
feet	30	centimeters
feet	0.303	meters
yards	0.91	meters
miles	1.6	kilometers
Area		
square inches	6.5	square centimeters
square feet	0.09	square meters
square yards	0.8	square meters
square miles	2.6	square kilometers
acres	0.4	hectares
Mass (Weight)		
ounces	28	grams
pounds	0.45	kilograms
short tons	0.9	metric tons
Volume		
teaspoons	5	milliliters
tablespoons	15	milliliters
fluid ounces	30	milliliters
cups	0.24	liters
pints	0.47	liters
quarts	0.95	liters
gallons	3.8	liters
cubic feet	0.03	cubic meters
cubic yards	0.76	cubic meters
Pressure		
pounds per square inch	0.069	bars
atmospheres	1.013	bars
atmospheres	1.033	kilograms per square centimeter
pounds per square inch	0.07	kilograms per square centimeter
Rates		
pounds per acre	1.12	kilograms per hectare
tons per acre	2.24	metric tons per hectare

Appendix C

English Units Conversion Chart		
Known	**Multiplier**	**Desired**
Length		
millimeters	0.04	inches
centimeters	0.4	inches
meters	3.3	feet
kilometers	0.62	miles
Area		
square centimeters	0.16	square inches
square meters	1.2	square yards
square kilometers	0.4	square miles
hectares	2.47	acres
Mass (Weight)		
grams	0.035	ounces
kilograms	2.2	pounds
metric tons	1.1	short tons
Volume		
milliliters	0.03	fluid ounces
liters	2.1	pints
liters	1.06	quarts
liters	0.26	gallons
cubic meters	35	cubic feet
cubic meters	1.3	cubic yards
Pressure		
bars	14.5	pounds per square inch
bars	0.987	atmospheres
kilograms per square centimeter	0.968	atmospheres
kilograms per square centimeter	14.22	pounds per square inch
Rates		
kilograms per hectare	0.892	pounds per acre
metric tons per hectare	0.445	tons per acre

Appendix D

COMMON AND SCIENTIFIC NAMES OF SELECTED PLANTS

Agave (*Agave* spp.)

Ageratum (*Ageratum houstonianum*)

Aglaonema (*Aglaonema simplex*)

Air plant (*Kalanchoe* spp.)

Almond (*Prunus amygdalus*)

Amaryllis (*Hippeastrum* spp.)

Apple (*Pyrus malus*)

Apricot (*Prunus armeniaca*)

Artichoke, Jerusalem (*Helianthus tuberosus*)

Ash (*Fraximus* spp.)

Aspen (*Populus tremuloides*)

Aster (*Callistephus chinensis*)

Avocado (*Persea americana*)

Azalea (*Rhododendron* spp.)

Baby's breath (*Gypsophila paniculata*)

Bamboo (*Bambusa* spp.)

Banana (*Musa paradisiaca*)

Basil (*Ocimum basilicum*)

Bay, sweet (*Laurus nobilis*)

Bean, broad (*Vicia faba*)

Bean, garden (*Phaseolus vulgaris*)

Bean, lima (*Phaseolus lunatus*)

Beech (*Fagus sylvatica*), *B. grandifolia*

Beet, sugar (*Beta vulgaris*)

Begonia (*Begonia socotrana*) and others.

Bentgrass, colonial (*Agrostis tenuis*)

Bentgrass, creeping (*Agrostis palustris*)

Birch (*Betula papyrifera*) and others.

Bird-of-paradise (*Strelitzia reginae*)

Blackberry (*Rubus* spp.)

Bleeding heart (*Dicentra* spp.)

Blueberry (*Vaccinium* spp.)

Bluegrass, Kentucky (*Poa pratensis*)

Bougainvillea (*Bougainvillea spectablis*)

Boxwood, common (*Buxus semperivirens*)

Broccoli (*Brassica oleracea* var. *botrytis*)

Browallia (*Browallia* spp.)

Brussels sprouts (*Brassica oleracea* var. *gemmifera*)

Buffalograss (*Buchloe dactyloides*)

Buttercup (*Ranunculus* spp.)

Cabbage (*Brassica oleracea*)

Camellia (*Camellia japonica*) and others.

Canna lily (*Canna indica*)

Canterbury bells (*Campanula medium*)

Carnation (*Dianthus caryophyllus*) and others.

Carpetgrass (*Axonopus affinis*)

Catnip (*Nepeta cataria*)

Caulifower (*Brassica oleracea* var. *botrytis*)

Celery (*Apium graveolens*)

Century plant (*Agave* spp.)

Cherry (*Prunus avium*) and others.

Chestnut (*Castanea* spp.)

Christmas cactus (*Zygocactus truncatus*); *Schlumbergera* hybrids

Christmas flower (*Euphorbia pulcherrima*)

Chrysanthemum (*Chrysanthemum* spp.)

Cinnamon (*Cinnamomum zeylanicum*)

Citrus (*Citrus* spp.)

Clematis (*Clematis hybrida*) and others.

Coleus (*Coleus blumei*) and others.

Columbine (*Aquilegia* spp.)

Compass plant (*Lactuca serriola*)

Corn plant (*Dracaena fragrans* 'Massangeana')

Corn (*Zea mays*)

Cosmos (*Cosmos* spp.)

Cotoneaster (*Cotoneaster congesta*)

Cottonwood (*Populus deltoides*) and others.

Crab apple (*Crataegus* spp.)

Cranberry (*Vaccinium macrocarpus*)

Crocus, autumn (*Cochicum autumnale*)

Croton (*Codiaeium variegatum*)

Cucumber (*Cucumis sativus*)

Cyclamen (*Cyclamen* spp.)

Cypress, bald (*Taxodium distichum*)

Daffodil (*Narcissus* spp.)

Dahlia (*Dahlia variabilis*) and others.

Daisy (*Dimorphotheca* spp.)

Dandelion (*Taraxacum officinale*)

Daphne (*Daphne odora*)

Daylily (*Hermerocallis* spp.)

Dill (*Anethum graveolens*)

Dogwood (*Cornus* spp.)

Dumbcane (*Dieffenbachia exotica*)

Echeveria (*Echeveria secunda*)

Eggplant (*Solanum melongena*)

Elephant ears (*Colocasia* spp.)

Elm (*Ulmus* spp.)

Fatshedera (*Fatshedera lizei*)

Fern, asparagus (*Asparagus densiflorus* 'Sprengeri')

Fern, Boston (*Nephrolepis exaltata*)

Fig, common (*Ficus carica*)

Fir, Douglas (*Pseudotsuga menziesii*)

Foxglove (*Digitalis purpurea*)

Gardenia (*Gardenia jasminoides*)

Geranium (*Geranium* spp.; *Pelargonium* spp.)

Ginger (*Zingiber officinale*) and others.

Gladiolia (*Gladiolus* spp.)

Globe amaranth (*Gomphrena globosa*)

Gloxinia (*Sinningia speciosa*)

Gooseberry (*Ribes* spp.)

Grape (*Vitis* spp.)

Grapefruit (*Citrus paradisi*)

Hazelnut (*Corylus* spp.)

Holly, American (*Ilex opaca*)

Horseradish (*Rorippa armoracia*)

Hosta (*Hosta undulata*)

Hyacinth (*Hyacinthus* spp.)

Impatiens (*Impatiens walleriana*)

Iris (*Iris* spp.)

Ivy, English (*Hedera helix*)

Jade plant (*Crassula argentea*)

Juniper (*Juniperus* spp.)

Kohlrabi (*Brassica oleracea* var. *caulorapa*)

Lantana (*Lantana camara*)

Larkspur (*Delphinium* spp.)

Lavender (*Lavandula officinalis*)

Lemon (*Citrus limon*)

Lettuce (*Lactuca sativa*)

Lily (*Lilium* spp.)

Lily of the valley (*Convallaria majalis*)

Lime (*Citrus aurantifolia*)

Lombardy poplar (*Poplus nigra*)

Magnolia (*Magnolia* spp.)

Maple (*Acer* spp.)

Marigold (*Tagetes* spp.)

Marjoram (*Majorana hortensis*)

Medicine plant (*Aloe vera*)

Melon (*Cucumis melo*)

Mistletoe (*Phoradendron* spp.)

Morning glory (*Ipomoea* spp.)

Moss, peat (*Sphagnum* spp.)

Moss rose (*Portulaca grandiflora*)

Mustard (*Brassica campestris*) and others.

Nasturtium (*Tropaeolum* spp.)

Norfolk Island pine (*Araucaria excelsa*)

Oak (*Quercus* spp.)

Olive (*Olea europaea*)

Onion (*Allium cepa*)

Orange (*Citrus sinensis*)

Orchid (*Cattleya* spp.) and others.

Oregano (*Origanum vulgare*) and others.
Palm, coconut (*Cocos nucifera*)
Palm, date (*Phoenix dactylifera*)
Pansy (*Viola tricolor*)
Papaya (*Carica papaya*)
Parsley (*Petroselinum sativum*)
Pea (*Pisum sativum*)
Peach (*Prunus persica*)
Peanut (*Arachis hypogaea*)
Pear (*Pyrus communis*)
Pecan (*Carya illinoensis*)
Peony (*Paeonia* spp.)
Peperomia (*Peperomia* spp.)
Pepper (*Capsicum annuum; C. frutescens*)
Peppermint (*Mentha piperita*)
Periwinkle (*Vinca minor*)
Petunia (*Petunia* spp.)
Phlox (*Phlox* spp.)
Pine (*Pinus* spp.)
Pineapple (*Ananas comosus*)
Pitcher plant (*Sarracenia* spp.) and others.
Pistachio (*Pistacia vera*)
Plum (*Prunus domestica*) and others.
Poinsettia (*Euphorbia pulcherrima*)
Ponytail palm (*Beaucarnea recurvata*)
Potato, Irish (*Solanum tuberosum*)
Potato, sweet (*Ipomea batatas*)
Pothos (*Scindapsus aureus*)
Prayer plant (*Maranta* spp.)
Primrose (*Primula* spp.)
Quince (*Cydonia oblonga*)
Radish (*Raphanus sativus*)
Raspberry (*Rubus* spp.)
Redbud (*Cercis* spp.)
Redwood, coastal (*Sequoia sempervirens*)
Redwood, giant (*Sequoiadendron giganteum*)
Rhododendron (*Rhododendron* spp.)
Rhubarb (*Rheum rhaponticum*)
Rose of Sharon (*Hibiscus syriacus*)

Rose (*Rosa* spp.)
Rosemary (*Rosmarinus officinalis*)
Rubber plant (*Ficus elastica*)
Rutabaga (*Rutabaga campestris* var. *napobrassica*)
Sage (*Salvia officinalis*)
Sedge (*Carex* spp.) and others.
Shasta daisy (*Chrysanthemum maximum*)
Snake plant (*Sansevieria trifasciata*)
Snapdragon (*Antirrhinum majus*)
Snowplant (*Sarcodes sanguinea*)
Spiderflower (*Cleome* spp.)
Spider plant (*Chlorophytum comosum*)
Spinach (*Spinacia oleracea*)
Spruce (*Picea* spp.)
Squash (*Cucurbita mixta; C. pepo*)
Saint augustinegrass (*Stenotaphtum secundatum*)
Statice (*Limonium* spp.)
Sugarcane (*Saccharum officinarum*) and others.
Sunflower (*Helianthus annuus*)
Sweet alyssum (*Lobularia* spp.)
Sycamore (*Platanus* spp.)
Thyme (*Thymus vulgaris*) and others.
Tomato (*Lycopersicon esculentum*)
Tulip (*Tulipa* spp.)
Turnip (*Brassica rapa*)
Velvet plant (*Gynura sarmentosa*)
Verbena (*Verbena hortensis*)
Violet, African (*Saintpaulia* spp.)
Walnut (*Juglans* spp.)
Wandering Jew (*Zebrina pendula*)
Watermelon (*Citrullus vulgaris*)
Willow (*Salix* spp.)
Wisteria (*Wisteria sinensis*)
Yam (*Dioscorea* spp.)
Yew (*Taxus* spp.)
Zebra plant (*Aphelandra squarrosa; Equus zebra*)
Zinnia (*Zinnia elegans*)
Zoysiagrass (*Zoysia japonica*)

Glossary

A

Abaxial. Turned away from the base or away from the axis (also dorsal).

Abscisic acid. Plant hormone that induces abscission and dormancy and inhibits seed germination, among other plant responses.

Abscission. The dropping of leaves, fruits, flowers, or other plant parts.

Absorption. The uptake of materials by a plant or seed, such as water or nutrients by the roots or pesticide and fertilizer through the leaves.

Accent plant. A plant used to create interest in a landscape by calling attention to a particular feature of an area.

Acid loving. Plants that grow best in acidic potting media or soil.

Active growth. A phase of the plant life cycle characterized by rapid stem lengthening and leaf production and sometimes flowering or fruiting.

Active ingredient. The ingredient in a pesticide that determines its effectiveness.

Adaptation. The process of change in structure or function of an individual or population due to environmental changes.

Adaxial. Turned toward the apex or the axis (also ventral).

Adhesion. The molecular attraction of liquids to solids, such as the attraction of water to soil particles.

Adventitious bud. A bud produced on a part of a plant where it is not expected, such as on a leaf vein or root.

Adventitious root. A root produced by a part of the plant other than seminal tissues, such as from the stem.

Aeration. As pertaining to the soil, it is the movement of air from the atmosphere into and through the soil.

Aggregation. The clinging together of soil particles to form larger units such as clods and clumps.

Agronomy. The art and science of producing crops and managing the soil.

Air layering. The rooting of a branch or top of a plant while it is still attached to the parent.

Algae. A single-cell plant that often grows in colonies.

Allelopathy. The injury of one plant of one species by another through excretion of toxic chemicals by the roots.

Amendment. A substance added to soil to alter its properties to make it more suitable for plant growth.

Anaerobic. An environment deficient in oxygen.

Anatomy. The internal structure of an organism.

Angiosperm. A flowering plant or a plant that produces its seeds in ovaries.

Annual. A plant that completes its life cycle in one year (grows, flowers, produces seed, and dies in one cropping season).

Anther. A saclike, pollen-producing portion of the stamen, usually found on top of the filament.

Anthocyanin. Found in vacuoles of the cell, this water-soluble pigment is responsible for a variety of red to blue colors found in plant parts (fruits, leaves, and flowers).

Apical dominance. Growth regulation in plants where auxins secreted by the terminal bud inhibit the growth of lateral buds on the same shoot.

Apomixis. The asexual production of seed without fertilization.

Arboretum. A place where trees are grown for educational and research purposes.

Arboriculture. The science of growing and caring for ornamental trees.

Asexual reproduction. The reproduction of a new plant without the union of male and female sex cells.

Autogamy. The pollination of a flower by its own pollen.

Auxin. A plant hormone that accelerates growth and is also involved in dormancy, abscission, rooting, tuber formation, and other activities.

Axil. The angle formed on the upper side of a leaf where a petiole joins a stem.

Axillary bud. A bud located in a leaf axil.

B

Backcross breeding. A cross of a hybrid with one of its parents.

Bactericide. A chemical used to control bacterial disease.

Balled and burlapped. A tree grown and sold with a ball of soil around its roots and wrapped in burlap.

Band application. Applying fertilizer in shallow trenches on each side of a row of plants.

Bare root. A tree (deciduous) sold dormant and without soil around its roots.

Basal plate. The base of a bulb.

Bed. A piece of land or area of soil prepared for planting.

Bench. A raised platform for growing plants.

Berm. A ridge of soil created around a newly planted tree to retain water.

Berry. A simple fleshy fruit derived from a single ovary in which the ovary wall is fleshy.

Biennial. A plant that completes its life cycle in two cropping seasons, the first season involving vegetative growth and the second flowering and death.

Binomial system. The system of naming plants developed by Carolus Linnaeus whereby a plant is given a two-part name representing the genus and species.

Biological control. The use of a natural enemy to control a pest.

Blanch. The practice of improving the quality of a vegetable by excluding light.

Blight. A bacterial or fungal disease of plants.

Borer. An insect that burrows into the stems, roots, leaves, or fruits of a plant.

Bract. A modified leaf from whose axil a flower or inflorescence may arise.

Brambles. Collectively, the fruits in the genus *Rubus*.

Breeder seed. Seed quantity increased by the plant breeder.

Broadcast. Scattering of seed or fertilizer uniformly over a given area.

Bud. A rudimentary structure consisting of meristematic tissue and a potential to develop into vegetative, reproductive, or a mixture of structures.

Budding. The grafting of a bud of one plant onto another plant so that the inserted bud develops into the new shoot.

Bud scales. Leaflike structures that cover the outside of a bud.

Bulb. An underground storage organ consisting of highly compressed leaves.

Bulblet. An immature bulb that develops at the base of a bulb.

C

Callus. A mass of parenchyma cells that forms around the wounded area of a plant to start the healing process. It can be artificially induced under tissue culture conditions.

Candle. The succulent new growth produced by a needle evergreen plant.

Cane. Long shoots or weeping stems such as those produced by grapes and brambles.

Canker. A disease that destroys the cambium and vascular tissue of a generally localized area of a plant.

Capillary action. The movement of water through micropores of a growing medium owing to the adhesion of water molecules to the medium.

Capillary water. The soil water held against gravitational force.

Carotene. A yellow plant pigment that is a precursor of vitamin A and gives color to plant parts such as orange fruits.

Caryopsis. A one-seeded dry fruit produced by grasses.

Cation-exchange capacity (CEC). A measure of the total amount of the ability of a soil to attract and hold nutrients or exchange cations.

Cell pack. A plastic container of seedlings in which each seedling has its own root packet.

Cellulose. A complex carbohydrate that forms a major part of plant cell walls and provides strength to stems and other parts.

Central leader. A training method for trees in which a central axis is maintained.

Certified seed. An increase in quantity of foundation seed certified by an approved certifying agency.

Chlorosis. A mineral deficiency symptom in which affected plants show yellowish or greenish-yellow coloration in their leaves.

Clay. One of the three primary soil particles; it has extremely small particles with colloidal properties.

Clod. A hard, artificially produced (e.g., by tillage) lump of soil that is difficult to break.

Clone. A plant that is asexually propagated from another; both plants are genetically identical.

Cold frame. An enclosed, unheated structure covered on top with glass or plastic used to harden off plants or protect tender plants during the early spring.

Cold hardiness. The minimum temperature at which a plant can survive (tolerance).

Cole crop. Vegetables of the genus *Brassica*.

Colloid. Very fine organic or inorganic particles that are interspersed.

Common name. The English name of a plant that may differ in various localities (e.g., maize and corn).

Compaction. Excessive packing or settling of a growing medium.

Companion crop. A crop planted along with another but harvested separately. The two crops positively impact each other.

Compatibility. Ability of different species to be sexually or physically united and grow together as one.

Complete fertilizer. A compound fertilizer containing a mixture of nitrogen, phosphorus, and potassium.

Complete flower. A flower with both male and female parts; sexually complete.

Compost. A soil amendment consisting primarily of decomposed organic matter.

Compound leaf. A leaf composed of more than one distinct leaflet.

Container growing. Growing plants to mature size in a container; also called pot gardening.

Cool-season plant. A plant species that is frost tolerant and prefers daytime temperatures ranging from 50 to 65°F (10 to 18°C).

Cork cambium. The layer of meristematic cells that produces bark on woody plants.

Corm. A short, swollen, and vertically growing underground stem that stores food.

Cormel. A young offshoot produced by a larger corm.

Cotyledon. Seed leaf; serves as a storage organ in some plants.

Cover crop. A fast-growing crop grown primarily for the purpose of covering a bare soil.

Crop rotation. Planting different species of crops in an area in a planned sequence, year after year, to prevent buildup of diseases or insects associated with particular crops.

Cross-pollination. The transfer of pollen from the stamen of one flower to the stigma of a flower on another plant.

Crown. The root-stem junction of woody plants; also the region at the base of the stem of herbaceous species from which branches or tillers arise.

Cultivar. Derived from the words *cultivated* and *variety,* often designating a product of plant breeding.

Cultural control. A method of pest control involving manipulation of the growing conditions of the plant.

Cut flower. A flower grown for the sole purpose of cutting and displaying in a vase or other arrangement.

Cuticle. An impermeable, waxy layer on the exterior of stems and leaves that retards water loss.

Cutting. A vegetative plant part used for propagation.

Cytokinin. A naturally occurring plant hormone involved in activities such as cell division, organ initiation, breaking dormancy, and other activities.

D

Damping-off. A disease in which a soilborne fungus attacks seedlings soon after germination at the soil line and kills the plants.

Day-neutral plant. A species that flowers regardless of the dark to light ratio.

Deciduous. A plant that sheds its leaves in fall.

Decomposition. The breakdown or rotting of plants or animals.

Defoliation. The dropping of most or all leaves from a plant.

Dehiscence. The splitting open of mature pods.

Desuckering. The removal of suckers from a plant.

Determinate. The growth pattern in which plants remain short (bush) and flower and ripen uniformly within a limited period.

Dethatching. Removal of thatch from a turf or lawn.

Dicot. A plant having two cotyledons (dicotyledons).

Dioecious. A species in which the male and female organs of reproduction are produced on separate plants.

Direct seeding. Sowing seeds in a permanent growing site.

Division. A method of asexual propagation requiring the cutting and dividing of plants.

Dormancy. A phase of a plant life cycle characterized by slowed or stopped growth.

Drip irrigation. A method of applying water to plants such that only the immediate vicinity becomes wet.

Drupe. A simple, fleshy fruit derived from one carpel, with a thin exocarp, fleshy mesocarp, and hard endocarp.

Dwarfing rootstock. A special rootstock that has the capacity to slow and limit the growth of the scion grafted onto it.

E

Embryo. The immature plant in a seed.

Endosperm. The part of a seed that serves as a food supply for the young, developing seedling.

Enzyme. A chemical that regulates an internal chemical reaction in an organism.

Epidermis. The outer layer of plant parts.

Epigeous germination. A type of seed germination, characteristic of dicots, in which cotyledons emerge above the soil surface.

Erosion. The wearing away of soil caused by agents such as water and wind.

Espalier. A method of training trees or shrubs to grow flat against a wall or trellis, sometimes in fancy patterns.

Essential elements. The micro- and macronutrients required for plant growth.

Etiolation. Tall, spindly growth induced by insufficient light conditions.

Everbearing. Strawberry and raspberry varieties that bear two crops annually.

Explant. A piece of tissue obtained from part of the plant of interest to be used in tissue culture.

F

Fertilizer analysis. The proportions of essential plant elements (usually nitrogen, phosphorus, and potassium) contained in a fertilizer.

Fibrous roots. A highly branched root system without a single dominant axis.

Field capacity. Moisture retained by a soil after free drainage under gravity has ceased.

Filament. The stalklike part of the stamen or male part of the flower that supports the anther.

Filler flowers. Flowers used to fill in the gaps to tie together a floral arrangement to give a complete appearance.

Filler material. An inert material added to a fertilizer to increase its volume.

Flat. A shallow, oblong-shaped wooden or plastic container used to start seedlings.

Fleshy roots. Modified plant roots that store carbohydrates or water.

Floral foam. A piece of dense, foamlike material with high water retention used to hold cut flowers in an arrangement.

Floriculture. The science and practice of cultivating and arranging ornamental flowering plants.

Flower stalk. The stem of the plant that supports the flower.

Foliar application. A method of supplying nutrients to plants by spraying the leaves with liquid fertilizer.

Forcing. A cultural practice of manipulating plants to flower at times other than their normal growing season.

Form. The overall shape of a plant.

Formal flower bed. A symmetrically designed bed for annual and perennial flowers.

Form flower. A type of flower with a unique shape such as bird-of-paradise, potea, or orchid.

Foundation planting. Planting shrubs or ground covers at the base of a house.

Freestanding bed. A bed designed to be accessible from all sides.

Frost hardy. Plants that are able to withstand frost.

Frost pocket. A low-lying area more susceptible to frost than the surrounding vicinity.

Frost tender. Plants that are killed at temperatures below freezing.

Fruit. A mature, swollen ovary.

Fungicide. A chemical used to control a fungus.

G

Gall. An abnormal swelling on a plant part.

Genetic engineering. The manipulation of the genome of an organism by incorporating genes from an alien source.

Genotype. The genetic makeup of an individual.

Germination rate. The percentage of seeds that will germinate under favorable conditions.

Gibberellin. A plant hormone that stimulates growth in stem and leaf by cell elongation and also influences vernalization.

Glazing. The clear or translucent material used to cover a greenhouse or related plant growing structure.

Grafting. The method of asexual propagation in which a branch or bud from one plant is implanted on another.

Gravitational water. The water that the soil is unable to hold against the force of gravity.

Greenhouse effect. The heating of the air in an enclosed translucent or transparent structure due to trapped energy from the sun.

Green manure crop. A fast-growing plant grown for the purpose of turning under the soil before it reaches maturity to provide organic matter in the soil.

Ground cover. Low-growing and spreading plants that form matlike growth over an area.

Growing medium. A material used for rooting and growing plants.

Guard cells. Specialized epidermal cells that open and close the stoma.

H

Hardening off. The cultural practice preceding transplanting whereby plants are acclimatized to field growing conditions by gradually subjecting them to decreasing temperature, moisture, and nutrition.

Hardpan. A layer of compacted soil that slows or stops the movement of water.

Hardwood cutting. A cutting made from the mature growth of a woody plant, usually taken in late fall or early spring.

Hardy plant. A plant adapted to adverse climatic conditions, such as cold temperature, prevailing in the growing area.

Heading back. A pruning strategy used to control the size and shape of shrubs by cutting back shoots to the parent branch to encourage new growth.

Heeling in. A method of temporary storage of bare-rooted trees and shrub seedlings before transplanting by covering the roots with soil or some organic matter in a shallow trench.

Herbaceous. A plant that does not normally produce woody growth.

Herbaceous cutting. A cutting derived from nonwoody plant parts.

Herbicide. A chemical used to control weeds.

Heterosis. The phenomenon whereby the hybrid of a cross exhibits greater vigor and growth than either parent (hybrid vigor).

Hill. A group of seeds planted close together.

Hormone. A growth-regulating substance produced in one part of the plant and used in another part.

Horticulture. The science and practice of growing, processing, and marketing fruits, nuts, vegetables, and ornamental plants.

Hot cap. A paper or plastic dome set over warm-season vegetables in early spring to protect them against frost and wind and to increase the daytime growing temperature.

Humidity. The amount of moisture in the atmosphere.

Hybrid. An offspring of two different varieties of one plant that possesses certain characteristics of either parent.

Hydroponics. A method of growing plants using water fortified with nutrients as the medium.

Hypogeous germination. A type of seed germination in which the cotyledon remains below the soil surface.

I

Incompatibility. The condition in which two plants will not cross-pollinate or a graft union is not successful in spite of appropriate conditions.

Indeterminate. A plant whose central stem or axis grows indefinitely while flowers are continually produced on lateral branches.

Integrated pest management (IPM). A system of pest control based on a good understanding of the life cycle of the pest as well as cultural and biological controls.

J

Juvenility. The first phase of plant development in which growth is vegetative.

L

Layering. A method of asexual propagation in which roots are developed on the stem of a plant while it is attached to the parent plant.

LD$_{50}$. The dose of a chemical at which 50 percent of the exposed test population dies.

Leaching. The washing out of the soil or medium of soluble material with water.

Leaf apex. The tip of a leaf.

Leaf base. The section where the leaf blade joins the petiole.

Leaf blade. The flattened portion of a leaf that extends from either side of the leaf stalk; also called lamina.

Leaf cutting. A cutting made from a leaf and its attached petiole used for propagation.

Leaflet. One of the expanded, small leaflike parts of a compound leaf.

Leaf margin. The edge of a leaf blade.

Lean-to greenhouse. A greenhouse that shares one wall in common with a house or other building.

Light intensity. The brightness of light.

Limbing up. The removal of the lower branches of a tree.

Limestone. A natural rock used to reduce soil acidity or raise pH.

Loam. A soil with approximately equal amounts of sand, silt, and clay.

Long-day plant. A species that flowers under conditions of long duration of light and short daily dark period.

M

Macronutrient. An essential element required in relatively large amounts for plant growth; also called major nutrient.

Macropore. A large space found between large particles or aggregations of particles.

Male sterility. A condition in which afflicted plants produce either no pollen or no viable pollen.

Medium. A material in which plants grow.

Meiosis. A nuclear cell division that occurs in reproductive organs resulting in the production of gametes with half the somatic chromosome number.

Meristem. A region of the plant consisting of undifferentiated and rapidly growing and dividing cells.

Microclimate. A small area with a climate differing from that of the surrounding area because of factors such as exposure, elevation, and sheltering structures.

Micronutrient. An essential element required in minute quantities for plant growth.

Micropore. A tiny space found between fine soil particles or aggregations (capillary pore).

Mineral soil. A soil consisting mainly of weathered rock particles such as sand or clay.

Miticide. A chemical used to control mites.

Mitosis. A nuclear cell division in which the products are genetically identical to the parent cell.

Modified central leader. A pruning method in which the tip of the central leader is cut back to the nearest scaffold branch.

Monocarp. A plant that grows vegetatively for more than one year but flowers only once and then dies.

Monocot. Dimunitive for monocotyledon; a plant having one cotyledon or seed leaf.

Mulch. A material used to cover the soil for purposes such as moisture conservation or weed control.

Mutation. A spontaneous change in the genetic constitution of a plant.

Mycoplasma. A microscopic organism intermediate between a virus and a bacterium that causes several plant diseases.

Mycorrhiza. Soil fungi that live in association with plant roots to the benefit of both roots and fungi.

N

Nematicide. A chemical used to control nematodes.

Node. The joint of a stem where leaves and buds are attached.

Nonselective herbicide. A herbicide that indiscriminately controls all kinds of plants.

O

Offset. A young plant produced at the base of a parent plant.

Olericulture. The science and practice of growing vegetables.

Organic matter. The decomposing bodies of dead plants and animals.

Organic soil. Soil composed primarily of decayed plant and animal remains.

Ornamental horticulture. The branch of horticulture that deals with the cultivation of plants for their aesthetic value.

Ovary. The lower part of the pistil in which the eggs are fertilized and develop.

P

Parent stock. The plant from which material is obtained for propagation.

Parthenocarpy. The development of a fruit without sexual fertilization.

Pathogen. A disease-causing organism.

Perennial. A plant that grows year after year without replanting.

Perlite. A coarse material made from expanded volcanic rock.

Pesticide. A chemical used to control undesirable organisms (pests).

Petiole. The stalk that attaches the lamina or leaf blade to the stem.

pH. A measure of the acidity or alkalinity of a medium or liquid.

Phloem. The portion of the vascular system in which photosynthates are transported throughout the plant.

Photoperiod. The length of day.

Photoperiodism. The response of plants to the relative length of light and darkness.

Photosynthesis. The chemical process by which green plants manufacture food using carbon dioxide and water in the presence of light.

Phototropism. The hormone-induced bending of a plant toward light.

Physiological disease. A diseaselike symptom produced by plants due to an improper or inadequate supply of essential nutrients.

Phytochrome. The chemical involved in plant response to photoperiod.

Pinching. The removal of the terminal bud of a plant to stimulate branching.

Pistil. The female reproductive part of the flower, consisting of the stigma, style, and ovary.

Plant breeding. The science of controlled pollination of plants to develop new cultivars.

Plugging. A method of lawn establishment whereby small cores of sod are transplanted.

Pollarding. A training method for deciduous trees whereby branches are severely pruned to leave stubs from which new shoots grow.

Pollen. The male sex cells borne in the anthers of flowering plants.

Pollination. The deposition of pollen on the flower stigma.

Pomology. The science and practice of fruit culture.

Pore spaces. The gaps between the particles of soil or other growing media.

Postemergence herbicide. A herbicide designed to kill weeds after they become established.

Pot-bound. A condition in which the roots of a plant growing in a restrictive container environment grow around the walls.

Preemergence herbicide. A herbicide designed to kill weeds as they germinate.

Profile. The vertical section of the soil showing horizons or layers.

Propagation. The reproduction or increase in the number of plants by sexual or vegetative methods.

Pruning. The removal of parts to control a plant's growth, size, or appearance.

Public area. The part of a landscape in front of a house and viewable from the street.

Pure line. A population of plants descended from a single homozygous individual.

R

Relative humidity. The amount of water vapor present in the air compared to the total amount the air could hold at its present temperature.

Resistance. The capacity of a plant to resist disease or insect attack.

Respiration. The breakdown of carbohydrates to yield energy for use by the cell.

Rhizome. The underground stem that grows horizontally and produces roots on the lower surface while producing shoots above the ground.

Rooting hormone. A chemical used in promoting rooting of cuttings.

Rootstock. The plant that functions as the root system in a grafted plant.

Runner. An above-ground stem that grows horizontally on the soil surface.

S

Scaffold branch. A main branch growing from the trunk of a tree.

Scarification. The mechanical scraping of a seed coat to facilitate germination.

Scientific name. The Latin name of a plant.

Scion. A piece of shoot or bud grafted onto the rootstock.

Seed coat (testa). The outer covering of the seed.

Seed leaves. The first leaves (cotyledons) produced by a seedling.

Selective herbicide. A herbicide designed to control certain species of plants but not others.

Self-incompatibility. The inability of pollen to fertilize eggs of the same plant.

Self-pollination. The transfer of pollen from the stamen to the stigma of the same flower or to the flowers of genetically alike plants.

Semihardwood cutting. A cutting made from the partially mature new growth of a woody plant.

Senescence. The process of physiological aging of the tissue of a plant or any of its parts.

Sepal. The usually green, leaflike, outermost part of a flower that encloses and protects the unopened flower bud.

Service area. The area of a home landscape designed for storage, garbage cans, and other such uses.

Short-day plant. A species that flowers under conditions of short duration of light and prolonged darkness.

Sidedressing. A method of fertilizer application in which fertilizer is placed in a narrow band along the sides of rows near plants.

Simple leaf. A leaf with a single blade or lamina.

Slow-release fertilizer. A fertilizer chemically formulated to release its nutrients over a long period of time.

Small fruits. The general terminology for plants that bear edible fruits of a relatively small size.

Softwood cutting. A cutting made from the new growth of a woody plant.

Soilless medium. A plant growing medium that contains no natural mineral soil.

Specimen plant. A showy ornamental plant grown solely for its unique beauty.

Spreader branch. A piece of wood inserted between the trunk and branch of a tree to widen the crotch.

Stamen. The male reproductive part of the flower consisting of the anther and filament.

Starter solution. A liquid fertilizer high in phosphorus content applied to seedlings to hasten growth.

Stigma. The top of the pistil that receives pollen.

Straight fertilizer. A single-element fertilizer that is formulated to supply only one of the elements essential for plant growth.

Stratification. The practice of exposing seeds to a low temperature to break dormancy.

Sucker. A vertically growing shoot arising from the base of an established plant.

Systemic pesticide. A pesticide absorbed into a plant, killing the organisms that prey on it.

T

Taproot. A single root that grows vertically into the soil.

Taxonomy. The science of identifying, naming, and classifying plants.

Terminal bud. A bud located at the tip of a stem.

Terrarium. A transparent glass or plastic container that retains high humidity; used for displaying and growing plants indoors.

Texture (plant). The visual impact of a plant due to the size of its leaves.

Thatch. A layer of grass clippings, stems, and roots that accumulates between the surface of a turf and the soil.

Thinning. The pruning technique for shrubs in which the oldest stems are removed to stimulate new growth; the selective removal of the excess fruits on a fruit tree to improve the size and quality of the remaining fruits. The term is also applied to the removal of excess seedlings from a row of dense planting for optimum growth.

Topdressing. A method of fertilizer application that involves the sprinkling of fertilizer over the soil surface.

Topiary. The formal training and pruning of a plant to grow in a controlled shape.

Topsoil. The uppermost layer of soil.

Trace element. A micronutrient.

Translocation. The movement of carbohydrates, water, minerals, and other materials within the vascular system of a plant.

Transpiration. The loss of water from a plant, usually through leaves, in vapor form.

Tuber. A thickened underground stem in which carbohydrates are stored.

Tunicate bulb. A bulb that has a dry membranous outer scale.

V

Variegation. The genetic patterning of leaves with white or yellow markings.

Vernalization. The practice of subjecting a plant to cool temperatures for the promotion or enhancement of growth or flowering.

X

Xylem. The portion of a plant vascular system in which water and minerals, taken in by the roots, move throughout the plant.

Index

common lamb's-quarter (*Chenopodium album*), 214
common milkweed (*Asclepias syriaca*), 214
communication, 705
community packs, 365, 366
companion cells, 36
compensatory root growth, 52
complementary DNA (cDNA), 174, 199, 200
complete dominance, 163
complete fertilizers, 89
complete flowers, 55
complete metamorphosis, 217
complete senescence, 150
complex leaves, 47
complex tissues, 33, 34–38
compost, 88, 674
compost activators, 682
composting, 679
  carbon-to-nitrogen ratio in, 681
  decomposers, 680–81
  maintaining piles in, 687
  materials in, 681
  principles of, 680, 683
  systems of, 682–86
  time factor, 682
compound lamina, 43
compound layering, 328
compressed-air tank spray, 262
compression wood, 53, 54
computer language, 703
computers, in horticulture, 702–10
concept-based searching, 708
condensation, 127
conducting tissues, 6, 36–38
conduction heat loss, 396
conifer cuttings, 309, 310
conifers
  for bonsai, 733, 734
  pruning, 559–60, 561
connected greenhouses, 384–85
conservation tillage, 109–110
contact herbicides, 268
contact poisons, 258
container composting, 683
container-grown trees, 493–94
container nursery, 470–71, 472
containers, 421
  for bonsai, 736, 740–41
  for cacti, 721
  for floral design, 449
  for houseplants, 345–47, 363–66
  for succulents, 714
  for terrariums, 725, 726
continual bloom garden, 481
contrast, in floral design, 449
controlled-environment horticulture. *See* greenhouses
controlled-release fertilizers (CRF), 89–90
convection heaters, 398, 399
conventional breeding, 177–85
  biological variation, 176–77
  duration of, 196
  for flowering plants, 175–76
  guidelines and classification, 185–86
  limitations of, 197
  methods of, 186–93
  mutations, 193–94
  parthenocarpy, 197
  polyploidy, 194–96
  wide crossing, 196
conventional tillage, 108, 109
Convolvulus, 213

cooling systems, in greenhouses, 401–2
cool-season grasses, 510, 512
cool-temperature plants, 67
copper, 85
coppicing, 550
cordon training system, 556, 557
core aerator, 526
cork cambium, 54
cormels, 332, 504
corms, 40, 41, 331, 332, 503–6
corn earworms, 221, 222
corn plant (*Dracaena fragrans*), 373
corner planting, 458
corolla, 55, 56
corolla tube, 55, 56
cortex, 39
cosmids, 198
cotyledons, function of, 42
coumarin, 144, 153
Crassula argenta, 371
Crassulacean acid metabolism (CAM), 135–36
creeping grasses, 511, 513
creeping plants, 11
crop modeling, 704
crop production, 284
crop rotation, 252, 254, 592–93, 678
cross-benching, 393, 394
crossing over, 164
cross-pollinated species, 185, 187–93
cross-pollination, 57, 147, 162, 176
croton (*Codiaeum variegatum*), 371
crown division, 327, 328, 333
crown lifting, 549
crown reduction, 549
crowns, 39, 326–27
crown thinning, 549
cryptogams, 57
cucumber beetles, 221
cucumbers (*Cucumis sativus*), 603–4
Cucumis spp.
  C. melo, 605–6
  C. sativus, 603–4
Cucurbita, 606–7
Cucurbitaceae, 24
cucurbits, 603–8
culinary herbs, 622, 624
cultivars, 4, 6–7, 8, 422
  adopted local, 176
  of grapes, 650
  resistant, 236, 254
  of small fruits, 642
  of strawberries, 644
cultivation, 110
cultivators, 111, 581
cultural pest control, 238, 252, 254–55
curtain wall, 381
cut flowers, 416
  arranging and designing, 694–97
  cultural procedures for, 691–93, 694
  greenhouse production of, 427–32
  industry, 427–28
  overview of, 689–90
  species of, 690, 691
  See also dried flowers
cuticle, 34
cutin, 34, 129
cuttings
  of cacti, 719, 720
  factors affecting rooting of, 311–13
  of succulents, 714
  transplanting, 314
  types of, 308–11
cutworms, 221, 222

cybrid, 338
cyclic lighting, 75, 406
cycocel, 153
cytokinins, 153
cytoplasm, 29
cytoplasmic DNA, 175
cytoplasmic factors, 190
cytoplasmic-genetic male sterility, 190
cytosine, 170, 171

D

daminozide, 153
damping-off, 276, 292
dandelions (*Taraxacum officinale*), 213
Dark Ages, horticultural history, xxxii–xxxiii
darkness, 75–76, 312
dark reaction, 133
databases, 703, 705
Daucus carota, 598–99
day-neutral plants, 75, 405
deadwood, 557
de Candole, Pyrame, 4
deciduous hardwood cuttings, 309
deciduous ornamentals, 19
deciduous perennials, 9
deciduous shrubs, 19
deciduous trees, 20, 499, 500, 733, 734
deciduous vines, 20
decomposers, 680–81
decomposition, 85
decorative lighting, 352
decreasing growth phase, 124
decumbent plants, stems of, 11
deep recirculating systems, 434
defoliating insects, 218–19
dehatching, 526–27
dehiscent fruits, 12, 14, 15
dehorning, 561
dehydrogenases, 129
deletions, 193
denaturation, 171
dendrochronology, 53
denitrification, 97
deoxyribonucleic acid (DNA), 29, 30, 129
  function of, 172–74
  structure of, 170–71
dermal toxicity, 244
desert cacti, 717
desert fan palms (*Washingtonia filifera*), 376
detached greenhouses, 383, 384
detached-scion grafting, 315, 318
determinate plants, 38, 145
developmental plasticity, 37
devernalization, 147
dibble, 301, 302
2,4-dichlorophenoxyacetic acid (2,4-D), 150
dicotyledons (dicots), 23–24, 39
  roots of, 50, 51
  seeds of, 57
  wood of, 53
dictyosomes, 32
Dieffenbachia, 375
DIF, 419
differentiation, 28, 125, 167
diffuse growth, 37
digestive glands, 35
Digitaria, 214
dihybrid cross, 160, 162
dioecious plants, 56
dioecy, 148, 176
diphenyl ethers, 269
diploid, 30, 162, 194

Diptera, 216
direct seeding, 295–96, 477
disaccharides, 127
disbudding, 431
disease cycle, 228
diseases
  bacteria, 224–25
  of blueberries, 656
  of bulbs, 506
  of cut flowers, 693
  defenses against, 230–31
  factors causing, 227–29
  fungi, 223–24
  in gardens, control of, 594
  genetic basis of, 232
  of grapes, 653
  in greenhouses, 275–76, 278–79, 413
  insecticides control strategies for, 265, 266
  in landscapes, 572–73
  of lawns, 525, 526
  mycoplasma-like organisms, 226
  of poinsettias, 423
  prevention through mulching, 677
  resistance to, 231
  of roses, 431
  of succulents, 714
  of terrarium plants, 729
  of trees, 493
  viruses, 225
disease triangle, 77, 228
disk harrow, 111
disk plow, 110
distribution, xi, 670
division, plants propagated by, 331, 333, 334
DNA. *See* deoxyribonucleic acid
domestication, 177
dominance, 160. *See also* partial dominance
dominance gene action, 192
dominance theory, 188
dominance variance, 184
dormancy, 69, 144
  of bulbs, 505
  of seeds, 144, 289–90
dorsoventral leaves, 43
double helix, 171
double-layer covering, 389
double pot production system, 471
double potting, 345
double sigmoid curve, 124
Downing, Andrew J., xxxiii
downloading, 706
Dracaena fragrans, 373
drainage, 102, 106–7, 466, 468
dried flowers, 697–700
drift, 260
drilled wells, 574
drill holes, 95
drip irrigation, 105–106, 571, 575–78
drip line, 570
driven wells, 574, 575
drop crotching, 557, 558
drought, 154
drum-mounted hose, 575, 576
drupe, 12, 13
dry fertilizers, 94–95
dry formulations, 260
dry fruits, 11, 12, 14–15
dry sets, 602
dug wells, 574
dumbcane (*Dieffenbachia*), 375
duplications, 193

Evergreen plants survive the harsh winter.

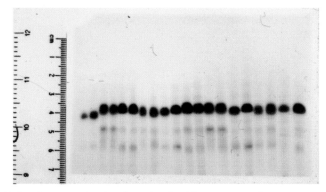

Using isozyme technology in plant improvement.

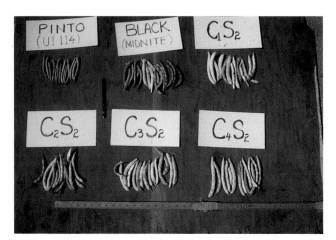

Genetic improvement of beans.

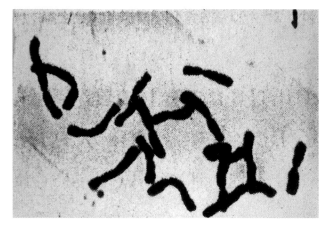

Chromosomes: contain DNA.

Decidous trees in landscape lose their leaves in fall.